Primitive Motile Systems in Cell Biology

PRIMITIVE MOTILE SYSTEMS IN CELL BIOLOGY

EDITED BY

Robert D. Allen

DEPARTMENT OF BIOLOGY
PRINCETON UNIVERSITY
PRINCETON, NEW JERSEY

Noburô Kamiya

FACULTY OF SCIENCE
OSAKA UNIVERSITY
OSAKA, JAPAN

The Proceedings of
A SYMPOSIUM ON THE MECHANISM OF
CYTOPLASMIC STREAMING, CELL MOVEMENT,
AND THE
SALTATORY MOTION OF SUBCELLULAR PARTICLES
Held at
Princeton University
April 2–5, 1963

1964

ACADEMIC PRESS New York and London

ACADEMIC PRESS INC.
111 Fifth Avenue, New York, New York 10003

United Kingdom Edition published by
ACADEMIC PRESS INC. (LONDON) LTD.
Berkeley Square House, London W.1

LIBRARY OF CONGRESS CATALOG CARD NUMBER: 64-15265

PRINTED IN THE UNITED STATES OF AMERICA

Symposium Participants

Numbers in parentheses indicate the pages on which the authors' contributions begin.

ABÉ, TOHRU H., *Laboratory of Biology, Hosei University, Tokyo, Japan* (221)

ALLEN, ROBERT D., *Department of Biology, Princeton University, Princeton, New Jersey and The Marine Biological Laboratory, Woods Hole, Massachusetts* (407)

ANDERSON, JOHN D., *Department of Physiology and Biophysics, University of Illinois, Urbana, Illinois* (125)

BERREND, ROBERT E., *Department of Zoology, University of Wisconsin, Madison, Wisconsin* (433)

BOUCK, G. BENJAMIN, *Biological Laboratories, Yale University, New Haven, Connecticut and Biological Laboratories, Harvard University, Cambridge, Massachusetts* (7)

BOVEE, EUGENE C., *Department of Zoology, University of California, Los Angeles, California* (189)

GOLDACRE, R. J., *Chester Beatty Research Institute, London, England* (237)

GRIFFIN, JOE L., *Department of Anatomy, Harvard Medical School, Boston, Massachusetts* (303)

GUSTAFSON, TRYGGVE, *The Wenner-Gren Institute, University of Stockholm, Sweden* (333)

HAYASHI, TOSHIO, *Biological Institute, College of General Education, Tokyo University, Komaba, Tokyo, Japan* (19)

HOFFMANN-BERLING, H., *Max-Planck-Institut für Physiologie, Heidelberg, West Germany* (365)

HONDA, S. I., *Department of Botany and Plant Biochemistry, University of California, Los Angeles, California* (485)

HONGLADAROM, T., *Department of Botany and Plant Biochemistry, University of California, Los Angeles, California* (485)

INOUÉ, SHINYA, *Department of Cytology, Dartmouth Medical School, Hanover, New Hampshire* (549)

JAHN, THEODORE L., *Department of Zoology, University of California, Los Angeles, California* (279)

JAROSCH, ROBERT, *Biological Research Division, Austrian Nitrate Works, Linz/Donau, Austria* (599)

KAMIYA, NOBURÔ, *Department of Biology, Faculty of Science, Osaka University, Osaka, Japan; Department of Biology, Princeton University, Princeton, New Jersey; and The Marine Biological Laboratory, Woods Hole, Massachusetts* (257)

KITCHING, J. A., *Department of Zoology, University of Bristol, Bristol, England* (445)

KURODA, KIYOKO, *Department of Biology, Faculty of Science, Osaka University, Osaka, Japan* (31)

MAHLBERG, PAUL G., *Department of Biological Sciences, University of Pittsburgh, Pittsburgh, Pennsylvania* (43)

MARSLAND, DOUGLAS, *Department of Biology, New York University, New York* (173, 331)

NAKAI, J., *Department of Anatomy, School of Medicine, University of Tokyo, Hongo, Tokyo, Japan* (377)

NAKAJIMA, HIROMICHI, *Department of Biology, Faculty of Science, and Institute for Protein Research, Osaka University, Osaka, Japan and Department of Biology, Princeton University, Princeton, New Jersey* (111)

O'NEILL, C. H., *Zoology Department, King's College, University of London, London, England* (143)

PARPART, ARTHUR K., *Department of Biology, Princeton University, Princeton, New Jersey and The Marine Biological Laboratory, Woods Hole, Massachusetts* (471)

REBHUN, LIONEL I., *Department of Biology, Princeton University, Princeton, New Jersey and The Marine Biological Laboratory, Woods Hole, Massachusetts* (503)

ROBINEAUX, ROGER, *Hôpital Saint-Antoine, Paris, France* (351)

ROTH, L. E., *Committee on Cell Biology and Department of Biochemistry and Biophysics, Iowa State University, Ames, Iowa* (527)

SHAFFER, B. M., *Department of Zoology, Cambridge University, Cambridge, England* (387)

STEWART, PETER A., *Department of Physiology, Emory University, Atlanta, Georgia and Biology Department, Brookhaven National Laboratory, Upton, New York* (69)

THOMPSON, C. M., *Zoology Department, King's College, University of London, London, England* (143)

WILDMAN, S. G., *Department of Botany and Plant Biochemistry, University of California, Los Angeles, California* (485)

WOHLFARTH-BOTTERMANN, K. E., *Zentral-Laboratorium für angewandte Übermikroskopie am Zoologischen Institut der Universität Bonn, Bonn, Germany* (79)

WOLPERT, L., *Zoology Department, King's College, University of London, London, England* (143)

Discussants

BISHOP, DAVID W., *Carnegie Institution of Washington, Baltimore, Maryland*

BURGERS, J. M., *Institute for Fluid Dynamics and Applied Mathematics, University of Maryland, College Park, Maryland*

DE BRUYN, PETER P. H., *Department of Anatomy, University of Chicago, Chicago, Illinois*

HAYES, WALLACE, *Department of Aeronautical Engineering, Princeton University, Princeton, New Jersey*

JAFFEE, LIONEL, *Department of Botany, University of Pennsylvania, Philadelphia, Pennsylvania*

KAUZMANN, WALTER, *Department of Chemistry, Princeton University, Princeton, New Jersey*

LING, GILBERT, *Pennsylvania Hospital, Philadelphia, Pennsylvania*

NOLAND, LOWELL, *Department of Zoology, University of Wisconsin, Madison, Wisconsin*

SZENT-GYÖRGYI, ANDREW G., *Dartmouth Medical School, Hanover, New Hampshire*

TAYLOR, EDWIN, *Committee on Biophysics, University of Chicago, Chicago, Illinois*

THIMANN, KENNETH V., *Department of Biology, Harvard University, Cambridge, Massachusetts*

Observers

CLARK, ELOISE E., *Department of Zoology, Columbia University, New York, New York*

COUILLARD, PIERRE, *Departement des Sciences Biologiques, Université de Montréal, Montréal, Canada*

COWDEN, RONALD R., *Department of Pathology, Medical Center, University of Florida, Gainesville, Florida*

DEVILLERS, CHARLES, *Willard Gibbs Laboratories, Yale University, New Haven, Connecticut*

DOUGHERTY, WILLIAM, *Department of Biology, Princeton University, Princeton, New Jersey*

FREUDENTHAL, H., *American Museum of Natural History, New York, New York*

GREEN, PAUL, *Division of Biological Sciences, University of Pennsylvania, Philadelphia, Pennsylvania*

GUTTES, E. S., *Department of Biology, Brown University, Providence, Rhode Island*

HACKETT, RAYMOND, *Department of Pathology, Medical Center, University of Florida, Gainesville, Florida*

HANSON, EARL, *Shanklin Laboratories of Biology, Wesleyan University, Middletown, Connecticut*

HAYASHI, TERU, *Laboratory of Biophysics, Pupin Hall, Columbia University, New York, New York*

HAYES, MRS. HELEN L., *Biology Branch, Office of Naval Research, Washington, D. C.*

HIRSHFIELD, HENRY, *Department of Biology, Washington Square College, New York University, New York, New York*

JACKSON, WILLIAM, *Department of Biology, Dartmouth College, Hanover, New Hampshire*

JACOBS, WILLIAM P., *Department of Biology, Princeton University, Princeton, New Jersey*

LEE, JOHN, *American Museum of Natural History, New York, New York*

RHEA, ROBERT PERRY, *Rockefeller Institute, New York, New York*

ROSENBERG, MURRAY, *Rockefeller Institute, New York, New York*

RUSTAD, RONALD, *Institute for Molecular Biophysics, Florida State University, Tallahassee, Florida*

SATO, HIDEMI, *Department of Cytology, Dartmouth Medical School, Hanover, New Hampshire*

SCOTT, THOMAS, *Department of Biology, Princeton University, Princeton, New Jersey*

TASAKI, ICHIJI, *Laboratory of Neurobiology, National Institutes of Mental Health, Bethesda, Maryland*

TAYLOR, A. CECIL, *Rockefeller Institute, New York, New York*

VISHNIAC, ROMAN, *New York, New York*

YANAGITA, T. M., *Department of Zoology, Ochanomizu University, Tokyo, Japan*

ZIMMERMAN, ARTHUR M., *Department of Pharmacology, Downstate Medical College, Brooklyn, New York*

Preface

This book is intended for scientists and students of the biological, biophysical, and medical sciences who are interested in the movements in and of living cells. In it are collected thirty papers presented at the *Symposium on the Mechanism of Cytoplasmic Streaming, Cell Movement, and the Saltatory Motion of Subcellular Particles,* held at Princeton University in April, 1963, together with the edited discussions.

At this Symposium nearly a hundred scientists, representing such disciplines as cell biology, plant physiology, protozoology, developmental biology, biophysics, physical chemistry, biochemistry, rheology, physics, engineering, and medicine, gathered to consider one of life's most elusive problems: How does movement occur at the cell level and below? Until quite recently, nearly all of these phenomena of movement, which we classify as "primitive motile systems," were so poorly understood that theories about them were almost as numerous as facts.

Within the past decade, however, research on motility has begun to bear fruit, due largely to the introduction of improved methods. Each contribution to the volume represents a sample of the best work being done in each area of the field. Each paper contains not only enough background material and bibliographic references to serve as an effective guide to the literature, but is followed by an edited version of the symposium discussion. The discussion should be a most valuable part of the volume for students and for new workers entering the field, for it points the way to the uncertainties and disagreements in each area of study.

"Free Discussion" sections contain remarks and comments by invited discussants, which by themselves would be worth publishing irrespective of the papers. For example, there are pertinent comments by Andrew G. Szent-Györgyi and G. Ling regarding the molecular mechanism of contraction and its control system, a discussion of wave motion by W. D. Hayes, a description of how mathematical models may be useful to biologists studying motility phenomena by J. M. Burgers, and a "running battle" among proponents of the various theories of ameboid movement.

It is over two decades since a similar volume was published in the motility field. The scope of the meeting and the resulting volume is so broad that its influence should doubtless be felt in many fields.

The editors would like to express their appreciation to a number of individuals who contributed to the success of the conference and speedy publication of its proceedings; to the other members of the organizing committee, Drs. Eugene Bovee, Douglas Marsland, and Lionel Rebhun; to the conference assistants, Mrs. Eleanor Benson Carver, Mrs. Prudence Jones Hall, Mr. Christopher D. Watters, and Mr. Konrad Bachmann; to Mrs. Olive Loria, stenotypist, and to Mrs. Mildred Nunziato and Mrs. Sarah Hayashi who assisted in the preparation of the discussions. We are also grateful to Drs. Lionel I. Rebhun, Walter Kauzmann, and Peter Stewart for performing important editorial tasks.

The conference was supported by a generous research grant [Nonr(G) 00023-63] from the Biology Division of the Office of Naval Research, United States Navy.

January, 1964

R. D. ALLEN
N. KAMIYA

Introduction

One might properly ask what "primitive motile systems" are and why they are of interest. If it had not been for the invention of the microscope, the study of motility might well have remained restricted to the study of muscular contraction. However, early microscopists saw and described the marvelous diversity of movements among protozoans and other lower organisms, and it was not long before hypotheses were advanced to explain the movements of these creatures. Each succeeding generation of biologists has seen the gliding of cells, protoplasmic streaming, pseudopodial movement, the beating of cilia and flagella, mitotic movements of chromosomes, saltatory motions of various cytoplasmic particles, contractions of myonemes and other structures, and various other "nonmuscular" movements. Despite two centuries or more of study with ever-improving methods of study, however, the basic problem as to the mechanism of these various movements has remained unsolved.

One point of view toward these "primitive motile systems" is that they may be slightly different manifestations of some single basic mechanism such as, for example, gel contraction. Such general viewpoints have been expressed from time to time and are very attractive from a theoretical point of view. However, before accepting any unifying theory at face value, it is desirable to classify motile systems into representative phenomenological groups, bearing in mind that the classification may be artificial, and then to test the predictions of any such unifying theory with representatives of each type of motile system. This approach has been applied only to a limited extent, with the result that unifying theories are few and rest on tenuous evidence.

The simplest kind of motion in cells is the *Brownian motion* of particles of micron dimensions, produced by the thermal agitation of neighboring molecules. These molecules are in motion in living and nonliving fluids alike, and therefore have little to do with motility, except insofar as any restrictions imposed on Brownian motion may contribute information regarding the structural properties of certain parts of the cytoplasm.

There are, however, other motions in cells that superficially resemble Brownian motion. These are the *"saltatory" (or jumping) motions,* which are sudden excursions of cytoplasmic particles over distances too extensive to be accounted for as Brownian motion. Such motions have been described in the plant literature as "Glitchbewegung" or "agitation." *Mitotic movements* of chromosomes are phenomenologically somewhat similar, except that they occur within a highly organized structure, the mitotic spindle.

Cytoplasmic streaming is a broad term applied to perhaps a dozen or more different kinds of phenomena in which visible particles move in groups in such a way as to indicate that they are carried by the streaming of cytoplasmic ground substance. The distinction between cytoplasmic streaming and saltatory motion must not be overemphasized, for there are apparently transitional states between the two situations. It needs to be established in many types of "cytoplasmic streaming" whether the flow of ground substance is solely responsible for the motion of particles. Cytoplasmic streaming occurs in cells of both plants and animals, as well as in such acellular organisms as slime molds. Within the plant kingdom alone, the diversity of streaming phenomena is impressive; it is possible to list perhaps a dozen types. Of these, it is now quite well established that two of the plant systems have very dissimilar aspects. This will be brought out in Part I. Little is known about the other types of streaming in plant cells.

Ameboid movement has usually been defined as "locomotion by means of pseudopodia," but it has often been considered by textbook writers as a special case of cytoplasmic streaming in which pseudopodia form and are used in locomotion. It is brought out in Part II that there may well be fundamental differences among two or three groups of amebae as to structure, details of movement, and mechanism of pseudopod formation. It seems abundantly clear, at least, that the cytoplasmic streaming which accompanies pseudopod formation has little in common, as far as mechanism at the cellular level is concerned, with the streaming which occurs in plant cells.

According to the definition of ameboid movement above, the Foraminifera, Radiolaria, and Testacea, among the Protista, and the numerous metazoan tissue cells that move by means of pseudopodia should be included in the same phenomenological grouping as the free-living amebae. However, when the details of movement are compared, the diversity found gives cause for concern whether the same basic mechanism could apply to all types of "ameboid movement."

At first sight, it would appear that the problems of mechanism might be more easily solved with the "less primitive" (i.e., more highly organized) motile systems, such as *ciliary and flagellar movement,* and of course, muscular contraction. However, the degree of structural organization in these systems is a mixed blessing, and here also the fundamental questions of molecular mechanics are still largely unsolved, although in the case of muscular contraction most authorities have settled on one of two leading theories, both of which lack decisive evidence.

What kind of information should we seek in studies of primitive motile systems? First and foremost, the observational details of each system must be recorded in as objective a manner as possible—free from interpretation in terms of any model. This is particularly true of the more complex movements of ameboid cells, which are highly dependent upon external environmental and internal physiological conditions. The advantages of recording observations on cine film are worth considering, especially if it is possible to publish the film and make it available at cost to other investigators. Second, since movements are produced by forces, it is important to identify the forces as contractile, electrical, osmotic, or whatever, and to localize within the cell the site at which the force is applied. Third, we must find out about the nature and availability of the chemical fuel; is it always ATP? How is the fuel withheld from the motile machinery during inactivity? Finally, we must find out as much as possible about the machinery which uses the fuel. Of what units is it composed? How are the forces generated within it and controlled?

In principle, most of these questions can now be posed with the aid of existing instrumentation. In fact, some of the more favorable experimental materials have been subjected to experiments, which in effect constitute the beginnings of such approaches. The next decade promises to be a very exciting one in the study of primitive motile systems, as one by one these systems emerge from the stage of descriptive analysis to the kind of approach outlined above.

THE EDITORS

Contents

Part IV Non-Brownian and Saltatory Motion of Subcellular Particles, and Mitotic Movements

PART I

Cytoplasmic Streaming in Plants and Myxomycetes (Mycetozoa)

Introduction

KENNETH V. THIMANN

Department of Biology, Harvard University, Cambridge, Massachusetts

I need not tell you that cytoplasmic movement is and has been over the years a region of ignorance. It is really remarkable that since Corti a great number of people have been making observations on a variety of plant and animal material, and yet basic ideas remain very scarce. Let me remind you that Corti's observations were made in 1774; this was the time of Lavoisier, and the area in which Lavoisier worked has developed fantastically into the fields which have become known to us as chemistry and biochemistry, including all of organic chemistry, of which there was practically nothing in 1774. Consider the elaborate development of synthetic organic chemistry, and, now, of electronic concepts in chemistry, and of the ramifications into biochemistry—oxidations, enzymes, and the current work on biosynthesis. All of these developments have constituted a real explosion of knowledge in fields other than the one which will be discussed in this symposium.

Another comment to be made about the date 1774 is that it was a year or two after Priestley's observations on photosynthesis (photosynthesis was first observed in 1771). Think what has happened in photosynthesis in the intervening period, the discovery of its anaerobic modification, the countless schemes of exchange of carbon atoms between one compound and another, and the growth, now, of some really basic understanding of this process.

It is even the sixtieth anniversary of the publication of Ewart's book on protoplasmic streaming. Sixty years is a perfectly tremendous time in almost any field and yet you can read Ewart today on many parts of the subject and find it reasonably up to date.

Of course, the explanation, I need not tell you, is that this problem comes very close to the heart of biology, and being so close to the heart, it is one of the most difficult in which to make definite advances.

Recently, Szent-Györgyi, Sr. (1960) made a very wise remark in one of the little books which he writes every few years[1]:

"We know life only by its symptoms, and what we call life is to a great extent the orderly interplay of the various forms of work. Since

[1] Quote has been very slightly paraphrased.

the dawn of mankind, death has been diagnosed mostly by the absence of one of these forms, the one that expresses itself in *motion*."

So we are confronted here, it might be said, with the basic problem of life and death.

Recently an accident took place on the highway. A man was run over and the driver claimed that he did not knock him down, but that the man fell in front of his car and was apparently pushed down there. So the judge, in charging the jury with their duty, said that they had three things to determine: First, did he fall? Second, if he did fall, was he pushed, and, if so, by whom? And, third, just how did this happen?

This is, I think, a perfect analog for the problems of our topic. When cytoplasm moves, or when individual particles move within it, are they pushed and, if so, by whom, and just how does this happen?

Now if we look back at what has been done (I made the point that it is not a great deal), of course the situation is not quite so gloomy. We do know a great many things, and it is perhaps worth taking the time to mention some of those that we do know. We do know that the cause of movement somehow involves adenosine triphosphate (ATP). At least in Myxomycetes we know that ATP is present, and we know when one injects ATP it produces very characteristic results. It liquefies Myxomycete plasmodium; it makes cytoplasm flow away from the point of injection in amebae; it reactivates sperm tails that have been stored in the cold; it greatly stimulates the streaming in *Acetabularia*.

Then again, we know, at least in Myxomycetes, that there is present a myosin-like protein which has many of the properties of myosin, and acts as an ATPase stimulated by calcium, with the calcium effect opposed by magnesium. This is one precious piece of knowledge; its exact application may not yet be clear, but at least it describes some part of what happens.

There is good evidence, also, that the streaming energy in Myxomycetes is derived from an anaerobic process, and we know from plant material, especially cambium, that streaming can continue for a very long time in nitrogen. We know a good deal also about the temperature relations of streaming and related movements. The Q_{10} is on the whole low enough to suggest that the limiting factor may well be a physical process rather than the provision of the metabolic energy (see the paper by Jarosch in this volume).

There is a great deal of information about the effects on movement of external influences, at least in plants. These include the effect of light, especially blue light, and its accompanying change in protoplasmic viscosity (Virgin, 1951, 1954). We are familiar too with many kinds of ionic effects and with the stimulation of streaming caused by physiological concentrations of auxin, and its interdependence with carbohydrate supply.

As to the mechanism, a survey of the literature indicates that there is a gradual crystallization of views around the importance of an interface between sol and gel. For this we have various supporting observations of which one of the clearest is the effect of pressure, which liquefies the gel and also stops streaming. The remarkable phenomena of independent movement of plastids and granules suggest that any protoplasmic surface can generate the forces leading to movement and this, of course, would strongly support the idea of some sort of a universally occurring type of interface. However, in such organisms as amebae, if there is a sol–gel interface it must of course be constantly changing or breaking down and reforming, so this seems to be a rather obscure part of the field, touched on by several contributors to this symposium.

One problem that comes very strongly to mind when one thinks of this field is the fact that the free movement of cytoplasm almost seems to preclude the presence in it of elaborate organization, and yet one sees under the electron microscope, even in motile cells, quite extensive structures. These structures are apparently still present even though the cytoplasm may be in active movement.

In the pollen mother cells of *Saintpaulia* (sectioned by my colleagues Leadbetter and Porter at Harvard), as is typical with vacuolated plant cells, there is very vigorous streaming. Nevertheless, although cytoplasm is evidently streaming in and out between the separated vacuoles, this highly mobile material contains elaborate endoplasmic reticular structures. That a small body like the Golgi could remain intact is not remarkable, for it will probably move intact like a completely enclosed structure; but that endoplasmic reticulum of various kinds including outgrowths of the nuclear membrane into the cytoplasm should remain there and move freely about seems to me a very remarkable thing.

Just recently, Kollman and Shumacher (1962) published pictures showing cytoplasmic structure in sieve tubes. I need not tell you that sieve tubes are the vessels through which sugar solution flows freely in mass flow. These sieve tubes have been reactivated in the spring. Fixation by permanganate or osmic acid shows similar structures in these long cells, namely the presence of elaborate endoplasmic reticular structures, in cytoplasm which is supposed to flow and give way to sugar solutions continuously. In the presence of such vigorous transporting activity as has been demonstrated to occur in trees, it is difficult to convince oneself that any such structure could possibly survive.

As I said at the outset, we are close to the heart of biology here, so that it is perhaps not surprising that there are a great many unsolved problems. After all, the presence of such puzzles is the principal stimulation of a symposium such as this.

References

Kollman, R., and Shumacher, W. (1962). *Planta* **57**, 195.

Szent-Györgyi, A. (1960). "Introduction to a Submolecular Biology," p. 10. Academic Press, New York.

Virgin, H. I. (1951). *Physiol. Plantarum* **4**, 255.

Virgin, H. I. (1954). *Physiol. Plantarum* **7**, 343.

Fine Structure in *Acetabularia* and Its Relation to Protoplasmic Streaming

G. BENJAMIN BOUCK

Biological Laboratories, Yale University, New Haven, Connecticut and Biological Laboratories, Harvard University, Cambridge, Massachusetts

Since the early 1930's, the classic works of Hämmerling and others (cf. review, Hämmerling, 1953) have demonstrated that the large uninucleate, unicellular, green alga, *Acetabularia*, is a unique experimental organism for the study of a variety of cellular problems. However, despite extensive physiological and morphogenetic investigations, an adequate description of cytoplasmic fine structure has not yet appeared. A study of the fine structure of experimentally treated plants is believed to be ultimately capable of revealing the extent of nuclear control of the form and distribution of cytoplasmic fine structure, and possibly the role of fine structure in morphogenetic expression. The present paper represents the results of an investigation of the "normal" cytoplasmic constituents and, it is hoped, will provide some basis for future experimental work in fine structure on enucleate and variously treated portions of *Acetabularia*.

Streaming in this plant has been described as occurring in parallel tracks or striations in the cortical gel, and running along the longitudinal axis of the plant. Kamiya (1959) has found that when the major cell components are centrifuged to the basal portion of the plant, they always return along the original cortical tracks—a fact which suggests a difference in fine structure in the two states of the cytoplasm (i.e., moving and stationary). Thus a study of the fine structure of *Acetabularia* promised also the possibility of determining the fine structural basis (if any) for active cyclosis.

Materials and Methods

Stock cultures of *Acetabularia* were generously supplied by Mr. John Terborgh of Harvard University. The plants were given continuous illumination at 25°C while growing in "Erdschreiber" salt water medium (Føyn, 1934). Two-centimeter plants were then removed from the culture flask with a glass needle and placed whole in a 2% osmium solution buffered with potassium phosphate (Millonig, 1961a) and adjusted to a final pH of 6.9 and a molarity of 0.49. Fixation was carried out at room

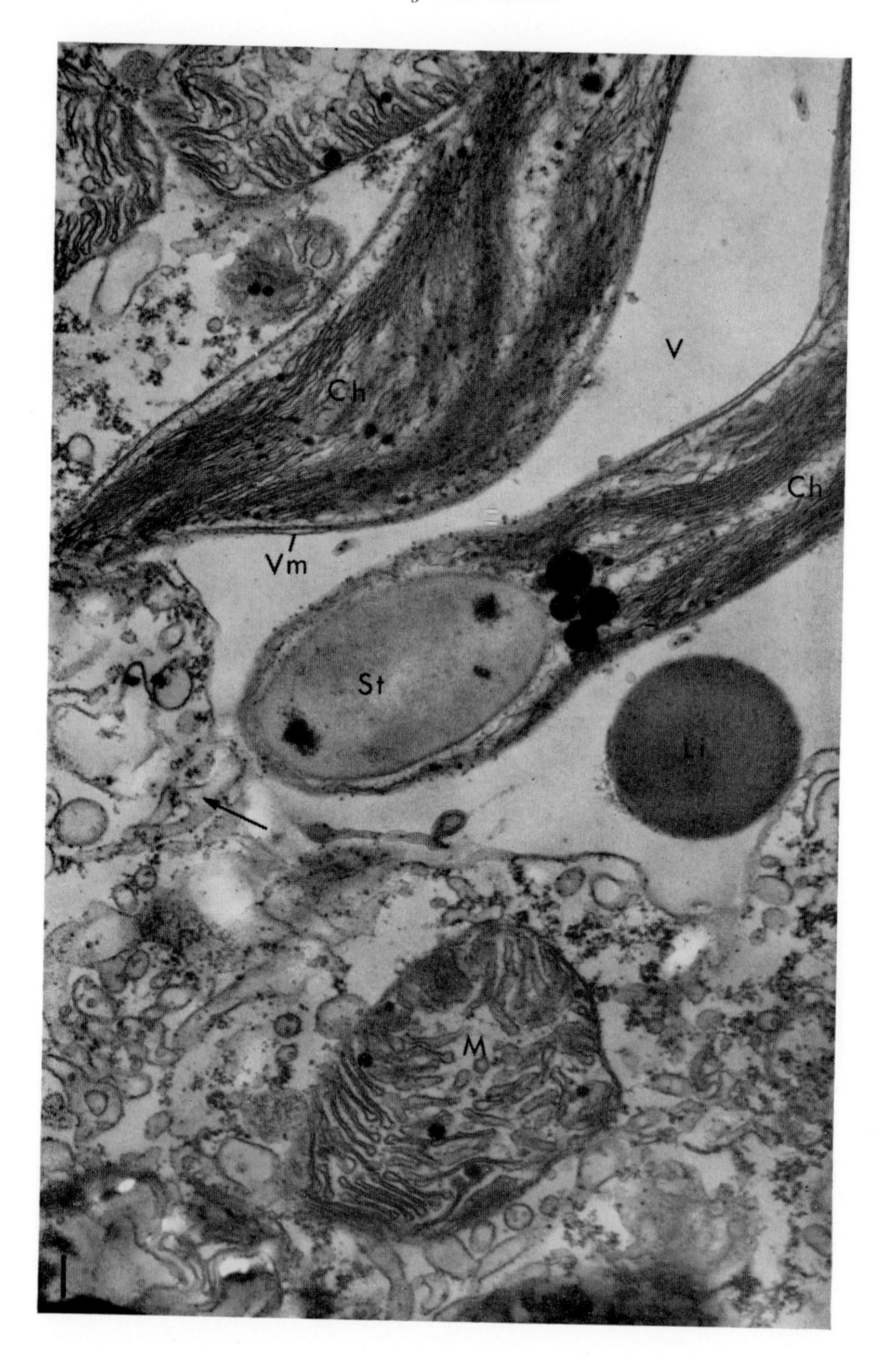
V
Ch
Ch
Vm
St
Li
M

temperature on a rotating drum for 42 hr. This was followed by rapid washing in distilled water, dehydration in methyl alcohol, and embedding in Epon (Luft, 1961). All dehydration steps were carried out gradually. The propylene oxide intermediate step before embedding (Luft, 1961) was found frequently to cause a severe bubbling within the plant, and was thus omitted from the embedding schedule. Plants were cut just before final embedding since infiltration was poor with whole plants or even with plants cut longer than a few millimeters in length. Sections of the polymerized blocks were cut with a diamond knife on either a Huxley or a Porter-Blum ultramicrotome, stained in Millonig's (1961b) lead stain, and photographed in either a Siemens Elmiskop I or a Japanese (JEM) electron microscope.

Results

Acetabularia presents special problems in electron microscopy due to difficulties with fixation and embedding. The thick wall, relatively small volume of cytoplasm, and extremely acid contents (pH 1–2) of the large vacuole make preparation difficult. Tandler (1962b) suggests the poor fixation usually achieved in *Acetabularia* may be the result of the rapid release of vacuolar oxalic acid into the cytoplasm when the plant is introduced into osmium solutions. The osmium being slow to penetrate cannot stabilize cytoplasmic structures before the oxalic acid has damaged or destroyed the membranous components. The use of phosphate buffers seems to reduce this latter problem somewhat.

Vacuole and Vacuolar Membrane

An examination of the cytoplasm of *Acetabularia* reveals that the relatively thin layer of cytoplasm adjacent to the cell wall is not uniformly cytoplasmic, but appears invaginated by portions of vacuole (Fig. 1). Thus, in a given section, strands of cytoplasm often seem unattached to other strands, but adjacent sections show them to be part of a continuum. The cytoplasm is separated from the vacuole by a clearly defined vacuolar membrane (tonoplast), consisting of the usual three layers (two electron-opaque bands separated by a less dense region). It is of some

---

Fig. 1. Section through the cytoplasm of *Acetabularia* illustrating the major cytoplasmic components. Note the vacuole (V) invaginates into the cytoplasmic region and the vacuolar membrane (Vm) does not touch directly on the chloroplast (Ch) envelope. A mitochondrion (M) appears surrounded by anastomosing tubular elements (endoplasmic reticulum). A branched tubule may be seen at the arrow. Li, lipid droplet. Magnification: $\times$ 33,000.

a
b

interest that the chloroplast envelope never touches on the vacuolar membrane but always leaves a narrow clear space separating the two membrane systems. The large lipid droplets, on the other hand, may come directly into contact with the vacuolar membrane.

Crystals

Within the vacuole there appears little evidence for preservable fine structure except for the occurrence of rather large crystals (Fig. 2b). These crystals described as high in bound indoles (Tandler, 1962a) have major spacings in one plane about 240 A and finer spacings separating these into approximately 120 A regions (Fig. 2a). The crystals are easily visible with the light microscope and sometimes appear in older cells in great abundance. Their significance in the intact cell is unknown. Under the culture conditions of these experiments the crystals tend to collect in the vacuole along the margin of the cytoplasm.

Chloroplasts

The chloroplasts consist of a limiting double membrane or envelope enclosing an internal matrix (Sager, 1959). This matrix is occupied in part by a series of roughly parallel discs or plates, characteristic "globuli," and frequently, starch grains. The discs may run parallel to one another throughout the length of the chloroplast (Fig. 3a), or the discs may be shortened and locally stacked to form granal regions (Fig. 3b). There appears to be little consistency as to whether a chloroplast will lack or possess grana, and apparently in a given cell chloroplasts of both types may be present. An older report (Mangenot, cited in Fritsch, 1945) that chloroplasts of *Acetabularia* probably lack starch is untenable, at least under the present culture conditions.

Mitochondria

Mitochondria are large and numerous with platelike cristae (often swollen at their tips) extending approximately half the width of the mitochondrion (Figs. 1 and 4). Small globuli may be seen in some sections of mitochondria, and these globuli appear similar to, but smaller than, the chloroplast globuli. There appears to be no special orientation

---

Fig. 2. (a) Portion of a vacuolar crystal cut in a plane to show 240 A spacings (between arrows). These spacings are bisected by less electron-opaque points. Magnification: × 58,000. (b) An entire crystal as seen suspended in the vacuole. Several different planes of regular spacings may be seen at the lower portion of the crystal. Note the margin of the crystal suggests that materials are being incorporated into or dissolved from its surface. Fixed in osmium vapors for ½ hr. Magnification: × 23,000.

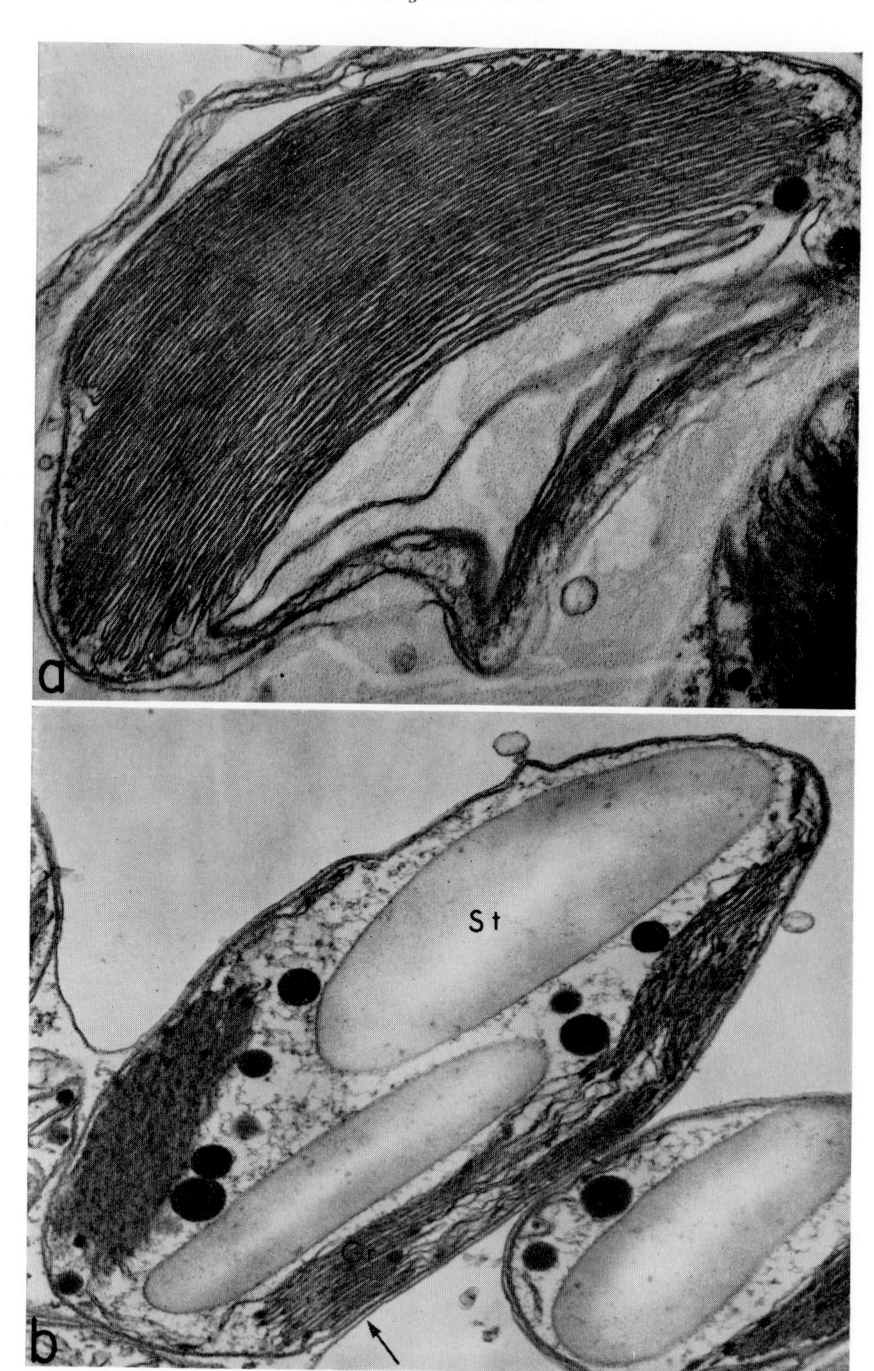
a
St
b

of mitochondria to a particular portion of the cytoplasm, and it is assumed they are randomly scattered.

Ground Substance (Hyaloplasm)

In the ground cytoplasm can be found two kinds of obvious structural organization: (1) Groups of ribonucleoprotein particles which seem unattached to a membrane system, and (2) an anastomosing system of tubules which possibly represent elements of a smooth endoplasmic reticulum (Figs. 1 and 4). These latter tubules are limited by a well-defined membrane which has about the same width as the plasma membrane and vacuolar membrane. The tubules vary considerably in diameter perhaps as a result of swelling or shrinkage during fixation procedures. Some areas of the cytoplasm seem crowded with these tubules whereas in other regions, profiles are widely scattered.

Lipid Droplets

Large, electron dense droplets which are spherical in outline and homogeneous in content are found throughout the cytoplasm. These droplets may have local "erosions" suggesting that material is being removed from their surface. They may represent the metaphosphate granules of Stich (1953).

Cell Wall

The thick wall of *Acetabularia* is composed of two structurally distinct regions. The inner, larger portion seems granular or possibly coarsely fibrous in nature, whereas the outer region appears homogeneous or possibly composed of finer and more closely packed fibrils (Fig. 5). In the innermost region of the wall there is an alternation of orientated layers, but this alternation is lost in older portions of the wall. The two-layered appearance of the wall agrees with Tandler's (1962a) conception of a wall composed of two regions, the outer of which is thin and stains metachromatically with toluidine blue. Frei and Preston (1961) find that the principal constituent of wall extracts of *Acetabularia* is mannose and not glucose (though there may be some glucose residues present) and conclude that cellulose is not the chief structural polysac-

---

Fig. 3. (a) A chloroplast in which the discs are not locally stacked into granal regions, but are closely packed and traverse the width of the chloroplast. Magnification: × 40,000. (b) A chloroplast containing granal stacks (Gr), starch grains (St), and electron opaque globuli. Note the vacuolar membrane (arrow) is apparently continuous around the chloroplast, but is separated from the chloroplast envelope by a narrow space. Another strand of cytoplasm occupied largely by another chloroplast may be seen in the lower right-hand corner. Magnification: × 37,000.

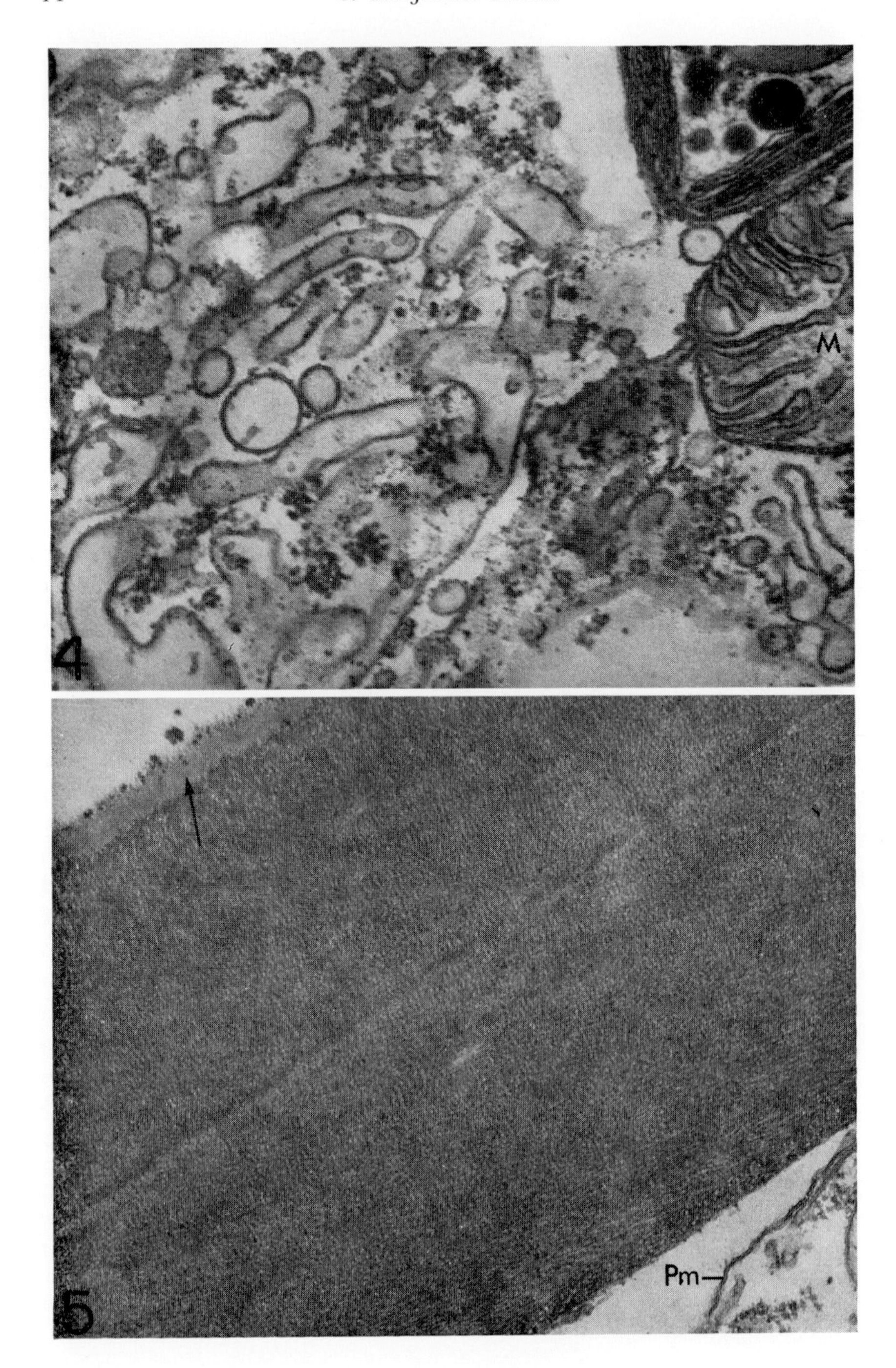
4
M
5
Pm

charide in the wall. The encrusting calcium carbonate found in walls of plants taken directly from the sea, but not found in the walls of plants grown in culture, presumably would be located within the outer thin mucilagenous layer (Fritsch, 1945).

Discussion

Although fibers have been identified in electron micrographs of several forms of moving cytoplasm (e.g., de Petris *et al.*, 1962; Wohlfarth-Botterman, 1962), their existence has not been confirmed by electron microscopy in the moving cytoplasm of green plants. Even the fibers seen in dark field microscopy in *Chara* (Jarosch, 1956) and *Nitella* (Kamiya, 1959) have not as yet been clarified by electron microscopy. Judging from the results of physiological and biochemical investigations (Takata, 1958, for example) it would seem likely that contractile actomyosin-like fibers do exist in green plants, but apparently present methods of preparation for electron microscopy do not preserve their structure. Or, possibly these fibers are present in such a diffuse form that their identification in electron micrographs will be most difficult. However, other cytoplasmic structure can be satisfactorily preserved and it is of interest to examine their possible role in streaming phenomena.

The anastomosing tubules seen in abundance in many regions of the cytoplasm of *Acetabularia* offer at least three alternatives for their possible function: (1) They may be elements of smooth endoplasmic reticulum somehow involved in cell wall deposition. Porter and Machado (1960, 1961) have shown the endoplasmic reticulum may be associated with the developing walls of higher plants, and the present writer's own unpublished examination of pea root cells has demonstrated that clear continuities may exist between the smooth endoplasmic reticulum and the expanding cell wall. However, in *Acetabularia* the anastomosing tubules are not confined to the outer margin of the cytoplasm where wall deposition is most active, nor have direct continuities of the tubular elements with the wall yet been established. (2) The quantity of anastomosing tubules in a given region may reflect the physical state of the

---

FIG. 4. A portion of the hyaloplasm showing the anastomosing tubules sectioned in various planes. Ribonucleoprotein particles show no affinity for the tubules. M, mitochondrion. Magnification: × 48,000.

FIG. 5. Part of the wall of a 2-cm *Acetabularia* showing two structurally distinct regions. The larger portion of the wall is coarsely fibrous in texture whereas the smaller portion (arrow) is homogeneous or composed of very fine fibers. Pm, plasma membrane. Magnification: × 40,000.

cytoplasm of that region. Thus large numbers of tubules may indicate a gel state whereas fewer tubules (or their absence) may characterize a sol condition. The extremely crowded appearance of the cytoplasmic tubules in some areas (Fig. 1, for example) suggests that organelles within this area are not free to move about with great ease. In other regions chloroplasts and mitochondria seem suspended in cytoplasm almost devoid of background structure, and would thus appear capable of unimpeded movement. Since it is known that the sol-gel interface (Kamiya and Kuroda, 1956) or sol–gel transformations (Abé, 1962) may be involved in the driving mechanisms for protoplasmic streaming, it would seem of prime importance to investigate further the structural basis for the sol and gel conditions. Unfortunately it has not been possible thus far to correlate positively the tracts of streaming cytoplasm seen in the living cell with a definite region of the cytoplasm after the cell has been fixed and embedded for electron microscopy. (3) The possibility that the tubules may also operate as a conveyor or communications network connecting various regions of the cytoplasm should also be considered. The occurrence of a system of anastomosing tubules in the sarcoplasm of striated muscle is well documented (cf. Porter, 1961), and this sarcoplasmic reticulum most probably serves to coordinate different regions of muscle fibers (either to relax the muscle or in some other capacity). Since contractile fibers may also be present in the moving cytoplasm of *Acetabularia*, it would not be too surprising to find a coordinating system provided to produce the uniform, unidirectional motion resulting from the action of many individual fibers.

Whatever their function it is clear that the anastomosing tubules (endoplasmic reticulum) from a major structural component of the hyaloplasm of *Acetabularia*. It is of some interest that tubular vesicles have also been seen in electron micrographs of the endoplasm of ameba (Schneider and Wohlfarth-Botterman, 1959). Obviously only continuing investigations of the fine structure of streaming cytoplasm will determine the universality of this kind of structure, and ultimately, its possible relation to the streaming process.

Summary

The fine structure of *Acetabularia* has been examined after osmium fixation and Epon embedding. Many of the cellular components seem similar to structures described elsewhere in other plant tissue. However, the presence of numerous anastomosing tubules in the ground cytoplasm (hyaloplasm) are noted, and theories for their possible function especially in relation to streaming are discussed. Large crystals in the vacuole are examined and are found to have regular spacings. Chloroplasts may or may not possess grana and often contain starch.

Acknowledgment

This work was begun during tenure of a post-doctoral fellowship from training grant USPHS 26707 to K. R. Porter. The author wishes to express his appreciation to Dr. Porter for suggesting this problem and for the use of his facilities in its initiation.

References

Abé, T. H. (1962). *Cytologia* **27**, 111-139.
de Petris, S., Karlsbad, G., and Pernis, B. (1962). *J. Ultrastruct. Res.* **7**, 39-55.
Føyn, B. (1934). *Arch. Protistenk.* **83**, 1-56.
Frei, E., and Preston, R. D. (1961). *Nature* **192**, 939-943.
Fritsch, F. E. (1945). "The Structure and Reproduction of the Algae," Vol. 2. Cambridge Univ. Press, London.
Hämmerling, J. (1953). *Intern. Rev. Cytol.* **2**, 475-498.
Jarosch, R. (1956). *Phyton (Buenos Aires)* **6**, 87-107.
Kamiya, N., and Kuroda, K. (1956). *Botan. Mag. (Tokyo)* **69**, 544-554.
Kamiya, N. (1959). *in* "Protoplasmatologia - Handbuch der Protoplasmaforschung" (L. V. Heilbrunn and F. Weber, eds.), Vol. VII, 3a. Springer, Berlin.
Luft, J. H. (1961). *J. Biophys. Biochem. Cytol.* **9**, 409-414.
Millonig, G. (1961a). *J. Biophys. Biochem. Cytol.* **11**, 736-739.
Millonig, G. (1961b). *J. Appl. Phys.* **32**, 1637.
Porter, K. R. (1961). *J. Cell Biol.* **10**, 219-226.
Porter, K. R., and Machado, R. D. (1960). *J. Biophys. Biochem. Cytol.* **7**, 167, 180.
Porter, K. R., and Machado, R. D. (1961). *Proc. European Regional Conf. Electron Microscopy, Delft, 1960*, **2**, 754-758.
Sager, R. (1959). *Brookhaven Symp. Biol.* **11**, 101.
Schneider, L., and Wohlfarth-Bottermann, K. E. (1959). *Protoplasma* **51**, 377-389.
Stich, H. (1953). *Z. Naturforsch.* **8b**, 36.
Takata, M. (1958). *Kagaku (Tokyo)* **28**, 142.
Tandler, C. J. (1962a). *Planta* **59**, 91-107.
Tandler, C. J. (1962b). *Naturwissenschaften* **49**, 112.
Wohlfarth-Bottermann, K. E. (1962). *Protoplasma* **54**, 514-539.

Discussion

Chairman Thimann: Now we are ready for discussion or questions.

Dr. Rebhun: Would it be possible to cut strips from the wall in *Acetabularia* before you fix the cell and, say, fix just part of this wall, with its associated ectoplasm?

Dr. Bouck: I tried this, but obtained extremely poor fixation. If successful, one might thus be able to reveal the structure of the endoplasm, without having materials from the vacuole affect fixation.

Chairman Thimann: The problem occurs not only with fixation but also with enzyme preparations, for example, from very acid cells such as the cells of leaves of *Kalanchoë* where there is a high concentration of malate. It has been found in these cases that one can get protection against the large amount of organic acid there by previously exposing the leaf tissue to ammonia. Ammonia penetrates very rapidly into the cells. Then, after a few minutes in ammonia, depending on the thickness of the cuticle, you can make your preparation of enzymes; the ammonia hasn't had time to be metabolized and the free acid is neutralized. I wonder if this is worthwhile trying in *Acetabularia*.

DR. BOUCK: It seems like a very good idea.

DR. ROTH: First of all, I would like to compliment you on your very nice work. I tried diligently for several weeks to fix *Acetabularia* and gave up. It is a very difficult cell with which to work; nevertheless, very interesting.

In the amebae there are areas seen in the cytoplasm which have no ribosomes, no vesicles, no mitochondria, but, nevertheless, look very compact. These are the areas that are frequently called "ectoplasm." Do you see any such concentrations of very finely filamentous material?

DR. BOUCK: Filaments, of course, are generally elusive in plant material. Spindle fibers are extremely hard to demonstrate by electron microscopy. I have not as yet seen fibers of any kind in the cytoplasm of *Acetabularia*.

DR. ALLEN: I would like to comment that it would be extremely useful for electron microscopists to study the relationship between the deployment of endoplasmic reticulum and regions of active cytoplasmic streaming in plant cells, in the light of the kinds of mechanisms of streaming Kavanau has recently proposed.

CHAIRMAN THIMANN: According to electron microscopists it (the ER) is less marked in regions of vigorous streaming.

DR. MAHLBERG: Were you able to see the nucleus in any of your sections?

DR. BOUCK: Yes. The nucleus has a typically pored membrane; but the contents were not preserved.

DR. WOLPERT: I presume you fixed your material in the cold.

DR. BOUCK: No.

DR. REBHUN: If I remember correctly, it was Dr. Takata who did work with glycerated *Acetabularia* using adenosine triphosphate to stimulate temporary streaming motion in this glycerated material. Assuming these are good repeatable experiments, one wonders if it wouldn't be worthwhile to do electron microscopy on this glycerated material. Have you tried this?

DR. BOUCK: No. I think it would be a good project.

DR. MAHLBERG: Is the tonoplast a "typical" unit membrane?

DR. BOUCK: Yes.

Role of the Cortical Gel Layer in Cytoplasmic Streaming

TOSHIO HAYASHI

Biological Institute, College of General Education, Tokyo University, Komaba, Tokyo, Japan

Although several types of cytoplasmic streaming are known in plant cells, rotational streaming will be dealt with here for the reason that it is simplest in dynamic organization.

The materials used in our experiments have been the internodal and rhizoid cells of the following Characeae: *Chara Braunii, Chara corallina,* and *Nitella flexilis.* These cells, some of which are truly enormous, measure from a few millimeters to 10 cm in length and several tenths of a millimeter in diameter. They exhibit vigorous and typical rotational streaming and are suitable for many kinds of experimental approaches to the cytoplasmic streaming problem.

First, we should like to describe briefly the general organization of a typical characeous cell, *C. Braunii.* Figure 1 shows a series of line drawings, and Fig. 1a shows the general location of the leaf, internodal, and rhizoid (or root) cells of the whole plant. The second drawing (Fig. 1b) shows a close-up of one end of an internodal cell in which the chloroplasts are arranged in spiral rows with a gentle slope. The streaming cytoplasm follows the path of these rows in a continuous belt around the cell. The cytoplasm is quiescent only at the indifferent or neutral zones marked by a lack of chloroplasts. Figure 1c, a longitudinal section, shows that the streaming endoplasm is bounded externally by a thin cortical layer in which most of the chloroplasts are found and internally by the tonoplast or vacuolar membrane. The cell wall and vacuole are nonliving parts of the cell. Between the two opposing streams there are two narrow zones, the so-called "indifferent zones," which are marked by a lack of chloroplasts. The cytoplasm is quiescent only at these zones. Each of the two opposed streams has a uniform speed over the circumference covering about 165° (Kamiya, 1962). Only in the narrow zones close to the indifferent lines the endoplasm streams with smaller rates.[1] The last drawing (Fig. 1d) is of a rhizoid cell, which is similar to internodal cells except

[1] The cross section of a *Nitella* or *Chara* internodal cell wall is circular. The contour of the tonoplast is also roughly circular because of local difference in endoplasmic thickness, but the endoplasm is thinner in the region of the indifferent zone.

for a lack of chloroplasts; the endoplasm streams as vigorously as that in the internodal cell.

The velocity of cytoplasmic streaming in adult internodal cells is very high as compared with that in other plant cells, and is kept almost constant under constant environmental conditions. It is usually 60 μ/sec at 20°C. But in young apical internodal cells, the cytoplasm streams much slower.

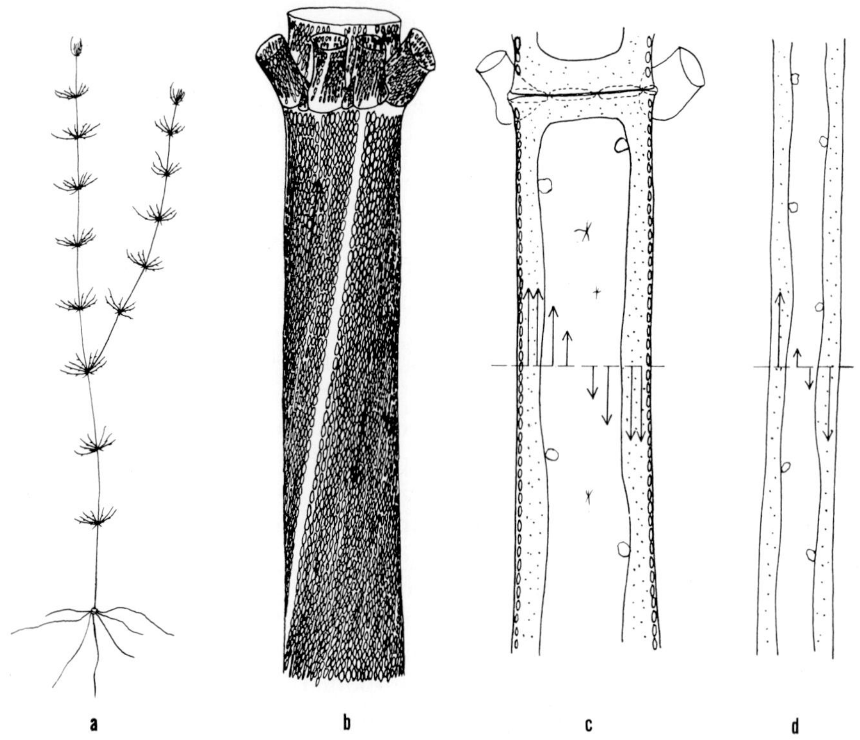

FIG. 1. (a) The whole plant of *Chara Braunii*. The internodal cells are about 0.5–2 cm long. (b) The closeup of the upper end of an internodal cell. (c) A longitudinal section of (b). (d) A longitudinal section of a part of the rhizoid. Arrows indicate the direction and the relative rate of the streaming in the cell.

Streaming velocity increases rapidly as the temperature rises. The acceleration of the streaming was found to be produced mainly by the decrease in viscosity of the cytoplasm and not by the increase in the motive force behind the streaming (Hayashi, 1960). These results substantially support the view of Lambers (1925) and are also in accord with the results obtained from myxomycete plasmodia (Kamiya, 1953).

There are two current theories about the site of motive force generation of rotational streaming in plant cells; one is to attribute it to the streaming cytoplasm itself, whereas the other takes the view that the motive force is generated at the boundary between the streaming endoplasm and the cortical gel plasm (cf. Kamiya, 1959). I should like to describe some recent experiments which may help establish the locus of motive force generation.

Exfoliation of Chloroplasts by Means of Centrifugation

A direct test for whether the cortical gel layer is of fundamental importance for the mechanism of the cytoplasmic streaming is to injure locally or totally destroy the cortical gel layer. In this case streaming should be impaired, modified, or prevented. On the other hand, if the whole mechanism of streaming should be located in the streaming endoplasm itself, then the endoplasm might be able to move after destruction of the cortical gel layer. It was from this viewpoint that the following experiment was conducted.

One way of damaging the cortical layer is by centrifugation. For these experiments we have used a home-made electrically driven centrifuge microscope of very simple design with a rotor only 4 cm in diameter (Hayashi, 1957a). With this it is possible to subject our material to accelerations up to 4000 *g* within 4 sec, yet stop the rotor quickly when the desired degree of cortical injury was attained.

The effects of centrifugation are as follows. First the endoplasm is displaced centrifugally and the vacuole centripetally without apparent injury to the cortical layer. At higher accelerations, from 400 to 1000 *g* depending on the species and season, the chloroplasts are exfoliated or torn off the cortex, sometimes singly, but more often in longitudinal rows. A substantial part of the cortical gel accompanies them.

Figure 2 indicates three examples of cells which suffered exfoliation of chloroplasts to an extent suitable for observation. The white area, in the cell in this figure is the site where the chloroplasts have been torn off and have fallen to the centrifugal end of the cell.

In Fig. 3 the cytoplasm is transported continuously from the left-hand side of the cell along the peninsula-like area where the chloroplasts are embedded normally, but it stagnates and piles up at the white area. Although the endoplasm can stream normally when in contact with the inner surface of chloroplast-containing, cortical, gel layer, the very same endoplasm immediately loses the ability to stream when it reaches the cortical layer from which the chloroplasts have been removed. The arrows in Fig. 3 show the direction and speed of the streaming at different loci which were obtained through cinematographic analysis.

We can infer that the cortical gel layer at the white area must have received rather severe injury, as is shown schematically in Fig. 4. Figure 4 shows the normal state of the cortical gel layer. Immediately after the centrifugal treatment there was no sign of any movement of the cytoplasm

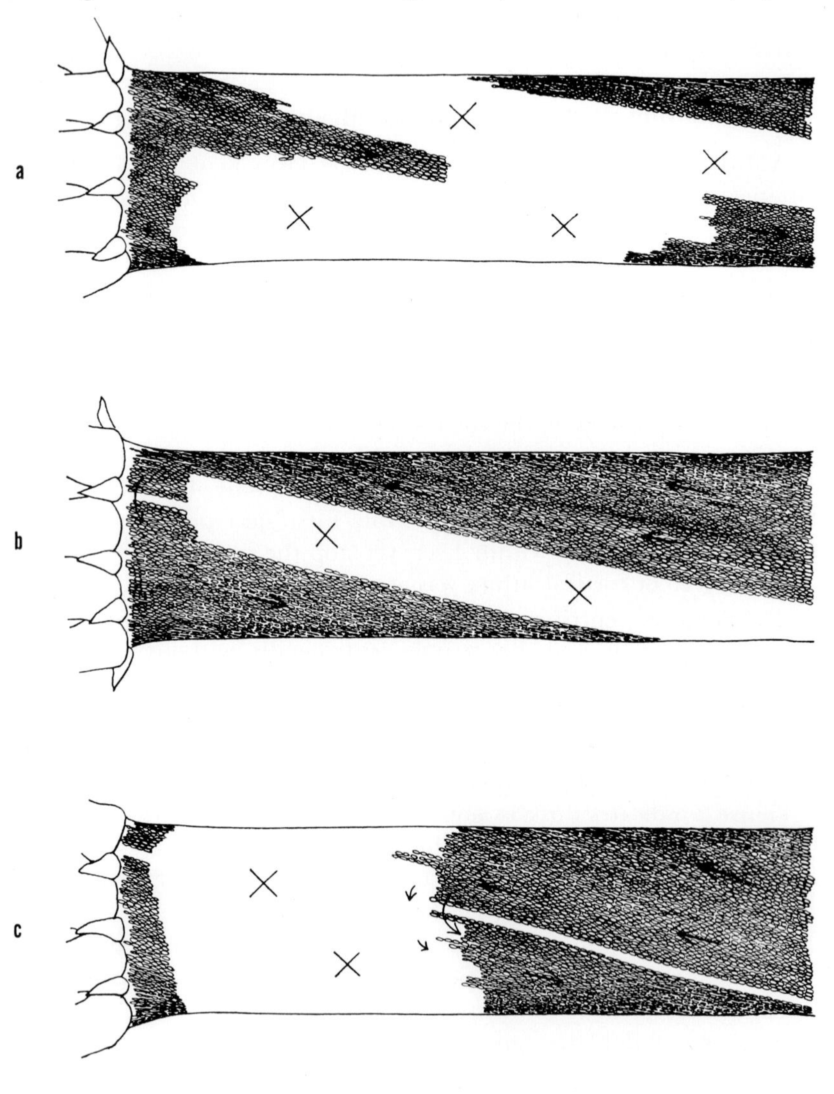

FIG. 2. Cells immediately after centrifugation at 400 *g* for 3 min (in winter). No cytoplasmic streaming is observable at the white area marked with X where the chloroplasts have been exfoliated, whereas normal streaming continues at the area where the chloroplasts have remained attached at their original positions. Arrows indicate the direction of the cytoplasmic streaming.

except for the rotation of the exfoliated chloroplasts and the normal Brownian motion of small particles, though the streaming continued at the part where the chloroplasts remained attached at their original positions (Fig. 4b). When the treatment of centrifugation is strong enough, the white area where the chloroplasts are exfoliated becomes too large for the cell to survive. This degree of cortical injury was achieved after about 3 min of centrifugation at an acceleration of 400 *g* in winter.

At any rate, the previously mentioned fact is a further confirmation of the findings of previous authors that, when the plasmagel layer of *Nitella* or *Chara* cells is injured locally by a needle or anesthetics, the

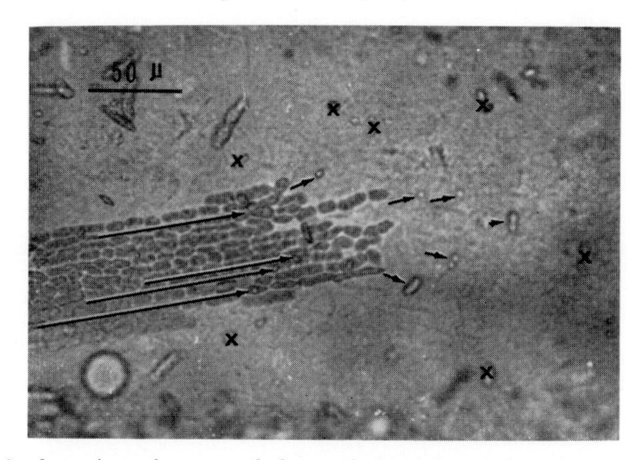

Fig. 3. Surface view of a part of the cortical gel layer of an internodal cell after centrifugation. Chloroplast layers are mostly exfoliated except at a peninsula-like area of the left. Arrows show the direction and speed of flow at different loci. The length of each arrow shows the distance covered by a particle in 5 sec.

endoplasmic flow no longer occurs at that spot but goes around it (Nichols, 1930; Linsbauer, 1929; Péterfi and Yamaha, 1931). These facts indicate the importance of the role played by the cortical gel layer in the mechanism of cytoplasmic streaming.

Recovery of Streaming

The cytoplasm at the white area where the chloroplasts have been exfoliated begins to move within 5 to 10 hr after the treatment, though the movement is feeble compared with that in other areas where chloroplasts have not come off (Fig. 4c). Prolonged observations on these injured cells showed first that the initially feeble movement became progressively stronger as time passed. Over a period of days the chloroplasts that had fallen to the centrifugal end of the cell returned to and settled down at the injured white area. Along with restitution of the chloroplasts to the white area, the cytoplasmic streaming at that area is recovered (Fig. 4d). The direction of flow in this case was always the same as before.

Although the cortical gel layer is indicated schematically in Fig. 4, it is difficult to demonstrate the layer microscopically in normal cells. This

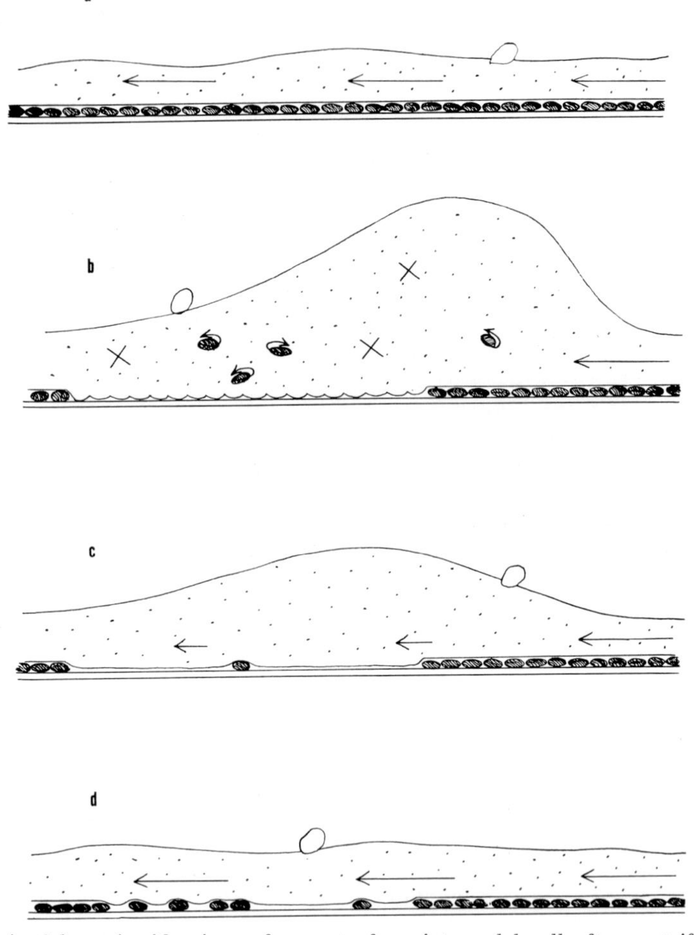

FIG. 4. Schematic side views of a part of an internodal cell after centrifugation showing the recovery of cytoplasmic streaming and restitution of the chloroplasts to the white area. (a) Before centrifugation; (b) 5–10 hr after centrifugation; (c) 3–5 days later; (d) about 2 weeks after the treatment.

is because the layer is so thin and attached so closely to the cell wall. Moreover, in the case of the internodal cell, the chloroplasts directly interfere with observations of the cortical gel layer.

From the evidence presented so far, one might conclude that either the cortical layer or the chloroplasts within it was required for streaming. However, from the fact that the rhizoid cells, which are devoid of chloroplasts, stream, we are inclined to believe that it is the cortical gel layer itself, and not the chloroplasts within it, which is essential to streaming.

Streaming in the Plasmolyzed Cell

Since the cortical layer is so difficult to see in internodal cells because of the chloroplasts, the ideal place to study the cortex is in rhizoid cells, which as we said, lack chloroplasts. Here it is possible to ascertain the presence or absence of this layer by plasmolysis (Hayashi, 1960). When the rhizoid cell of *C. corallina* is plasmolyzed by a 0.2 *M* solution of calcium chloride (Hayashi and Kamitsubo, 1959) the endoplasm streams normally along the inner side of the cortical layer which is microscopically visible and about 1 μ thick. On the other hand, at the site where no cortical gel as such is observable, the endoplasm does not stream.

In the protoplast plasmolyzed with calcium chloride it occasionally happens that the gel layer is found not at the periphery but along the tonoplast. In such a case the cytoplasmic movement is most active in the immediate proximity to the tonoplast and becomes weaker the greater the distance from the vacuole. In normal cells, however, the endoplasm streams, together with the tonoplast, with almost the same velocity throughout its thickness (Kamiya and Kuroda, 1956).

All the above facts show clearly that the boundary between organized cytoplasmic gel layer and the endoplasm plays an essential role in generating the motive force of the cytoplasmic streaming in these cells.

Behavior of the Endoplasm under Moderate Centrifugation

Next we shall talk briefly about the behavior of the endoplasm when the internodal cell is centrifuged at low accelerations, at which chloroplasts do not come off from the cortical gel layer. Instead, at these low accelerations the endoplasm accumulates at the centrifugal end of the cell. Most of the chloroplasts usually remain in position when the centrifugal acceleration is below 400 *g*. With the aid of a centrifuge microscope, we can observe directly the displacement of the endoplasm while the cell is being centrifuged (Fig. 5). When the centrifugal acceleration is decreased below 40 *g*, the sedimented endoplasm can go back toward the centripetal end in the form shown at the bottom of Fig. 5. The restitutional streaming toward the centripetal end of the cell always takes place only along one side of the cylindrical cell wall, the side along which the endoplasm flowed in that same direction before the cell was centrifuged. This fact was established by Breckheimer-Beyrich (1954).

Kamiya and Kuroda (1958), who calculated the motive force responsible for rotational streaming from measurements of the balance-acceleration required to stop centripetal streaming, found that the thinner the endoplasmic layer, the greater the centrifugal acceleration required to stop streaming. This constitutes further support for the conclusion that the site of the motive force of rotational streaming is not distributed throughout the streaming endoplasm but localized at the boundary be-

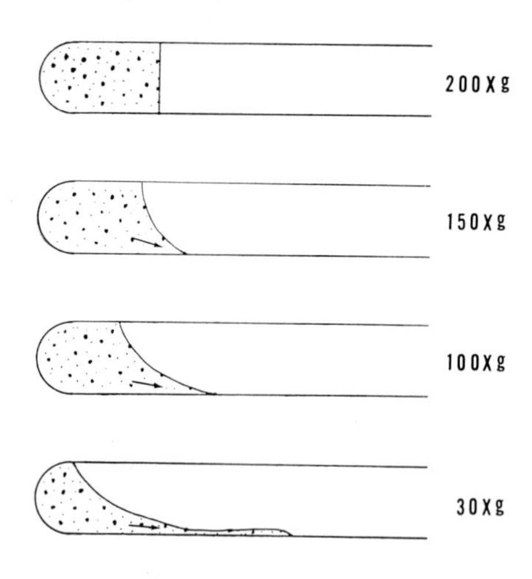

FIG. 5. Schematic profiles of the endoplasm collected at the centrifugal end of the internodal cell under various accelerations. At 200 *g*, the endoplasm accumulates at the centrifugal end of the cell, where no cytoplasmic streaming toward the centripetal end is visible. Below 200 *g* endoplasm starts streaming against the centrifugal force in an extremely thin layer. When the centrifugal acceleration is decreased down to 40 *g*, the restitutional streaming of endoplasm becomes more conspicuous.

tween the outer edge of the endoplasm and the inner surface of the cortical gel layer.

Local Chemical Fixation

At this point it would be pertinent to touch upon a recent experiment (Hayashi, 1960) in which the cell was fixed locally by bringing a micropipette filled with various fixatives in contact with the cell. Under such a situation, cytoplasmic streaming could not be observed for several hours at the spot where the cortical gel was fixed. The chloroplasts at that spot fell off together with the endoplasm into the vacuole, while the normal streaming kept going elsewhere. By plasmolyzing the cell, it was clearly ascertained that the cortical gel layer disappeared soon after the treatment but appeared again several hours later followed by the recovery of the cytoplasmic streaming there. This may also be interpreted as evidence supporting the significance of the cortical gel layer in the rotational streaming typical of this plant.

Behavior in a Pathological Cell

To conclude, I would like to add one more observation on an interesting abnormality of streaming in internodal cells of *C. corallina* infected by an unknown species of the oribatid mite (Hayashi, 1957b). No rotational streaming can take place in these cells, but the chloroplasts rotate independently around their own axes (Fig. 6b). This phenomenon may

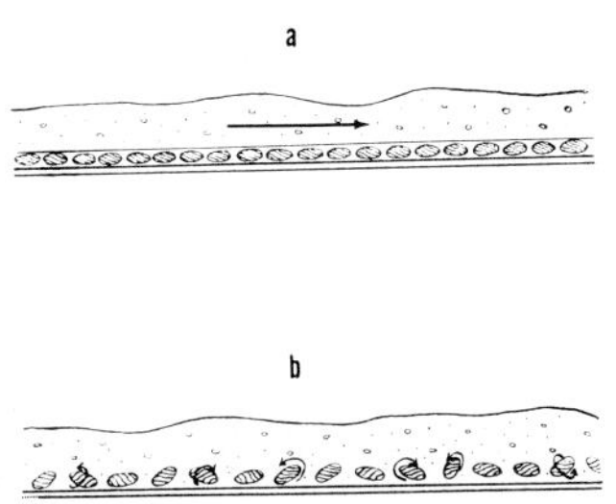

FIG. 6. Schematic representation of longitudinal sections of a normal (a) and an abnormal cortical gel layer of the internodal cell infected by an oribatid mite (b).

imply that the cortical gel layer of the cell in question can no longer hold the chloroplasts firmly enough and such an abnormal gel layer cannot drive the endoplasm along that surface.

Summary

To sum up, we can say that rotational streaming of cytoplasm found in the cells of the Characeae stops as soon as the cortical gel layer is destroyed by any means whatever, and that streaming is resumed as the cortical gel layer is reformed. The motive force for rotational streaming is an active shearing force generated at the boundary between the cortical gel and endoplasm. What comes into focus concerning the mechanism of cytoplasmic streaming in this case is, therefore, the key problem of how the active shearing force is produced at this boundary.

REFERENCES

Breckheimer-Beyrich, H. (1954). *Ber. Deut. Botan. Ges.* **67**, 86.
Hayashi, T. (1957a). *Botan. Mag. (Tokyo)* **70**, 168.
Hayashi, T. (1957b). *J. Japan. Botan.* **37**, 82.
Hayashi, T. (1960). *Sci. Papers Coll. Gen. Educ. Univ. Tokyo* **10**, 245.
Hayashi, T., and Kamitsubo, E. (1959). *Botan. Mag. (Tokyo)* **72**, 309.
Kamiya, N. (1953). *Ann. Rep. Sci. Works Fac. Sci. Osaka Univ.* **1**, 53.

Kamiya, N. (1959). *Protoplasmatologia* **8**, 3a, 1.
Kamiya, N. (1962). *In* "Handbuch der Pflanzenphysiologie" (W. Ruhland, ed.), Vol. XVII, Part 2, p. 980. Springer, Berlin.
Kamiya, N., and Kuroda, K. (1956). *Botan. Mag. (Tokyo)* **69**, 544.
Kamiya, N., and Kuroda, K. (1958). *Protoplasma* **50**, 144.
Lambers, M. H. R. (1925). *Proc. Koninkl. Ned. Akad. Wetenschap.* **28**, 340.
Linsbauer, K. (1929). *Protoplasma* **5**, 563.
Nichols, S. P. (1930). *Bull. Torrey Botan. Club* **57**, 153.
Péterfi, T., and Yamaha, G. (1931). *Protoplasma* **12**, 279.

Discussion

Chairman Thimann: May I ask, sir, how it is that the chloroplasts get back into the white areas? There is no streaming to carry them there. Do they get back under their own motility?

Dr. Toshio Hayashi: The chloroplasts come back very slowly. The endoplasm at the white area does not stream, but is somehow "pushed" into the white area.

Dr. Zimmerman: Dr. Hayashi, is there any relationship to the thickness of the cortical gel layer and the rate of protoplasmic streaming?

Dr. Toshio Hayashi: I cannot give you a definite answer for your question, since the cortical gel layer is so thin that it is difficult to determine its exact thickness with an optical microscope. Probably what is important for the streaming is an organized inner surface of the cortical gel layer and not its thickness.

Dr. Burgers: Would you repeat your conclusion? You said the movement is localized at the boundary between the cortical gel layer—and what followed?

Dr. Toshio Hayashi: The motive force for rotational streaming is generated at the boundary between the cortical gel and endoplasm.

Dr. Marsland: I would like to ask how you differentiate between the cortical gel and the endoplasm. How do you find the difference?

Dr. Toshio Hayashi: The cortical gel is stationary, and the endoplasm is moving.

Dr. Jacobs: When you exfoliate the chloroplasts with centrifugal acceleration, do some chloroplasts "diffuse" in and then divide in the exfoliated area or are they all moved in from other areas?

Dr. Toshio Hayashi: They grow, but I am not quite certain about whether they divide.

Dr. Marsland: I think these observations fit very well with ones I made many years ago on *Elodea,* which showed that the application of high pressure caused solation, presumably of this gel layer. The rate of movement dropped off in proportion to the degree of solation. This would also certainly indicate the importance of the gel structure.

Dr. Ling: First, in one of the movies you showed, when the chloroplasts were more or less randomly distributed, the cytoplasm seemed to accelerate along the path where there was a string of these chloroplasts. Is this observation correct? Second, if this is the case, is it possible that the chloroplasts may still add an effect onto the cortical cytoplasm in spite of the fact that in rhizoid cells lacking chloroplasts streaming still takes place.

Dr. Toshio Hayashi: Cytoplasm sometimes streams faster along the path formed by aligned chloroplasts. This, I suppose, is probably because these chloroplasts are connected to one another by cytoplasmic fibrils. The fibrils are probably laid on the inner surface of the chloroplast rows in the normal cell and are very likely responsible for driving the endoplasm. What is important for the mechanism of endoplasmic streaming is not the chloroplast per se but these fibrils adhering to them.

CHAIRMAN THIMANN: We are going to hear this from Dr. Kuroda. I understand there is only rotation and not translation.

DR. ALLEN: Single chloroplasts can, however, be translated under some circumstances.

CHAIRMAN THIMANN: That is important.

DR. GRIFFIN: In the rhizoid can you detect a difference between the cortical gel and the endoplasm?

DR. TOSHIO HAYASHI: No, I cannot, with any certainty, in the normal rhizoid, but if it is plasmolyzed we can detect a thin cortical layer which has a refractive index higher than that of the endoplasm.

DR. ANDREW G. SZENT-GYÖRGYI: What is the difference between the cortical gel and the endoplasm? Is it a matter of aggregation? Are they composed of different materials? Is there an equilibrium between the endoplasm and cortical gel, or are they both structurally stable?

DR. TOSHIO HAYASHI: They are structurally stable and different in function, but probably there is a transformation of material from one to the other.

DR. INOUÉ: I would like to put the same question in a little more specific form. What happens to the cortical gel layer when you centrifuge the cell and the cortical gel is peeled off? I gather that is what you are saying does happen?

DR. TOSHIO HAYASHI: When chloroplasts are exfoliated, there still remains a damaged cortical layer. Once cortical gel layer is injured, however, it is no longer capable of producing the streaming. What is essential, I believe, is the organized inner surface of the cortical gel layer. Whether it contains chloroplasts or not is immaterial.

DR. JAROSCH: We observe frequently that a piece of a chloroplast chain is detached from the cortical layer. At the moment this occurs, the piece of the chloroplast chain moves independently in the direction opposite to that of the endoplasmic streaming. What is inferred from this fact is that cortical gel structure responsible for producing the active shifting force is attached to the surface of the detached chloroplasts.

Behavior of Naked Cytoplasmic Drops Isolated from Plant Cells[1]

KIYOKO KURODA

Department of Biology, Faculty of Science, Osaka University, Osaka, Japan

A number of studies (Donné, 1838; Dutrochet and Brongniart, 1838; Velten, 1876; Stålberg, 1927; Valkanov, 1934; Yotsuyanagi, 1953a,b; Jarosch, 1956a,b, 1958, 1960; Kamiya and Kuroda, 1957b) have been made on the movement in fragments of cytoplasm isolated from cells of Characeae. Most previous workers have squeezed out the cell contents mechanically in order to obtain cytoplasmic drops *in vitro*. These cytoplasmic fragments naturally contain both streaming endoplasm and cortical gel plasm. In the light of our knowledge about the role played by the cortical gel plasm in the mechanism of cytoplasmic streaming, however, it would seem especially significant to study how the endoplasm behaves when it is isolated from the cell free from the cortical plasm.

Effusion of Endoplasm

Kamiya and Kuroda (1957a) have found that amputation of a portion of an internodal cell of *Nitella* is not fatal, provided the cell is somehow protected from collapsing after the cell is cut.

For this purpose, Kamiya and Kuroda (1957a) prepared a special glass chamber. As shown in Fig. 1, an internodal cell (c) is placed through a narrow groove in the wall of the chamber (A) so that a greater part of the cell remains inside and the smaller, lower end is left outside. The chamber is filled with a medium which is not only isotonic but also similar in its cationic constitution to the natural cell sap. It contains 0.08 *M* KNO_3, 0.05 *M* NaCl, and 0.004 *M* $Ca(NO_3)_2$. The portion of the cell sticking out of the chamber is put in a cuvette (B) filled with the same solution. In order to prevent the cell from collapsing when it is cut and open, we applied weak negative hydrostatic pressure of about 2 cm of mercury to the interior of the chamber. Under these conditions the cell was cut in the solution with sharp scissors at the site indicated with the dash–dot line. Owing to the

[1] This work has been supported partly by a grant for Fundamental Scientific Research from the Japanese Ministry of Education.

negative pressure in the chamber, the cell maintained its original form even when it was cut and left open.

After the cell is cut the endoplasm continues to stream normally along the cortical gel layer as far as the cell opening, where it comes

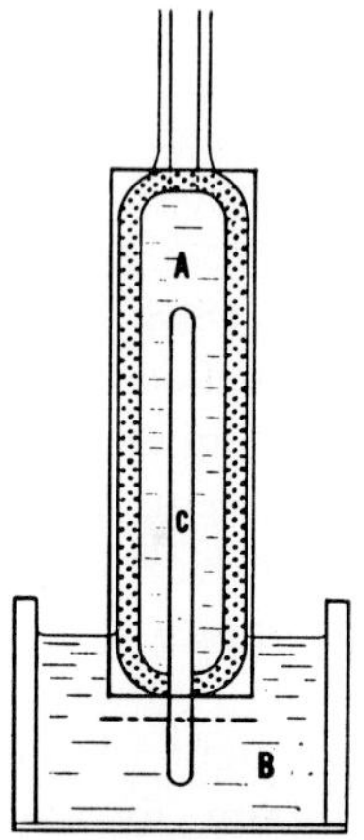

FIG. 1. Operation setup for amputating one end of a *Nitella* cell. While applying negative hydrostatic pressure in chamber (A), the cell (c) is cut in the cuvette (B) at the site indicated with the dash–dot line.

FIG. 2. Effusion of the endoplasm from the cell opening about 5 min after amputation of one end.

out gradually into the medium and falls down continuously onto the bottom of the cuvette without being dispersed into the medium (Fig. 2). The effusion continues as long as 20–40 min without interruption; when the supply of endoplasm becomes scanty, a continuous strand of endoplasm is broken to form intermittent pendent drops, each of which is detached from the cell by gravity, and falls onto the bottom where it forms a sessile drop.

The cytoplasm isolated in this manner consists of endoplasm alone and suffers little injury, if any. These endoplasmic drops can survive usually for 10–50 hr in the medium (Kamiya and Kuroda, 1957a).

Movement in an Isolated Endoplasmic Drop

When the endoplasm comes out of the cell in the form of a continuous strand, we notice that the entire strand falls down as a complete unit, there being no velocity gradient within it. What is noted here, however, is that the chloroplasts in the falling endoplasm rotate vigorously around their own axes, just as the few chloroplasts in the streaming endoplasm do in the intact cell.

When the endoplasm has settled on the bottom in the form of a sessile drop, both the nuclei and chloroplasts tend to sink toward the bottom, since they have a greater specific gravity than the cytoplasmic matrix. For this reason, the observation of their motion was best carried out with an inverted microscope with which the moving chloroplasts and nuclei can be seen most clearly in the same focal plane.

Endoplasm which streamed when in the intact cell no longer exhibits any mass streaming when isolated *in vitro* by this procedure. This does not mean, however, that there is no motion inside the drop. Besides Brownian motion, what is noticed here is vigorous rotation of chloroplasts and nuclei around their own axes (Kamiya and Kuroda, 1957b). The rate and direction of the rotation vary with each individual chloroplast and nucleus; the rate is usually within the range of 0.5–2 rps for the chloroplasts and 2 rpm for the nuclei.

An important question arises here as to why there is no mass streaming in the isolated endoplasm that was once streaming in the cell. Certainly injury caused by the operation of cutting the cell is not the cause of this, since chloroplasts and nuclei continue to rotate.

Through analysis of the intracellular velocity distribution of cytoplasmic streaming in the internodal cell of *Nitella,* we came to the conclusion that the motive force of the streaming is an active shifting force generated at the boundary between the endoplasm and cortical gel layer (Kamiya and Kuroda, 1956). This is to say that the endoplasmic sol alone is not capable of streaming. Evidence presented by Dr. Hayashi in this Symposium provides further support for this conclusion. The fact that there is no velocity gradient within the falling endoplasm also may be understood from this viewpoint. Thus, it is reasonable to believe that the absence of cortical gel is the reason that no mass streaming occurs in the isolated endoplasmic drop.

This same idea can also explain the fact that chloroplasts and nuclei continue to rotate in the isolated endoplasm, if we assume that the boundary between these organelles and the ground cytoplasm generates the same active shifting force as the corticoendoplasmic boundary. Thus we can say that when the solid phase is fixed (e.g., the cortical layer), cytoplasmic streaming takes place as observed in the intact cell, whereas when it is suspended (e.g., rotating organelles), the solid phase itself moves around as in the case of rotation of chloroplasts or nuclei.

When a chloroplast rotates slowly in the isolated endoplasm, it was observed, sometimes, that the endoplasm just around the chloroplast moves in the opposite direction. In order to clarify the relation of the movement of the chloroplast to the surrounding endoplasm, we tried to stop a chloroplast from rotating with a microneedle. As soon as rotation of the chloroplast was stopped, a local streaming of endoplasm was observed to take place along the surface of the chloroplast in a direction opposite to the original chloroplast rotation. If the microneedle was removed, the chloroplast started rotating again in the original direction, and there was no observable countercurrent of the matrix around the chloroplast.

These facts indicate that the rotation of chloroplasts observed in the isolated endoplasm is essentially the same phenomenon as the cytoplasmic streaming found in intact cells. Both are believed to be caused by the sliding force generated at the boundary between the plasmasol and plasmagel. The nature of this force is, however, still unknown.

Moving Cytoplasmic Fibrils

Recently, it was reported by Jarosch (1956a,b) that isolated protoplasmic drops squeezed out of an internodal cell of *Chara foetida* contain many moving cytoplasmic fibrils. According to Jarosch these moving fibrils are so thin that only a well trained eye can recognize them in the dark field under high magnification. Both Kamiya and I confirmed many of his observations with *Nitella flexilis.* It was postulated by Jarosch (1956b, 1958, 1960) that these moving fibrils are oriented in an organized fashion on the inner surface of the cortical gel layer and that the protoplasmic streaming is brought about by the interaction between the endoplasm and those cytoplasmic fibrils. Moving fibrils, found in the cytoplasmic drops obtained from *Chara* or *Nitella,* may be considered as having detached themselves from the cortical gel plasm which is generally sloughed off when a drop of cytoplasm is squeezed out of the cell mechanically. However, in order to account for the rotation of the nuclei in the cytoplasm streaming in the intact cell, the presence of the fibrous elements is to be expected also in the endoplasm. As a matter of fact, the moving fibrils could be found in the isolated drop of endoplasm without containing cortical gel plasm. They

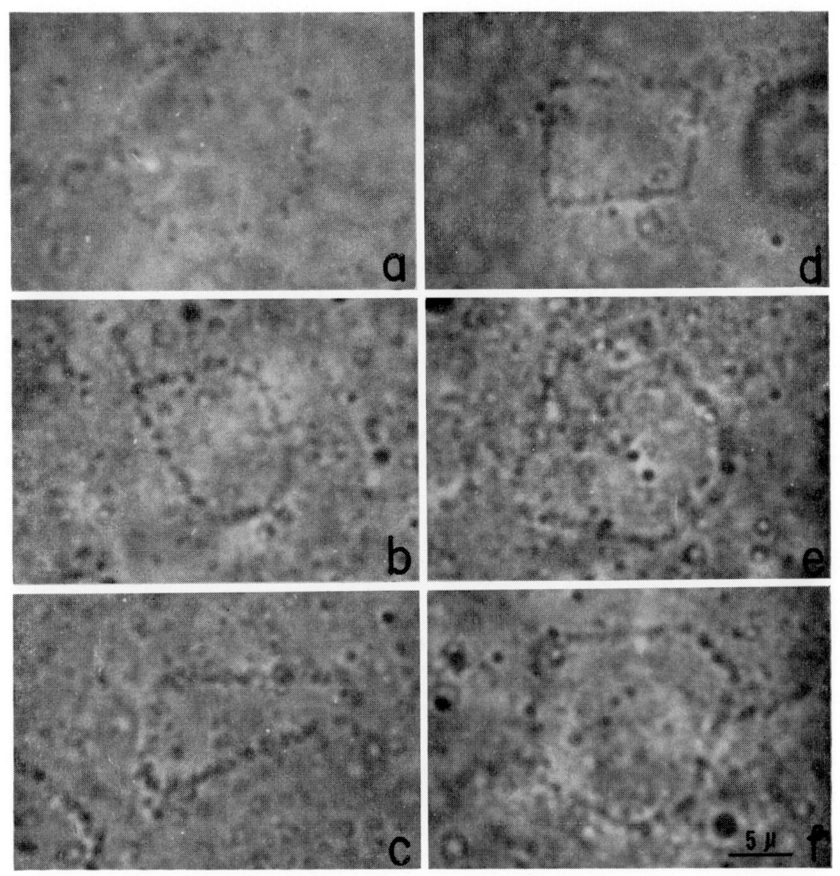

FIG. 3. Circular and polygonal loops of cytoplasmic fibrils in the isolated endoplasm observed with an inverted microscope in bright field. Moving fibrils first appear in the form of ring (a). Later, they assume different shapes (b–f).

can be seen also in the bright field fairly well. Since the fibrils have a tendency to appear along the surface of the drop, they are observed clearly in one focal plane with the aid of an inverted microscope.

The isolated endoplasm contains only a small number of moving fibrils right after isolation; but the number increases gradually. Many undulating polygons of fibrils appeared in the same drop in a few hours.

This seems to indicate that the streaming endoplasm in the intact cell contains only a few visible moving fibrils, but contains a large number of elements capable of forming moving fibrils.

In the isolated endoplasm, moving fibrils first appear in the form of an extremely fine filament or loop which is in active motion. The

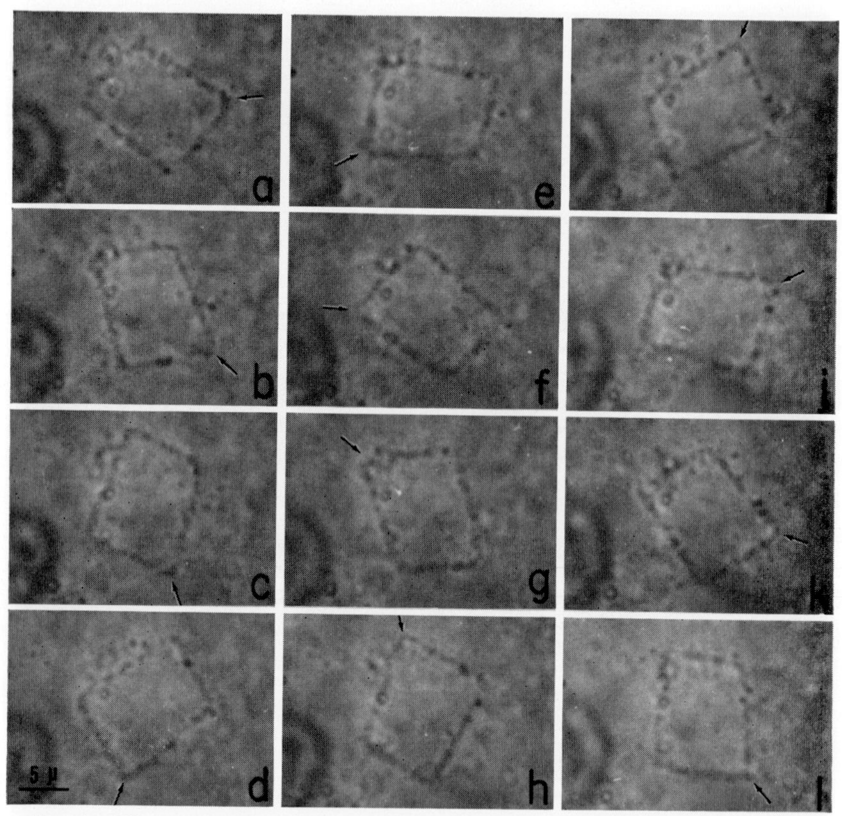

FIG. 4. The propagation of angles of a polygon as waves. Particles attaching to the fibril do not change their positions. Intervals of successive pictures: ½ sec. Arrows show a corner which propagates as a wave. (Observed with an inverted bright-field microscope.)

behavior of moving fibrils in the endoplasmic drops is quite similar to that in the squeezed out, cytoplasmic drop described by Jarosch (1956b). Later these fibrils assume most fascinating shapes, such as semicircles, triangles, squares, pentagons, hexagons, and other polygons (Fig. 3). It must be stressed that these thick fibrils display a remarkable undulating motion (Fig. 4); these undulations are observed only in thick fibrils which do not rotate vigorously as Jarosch pointed out. However, the undulating motion propagates a wave in the direction usually opposite to that of rotation. The polygons in which corners propagate as waves counterclockwise is about the same in number as those in which corners travel clockwise. By means of a cinematographic technique, we determined the rate of propagation of the undulating motion to be 9.8 μ/sec in an average of 15 polygons at 19–20°C and 13.1 μ/sec in an

TABLE I

Rate of Wave Transmission in the Loops of Cytoplasmic Fibrils

Temp. (°C)	Shape of polygon	Length of fibril (μ)	Rate of wave transmission (μ/sec)
19	Semicircle	37	9.3
19	Fan-shaped	37	10.4
19	Square	27	10.0
19	Pentagon	52	5.8
19	Pentagon	50	11.0
19	Pentagon	40	11.0
19	Pentagon	60	12.0
19	Pentagon	62	13.0
19	Hexagon	51	10.5
19	Hexagon	41	10.7
19.5	Pentagon	45	10.8
20	Fan-shaped	50	8.0
20	Triangle	41	5.3
20	Pentagon	43	9.0
20	Pentagon	57	10.8
24	Triangle	25	11.5
24	Square	37	15.5
24	Pentagon	48	9.6
24	Pentagon	40	13.2
24	Pentagon	72	13.0
24	Pentagon	73	14.6
25	Semicircle	39	12.9
25	Pentagon	49	14.7

average of 8 polygons at 24–25°C. The rate of propagation increases with temperature, but it is independent of the number of angles, the shape and the size of the polygons; the angular velocity of wave propagation is greater the smaller the polygons (Table I).

Since moving fibrils in the isolated endoplasm could be observed in bright field, several micromanipulations could be made in order to investigate the mechanical properties of fibrils. We first tried to disturb the polygons mechanically when they were in an early stage of formation. When one of these polygons was dragged by a microneedle, it was split into a number of fine fibrils, each of which rotated very furiously without any sign of undulation. The fact to be pointed out in this

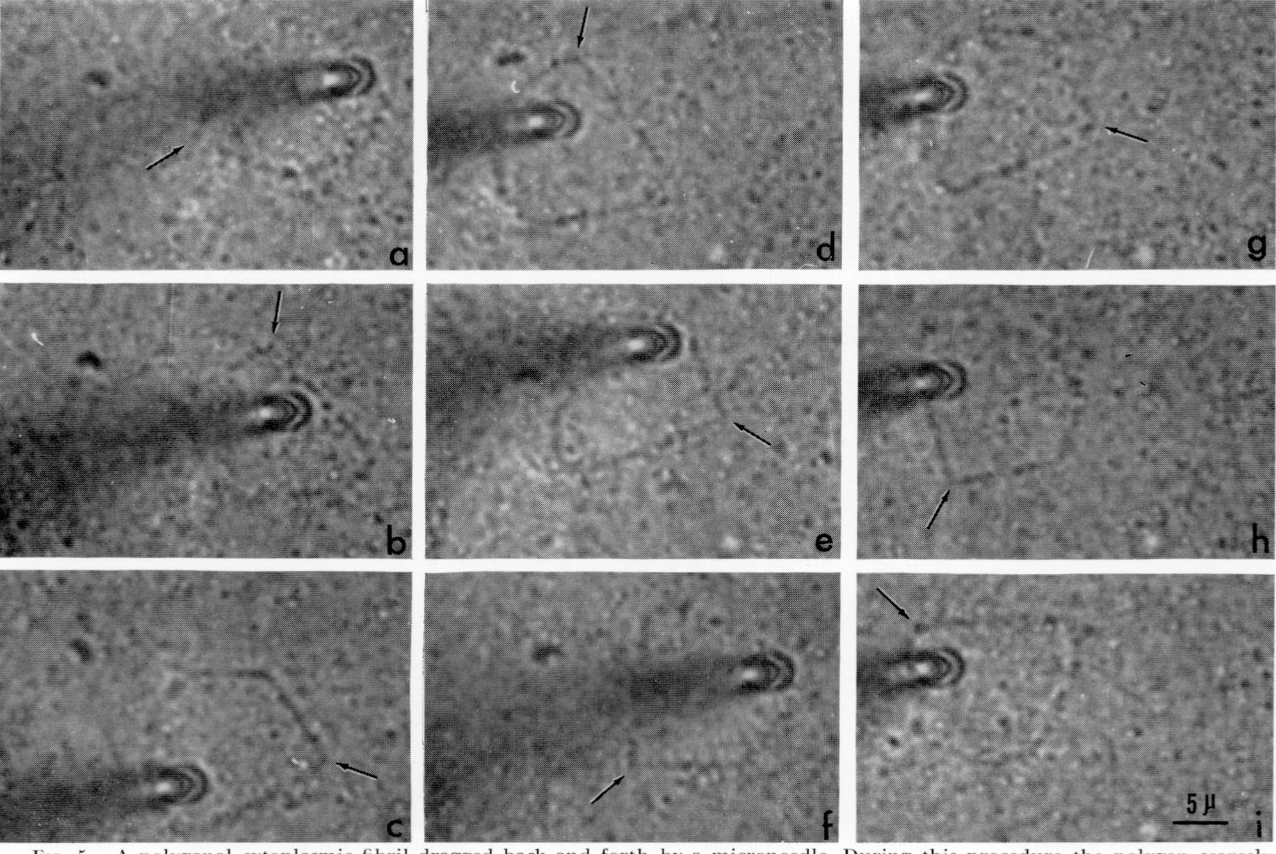

FIG. 5. A polygonal cytoplasmic fibril dragged back and forth by a microneedle. During this procedure the polygon scarcely changed its form and continued wave transmission just as before the dragging. Arrows show a corner which propagated as a wave. Intervals between a–b, b–c, d–e, e–f, g–h, h–i: 2 seconds; c–d: 5 seconds; f–g: 24 seconds. (Observed with an inverted microscope in bright field.)

case is that the newly formed fibrils always rotate in the direction opposite to that of the transmission of the corners in the original polygon.

Next, the same operations were performed on polygons a few hours later when they had become thicker. Unlike the newly formed polygons, polygons at this later stage were easily dragged back and forth by the microneedle. During this procedure the polygons scarcely changed their form, and the corners continued to travel along the fibrils just as if they were free from dragging (Fig. 5).

The facts mentioned previously indicate that the polygon is composed of many finer fibrils, each of which is capable of moving rapidly. During the formation of a polygon, these fine fibrils apparently form thicker fibrils by accretion until eventually they become strong enough to continue wave transmission even after some mechanical disturbance.

It was shown that the isolated endoplasm contains fibrous elements which could form moving fibrils visible at the microscopic level. In spite of this, no mass streaming takes place in the isolated endoplasm. This may be because the moving fibrils are scattered around and have no organized orientation and coordination. If the moving fibrils are gradually arranged in an orderly fashion during the process of incubation, mass streaming could begin to occur also in the isolated endoplasm. A few experiments were performed under aseptic conditions with the purpose of prolonging the survival time of the isolated drop of the endoplasm. The incubation medium consisted of a $\frac{3}{4}$ concentrated solution mentioned previously to which was added glucose, biotin, vitamin B_1, indoleacetic acid, the microelements of Heller, and a $\frac{1}{2}$ concentrated Knop's solution. After 1 day's incubation, some of the incubated endoplasm began to show local cytoplasmic streaming. With the present technique the isolated endoplasm could survive about 3 days outside the cell. Naturally, experiments along this line await further improvement in the technique, but the results obtained so far are rather encouraging. Studies on a fragment of cytoplasm *in vitro* and the analysis of its behavior have been conducted by many workers in the past. Systematic work in this field will without doubt help us gain an insight into the mechanism of intracellular motions as well as many other important physiological phenomena.

References

Donné, A. (1838). *Ann. Sci. Nat., Bot.* [2]**10**, 346.

Dutrochet, M., and Brongniart, A. (1838). *Ann. Sci. Nat., Bot.* [2]**10**, 349.

Jarosch, R. (1956a). *Oesterr. Akad. Wiss. Math.-Naturw. Kl., Anz.* No. **6**, 58.

Jarosch, R. (1956b). *Phyton (Buenos Aires)* **6**, 87.

Jarosch, R. (1958). *Protoplasma* **50**, 93.

Jarosch, R. (1960). *Phyton (Buenos Aires)* **15**, 43.

Kamiya, N., and Kuroda, K. (1956). *Botan. Mag. (Tokyo)* **69**, 544.
Kamiya, N., and Kuroda, K. (1957a). *Proc. Japan Acad.* **33**, 149.
Kamiya, N., and Kuroda, K. (1957b). *Proc. Japan Acad.* **33**, 201.
Stålberg, N. (1927). *Botan. Notiser.* p. 305.
Valkanov, A. (1934). *Protoplasma* **20**, 20.
Velten, W. (1876). *Oesterr. Botan. Z.* **26**, 77.
Yotsuyanagi, Y. (1953a). *Cytologia* **18**, 146.
Yotsuyanagi, Y. (1953b). *Cytologia* **18**, 202.

Discussion

CHAIRMAN THIMANN: It is a fascinating picture. I am sure there are a lot of comments.

DR. REBHUN: Your pictures seem to indicate two kinds of phenomena concerned with motion of the fibrils in the cytoplasm. When these fibrils move, they appear to have a countercurrent stream moving in the opposite direction. When they are aggregated into polygons, I have looked to see if there was a countercurrent stream and it appeared to me there was not. That is, it appeared when these fibrils formed polygons, whatever the process deforming the shape of the polygons was entirely within the fibril and not coupled to motion of the surrounding cytoplasm. Does that agree with your observations?

DR. KURODA: I think so.

DR. COUILLARD: Your polygons appear remarkably coplanar all the time. Have you considered the possibility that you might be observing an optical section of a three-dimensional body such as a polyhedron?

DR. KURODA: The polygons we observed are not an optical section of a three-dimensional body. The reason why they lie in the same plane is because they tend to appear in the peripheral layer of the drop. The polygons which are found in the inner space of the drop can take all possible orientations. In this case we observe only a fraction of them in one plane of focus.

DR. MAHLBERG: Do the polygons ever stop moving; and, if so, do they reverse direction?

DR. KURODA: They continue moving without interruption for hours. Sometimes the motion of the polygons lasts more than 24 hr. The direction of wave transmission never reverses.

DR. MAHLBERG: It appeared some of them slowed down during the filming. I wondered if the organelles continued moving independently, that is, in a wave action.

DR. KURODA: After the polygons have been formed, wave propagation generally continues at nearly a constant speed for a few hours.

DR. MAHLBERG: Does the wave propagation ever stop?

DR. KURODA: I have never seen polygons that stopped completely. Only in a degenerating drop, cytoplasmic fibrils accrete in a needlelike structure which is stationary.

DR. GOLDACRE: Do you ever see short lengths of the polygon material moving in a straight line?

DR. KURODA: There are no such short pieces of the polygon material observable.

DR. GOLDACRE: Can you not snap the loop by pulling on two needles?

DR. KURODA: I tried to pull the loop using two microneedles. If it is stretched, the polygon is split into finer fibrils which are scarcely visible. I suppose they are motile, but they are almost beyond the resolution of the optical microscope and it is hard to observe their motion.

DR. BOUCK: First, do you ever see fibrils in an intact cell or are they only in these isolated drops? Second, what are the particles?

DR. KURODA: The polygons have been also found in a centrifuged cell of *Nitella* by Dr. Kamitsubo (unpublished) quite recently. Naturally this is not an intact cell but it is still living. My answer to the second question is that these granules are not identified cytologically or electron microscopically. According to Dr. Jarosch's classification of the cytoplasmic inclusions, they are probably prespherosomes.

DR. JAROSCH: As to the first question of Dr. Bouck, I have seen moving fibrillar rings showing undulatory movement in the intact cell also. In a few cases it was possible to observe them with dark-field illumination lying in the stationary protoplasm along the indifferent line. But this seems to be an exception. In most cases, there are no visible fibrillar structures in the intact cell.

DR. DEBRUYN: One ought to get a motion something like that observed in your polygons if you could take a flagellum and tie the tail to the other end. Is anything known of the fine structure of these polygons?

DR. KURODA: No. Nothing is known electron microscopically of the fine structure of these polygons. Dr. Terada studied both endoplasmic drops and internodal cells of *Nitella,* with an electron microscope, but he has not been successful in finding fibrous structures or any structure which seems to have a bearing on the polygons.

DR. INOUÉ: I would like to point out the possibility that the polygonal movement could be explained by a twisting motion as seen in the axostyle of certain protozoa. If this is so, I would make two predictions which one could verify with the electron microscope; one, the cross section of the so-called fibrils ought to be asymmetrical instead of circular, and two, one should see not a regular sine wave but a three-dimensional saw-tooth wave. Judging from your photographs, I think there is a good likelihood that you are observing the propagation of such a twisting motion.

DR. ALLEN: One idea that comes to mind is the possible relationship between the movement of these fibrils and streaming of the intact cell. These fibrils might be oriented along the cortex in such a manner that some kind of waves passing along the fibrils (such as the straightening or stiffening waves) could drive the endoplasm.

DR. GOLDACRE: I wonder about the membrane that appears to be around the isolated drops of cytoplasm. You say this is permeable to the kind of substances that will not enter the living cell, such as certain vital dyes.

DR. KURODA: There is a definite membrane at the surface of the isolated endoplasmic drop. This membrane, I suppose, is more like a tonoplast than the protoplasmic surface membrane. It has a weak tension amounting to 0.001–0.006 dyne/cm. It is semipermeable to solutes. This drop can be vitally stained with some dyes, such as erythrosine, chrysoidine, and eosine, but details of permeability characteristics of this membrane are still unknown. When the drop degenerates, there is often a "ghost" left.

Rates of Organelle Movement in Streaming Cytoplasm of Plant Tissue Culture Cells[1]

PAUL G. MAHLBERG

Department of Biological Sciences, University of Pittsburgh, Pittsburgh, Pennsylvania

Introduction

Streaming is usually evaluated in terms of the velocity of organelles of various sizes embedded in the cytoplasm. In the absence of microscopically visible organelles or other particles, it is very difficult to determine motion in the groundplasm. Currently, emphasis on the streaming process is centered upon so-called lower plants such as myxomycetes (e.g., *Physarum*) and certain of the Characeae, exemplified by *Nitella* and *Chara* (Kamiya, 1959, 1962) and, indeed, these organisms are excellent subjects in which to study streaming.

The streaming pattern may be correlated with the differentiation process of individual cells. This was suggested by Denham (1923) and Martens (1940) for *Tradescantia* stamen hairs where the streaming pattern appears to follow the wall ridges, and by Probine and Preston (1958) in *Nitella* where the helical pattern of streaming along the wall appears to be correlated with the orientation of the cellulose microfibrils of the wall. More effort, I believe, should be directed toward an analysis of streaming in cells of higher plants to determine the influence of streaming on differentiation. A possible point for departure can be the use of individual cells under continuous culture conditions where the cells are capable of division and possess the capacity to differentiate.

Various studies have been made of streaming in higher plant cells with emphasis on how different physical and chemical factors affect the rates of organelle movement. Several excellent reviews of these data have been published (Kamiya, 1959, 1962).

At present I wish to emphasize the variations in the streaming rate and pattern which can exist in one cell over a given interval of time. Convenient materials for these investigations are the individual cells of callus origin which can be maintained under tissue culture conditions. Such cells are parenchymatous in character and possess a large central

1 This work was supported in part by a grant (C-5714) from the National Institutes of Health, Department of Health, Education and Welfare.

vacuole. A peripheral layer of cytoplasm is in contact with the inner surface or the thin cell wall, and cytoplasmic strands traverse the vacuole, extending in different directions across the vacuole. Individual cells were chosen for analysis to avoid any intercellular influences upon the streaming process.

Very precise measurements of organelle speed can be performed by analyzing cine-film recordings of their movement. By this method it was possible to measure simultaneously the rates of movement for several organelles at different locations within the same cell. In this manner the streaming rates were analyzed in the peripheral cytoplasm, in transvacuolar strands of different diameters, in transvacuolar strands of diverse configurations, in cells of different genera, and for cells maintained in the culture chambers for varying durations of time.

Materials and Methods

The materials employed in this study consist of individual cells of *Euphorbia marginata* and genetically pure strains of *Nicotiana tabacum* grown as callus in tissue culture. These strains included a pigmented form, homozygous dominant for one of two factors controlling chlorophyll synthesis, and an albino form where both factors controlling chlorophyll synthesis are homozygous recessive. The callus was derived from the hypocotyl of the germinated embryo and cultured on an agar-base medium containing mineral salts, sucrose, and specific growth regulators.

The culturing procedures have been described elsewhere for *Euphorbia* (Mahlberg, 1962) and for *Nicotiana* grown on a synthetic medium (Venketeswaran and Mahlberg, 1962). All cultures were maintained under continuous fluorescent lighting (250 ft-c) at 24–25°C.

To secure individual cells a small piece of callus was introduced into the liquid form of the culture medium and shaken (170 rpm) for a period of 2–6 days to dissociate cells from the callus. Small drops of the liquid medium containing individual cells were micropipetted into aseptic chambers for viewing. A chamber consisted of a ring of sterile petroleum jelly separating a standard slide and a 7/8-in. coverslip. The depth and volume of the chambers were approximately 0.2 mm and 0.03 ml, respectively. The chambers were maintained and viewed at a temperature of 24–25°C.

The cells selected for observation were photographed on 16-mm cine film at 16 frames per second under phase contrast (40 ×, N. A. 1.0) within 8 hr after mounting in the microchamber unless otherwise indicated. A strand selected for photographing occupied most of the film frame and was in focus along its entire length. Measurements were performed with a film analyzing projector by projecting the film onto a 3 × 4 ft piece of white paper. The position of individual organelles was followed on successive frames and recorded at 0.5-sec intervals. The magnification was such that positional changes corresponding to 0.15–0.25 μ could be detected. Measurements of transvacuolar strand diameter were approximate but adequately meaningful for comparative purposes. Filmed and measured in this way, the rate and direction of movement of several organelles in different portions of the cytoplasm could be determined simultaneously.

Observations

The materials selected for this investigation are those which I am currently employing in studies on cellular differentiation. Individual cells in suspension cultures vary in size and shape from spherical cells 40 μ in diameter to elongated cells 1500 μ in length and about 50 μ in diameter. The cells possess a large population of organelles showing striking uniformity in size (ca. 0.2–0.4 μ) and spherical in form. Initial electron microscope studies on these organelles in *Euphorbia* have revealed the fine structure characteristic of mitochondria. These studies will be expanded and extended to the tobacco strains in the future.

Larger organelles, at present assumed to be plastids, were distributed throughout the cytoplasm of the *Euphorbia* and green strain of tobacco; however, in the latter they were few in number. The albino tobacco cells contained no large organelles which could be interpreted at this time as plastids.

The uniformity in size of the organelles was advantageous for the comparison of their speeds. Large organelles did not interfere with the streaming movements being plotted for organelles of small size. Since no data are presently available on the mass of the small organelles, it will be assumed for the present purposes of comparison that they are similar in mass. The transvacuolar strands are a very convenient and even ideal location within the protoplast to record the rate of streaming. Organelles can be easily and accurately followed in such strands.

The comparisons of speed for organelles were made in three, more or less spherical, albino tobacco cells with the largest dimensions of 120, 140, and 174 μ; a spherical green tobacco cell with a maximum diameter of 125 μ; and two *Euphorbia* cells 180 and 200 μ along the longest axis.

Thin Transvacuolar Strands

Organelles in a very thin transvacuolar strand approximately 0.5 μ in diameter were observed to move at a rather uniform speed as illustrated for a short strand in tobacco (Fig. 1) or a longer strand in *Euphorbia*

(Fig. 2). In both cells the rate of movement for each organelle was linear with little change in speed evident for an organelle over a maximum period of 7 sec. These observations represent the ideal condition. It should be emphasized that important differences in average speeds did exist between the organelles. In the tobacco cell (Fig. 1), the organelles designated by the letters B and D were the first to appear on the strand separated from each other by distance of 4.4 μ. Each organelle moved at a different speed, one, B, moving at an average speed of 8.8 μ/sec or 1.4 μ/sec faster than the other, D. The organelle A appeared on the strand 13

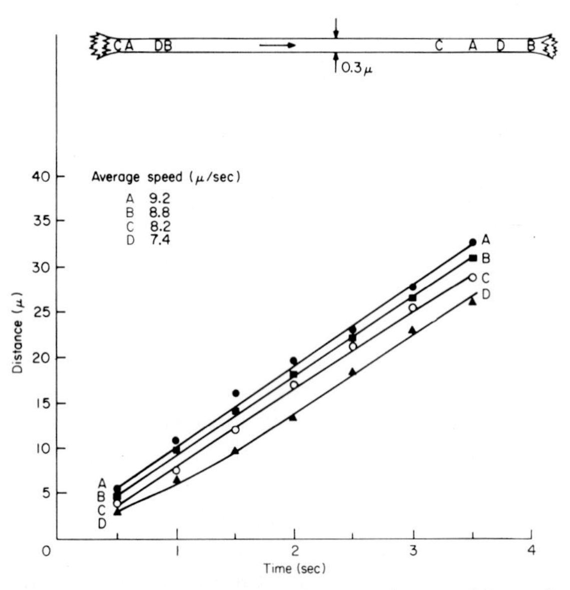

FIG. 1. Transvacuolar strand, 0.3 μ in diameter, in an albino tobacco cell 120 μ in diameter. The four organelles moved with a uniform velocity along a transvacuolar strand approximately 30 μ in length. (Interval = ½ sec.) Top diagram—relative positions of organelles at beginning and end of plotting.

sec after the appearance of D, moving linearly at an average speed of 9.2 μ/sec. The fourth organelle, C, appeared 2 sec later moving at the somewhat slower speed of 8.2 μ/sec.

In thin strands of considerable length the movement of organelles follows a somewhat similar pattern as represented by *Euphorbia* (Fig. 2). The organelle C moved at a nearly uniform speed for a distance of 56 μ. A second organelle, A, accelerated over a distance of 52 μ, whereas the third organelle, B, 5 μ behind A, progressively decreased in velocity over a distance of 58 μ.

It should be noted that the speed of an organelle during the first

1/2–sec interval upon entering the strand can vary considerably (note time period at 1/2 sec in Fig. 1). The organelle C moved into the strand at a speed of 1.6 μ/sec, the other two organelles entered with speeds of 9.2 and 10.2 μ/sec. The organelle C which entered the strand at the slower speed was 1 and 6 sec, respectively, in advance of A and B. Other

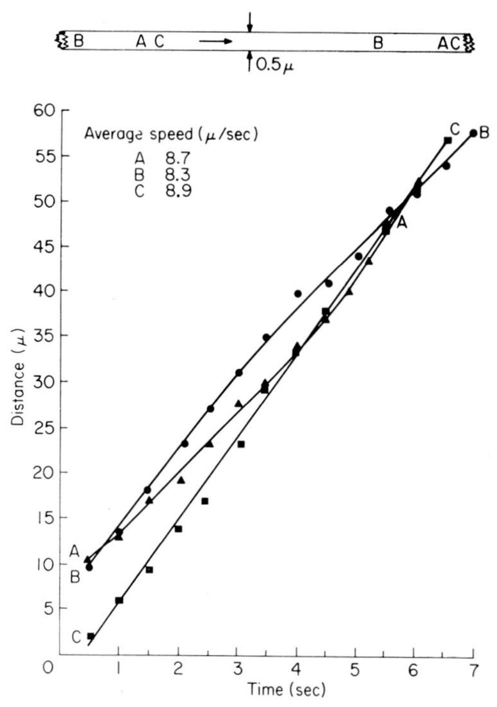

FIG. 2. Transvacuolar strand 0.5 μ in diameter in a *Euphorbia* cell 200 μ in length. Organelles entered the strand at different speeds and moved either linearly with time (C), accelerated (A), or decelerated (B) slowly along the 50–60 μ course during plotting. (Interval = ½ sec.) Top diagram—relative positions of organelles at beginning and end of plotting.

organelles followed similar patterns of movement when entering the strand.

The movement of an organelle in a thin strand may not be linear with time (Fig. 3). In a strand 0.8 μ in diameter the two organelles B and A separated by a distance of 6.5 μ decelerated during the first 4 sec and subsequently accelerated during their movement along the strand. Both organelles followed a similar pattern which was quite irregular when compared with the previous graphs. Eight seconds after their appearance, two new organelles, C and D, appeared on the strand. Their velocities were quite different as noted by the curves. These organelles progressively accelerated for 3 sec whereupon their velocities became more or less constant along the strand. In contrast to the first pair of organelles, the pattern of movement for organelles C and D was not synchronized along the strand.

The rate of streaming along a thin strand can be subject to even greater variations than indicated in the previous figure when movements

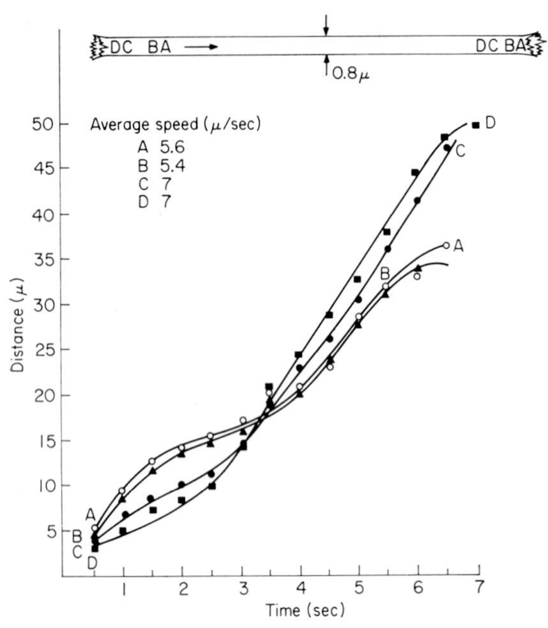

FIG. 3. Transvacuolar strand 0.8 μ in diameter in an albino cell 174 μ in diameter. The four organelles followed two curve patterns. See text. (Interval = ½ sec.) Top diagram—relative positions of organelles at beginning and end of plotting.

in a strand are followed for extended periods of time. This is exemplified in Fig. 4 where the appearance of two sets of organelles, A, B and C, D, in a strand 1 μ in diameter is separated by a 9-sec interval. The organelles appearing first, A and B, traversed the strand during a period of 8.5 sec, slowly accelerating during their movement. Their average speeds were 2.7 and 2.9 μ/sec, respectively. Note in the diagram that organelle B accelerated to a greater extent than A, passing the latter along the strand.

In Fig. 4, the organelles C and D separated by a distance of 1.5 μ entered the strand 9 sec after the first set. The second set moved at different average speeds of 0.7 and 1.7 μ/sec, respectively. After these organelles had traveled for a period of 4 sec, they were separated by a distance of 10.5 μ. Both organelles then underwent rapid deceleration to zero velocity for several seconds, temporarily reversed direction, and then resumed movement in their original direction. The organelles entered and moved along the strand with different initial velocities but note that their deceleration and reversal in directions of movement were synchronized through time. This pattern of movement for these organelles separated from each

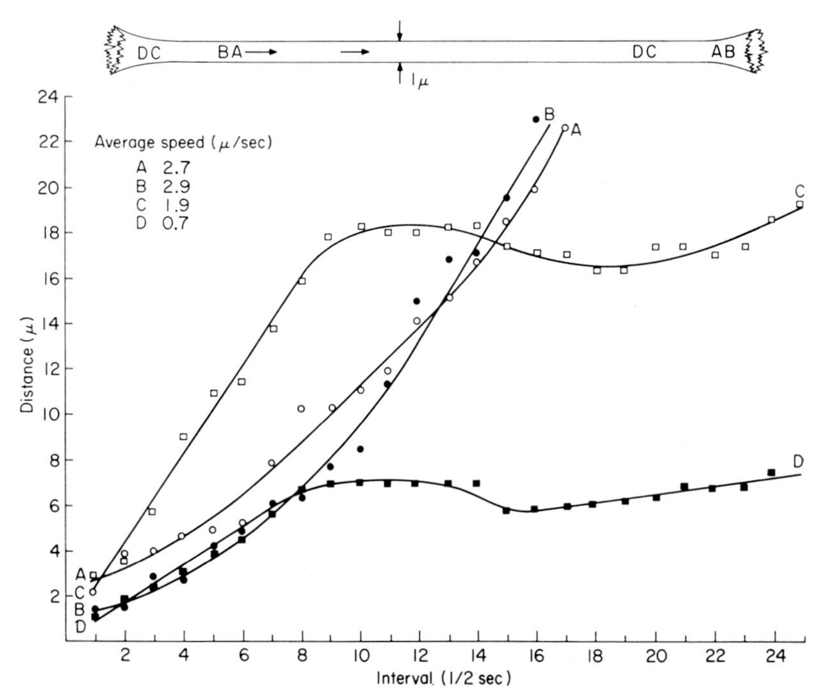

Fig. 4. Transvacuolar strand 1 μ in diameter in an albino cell 120 μ in diameter. Organelles A and B accelerate along the course, B passing A. Organelles C and D which appear 9 sec after A and B, moved at different speeds but experience similar degrees of impedance along the strand although they are separated by 10.5 μ. Top diagram—relative positions of organelles at beginning and end of plotting.

other by 10.5 μ is suggestive that a unit of groundplasm at least 10.5 μ in length stopped, reversed direction, and then proceeded forward again.

Strand of Large Diameter

The streaming pattern of organelles was followed in a thick strand approximately 5.7 μ in diameter in an albino cell (Fig. 5). Initially I attempted to follow the movement of several organelles both along the surface and in the center of the strand to detect positional differences in speed for the organelles, as one would expect for laminar flow in a physical system. Lateral movement of the organelles, from the surface to a more central position in the strand and vice versa, prevented me from gaining adequate data on this point.

Numerous organelles were moving in the thick strand, all moving in the direction indicated by the arrow. The recorded variations in velocity characterize the movement of the organelles. Those near the surface of the strand as well as those within moved readily in a lateral direction.

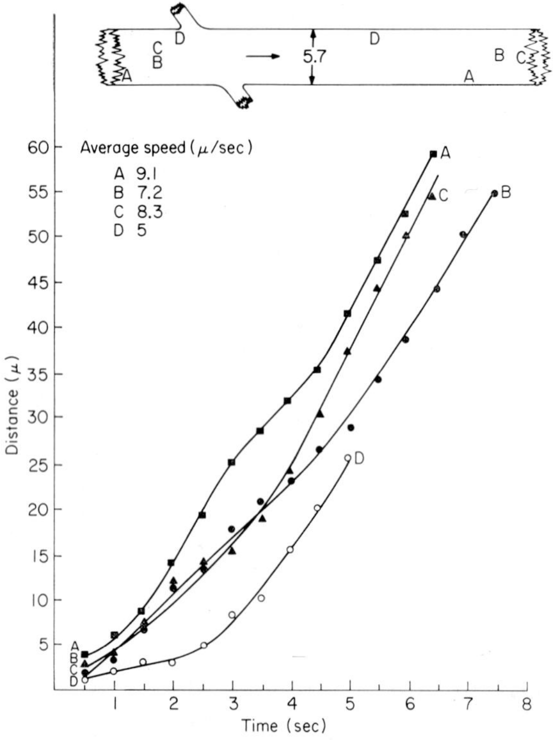

FIG. 5. Transvacuolar strand 5.7 μ in diameter in an albino cell 174 μ in diameter. Organelles were plotted simultaneously at different positions in transvacuolar strand. (Interval = ½ sec.) All organelles moved with different average speeds. Top diagram—relative positions of organelles at beginning and end of plotting.

Organelles A and D emphasized recorded ranges in velocities. Organelles B and C remained adjacent to each other for 3.5 sec, whereupon C increased in velocity and moved ahead of B. There was little to suggest synchronization of organelle movement in strands of large diameter.

THIN-THICK TRANSVACUOLAR STRAND COMBINATION

Upon viewing the transvacuolar strand reticulum in the cell, one can observe combinations of strands which vary in their diameters. A typical pattern is the form of the letter "Y" as in Fig. 6 consisting of a strand 4 μ in diameter and a divergent thin strand 0.5 μ in diameter.

Organelle movement was traced in the Y strand combination to determine the velocity of organelles when moving from the strand of large diameter into the small strand. The rate of flow of a Newtonian sol in a Y tube should be greater in the thinner arm than in the base. Actually such a change in velocity did not occur. The speed of the organelles A and B along the thick strand averaged 9.3 and 8.9 μ, respectively. The

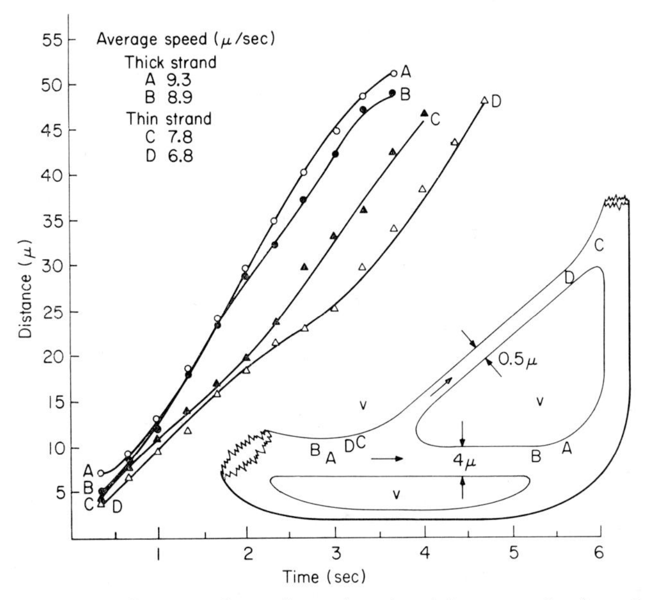

Fig. 6. Transvacuolar strand configuration involving strands 4 and 0.5 μ in diameters. Organelles either remain in thick strand or enter the thin strand. Average speeds of organelles entering the thin strand were somewhat less than those for the organelles remaining in the thick strand. (Interval = ½ sec.) Diagram at right shows relative positions of organelles at beginning and end of plotting. Vacuole, v.

organelles C and D which moved from the thick strand into the thin strand exhibited no great acceleration. The velocity of these organelles, having average speeds of 6.8 and 7.8 μ, respectively, was actually less than that recorded for the organelles in the thick strand. Similarly, the average speeds for two additional organelles which were plotted along the thin strand were lower than the speeds of the organelles along the thick strand.

Peripheral Cytoplasm–Transvacuolar Strand Combination

The simultaneous comparison of organelle movements was prepared for the peripheral cytoplasm and a transvacuolar strand 1 μ in diameter

(Fig. 7). Each organelle exhibited somewhat different average speeds and curve patterns. The average speed of two organelles (A = 5.9 μ/sec) and (B = 5.8 μ/sec) exceeded the speed of the organelles along the transvacuolar strand.

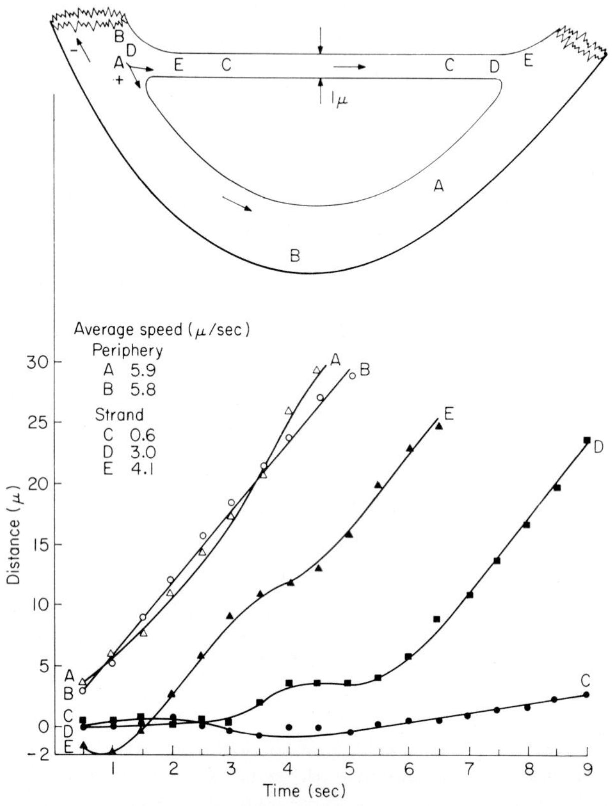

FIG. 7. Combination of a transvacuolar strand 1 μ in diameter and the peripheral cytoplasm in an albino cell 120 μ in diameter. Certain organelles in strand were impeded or reversed their direction during their movement. (Interval = ½ sec.) Top diagram—relative positions of organelles at beginning and end of plotting; (+) and (—) indicate changes in direction of movement.

The movement of organelles along the strand was subject to interference. Organelle C, the first to appear in the strand, moved at a slow speed averaging 0.6 μ/sec during a 9-sec period of time. Organelle E, appearing 3 sec after C, initially moved in the direction indicated as (—) for 1 sec, thereafter it reversed, to move in the (+) direction along the strand with a rather constant velocity. The path of organelle D which entered the strand 4 sec after C was impeded along its course for 6 sec after which it followed a linear time–distance relationship. The average speed of D was 3 μ/sec, somewhat slower than the 4.1 μ/sec recorded for E.

A most interesting phenomenon occurred when organelles D and E entered the strand (Fig. 8). Organelle E passed the impeded C after 3 sec and continued unimpeded along the strand. The distance between these organelles increased initially when E moved in the (—) direction (—2 on the graph) and thereafter altered its course to move in the (+) direction.

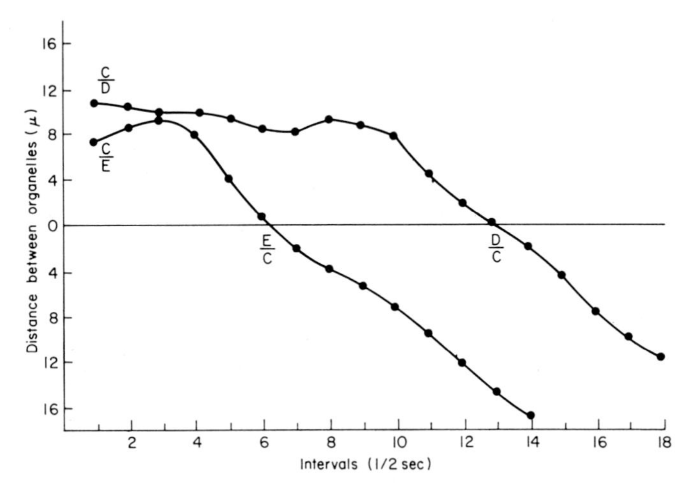

FIG. 8. Passing of organelles within a transvacuolar strand 1 μ in diameter (as in Fig. 7). Since organelles C, D, and E moved at different average speed, interceptions occurred along the transvacuolar strand. Symbols C/D and C/E indicate that the organelle C initially was ahead of either D or E. Positions were transposed D/C and E/C at the time interval when the organelles D and E passed C.

tion. Organelle D, at first moving slowly along the strand, increased in speed to converge upon and pass organelle C at interval 13, and thereafter, continued to move unimpeded along the strand.

The average speeds for various organelles, both in broadly grouped categories of transvacuolar strands with diameters of two magnitudes and in the peripheral cytoplasm, are presented in Table I.

Two-Directional Movements along One Strand

Frequently movement can occur in two directions along one strand as in a strand 1.7 μ in diameter in an albino cell (Fig. 9a). The first organelle, A, plotted along the strand at an average speed of 2.2 μ/sec, decelerated and at the time intervals 2 and 4 sec reversed the direction of movement for very short periods of time. The second organelle, B, moved along the strand linearly for a distance of 26 μ at an average speed of 6.5 μ/sec. This organelle B passed organelle A during its course along the strand.

Organelle C moving in the opposite direction progressed linearly for a distance of 14 μ during the first 2.5 sec of movement, whereupon it decelerated to zero speed and oscillated during the successive 3 sec while measurements were being made.

TABLE I

AVERAGE AND RANGE IN SPEEDS FOR ORGANELLES IN DIFFERENT POSITIONS WITHIN THE CELLS, AND IN CELLS MAINTAINED IN MICROCHAMBERS FOR VARYING LENGTHS OF TIME

Position or condition	Diameter (μ)	No. of organelles plotted	Speed range in any one interval (μ/sec)	Average speed (μ/sec)
Transvacuolar	0.3–1.0	33	0–19	6.0
strand	1.1–6.0	29	0–19	5.8
Peripheral cytoplasm	—	7	2.8–7.6	6.3
Constriction of				
transvacuolar	1	4	1.4–10.8	7.9
strand	6	6	2.2–14.2	6.5
Age:				
8 hr or less (from				
above)	—	79	0–19	6.5
6 days	—	6	1.0–8.4	4.7
10 days	—	4	3.0–9.0	6.7

Organelles C and A traveling in opposite directions were subjected to deceleration and reversal in direction at two positions during their course along the strand (Fig. 9b). These positions and times at which the reversal occurred could not be correlated.

These movements in opposite directions and at different velocities is indicative that the streaming cytoplasm possesses a structured character consisting of numerous streams or possibly fibrils along which the organelles are transported.

REVERSAL IN DIRECTION OF MOVEMENT

Organelles traveling in the same strand can be observed to change their direction more dramatically than illustrated previously. This change may persist for periods of time during which the organelle can move a considerable distance (Fig. 10). Each organelle appeared to respond independently of other organelles along the strand. The average speed of the organelles was plotted for each interval of time to illustrate the magnitudes of movement. Organelle B moving at an average speed of 2.4 μ/sec traveled the length of the thick strand in about 10 sec. Organelle C, after moving along the thick strand a distance of 11 μ, altered its direction to enter the thin strand. There was no increase in the velocity of this organelle upon entering the thin strand 1 μ in diameter.

The path of organelle A was subject to change. During the initial 2.5 sec it moved in the same direction as the other organelles, whereupon

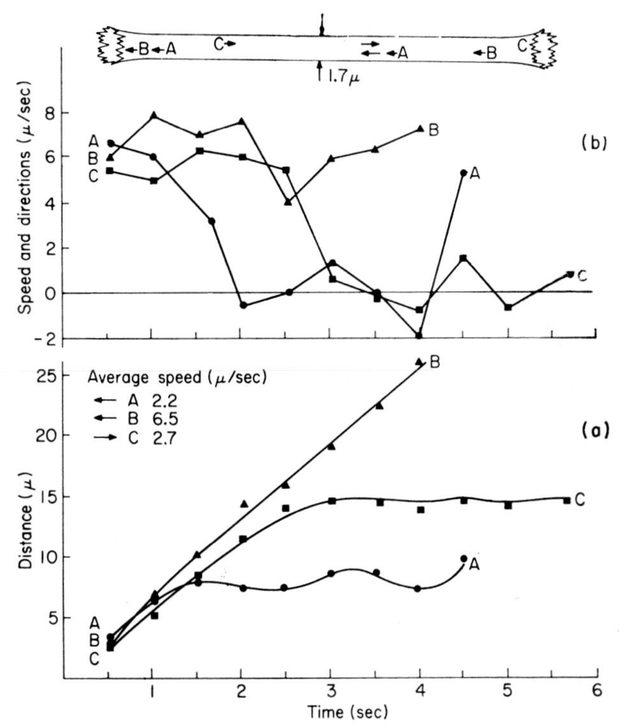

Fig. 9. Transvacuolar strand 1.7 μ in diameter in an albino cell 120 μ in diameter. Movement of organelles is evident in opposite directions along strand—(a) the time-distance curve and average speed of the organelles; (b) the magnitudes of speed for an organelle during an interval. (Interval = ½ sec.) Arrows indicate direction of movement. Note that organelles A and C, moving in opposite directions, underwent periodic reversals in direction (+) or (—) as they moved. Top diagram—relative positions of organelles at beginning and end of plotting.

A reversed direction to travel a distance in excess of 8 μ along the thick strand. Again, this organelle reversed direction and after traveling a distance of 19 μ in the thick strand, it altered its direction again to enter the thin strand. The magnitude of interval speed per second for organelle A varied considerably attaining a maximum value of nearly 7 μ/sec in either direction and with average speeds of (+) 3.4 and (—) 2.5 μ/sec.

One interpretation of this strand complex is to suggest that several or many cytoplasmic streams exist within the thick strand. The stream or streams transporting organelle B move it along the large strand in the (+) direction, whereas other streams diverge from the thick into the thin strand (organelle C). The movement of organelle A may well represent shifting in position from one stream to another, these moving in different directions and at different velocities.

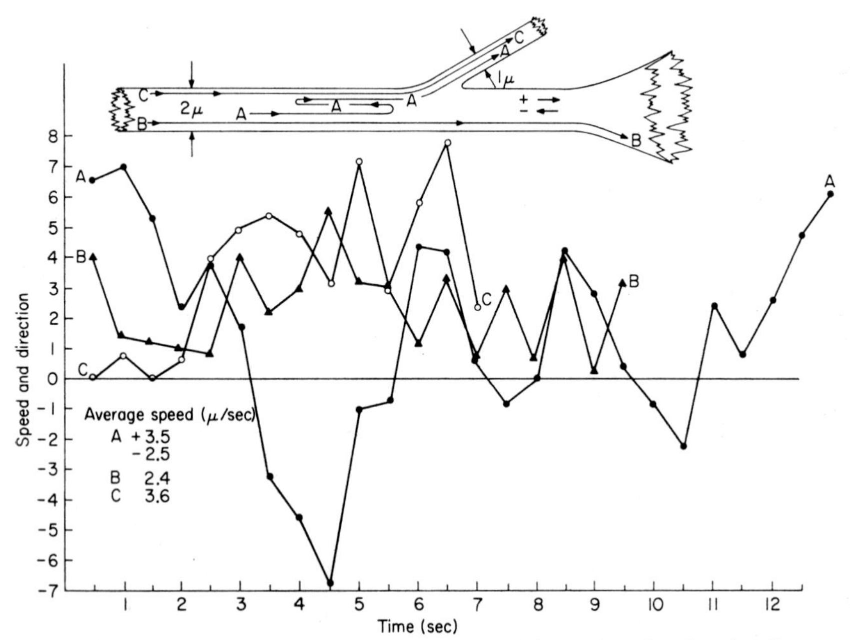

FIG. 10. Transvacuolar strand configuration involving a 2 and a 1 μ in diameter strands in a chlorophyllous cell 125 μ in diameter. The three organelles followed different patterns of movement: B along the thicker strand; C entering into the thin strand from the thick strand; and A reversing directions at several points while following the general path indicated in diagram. The maximum speed of A in (+) direction during a given interval was as great as that attained in the (—) direction during the plotting. (Interval = ½ sec.) Top diagram—relative positions of organelles at beginning and end of plotting.

CONSTRICTION OF A TRANSVACUOLAR STRAND

The form of the transvacuolar strand can be exceedingly variable. One form which was most instructive is represented in Fig. 11 in which a thick strand 6 μ in diameter suddenly constricted to a thin strand 1 μ in diameter. Organelle positions and period of movement relative to the constriction are represented on the abscissa; the dashed vertical line between interval 10–11 represents the time interval when organelles C and D entered the strand constriction. The other organelles (A and B) were plotted either for the thick region of the strand (left of the dashed vertical line) or for organelles (E and F) in that portion of the strand which was small in diameter (right of dashed vertical line).

The rates of movement for organelles in the strand 6 μ in diameter were plotted simultaneously. The organelle B moved linearly at an aver-

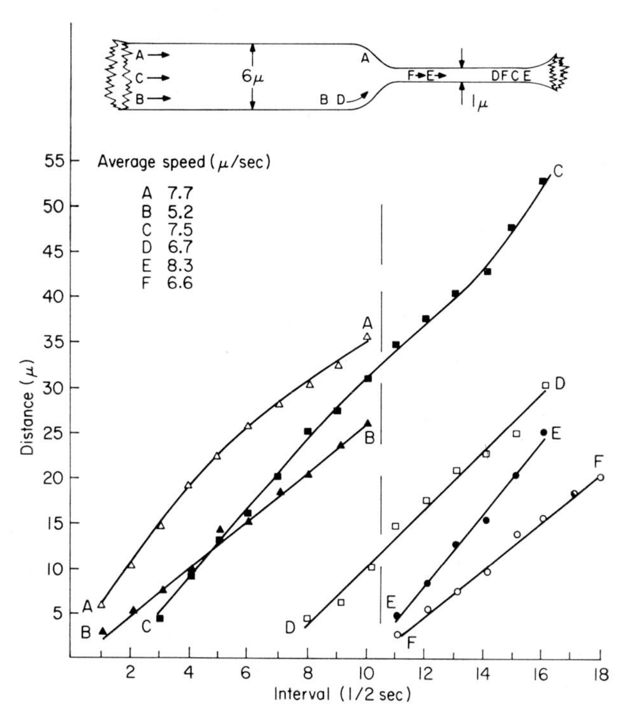

Fig. 11. Constricted strand in a *Euphorbia* cell 180 μ in diameter. The vertical line at interval 10–11 represents interval when an organelle entered constriction (1 μ diameter) from the thick portion (6 μ diameter) along the strand. Some variation in average speeds are evident for different organelles. No significant change in speed is evident for organelles C and D entering the constriction from the thick portion of the transvacuolar strand. Top diagram—relative position of organelles at beginning and end of plotting.

age speed of 5.2 μ/sec whereas organelle A progressively decelerated along the strand and was recorded to average a higher speed of 7.7 μ/sec. These organelles could not be followed into the constriction.

The organelles F and E, plotted simultaneously along the constricted portion of the strand 1 μ in diameter, moved at a rather uniform rate averaging 6.6 and 8.3 μ/sec, respectively. Note that both organelles moved at different rates of speed in the thin strand.

The movement of the two organelles C and D was traced as they moved both in the thick strand and in the thin strand. No appreciable increase in speed is evident for either organelle. The interval distances for organelle D plotted as a straight line with an average speed of 6.7 μ/sec. Organelle C decelerated at the neck of the constriction and subsequently accelerated at a low rate of speed in the constricted strand, averaging 7.5 μ/sec along the entire strand. When the average speeds of these organelles, C and D, were examined relative to the region of large diameter (6 μ), it was found that they moved with an average speed of 7 and 7.7 μ/sec; their average speeds in the constricted strand were 6.6 and 7.2 μ/sec, respectively. Note that they decreased rather than increased in speed when entering the thin strand. The average speeds of all organelles plotted along the strand revealed only a slight increase in speed in the 1 μ sector compared to the portion 6 μ in diameter (Table I).

In a physical system the velocity of a pressure-induced laminar flow for a liquid will increase when passing from a tube of large diameter to one of small diameter. The change in velocity within a fluid can be expressed by the following equation:

$$V_1 = V_2 \left(\frac{d_2}{d_1}\right)^2 \qquad (1)$$

Where V_1 and V_2 represent mean velocities, and d_1 and d_2 represent diameters along the large and small diameter sections of the tube.

This law of classic fluid dynamics is not applicable to the flow of cytoplasm. The organelles when entering into the constricted portion of the strand 1 μ in diameter should increase in velocity nearly 30 times the recorded values for the organelles in the strand 6 μ in diameter.

The velocities for the organelles in this as well as other strands emphasize that the cytoplasm possesses a structural character, possibly consisting of numerous fibrils or streams. The minor changes in velocity detected in the constricted strand emphasize that streaming of cytoplasm is not a pressure-induced flow.

Effect of Time on Rate of Streaming

The previous observations were made upon cells within 8 hr after preparation of the microchambers. Observations were continued on the individual cells daily for a period of 10 days and irregularly thereafter for 30 days to determine any changes in the pattern or rate of streaming. Recordings on film were restricted to 6 and 10 days after chamber preparation. All the tobacco cells mentioned previously were streaming after 6 days.

The analyses of organelle movement along a thin strand, 0.5 μ in diameter, for an albino cell maintained in the microchamber for the 6-day interval is indicated in Fig. 12. The three organelles, A, B, and C, illustrated for the thin strand were plotted at varying velocities along the strand. The organelle A moved at an average speed, 6.4 μ/sec, considerably greater than the other organelles. Note that this organelle A is positioned between the two slow moving organelles.

The same strand, after 11 sec, increased to 1 μ in diameter (Fig. 13). The change in diameter did not appreciably affect the velocity of the

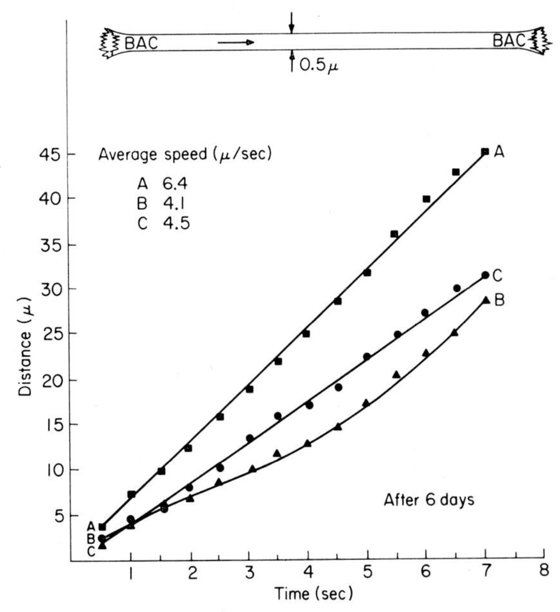

FIG. 12. Transvacuolar strand 0.5 μ in diameter in an albino cell 174 μ in diameter after 6 days in microchamber. Average speeds of the organelles were somewhat variable. (Interval = ½ sec.) Top diagram—relative positions of organelles at beginning and end of plotting.

organelles. The average speed recorded over a distance of approximately 35 μ for the three organelles A, B, and C was 4.2, 4.6, and 4.7 μ/sec, respectively, with only minor deviations from a linear time–distance relationship.

Active streaming was evident in the cells after 10 days enclosure in the microchambers (Fig. 14). The velocities for the organelles in the peripheral cytoplasm did not decrease appreciably after 10 days in the chamber from those speeds registered for organelles during the first 8 hr. Whereas the organelles A, B, and C moved at a rather uniform velocity, others were observed to move more irregularly along a course.

The average speeds for all the organelles plotted in the cells after 6 and 10 days in the microchambers are indicated in Table I.

Streaming continued in all the cells employed for these measurements until all cells plasmolyzed and appeared to die at the end of 18 days. The cultured cells appeared to be a very tolerant form of cell since the streaming rates of the above magnitude were evident in other cells 30 days after being mounted. After this period of time, the observations were discontinued. The oxygen and carbon dioxide gas tensions, both gases known

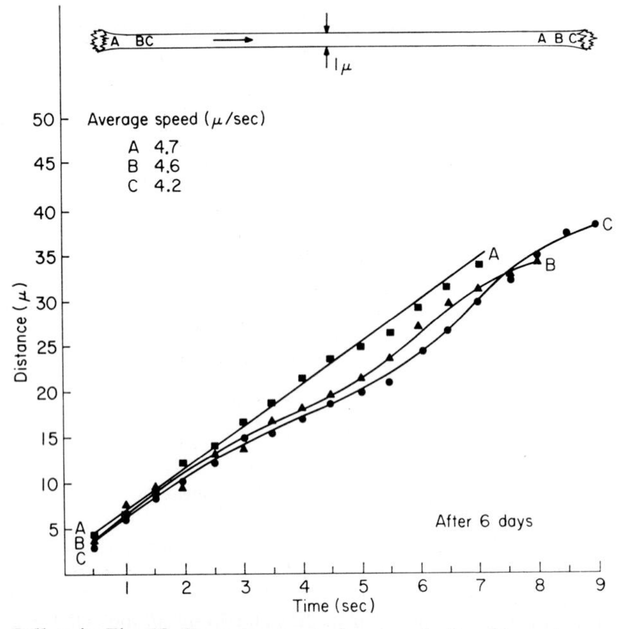

FIG. 13. Cell as in Fig. 12. Same transvacuolar strand after 11 sec increased to 1 μ in diameter. Average speeds of the organelles corresponded closely with the slow moving organelle in Fig. 12. (Interval = ½ sec.) Top diagram—relative positions of organelles at beginning and end of plotting.

to measurably affect the streaming rate in other materials, were not known for the sealed microchambers.

Alterations in the form of the cytoplasm in the cultured cells became evident over prolonged periods of observation. Initially, the peripheral cytoplasm and that along a transvacuolar strand appeared optically homogeneous except for the organelles and the vacuolar membrane. After 10 days in culture, however, other cells were observed to develop an extensive system of phases or membranes which were arranged more or less parallel to each other and at irregular intervals were joined together into an anastomosing network. This system was very mobile and portions of membranes could be observed to undergo contortions during movement in the cell. Segments of the new membrane varying from 5–40 μ in length were observed to separate from one position, move for some distance, and reattach to the membrane or phase system at another position within the cell.

Further observations indicated that each membranelike structure may be tubular in form. As viewed on film, a membrane periodically formed dilations which contained a volume of so-called groundplasm and organelles. These dilations, too, were subject to change and would rapidly

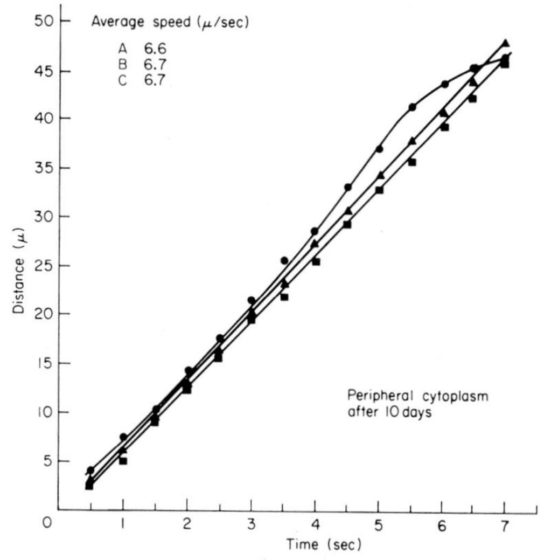

FIG. 14. Peripheral cytoplasm in chlorophyllous cell 125 μ in diameter after 10 days in the microchamber. The period of time that the cell has been maintained in the culture chamber has not noticeably decreased the average speed of the organelles when their speeds are compared with organelles in many of the other graphs either for the transvacuolar strands or for the peripheral cytoplasm. (Interval = ½ sec.)

constrict to assume the form of the tubular membrane structure. The organelles were not necessarily confined to the groundplasm within these tubelike structures, but often were observed between the parallel tubes. At other locations in the cell, the organelles were observed to cross the membrane system at oblique angles.

Discussion

Heilbrunn (1952), Kamiya (1959), and Lundegårdh (1922) have demonstrated the presence of a cortical gel layer of cytoplasm in the internodal cell of *Nitella* and *Chara* as well as the plasmodium of myxomycetes. One interpretation of the streaming mechanism centers upon the physical differences between the gel and sol phases and that forces generated at the interface between these two phases are responsible for the movement of the sol phase. The observations of Kamiya and Kuroda (1957a,b, 1958) upon isolated drops of *Nitella* emphasize the importance of a gel phase. It has been difficult to demonstrate a gel layer in the cells of higher plants and, indeed, the streaming activity in the cultured cells viewed on film indicate that no conspicuous gel layer was detectable to the inside of the plasma membrane (ectoplast), and none to the inside of the vacuolar membrane (tonoplast) in a transvacuolar strand.

It is not probable, in my interpretation, that the diverse movements of the organelles are to be explained by changes in cytoplasmic viscosity. The rates of movement of organelles can be similar in transvacuolar strands of varying diameter. The cytoplasm in transvacuolar strands of small diameter does not appear to be more viscous than that in strands of large diameter when one evaluates the average speed of organelles along the linear part of the curves (compare Figs. 1, 2, 6, and 11). Two-directional movement and passing of organelles along a transvacuolar strand are most difficult phenomena to explain relative to changes in viscosity along the length or across the diameter of a transvacuolar strand.

The occurrence of a rhythm in the streaming cytoplasm of *Physarum polycephalum* has been demonstrated repeatedly in many excellent experiments by Kamiya (1962) and his co-workers. Indeed, Kamiya in analyzing plasmograms has indicated that a series of rhythms coexist within the protoplasmic system of *Physarum*. My initial efforts in analyzing the movement of organelles in cultured cells were directed in part toward determining whether a rhythm could be detected in the cytoplasm of the higher plant cells.

In cultured cells the pattern of cytoplasmic strands along the wall is constantly changing and the cytoplasm does not exhibit a rhythmic reversal in direction of flow such as observed in *P. polycephalum*. Similarly the transvacuolar strands *in toto* are in motion, some disappearing from the strand reticulum, others appearing from the peripheral cytoplasm (Küster, 1956; Mahlberg, 1961; Mahlberg and Venketeswaran, 1962, 1963). Superimposed upon these changes are the movements of the cytoplasm and the organelles within a strand. To complicate the problem further, quantities of cytoplasm in cultured cells are observed to aggregate and move en mass and later disperse, the entire phenomenon appearing to lack periodicity (Mahlberg, 1961).

Comparison of organelle velocities revealed that no apparent rhythm existed in the cytoplasm which involved 1/2–sec intervals of time or multiples of this unit of time. The time–distance relationships plotted for many organelles in different positions within the cell demonstrated that an organelle during any one interval of time can move at different rates of speed varying from 0–19 μ/sec. Some organelles were observed to decelerate to a speed of 0 μ/sec. Other organelles accelerated along their path even when confined to strands of small diameter. The speed of several organelles was evident as a sigmoid curve. Many organelles moved with a nearly constant velocity—perhaps surprisingly so—as represented by their linear time–distance relationships in several transvacuolar strands.

As noted for nearly all curves, variations in velocity for a single organelle can occur for each 1/2 sec. This nonconstant speed for the organelle is emphasized in Figs. 9 and 10 in particular. The changes in speed appear to be quite irregular, and my initial attempts to detect a synchronism for organelle movements have not been successful.

Organelles moving with a constant velocity would represent most accurately the streaming rate of the cytoplasm. However, each of several organelles can move with constant but different speeds along the same strand, making it difficult to select the speed representing the true rate of streaming. Attempts to correlate movement of different organelles along the same strand over a period of 20 sec and more did not reveal a rhythmic or pulsative character for the cytoplasm. For example, two organelles can move at different speeds when crossing the same zone in thin strands, either at the same time or at different times. As noted in Fig. 4, however, there were instances along a strand when organelles (such as C, D) separated from each other by a considerable distance responded with very similar patterns in movement. These nearly identical movements suggested that a filamentous mass of cytoplasm of considerable length was undergoing changes in the rate of flow.

The most difficult patterns of movement to interpret are those where organelles move in opposite directions, where organelles pass each other, or where individual organelles reverse directions during their movement along the strand. These variations suggest that the cytoplasm within a strand is not moving at a similar velocity; rather that numerous individual streams (microstreams) of flow, or possibly that fibrils (Jarosch, 1958) constitute the cytoplasm of a strand. The cytoplasm, thereby, can be interpreted as a form of structured fluid as suggested by Allen (1962) for *Amoeba*. This is supported by the observations upon organelle velocity in constricted strands and those of varying diameters. The flow rate does not correspond to predictable changes in velocity, as indicated in Eq. (1), when the cytoplasm moves through strands of different diameters. The cytoplasm, therefore, does not appear to be a homogeneous or nonstructured matrix, nor does it appear to possess an internal pressure which controls the movement of cytoplasm. It should be emphasized that the interpretation of the structured character of cytoplasm as microstreams or fibrils must be sufficiently flexible to be definable at the molecular level as more data are accumulated on the energetics of streaming.

The microstreams are interpreted to be of various lengths and are interdigitated with each other; they represent the bulk flow of cytoplasm, are capable of sliding over one another, and move at varying rates in different directions.

Some impression of the maximal diameters of microstreams can be gained from the following observed movements of organelles. Tentatively, the diameter of a microstream, one or more moving an organelle, can be related with the varying velocities for organelles along a transvacuolar strand. In Fig. 4, the organelle B passed A along a strand 1 μ in diameter. Since the speed of both organelles was nearly linear, at least two microstreams about 0.5 μ in diameter should be present in that strand. The three organelles (Fig. 7) in the thin strand, 1 μ in diameter, moved at different rates, two of the organelles crossing the path of the third. Their movement represented at least three different microstreams approximately 0.3–0.4 μ in diameter, since all organelles were moving at different rates of speed.

In a constricted transvacuolar strand (Fig. 11) the thick portion of the strand is interpreted to consist of numerous interdigitated microstreams moving at different velocities in both directions whereas the constricted portion contained fewer such interdigitated microstreams. Various microstreams are interpreted to terminate or become folded in the enlarged portion of the strand. This interpretation is supported from observations which indicate that some organelles decelerate appreciably upon approaching the zone of constriction and reverse directions. It was too difficult to plot the course of these organelles for any distance since their path was quickly obscured by other organelles. Microstreams in the thick portion of the strand were interpreted to move individually into the constricted portion of the strand. Moving as interdigitated units the microstreams would not increase appreciably in speed when they became components of the cytoplasm of the constriction.

The movement of organelles is associated primarily with the microstreams, one or more carrying the organelles at different velocities or in different directions. The most generally accepted interpretation of organelle and particle movement emphasizes the role of contractile or fibrous proteins which effect the movement of all cytoplasm and which carry the particles and organelles around the cell (Frey-Wyssling, 1955; Pfeiffer, 1953; Schmidt, 1941; Seifriz, 1944). However, the capacity of independent motility for metabolizing organelles must not be disregarded although direct evidence for the mechanism of self-propulsion is presently unavailable (Bünning, 1935; duBuy and Olson, 1940).

As interpreted here the sudden changes in direction or in velocity of an organelle reflect a shift in positions between different microstreams, these moving at different rates or in different directions. The metabolizing organelles are being carried within or upon, but not affixed to, the microstreams. Organelle transfer is effected by its capacity of self-motility whereby it moves from one microstream to another.

A streaming pattern observed in *Acetabularia,* described by Kamiya (1962) as multistriated streaming, is also apparent in higher plant cells. The plasmic network, representing either phase differences or membranelike structures, which became apparent in some of the cultured cells after 10 days, may be similar to the membranes observed by other investigators in different cells (Boresch, 1914; Honda *et al.*, 1962; Küster, 1956; Scarth, 1942) or the striated tracks in *Phycomyces* and *Acetabularia* (Kamiya, 1959). Possibly the fibrils identified by Jarosch (1958) in *Chara* may be identical to the membranous structures observed in the cultured cells.

The movement and contortions exhibited by segments of this membranelike system was associated with the streaming of the cytoplasm in general. During the initial phases of development numerous dilations were evident along the membranelike complex, but these became less conspicuous with time. The groundplasm confined within the compartmentalized cytoplasm continued to stream as indicated by the movements of the organelles. The apparent segregation of the cytoplasm did not significantly alter the average speed of the organelles even after 10 days in the microchambers. Characterization of the membranelike structures in the cytoplasm will require further investigation.

Summary

The velocity of an organelle within streaming cytoplasm was recorded to vary from 0–19 μ/sec over a period of several seconds when plotted at 1/2-sec intervals. In this initial study it was not possible to detect a pulsative or rhythmic pattern for such variations in velocity.

Each of several organelles can move at different velocities along a strand, frequently passing each other as they move in the same direction. Bidirectional streaming often was detected in strands of different diameters; in several instances individual organelles were observed and recorded to alter their direction of movement along a strand.

Any two organelles can follow very similar, essentially synchronized, patterns of movement along a strand. Such patterns can be complex in that they may include variations in velocities as well as changes in direction of movement.

In a constricted strand the velocity of an organelle does not increase appreciably when the organelle enters the constriction, indicating that streaming cytoplasm is not under pressure.

The cytoplasm appears to possess structural characteristics, possibly consisting of microstreams of unknown dimensions. The numerous microstreams interdigitate with each other and are interpreted to move at varying velocities and in different directions.

Streaming may represent a dual problem involving motion at two different levels: movement of the ground cytoplasm and movement of metabolizing organelles. Movements of the organelles represent, primarily, transport by the microstreams.

Several organelles being carried by one or more microstreams moving at similar velocities would exhibit a synchronized pattern of movement. It is presently interpreted that metabolizing organelles may possess the capacity for self-motility. Their rapid changes in velocities and in directions may reflect their capacity to utilize energy to shift their position from one microstream to another which is moving at a different speed or in a different direction.

Acknowledgment

The author is indebted to Dr. S. Venketeswaran and Dr. E. de Fulvio for assistance in preparation of portions of the film.

References

Allen, R. D. (1962). *Sci. Am.* **206**, 112-122.

Boresch, K. (1914). *Z. Botan.* **6**, 97-156.

Bünning, E. (1935). *Z. Wiss. Mikroskopie* **52**, 166-172.

Denham, H. J. (1923). *J. Textile Inst. Trans.* **14**, T85-T113.

duBuy, H. G., and Olson, R. A. (1940). *Am. J. Botan.* **27**, 401-413.

Frey-Wyssling, A. (1955). *Protoplasmatologia* **2**, A2, 1-244.

Heilbrunn, L. V. (1952). "An Outline of General Physiology." Saunders, Philadelphia, Pennsylvania.

Honda, S. I., Hongladarom, T., and Wildman, S. G. (1962). *In* "Organelles in Living Plant Cells." Film. Educational Film Sales, University of California, Berkeley, California.

Jarosch, R. (1958). *Protoplasma* **50**, 93-108.

Kamiya, N. (1959). *Protoplasmatologia* **8**, 3a, 1-199.

Kamiya, N. (1962). *In* "Encyclopedia of Plant Physiology" (W. Ruhland, ed.), Vol. XVII, pp. 979-1035. Springer, Berlin.

Kamiya, N., and Kuroda, K. (1957a). *Proc. Japan Acad.* **33**, 149-152.

Kamiya, N., and Kuroda, K. (1957b). *Proc. Japan Acad.* **33**, 201-205.

Kamiya, N., and Kuroda, K. (1958). *Proc. Japan Acad.* **34**, 435-438.

Küster, E. (1956). "Die Pflanzenzelle." Fischer, Jena, Germany.

Lundegårdh, H. (1922). "Linsbauers Handbuch der Pflanzenanatomie," Vol. 1, pp. 1-402. Gebruder Borntraeger, Berlin.

Mahlberg, P. (1961). *In* "Streaming in Cultured Cells of *Euphorbia marginata.*" Film. W. R. Smith, Pittsburgh, Pennsylvania.

Mahlberg, P. (1962). *Exptl. Cell. Res.* **26**, 290-295.

Mahlberg, P., and Venketeswaran, S. (1962). *In* "Formation of Transvacuolar Strands in Cultured Cells of *Euphorbia marginata.*" Film. W. R. Smith, Pittsburgh, Pennsylvania.

Mahlberg, P., and Venketeswaran, S. (1963). *Am. J. Botan.* **50**, 507–513.

Martens, P. (1940). *Cellule Rec. Cytol. Histol.* **48**, 247-306.

Pfeiffer, W. (1953). *Ber. Deut. Botan. Ges.* **66**, 2-5.

Probine, M. C., and Preston, R. D. (1958). *Nature* **182**, 1657-1658.

Scarth, C. W. (1942). *In* "Structure of Protoplasm" (W. Seifriz, ed.), pp. 99-107. Iowa State Univer. Press, Ames, Iowa.

Schmidt, W. J. (1941). *Ergeb. Physiol. Biol. Chem. Exptl. Pharmakol.* **44**, 27-95.

Seifriz, W. (1944). *In* "Medical Physics" (O. Glasser, ed.), pp. 1127-1145. Year Book Publ., Chicago, Illinois.

Venketeswaran, S., and Mahlberg, P. (1962). *Physiol. Plantarum* **15**, 639-648.

Discussion

Dr. Allen: I must say I am quite struck by the similarity of your material to Foraminifera which I have studied and will report on later. One thing I have observed both in your moving pictures and in Foraminifera is that particles, if you analyze them over periods of time such as a few seconds, appear to be going at more or less uniform velocities; but, if you analyze them over shorter intervals, they move in jerks.

I got the definite impression, that in your film many or most of the particles move in a more or less jerky fashion. I think this shows in some of your graphs in which the points lie first above and then below the line.

Dr. Mahlberg: Yes, I agree that the motions are somewhat irregular. At first we thought there might be a periodicity in the motion, but we were not able to demonstrate this.

Dr. Nakai: I am very much interested to see that your film shows bidirectional streaming. As I will show in my film, neuronal pseudopodia also show bidirectional movement, however, different kinds of particles are involved and the character of the movement is different. I noticed that some of the fibers have a thickness of about 1 μ, and that the particles are hardly smaller. Therefore, I wonder whether there is enough room in a cytoplasmic strand 1 μ thick to generate separate streams, each carrying particles and sometimes going in opposite directions. Would you consider the hypothesis that the particles are actively metabolizing cell organelles which can interact with the immediate cytoplasmic surroundings and can generate a local streaming rather than simply following predetermined pathways?

Dr. De Bruyn: One could certainly imagine that a metabolizing organelle must take in molecules of very small size for synthesis, then must allow other molecules to diffuse out.

There are apparently in mitochondria little holes, or sieves, which may pass the large molecules unable to go through the membrane proper. If this were so, then you might get surface tension differences which then would shoot these organelles off under self-motility. This is one possibility.

Dr. Bishop: Are you identifying these particles? Are they metabolic particles or are they uncertain large granules? You didn't say much about them.

Dr. Mahlberg: The preliminary work on *Euphorbia* indicates that these are very likely mitochondria. Assuming all of the particles are as similar as they appear, we could call them metabolizing organelles. We can also see other dissimilar granules. I didn't point any out, but there are small granules in these cultured cells that exhibit very, very little motion at all.

DR. REBHUN: We have been studying so-called independent movements in the cytoplasm and their organization during mitosis. It is rather interesting that the class of particles which can participate in movement such as this, i.e., moving in one direction up to 30 μ and suddenly reversing or stopping, is extremely wide. It may be melanin granules in melanocytes and granules that have been identified as crystalline heme in certain eggs, and a variety of other particles in other cells. However, perhaps the most telling results in this direction were gotten by Andrews in *Stentor* and various other ciliates in which he observed carmine particles, India ink, and starch granules participate in these peculiar movements. Chambers saw these movements in the oil droplets "sliding" along astral rays; movements very similar to those of other naturally included particles. In such cases it would be, of course, impossible to ascribe this to metabolic processes within the particle.

DR. MAHLBERG: The cytoplasm is definitely moving. All that I am suggesting for self-motility of organelles is that they are able to transfer themselves from one stream to another. I am not sure how the inert granules do it. Perhaps they are in the boundary between.

On the other film, one can see the development of what appeared to be an entire membrane system within these cells. It was possible to see along one strand a series of membranes with movement going in opposite directions.

DR. DE BRUYN: It looks as if you might have something like elastic ropes moving sometimes in one direction and sometimes going back. They may not be really streams in the hydrodynamic sense.

CHAIRMAN THIMANN: Perhaps something like Miss Kuroda's polygons.

DR. MAHLBERG: It could be something of that type. I am suggesting here there is an end, so to speak, to a stream.

DR. JAHN: I would like to emphasize the point made by Dr. Allen, that the phenomena you described are very similar to those in Foraminifera. Furthermore, in forams the "ropes" not only are elastic but they are on the surface so that insoluble dye granules can attach to the elastic rope and be pulled along, then become transferred to another "rope," and return.

DR. MAHLBERG: The organelle itself does not appear to be attached to the kind of membrane system I described; they are quite able to move around indiscriminately. They can aggregate for a while and then disperse.

The Organization of Movement in Slime Mold Plasmodia[1]

PETER A. STEWART

Department of Physiology, Emory University, Atlanta, Georgia and Biology Department, Brookhaven National Laboratory, Upton, New York

Plasmodial Characteristics

At first sight, plasmodia of the acellular slime molds, such as *Physarum polycephalum* (Fig. 1) appear to be ideal organisms in which to study protoplasmic movement. They are large, easily manipulated, and display high speeds of protoplasmic streaming, up to 1300 μ/sec (Kamiya, 1959). On the other hand, the patterns of movement observed are labile and extremely complex. The streaming fluid protoplasm is non-Newtonian, and the gel-like structure through which it streams possesses poorly defined rheological properties (Frey-Wyssling, 1952). Detailed quantitative application of hydrodynamic theory therefore appears to be fruitless (Stewart and Stewart, 1959). Nonetheless, a model which explains the complexities of the movement patterns can be constructed on the basis of four assumptions about the properties of the nonstreaming protoplasm, and is described later. First, we consider the major characteristics of plasmodial movement which such a model must explain.

1. The most active streaming in a plasmodium appears in rather clear-cut channels, ranging up to a millimeter in diameter, but smaller streams can also be seen throughout the substance of the plasmodium, even in the walls of the larger channels.

2. In any particular channel, flow reverses direction repeatedly, but not with a precise periodicity.

3. When reversal of streaming occurs, streaming does *not* stop all over the plasmodium and then begin again in the opposite direction everywhere. Reversals occur in different regions at different times, and the relationships between the timing of reversals at different places are constantly changing.

4. Two types of reversal of streaming direction occur and alternate at any point. Afferent reversals are those in which the streaming on both

[1] The work on which this article is based was supported in part by USPHS grant #E-1433 to Emory University and partly carried out at Brookhaven National Laboratory under the auspices of the U. S. Atomic Energy Commission.

sides of the reversing region is toward that region, whereas in efferent reversals streaming is away from the region of reversal.

5. Both types of reversal travel across the plasmodium, with speeds of the order of 10 μ/sec.

6. Local streaming movements occasionally occur for short periods of time in small regions, more or less independently of surrounding streaming patterns.

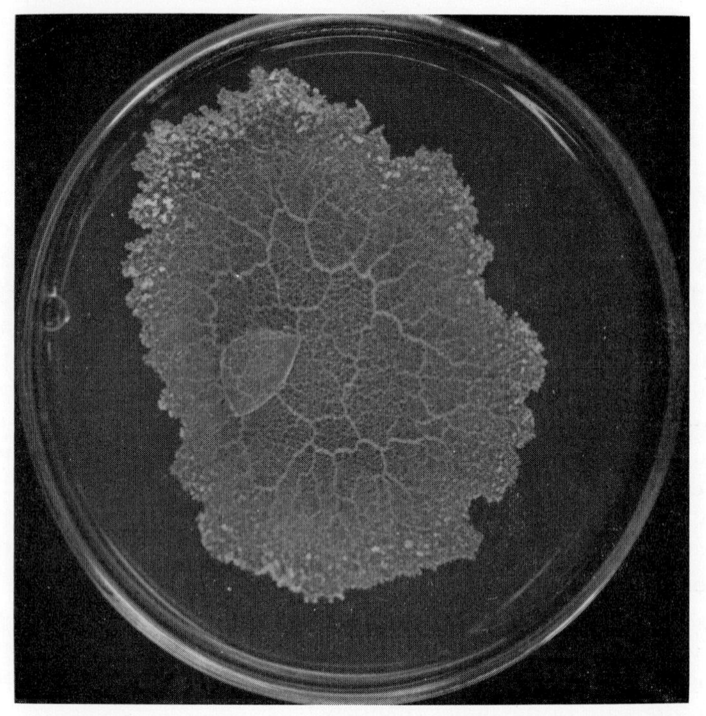

FIG. 1. A plasmodium of *Physarum polycephalum* growing on nutrient agar in a 14-cm petri dish.

7. Under high magnification, many particles in a streaming plasmodium are seen to undergo occasional sudden movements over distances many times their own diameters and in directions apparently unrelated to any nearby streaming movements. These movements have been called "independent movements" (Stewart and Stewart, 1959), but are now usually referred to as saltatory movements (see Chapters by Rebhun and Parpart in this volume). Analysis of cine films has shown that the particles often move at constant speeds over most of their paths. Since they are moving through a viscous medium, this fact implies the continuous production of motive force along the paths. The mechanism of these

movements in plasmodia, and their relationship, if any, to streaming movements are both obscure at present. The explanatory model presented later provides no explanation for them.

8. In an actively streaming plasmodium, there are two different phases or states of protoplasm, one more fluid and streaming, the other nonstreaming and apparently more gel-like. The two states are not sharply distinguishable, either in space or in time. They are constantly being interconverted, and no clear boundaries between them can be drawn for more than a few seconds. The fluid state interpenetrates the gel-like state in a variety of morphological patterns, ranging from large channels to very small local streamlets. The structure of a streaming plasmodium at the light microscope level is, therefore, best described as a labile meshwork of nonstreaming gel-like protoplasm with a fluid protoplasmic phase coursing throughout its interstices and constantly interchanging with it (Stewart and Stewart, 1959). Microscopically visible particles display vigorous Brownian movement in both states of the protoplasm, suggesting that the submicroscopic structural differences between them are slight, as is also suggested by their continuous interconversion.

A dramatic rapid change in this structure occurs when a streaming plasmodium is suddenly treated with a variety of agents such as carbon dioxide, ether vapor, or even mechanical shock (Seifriz, 1941). Streaming ceases abruptly, and the whole plasmodium converts to a firm gel in which no Brownian movement is seen, except inside of vacuoles. After a time lasting from several seconds to many minutes, and often in the continued presence of the applied agent, Brownian movement suddenly reappears throughout the plasmodium and small local streaming movements begin. The streams gradually increase in size and fuse, finally forming an overall movement pattern similar to that occurring before the stimulus. How similar the new pattern is to the old depends on how long the induced gelation lasts.

The sudden transition from a streaming plasmodium containing fluid and nonfluid states of protoplasm, both showing Brownian movement, to a uniformly gelled protoplasmic mass incompatible with streaming, indicates that there are two quite different nonfluid states of living plasmodial protoplasm. At present, nothing is known about the structural bases for the differences between these two "gels," or the nature of the transition from one to the other.

9. The fluid protoplasm in a streaming plasmodium can be moved in its channels by an externally applied pressure gradient, for example by pressing unevenly on a cover slip. The nonfluid gel-like meshwork responds elastically to such forces. If properly timed and oriented, external pressure gradients can be applied so as to prevent streaming in any

chosen channel for considerable periods of time. Streaming continues in nearby channels, although the pattern is distorted by the applied forces. Kamiya and his co-workers have used this fact as the basis of an ingenious double-chamber technique, with which they have made extensive studies of changes in local motive force in response to a variety of agents (Kamiya, 1959).

10. Time-lapse studies of plasmodia reveal a different aspect of the movement patterns. What is seen in cine films with the time scale speeded up fiftyfold is most simply described as a succession of waves of local swelling or expansion moving across the plasmodium. These waves do not follow a constant pattern and they do not generally seem to originate from a single or constant source. They sweep across the plasmodium with speeds on the order of 10 μ/sec. The peak of each wave is seen to be a region in which an afferent reversal of streaming is occurring, and between any two peaks is a region in which an efferent reversal takes place. Once these waves have been seen in time-lapse movies, they can easily be seen by the naked eye when a streaming plasmodium is carefully examined at intervals of 20 or 30 sec. Their profiles have not yet been carefully mapped, but they are clearly nonsinusoidal.

Explanatory Model

To explain many of these aspects of streaming in plasmodia, we propose a model based on the following four assumptions:

Assumption 1. The gel-like meshwork of nonfluid protoplasm throughout the plasmodium is in a state of tension, so that the interstitial fluid in the channels and streamlets is under slight hydrostatic pressure.

The existence of such an internal pressure is easily demonstrated by tearing open the surface of a channel with a sharp point. A drop of fluid protoplasm spurts out through the wound, and then gels in a few seconds, thereby preventing further outflow.

Assumption 2. A local weakening of the nonfluid meshwork occurs, such that it loses its tension, or relaxes, or perhaps simply converts to fluid to a large extent.

Because of the internal pressure, fluid protoplasm from adjacent regions should then stream into this region, causing it to swell. Flow will occur through whatever channels are available. The adjacent regions will thus lose interstitial fluid and shrink in volume. Wherever this happens, a rearrangement or contraction of the gel-like meshwork must occur, since the tension in it is maintained, according to assumption 1.

Assumption 3. The process of local weakening is reversible, so that after a short time, on the order of a minute, the gel-like meshwork regains its original strength locally.

This could occur by conversion of the inflowing fluid protoplasm from adjacent regions into nonfluid meshwork. Inflow will then cease. Note that it is not required that any significant volume change accompany this conversion.

Assumption 4. This cycle of changes in the gel-like meshwork is contagious, so that it spreads from the initial locale to adjoining regions, and thus propagates through the plasmodium.

Our four assumptions add up to a picture of waves of temporary, reversible, local weakening of the gel-like meshwork, spreading through the plasmodium. Streaming then results from the response of the fluid protoplasm to the pattern of pressure gradients accompanying these waves of local weakening and recovery.

Discussion

The model described by these four assumptions accounts for the complexities of plasmodial movement patterns in simple terms. The waves of expansion seen in the time-lapse pictures are interpreted as the visible expression of the waves of local weakening of the meshwork. Regions of afferent reversal are regions in which a wave of weakening is at its peak. Efferent reversals occur in regions between these peaks. Reversals travel across the plasmodium because the waves of weakening do, and at the same speed. The inconstant relationship between time of reversal and time of maximum channel diameter at a point (Kamiya, 1950) can be understood as a result of interaction between the direction of the channel axis and the direction of travel of the waves. Only in case a wave of expansion moves along a path parallel to the channel axis should a direct correlation between diameter and reversals be expected. In other geometrical situations the relationship will be complex.

Furthermore, since the fluid and gel-like states of a streaming plasmodium differ so subtly and are so readily interconverted, stable patterns for the waves of weakening are not to be expected. The pattern followed by the sequences of reversals should therefore be very changeable. Occasional small local changes in gel strength may occur and not be propogated, leading to restricted local movements, as actually observed. When a drop of fluid protoplasm spurts out at a tear in the plasmodial surface, local streaming patterns will be distorted until the drop gels, when the over-all streaming pattern will again take over in response to the sequence of waves in the meshwork. This is precisely what is seen in this situation.

Pressing on the cover slip over a plasmodium moves the fluid protoplasm about in its channels simply by swamping the local pressure gradients normally produced by the wave sequence. Motive force measurements by the double-chamber technique are, therefore, interpretable as measurements of the interaction between waves of weakening in the two plasmodial segments and the externally applied pressure difference between the two chambers.

Locomotion of a plasmodium, in terms of this model, is the result of three interacting factors: the detailed wave forms of the waves of weakening, the pattern they follow as they spread over the plasmodium, and the locations of their sources. The time-lapse records suggest that any point of a plasmodium may be able to act as a source, depending on external conditions. The detailed patterns followed by the waves are not stable, as already mentioned, but for locomotion to occur a statistical kind of stability must be present, and is observed. More than one source is often apparent in a single plasmodium, further increasing the complexity of the patterns of waves and, therefore, of streaming movements. The detailed wave forms will affect locomotion, since they will determine whether, and in which direction, net displacement of fluid protoplasm results from the passage of the waves through a local region.

One more factor must be added to this picture in order to explain purposeful and integrated locomotor activity, namely, some sort of overall integrative control system. A plasmodium is an organism, and displays purposeful as well as apparently spontaneous behavior. An explanation of the behavior-controlling system seems to be beyond the capabilities of the model presented here, although it is tempting to suggest that if the gel-like meshwork can transmit the necessary signals for the propagation of the waves of weakening, it may also be able to transmit controlling signals. On the other hand, Kamiya's observation that the phasing of dynamoplasmograms is maintained through short periods of complete gelation induced by carbon dioxide (Kamiya, 1959) suggests the operation of a control system which is independent of actual streaming and of the gel-like meshwork.

The movement and morphological changes of *Amoeba striata* has been analyzed in somewhat similar terms by Abé (1962). His precise description of the events associated with movement in living amebae is a beautiful example of how much can still be learned from careful observation with the light microscope. In this ameba, the regions of breakdown and reconstitution of "gel" are separated in space, and maintained that way by the gel structure, whereas in a plasmodium both processes occur everywhere, but are separated in time in a consistent way so as to result in streaming, reversals, and locomotion.

At the end of the scale of levels of explanation lies the question of a macromolecular basis for the model proposed here. In view of the studies of several workers on an actomyosin-like protein extracted from plasmodia

(Loewy, 1952; Ts'o *et al.*, 1957; Nakajima, 1960), it is intriguing to speculate that the gel-like meshwork may consist primarily of this protein system. The cause of local weakening of structure might then be as simple as local changes in adenosine triphosphate (ATP)—adenosine monophosphate (AMP) ratios, and the ATPase activity associated with this system could provide the necessary reversibility of the effect. Added ATP has been shown to increase the proportion of fluid to nonfluid protoplasm in plasmodia (Ts'o *et al.*, 1956) and to increase the motive force (Kamiya *et al.*, 1957). It is also possible that the change in ATP–AMP ratio might involve some sort of chain reaction in the meshwork, which would account for the propagation of this change and, therefore, of the wave of weakening.

Unfortunately, any attempt to fix plasmodial protoplasm results in its immediate conversion to the completely gelled state, in which streaming and the gel-like meshwork do not occur. Interpretation of electron or light micrographs of conventionally fixed specimens is therefore extremely difficult. Homogenization and extraction procedures preliminary to biochemical studies also wreak complete havoc with the delicately balanced fluid–gel relationship. Even simple squashing of a plasmodium under a cover slip leads to a third type of gel-like structure and the postmortem appearance of numerous fibers (Stewart and Stewart, 1959). Analysis of the streaming problem at the molecular level is thus rendered inordinately difficult by the lability of the systems involved and by the fact that streaming cannot be isolated for study. For the resolution of these problems, more refined techniques of observation applicable to living material must be developed. Methods which require the death of organism can be of only limited value.

Whatever the facts at the molecular level, the four assumptions presented here provide a descriptive explanation at the supramolecular level, in terms of which the complex phenomena of protoplasmic streaming in slime mold plasmodia form a consistent set of interrelated events. The nature of the control systems that operate on this model so as to integrate streaming movements into behavior remains unknown.

Acknowledgments

I am indebted to Mr. Robert Smith and Mr. A. P. Christoffersen of the Photography and Graphic Arts Division of Brookhaven National Laboratory for their proficient assistance and cooperation in the preparation of the film shown at this symposium.

References

Abé, T. H. (1962). *Cytologia (Tokyo)* **27**, 111.

Frey-Wyssling, A. (1952). "Deformation and Flow in Biological Systems." North-Holland, Amsterdam.

Kamiya, N. (1950). *Cytologia (Tokyo)* **15**, 194.
Kamiya, N. (1959). *Protoplasmatologia* **8**, 1.
Kamiya, N., Nakajima, H., and Abé, S. (1957). *Protoplasma* **48**, 94.
Loewy, A. G. (1952). *J. Cellular Comp. Physiol.* **40**, 127.
Nakajima, H. (1960). *Protoplasma* **52**, 413.
Seifriz, W. (1941). *Anesthesiology* **2**, 300.
Stewart, P. A., and Stewart, B. T. (1959). *Exptl. Cell Res.* **17**, 44.
Ts'o, P. O. P., Bonner, J., Eggman, L., and Vinograd, J. (1956). *J. Gen. Physiol.* **39**, 325.
Ts'o, P. O. P., Eggman, L., and Vinograd, J. (1957). *Biochim. Biophys. Acta* **25**, 532.

Discussion

Dr. Allen: I would like to ask if you assume that the motive force generation, whatever it is, is uniform over the mold or whether this might change in different parts. For example, suppose you have a double-chamber preparation, as Dr. Kamiya has, with cytoplasm streaming from one side to the other, do you expect contraction takes place everywhere and then a wave of weakening simply spreads from one side to the other, or do the contraction and weakening have to vary independently?

Dr. Stewart: When you look at what is happening in the channel separating the blobs in the double-chamber technique, you don't know what is happening out in the blobs. The fact is when you stop streaming in the channel, streaming continues independently in each blob. I haven't yet done this, but I would guess that if you took time-lapse pictures of such a preparation you would see similar kinds of waves moving across the two separated segments of the plasmodium. What is happening in the channel connecting them is the interaction between these waves.

Dr. Allen: Suppose the mass of the blobs were reduced almost to zero?

Dr. Stewart: I think that is a very interesting experiment. I wonder if you have done that, Dr. Kamiya? What happens when you have a system like this in the double-chamber situation?

Dr. Kamiya: Some time ago, Dr. Ohta cut off one blob on the dumbbell-shaped plasmodium while the other blob was left as it was. In that case there was no detectable change in the motive force. Generally speaking, there is no direct relationship between the motive force and the size of the plasmodium. This is somewhat comparable to the fact that the electromotive force is independent of the number of batteries as far as they are connected in parallel. Increasing the mass of the blobs may correspond, as it were, to the parallel connection, rather than the serial connection, of minute parts generating the motive force.

Dr. Marsland: You were speaking of a wave of relaxation. How can you differentiate between a wave of relaxation and contraction?

Dr. Stewart: I am not really sure whether one can or not. I would be interested in any further comments on that. It seems to me that when one looks at the over-all streaming patterns in the plasmodium, it is much easier to understand these, on a "hydrodynamically intuitive" basis, if one assumes that something is "giving way" and cytoplasm is streaming in from all available sources. On the other hand, if you look at it from the point of view that the ectoplasm is squeezing endoplasm out of a certain region my feeling is that you should observe a slightly different pattern from what is seen.

Dr. Allen: We have asked this question actually in another way. We were interested in making a direct test of whether endoplasm might be pulled in one direction by an endoplasmic contraction or pushed in the other direction by an ectoplasm contraction of the type Goldacre has assumed for the ameba. Two of my students, Reid Pitts and David Speir, have done some rather interesting experiments using very small blobs of *Physarum* connected by a single strand. Below the two blobs are placed two thermistors. By means of a sensitive lock-in amplifier technique, they have been able to measure a temperature difference cycle. We were, of course, interested in finding out which of the two ends became hotter during a single streaming cycle because the expectation was that contraction on one side or the other (front or rear) would produce a certain amount of heat since contractile processes are in general about only 20% efficient. We did find a cycle, and the polarity was such that the heating always occurred at the origin of the stream. So this kind of streaming, indeed, appears to be a pressure-induced flow.

DR. TERU HAYASHI: In the light of those comments, it seems to me the idea of a meshwork under tension which depends for the weakness of the meshwork somewhere to generate flow is equivalent to a passive kind of pushing. It is passive in the sense you have first to stretch the meshwork to this point of tension.

DR. STEWART: I believe it is passive locally.

DR. PARPART: I should like to rise in defense of the mouse, which is a complex organism. For sometime we have been examining the circulation in the spleen in normal and anesthetized animals and found exactly the same sort of picture you have in *Physarum*: reversals, open areas, all sorts of complexes. Yet this is a pulsed pressure system.

DR. STEWART: This certainly is a pressure system. So, I believe, is this picture that I have presented. The driving force is hydrostatic pressure.

DR. KITCHING: I should like to ask how the organism reacts to external stimuli; is one to suppose the basis of the streaming cycling resides in the plasmagel or might it reside in the plasma surface?

DR. STEWART: From a theoretical point of view, I would like to think it resides in the surface. I have suggested in the manuscript that if it is true that the meshwork carries a traveling wave of weakening, this suggests it can transmit "information" in some vague way. This idea is so vague as to be almost worthless, but perhaps not quite.

If the surface can transmit information, it must be in the form of some kind of "signals." As soon as one says that, one has introduced the idea of a nervous system. The problem is that the slime mold doesn't *appear* to have any stable organization, and yet it *does* have some organization; it is an organism. I think your question is important, but I have no idea how to answer it.

DR. WOLPERT: If I understand correctly, as you say, it is a hydrostatic pressure mechanism. As I understand in some of your early papers, you mentioned streams can go past each other. Am I correct?

DR. STEWART: Not in this organism, but such movements have been described for *Reticulomyxa.*

DR. WOLPERT: In other words, in one cross section of a *Physarum* channel streaming is all in a given direction.

DR. STEWART: Yes; in a given channel at a given time.

DR. WOLPERT: Isn't *Physarum* the organism in which you reported circular streaming?

DR. STEWART: Yes, if you mean, a plasmodium with a circular vein around it. This is one of the situations where I am reminded of a comment of Dr. Allen's, to the effect that whenever working with amebae or slime molds, one should never put one under the microscope unless one is prepared to take moving pictures. This observation was made and published several years ago, but I am no longer prepared to stand up in public and say I know for sure streaming was going on in the same direction at the same time throughout the circuit. I was quite sure at the time, but I have never seen it happen again, and I don't have a film record of it.

Dr. Rebhun: If I remember correctly, a film Dr. Seifriz produced a number of years ago indicated carbon dioxide would gel *Physarum* cytoplasm. When the CO_2 was removed, the streaming resumed with essentially the same pattern as before its application.

Dr. Stewart: Depending on how long it stays gelled.

Dr. Rebhun: I wondered how long this persisted, and if this would not indicate there is considerable long-term organization even though temporarily labile.

Dr. Stewart: I think Dr. Kamiya has something to add to this, because he has done double-chamber experiments showing that if you measure the motive force over a period in which CO_2 gelation is induced, the motive force cycle will pick up in just the phase it would have had if it had never stopped. This is only true if it doesn't stay gelated too long. There are also electrical data that go along with this.

What this suggests to me is that this meshwork–fluid system has some short-term stability. Also, the gelation which occurs with CO_2 or mechanical shock, is quite a different structure from the gel which I am talking about as being concerned with movement.

Dr. Allen: How long can this stability last, Dr. Kamiya?

Dr. Kamiya: One or 2 min.

Dr. Stewart: This short a duration would make sense, since the characteristic period of these waves is on the order of 2 min.

Differentiations of the Ground Cytoplasm and Their Significance for the Generation of the Motive Force of Ameboid Movement[1]

K. E. Wohlfarth-Bottermann

Zentral-Laboratorium für angewandte Übermikroskopie am Zoologischen Institut der Universität Bonn, Bonn, Germany

Throughout the history of the investigation of ameboid movement the relevant theories have often been closely tied to current views of cytoplasmic structure (cf. De Bruyn, 1957; Noland, 1957). The morphologist, dealing with protoplasmic streaming, has primarily to answer the question of *which cytoplasmic structures furnish the motive force* for this interesting and important phenomenon of living matter. Furthermore, it is a problem of morphological interest to see how these structures might be ordered to give a *functional pattern,* i.e., *a mode of action resulting in protoplasmic movement.* Such knowledge would be a highly desirable basis for physiological and biochemical studies.

The purpose of this paper is to show that today, for the first time, modern cell morphology seems able to demonstrate by means of the phase-contrast and electron microscopes differentiations of the groundplasm or ground cytoplasm which may be responsible for the generation of the motive force. We studied the classic objects for this kind of research, namely, amebae and slime molds.

Thus far electron microscopy has not been very successful in revealing the structural organization of the groundplasm. In the past decade there has emerged an image of the typical cell structure, which is so generally well understood today that it is unnecessary to give details here. The spatial relationships of the newly discovered membrane systems are surely a major contribution to animal and plant cytology. These membranes provide an enormous expanse of inner surface and achieve a separation of different spaces within the cytoplasm (cf. Palade, 1956; Danielli, 1958; Ruska, 1962). In most cells, the groundplasm is the most significant of these spaces as far as the volume is concerned. It is the groundplasm in which the membrane structures of the cell are embedded, and therefore it has been called the cytoplasmic matrix

[1] Support from the Deutsche Forschungsgemeinschaft for the experimental work is gratefully acknowledged.

(cf. Wohlfarth-Bottermann, 1960, 1961a,b,c). It is reasonable to hold groundplasm responsible for the generation of the motive force. However, two uncertainties must be considered in connection with this assumption:

1. It is not clear in all cases, whether protoplasmic streaming is set in motion by contraction phenomena (cf. Stewart and Stewart, 1959a; Allen, 1961b; Abé, 1962; Kavanau, 1962a,b).

2. Because of the artifact problem, it is more difficult to obtain reliable morphological information about the groundplasm than about membranes.

If the cytoplasm is thought of as the source of motive force resulting from contraction phenomena (cf. De Bary, 1864; Schulze, 1875; Mast, 1926, 1931), then the morphologist immediately wants to know whether there is *any evidence of threadlike or reticular structures.* Such structures have been postulated since the beginning of protoplasmic investigation (cf. Zeiger, 1943; Haas, 1955; Oberling, 1959). In most electron microscope studies it has not been possible to distinguish clearly the ultrastructure of the cytoplasmic ground substance: a survey of the existing literature reveals mostly negative findings and so the general opinion has been that the groundplasm, even after the best fixation, appears structureless under the electron microscope.

A promising approach to the study of groundplasm has seemed to us to be to work with cells in which large and defined areas contain only this material. The ameba *Hyalodiscus simplex* (Wohlfarth-Bottermann, 1960), the normal form of which always shows a clear separation of hyaline ectoplasm and granular endoplasm, has proved particularly suitable for this purpose. The large front lobe consists of pure groundplasm, and contains no membrane structures such as endoplasmic reticulum or vacuoles and is intimately connected with the motive mechanism of this cell. The electron microscope has revealed, independently of the fixation and electron-staining methods used, three structural aspects: the groundplasm is *either structureless, or it is composed either of globular or threadlike structures* (Wohlfarth-Bottermann, 1960, 1961a). We found pleomorphism of the groundplasm (even in single cells), which may be correlated with different physiological states. Though threadlike structures have been postulated for a long time as components of a "contractile gel reticulum" we must ask to what degree they represent real, vital components of the cytoplasmic matrix.

In living amebae, fibrils have been described several times in the light microscope (Wittmann, 1950, 1951; Goldacre, 1961; Käppner, 1961). Also by means of the electron microscope similar fibrils are detected in

Hyalodiscus: they represent bundles of filamentous structures of the groundplasm (Schneider and Wohlfarth-Bottermann, 1959, p. 379, Fig. 1; cf. Wohlfarth-Bottermann, 1963a); we can thus assume that they are not artifacts. The evidence for this will now be given in connection with another "classic" object for studying protoplasmic streaming, i.e., the slime mold, *Physarum polycephalum.*

This organism is particularly suitable for morphological investigations, and the physiology of this type of protoplasmic streaming is well known (cf. Kamiya, 1959, 1960a,b). The observation of living microplasmodia of *Physarum* suggests strongly that contraction mechanisms are involved. We know from many studies some of the chemical processes which underlie the generation of the motive force in *Physarum,* but until now no one has demonstrated a physical basis for this type of streaming.

I need not cite here the numerous physiological studies that have been made; I only want to recall that they seem to suggest a pressure-flow mechanism. Similarly, biochemical methods in general (Loewy, 1952; Ts'o *et al.,* 1956, 1957; Nakajima, 1956, 1957, 1960) demonstrate that it is possible to prepare an adenosine triphosphate (ATP)-sensitive, "contractile protein" from the plasmodium. These proteins behave much like actomyosin and myosin B of muscle and were, therefore, designated "myxomyosin" and "plasmodial myosin B." The myxomyosin molecules are threadlike structures of 70 A in width.

Kamiya *et al.* (1957) were able to show that the motive force of this streaming (which alternates its direction rhythmically) is significantly increased by the addition of ATP. All these results indicate a mechanism for the conversion of chemical to mechanical energy, which is tied to an ATP-sensitive protein, and which thus may be related to the mechanism of muscle contraction. Unfortunately, until the present, morphologists have been unable to detect relevant filamentous structures in the cytoplasm of *Physarum* (Stewart and Stewart, 1959b), i.e., a substrate which might be structurally capable of contracting and whose functional disposition might speak for a pressure-flow mechanism indicated by the physiological results cited previously.

The plasmodial channels in which the endoplasm flows back and forth have an outer ectoplasmic gel layer. The electron microscope has revealed that this ectoplasmic gel layer is substantially *groundplasm* (Wohlfarth-Bottermann, 1962) *containing threadlike structures* morphologically identical with those found in the groundplasm of amebae. We shall henceforth designate these filamentous structures as "plasma filaments" (Wohlfarth-Bottermann, 1962), and we believe that they rep-

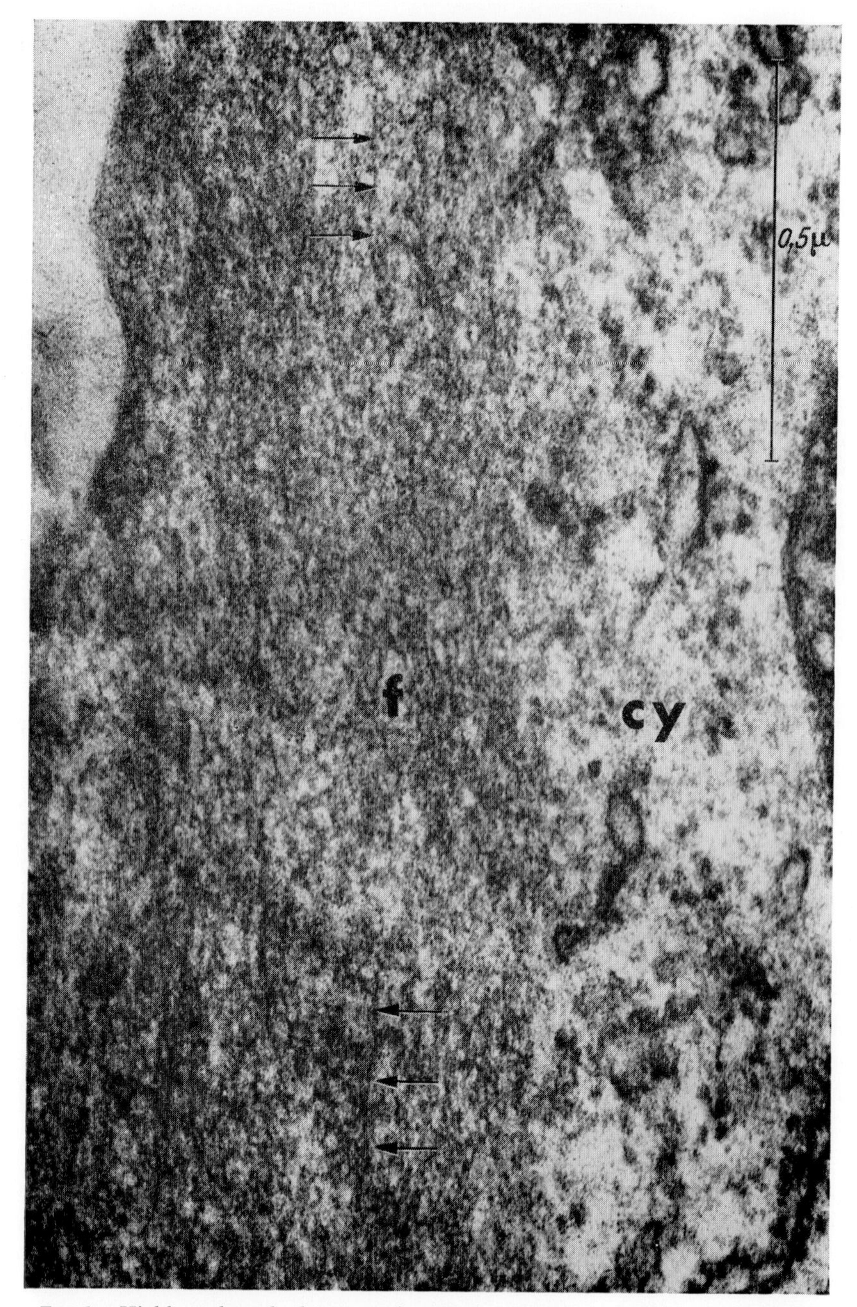

FIG. 1. Highly enlarged electron micrograph (ultrathin section) of a part of a protoplasmic fibril from the outer region of a protoplasmic thread. Normal plasmodium of *Physarum polycephalum*. The fibril (f) is formed by very narrow, longitudinally

resent an interesting and important differentiation of the cytoplasmic matrix. Let us see whether we can gain more information about their function.

Individual plasma filaments of the *Physarum* groundplasm (Fig. 1) can be seen electron optically with approximately the same fidelity as the thin myofilaments of smooth muscle cells. In the course of our electron microscopic studies on slime mold plasmodia we very soon found that the *plasma filaments can form compact fibrils* (Figs. 1 to 3), in part by means of parallel arrangement (Wohlfarth-Bottermann, 1962). This finding supports the similar observation on amebae, where the groundplasm may likewise contain plasma filaments (Wohlfarth-Bottermann, 1961a), which can orientate themselves to form fibrils (Schneider and Wohlfarth-Bottermann, 1959, Part 1; cf. Wohlfarth-Bottermann, 1963a). In contrast with amebae, one finds the fibrils in plasmodia in great numbers and a thorough electron microscopic analysis has revealed that the fibrils are often intricately interconnected at "nodal points" (Fig. 3). In the cytoplasm of *Physarum* plasmodia there is a more or less *coherent network of fibrils,* and these fibrils are simply a conspicuous differentiation of the groundplasm. It must be pointed out that not all fibrils possess the same fine structure. Many of them are built up by structural units showing no parallel arrangement (Fig. 2). It is likely that the variations in the fine structure can be interpreted as functional differences, but until now it has not been possible to show a clear correlation, say, for instance, with respect to their dynamic behavior.

The electron microscope pictures of sections through fibrils show that the dimension of many fibrils reaches a size visible in the light microscope (Fig. 3). For the study of the functional arrangement of the fibrillar network within the larger plasmodia, the electron microscope is for several reasons quite unsuitable: thin-sectioning techniques providing preparations of 200–600 A thickness and of limited area do not allow a quick and quantitative survey. For this purpose we had to find a light microscopic method, which enabled us to cut plasmodial strands and protoplasmic drops in serial sections according to conditions of fix·ation and embedding that make it possible to see the fibrils in the light microscope. We succeeded in developing a quite simple procedure (Wohlfarth-Bottermann, 1963b): after fixation as used for electron microscopy (Wohlfarth-Bottermann, 1957), we embedded the plasmodia in

---

arranged *plasma filaments* (arrows) of the groundplasm. On the right side of the picture —normal cytoplasm (cy) with ribosomes and membrane structures. Magnification: × 107,000. (Wohlfarth-Bottermann, 1962.)

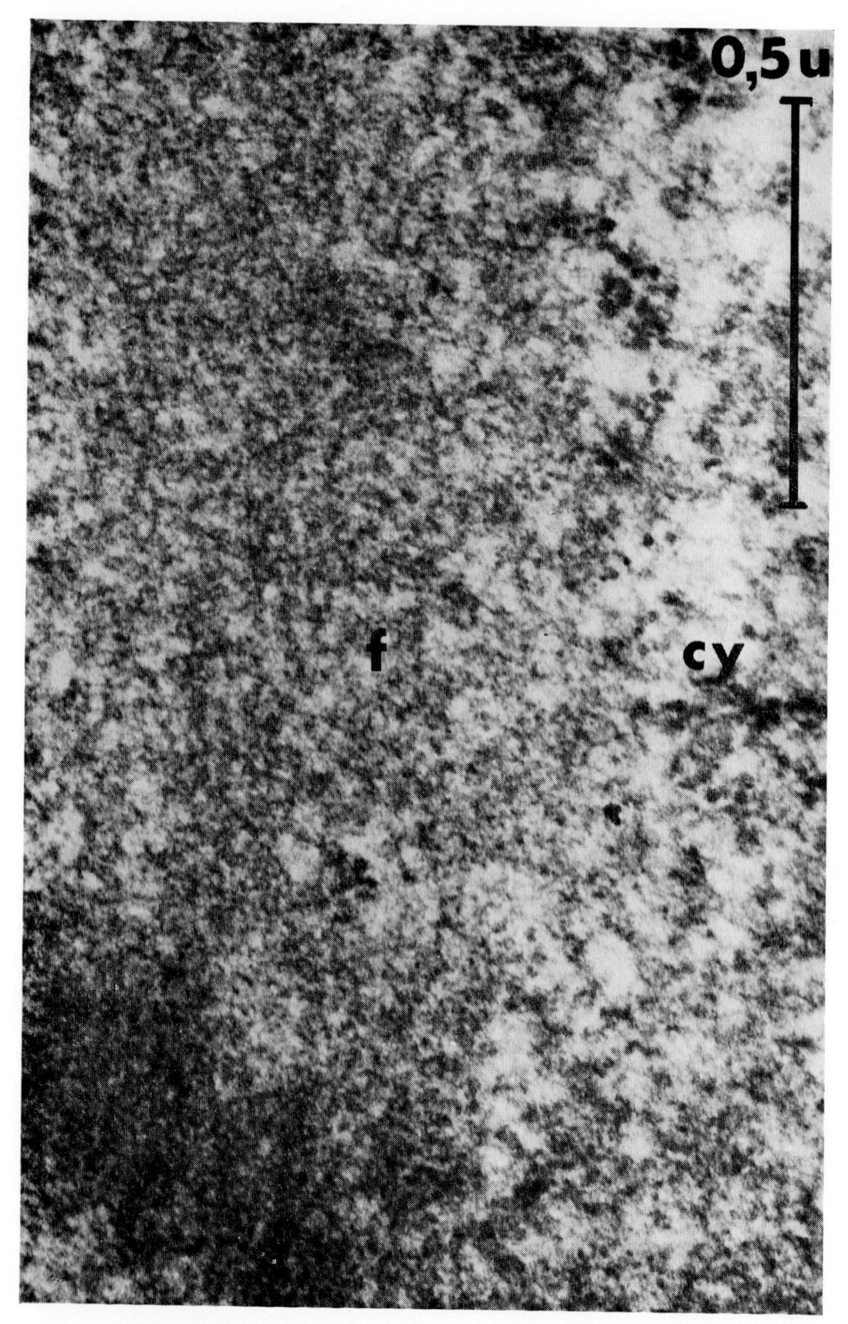

FIG. 2. Highly enlarged part of a protoplasmic fibril from the outer region of a protoplasmic drop (10 min old) (*Physarum polycephalum*). The fibril (f) is formed by plasma filaments showing no parallel arrangement. On the right side of the picture —normal cytoplasm (cy) with ribosomes. Magnification: × 107,000 (unpublished).

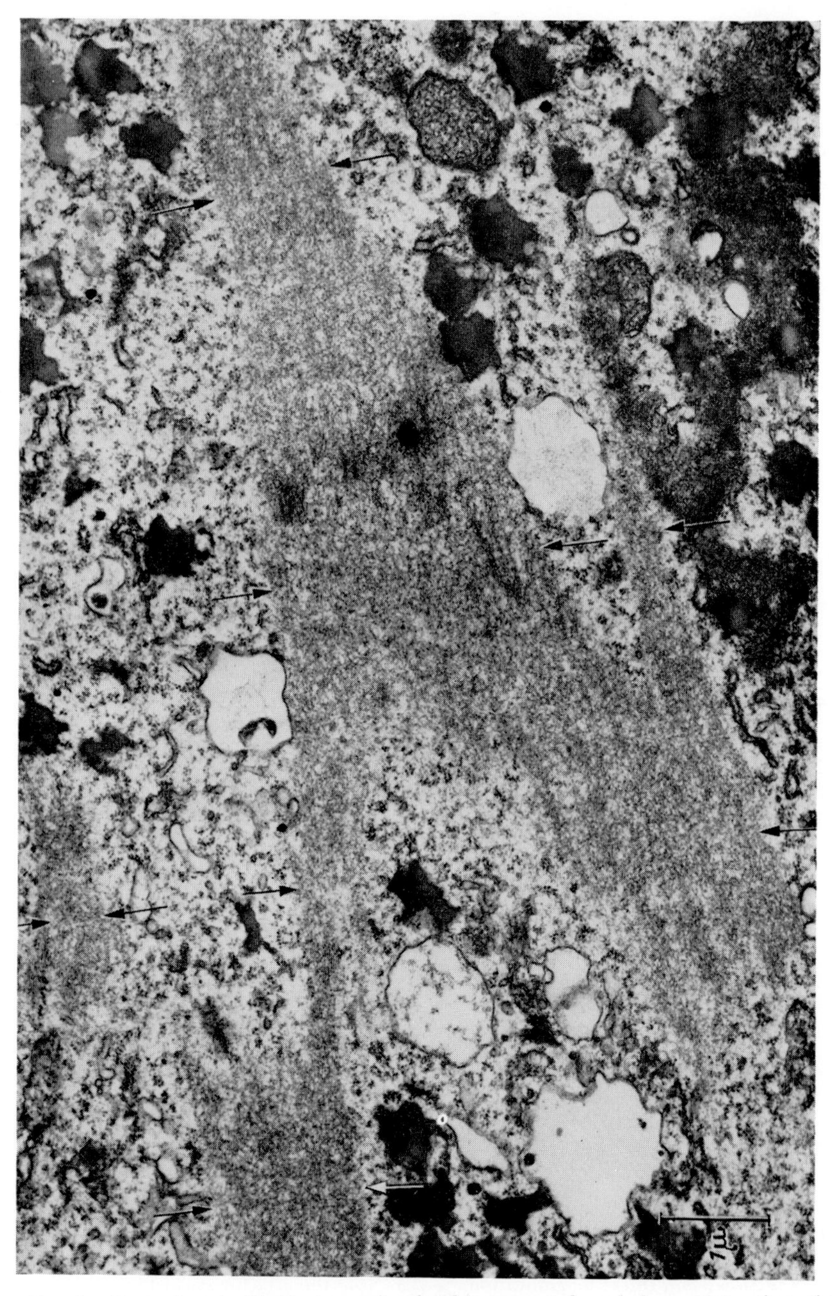

Fig. 3. Protoplasmic fibrils (arrows) of *Physarum polycephalum*. Note that the fibril branches out at different "ramification points." Electron micrograph of an ultrathin section. Magnification: $\times$ 15,000. (Wohlfarth-Bottermann, 1963c.)

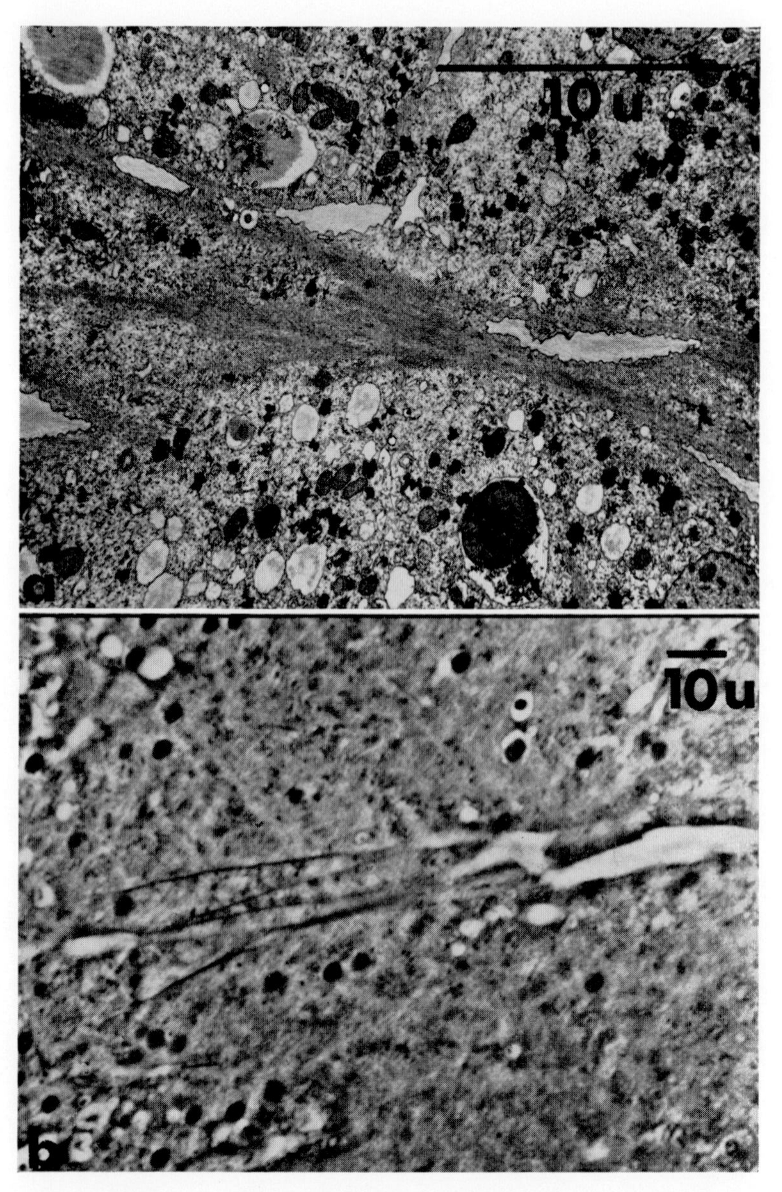

FIG. 4. Electron and phase-contrast micrographs of the protoplasmic fibrils of *Physarum polycephalum* for comparison. (a) Low-power electron micrograph; ultrathin section; magnification: $\times$ 4000. (b) Phase-contrast microscopic picture; paraffin wax section 3–5 μ thick; magnification: $\times$ 725. Note the attachment of fibrils to vacuoles in both cases and the deformation of the vacuoles (unpublished).

paraffin wax ("Tissuemat") and cut serial sections of 3–5 μ thickness on the LKB ultramicrotome (Schneider, 1962). Without any staining, phase-contrast microscopic examination of the sections is possible after the embedding medium has been removed. In Fig. 4 one can compare the electron and phase-contrast microscopic images. It is absolutely clear that these microscopes reveal identical structures. There are many reasons for this statement, one of them being that the filamentous struc-

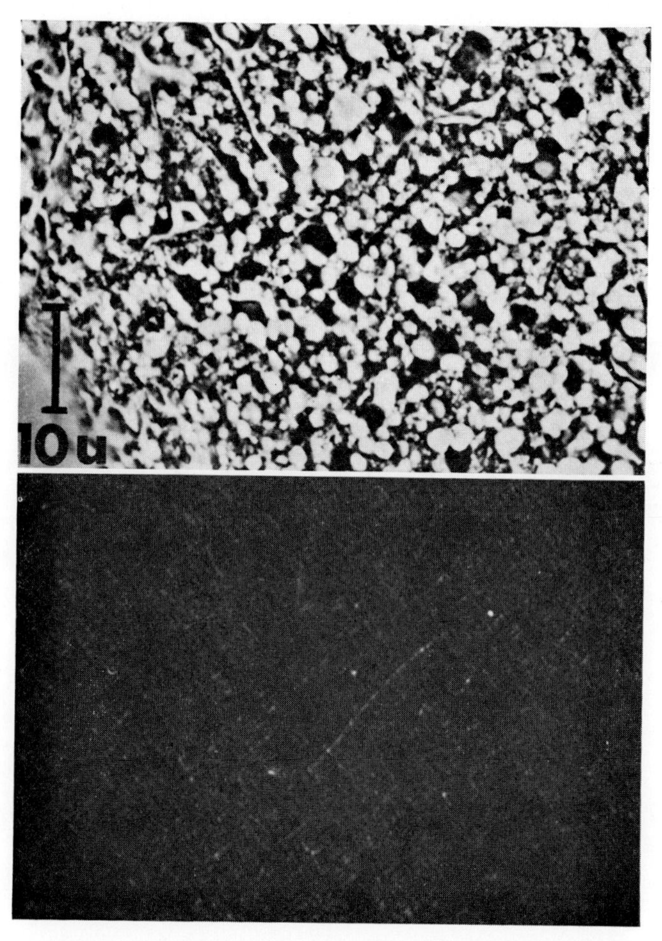

FIG. 5. Protoplasmic fibril of *Physarum polycephalum*. Osmium-chromium fixation; paraffin wax embedding; section of 3–5 μ thickness. *Top:* phase-contrast micrograph. *Bottom:* polarizing micrograph, revealing the weak birefringence of the fibril. Magnification: × 1250. Photograph, W. Cebulla, Ultraphot Carl Zeiss, Oberkochen (unpublished).

tures seen in the phase-contrast microscope are birefringent in the polarizing microscope (Fig. 5). This finding is in accordance with the fine structure of certain fibrils revealed by the electron microscope, namely, their composition of longitudinally arranged plasma filaments.

After a suitable method had been developed for examining the fibrils in the light microscope, we found it was possible to isolate them by means of quite simple procedures developed by muscle physiologists: the *protoplasmic fibrils survive for many days after glycerine extraction* of the plasmodia and protoplasmic drops. Figure 6 shows fibrils after glycerine extraction, comparing results achieved by phase contrast (Fig. 6a) and electron microscopy (Fig. 6b). This means that they could be used as "models" in the sense of the muscle physiologists. Indeed we have tried to test their contractility by the methods worked out by Hoffmann-Berling (1958), but until now we have not been able to find clear evidence of shortening. This negative finding, however, is not evidence against their being contractile, because our control experiments with amebae and muscles only gave positive results on muscles

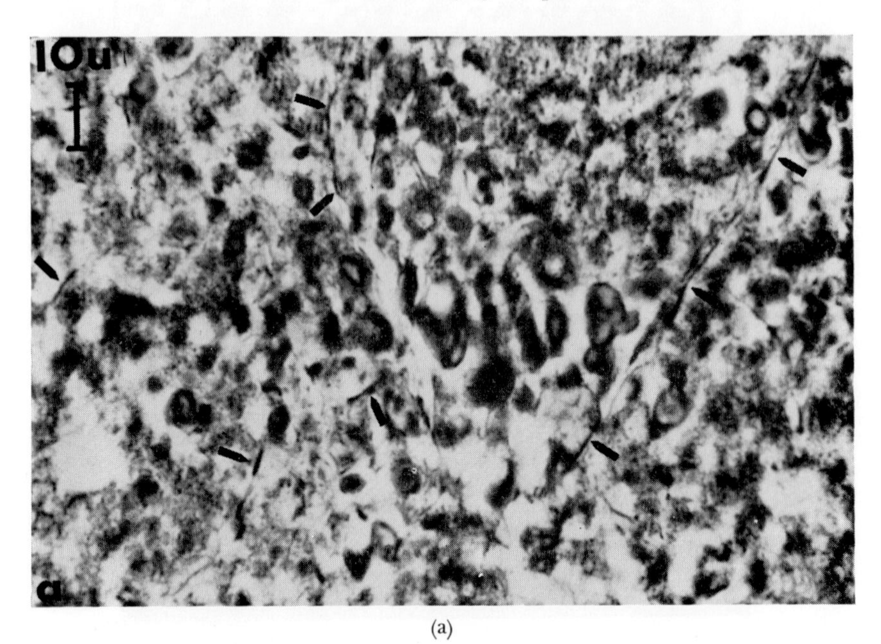

(a)

FIG. 6. Protoplasmic fibrils (arrows) of *Physarum polycephalum* after glycerine extraction (72 hr) of a protoplasmic drop, followed by osmium-chromium fixation. (a) Paraffin wax embedding; section of 3–5 μ thickness; phase-contrast microscope; magnification: × 800 (unpublished). (b) Vestopal-W embedding; ultrathin section; electron microscope; magnification: × 25,000 (unpublished). Note the clear representation of the plasma filaments (b) after glycerine extraction.

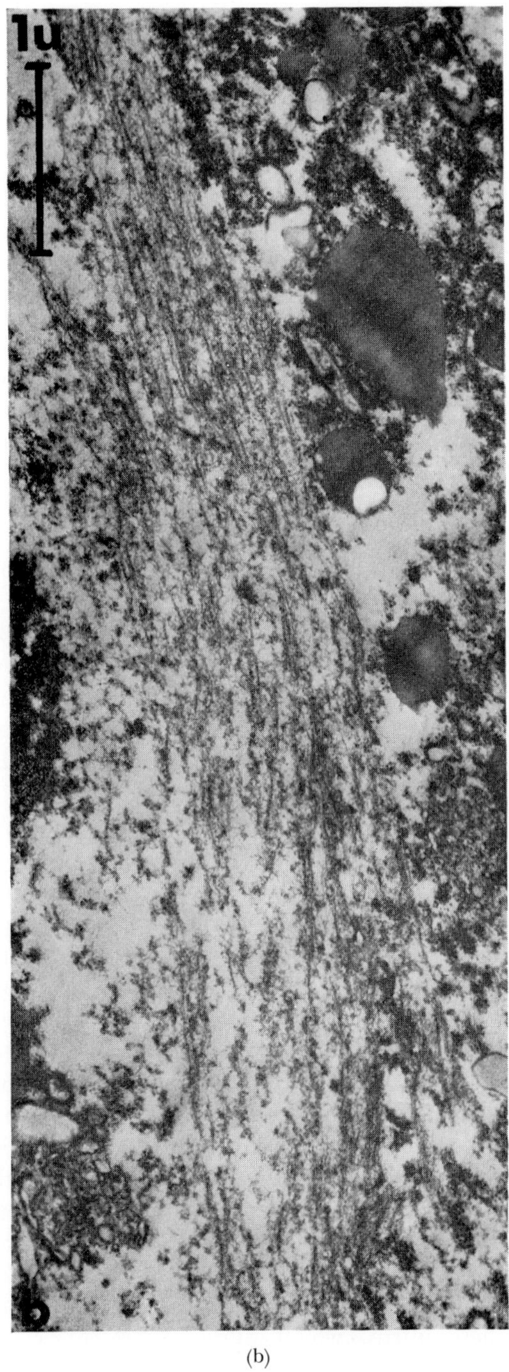

(b)

and never on amebae (*Amoeba proteus*), whereas Hoffmann-Berling described contraction phenomena on models in both cases. Beyond this it must be considered [though the plasma filaments after glycerine extraction are revealed particularly well (Fig. 6b)] that the electron microscopic pictures point to disintegration processes of the fibrils, possibly preventing a contraction of the model. Yet the extraction experiments with glycerine show that the isolation of the fibrils for biochemical and further physiological studies is possible.

The occurrence and the functional arrangement of the fibrils in protoplasmic drops from the plasmodial strands of *Physarum* may give some further information about their function. When a plasmodial channel is punctured with a glass needle, the endoplasm flows out forming a drop which is reabsorbed after a certain time by the plasmodial network. We can assume that this is a process that occurs quite often in nature, too, when the plasmodium, living in the soil gets injured. Before the reabsorption of the endoplasmic drop by the plasmodial channel can take place, the protoplasm must in any case be differentiated into an "outer gel layer" and an inner, "more fluid portion." It is well known that this process of differentiation is an important one for ameboid movement. It can be studied quite well in protoplasmic drops which are from 0 to 10 min old. In this connection it is only of interest that the occurrence of fibrils in drops of different ages and the functional arrangement of the fibrils give us further information about their function. The protoplasmic drops are objects yielding easily obtainable and readily reproducible results.

Drops fixed at the moment of formation ("0 min drops") contain no fibrils visible in the phase-contrast or electron microscope. Several minutes later, the formation of the fibrils and a differentiation of the protoplasm into an outer layer and an inner core begins. Drops 10 min old contain in their outer plasmic gel layer many large and interconnected fibrils (Wohlfarth-Bottermann, 1962), whose thickness and configuration are characterized in Figs. 12b and 13B. Paraffin sections 3–5 μ thick, examined under the phase-contrast microscope have thus far revealed maximum lengths of 1/3 mm (!) and maximum widths of 8 μ (Wohlfarth-Bottermann, 1963b). It is a curious fact that these structures have not been seen earlier. They are often interconnected by ramifications and thus form a coherent network. In interpreting their function it is important to remember that the fibrils are not present in "0 min drops," but appear for the first time some minutes later, that is, precisely *when the drop begins to be reabsorbed* by the plasmodial channel. A simple explanation for this morphological finding would be the assumption that these fibrils, which simply represent a big threadlike differentiation of the groundplasm, supply the motive force needed to bring the drop of protoplasm back into the plasmodial channels, even against a pressure within the strands.

Let us see whether we can find further support for this view. We have already cited the myxomyosin molecules as threadlike structures 70 A in diameter according to Ts'o *et al.* (1957). We found that *the plasma filaments have the same size.* If the hypothesis is correct that myxomyosin and plasma filaments are identical structures, then one could make the prediction that our fibrils should show ATPase activity, because myxomyosin is an ATP-sensitive, contractile protein as found by biochemical methods (Loewy, 1952; Ts'o *et al.,* 1956, 1957; Nakajima, 1956, 1957, 1960).

Indeed, *ATPase activity in the protoplasmic fibrils can be demonstrated cytochemically* both in the light and in the electron microscope (Wohlfarth-Bottermann and Komnick, see Wohlfarth-Bottermann, 1963a). We used frozen sections of unfixed, 10-min-old drops of *Physarum* protoplasm. The control preparations for these experiments were the usual ones: incubation without ATP, as well as normal incubation after thermal inactivation of the enzymes. Neither control showed any positive reaction. The greatest amount of ATPase activity is restricted to the outer region of the drops (cf. OR Fig. 13B), whereas the inner region (cf. IR Fig. 13B) shows no important reaction. The fibrils in the outer region of the drop show a heavy positive reaction. To be sure, this ATPase activity is not restricted to the fibrils but is also found, though to a lesser extent, in the whole cytoplasm of the drop's outer region. But this finding is in accordance with our view that the fibrils are only a specialized form of the groundplasm and that, of course, this matrix exists in a less differentiated form all over the cytoplasm.

The demonstration by light microscope of ATPase activity could, with suitable modifications of method, be supplemented by relevant electron microscopic investigations (cf. Wohlfarth-Bottermann, 1963a). There is an amassing of the reaction product from the cytochemical ATPase reaction in the protoplasmic fibrils, and higher magnification revealed the reaction product adhering to the plasma filaments, i.e., the units of the fibrils. One should point out here, however, that demonstrations of enzyme activity by means of the electron microscope are still problematical with respect to certainty and reproducibility. The significance of such findings depends on the extent to which they can be corroborated by other methods. With regard to this finding, however, we should therefore mention that the electron microscopic result is confirmed by the classic light microscopic technique which also reveals

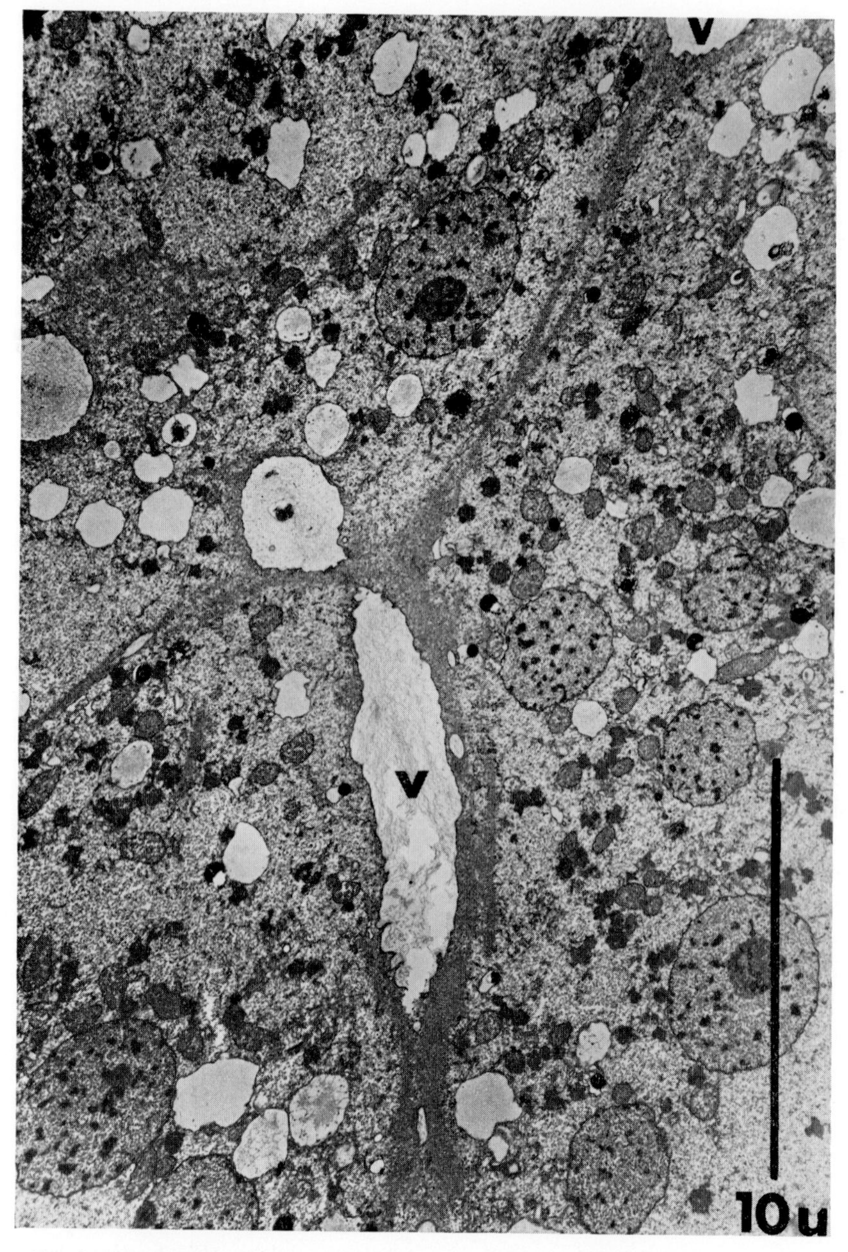

FIG. 7. Low-power electron micrograph revealing the characteristic deformation of vacuoles (V) with fibril attachments; (*Physarum polycephalum*). Note the deformation of the vacuole in the longitudinal direction of the fibril. Magnification: $\times$ 5500 (unpublished).

the ATPase activity of the fibrils. This cytochemical *result supports our assumption that myxomyosin and plasma filaments are identical.*

After this excursion into the field of cytochemistry let us return to the solid basis of morphological facts. Both light microscopic and electron microscopic investigation of drops, strands, or compact layers of *Physarum* plasmodia prove that many of the fibrils found have contact with vacuoles of different sizes. Such vacuoles occur abundantly in the *Physarum* protoplasm (a further observation that is very easy

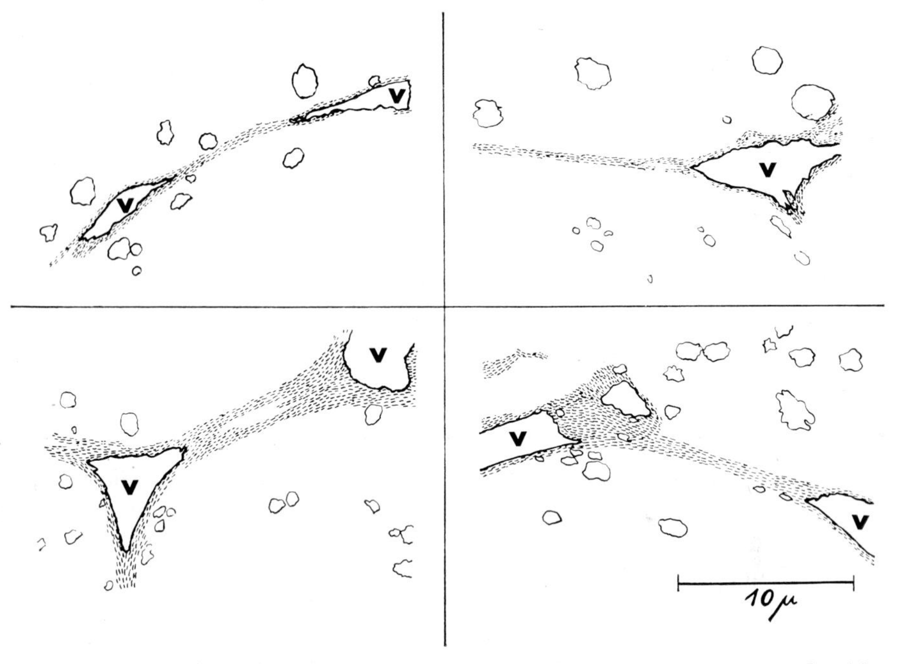

FIG. 8. Drawings of electron micrographs of four fibrils deforming vacuoles (v) in the direction of the longitudinal axis of the fibrils (unpublished).

to make is the frequent attachment of fibrils on the plasmalemma of drops, channel walls, and the limiting membrane of plasmodial protoplasm). Figure 7 shows this finding in a low-power electron microscopic survey picture. It is nearly impossible to find vacuoles with fibrils without the vacuoles seeming to be considerably deformed in a characteristic way. Figure 8 presents four drawings from electron micrographs, which illustrate that the vacuoles with attached fibrils are always deformed in the longitudinal direction of the fibrils. The morphologist can scarcely interpret this phenomenon in any other way than that the *fibrils impose a tension on the vacuole membrane* to which they are attached.

We planned and carried out the following experiments with the hypothesis in mind that the protoplasmic fibrils of *Physarum* are rather ephemeral structures. It is not appropriate to give extensive support for this assumption here. It will be enough to mention that the very rapid formation of the fibrils in protoplasmic drops and general and special experiences with other cellular ultrastructures as well (Wohlfarth-Bottermann, 1959a, 1963a,c) demonstrated a remarkably dynamic capacity of the cytoplasmic building elements. Our idea was that, provided the fibrils were indeed short-lived, we could find *differences in their quantitative occurrence* in small plasmodial threads with defined protoplasmic streaming activity. For this purpose, we used the following experimental arrangement (Fig. 9): protoplasmic strands of convenient thickness (0.1–0.5 mm) and length (15 mm) were cut from the plasmodium of *Physarum* grown on filter paper (Camp, 1936). In a moist chamber, which permitted observation by the aid of a stereomicroscope, we mounted the strands in two different ways. One position was to hang them up, so that the longitudinal axis of the strand was oriented *vertically*[2]; the other was to lay the strands on glass slides so that the threads were orientated *horizontally*. It is well known that a short time after isolation the protoplasmic streaming reappears under such conditions and is quite normal.

We now watched by means of a stereomicroscope the direction of the moving protoplasm in the plasmodial capillary and the rhythmic reversal of the flow. Sometimes it was possible to make the plasmic flow directly visible (this is the case with very thin strands lying on glass), but we often had to depend on measuring the lengthening or shortening of the ends of the strands as a sign of influx or efflux, respectively, of protoplasm. The objects were fixed just at the moment when either the influx or the efflux of protoplasm had reached its maximum speed. After fixation we cut off the ends we had observed, embedded them in paraffin wax, and divided them in serial sections of 3 μ thickness. In this manner we examined 80 plasmodial strands with different positions and directions of plasmic flow. Figures 10 and 11, which show the results of this investigation, are drawings of unstained sections evaluated in the phase-contrast microscope. For each drawing, the position of the strand, the plane of the section, and the direction of the protoplasmic flow at the moment of fixation are recorded in a diagram.

Transverse sections through the front of horizontally lying proto-

---

[2] In a further group of experiments the "twisting movement" of free hanging threads (Kamiya and Seifriz, 1954) was avoided by adhering the strands to glass slides in the same vertical orientation.

plasmic strands often reveal no very distinct differences between threads with different streaming directions. Counting the fibrils in terms of influx and efflux, however, proved that those endpieces, which squeezed out the protoplasm, contained on an average nearly twice as many fibrils as those threads with an influx of protoplasm. Longitudinal sections, especially in the tangential region of the threads (Fig. 10A and B), show this finding more clearly.

In another group of experiments (cf. Fig. 9) we hung up threads, with the idea in mind that here in a vertically oriented thread, the efflux of

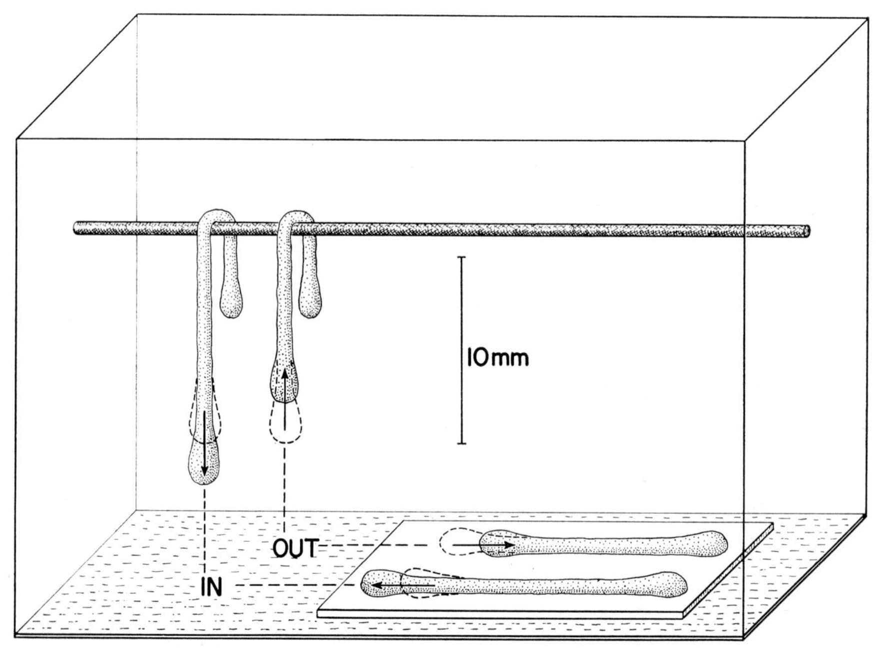

FIG. 9. Schematic representation of the experimental arrangement using protoplasmic threads of *Physarum* with defined direction of protoplasmic flow. Drawing, W. P. Fischer (unpublished).

protoplasm had to be effected *against the force of gravity,* and, in contrast to this, the influx should be *assisted by gravitation.* Provided that our view of the function of the fibrils was correct and, also, provided that they are rather ephemeral structures, the quantitative differences in the number of the fibrils should be clearer in this experimental arrangement. If the fibrils have a direct function for the generation of the motive force we should expect to see *the fibrils in large numbers in those threads pumping the protoplasm against gravity.* This prediction proved to be

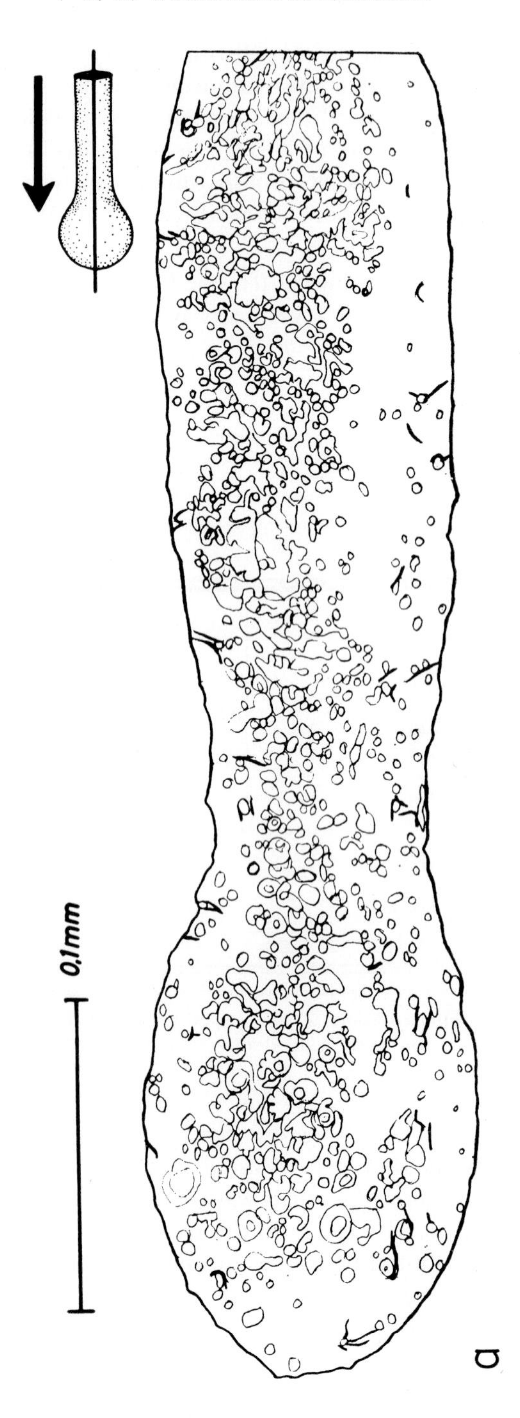
0,1mm
a

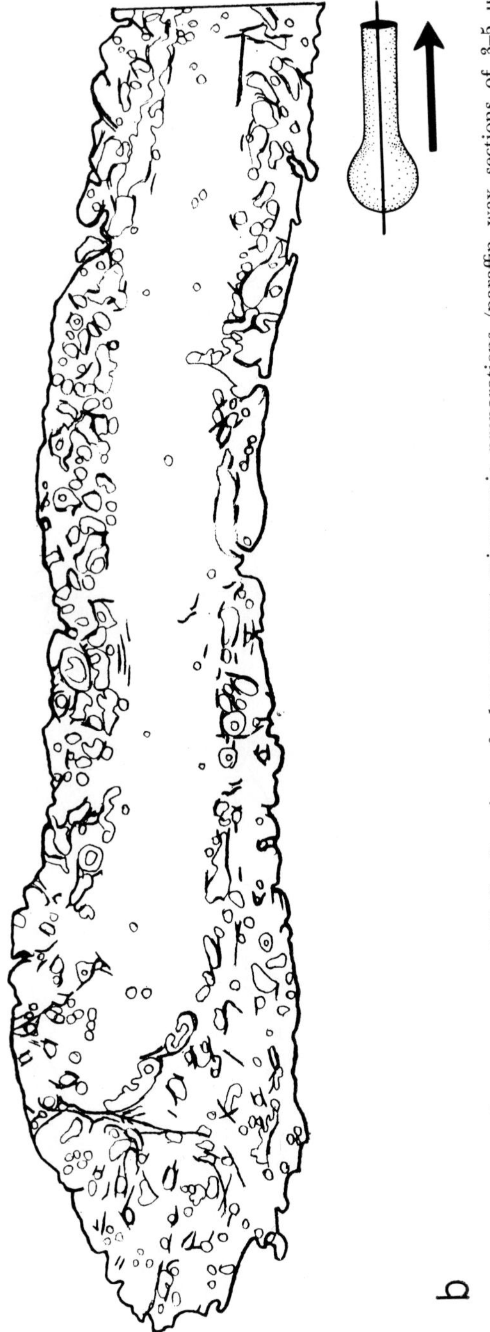

FIG. 10. Functional arrangement of fibrils. Drawings of phase-contrast microscopic preparations (paraffin wax sections of 3–5 μ thickness) of *horizontally lying* protoplasmic strands of *Physarum*, fixed while streaming in a defined direction. Orientation of section and direction of streaming is shown by arrow. Drawing, B. Koeppen-Lesche. (A) Fixation at the moment of *influx* of protoplasm; (B) fixation at the moment of *efflux* of protoplasm (unpublished).

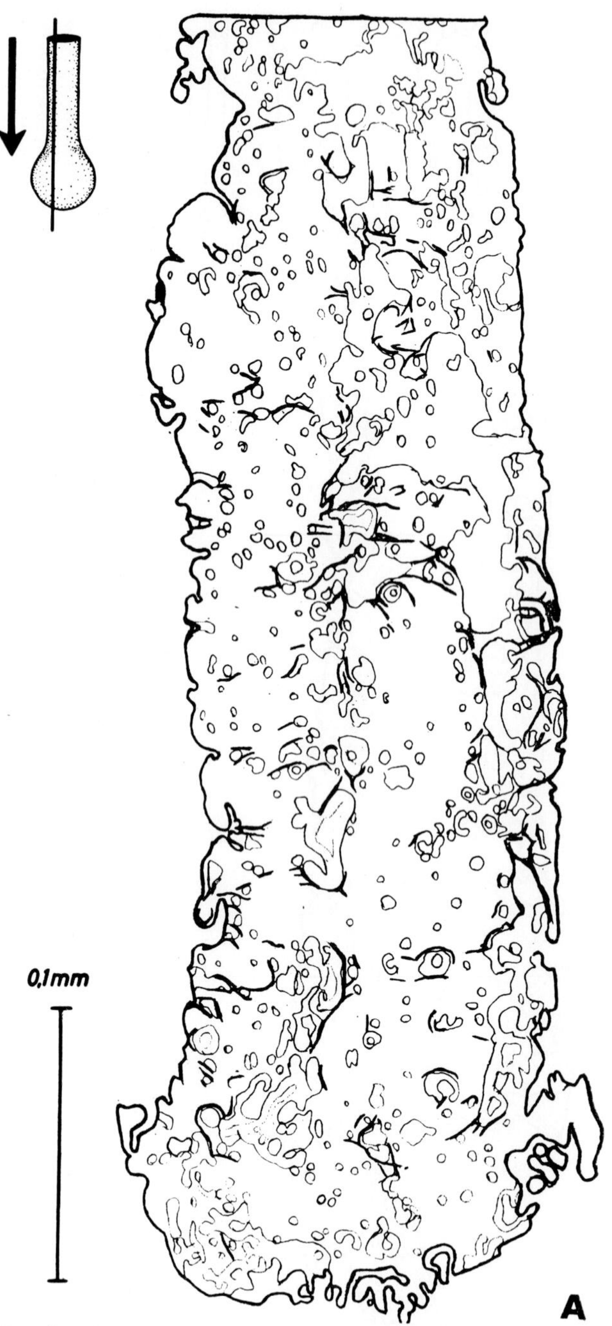

FIG. 11. Functional arrangement of fibrils. Drawings of phase-contrast microscopic preparations of *vertically suspended* protoplasmic strands of *Physarum*. Drawing, B.

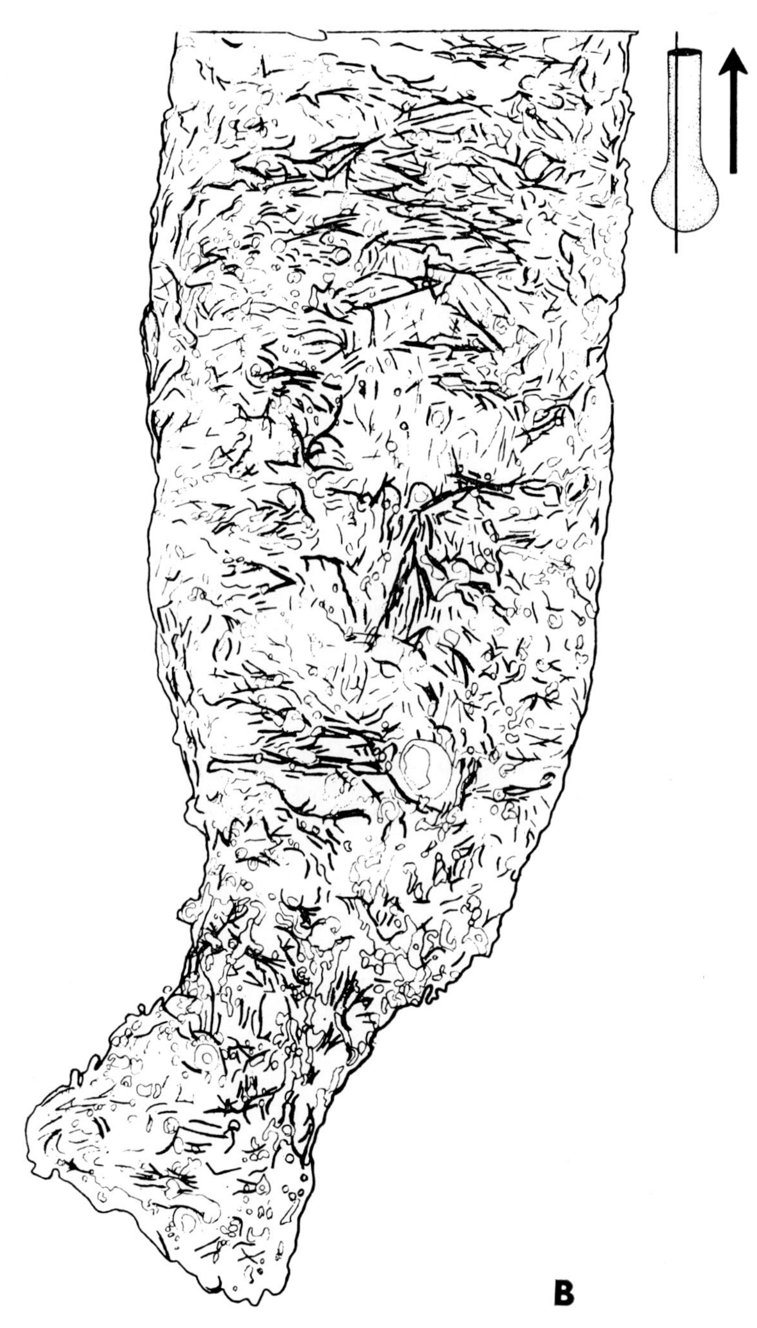

Koeppen-Lesche. (A) Influx of protoplasm; (B) efflux of protoplasm.

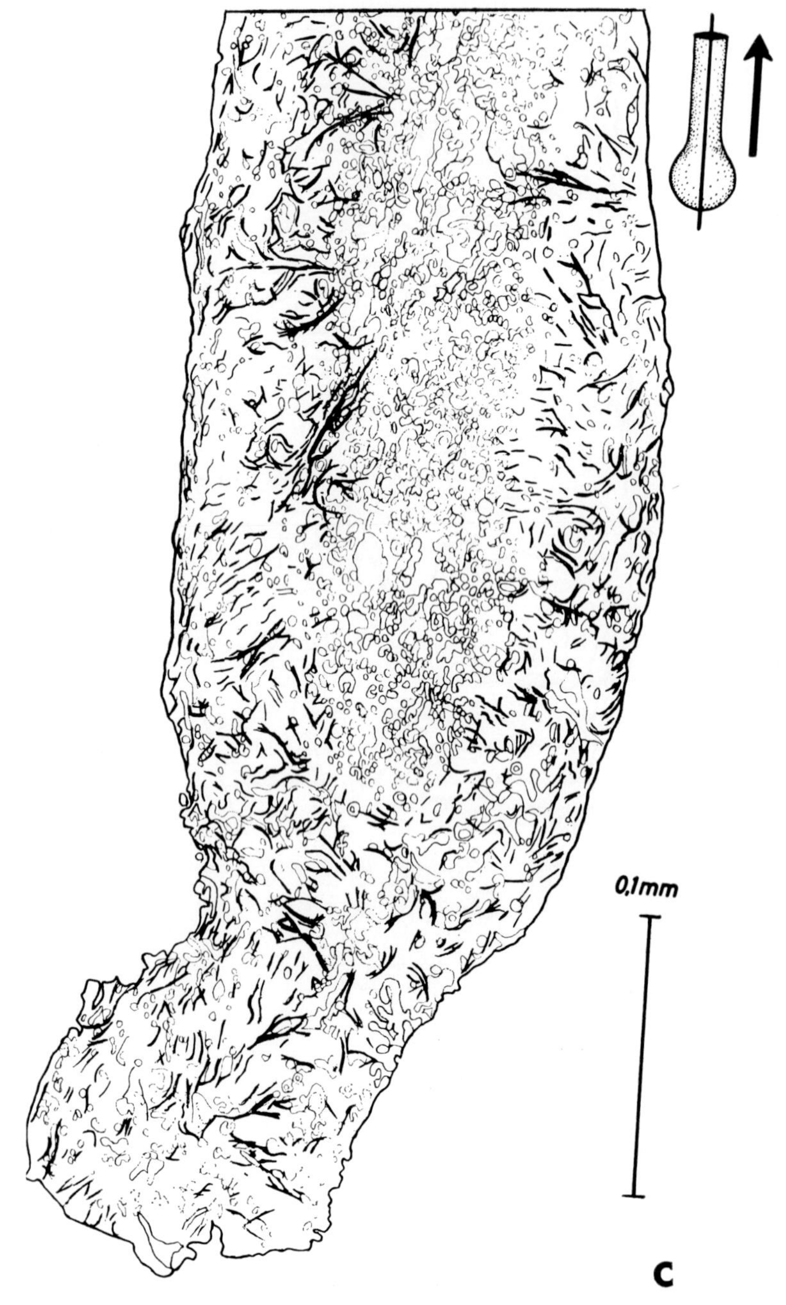

FIG. 11. (C) Median section through the thread presented in B (unpublished).

correct. In both transverse and longitudinal sections, there was an accumulation of fibrils in those specimens where the motive force had to work against gravity[3] (Fig. 11A–C). Indeed the threads coming down also contain fibrils, but at first sight they seem far fewer. There seems to be a *direct relationship between the amount of motive force needed and the number of fibrils.* The fibrils show many interconnections by branching (Fig. 11B and C), so we can be sure they frequently form a cohesive network. This network obviously represents a highly dynamic organization.

Figure 11C shows a median section of the thread in Fig. 11B. This illustration (Fig. 11C) demonstrates the occurrence of the fibrils *exclusively in an outer zone* (probably equivalent to the channel wall) of the thread whereas they cannot be found in the inner core.

Equally interesting and convincing results can be obtained by a comparative investigation of protoplasmic drops fixed at different times after their formation. Drops fixed immediately after formation (0 min drops) have a very homogeneous cytoplasm without fibrils (Fig. 12a); 10 min afterward, when the reabsorption of the drop is in full progress, the fibrillar network is present (Fig. 12b).

Figure 13 shows drawings of sections through drops of different ages. The diagrams explain the ages of the drops and the direction of the preponderant protoplasmic flow. The morphologist, without knowing the physiological and biochemical findings that support a pressure-flow mechanism by contractility phenomena tends to interpret the results illustrated in Fig. 13A and B in the sense cited previously. We believe that the *fibrillar network revealed is one site of the force causing streaming.*

Admittedly, the mode of action of these fibrillar elements remains to be analyzed. One should remember, however, that there is no direct proof of contractility in the myofilaments of smooth muscle cells; nevertheless, nobody doubts their capacity for contraction. Even so we shall shortly review and discuss the facts presented to relate them to the physiological and biochemical results and to some current theoretical interpretations. In both amebae and slime molds, the *groundplasm* shows filamentous structures, the plasma filaments, which can form compact fibrils in part

---

[3] It seems probable that a certain number of the fibrils are involved in delivering the motive force for the "twisting movement" observable on free hanging threads (Kamiya and Seifriz, 1954). We avoided as far as possible fixing threads which showed greater activity in this respect. In any case this does not put in doubt the validity of the result because, as demonstrated by Kamiya and Seifriz (1954), this twisting motion is independent of the direction of protoplasmic streaming. But if this were not so, the abundant occurrence of fibrils in twisting threads would speak for the dynamic function of the fibrillar network.

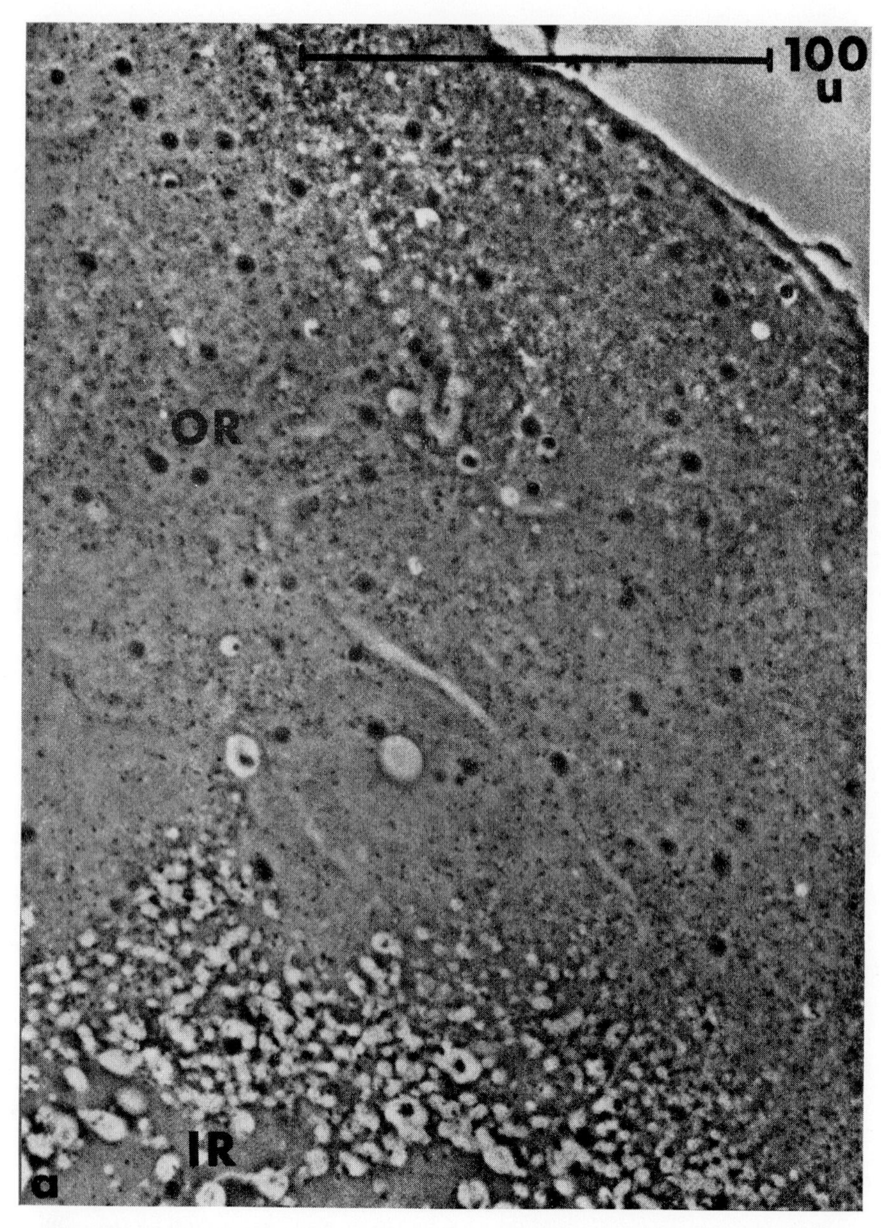

FIG. 12. Phase-contrast micrographs of 3–5 μ paraffin wax sections (median section through protoplasmic drops of different ages). (a) "0 min drop" (fixed as soon as it had emerged from the plasmodial tube network). The homogeneous protoplasm of the outer region (OR) contains no fibrillar structures visible in the light microscope. IR

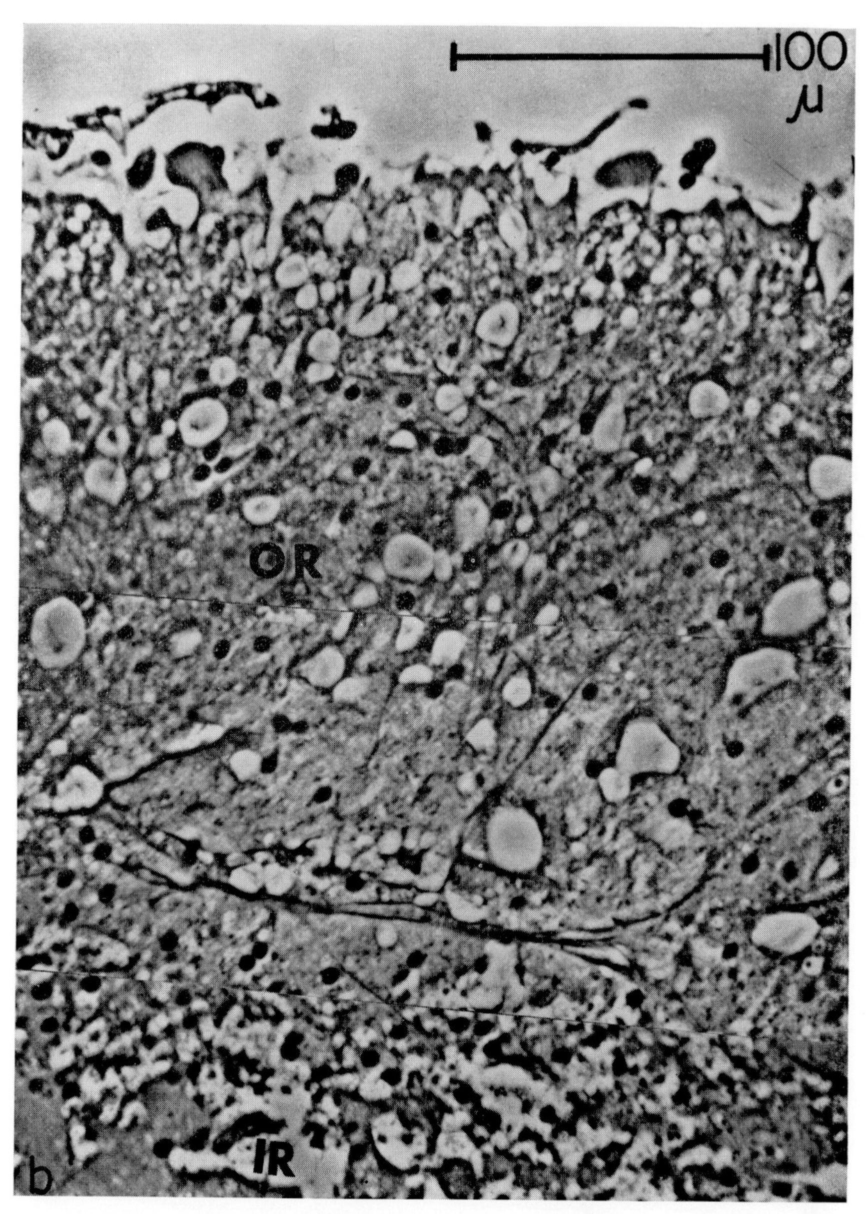

= inner region of the drop. Magnification: × 620 (unpublished). (b) "10 min drop" (fixed 10 min after it had emerged from the plasmodial tube network). Note that the outer region (OR) of the drop now contains large fibrils joined into a network through ramification points. IR = inner region of the drop. Magnification: × 420 (unpublished).

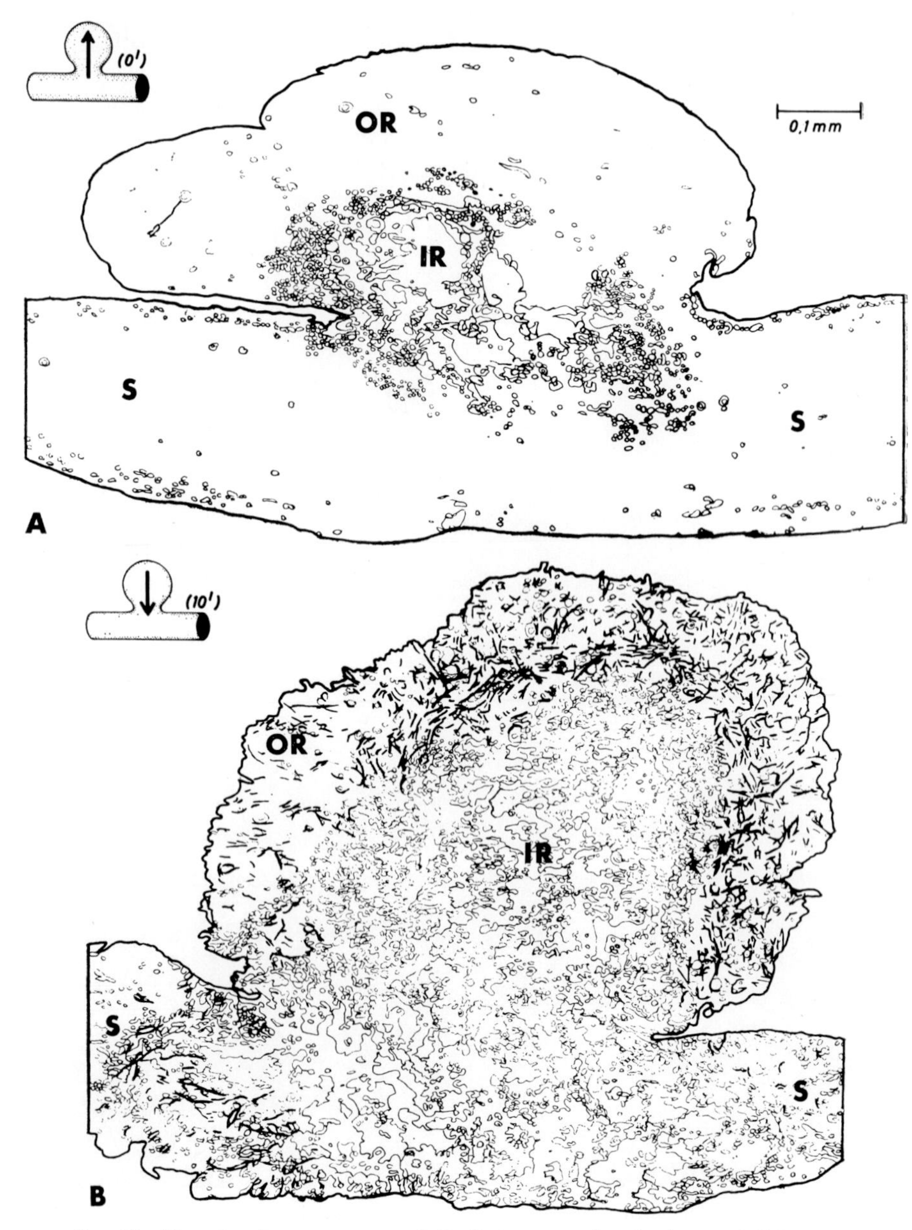

FIG. 13. Functional arrangement of fibrils in protoplasmic drops of different ages (*Physarum polycephalum*). Drawings of phase-contrast microscopic preparations (median paraffin wax sections of 3–5 μ thickness). Osmium-chromium fixation, no staining. Drawing, B. Koeppen-Lesche. (A) Drop fixed just after it had emerged from the strand (S). The arrow in the diagram represents the direction of the main streaming tendency

by means of parallel arrangement.[4] The fibrils possess ATPase activity, and have a functional disposition in protoplasmic threads of *Physarum* justifying the assumption that the cohesive network built up by them in the plasmagel wall of the capillary tubes represents the physical structure responsible for the generation of the motive force of the streaming. This thesis is in accordance with physiological and biochemical results indicating that the enzymatic and mechanochemical system of protoplasmic streaming in *Physarum* has much in common with the basic mechanisms of muscle contraction (cf. Kamiya, 1959; Hasselbach, 1962). The possibility of making this long postulated "gel-reticulum" visible perhaps helps us to throw further light on the dynamic organization producing the motive force.

These conspicuous structures now revealed, which *probably produce contraction of a cortical gel layer,* seem to corroborate from the morphological view the biochemical and physiological results of Kamiya and his group (cf. 1959, 1962) according to which the streaming in *Physarum* (at least in part) "is caused passively by a difference in internal pressure" (cf. De Bary, 1864). However, it should be mentioned that not all movement phenomena of this organism can be explained by a "simple" pressure-flow theory alone, because the facts are more complicated (Stewart and Stewart, 1957, 1959a). Though the microscopic observation of living microplasmodia seems to reveal active contraction phenomena and a squeezing out of protoplasm in the sense of a pressure flow at the moment of the efflux of protoplasm, it is unlikely also with *Physarum* that such "a simple mechanism" alone deals with all aspects. Here, however, we should bear in mind that the fibrillar network now demonstrated is nothing but a particularly high *differentiation of the cytoplasmic matrix*. This "undifferentiated" matrix, existing everywhere in both a stiff protoplasmic gel and the flowing endoplasm, should be capable of producing motive force at any point where it might be necessary. In view of this, both localized and limited counterstreaming and a movement of individual

[4] Last year, De Petris *et al.* (1962) described in normal mononuclear phagocytes filamentous structures with a diameter of 40–60 A, often forming bundles. The authors discuss the hypothesis "that these filaments are contractile in nature and are related in some way to the dynamic activity of the cells, either to the movements of the cell as a whole, or to internal shifts of organelles." Joyon and Chawet (1962) found filaments of unknown nature in the cytoplasm of the ameba *Hyalosphenia papilio.*

---

of the protoplasm at the moment of fixation. IR = inner region of the drop. (B) Drop fixed 10 min after it had emerged from the strand (S). Note the amassment of fibrils in the outer region (OR) of the drop. The arrow in the diagram represents the direction of the main streaming tendency of the protoplasm at the moment of fixation. IR = inner region of the drop (unpublished).

particles against the main stream (Stewart and Stewart, 1959a) are not fully unintelligible.[5]

In view of our results both in amebae and in slime molds contractility phenomena (which are perhaps comparable to a certain extent with muscle contraction) remain the most satisfactory explanations regarding the source of motive force.

At the present time there seem to be no compelling reasons to abandon the contractility theories in favor of surface tension theories (Kavanau, 1962a,b) or other modern explanations (Jahn and Rinaldi, 1959; Stewart and Stewart, 1959a; Bingley and Thompson, 1962). The front contraction theory of Allen (1961a,b), postulated for amebae contractility of the endoplasm, can also exert a fruitful influence on further morphological research with *Physarum*.

Though modern cell morphology is now able to reveal the *gel reticulum of the protoplasm,* the morphologist today does not yet seem to be in a position to decide between the different modern contraction theories—of which one should also mention the interesting concepts of Goldacre (1961) and Yagi (1961)—but perhaps this is only a question of time and effort.

It remains to be tested whether the ideas of "sol-gel transformation" are not intimately correlated with the conceptions of contractility (cf. Wohlfarth-Bottermann and Schneider, 1958; Abé, 1961, 1962; Wohlfarth-Bottermann, 1963a). Undoubtedly, the streaming in a plasmodial tube of *Physarum* requires that an outer, gel-like zone should surround the endoplasm flowing within. Furthermore, it should be noted that sol-gel transformation theories cannot be sharply differentiated from contraction theories, since essential elements of the sol-gel transformation theory are contained in the various contractility theories, or to state it more accurately, sol-gel transformation theories have nearly always assumed a

[5] The enigmatic counterstreaming in *Foraminifera rhizopodia* observable in the light microscope cannot be definitely demonstrated with this instrument: the electron microscope clearly revealed that a small rhizopod, appearing in the light microscope as *one* thread, really consists of a conglomeration of *many* minute strands of submicroscopic size (Wohlfarth-Bottermann, 1961b). This finding makes it conceivable that within a single strand there is a *one-way* streaming, but within different, closely neighboring strands the protoplasm streams in opposite directions, thus simulating a counterstreaming in the light microscope, whose resolution power is unable to visualize the individual threads composing a rhizopod. The lack of proof of a counterstreaming in *Foraminifera,* the fact that each minute thread is limited individually by a definite membrane, and the morphological finding that one rhizopod is simply an open bundle of many individual threads make it difficult at present to explain this type of streaming with various kinds of interfacial forces in the sense of Jahn and Rinaldi (1959) and of Kavanau (1962b).

contractility of the ectoplasm or "plasmagel" (cf. Landau *et al.*, 1954; Marsland, 1956a,b; Hirshfield *et al.*, 1958; Landau, 1959; Yagi, 1961; Landau and Thibodeau, 1962).

The purpose of this paper was to point out that modern cell morphology is able today to demonstrate *the long-postulated cytoplasmic gel reticulum* and its interesting *functional pattern* in *Physarum*. Therefore, morphology is in a better position than ever before to cooperate with cell physiology and promises to make significant contributions toward the goal of elucidating the secret of protoplasmic streaming.

References

Abé, T. H. (1961). *Cytologia* **26**, 378.

Abé, T. H. (1962). *Cytologia* **27**, 111.

Allen, R. D. (1961a). *Exptl. Cell Res. Suppl.* **8**, 17.

Allen, R. D. (1961b). *In* "The Cell" (J. Brachet and A. E. Mirsky, eds.), Vol. II, pp. 135-216. Academic Press, New York.

Bingley, M. S., and Thompson, C. M. (1962). *J. Theoret. Biol.* **2**, 16.

Camp, W. G. (1936). *Bull. Torrey Botan. Club* **63**, 205.

Danielli, J. F. (1951). *In* "Cytology and Cell Physiology" (G. H. Bourne, ed.), pp. 150-182. Clarendon Press, Oxford, England.

Danielli, J. F. (1958). *In* "Surface Phenomena in Chemistry and Biology," pp. 246-265. Pergamon Press, New York.

Davson, H., and Danielli, J. F. (1943). *In* "The Permeability of Natural Membranes." Cambridge Univ. Press, Cambridge, England.

De Bary, E. (1864). "Die Mycetozoen (Schleimpilze). Ein Beitrag zur Kenntnis der niedersten Organismen," 2nd ed. Engelmann, Leipzig, Germany.

De Bruyn, P. P. H. (1957). *Quart. Rev. Biol.* **22**, 1.

De Petris, S., Karlsbad, G., and Pernis, B. (1962). *J. Ultrastruct. Res.* **7**, 39.

Goldacre, R. J. (1961). *Exptl. Cell Res. Suppl.* **8**, 1.

Haas, J. (1955). "Physiologie der Zelle." Borntraeger, Berlin.

Hasselbach, W. (1962). *Fortschr. Zool.* **15**, 1.

Hirshfield, H. I., Zimmerman, A. M., and Marsland, D. (1958). *J. Cellular Comp. Physiol.* **52**, 269.

Hoffmann-Berling, H. (1958). *Fortschr. Zool.* **11**, 142.

Jahn, T. L., and Rinaldi, R. (1959). *Biol. Bull.* **117**, 100.

Joyon, L., and Chawet, R. (1962). *Compt. Rend. Acad. Sci.* **255**, 2661.

Käppner, W. (1961). *Protoplasma* **53**, 504.

Kamiya, N. (1959). *Protoplasmatologia* **8**, 3a.

Kamiya, N. (1960a). *Ann. Rev. Plant Physiol.* **11**, 323.

Kamiya, N. (1960b). *Ann. Rep. Sci. Works, Fac. Sci. Osaka Univ.* **8**, 13.

Kamiya, N. (1962). *In* "Handbuch der Pflanzenphysiologie" (W. Ruhland, ed.), Vol. XVII, part 2, pp. 979-1035. Springer, Berlin.

Kamiya, N., and Seifriz, W. (1954). *Exptl. Cell Res.* **6**, 1.

Kamiya, N., Nakajima, H., and Abé, S. (1957). *Protoplasma* **48**, 94.

Kavanau, J. L. (1962a). *Life Sciences* **5**, 177.

Kavanau, J. L. (1962b). *Exptl. Cell Res.* **27**, 595.

Landau, J. V. (1959). *Ann. N. Y. Acad. Sci.* **78**, 487.

Landau, J. V., and Thibodeau, L. (1962). *Exptl. Cell Res.* **27**, 591.

Landau, J. V., Zimmerman, A. M., and Marsland, D. A. (1954). *J. Cellular Comp. Physiol.* **44**, 211.
Loewy, A. G. (1952). *J. Cellular Comp. Physiol.* **40**, 127.
Marsland, D. A. (1956a). *Pubbl. Staz. Zool. Napoli* **33**, 182.
Marsland, D. A. (1956b). *Intern. Rev. Cytol.* **5**, 199.
Mast, S. O. (1926). *J. Morphol.* **41**, 347.
Mast, S. O. (1931). *Protoplasma* **14**, 321.
Nakajima, H. (1956). *Seita No Kagaku* **7**, 256.
Nakajima, H. (1957). *22nd. Ann. Meeting Botan. Soc. Japan.*
Nakajima, H. (1960). *Protoplasma* **52**, 413.
Noland, L. E. (1957). *J. Protozool.* **4**, 1.
Oberling, Ch. (1959). *Intern. Rev. Cytol.* **8**, 1.
Palade, G. E. (1956). *J. Biophys. Biochem. Cytol.* **2**, 85.
Robertson, J. D. (1960a). *Progr. Biophys. Biophys. Chem.* **10**, 343.
Robertson, J. D. (1960b). *Proc. Intern. Conf. Electron Microscopy, 4th, Berlin, 1958,* **2**, 159-171.
Ruska, H. (1962). *Proc. Intern. Congr. Neuropathol., 4th.* **2**, 42-49.
Schneider, L. (1962). *Sci. Tools* **9**, 16.
Schneider, L., and Wohlfarth-Bottermann, K. E. (1959). *Protoplasma* **51**, 377.
Schulze, F. E. (1875). *Arch. Mikroskop. Anat. Entwicklungsmech.* **11**, 329.
Stewart, P. A., and Stewart, B. T. (1957). *Federation Proc.* **16**, 125.
Stewart, P. A., and Stewart, B. T. (1959a). *Exptl. Cell Res.* **17**, 44.
Stewart, P. A., and Stewart, B. T. (1959b). *Exptl. Cell Res.* **18**, 374.
Ts'o, P. O. P., Eggman, L., and Vinograd, J. (1956). *J. Gen. Physiol.* **39**, 801.
Ts'o, P. O. P., Eggman, L., and Vinograd, J. (1957). *Biochim. Biophys. Acta* **25**, 532.
Wittmann, H. (1950). *Protoplasma* **39**, 450.
Wittmann, H. (1951). *Protoplasma* **40**, 23.
Wohlfarth-Bottermann, K. E. (1957). *Naturwissenschaften* **44**, 287.
Wohlfarth-Bottermann, K. E. (1959a). *Zool. Anz.* **23**, Suppl., 393.
Wohlfarth-Bottermann, K. E. (1959b). *Z. Zellforsch. Mikroskop. Anat.* **50**, 1.
Wohlfarth-Bottermann, K. E. (1960). *Protoplasma* **52**, 58.
Wohlfarth-Bottermann, K. E. (1961a). *Protoplasma* **53**, 259.
Wohlfarth-Bottermann, K. E. (1961b). *Protoplasma* **54**, 1.
Wohlfarth-Bottermann, K. E. (1961c). *Protoplasma* **54**, 307.
Wohlfarth-Bottermann, K. E. (1962). *Protoplasma* **54**, 514.
Wohlfarth-Bottermann, K. E. (1963a). *Intern. Rev. Cytol.* **16**, pp. 61-132.
Wohlfarth-Bottermann, K. E. (1963b). *Protoplasma* **54**, 514.
Wohlfarth-Bottermann, K. E. (1963c). *Naturwissenschaften,* **50**, 237.
Wohlfarth-Bottermann, K. E. (1963c). *Naturwissenschaften* **50**, 237.
Yagi, K. (1961). *Comp. Biochem. Physiol.* **3**, 73.
Zeiger, K. (1943). *Klin. Wochschr.* **22**, 201.

Discussion

Chairman Thimann: Would you care to guess the fraction of the slime mold material which is occupied by the fibers?

Dr. Wohlfarth-Bottermann: Yes, I think the number of fibrils depends on the physiological state and is very different in different areas of the plasmodium. If you have a very thin microplasmodium, then there are not many fibers; but if you ask for the total mass of fibril-forming material, then the whole mass of groundplasm has to be considered. In that case, the value would be rather high.

DR. BOUCK: What is the nature of the endoplasmic reticulum?

DR. WOHLFARTH-BOTTERMANN: I think mostly one doesn't find endoplasmic reticulum but rather vacuoles. Most people, I think, will call them regular endoplasmic reticulum; but I would not. I never understood why a vacuole should be called a reticulum.

DR. STEWART: Have you ever seen the fibers in a living plasmodium?

DR. WOHLFARTH-BOTTERMANN: Yes. We have occasionally seen them in living microplasmodia with the phase-contrast microscope. Also, I saw polarization microscopic pictures made by Dr. Nakajima which were very convincing.[6]

DR. GOLDACRE: I demonstrated fibrils that can be seen in the light microscope by the technique of allowing an ameba to lose half its water.

DR. WOHLFARTH-BOTTERMAN: I know. We intended to try using your technique before we saw fibers in our ameba after mere fixation. However, I have referred to your paper.

---

[6] *Note added in proof:* Using the *freeze-substitution techniques,* G. S. Omenn saw "strand-like material which might possibly resemble endoplasmic reticulum" in plasmodia of *Physarum* in the electron microscope without recognizing the importance of these pictures (Plates 8 and 9, G. S. Omenn: The myxomycete *Physarum polycephalum.* Cultivation, Ultrastructure, Molecular Model for Contractility. Unpublished Senior Thesis carried out under the direction of L. I. Rebhun, Department of Biology, Princeton University, 1961). There is no doubt that the "strands" of Omenn are identical with the fibrils discussed here. That means that these structures can also be demonstrated by freeze-substitution techniques.

The Mechanochemical System behind Streaming in *Physarum*

HIROMICHI NAKAJIMA

Department of Biology, Faculty of Science, and Institute for Protein Research, Osaka University, Osaka, Japan, and Department of Biology, Princeton University, Princeton, New Jersey

Plasmodia of the myxomycete, *Physarum polycephalum,* exhibit vigorous streaming of endoplasm which rhythmically reverses its direction. The streaming of endoplasm is also responsible for changes in the form of the organism and causes locomotion in general.

Just as in the case of muscle, streaming protoplasm is a mechanochemical system where the chemical energy acquired by metabolism is converted into the mechanical work of streaming. The nature of the mechanochemical system behind streaming, therefore, is certainly one of the essential problems in considering the mechanism of streaming.

In our earlier studies we showed that the immediate energy source for the flow in *Physarum* is adenosine triphosphate (ATP) provided by a fermentation process (Kamiya *et al.,* 1957). The first problem we shall consider is the nature of the mechanochemical system which responds to ATP. What draws our attention to this subject is a contractile protein like actomyosin.

Actomyosin-Like Protein in Plasmodia of Physarum

Loewy (1952) first discovered an actomyosin-like protein system in *Physarum* plasmodia. Later, Ts'o *et al.* (1956a, b, 1957a, b) obtained an ATP-sensitive protein in this organism. This they purified by successive salt fractionation and differential centrifugation and named it "myxomyosin." They studied especially its physicochemical properties. Recently, detailed examinations of the so-called "contractile protein" of *Physarum* have been made by Hatano (1962) and Rebhun (1962).

We (Nakajima, 1956, 1960) also isolated a structural protein fraction from the same source by the procedure used for the preparation of muscle myosin B. Thus, the protein fraction was extracted with Weber-Edsall solution (0.6 *M* KCl, 0.01 *M* Na_2CO_3, 0.04 *M* $NaHCO_3$) and was precipitated by dilution. It was yellowish-white and very turbid. We have called this preparation "plasmodial myosin B" to denote its method of preparation. Some of its properties will now be discussed.

Viscous Properties

Figure 1 shows the change in viscosity of plasmodial myosin B on the addition of ATP. When ATP is added, the viscosity drops rapidly at first, then slowly increases to the original level. The time course of this change in viscosity is practically identical with that of muscle myosin B.

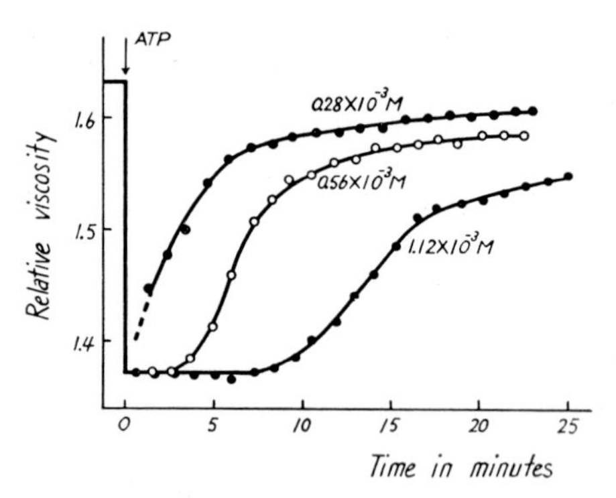

Fig. 1. Effect of ATP on the viscosity of plasmodial myosin B; pH 6.6, 15°C; K^+, 0.55 *M*. The final concentrations of ATP are indicated on the figure (Nakajima, 1960).

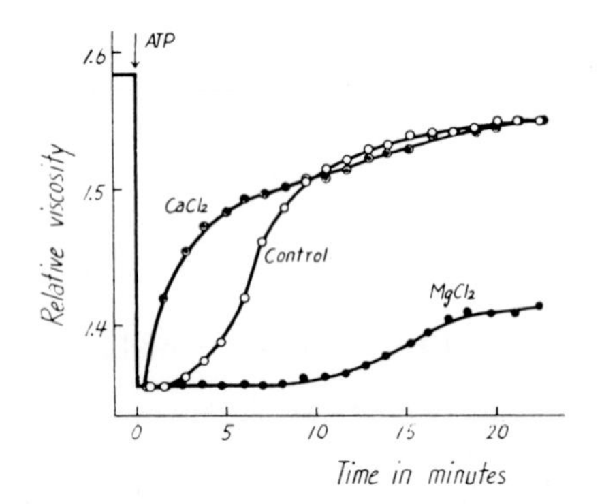

Fig. 2. Effect of $CaCl_2$ and $MgCl_2$ on the viscosity of plasmodial myosin B; pH 6.6, 15°C; K^+, 0.55 *M*; ATP, 0.56×10^{-3} *M*; $CaCl_2$, 10^{-2} *M*; $MgCl_2$, 10^{-2} *M* (Nakajima, 1960).

The degree of change in viscosity, i.e., ATP sensitivity as expressed by Portzehl *et al.* (1950), is 40–60%. These values are smaller than those for myosin B from skeletal muscle but similar to myosin B from smooth muscle (Portzehl *et al.* 1950; Needham and Cawkwell, 1956; Tonomura *et al.*, 1955; Tonomura and Sasaki, 1957).

In Fig. 2 is shown the effect of $CaCl_2$ and $MgCl_2$ on the viscosity of plasmodial myosin B. When added along with ATP, $CaCl_2$ shortens the duration of the viscosity drop, whereas $MgCl_2$ lengthens it. On the other hand, the magnitude of the ATP-induced viscosity drop is not affected by these cations.

We have obtained the following additional results:

1. Adenosine monophosphate (AMP) produces no viscosity change such as does ATP in plasmodial myosin B.

2. Inorganic pyrophosphate does cause a decrease in viscosity and its effect is considerably enhanced by $MgCl_2$. However, unlike the case of $MgCl_2$, $CaCl_2$ nearly returns the pyrophosphate-induced viscosity drop to the control value.

3. $MgCl_2$ or $CaCl_2$ alone hardly causes any change in viscosity.

4. When added with ATP, ethylenediaminetetraacetic acid (EDTA) inhibits both the initial drop and subsequent return to the original viscosity.

Comparing the viscous properties and enzymatic activity which will be discussed later, it is inferred that, as with muscle myosin B, the initial drop in viscosity induced by ATP is a nonenzymatic process which is unrelated to the dephosphorylation of ATP, whereas the recovery process is enzymatic and associated with the hydrolysis of ATP.

Superprecipitation

We have also observed the superprecipitation of plasmodial myosin B gels on the addition of ATP (Fig. 3). Superprecipitation of this protein, however, proceeds rather slowly under the present experimental conditions.

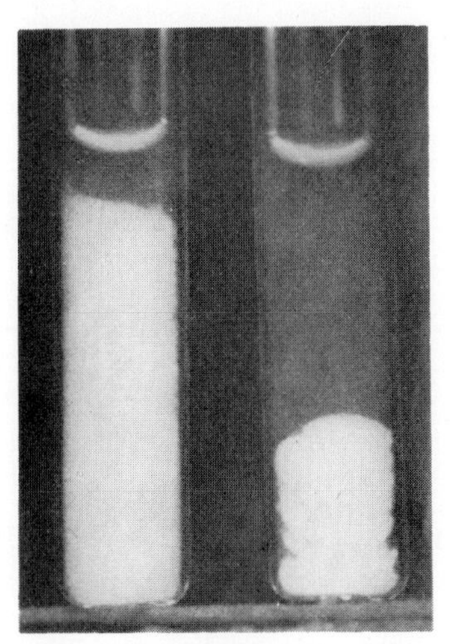

Fig. 3. Superprecipitation of plasmodial myosin B; pH 6.6, 19°C; K^+, 0.11 *M;* ATP, 0.47×10^{-3} *M*; plasmodial myosin B, 1.31 mg/ml. Left, control; right, ATP treatment. Photographed 70 min after addition of ATP (Nakajima, 1960).

Enzymatic Properties

As mentioned previously, plasmodial myosin B either shows viscosity changes or superprecipitation on the addition of ATP. Which of these changes takes place depends on whether the plasmodial myosin B is in the sol or gel state; this in turn depends of the K^+ concentration. Therefore, we studied the enzymatic properties on varying the concentration of K^+.

Figures 4a and b show the time courses of orthophosphate liberation when ATP, AMP, and inorganic pyrophosphate were added to plasmodial myosin B. Figure 4a is the case at a high K^+ concentration (0.5 *M*) and Fig. 4b at a low K^+ concentration (0.1 *M*). The broken line indicates the

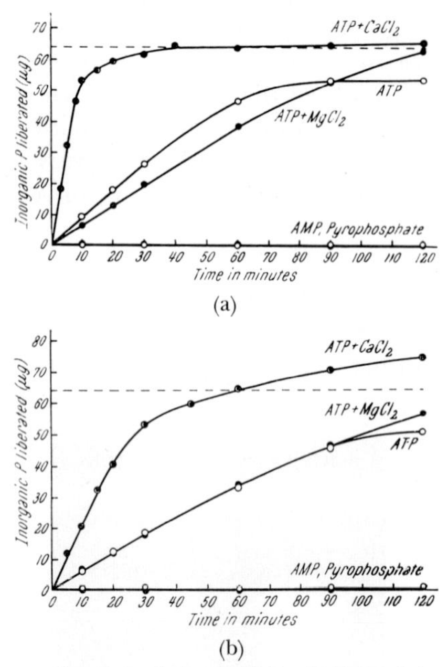

Fig. 4a. Time courses for hydrolysis of ATP, AMP, and inorganic pyrophosphate at a high K^+ concentration; pH 6.6, 28°C; K^+, 0.5 *M;* ATP, 0.69×10^{-3} *M;* AMP, 0.83×10^{-3} *M*; pyrophosphate, 0.85×10^{-3} *M*; $CaCl_2$, 3.3×10^{-3} *M*; $MgCl_2$, 3.3×10^{-3} *M*; plasmodial myosin B; 0.141 mg of N (Nakajima, 1960).

Fig. 4b. Time courses for hydrolysis of ATP, AMP, and inorganic pyrophosphate at a low K^+ concentration. The conditions are the same as in Fig. 4a, except that the K^+ concentration is 0.1 *M* (Nakajima, 1960).

level corresponding to half the labile phosphate of ATP. It is clearly shown in these figures that plasmodial myosin B has ATPase activity, and that it hydrolyzes neither AMP nor inorganic pyrophosphate. In the presence of a high K^+ concentration, the ATPase activity is stimulated by $CaCl_2$, whereas it is inhibited by $MgCl_2$ (see also Table I). At a low K^+ concentration, $CaCl_2$ accelerates the activity, whereas $MgCl_2$ has no effect on it. The specific activities calculated from these figures are 1.44 and 0.50 μmoles of P/mg of N/min at K^+ concentrations of 0.5 and 0.1 *M*, respectively (in the presence of 3.3×10^{-3} *M* $CaCl_2$). These values are smaller than those of skeletal muscle myosin B but higher than those of smooth muscle myosin B (Needham and Cawkwell, 1956, 1958).

As shown in these figures, when the K^+ concentration is low, or when $MgCl_2$ is present in the reaction mixture, more than half the labile phosphate of ATP is split during longer incubation periods. This may be due to myokinase which may be present in this preparation.

Table I summarizes the effect of the K^+ concentration on the ATPase activity with and without $CaCl_2$ and $MgCl_2$. The ATPase activity is

TABLE I

EFFECT OF K^+ CONCENTRATION ON ATPASE ACTIVITY WITH AND WITHOUT $CaCl_2$ AND $MgCl_2$

Molar concentration K^+	μmoles of P/mg of N/min		
	None	$CaCl_2$	$MgCl_2$
0.08	0.11	0.24	0.15
0.15	0.21	0.76	0.18
0.30	0.23	1.46	0.14
0.45	0.25	1.46	0.12
0.60	0.25	1.34	0.10
0.80	0.24	1.34	0.09

[a] pH 6.6, 28°C; ATP, 0.74×10^{-3} *M*; $CaCl_2$, 3.3×10^{-3} *M*; $MgCl_2$, 3.3×10^{-3} *M*; plasmodial myosin B; 0.101 mg of N (Nakajima, 1960).

stimulated by an appropriate amount of K^+. $CaCl_2$ accelerates the activity over a wide range of K^+. On the other hand, $MgCl_2$ increases ATPase activity at low K^+ concentrations of less than 0.1 *M* but inhibits it at high K^+ concentrations.

In addition to these facts, we found the following enzymatic properties:

1. There is antagonism between Ca^{++} and Mg^{++} both at high and low K^+ concentrations.[1]

2. The pH optima are at about 6.2 and 5.4 in the presence of high and low K^+ concentrations, respectively. On the alkaline side, the activity increases with pH.

3. 2,4-Dinitrophenol (DNP) stimulates the enzymatic activity at both high and low K^+ concentrations.

[1] High and low K^+ concentrations correspond to around 0.5 and 0.1 *M*, respectively.

4. Parachloromercuribenzoic acid (PCMB) inhibits the ATPase activity. This inhibition is partially reversed by cysteine.

5. Monoiodoacetic acid (MIA) has no effect on the ATPase activity at either high or low K^+ concentration.

6. EDTA inhibits the enzyme both with high and low K^+ concentrations.

These results demonstrate that the major part of the enzymatic properties of plasmodial myosin B coincide with the distinctive features of myosin B in muscle.

As mentioned previously, the ATP sensitivity of the viscosity drop is small in plasmodial myosin B. Besides, the ATPase activity is stimulated by DNP at low K^+ concentrations. These facts suggest a low actin content in this protein preparation (cf. Szent-Györgyi, 1951; Chappell and Perry, 1955; Greville and Needham, 1955; Greville, 1956; Yagi, 1957).

To sum up, data represented here indicate that there is in myxomycete plasmodia a protein fraction which has many physicochemical and enzymatic similarities to the properties of muscle myosin B (or actomyosin). On the other hand, we have demonstrated already that when ATP is added externally to plasmodia of *Physarum* (Kamiya *et al.*, 1957), or when ATP is injected into the organism (Takata, 1957), the motive force of streaming is strongly increased. In addition, it is known that *Physarum* plasmodia contain a large amount of ATP (Hatano and Takeuchi, 1960). As the protein in question reacts specifically with ATP *in vitro*, we may assume that this protein is the active principle that responds to ATP in a characteristic way *in vivo*. It is possible to say that plasmodial myosin B is a substance playing a crucial part in mechanochemical phenomena underlying protoplasmic streaming in *Physarum*.

Birefringence in Plasmodia

If actomyosin-like protein is a substance corresponding to the mechanochemical system behind streaming in *Physarum*, it would be expected that the deployment of molecules or micelles of this protein could be studied in the living plasmodia by means of the polarizing microscope. A study of the pattern of birefringence in *Physarum* plasmodia has been in progress with Dr. Allen (Princeton University); the analysis is so far incomplete, but we shall present a brief progress report at this time, pending a more complete report to be published later. The photographs (Figs. 5–11) are of flattened portions of plasmodia suspended in a sucrose solution (0.1 and 0.05 M) to prevent osmotic damage.

One of the most striking features of plasmodial ultrastructure visible with polarized light is shown in Fig. 5. The arrow in the pictures indi-

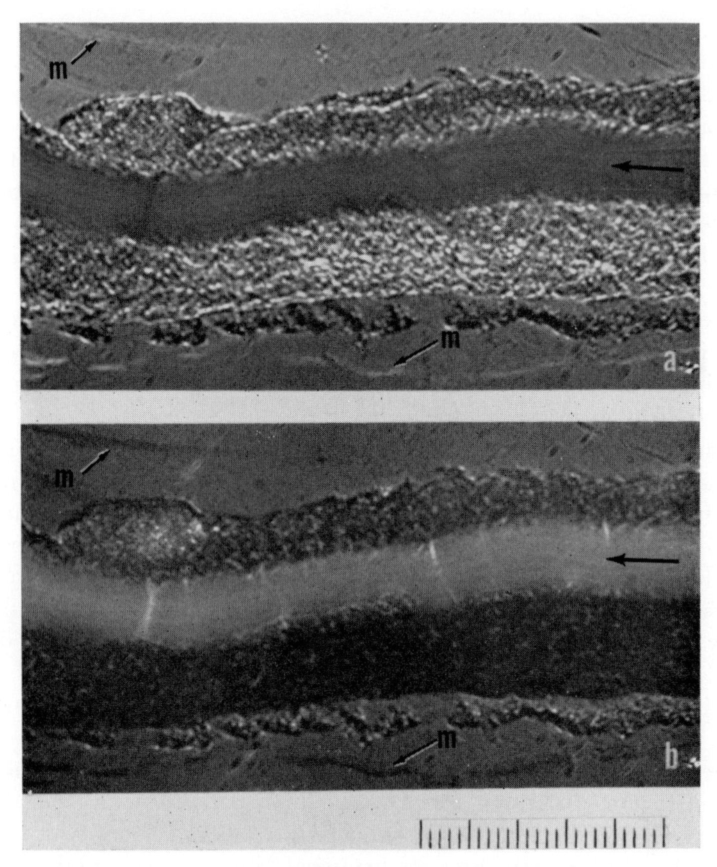

FIG. 5. Birefringence in a plasmodial strand. Angle settings of mica compensator ($\lambda/23$): (a) 3°; (b) —2°. m, Mucus secreted by plasmodia. Sucrose, 0.05 *M*. Scale, 10 μ interval.

cates the streaming endoplasm and its direction of flow. Portions surrounding endoplasm are stationary ectoplasm. The upper and lower pictures are taken at opposite compensator settings to show reversal of contrast in the two pictures. The compensator settings indicate that the signs of birefringence with respect to the axis of streaming are opposite for the endoplasm and ectoplasm; the endoplasm is negative, whereas the ectoplasm is positive.

If proteins are responsible for birefringence, positive axial birefringence in the ectoplasm is interpreted as orientation of proteins at the micellar or filament level along the axis of the strand. In the case of the endoplasm, which presumably contains the same molecular species as the ectoplasm, and differs supposedly only in physical state, the sign is negative, indicating that the alignment of proteins would have to be perpendicular to the axis of streaming. However, we cannot say with any assurance what kind of molecules and what kind of deployment might be responsible for these birefringence effects.

Concerning the endoplasmic birefringence, we further observed that a substantial part of it remains not only when the flow is stopped during the streaming cycle, but even after fixation and preservation in 50% glycerol.[2] These facts suggest that the endoplasmic birefringence is not simple flow birefringence alone.

It is also noticed in these pictures that there are birefringent structures either in or around the endoplasm and oriented orthogonally to the streaming. The mucus secreted by plasmodia has positive birefringence (designated by m in the pictures).

Figures 6–8 show very clearly the presence of birefringent fibrillar structures in the cytoplasm as well as diffuse birefringence surrounding them. In these specimens, scarcely any endoplasmic streaming occurs. In the top pictures are shown the advancing front; in the middle frames, the expanded inner portion; and in the lower pictures, a portion of an area with ramifying strand. In the case of active plasmodia, the endoplasm usually streams through such a strand with rhythmical changes in direction, then diverges into smaller and smaller branches through the expanded portion, and finally into very small streamlets at the advancing front.

These pictures show that in the expanded portion (Figs. 6 and 7), the birefringent structures are arranged orthogonally, with one almost perpendicular to the edge of the spreading tip. On the ramifying strand (Fig. 8), the fibrillar structures are arranged along the axis of the strand. These fibrillar structures have positive birefringence.

In general, birefringent fibrillar structures are usually not visible in many plasmodia in which the endoplasm exhibits the most active streaming. In other words, they are induced structures, which may be thought of as having crystallized out of pre-existing organization of lower order.

We can also observe the belt-shape birefringent structures (Figs. 9–11). They are clearly visible in this specimen, in which no streaming was discernible. The orientations of belt-shape structures are very similar to those of the fibrillar structures; in the expanded portion (Figs. 9 and 11) the birefringent structures are arranged orthogonally, and in the ramifying strand (Fig. 10) they are oriented parallel to the axis of the strand. Therefore, both types of structures may have originated from the same source.

---

[2] The specimen was washed three times with water and suspended in water before observation.

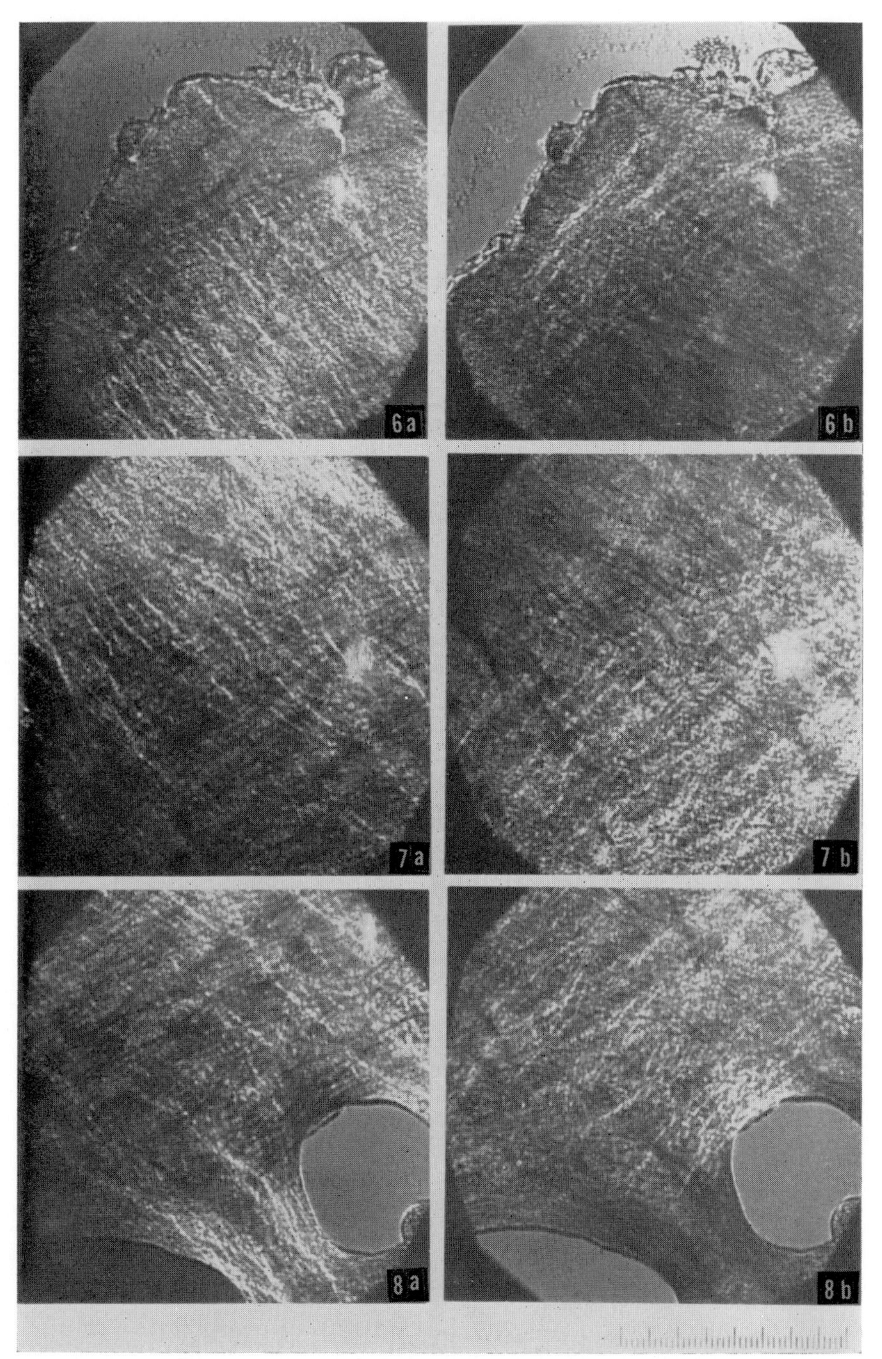

FIGS. 6–8. Fibrillar birefringent structures in plasmodia. Angle settings of mica compensator ($\lambda/23$), Figs. 6a, 7a, 8a; 3°; Figs. 6b, 7b, 8b; —3°. Sucrose, 0.1 *M*. Scale, 10 μ interval.

To summarize the polarized-light results obtained so far, it can be said that the endoplasm shows negative birefringence, whereas the ectoplasm shows positive birefringence owing both to visible fibrils and to unresolved structural elements. At the present time, we are unable to give a detailed interpretation at either the submicroscopic or molecular levels of this birefringence.

We have shown the beginnings of two different experimental approaches to the molecular basis of protoplasmic streaming in *Physarum*. The first is chemical; it is obviously necessary to learn what molecules are present in plasmodia and how they interact under well-defined *in vitro* conditions. The second method is physical; without identifying the molecules, it attempts to delineate their deployment and alignment within the living plasmodia. Living motile systems are obviously dynamically organized societies of molecules. Therefore, both approaches are required if we are to understand the whole process.

References

Chappell, J. B., and Perry, S. V. (1955). *Biochim. Biophys. Acta* **16**, 285.
Greville, G. D. (1956). *Biochim. Biophys. Acta* **20**, 440.
Greville, G. D., and Needham, D. M. (1955). *Biochim. Biophys. Acta* **16**, 284.
Hatano, S. (1962). Personal communication.
Hatano, S., and Takeuchi, I. (1960). *Protoplasma* **52**, 169.
Kamiya, N., Nakajima, H., and Abé, S. (1957). *Protoplasma* **48**, 94.
Loewy, A. G. (1952). *J. Cellular Comp. Physiol.* **40**, 127.
Nakajima, H. (1956). *Seitai no Kagaku* **7**, 256 (in Japanese).
Nakajima, H. (1960). *Protoplasma* **52**, 413.
Needham, D. M., and Cawkwell, J. M. (1956). *Biochem. J.* **63**, 337.
Needham, D. M., and Cawkwell, J. M. (1958). *Biochem. J.* **68**, 31p.
Portzehl, H., Schramm, G., and Weber, H. H. (1950). *Z. Naturforsch.* **5b**, 61.
Rebhun, L. (1962). Personal communication.
Szent-Györgyi, A. (1951). *J. Biol. Chem.* **192**, 361.
Takata, M. (1957). *22nd Ann. Meeting Botan. Soc. Japan.*
Tonomura, Y., and Sasaki, A. T. (1957). *Enzymologia* **18**, 111.
Tonomura, Y., Yagi, K., and Matsumiya, H. (1955). *Arch. Biochem. Biophys.* **59**, 76.
Ts'o, P. O. P., Bonner, J., Eggman, L., and Vinograd, J. (1956a). *J. Gen. Physiol.* **39**, 325.
Ts'o, P. O. P., Eggman, L., and Vinograd, J. (1956b). *J. Gen. Physiol.* **39**, 801.
Ts'o, P. O. P., Eggman, L., and Vinograd, J. (1957a). *Arch. Biochem. Biophys.* **66**, 64.
Ts'o, P. O. P., Eggman, L., and Vinograd, J. (1957b). *Biochim. Biophys. Acta* **25**, 532.
Yagi, K. (1957). *J. Biochem.* **44**, 337.

Discussion

DR. REBHUN: I recently had an opportunity to examine the protein that has been extracted by Dr. Nakajima's technique. I think it is rather fortunate you use different terms for your protein than Dr. Ts'o and his collaborators used for theirs. I think the evidence is they are different proteins. Specifically, we started out using Dr. Nakajima's techniques on frozen plasmodia. We were unsuccessful after six or seven tries in ex-

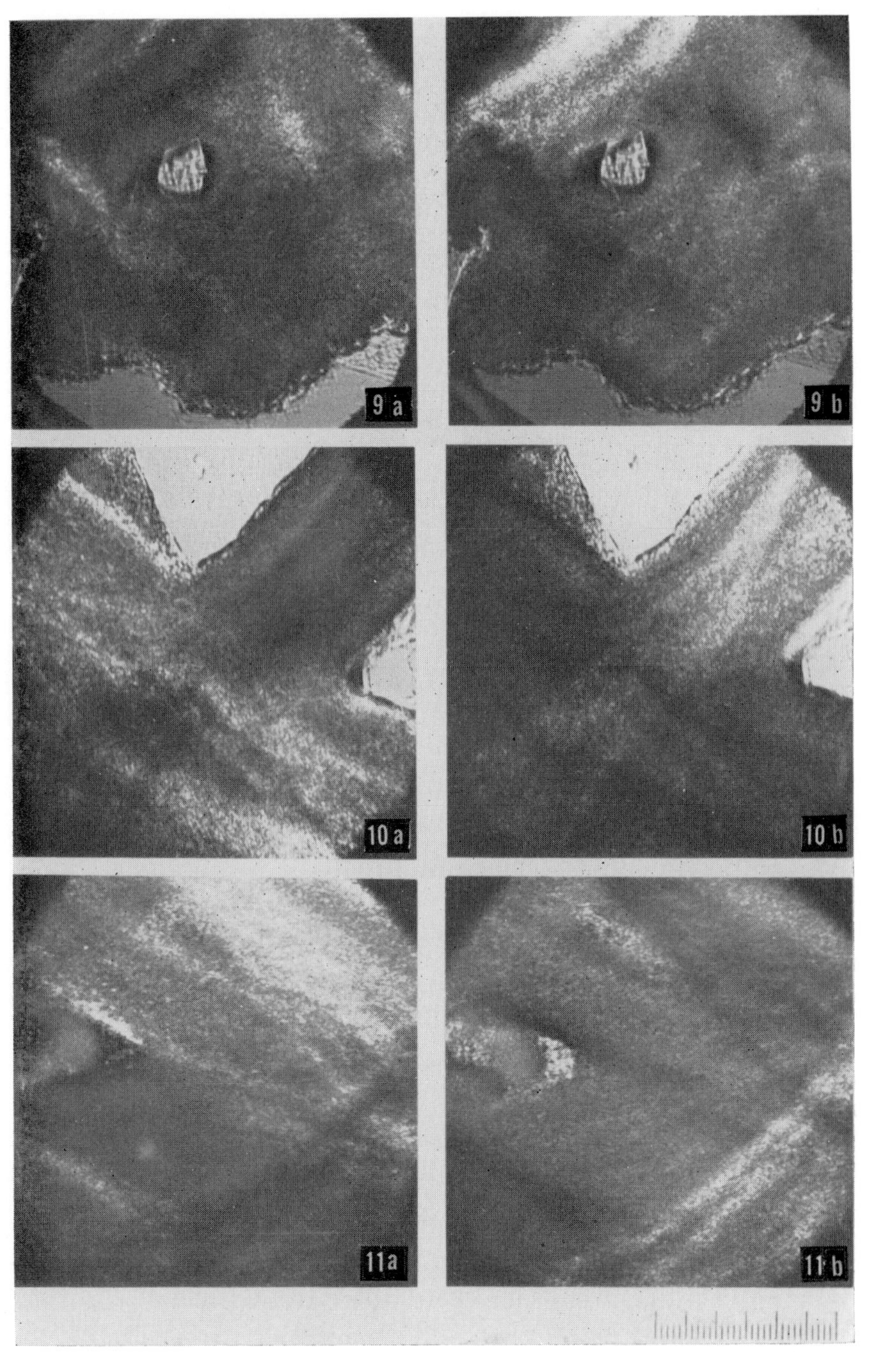

FIGS. 9–11. Belt-shape birefringent structures in plasmodia. Angle settings of mica compensator ($\lambda/23$): Fig. 9a, 8°; Fig. 9b, —6°; Fig. 10a, 9°; Fig. 10b, —10°; Fig. 11a, 6°; Fig. 11b, —7°. Sucrose, 0.1 *M*. Scale, 10 μ interval.

tracting the so-called plasmodial myosin B, using Dr. Nakajima's technique on the frozen mold. However, we went to fresh mold and it came out perfectly well.

We also found that if we varied the pH of the extracting solution—and we used a variety of solutions—if we got down below approximately pH 8.0, we were not able to extract the protein. Similarly, in diluting the extract, the pH had to be below pH 6.6 or a precipitate would not form.

Thus, myxomyosin (Dr. Ts'o) and plasmodial myosin B (Dr. Nakajima) differ on several counts: first, plasmodial myosin B is not extractable, under any conditions that we have tried, from frozen mold, whereas myxomyosin is; second, plasmodial myosin B is not extractable at pH's below approximately 8, whereas myxomyosin is extractable in KCl in distilled water, presumably below pH 7.0; and last, plasmodial myosin B precipitates at low ionic strength, whereas myxomyosin not only is soluble but appears to dissociate irreversibly into smaller units. These proteins are undoubtedly different. They may very well be related, however, and it may be that myxomyosin is a degeneration product of plasmodial myosin B. At least I presume degeneration goes in that direction. I also presume that fresh mold is a better material to use for extraction than slowly frozen mold.

I think it is rather interesting that one can prepare at least two proteins, both of which have mechanochemical properties in the sense of an ATPase which shows reversible change in viscosity on addition of adenosine triphosphate (ATP).

Chairman Thimann: Do you want to make comment on that? Have you ever extracted frozen mold?

Dr. Nakajima: No. I have used only fresh mold.

Dr. Hoffmann-Berling: Judging from Dr. Rebhun's comment, the contractile protein of slime mold plasmodia behaves somewhat like the contractile protein from smooth muscle. The contractile complex can be extracted only from fresh muscle. With the onset of rigor mortis, the actin component becomes inextractable, and the muscle delivers a myosin which is devoid of actin and which, of course, is noncontractile. Furthermore, actin can be extracted more completely from fresh muscle, if the pH is kept well above 8. I would guess that plasmodial myosin B (Dr. Nakajima) corresponds to the contractile complex protein of the slime mold plasmodia, whereas myxomyosin (Dr. Ts'o) represents the plasmodial myosin deficient in actin.

In Dr. Nakajima's paper it struck me that the rate of ATP dephosphorylation is enhanced at ionic strengths of 0.2. The ATP-splitting activity of muscle actomyosin decreases sharply in concentrated salt solutions. However if activation is brought about, not in the usual way by 10^{-3} mole liter Mg^{++} but by adding Ca^{++}, there is an augmentation of enzymatic dephosphorylation of muscular actomyosin at high salt concentrations. Have your splitting experiments been done in the presence of high concentrations of Ca^{++}?

Dr. Nakajima: Yes. These are the values obtained in the presence of 3.3×10^{-3} *M* $CaCl_2$ (28°C, pH 6.6). As you pointed out, in the case of skeletal muscle myosin B, the ATPase activity is maximal when potassium is at the physiological concentration, that is, at about 0.1 *M*, in the presence of $CaCl_2$ and decreases with increasing potassium concentration. However, in the case of smooth muscle myosin B, the activity increases with increasing potassium concentration until the maximum at about 0.6 *M* is reached. Therefore, plasmodial myosin B ATPase is affected by the K^+ concentration in a manner more like smooth muscle than skeletal muscle.

Dr. Hoffmann-Berling: Since the actin component of smooth muscle actomyosin and of fibroblast cell actomyosin is much less soluble in concentrated salt solutions than the myosin component, it is a problem to get out with the myosin an appropriate amount of actin. Since the addition of Ca^{++} activates the ATP-splitting activity only of free muscular myosin (L-myosin) but not of the actomyosin complex, and since Ca^{++} increases the ATP-splitting rate in your preparations, I would guess that your preparations contain an excess of free myosin and are deficient in actin. A deficiency in actin can roughly be estimated by measuring the mechanical activity of the protein, i.e., by centrifuging the gel in graduated tubes and measuring the amount of volume shrinkage after the incubation of the gel with ATP. Can you say anything about this?

DR. NAKAJIMA: I did not measure the volume shrinkage. However, the superprecipitation of plasmodial myosin B proceeds rather slowly under the present experimental conditions. In addition, the ATP sensitivity of the viscosity drop is small and the ATPase activity is stimulated by DNP at low K^+ concentrations. From these facts, I suggest that actin content in plasmodial myosin B is low.

DR. INOUÉ: May I speak concerning this negative birefringence of the endoplasm? Did I notice that the contrast of the endoplasm was similar to that portion of the plasma surface which was parallel to the endoplasm?

DR. NAKAJIMA: In the case of the mucus layer, surface membrane, and ectoplasm, the slow axis is parallel to the strand axis. In the case of the endoplasm, the slow axis is perpendicular to the strand axis.

DR. INOUÉ: In cell division when asters first develop, they appear with a positive birefringence. When the fibers are about to disappear in telophase, they sometimes show negative birefringence. We do not know the explanation, but it may be related to your observation.

DR. SATO: Considering the negative birefringent nature of endoplasmic streaming, which I believe would be based on both flow and form birefringence, did you notice any changes of the sign and the order of magnitude of birefringence during the whole process of flow reversal?

DR. NAKAJIMA: We have not yet seen such changes.

DR. ALLEN: I think it is probably safe to assume that the fibers Dr. Nakajima has been looking at in polarized light are either similar to or identical with the structures that Dr. Wohlfarth-Bottermann described. We have been rather encouraged to find that he found evidence of the ATPase system being there, and I am sure he is somewhat relieved to find we see the same thing in the living state. I think these studies complement one another well.

Regional Differences in Ion Concentration in Migrating Plasmodia

JOHN D. ANDERSON

Department of Physiology and Biophysics, University of Illinois, Urbana, Illinois

Introduction

Myxomycete plasmodia migrate toward the cathode when weak currents are applied continuously (Watanabe *et al.,* 1938). Anderson (1951) reported that the effect of the current was to inhibit migration toward the anode rather than to stimulate movement to the cathode. He could not find any appreciable differences in the rates of migration of equal-sized plasmodia when subjected to a range of current densities from about 1.0 to 8.0 $\mu a/mm^2$ in the agar substratum. In these experiments the durations were from 2 to 6 hr, and the distances traversed by the plasmodia were only a few centimeters.

A series of studies were undertaken to determine whether cataphoretic movement of the common cations—sodium, potassium, and calcium—could be detected. Calcium concentrations of about 25 meq/kg (wet weight) in the anterior regions and 15 meq/kg in the posterior were found, but there was no difference between the experimental samples and the controls (Butkiewicz, 1953). However, Roter (1953) found that, whereas sodium was usually higher in the posterior regions of both experimental samples and controls, potassium concentrations were definitely higher in the anterior regions of only those plasmodia whose migration had been oriented by direct current. Although these findings ruled out the possibility that the effect of the current was a simple cataphoretic one, the magnitude of the differences found was surprising since one would expect the vigorous shuttling type of protoplasmic streaming in this organism to equalize ion concentrations in all parts of the plasmodium.

In more extensive studies (Anderson, 1962), plasmodia which were subjected to direct currents had potassium concentrations of about 30 meq/kg in the anterior (cathodal) region and about 22 meq/kg in the posterior (anodal) regions. Thus there was a loss of about 30% of potassium in the posterior or anodal regions. No significant differences in protein concentration between the two regions could be detected by the analytical method used, nor were there any obvious differences in the gross morphology of the electrically oriented experimental samples and the controls.

The potassium concentration in the controls tended to be less in the posterior regions, but the difference between the two ends was less than 5% and not significantly different. However, the question arose as to whether this slight difference would become larger if the control plasmodia were allowed to migrate for a much greater length of time. In preliminary experiments differences greater than 5% in potassium concentrations were detectable between the advancing nonchanneled areas and the more posterior channeled regions of plasmodia. These plasmodia migrated for more than 12 hr over large sheets of non-nutrient agar and established a definite and consistent orientation, i.e., the trailing region did not produce an advancing front. These findings led to the present work, i.e., the question whether differences in potassium concentration might be associated with oriented migration generally rather than with just electrically imposed orientation.

Materials and Methods

Two kinds of culturing techniques were used. Axenic cultures were grown in the dark at 22–26°C on a medium slightly modified from that reported by Daniel and Rusch (1961). The modification consisted of the substitution of phosphate buffer at pH 6.0 for $CaCO_3$, hemin (Kelley *et al.*, 1960) (10 mg/liter) for chick embryo extract, and the addition of 2% agar. Nonaxenic, oatmeal-fed cultures were maintained by the method described by Camp (1936). No significant differences in potassium concentrations could be found between migrating plasmodia derived from the two types of culture.

In the experiments using radioactive potassium (K^{42}), the isotope as KCl was mixed with oatmeal. The cultures were fed this mixture 12 hr before transplants were made. In these experiments only transplants free of whole grains of rolled oats were used.

Sodium and potassium analyses were performed by standard flame photometric techniques using the internal standard method. Boiling the sample in water for 5 min released all the sodium and potassium from the plasmodium.

In most of the experiments the transplants were placed at one end of a Lucite tray (90 cm long, 30 cm wide, and 5 cm deep) on 3% agar (made with tap water) which had been poured to a thickness of 4 mm. Since the plasmodium will not migrate any appreciable distance over a paraffin surface, sampling was facilitated by laying strips of Parafilm (Marathon Company, Division of American Can Company, Menasha,

Wisconsin) on the agar along the edges of the tray, thus keeping the plasmodium migrating down the center portion of the tray.

Samples were weighed to the nearest 0.1 mg. Minimum sample size was 10 mg.

Results

In these experiments migration from the site of transplantation is unidirectional in spite of the fact that periodic shuttle streaming occurs. Once a definite orientation has been established, i.e., a leading non-

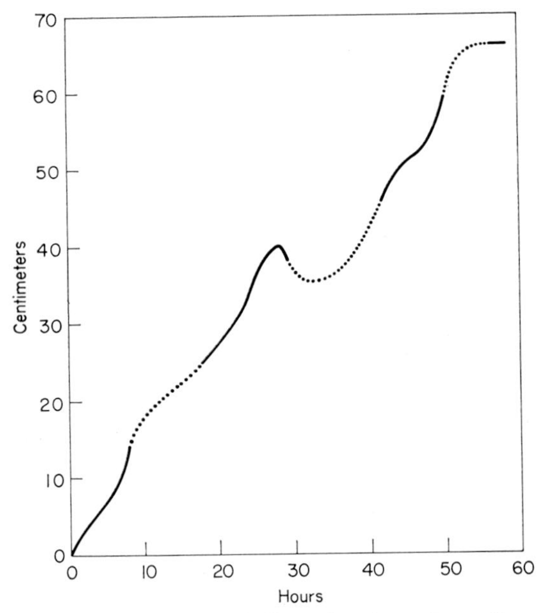

FIG. 1. Distance traversed by vertically migrating plasmodium as a function of time. Dotted portion of curve denotes periods during which no readings were taken. Sporulation occurred at 57 hr. Pauses in rate of vertical migration may be associated with nuclear division. Guttes and Guttes recently reported that plasmodial motility is arrested during mitosis (Guttes and Guttes, 1963).

channeled front followed by a network of channels, any change in direction of migration arises only from the leading nonchanneled area. Provided they are not fed and regardless of the length of time or the distance migrated, these migrating plasmodia never have been observed to reverse their direction of migration back toward the original site of transplantation. Even if forced to migrate vertically, the plasmodium does not recede back to the site of transplantation and it does not establish a new advancing front. The time course of one such vertically migrating plasmodium is plotted in Fig. 1. This particular plasmodium reached a height of 66 cm above the site of transplantation before sporulation occurred. No analyses for sodium and potassium have been made on the vertically migrating plasmodia.

The potassium concentration of plasmodia growing on nutrient agar medium is about 50 meq/kg. Transplants from such cultures placed in petri dishes on nonnutrient agar steadily lose potassium until concentrations of about 25 to 30 meq/kg are reached (Fig. 2). The

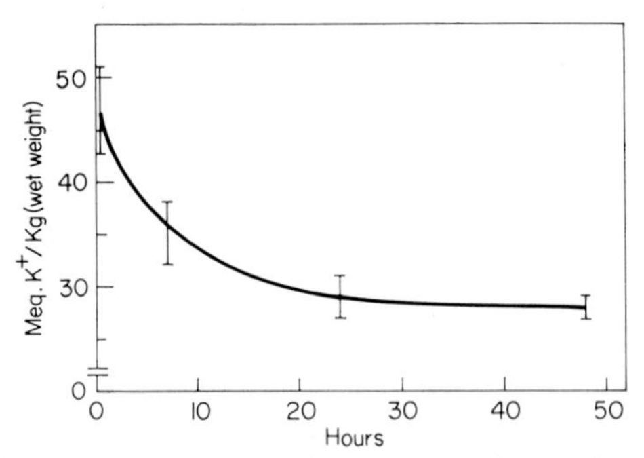

FIG. 2. Loss of potassium in transplants taken from plasmodia growing on nutrient agar. Each point represents values from 5 transplants. The sample analyzed included all the plasmodium.

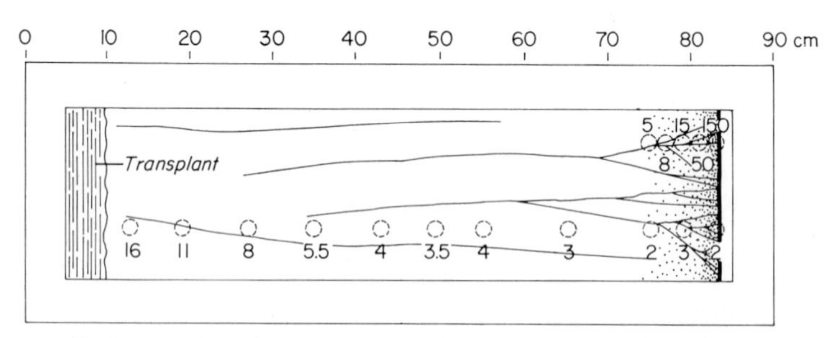

FIG. 3. Diagrammatic representation of the location and amount of radioactivity (K^{42}) in the nonnutrient agar substratum over which a plasmodium had migrated. Samples taken at 26 hr. Surface area of sample was 60 mm^2. Numbers by circles are counts per minute per square millimeter of agar surface. Before the bottom row of samples were taken, all the plasmodium was removed and the surface was wiped with Kleenex wetted with distilled water. Samples in the upper right-hand corner were removed and counted with the mold intact.

entire plasmodium was analyzed in obtaining these values. With the use of radioactive potassium it was possible to monitor the loss of potassium from a migrating plasmodium to the agar. Shown diagrammatically in Fig. 3 are the results obtained from one plasmodium that had traversed a distance of 80 cm in 26 hr. The loss is continuous; it is greater at the beginning of the experiment and then tends to reach a steady value.

The magnitude of the loss of potassium into the agar substratum from plasmodia subjected to direct current was surprisingly large (Table I). In this preliminary experiment it was found that the isotope

TABLE I

K^{42} IN AGAR SUBSTRATUM[a]

	Experimental		Control	
	Anterior	Posterior	Anterior	Posterior
	42	70	12	14
	70	60	1	1
	60	130	4	3
Average	57	87	6	6

[a] Counts per minute per square millimeter of surface of agar below plasmodia oriented by direct current. Current density, 3.5 $\mu a/mm^2$.

TABLE II

K^{42} COUNTS IN MIGRATING PLASMODIUM OF *P. polycephalum*[a]

Sampling region	Sample No.		Cpm/mg dry weight
A. Leading edge	34		384
	35		381
	54		415
	58		386
	59		378
		Average	389
B. No visible channels, 4 mm back from leading edge	36		390
	37		381
	52		392
	57		380
	60		382
		Average	385
C. Visible channels in thin protoplasmic sheet	38		360
	39		350
	53		400
	56		381
	61		354
		Average	369
D. Channels	40		450
	41		295
	54		356
	55		274
	62		307
		Average	336

[a] Radioactivity in various regions of plasmodium that had migrated for 26 hours and 80 cm from site of transplantation. See Fig. 4 for pictorial representation of location of samples.

count was about ten times greater in the agar under the experimental plasmodia than under the controls.

In molds which have migrated for more than 10 hr and over distances of 20 cm or more, higher concentrations of potassium have been

FIG. 4. Illustration of sampling regions. A, advancing front; B, nonchanneled area, i.e., contains no macroscopically visible channels; C, channels visible in the thin protoplasmic sheet; D, channeled area. Approximately equal-sized samples can be obtained from each area. Faint vertical lines are wax pencil marks, 10 cm apart, on bottom of Lucite tray.

found in the leading nonchanneled regions than in the more posterior channeled regions. (Figure 4 illustrates the sampling regions.) Sodium concentrations, although more variable, are just the reverse. Potassium concentrations of about 30 meq/kg and sodium concentrations of 5 to 8 meq/kg are consistently found in the leading edge (area A in

Fig. 4) whereas values of about 20 meq/kg for potassium and 10 to 15 meq/kg for sodium are found in the posterior channeled regions (area D in Fig. 4). The results from preliminary experiments in which radioactive potassium was used show the same trend as the flame photometric analyses (Table II). Radioactive sodium has not yet been used.

Much better agreement between "duplicate" samples, i.e., two or more samples taken in area A, is obtained if the time of sampling is correlated with the advent of an advancing wave or the surging forward of the leading edge. The results of potassium and sodium analyses of duplicate samples taken at the height of a forward surge of the leading

TABLE III

POTASSIUM AND SODIUM ANALYSES OF PLASMODIUM OF *P. polycephalum*[a]

Sampling area	Potassium (meq/kg)	Sodium (meq/kg)
A	29.1	5.8
	30.2	7.5
B	27.8	8.5
	24.6	8.2
C	20.3	12.0
	21.7	15.5
D	19.2	11.6
	20.3	11.8

[a] Duplicate samples taken from different regions of migrating plasmodium just as forward surging of the leading edge was at its height. See Fig. 4 for pictorial presentation of samples.

edge are presented in Table III. The potassium concentration decreases from front to rear whereas the sodium concentrations follow the reverse pattern.

Discussion

Physarum polycephalum survives over periods of stress by at least three commonly observed mechanisms: migration to a new food supply, sclerotization (Jump, 1954), and sporulation (Daniel and Rusch, 1962, and others). In this study we are primarily concerned with the migration phenomenon. As long as nutritional conditions are favorable, no appreciable locomotion occurs; rapid streaming and movement of the plasmodium begins when the food supply has been at least partially depleted (Guttes *et al.,* 1961). Migrating plasmodia gradually become smaller and smaller; the energy must come from stored reserves and autodigestion. The products are excreted.

Therefore, it is plausible that the analyses presented here are in part an artifact since it is impossible to take a sample from the plasmodium without including some of its excretory products. Thus the values obtained might more properly describe conditions in the active protoplasm at some time interval prior to sampling. We have attempted to estimate the amount of external salts by cutting out agar blocks with the mold intact and gently flushing distilled water over the plasmodium before taking samples. Most of the sodium (60%) but little of the potassium (15%) in the channeled areas is removed by flushing water over the plasmodium. These studies are preliminary and incomplete but suggest that sodium is being excreted much more freely than potassium from the channeled regions. However, it is to be noted that a major loss of radioactive potassium occurs immediately under the advancing front, and further migration of the mold over this point does not increase the amount of potassium in the agar (see Fig. 3). Future studies using radioactive sodium are clearly indicated. Also the rather abrupt change in ion concentration from the leading front to the channeled regions must be considered to indicate some special activity in the region where channels are being formed. Work on this is in progress.

These studies, which were originally initiated by the rather simple discovery that there was unequal distribution of sodium and potassium in plasmodia oriented by direct current, have not yet fully explained the role played by either of these ions. The variability found in sodium concentrations and the fact that we have had cultures very low in sodium (Anderson, 1962) leads us to conclude tentatively that sodium does not play an essential role and is indicative of the plant nature of slime molds. An essential role for potassium is indicated by the findings that the potassium concentration in the advancing nonchanneled front remains remarkably constant during long periods of migration and that there is a loss of potassium on the anodal side of plasmodia in direct current. It is suggested that anodal inhibition of migration is the result of this loss of potassium.

Nakajima (1960) has reported that crude preparations of myxomyosin extracted with KCl-free Weber-Edsall solution did not respond to adenosine triphosphate whereas those prepared in the presence of KCl did.

We have not measured potential differences between different regions of the migrating plasmodium and, therefore, can only speculate that these differences in potassium and sodium concentrations may have some relationship to the potential changes associated with streaming (Kamiya and Abé, 1950). Our efforts to measure true transmembrane potentials in these plasmodia have been complicated by the apparent formation of a membrane around the electrode tip, and, therefore, we have been unable to apply the techniques that Bingley and Thompson

(1962) used on amebae to measure effects of external potassium and sequential changes in potential gradients during movement.

Conclusions

1. Migrating plasmodia of the slime mold, *Physarum polycephalum,* maintain a potassium concentration of about 30 meq/kg (wet weight) in the advancing nonchanneled region and about 20 meq/kg in the posterior channeled regions.
2. Sodium concentrations, in contrast to potassium, are higher in the posterior channeled regions (12–15 meq/kg) than in the advancing nonchanneled front (5–8 meq/kg). Sodium concentrations are more variable than potassium.
3. Potassium content of plasmodia grown on nutrient agar is about 50 meq/kg. Plasmodia transferred from nutrient agar to nonnutrient agar lose potassium until a concentration of 25–30 meq/kg is reached.
4. Migrating plasmodia constantly lose potassium to the agar substratum. After about 10 hr of migration the rate of loss remains uniform.
5. Plasmodia whose direction of migration has been oriented by passing direct current through the agar lose about ten times as much potassium as those not subjected to the action of the current.
6. It is considered that potassium, but not sodium, has an essential function in migration in *P. polycephalum* and that the loss of potassium from the mold in direct current may be a primary aspect of the ability of current to orient migration.

Acknowledgments

I wish to thank Mr. Donald M. Miller for his assistance in these investigations and Dr. Howard S. Ducoff for his assistance and the use of equipment and facilities for the isotope work.

The investigations on plasmodia grown in axenic culture were carried out in the laboratories of the Department of Physiology, Emory University, Atlanta, Georgia, and were supported in part by research grant No. E-1433C5 from the National Institute of Allergy and Infectious Diseases, United States Health Service. The author is indebted to Dr. Peter A. Stewart and Dr. Babette T. Stewart, Emory University, for the use of their laboratories and facilities, their cultures of *P. polycephalum,* and for their interest and assistance.

References

Anderson, J. D. (1951). *J. Gen. Physiol.* **35**, 1-16.

Anderson, J. D. (1962). *J. Gen. Physiol.* **45**, 567-574.

Bingley, M. S., and Thompson, C. M. (1962). *J. Theoret. Biol.* **2**, 16-32.

Butkiewicz, J. V. (1953). "The effect of specific electric currents on the anterior and posterior calcium concentrations of migrating *Physarum polycephalum.*" M.S. Thesis, University of Illinois.

Camp, W. G. (1936). *Bull. Torrey Botan. Club* **63**, 205-210.
Daniel, J. W., and Rusch, H. P. (1961). *J. Gen. Microbiol.* **25**, 47-59.
Daniel, J. W., and Rusch, H. P. (1962). *J. Bacteriol.* **83**, 234-240.
Guttes, E., and Guttes, S. (1963). *Exptl. Cell. Res.* **30**, 242-244.
Guttes, E., Guttes, S., and Rusch, H. P. (1961). *Develop. Biol.* **3**, 588-614.
Jump, J. A. (1954). *Am. J. Botany* **41**, 561-567.
Kamiya, N., and Abé, S. (1950). *J. Colloid Sci.* **5**, 149-163.
Kelley, J., Daniel, J. W., and Rusch, H. P. (1960). *Federation Proc.* **19**, 243.
Nakajima, H. (1960). *Protoplasma* **52**, 413-436.
Roter, E. C. (1953). "Na and K concentrations in the cathodal and anodal portions of *Physarum polycephalum* during galvanotoxis." M.S. Thesis, University of Illinois.
Watanabe, A., Kadati, M., and Kinoshita, S. (1938). *Botan. Mag. (Tokyo)* **52**, 441-445.

Discussion

Dr. Mahlberg: Have you worked at all with phosphorus?

Dr. Anderson: We are now beginning to do so.

Dr. Kitching: Dr. Carter demonstrated a high level of potassium in the ciliate *Spirostomum*. The level maintained depends on the state of nutrition.

Have you made any attempt to relate the potential difference across the surface of this organism with the potassium levels?

Dr. Anderson: We have been unsuccessful in obtaining a stable transmembrane potential. When a microelectrode is inserted, a potential of 60–70 mv is measured, but immediately begins to decrease to zero.

Asymmetry potentials (between different regions of the plasmodium) can be measured by externally applied electrodes. The work of Kamiya and S. Abé are excellent examples. I would like to think that the changes in potassium concentrations might somehow be related to potential changes described by them, but I have no direct evidence.

Dr. Jahn: Did I understand correctly that when you apply the current, an accumulation of potassium takes place at the cathode?

Dr. Anderson: No. Potassium concentrations in the advancing regions of migrating plasmodia subjected to dc orientation are the same as those of control plasmodia. It is posterior to these advancing regions where the 30% loss in potassium occurs when the movement is oriented by direct current.

Dr. Jahn: How about sodium?

Dr. Anderson: The concentration of sodium is very variable, but always higher at the posterior.

Dr. Jahn: How about calcium?

Dr. Anderson: Calcium concentrations are found to be higher in the advancing region of both experimental and control plasmodia.

Dr. Ling: In your sampling technique, how wide were the strips of slime mold pieces which you cut? You mentioned that they were of equal length.

Dr. Anderson: Anywhere from 4 to 10 mm, depending on the experiment.

FREE DISCUSSION

Dr. Ling: In connection with possible mechanisms for the observed ionic gradients within a single cell, which Dr. Anderson just reported, there are two hypotheses one can examine. In the first one, one may assume that the cytoplasm is essentially a dilute solution of salts. In this case, the potassium ion concentration has a diffusion coefficient approximately equal to 2×10^{-6} cm^2/sec. Left alone, the potassium ion in the different regions of slime mold, which Dr. Anderson examined, will mix and reach equal concentrations in a matter of minutes. In order to maintain the gradient observed, there must be an ionic pump mechanism. However, the rate of pumping will be many times higher than that calculated for pumping across cell surfaces. This large amount of energy is not compatible with the maximum rate of energy delivery in living cells.

Let us examine a second possibility. In this, the potassium ion is considered intimately associated with ionic groups on proteins and protein complexes within the protoplasm. Because of the differences between the association energies of potassium and of sodium, such a system will select one ion over the other one. Examples of this type of ion-selecting system are, of course, now well known. For example, the exchange resins select potassium ion over sodium ion or sodium ion over potassium ion, depending on the nature of the charged groups that the resins bear. Selective ionic accumulation may be different in different parts of a cell without the need of continuous energy expenditure.

Dr. Anderson's observation of a change of the selectivity ratio with sol-gel transformation also agrees with this model. It has been shown that within the living cell ions can exist in two states: adsorbed on fixed ionic sites or interstitial occupying space between sites. The degree of association is governed by an energy term and an entropy term. Entropy, as is well known, exists in several forms: translational, rotational, vibrational, etc. An estimation can be made to show that the majoı contribution determining ionic association in aqueous solutions is predominantly that due to the rotational entropy. Within the protein matrix, which carries a large number of fixed ionic and H-bonding sites, water molecules cannot rotate themselves as freely as in pure water. Evidence for this has recently been produced by the measurement of nuclear magnetic resonance of the water in collagen (Dr. Berenblum). Since potassium and sodium ions are hydrated, they have a layer of rigidly attached water molecules around each ion. Like water they will also suffer a loss of rotational freedom within the protoplasmic network. This loss of rotational entropy will be greater in a gel where the H-bonding sites are more fixed than in the sol form. It may be anticipated, therefore, that there will be a larger proportion of ions associated in the gel state than in the sol form. Since it is only in the associated state that the association energy differences can determine potassium over sodium selectivity, one can anticipate that as a gel transforms to a sol, there is a tendency for the ionic distribution to approach that in the surrounding medium.

Chairman Thimann: I wanted to ask Dr. Kuroda a question about rotating chloroplasts. You have to have light in order to look at them, so they could conceivably be carrying out the chemical reactions of chloroplasts. Have you ever looked at them in green light? You see, in green light the chloroplasts would not be able to be carrying out a photosynthetic reaction, so it would have no source of energy for that. I wondered whether they continue to rotate in green light.

Dr. Kuroda: Streaming is capable of continuing in the dark.

Chairman Thimann: I am talking about the rotation of chloroplasts; not streaming.

Dr. Kuroda: Chloroplasts rotated 3 days in the dark. Therefore, photosynthesis has probably little to do with the chloroplast rotation. Nuclei also rotate.

Dr. Rebhun: We have looked at amebae in the Harvey-Loomis centrifuge microscope, and have been struck by the existence of particles persisting in the hyaline zone after most of the particles have been stratified in a centrifugal direction. In watching these particles, it is clear that many of them are moving in loops coming from the centrifugal mass of granules, going counter to the direction of centrifugal force and then turning to re-enter the particle mass. Since there are many dark granules in these hairpin loops, we assume they are Dr. Griffin's carbonyl diurea crystals which have a specific gravity of about 1.6. No obvious differences in velocity of particles could be observed with, as opposed to, against the direction of centrifugal force in these streams. Since the cells were examined at forces up to 1100 *g* it would appear that local cytoplasmic strands can have very considerable structural rigidity.

PART II

Cytoplasmic Streaming and Locomotion in the Free-Living Amebae

Introduction

P. P. H. De Bruyn

University of Chicago, Chicago, Illinois

It has been more than 15, rather closer to 17, years since I have had any personal contact with the work that is being presented in this Symposium. During this time many developments have occurred and many contributions have been made about which I am not properly informed. If it were not for the fact that some of you have been so kind as to send me reprints from time to time, I would be almost totally ignorant of the recent developments in this field.

Notwithstanding my remoteness to the problem, I have tried to examine how the recent work has changed, modified, or resulted in progress in the understanding of cytoplasmic movement. I have even attempted to speculate a bit about what progress can be expected or what possible new approaches may be fruitful. It may seem to many of you that in the last decades little basic information has been obtained which has led to adequate concepts explaining the movement of cytoplasm. However, although we are still very far removed from knowing how an organism without visible locomotive organelles moves, progress has been considerable, and the definition of the problem has become more circumscribed. We have long since rejected the idea that surface tension was responsible for ameboid movement, that movement is caused by an elastic recoil from a tension resulting in differences in tonicity of the cytoplasmic constituents of the cell, that gelation itself is the cause for cytoplasmic contraction, and other notions.

The influence of the terminology borrowed from colloid chemistry has been notable and, in some respects, detrimental. The terms *sol* and *gel, solation* and *gelation,* merely compare the visible and measurable changes in the ameboid cytoplasm with changes in state in colloids of a very diverse and different nature but do not tell us very much about the changes in the proteins of the cytoplasm that are responsible for the cytoplasmic movement. The descriptive use of colloid chemical terminology brings to mind structural concepts and processes for the existence of which, in cytoplasmic movement, there is no evidence. For this reason, I believe that the return of the terms *endoplasm* instead of *plasmasol* and *ectoplasm* instead of *plasmagel* will be helpful.

Contractility is now the most generally used term for the motor force in cytoplasmic movement, and the arguments have centered mostly on the site of the contraction. We must keep in mind, however, that the contractile material is not localized. Because of the internal cytoplasmic movement, this material is generally distributed throughout the cell. Its contractile action is localized.

Contraction brings to mind a directionally polarized change of shape as occurs in fibrillar structures and which presumably is based, as in contraction of muscle, on a unilateral change in the configuration or position of macromolecular components. In ameboid organisms, there is now morphological evidence for fibrillar organization, too (Wohlfarth-Bottermann). The question is now whether contraction can be localized in the fibrillar components. Are they always present where contraction in the ameboid organism occurs? Or are they formed when the need arises from a three-dimensional molecular reticulum? A macromolecular reticular continuum has been and still seems to be an attractive substrate for the explanation of ameboid activity. Such a molecular reticulum can produce changes in shape not only by contraction of linear molecules but also by changes in the linkages or side bonds, and, which may, as a consequence of this, cause changes in viscosity or apparent viscosity. The motor force can be located in the part of the cytoplasm that streams or rather appears to be streaming. Allen's fountain-zone contraction theory is in conformation with this notion.

The term "morphoplasticity" seems appropriate in this instance, although its original definition applies to a system, or substance, which undergoes structural changes under the influence of adenosine triphosphate. That some such morphoplastic substance is present in ameboid organisms has become clear from the preparation (from the slime mold) of extracts which respond reversibly with viscosity changes to physiological quantities of adenosine triphosphate. Furthermore, Hoffmann-Berling has given evidence that similar contractile proteins, sensitive to ATP, are present in most typically ameboid cells, like fibroblasts. It appears that the fact is emerging that there is a common chemical basis for contraction and, consequently, cell movement in widely different cell types.

The membrane of an ameboid organism has in the past been assigned a variety of roles. Its function as a locomotor force has been essentially discarded. It is true that cytoplasmic movement of some sort may occur in unattached cells, but membrane contact with the substrate produces an organized locomotory progression. This is particularly clear in the ameboid moving leucocytes. When these cells are unattached and are suspended in a liquid medium, the pseudopods arise from all sides of the cell body and the cells exhibit a random activity such as occurs in the depolarized phase of the lymphocytes. Contact with a solid substrate changes this immediately into a polarized shape of the cell and a polarized progression. The membrane has again received attention from Goldacre, who assigned to it a role in ameboid movement in his feedback regulatory process resulting from membrane-plasmagel contact. This idea is perhaps speculative, but it is not, therefore, necessarily wrong. Weiss's ideas that the molecular configuration of the cell surface and its interaction with the substrate may affect internal molecular arrangements may provide, perhaps, new ways for the exploration of ameboid movement.

The study of locomotory function of blood cells offers some unusual opportunities for observations on specific cellular characteristics of ameboid movement and cellular function. The most primitive blood cell, the lymphocyte, moves like a monopodial ameba. As this cell develops into either a granulocyte or a macrophage, this type of movement of the lymphocyte changes into the more depolarized type of the mature blood cell or macrophage. This is accompanied by the acquisition of the properties of phagocytosis and pinocytosis. The development of these membrane functions with the accompanying changes in locomotory characteristics seems to point to a role of the membrane in cytoplasmic movement and an effect on internal molecular arrangement. Later, the macrophage may develop into a fibroblast, which is a more or less sessile cell (or at least one which does not possess a motility such as occurs in its predecessors) and which has at least one function which its predecessors do not have: the ability to form fibers. During the progressive development from ameboid cell to fibroblast, profound changes must take place in the morphoplastic or contractile proteins, which are worthy of attention. The fact itself that specific locomotory characteristics exist in various types of blood cells has been a very useful tool in problems connected to blood formation. The work of Robineaux is one of the latest examples of this.

The studies of ameboid movement have almost entirely been limited to direct observations or measurements with physical tools. With few exceptions, such as mentioned earlier, chemical approaches have been lacking. Of course, exploration by direct chemical means will generally cause an intervention with the very process one is examining. Different metabolic and different structural processes can be expected to occur at different sites in an actively moving cell. To be able to fix a moving cell, leaving undisturbed this differential localization, of which the general shape of the cell is an expression, will open up possibilities of cytochemical studies, which in themselves may provide fruitful insight into the mechanism of ameboid movement. And it may be that if an ameba is fixed in such a way, fine-structure observations will be more revealing with respect to the study of ameboid movement than they have been thus far. Rapid freezing, freeze-drying, or freeze substitution are methods that immediately come to mind.

It is unfortunate that the hypotheses that have been advanced to explain ameboid movement cannot be put to experimental verification. Many have been proposed, but most have rested on conviction rather than on adequate observational foundations.

In this connection, I am reminded of a passage which occurs in the memoirs of Jacques Casanova, in which he describes a visit with his friends to a cabinet of Natural History. He writes as follows:

"The caretaker showed us a packet bound in straw that he told us contained the skeleton of a dragon; a proof, added he, that the dragon is not a fabulous animal."

Studies on the Isolated Membrane and Cytoplasm of *Amoeba proteus* in Relation to Ameboid Movement

L. Wolpert, C. M. Thompson, and C. H. O'Neill

Zoology Department, King's College, University of London, London, England

Introduction

It is a standard approach in experimental biology to try and isolate functional and structural systems from the intact organism so as to make them more amenable to study. In cell biology the isolation of subcellular constituents has been particularly fruitful; this is evident from the advances following from the isolation of mitochondria, mitotic apparatus, and muscle proteins. It is thus surprising that, despite notable exceptions, so little attention should have been paid to the isolation of the subcellular components in the study of ameboid movement. One of the difficulties peculiar to the planning of an isolation of the systems involved in ameboid movement is that one does not have well-defined structural or chemical criteria on which to base the isolation. There is nothing in the ameba to which one can point with confidence and say: "If we isolate that structure or that biochemical system we shall have the components responsible for movement." There is no compelling reason to think that most of the protein is involved in movement as in muscle; nor are there clearly recognizable structures as in cilia, or the mitotic apparatus. Worse still, we do not even know whether the site of the motive force resides within the surface membrane or cytoplasm. This point is particularly important since movement involves change in form of both cytoplasm and surface membrane. For such reasons one must rely largely on isolating a system that will show some of the phenomena seen in the living cell; for example, isolated surface membranes must change their form or a cytoplasmic fraction must show streaming.

There is another approach which has received considerable support, and that is to base the isolation procedure on the assumption that cell movement involves a system very similar to that responsible for muscular contraction. A persuasive analogy may be drawn, and this has been strengthened by the pioneer studies of Loewy (1952) and later workers who isolated an actomyosin-like protein from slime molds, and also of Hoffmann-Berling (1960) who prepared glycerol-extracted models of fibroblasts which resembled in many ways those of glycerol-extracted muscle, and an actomyosin-like protein fraction from tissue cells. However, such studies and their extension have not really established that the ideas on muscular contraction can really be applied to problems of cell movement. Suggestive though it may be, the isolation of a system showing viscosity changes with adenosine triphosphate (ATP) in no way guarantees that such a system is responsible for cell movement.

We have, therefore, adopted a somewhat different approach. At the time our studies began, about 4 years ago, the most satisfactory theory of ameboid movement was the extension and modification of Mast's (1926) ideas by Goldacre (1952). This had two main postulates: (*a*) a contractile gel capable of sol-gel transformations, contraction being accompanied by solation; and (*b*) the surface membrane initiates contraction in the gel, is resorbed at the tail of the ameba, and is re-formed at the advancing end, the whole membrane being formed anew each time the cell passed through its own length. We, therefore, set out to isolate the contractile gel and the surface membrane. For the gel, the criterion would be its ability to contract or at least to undergo sol-gel transformations. With respect to the membrane, we felt that if this could be isolated, one might be better able to study its dynamics, particularly the proposed rapid turnover and the mechanism of its resorption and re-formation.

Dynamics of the Surface Membrane during Cell Movement

The idea that the cell membrane plays an active part in ameboid movement goes back a long time and is, in many ways, attractive. Early theories suggesting that changes in surface tension would provide the basis for ameboid movement were discounted when it was realized that at the cell surface there was a structural membrane that showed elastic properties. Electron microscope studies of the membrane of *Amoeba proteus* suggest that it is made up of a typical plasma membrane with a thick outer coat (Pappas, 1959; Mercer, 1959). The plasma membrane appears as a double layer about 80 A thick which is consistent with the bimolecular lipid leaflet with protein on either side, postulated by Davson and Danielli (1942), and termed the "unit membrane" by Robertson (1959). The surface coat appears as a fibrillar layer about 0.2 μ thick and is probably composed of mucoprotein (O'Neill, 1963). These structures can be clearly seen in membranes of *A. proteus* isolated in bulk (Fig. 1) (O'Neill and Wolpert, 1961).

The membrane may play both physiological and mechanical roles in ameboid movement. A most significant example of the former is the possibility that changes in membrane potential may control the initiation of pseudopods and the rate and direction of streaming (Bingley and

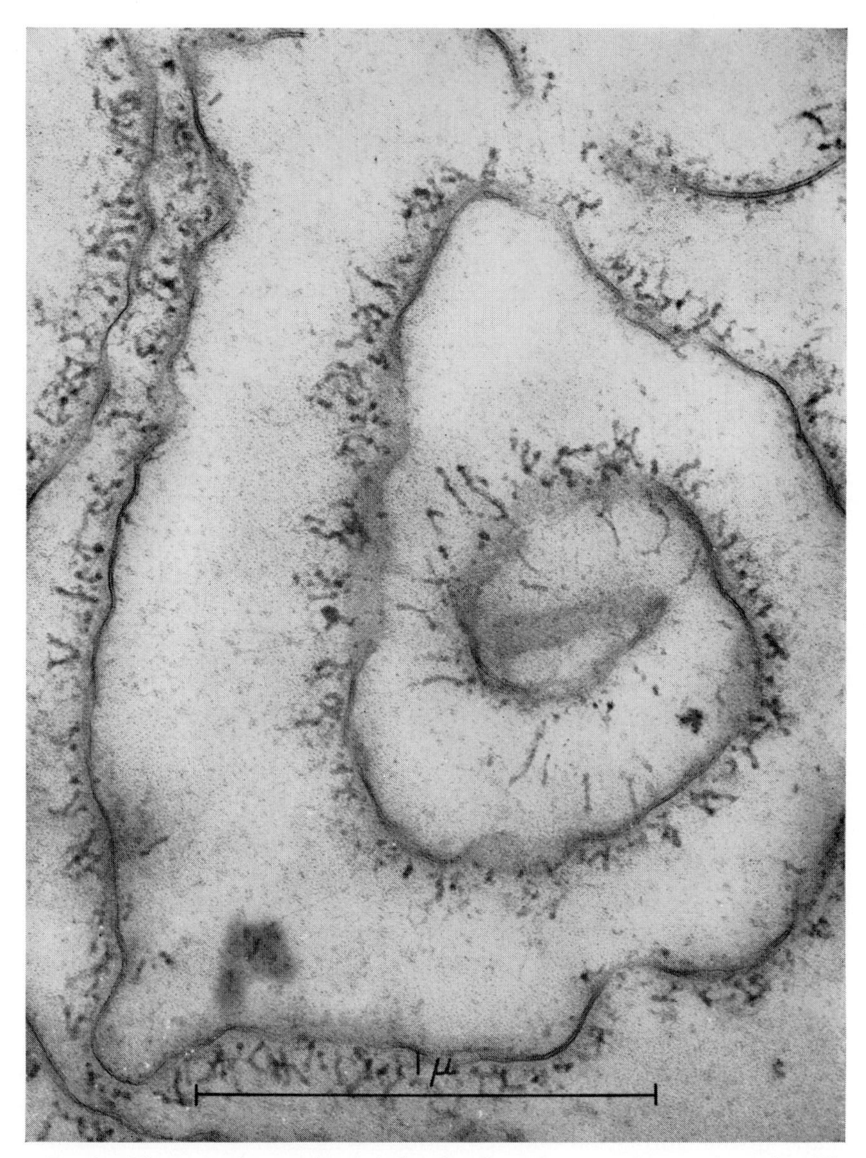

FIG. 1. Electron micrograph of isolated cell surface membrane of ameba. The fibrillar surface coat can be seen overlying the double-layered unit membrane (80 A thick) characteristic of the surfaces of cells. No other components of the cell can be detected. (Photograph by Dr. E. H. Mercer.)

Thompson, 1962). Here we are mainly concerned with the mechanical role of the membrane.

It has been suggested that the apparent mechanical properties of the membrane are not due to the plasma membrane but to a relatively thick (1–2 μ) gel-like cortex beneath the plasma membrane (Marsland, 1956; Mitchison, 1952, 1956), although this has been questioned (Wolpert, 1960b; Mercer and Wolpert, 1962). No evidence can be found for a cortex of this type in *A. proteus* either in intact cells or in isolated membranes.

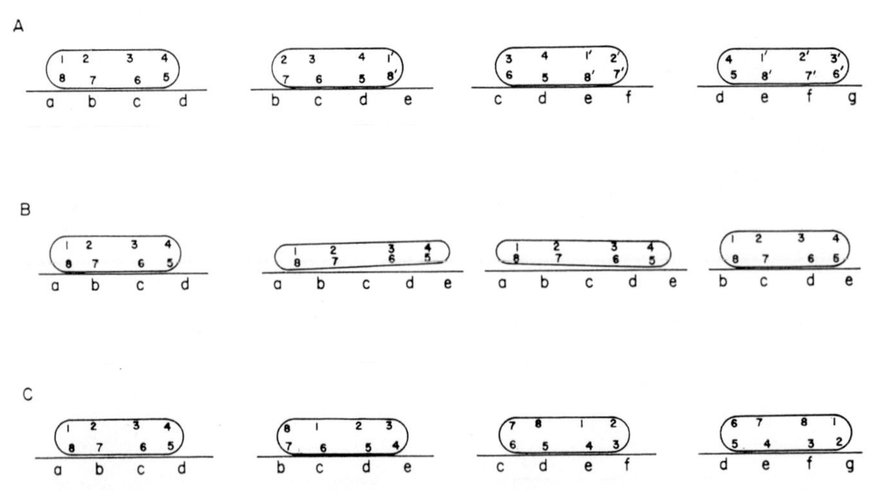

FIG. 2. Diagram to show possible ways the membrane can change its shape during cell movement. Sites on the membrane are distinguished by numbers, and sites on the substratum by letters. Three possibilities arise from geometrical considerations. (A) Membrane is supplied to the surface at the front, and withdrawn from the surface at the rear; the surface elsewhere is stationary. (B) The membrane is alternately extended and contracted. Expansion occurs while the cell is in contact with the substratum at the rear, and contraction occurs while the cell is in contact with the substratum at the front. (C) The membrane flows to accommodate changes in shape. The membrane flows forward over the upper surface of the cell, while it remains stationary at the points of contact with the substratum.

There is a well-defined problem as to how the membrane changes its form as the cell moves forward. Three main possibilities exist, which may be derived from consideration of the geometry of the moving cell (Fig. 2).

(*a*) The membrane is stationary over most of the cell and change in shape involes formation and resorption of membrane (Fig. 2A). This is the requirement of the theories of Goldacre (1952, 1961) and apparently Bell (1961). The cell contents move through a stationary tube of membrane, new surface being formed at the front, while there is withdrawal of membrane from the surface at the rear. Goldacre suggests that the force for movement is provided by contraction of the ectoplasmic gel at the rear, whereas Bell suggests that the formation of new surface itself provides the force. It is a consequence of the theory of Goldacre at least, that the rate of turnover of the membrane between the surface and the interior must be high during movement and the cell may be expected to renew its membrane completely each time it passes through its own length, namely about every 5 min.

(*b*) The membrane is reversibly extensible (possibly elastic) and localized changes in shape may take place by localized increase or decrease in surface area without the formation or removal of surface material (Fig. 2B). This is the basis of the theory for the movement of amphibian embryonic cells put forward by Holtfreter (1949) and for fibroblast movement by Ambrose (1961). No such theory has been put forward for ameboid movement.

(*c*) The membrane of the cell flows freely, and this will often be seen as forward motion over the upper surface when the cell is attached to the substratum over the whole of its length (Fig. 2C) (Mast, 1926; Griffin and Allen, 1960). In this case membrane turnover will be low.

These examples illustrate the main ways that the membrane may change in form during locomotion, and numerous variations are, of course, possible. For example, it is important to note that cell movement by pseudopod extension and shortening (cf. Gustafson and Wolpert, 1963) may involve any of the three possibilities.

From these considerations three experimentally testable questions concerning the behavior of the surface during locomotion can be posed. (1) Is there localized rapid membrane turnover? (2) Does the surface extend reversibly? (3) Does the surface flow forward? There is remarkably little unequivocal evidence that allows us to answer these questions. With respect to ameboid movement, only possibilities (1) and (3) have been put forward. On the one hand, Goldacre (1952, 1961), from observations on the behavior of large obstacles such as oil drops attached to the membrane and from the accumulation of neutral red at the tail, has claimed that turnover is rapid and that membrane is resorbed at the tail and formed anew at the front. But these observations are open to other interpretations. For example, the fact that an oil drop attached to the surface does not move forward may only show that membrane flow in the region of attachment has been stopped (Wolpert and O'Neill, 1962). On the other hand, from studies of particles attached to the surface, Mast (1926) and Griffin and Allen (1960) have suggested that the membrane is fluid and flows forward. However, Goldacre (1961) has suggested that this is open to other interpretations.

In order to try and resolve this controversial issue we considered it highly desirable to mark the cell surface by a method not subject to the uncertainties inherent in the observation of the movement of adhering particles or the accumulation of dyes, in order to follow the behavior during movement. Such a marker is provided by antibody specific to the cell surface in combination with fluorescent dye.

Fluorescent-labeled antibody, specific to the cell surface of *A. proteus,* has been prepared from cell membranes isolated in bulk. The gross chemical analysis of the membranes indicates the presence of 32% lipid, 25% protein, and 15% polysaccharide (O'Neill, 1963). Antisera to this fraction contain a single major antibody, together with two minor ones. This major antibody gives exceptionally intense lines in double diffusion tests, and is almost certainly directed against the mucoprotein coat (O'Neill, 1963; Wolpert and O'Neill, 1962).

Experiments with Antibody-Labeled Cells

The labeled antibody may be applied to living amebae in a concentration of 0.05% globulin, without damage, for at least 5 min, and after repeated washing the cell surface alone can be seen in the fluorescence microscope to be brilliantly stained. Such cells soon begin to move normally and are capable of pinocytosis and phagocytosis. The fluorescence at the surface remained very strong during the first hour after staining, but the intensity at the surface slowly decreased during the subsequent hours, eventually no longer being detectable after 12 hr had elapsed. The half-life of the label at the surface was judged to be 5 hr. This loss of fluorescence from the surface was paralleled by the appearance of fluorescence within the cell, where it was initially confined to the surface of small vesicles. This suggested that it was entering on the surface of pinocytic vesicles' and on occasion pinocytic channels, brilliantly fluorescent, were seen at the tail. With time, all the fluorescence accumulated in the large vacuoles which were defecated (Holter, 1961). The cells tolerate this treatment well, and have been relabeled at daily intervals for 6 days.

Careful examination with N.A. 0.95 objectives of the whole periphery of the motile cell showed substantially uniform staining at all stages, and, in particular, the tips of advancing pseudopods could always be clearly seen to bear a flourescent surface film.

Local areas of the surface of cells could also be stained by drawing most of the cell into a micropipette and placing the exposed end in 0.5% concentrated fluorescent antibody. After less than 1 min, the cell was expelled into a large volume of fresh culture medium. Cells treated in such a way in all cases showed brilliantly fluorescent tails, while the major part of the cell surface remained unlabeled. The stained tail surfaces remained evident for an hour or more. Intermittently during this period, islets of labeled surface could be observed to break off from the main body of the surface at the tail, and these rapidly dispersed while moving forward over the upper surface of the cell.

The four main conclusions that can be drawn from such studies (Wolpert and O'Neill, 1962) are:

(*a*) The surface of the ameba behaves like a fluid layer, flowing forward over the hyaline layer to accommodate the changing shape of the cell as described in Fig. 2C, and this in in agreement with observations on particles attached to the surface.

(*b*) There is no evidence for rapid turnover of the surface or for formation of new surface at the tip of advancing pseudopods as suggested by the theories of Goldacre (1961) and Bell (1961). These two conclusions are strictly applicable to the surface coat only, since the antibody probably attaches to it and not to the plasma membrane. We cannot exclude the possibility that the surface coat is flowing over a stationary plasma membrane which is being resorbed at the rear and formed at the front. Such a possibility seems highly improbable since it requires shear in the surface coat which has a well-defined structure, orientated perpendicular to the surface. The viscous drag in this case may be many times greater than if the whole membrane flows forward and the zone of shear is in the fluid hyaline layer between the membrane and the ectoplasmic gel. (For Newtonian flow, the total force required will be inversely proportional to the thickness of the layer, and proportional to the viscosity. The coat is at least 10 times thinner than the hyaline layer. The hyaline layer is very fluid, as shown by Brownian movement, and while no figure is available for the coat, its structure and composition suggest a higher viscosity.) Furthermore, it would be necessary to assume that the plasma membrane separates from the fibrillar coat at the rear and is withdrawn into the cell interior by a mechanism different from that of pinocytosis as observed in the ameba, and that new plasma membrane is formed by intussusception at the front. There is no evidence or plausible mechanism for this at present. Only if these improbable postulates are accepted do mechanisms involving rapid membrane turnover (Fig. 2A) remain tenable in their essence. Such additional complexity is difficult to justify and we thus prefer, at this stage, the simpler hypothesis that both coat and plasma membrane move in unison and flow forward.

(*c*) There is a slow turnover of the membrane, and its half-life is about 5 hr. This would correspond to a time of 8 hr for replacing an area of surface equal to the total area of the cell, which is equivalent to a turnover of 0.2%/min of the total surface area. This is consistent with observations on the time required to recover from pinocytosis (Chapman-Andresen, 1963).

(*d*) The tail behaves differently from the rest of the cell, and the surface there does not flow forward. Pinocytosis probably occurs there and this accounts for the slow turnover of the surface. The state of the membrane at the tail is similar to that of the whole membrane of a pinocytosing cell. This view of the tail is in marked contrast to that of Goldacre (1952) who suggested that it is a site of membrane contraction and resorption. It should be noted that the pinocytosis at the tail might have been induced by the application of the antibody, and when cells are left in the antibody solution for more than 20 min pinocytosis commences, although the total concentration of protein is some 10 times lower than that normally required to elicit pinocytosis. Thus, in unlabeled cells the turnover may be even slower. The differentiation of the membrane at the tail may be similar to that at points of contact with the substratum. The membrane must not flow in such regions if a mechanical force is to be exerted on the substratum so as to bring about movement. It is, of course, possible that the flow is prevented in such regions by adhesion to the substratum or the underlying gel.

These conclusions, emphasizing the fluidity and slow turnover of the membrane, relate to our experiments on *A. proteus*, and it now becomes necessary to consider briefly the problem in more general terms and in relation to other cells.

Changes in Area of the Cell Membrane

One of the problems outlined previously is whether the cell membrane, during movement, undergoes reversible changes in area as illustrated in Fig. 2B, and, particularly, whether this could provide the force for cell movement. One of the best studied systems is the change in form at cell cleavage, and although there is evidence that a contraction occurs in the region of the cell surface, particularly in the furrow, without necessarily a change in area, it has been argued that the origin of this force is probably not in the membrane but in the cytoplasm just beneath it (Wolpert, 1963a). There is quite good evidence that the membrane of the cell is elastic, can resist tension, and can form quite a tough mechanical barrier and yet have some fluid properties (reviewed Wolpert, 1960a), although little is known about the mechanical properties of the ameba membrane. That the membrane is elastic means only that the deformation produced by an applied force is proportional to the force and not that a deformation involves a change in surface area. There is, in fact, very little evidence to show that the cell membrane can alter its area without the removal or introduction of surface material. For the red blood cell the change in form between disc and sphere occurs without change in area, any swelling of the spherical form leading to hemolysis (Ponder, 1948). Apparent increases in surface area accompanying swelling of a rounded cell, e.g., the sea urchin egg, do not necessarily involve a real increase in area of the surface, since the membrane of the sea urchin egg and other cells may be highly convoluted. Again, the appearance of moving cells certainly suggests that changes in area are taking place, but there is surprisingly little evidence that this is true.

Consideration of the molecular structure of the plasma membrane forces one to a similar conclusion. Haydon and Taylor (1963), discussing the stability of bimolecular leaflets, have emphasized that the lipid molecules must be tightly packed. From their considerations it also seems that the plasma membrane could not change its area significantly without disruption or introduction or removal of further lipid.

Formation of New Surface Membrane

Although our studies suggest that during ameboid movement surface membrane turnover is slow and that changes in surface area would also require turnover, it is relevant to ask if there is evidence from other cells showing that membrane turnover is ever rapid. Turnover of cell membrane, i.e., the formation or resorption of membrane, occurs in a variety of cellular processes such as growth, cell cleavage, pinocytosis, phagocytosis, and vacuolar secretion. Although quantitative measurements are, in the main, lacking, calculations from some of the available data are summarized in Table I. From this table it can be seen that the highest rate of formation is 6%/min for cleavage of the sea urchin egg, which is a factor of 3 less than the 20%/min required on the theory that the ameba renews its surface each time it passes through its own length, and 30 times more than the observed rate. However, even a 1%/min rate of increase does show that, in principle, membrane formation could be rapid enough to provide new membrane for cell movement requiring slower rates of locomotion or pseudopod formation.

There are two special cases where it does seem that rapid formation of new surface occurs. The first is the so-called surface precipitation reaction (Chambers and Chambers, 1961) that results in new surface formation when the surface is torn apart. However, for the ameba at least, the evidence is by no means conclusive that new surface is, in fact, being formed rather than that the existing surfaces come together and fuse. The second is the observation that amebae and tissue cells become rounded, with apparently unconvoluted surfaces, when subjected to high hydrostatic pressures, and that on release of the pressure the cells rapidly resume their ameboid forms. As Landau (1961; Landau and Thibodeau, 1962)

TABLE I

MAXIMUM RATES OF MEMBRANE FORMATION IN VARIOUS SYSTEMS[a]

Process	Reference	Rate of formation relative to surface area of cell (%/min)	Rate of formation relative to volume of cell ($\mu^2/100\ \mu^3$/min)
1. Yeast cell (*Schizosaccharomyces pombei*) growth (complete cycle between divisions)	Mitchison (1957)	0.7	0.5
2. Newt egg: first cleavage. Approximate figure obtained by the asumption that the egg is a perfect sphere and that the cleavage furrow divides the cell into two hemispheres	Selman and Waddington (1955)	1.5	0.006
3. Sea urchin egg: first cleavage, assuming membrane formation to take place during the formation of the cleavage furrow only	Wolpert (1963b)	6.0	0.3
4. Myelinization: average rate of growth of myelin spiral of rabbit peripheral nerve Schwann cell, relative to calculated area of cell at start of myelinization	Williams (1963)	0.03	—
5. *Amoeba proteus:*			
(*a*) Movement: observation of loss of material from surface during movement when labeled with membrane specific antibody	Wolpert and O'Neill (1962)	0.2	0.01
(*b*) Observation of time taken to recover from pinocytosis	Chapman-Andresen (1963)	0.2	0.01
(*c*) Contractile vacuole: assuming the vacuole to fuse with the cell surface in its entirety at the end of each cycle	Adolph (1926)	1.0	0.2
Minimum rate of turnover required by the postulate that the ameba completely renews its surface when it moves a distance equal to its own length	—	20	1.0

[a] The rates have been calculated from the data of the authors quoted, making assumptions as indicated in the table. Where several alternative figures are available, the greatest has been chosen.

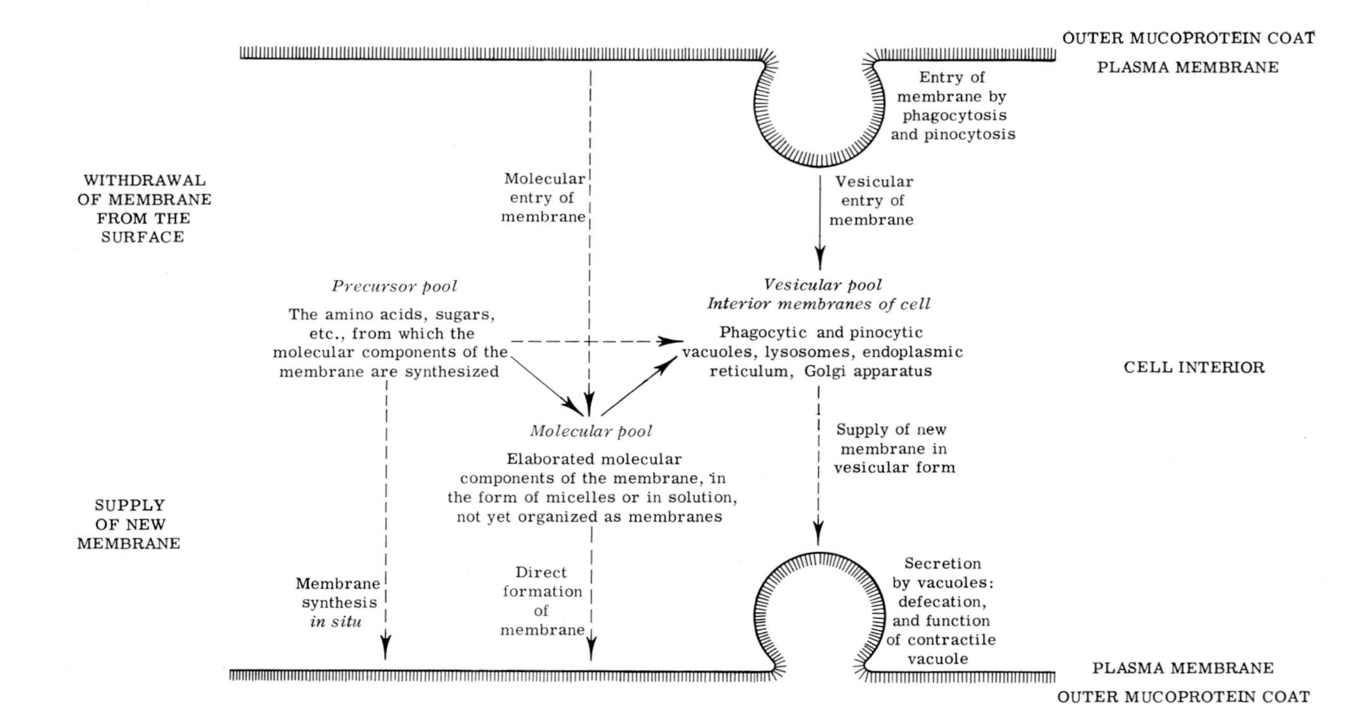

Fig. 3. Diagram of interchange between cell surface and cell interior. Continuous lines represent pathways whose existence is believed to be established; dotted lines represent possible pathways. The diagram shows the entry of membrane in vesicular form, and the possible supply of new plasma membrane to the surface in the form of vesicles. Three pools are postulated in the cytoplasm: (a) *precursors* providing the material for the elaboration of (b) the *molecular components* of the membrane, which themselves become organized (c) in *vesicles*. For simplicity arrows have been drawn in one direction only.

points out, this certainly suggestive of rapid resorption and re-formation of the membrane.

Our knowledge of the mechanisms by which new membrane is formed or resorbed is certainly poor, but various possibilities are illustrated in Fig. 3 which shows that membrane can leave or enter the surface in the form of vacuoles (as in pinocytosis and vacuolar secretion). Entry of membrane during pinocytosis is well established (Nachmias and Marshall, 1961) and so is formation of new surface during vacuolar secretion (Palade, 1959). It is not clear whether this mode of formation of new membrane occurs in the ameba, but a pathway by which the contents of pinocytosis vacuoles ultimately appear in defecation vacuoles has been described by Brandt and Pappas (1962). Our experiments with the fluorescent antibody indicate a similar pathway. The possibility of entry (or exit) in molecular or micellar units is also illustrated, although there is as yet little evidence for such a mechanism. It is also possible that the membrane may be synthesized *in situ* by enzymatic sites forming part of the membrane itself, particularly as the internal membranes of the cell have many synthetic functions. The problem of membrane formation is one of general importance, which could have significant implications for cell movement. Now that we can obtain isolated membranes we are hopeful that we may be able to investigate this problem using radioisotopes, and particularly to determine the turnover rate of the plasma membrane.

The available evidence leads us to the conclusion that the surface membrane of the ameba flows freely during movement. It probably is also elastic, but cannot undergo large changes in surface area without addition or removal of membrane material, both of which are relatively slow processes. It seems that, for the moment, the membrane must be assigned a relatively passive mechanical role in movement, although it probably plays a very important role in the control of movement. We must thus look to the interior of the cell for the motive force.

Studies of the Isolated Cytoplasm

One of the earliest observed characteristics of protoplasm was its apparent contractility. Since Mast's classic studies on amebae, "contractile" processes in the cytoplasm have been emphasized as the most probable basis for cell movement, and particular attention has been focussed on sol-gel transformations. It has more recently been established that cytoplasm is capable of autonomous streaming, for which the crucial evidence is the demonstration by Allen *et al.* (1960) that naked cytoplasm from ruptured amebae continues to stream.

As already pointed out, the analogy between muscular contraction and cytoplasmic motility has dominated attempts to isolate the active components. Table II summarizes some of the features of the systems extracted from nonmuscular cells for comparison with muscle. It should be noted that all the extracts are made in high salt concentrations. Although some of these results support the muscle protein analogy, namely, the changes of viscosity of myxomyosin and the superprecipitation of sarcoma cell protein (both ATP dependent), there is no compelling reason to believe that the motile system has yet been isolated in a form which can be related to what is seen in the intact cell. The results of Sakai (1962) indicate that the contraction of reconstituted threads may not require an ATP-dependent system.

A further difficulty, and one which also weakens the analogy, is that studies on the fine structure of ameboid cells have not, so far, brought to light any structures that can with confidence be assigned a role in motility and compared with myofilaments. The only possible exception to this is the recent demonstration of fibers in electron micrographs of the slime mold by Wohlfarth-Bottermann (1962). Amebae contain a great quantity of vesicular material resembling smooth endoplasmic reticulum (Mercer, 1959; Pappas, 1959; Daniels and Roth, 1961), but its function is not known. The electron microscope morphology of the cytoplasm is perplexing in that there is no detectable difference even between different regions of the cytoplasm. For example, there is still no clear indication of the nature of the boundary between the hyaline layer and the ectoplasm, or between endoplasm and ectoplasm.

Isolation of a Motile Cytoplasmic Fraction

In our own attempts to isolate the motile system of *A. proteus* we tried, in the first instance, to obtain an actomyosin-like fraction using variations on the extraction methods with strong salt solutions found successful with muscle, sarcoma, and slime mold. In no case, however, did we obtain a preparation which fulfilled the necessary criterion, namely, unequivocal reversible change of viscosity with added ATP. These early experiments led us to conclude that if present, such a system was highly labile. We also reasoned that even if we brought an extraction procedure to the point of duplicating the results achieved with, say, the slime mold (Ts'o *et al.*, 1956a,b, 1957a,b), we should be little nearer to the molecular basis of motility in the intact cytoplasm. For these two reasons, we began to look for a way of isolating the system not by a chemical extraction procedure but by a primarily physical one which would yield cytoplasm with as many as possible of its intracellular characteristics intact, that is, the power to stream, to contract, and to undergo

TABLE II
SUMMARY OF RESULTS OF SALT EXTRACTIONS OF MOTILE CELLS FOR COMPARISON WITH MUSCLE[a]

1 Tissue and name of system	2 Reference	3 Extraction medium	4 Effect of ATP on extract	5 Reprecipitation	6 Effect of ATP on insoluble form
Muscle, actomyosin	Weber and Portzehl (1952)	0.6 *M* KCl, pH 8–9	Fall of viscosity; reversed as ATP is split	As gel or load bearing fibers below 0.2 *M* KCl	Rapid contraction of fibers. Superprecipitation of gel
Sarcoma cells	Hoffmann-Berling (1956)	0.6 *M* KCl + ATP	Fall of viscosity; reversed as ATP is split	Flocculent precipitate 0.05–0.1 *M* KCl	Slow shrinkage of pieces of gel
Slime mold, myxomyosin	Ts'o *et al.* (1956a,b)	1.4 *M* KCl	Fall of viscosity; reversed as ATP is split	25–40% saturated ammonium sulfate	—
Slime mold, myosin B	Nakajima (1960)	0.6 *M* KCl, pH 8–9	Fall of viscosity; reversed as ATP is split	Below 0.1 *M* KCl	Slow superprecipitation
Sperm tails	Pautard (1962)	0.5 *M* KCl, pH 8–9	—	0.05 *M* KCl, pH 5.2	Formation of streaming network of gel; contraction, rhythmic in places

TABLE II (*Continued*)

1 Tissue and name of system	2 Reference	3 Extraction medium	4 Effect of ATP on extract	5 Reprecipitation medium	6 Effect of ATP on insoluble form
Sea urchin eggs	Sakai (1962)	0.6 *M* KCl on water-insoluble residue	—	Cold acetone or distilled water	None. But 40% contraction with dior trivalent ions, reversed by EDTA,[b] also with oxidizing agents, reversed by reducing agents
A. proteus	Simard-Duquesne and Couillard (1962a,b)	0.6 *M* KCl on water-insoluble residue	ATP splitting	—	—

[a] Column 4 shows the similarity of slime mold and sarcoma cell extracts to actomyosin in their viscosity responses to ATP. Column 5 and 6 show that some of these extracts can be precipitated at low ionic strength to form contractile gels, also similar to actomyosin. The results of Sakai (1962) show that ATP is not necessarily involved in the contraction. It should be noted that insoluble contractile models have been prepared by glycerol extraction from tissue cells (Hoffmann-Berling, 1956), sperm tails (Bishop, 1958), and amebae (Simard-Duquesne, 1962b).

[b] Ethylenediaminetetraacetate.

sol-gel changes. These we regarded as better criteria for recognizing the active component than viscosity changes or ATPase activities.

In general, the cytoplasm of an ameba, burst in a watery medium, either coagulates or disperses rapidly, destroying the possibility of delicate sol-gel changes. Our first objective was, therefore, to find a medium in which the cytoplasm remained undamaged as regards motility. We set out to test the effect of various media on the cytoplasm of amebae before, during, and after rupture. This was done by compressing a dense suspension of cells with a cover slip and watching the changes in the behavior and appearance of the cytoplasm. Although crude, this method gave relatively consistent and quite valuable results. Of particular interest was the observation that sometimes the cytoplasm from several cells would run together on the slide and remain without dispersing for several minutes. During this time, granules could be observed to move about in different directions by sudden displacements in such a way as to suggest a twitching network of oppositely directed streams. Sometimes whole areas would contract and tear in jerks from other areas. This behavior was taken as a sign of the survival, in disorganized form, of the motile mechanism outside the cell. As a rule addition of salts, especially calcium, prevented the twitching, whereas ATP and SH compounds tended to favor it. Cysteine reversed, to some extent, the dispersing effect of high potassium chloride concentration. Distilled water was as effective in prolonging twitching as ATP, but the contractions appeared less vigorous. The conclusion was that the most useful conditions for preparing ameba homogenates would be in the presence of ATP and SH compounds, and the exclusion of salts; but addition of either or both, before homogenization, did not yield bulk homogenates that showed detectable motility. Homogenization in total absence of medium and subsequent addition to ATP was also ineffective under the conditions tried. We, therefore, decided to dispense with homogenization altogether, a possibility hinted at by Marshall *et al.* (1959) and the experiments of Allen *et al.* (1960). After many trials we found that an active preparation showing dramatic streaming and syneresis can be obtained if the cells are first cooled at 4°C for 24 hr and then centrifuged whole at 35,000 *g* for 10 min. All the larger inclusions such as nuclei, food vacuoles and crystals are pinched off at the heavy pole; they pass to the bottom of the tube and are discarded. The cells remain tightly packed as smooth bags with a stratified contents of fat droplets, mitochondria, and various small vesicles. The membranous bags are emptied of their contents by gentle homogenization and spun out in the cold at 1200 *g*. The resulting extract, whose preparation is described in greater detail elsewhere (Thompson and Wolpert, 1963) is called extract I.

MOTILITY OF EXTRACTS

Extract I is heterogeneous and contains fat droplets, granules, and vesicles visible under the phase-contrast microscope. It is difficult to avoid slight contamination by surface membranes (10–100 membranes/ml of extract). Motility is judged by observing the behavior of granules in a drop of extract in a sealable chamber 50 to 100 μ deep. At 4°C or at room temperature only Brownian motion of granules can be seen. There are no signs of displacements of granules which could be distinguished from those of a colloidal suspension of inert particles of comparable size. However, the addition of neutralized ATP under appropriate conditions produces striking effects which have been observed and filmed many times. The sequence of events may be described as follows. The first signs of movement, appearing after about 30 sec, are sudden "saltatory" displacements of individual granules in any direction. The distance covered may be only a few microns, but the displacements are quite distinct from Brownian motion. Displacements continue in any direction but gradually become more widespread, whole areas moving as single blocks. The entire field appears to twitch. The moving blocks of cytoplasm appear to become connected to each other, so that the motion of one area is transmitted to others through a constantly changing network of oppositely directed streams of granules. The movement appears to become more definitely orientated into one of two alternative patterns:

(*a*) Streams of granules become arranged in parallel lines moving in opposite directions through the bulk of the cytoplasm. The streams persist for up to 10 min and die out slowly, leaving a uniform distribution of granular material over the whole area of the chamber. There is no net transfer of material in any direction, although streams of particles may be moving in opposite directions to each other at 80 μ/sec for distances of 650 μ. In general these movements are very similar to those observed by Allen *et al.* (1960) in single ruptured cells.

(*b*) Granules cease to move in oppositely directed streams. Instead, large areas of the field appear to gel into sheets which contract quickly to a small fraction of their original area (Fig. 4). This type of movement does not persist as long as the streaming. As gelled areas form and contract, many larger inclusions are trapped and drawn together, including any membrane fragments. In one particular case, at the edge of a contracting gel area, particles were seen to be moving toward the center at a speed of 150 μ/sec relative to the slide. At the same time, other granules were moving in the opposite direction at 50–100 μ/sec. It seemed they were being squeezed out from the interstices of the contracting gel network. In this pattern of movement there is always a net transfer of material as dense localized masses are formed. When two sheets of gel contract in different directions and tear apart from each other, the connections between them become drawn out and sometimes clearly visible strands (see Fig. 4) are formed which can extend for 500 μ. These thin strands have been observed to shorten and pull in smaller clumps of contracted material toward the main mass over a distance of several hundred microns. If a thin strand breaks, the two ends snap apart like rubber.

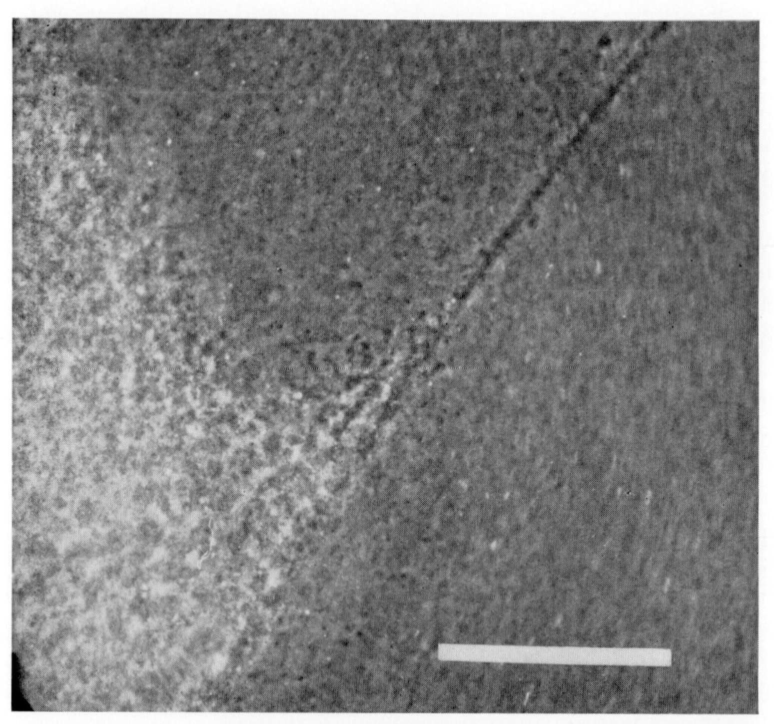

FIG. 4. The edge of a contracted gel mass formed from extract I in the presence of 2.5 m*M* ATP. The preparation was photographed in a chamber and shows part of a gel strand connected to another gelled region. Scale mark is 100 μ.

The second type of movement is seen particularly well when a mixture of extract and ATP is introduced into a capillary, of diameter about 1000 μ. The central mass then takes the shape of a thread which can conveniently be blown out for examination or fixed for electron microscopy (discussed later).

CONDITIONS FOR MOTILITY AND DEPENDENCE ON ATP

The types of motility described have been induced so far only by the addition of ATP or adenosine diphosphate (ADP), ADP showing less activity. In the absence of ATP or ADP no motility has been observed. Formation of contracted masses in capillary tubes occurs when the final concentration of ATP lies in the range 1–5 m*M*; 10 m*M* is inhibitory. Adenosine monophosphate (AMP) and all other compounds so far tested are without effect (pyrophosphate, orthophosphate, SH compounds, and salts). Salts, particularly potassium chloride and traces of calcium, abolish the effect of ATP, and 1–5 m*M* calcium chloride causes irreversible clumping of the vesicles. The induction of motility is dependent upon temperature. The capacity of extracts stored at 0°C to respond to added ATP gradually diminishes and disappears after 1–2 hr, and ATP does not have any effect on extracts kept at 4° or 22°C. Motility is observed only if extracts previously mixed with ATP in the cold are allowed to come to room temperature.

Partial Purification of Active Component

Spinning extract I (Thompson and Wolpert, 1963) for 10 min at 9000 *g* in the cold removes a large proportion of the granules. The resulting extract II shows almost identical behavior with ATP to that of extract I. The preparation is, however, more stable (up to 6 hr). Further spinning at 35,000 *g* for 10 min results in the disappearance of ATP-induced motility from the supernatant. Surprisingly, the pellet is also inactive provided that care has been taken to wash traces of the supernatant away from the sides of the tube with cold distilled water. The supernatant is largely free of granules and vesicles, but still contains drops of coalesced lipid which make it possible to detect motility. The pellet contains many vesicles and granules, and when resuspended in distilled water looks like extract II. It is found that when equivalent quantities of resuspended pellet and supernatant are mixed with cold ATP the usual response is restored, although neither are active alone. Spinning the 35,000 *g* supernatant a further 80 min at 150,000 *g* removes very little extra material and leaves a supernatant that will also give ATP responses when combined with the 9–35,000 *g* pellet. The response to ATP is lost if either pellet or supernatant are warmed to 100°C. The pellet and supernatant can be stored separately in the deep freeze and retain some activity after a week, but the pellet does not survive a second freezing.

Electron Microscopy of the Contracted Gel

Pieces of the thread formed by contraction of extract I in capillaries were prepared for electron microscopy, in collaboration with Dr. E. H. Mercer, by fixing in osmium (Palade's fixative) and embedding in Araldite (Mercer and Birbeck, 1961). The contracted mass is very heterogeneous and contains fat droplets, vesicles both small and large, mitochondria, and occasional fragments of cell membranes. The most striking features are arrays of well-defined and partially orientated fibers and a more or less homogeneous region with a spongy appearance (Fig. 5). Such spongy regions are continuous with the fibers (Fig. 6). The fibers are about 120 A thick and about 0.5 μ long. There is no evidence of

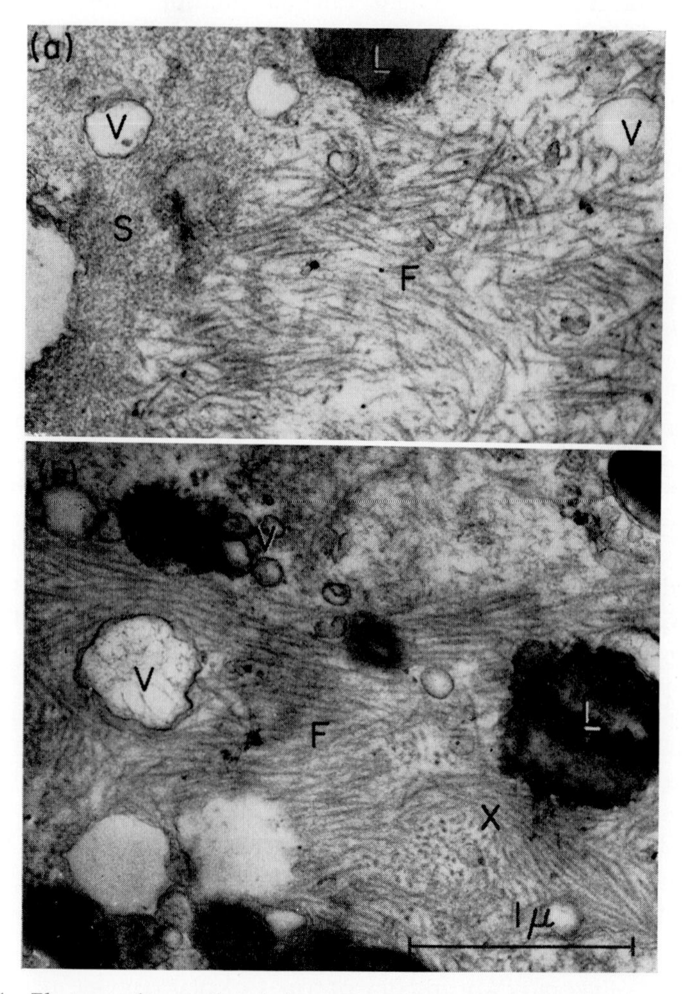

FIG. 5. Electron micrographs of the contracted gel obtained from extract I after the addition of ATP. The gel is heterogeneous and contains vesicles (V), lipoid droplets (L), and fibers (F). The fibers are about 120 A wide and 0.5 μ long. In (*a*) a spongy region (S) can be seen to be continuous with the fibers. Note in (*b*) the parallel alignment of the fibers. Fibers appear in cross section at (X) in (*b*). Magnification: × 30,000. (Photograph by Dr. E. H. Mercer.)

banding. Under high power the spongy region appears to contain some small particulates and small fibrils. The fibers seem to fuse with the spongy material, which suggests that it might be the precursor of the fibers.

Electron micrographs of gel prepared from extract II show it to be less heterogeneous than that from extract I, many of the vesicles having

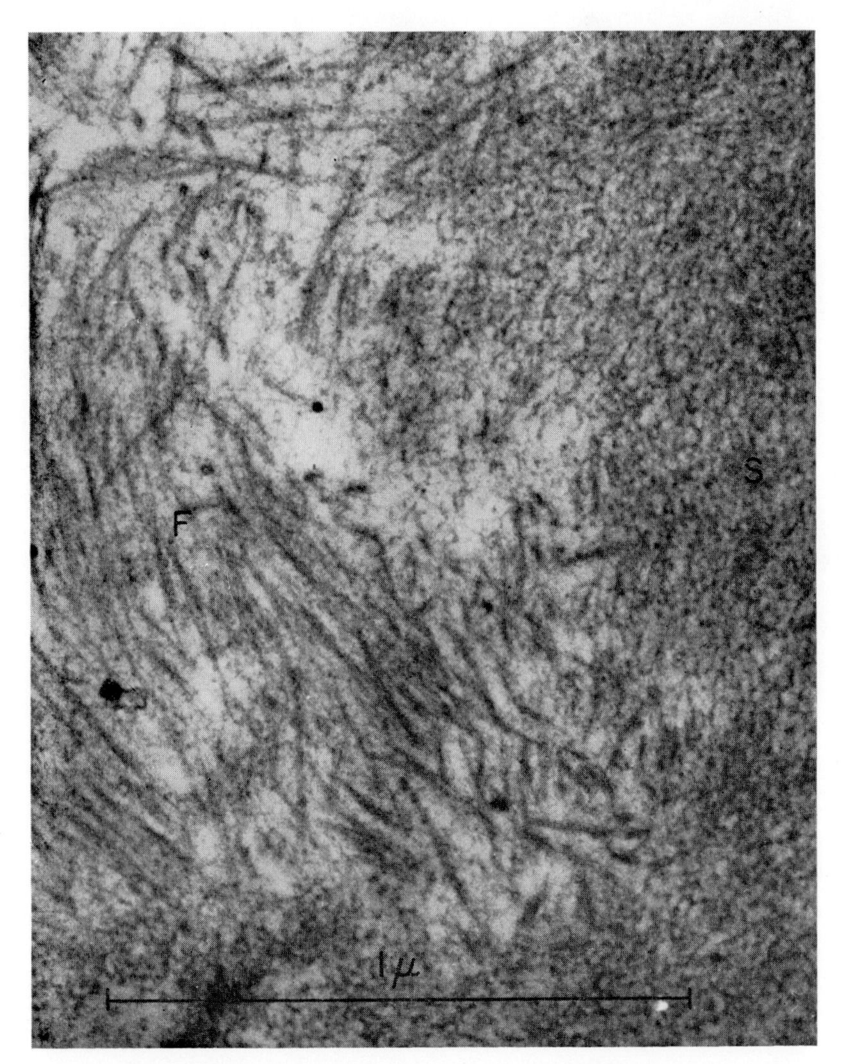

FIG. 6. Electron micrograph of spongy region (S) and fibers (F) of gel prepared as in Fig. 5. The fibers seem to be continuous with the spongy region which contains small fibrils and particles. Magnification: × 80,000. (Photograph by Dr. E. H. Mercer.)

been spun out. The fibrils now appear somewhat different (Fig. 7). They are about 80–90 A thick and, although they do not show banding, they do appear denser in some regions than in others. They also, on occasion, give the impression of being coiled.

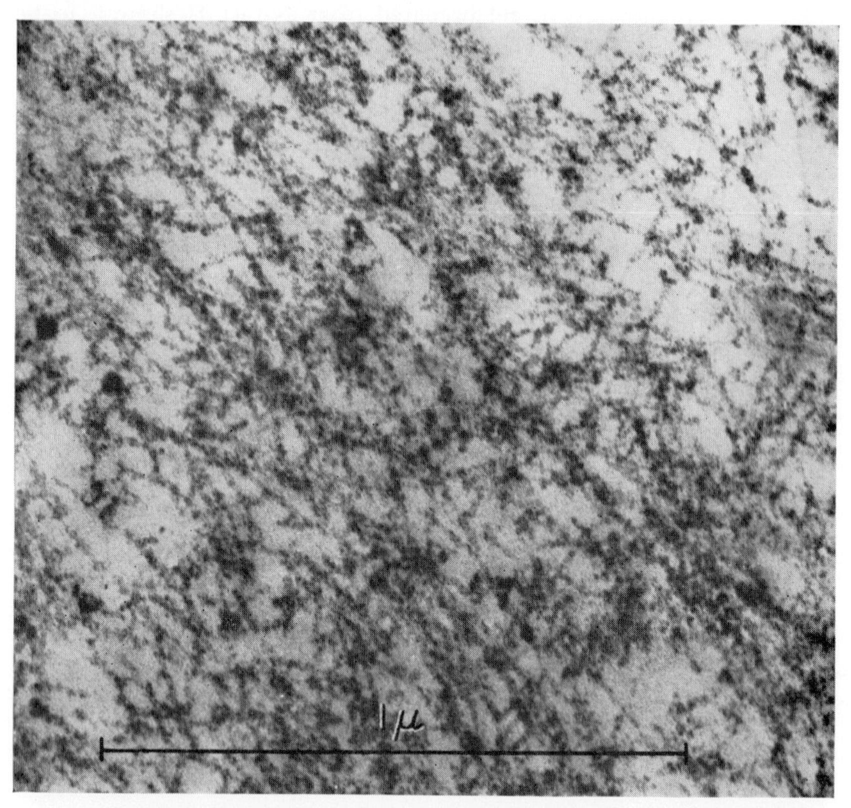

FIG. 7. Electron micrograph of fibers in gel obtained from extract II by the addition of ATP. The fibers are about 90 A wide. (Photograph by Dr. E. H. Mercer.)

Discussion

The results obtained so far should be regarded as preliminary, and detailed interpretation postponed until more data have been accumulated. There are, nevertheless, some general points that merit attention. The crude cytoplasmic extracts exhibit four main properties when brought to room temperature in the presence of ATP: (1) "gelation"; (2) streaming of particles often in localized regions and sometimes for long distances; (3) contraction and the ability to transmit tension; (4) syneresis. All these properties have been ascribed to the cytoplasm of whole amebae or are demanded by current theories of movement. The localized streaming is very similar to that of dissociated *Chaos chaos* cytoplasm (Allen *et al.*, 1960). The contraction and syneresis may be compared to the formation of the hyaline cap at the tip of a pseudopod (Allen *et al.*, 1962) and to the rapid cytoplasmic contraction seen in amebae immediately after decompression (Marsland, 1956; Landau *et al.*, 1954). The effect of temperature is also completely in line with the evidence that low temperature favors a breakdown in the structures responsible for movement (Marsland, 1956; Landau, 1959). Whole amebae do not stream at 4°C and the resumption of streaming at room temperature follows a similar time course to the development of motility in extracts. Inoué (1959) has shown that lowered temperatures cause a reversible disruption of the mitotic spindle in marine worm eggs which is almost certainly analogous.

The gentle, stepwise preparation of extracts, the wide variety of patterns of motility displayed, and their high lability suggest that we really are dealing with the same system as in the cell, and not an artifact of temperature, pH, or the addition of polyvalent ions. It should, however, be pointed out that some proteins, which are unlikely to be connected with motility and yet show some analogous behavior, have been extracted from other tissues. Examples are the polymerization of collagen (Hodge and Schmitt, 1961) and the contraction of threads of histone from human, liver cell nuclei (Zbarskii, 1951).

The general impression we have of movement in extracts is that it is the result of gelation followed by syneresis. Gelation is to be understood as a polymerization or linkage of subunits so as to form a structural network capable of transmitting tension. It occurs primarily while the temperature is rising. As soon as gelation has occurred the system seems capable also of generating tension. Although we consider it likely that the same mechanochemical activity of the linked subunits brings about both streaming and syneresis, the streaming patterns may vary, as for example when (*a*) granules are expelled from the gel with the syneretic fluid, presumably by pressure; (*b*) partially aligned rows of granules move past each other as two gelled regions tear apart; (*c*) more prolonged streaming analogous to frontal contraction takes place. We conclude that for this last to happen a delicate equilibrium must be established which allows in some regions a reversal of the changes accompanying gelation and contraction, so that the same subunits can go through repeated cycles. This is presumably what occurs in the intact cell.

The behavior of the extracts cannot at present be used to distinguish between the current contractile theories of ameboid movement. In general terms Goldacre (1961) supports a theory of contraction and syneresis at the tail followed by "solation" and the generation of hydrostatic pressure, whereas Allen (1961) suggests contraction and syneresis at the front followed by "gelation" and the transmission of tension. In the extracts both these sets of phenomena occur, although the latter is more common. However, the autonomous motility of the cytoplasm in the absence of the surface membrane argues against Goldacre's suggestion that streaming is produced by the generation of hydrostatic pressure in a closed system.

The molecular basis for the motility of extracts remains obscure. ATP is necessary but its role is uncertain. The presence of fibers in the contracted gel suggests that they play a role by sliding or coiling, but again much more evidence is needed. We have not, as yet, established whether the fibers exist in the extracts before the addition of ATP. Although the published electron micrographs of amebae do not show these fibers, they can be found under certain conditions (Mercer, 1963) and the demonstration of fibers in slime mold (Wohlfarth-Bottermann, 1962) suggests that with modified fixation methods they may be found more widely. In the electron micrographs illustrated here, the close juxtaposition of fibers with the spongy material suggests that some interconversion may occur.

Whatever the answers to the many outstanding problems, we are confident that the extracts provide a valuable starting material for further work on the molecular basis of cytoplasmic streaming.

Summary

A specific antibody has been prepared against isolated ameba surface membrane. This antibody labeled with fluorescein has been used to mark the surface and observe its behavior during locomotion. It has been concluded that the surface flows forward during movement and that the turnover of the surface is slow. The surface membrane appears to play a relatively passive role in locomotion and the forces causing movement must be sought for in the cytoplasm.

A cytoplasmic fraction has been isolated in bulk, which shows motility very similar to that seen in intact cells. Induction of motility appears to be ATP dependent and sensitive to temperature. The fraction is capable of gelation, streaming, and syneresis. Electron micrographs of the gel resulting from syneresis show orientated fibers that may be responsible for the motility.

Acknowledgments

We are grateful to Professor J. F. Danielli for his advice and encouragement, to Dr. E. H. Mercer of the Chester Beatty Research Institute for the electron microscopy and helpful discussions, to Dr. D. A. Haydon of the Department of Colloid Science,

Cambridge, for discussion on certain aspects of membrane properties, and to Mrs. J. L. Stewart for culturing the amebae. This work has been supported by the Nuffield Foundation, and the Agricultural Research Council have provided a grant for a preparative ultracentrifuge.

References

Adolph, E. F. (1926). *J. Exptl. Zool.* **44**, 355.
Allen, R. D. (1961). *Exptl. Cell Res. Suppl.* **8**, 17.
Allen, R. D., Cooledge, J. W., and Hall, P. J. (1960). *Nature* **187**, 896.
Allen, R. D., Cowden, R. R., and Hall, P. J. (1962). *J. Cell. Biol.* **12**, 185.
Ambrose, E. J. (1961). *Exptl. Cell Res. Suppl* **8**, 54.
Bell, L. G. E. (1961). *J. Theoret. Biol.* **1**, 104.
Bingley, M. S., and Thompson, C. M. (1962). *J. Theoret. Biol.* **2**, 16.
Bishop, D. W. (1958). *Anat. Record* **131**, 533.
Brandt, P. W., and Pappas, G. D. (1962). *J. Cell. Biol.* **15**, 55.
Chambers, R., and Chambers, E. L. (1961). "Exploration into the Nature of the Living Cell," p. 100. Harvard Univ. Press, Cambridge, Massachusetts.
Chapman-Andresen, C. (1963). *Proc. 1st Intern. Cong. Protozool.* In press.
Daniels, E. W., and Roth, L. E. (1961). *Radiation Res.* **14**, 66.
Davson, H., and Danielli, J. F. (1942). "The Permeability of Natural Membranes." Cambridge Univ. Press, London.
Goldacre, R. J. (1952). *Intern. Rev. Cytol.* **1**, 135.
Goldacre, R. J. (1961). *Exptl. Cell Res. Suppl.* **8**, 1.
Griffin, J. L., and Allen, R. D. (1960). *Exptl. Cell Res.* **20**, 619.
Gustafson, T., and Wolpert, L. (1963). *Intern. Rev. Cytol.* **15**, 139.
Haydon, D. A., and Taylor, F. H. (1963). *J. Theoret. Biol.* **4**, 281.
Hodge, A. J., and Schmitt, F. O. (1961). *In* "Macromolecular Complexes" (M. V. Edds, ed.), p. 19. Ronald Press, New York.
Hoffmann-Berling, H. (1956). *Biochim. Biophys. Acta* **19**, 453.
Hoffmann-Berling, H. (1960). *In* "Comparative Biochemistry" (M. Florkin and H. S. Mason, eds.), Vol. II, p. 341. Academic Press, New York.
Holter, H. (1961). *In* "Biological Structure and Function" (T. W. Goodwin and O. Lindberg, eds.), Vol. 1, p. 605. Academic Press, New York.
Holtfreter, J. (1949). *Exptl. Cell Res. Suppl.* **1**, 497.
Inoué, S. (1959). *Rev. Mod. Phys.* **31**, 402.
Landau, J. V. (1959). *Ann. N. Y. Acad. Sci.* **78**, 487.
Landau, J. V. (1961). *Exptl. Cell Res.* **23**, 538.
Landau, J. V., and Thibodeau, L. (1962). *Exptl. Cell Res.* **27**, 591.
Landau, J. V., Zimmerman, A. M., and Marsland, D. A. (1954). *J. Cellular Comp. Physiol.* **44**, 211.
Loewy, A. G. (1952). *J. Cellular Comp. Physiol.* **40**, 127.
Marshall, J. M., Schumaker, V. N., and Brandt, P. W. (1959). *Ann. N. Y. Acad. Sci.* **78**, 515.
Marsland, D. A. (1956). *Intern. Rev. Cytol.* **5**, 199.
Mast, S. O. (1926). *J. Morphol.* **41**, 347.
Mercer, E. H. (1959). *Proc. Roy. Soc.* **B150**, 216.
Mercer, E. H. (1963). Private communication.
Mercer, E. H., and Birbeck, M. S. C. (1961). "Electron Microscopy." Blackwell, Oxford, England.
Mercer, E. H., and Wolpert, L. (1962). *Exptl. Cell Res.* **27**, 1.

Mitchison, J. M. (1952). *Symposia Soc. Exptl. Biol.* **6**, 105.
Mitchison, J. M. (1956). *Quart. J. Microscop. Sci.* **97**, 109.
Mitchison, J. M. (1957). *Exptl. Cell Res.* **13**, 244.
Nachmias, V. T., and Marshall, J. M. (1961). *In* "Biological Structure and Function" (T. W. Goodwin and O. Lindberg, eds.), Vol. 2, p. 605. Academic Press, New York.
Nakajima, H. (1960). *Protoplasma* **52**, 413.
O'Neill, C. H. (1963). In preparation.
O'Neill, C. H., and Wolpert, L. (1961). *Exptl. Cell Res.* **24**, 593.
Palade, C. E. (1959). *In* "Subcellular Particles" (Tero Hayashi, ed.), p. 64. Ronald Press, New York.
Pappas, C. D. (1959). *Ann. N. Y. Acad. Sci.* **78**, 448.
Pautard, F. G. E. (1962). *In* "Spermatozoon Motility." p. 189. Am. Assoc. Advancement Sci., Washington, D. C.
Ponder, E. (1948). "Haemolysis and Related Phenomena." Grune and Stratton, New York.
Robertson, J. D. (1959). *Biochim. Soc. Symp. (Cambridge, Engl.)* **16**, 3.
Sakai, H. (1962). *J. Gen. Physiol.* **45**, 411.
Selman, G. G., and Waddington, C. H. (1955). *J. Exptl. Biol.* **32**, 700.
Simard-Duquesne, N., and Couillard, P. (1962a). *Exptl. Cell Res.* **28**, 85.
Simard-Duquesne, N., and Couillard, P. (1962b). *Exptl. Cell Res.* **28**, 92.
Thompson, C. M., and Wolpert, L. (1963). *Exptl. Cell Res.* **32**, 156.
Ts'o, P. O. P., Bonner, J., Eggman, L., and Vinograd, J. (1956a). *J. Gen. Physiol.* **39**, 325.
Ts'o, P. O. P., Eggman, L., and Vinograd, J. (1956b). *J. Gen. Physiol.* **39**, 801.
Ts'o, P. O. P., Eggman, L., and Vinograd, J. (1957a). *Arch. Biochem. Biophys.* **66**, 64.
Ts'o, P. O. P., Eggman, L., and Vinograd, J. (1957b). *Biochem. Biophys. Acta* **25**, 532.
Weber, H. H., and Portzehl, H. (1952). *Advan. Protein Chem.* **7**, 162.
Williams, R. J. P. (1963). Private communication.
Wohlfarth-Bottermann, K. E. (1962). *Protoplasma* **54**, 514.
Wolpert, L. (1960a). *Intern. Rev. Cytol.* **10**, 163.
Wolpert, L. (1960b). *Proc. Roy. Phys. Soc. Edinburgh* **28**, 107.
Wolpert, L. (1963a). *In* "Symposia of the International Society for Cell Biology." Vol. 2. "Cell Growth and Cell Division" (R. J. C. Harris, ed.), Academic Press, New York.
Wolpert, L. (1963b). In preparation.
Wolpert, L., and O'Neill, C. H. (1962). *Nature* **196**, 1261.
Zbarskii, I. B. (1951). *Biokhimiya* **16**, 112; from *Chem. Abstr.* (1951) **45**, 7613d.

Discussion

Dr. Goldacre: There are a number of points I would like to make, if I might. First, I would like to congratulate Dr. Wolpert and his colleagues in isolating membrane from cytoplasm. This brings a bit closer to the possibility of testing various postulated cell mechanisms by putting cells together from their components.

The first point I would like to make is this: if the movements of the membranes referred to by Dr. Wolpert refer only to the mucous coat, there seems to be no conflict between our views, because those that I put forward refer to the movement of the plasma membrane which appears to move backward as shown by the movements of attached hemispherical oil droplets.

The carmine particles move forward possibly because they are attached to the mucous coat which is pushed along passively by the ameba. Bell and Jeon have shown that the mucus with attached fluorescent label can be scraped off the cell, whereas the plasma membrane is left intact.

Now, another point. The membrane is capable of sudden changes in area. I think perhaps the experiments of Dr. Marsland indicate that under pressure amebae may become spherical, yet on release of pressure they can put out pseudopods, showing rapid change in membrane area.

DR. WOLPERT: May I deal with the membrane area problem first? I think it is possible to appear to increase the area of membrane by flattening a cell; but until you can measure the surface area, you don't have much basis for argument. We have compressed an ameba to a known thickness and watched it move, and from films measured the surface area; we found that, although the ameba changes *shape* considerably during movement, the *area* was constant within experimental error.

DR. GOLDACRE: What about Dr. Marsland's experiment?

DR. WOLPERT: I think you have to make sure there is no water entering or leaving the ameba, otherwise it may change its volume. This experiment is very easily done.

DR. MARSLAND: This experiment has been done by Landau. He draws the ameba into a capillary where it is a perfectly cylindrical shape with rounded ends. Under hydrostatic pressure there is no change in volume. The same holds true on release of pressure.

DR. WOLPERT: If this is true, then this seems to be one situation where you do get rapid membrane formation. I think Dr. Landau has also shown that, in the pressurized cell, the smooth, round surface is not convoluted. One has to be very cautious with this type of observation, because some cells, such as sea-urchin eggs which look perfectly round under the light microscope, can be seen to have a greatly convoluted surface when viewed in the electron microscope.

Your other point, on membrane contamination, is that one can never be absolutely sure one has not got membrane contaminants. However, I think it is very unlikely that they play a role in initiating motility in the cytoplasm because we separate the membranes by very gentle homogenization so as not to break them up. The membranes are, in fact, quite hard to break up. In homogenates they appear as visibly distinct bags which are centrifuged out. Our observations and photography are made with a phase-contrast microscope, in which it is almost impossible to miss a membrane. It stands out like a crumpled paper bag, and it is quite clear that movement can occur in regions in which there is no evidence of any membrane whatsoever.

I would also say that the fact that one can get the sort of streaming movements that we observe in the isolated cytoplasm, which is not in a closed hydraulic system, would suggest that pressure types of hypotheses for ameboid movement must be questioned; but this is perhaps another point we shall discuss later.

DR. WOLPERT: This relates to Dr. Goldacre's point. If, as he argues, the surface coat flows over the plasma membrane, then the plasma membrane must be renewed each time the cell passes through its own length. We find it very difficult to see how this could happen, especially since it means that the plasma membrane must be separated from the coat and enter the cell in the tail region. I think we should be aware of the fact that there is no evidence that the plasma membrane is being turned over. I do not say it is not, just that it seems very unlikely.

DR. GOLDACRE: Have you read my paper?

DR. WOLPERT: Yes. You speak of the oil-droplet experiments.

DR. GOLDACRE: And six others.

DR. WOLPERT: Let's take the oil-droplet one. The oil drop that sticks to the surface doesn't, as you would say, move back. The ameba moves past the oil droplet. If the membrane is fluid, this may simply mean you are preventing the membrane from flowing forward in this region, and the rest of the membrane simply flows past it.

DR. BISHOP: Dr. Wolpert, may I ask you a simple question? Is there any membrane antibody preparation that has been tested by gel diffusion?

DR. WOLPERT: Yes, there is.

DR. BISHOP: Is there only one band?

DR. WOLPERT: It is a complicated story. With a gel-diffusion test nothing is obtained with an intact membrane—the surface coat has to be dispersed. If this is done, then one obtains one main line with a couple of auxiliaries. We also have evidence for specificity. If the labeled antibody is put on a cell section, then it stains only the surface and certain vesicles within the cell.

DR. REBHUN: Dr. Pappas has made a very good observation on the effect of thorium oxide on the surface layers of amebae. When he kept the ameba in thorium oxide for some time, the entire polysaccharide layer peeled off the ameba. Electron micrographs showed that the only thing left was the plasmalemma. In this particular case you would have a naked ameba which regenerates its coat only after several hours. I wonder if you could use your membrane antibody in this case? There would be very little possibility of any cross reaction with the polysaccharide.

DR. WOLPERT: We have had no success with this method. We will have to try again. One of our aims is, in fact, to get a naked ameba. This may enable us to label specifically the plasma membrane. So far we have not been able to find anything that will take off the coat from the intact cell.

DR. REBHUN: One point with respect to the possibility of forming large amounts of membrane rapidly. What do you think occurs in the surface precipitation reaction?

DR. WOLPERT: I am not convinced the surface-precipitation reaction as described really occurs.

DR. REBHUN: Well, I think there is little doubt it does, especially in the case of the large number of vacuoles which can form rapidly within a cell. Do you believe there are no membranes bounding these?

DR. WOLPERT: I think there is some kind of interface, but I do not believe such an interface has ever been observed critically enough for us to know how closely it resembles a cell membrane particularly in the case of the ameba. These surfaces may or may not be similar. Perhaps Dr. Stewart could tell his story about formation of new membrane in slime mold.

DR. STEWART: If you have a large channel, and cause a droplet of endoplasm to flow out, then the droplet presumably has a new surface. This can be done under water or in a variety of solutions. If you fix the droplet quickly and look at it in the electron microscope, the surface layers appear to lack vesicles. We have assumed that this is because the way it forms its new membrane is to bring vesicles to the surface and spread the vesicular membranes over the surface.

DR. WOLPERT: Does it happen quickly?

DR. STEWART: It happens very quickly. You can do the same operation repeatedly until you get a string of six or seven droplets in a row. So, obviously, if there is a membrane there, it must have been formed; but we feel it is formed from stores of membranous vesicles and not simply by precipitation. The electron microscope evidence is, of course, not the final word, but at least it is consistent with the picture I am presenting.

DR. WOHLFARTH-BOTTERMANN: We have fixed protoplasmic drops at the moment of their generation and found that these drops possess as a boundary to the external medium a normal cytoplasmic membrane. I frankly doubt that this new cytoplasmic surface is formed at the expense of vesicular membranes

DR. HOFFMANN-BERLING: I would like to comment on the experiment, in which you stopped motion by centrifuging the isolated protoplasm. In muscle and in the non-muscular cells which have been analyzed, contraction is reversible. The agent which inhibits contraction and induces the contractile structures to return to the relaxed state are particles or tubules inside the contractile system. The particles actively store calcium. The depletion of this ion from the contractile structures seems to be the main cause for the cessation of contraction. The relaxing particles are rather large and can be centrifuged out of muscle homogenates at 20,000 *g* in an hour.

I do not know how long you centrifuged the particles. However one of the explanations of your experiment could be that the movements observed resulted from local, reversible contractions and that, by centrifuging, you eliminated some factor which was indispensable for relaxation. If, in your system too, relaxation were the result of a reduction of free calcium, then it would be interesting to know what happens if you add calcium to concentrations of, let us say, $>10^{-4}$ M.

DR. WOLPERT: It is a very nice possibility that Dr. Hoffmann-Berling suggests. Traces of calcium in the order of 1 to 2 mmoles bring about irreversible gelation.

Pressure-Temperature Studies on Ameboid Movement and Related Phenomena: An Analysis of the Effects of Heavy Water (D_2O) on the Form, Movement, and Gel Structure of *Amoeba proteus*[1,2]

DOUGLAS MARSLAND[3]

Department of Biology, New York University, New York

Introduction

Studies on the biological effects of D_2O began shortly after Urey (1933) first reported on the identification and isolation of deuterium. These early studies, though handicapped by the exceedingly small amounts of heavy water which were then available, soon indicated that surprisingly large effects may result when D_2O is partially substituted for H_2O in protoplasmic systems (cf. Lewis, 1934). The mechanisms of mitosis proved to be particularly susceptible to inhibition by this type of isotopic substitution, as was shown, very early, by Lucké and Harvey (1935) and Ussing (1935) and, considerably later, by Gross and Spindel (1960) and Marsland and Zimmerman (1963).

It was noted by Marsland and Zimmerman that the mechanism of karyokinesis, namely, the spindle-aster complex, or mitotic apparatus, is much more susceptible to D_2O blockage than is the mechanism of furrowing, by which cytokinesis is achieved. A 70% substitution of D_2O for H_2O in the environing sea water sufficed to "freeze" the structure and activity of the mitotic apparatus and to block the movement of chromosomes in the dividing eggs of *Arbacia punctulata*. But even greater substitutions, up to 95%, were ineffective in blocking the prog-

[1] Based on experiments performed partly at the Bermuda Biological Station and partly at the Marine Biological Laboratory, Woods Hole, Massachusetts.

[2] As originally planned this presentation was to have included an evaluation of current contractile hypotheses as to the mechanisms of ameboid movement. Time and space precluded this. Instead, a broad interpretation of the tube-wall contraction hypothesis, upon which the evaluation was to have been based, is presented in my introductory statement as chairman of the third part of this Symposium.

[3] Work supported by grant series CA 00807, from the National Cancer Institute of the National Institutes of Health, U. S. Public Health Service.

ress of the cleavage furrows. In deuterated eggs, moreover, the cortical cytoplasm, which is generally acknowledged to be the site where the energy of furrowing is developed, displayed a distinctly firmer, or stronger, gelational state, when tested by the pressure-centrifuge technique (Marsland and Zimmerman, 1963).

It has been proposed (cf. Marsland, 1956) that the mechanisms of furrowing and of ameboid movement are fundamentally similar. Both represent contractile gel systems and both are activated by a continuing cycle of sol $\rightleftarrows$ gel reactions. Consequently, it is of interest to compare the effects of deuteration upon the two systems. Moreover, the equilibria of gel systems generally are known to be particularly sensitive to the pressure factor, and the effects of pressure upon the gel systems of both the ameba and the dividing cell have been studied quite extensively (cf. Marsland, 1956). Accordingly, pressure has been utilized as a tool in this analysis of the effects of deuteration on the form and movement of the ameba.

The progressive solation imposed upon protoplasmic gels by increasing pressure and by decreasing temperature (Marsland, 1956) clearly indicates that a positive volume change ($+\Delta V$) must be inherent in the formation of such gel structures and that such gelations are endergonic processes. Moreover, it seems likely that hydrogen bonding plays a significant role in the formation and maintenance of protoplasmic gel structures (Marsland *et al.*, 1962). Consequently it is to be expected that the substitution of deuterium for ordinary hydrogen in a significant fraction of the water in and around the cell, should introduce significant changes, not only in the plasmagel structure of the ameba, but also in the cycle of sol $\rightleftarrows$ gel reactions which is presumed to energize the movement.

Materials and Method

Deuterated Media

Heavy water, having a D_2O content of 99.8 mole % was obtained from the Bio-Rad Laboratories, Richmond, California. This was utilized in the preparation of dilute Brandwein solution (Bandwein, 1935), here specified as D_2O-Brandwein. Experimental immersion media, in which the D_2O content was 30, 50, and 70%, respectively, were prepared by mixing 1.5, 2.5, and 3.5 ml of D_2O-Brandwein with 3.0, 2.0 and 1.0 ml, respectively, of ordinary H_2O-Brandwein, and then bringing the final volume of each experimental solution up to 5 ml by the addition of 0.5 ml of culture fluid in which 50–100 amebae were suspended. For the 90% D_2O medium, 0.5 ml of ameba suspension was added to 4.5 ml of D_2O-Brandwein; and for the 98.8% medium, 0.5 ml of the 90% medium, containing the amebae in suspension, was transferred to a fresh

4.5 ml sample of pure D_2O-Brandwein. For the sake of brevity a Brandwein solution in which a certain percentage, say 90%, of the H_2O content has been replaced by D_2O will be designated as 90% D_2O-Brandwein.

Amebae

Excellent cultures of *Amoeba proteus* were obtained from Professor J. A. Dawson of the College of the City of New York. These were maintained in an incubator at 18°C. Specimens from the original cultures continued to yield consistent results for about 4 weeks and the same was true for amebae from subcultures, prepared every 2 weeks by the method of Brandwein (1935).

Pressure Apparatus and Techniques

The microscope-pressure chamber equipment has been described previously by Marsland (1950). This equipment permitted the amebae to be observed at magnifications up to 600 diameters, while exposed to pressures, which in these experiments, ranged up to 12,000 psi. The pressure pump, a modified hydraulic jack, permitted the pressure to be built up at the rate of 4000 psi/sec. The period of compression, at the end of which a count was made to determine the number of specimens which had maintained a plainly lobose form, was standardized at 20 min. Each count, made at a magnification $\times$ 30, included some 80 to 100 specimens and required about 2 min. Decompression, achieved by the release of a needle valve, was virtually instantaneous.

The specimens in the deuterated media were confined within a small cylindrical glass chamber (6 $\times$ 6 mm), placed within a larger plastic cylinder (2 cm diameter $\times$ 1 cm height) which was filled with D_2O-Brandwein and covered with a plastic diaphragm. This arrangement prevented the amebae from migrating too far from the center of the window of the pressure chamber, where they would be outside the microscopic field. Also it prevented dilution of the experimental media by the distilled water which filled the main volume of the pressure chamber. The experiments were performed at room temperature, which was kept at 21° ± 1°C.

Results

Preliminary Observations

Amebae immersed in even the most heavily deuterated media displayed a remarkable tolerance to the treatment (Fig. 1). At the maximum, which was 98.8% D_2O-Brandwein, active locomotion continued for 8 days, beyond which it did not seem profitable to extend the observations. At the end of the eighth day, 94% of more than a hundred specimens originally isolated in the medium still displayed an actively extended form and vigorous locomotion. Apparently no divisions occurred during the period since the number of specimens found at the end was equal, at least approximately, to the number originally isolated.

Probably some ingestion occurred in the maximally deuterated medium, although this was not observed directly. Food vacuoles were ob-

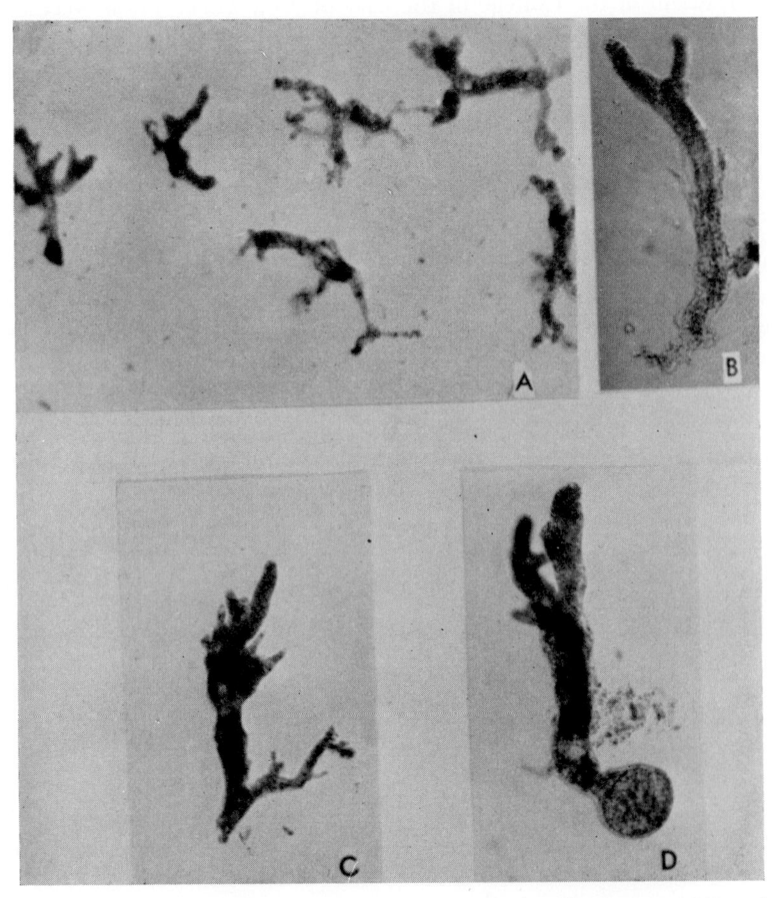

FIG. 1. Form and movement of *Amoeba proteus* in heavily deuterated Brandwein medium. (A) Low-power field showing typical form of amebae on the eighth day of immersion in 98.8% D_2O-Brandwein solution (98.8% of H_2O replaced by D_2O). All the specimens showed vigorous ameboid streaming. (B) Single specimen, same day, same conditions; locomotion very active. (C) Branching form of specimen after 6 days in 90% D_2O-Brandwein. (D) Newly formed food vacuole observed in specimen on fourth day of immersion in 90% D_2O-Brandwein. The large food vacuole at the posterior extremity contained a polychaetous worm, the cilia and bristles of which were still very actively motile. Magnifications: $\times$ 30 for A; $\times$ 100 for B, C, and D.

served during the first 3 days of isolation. It must be noted, however, that very few food organisms (mainly *Chilomonas* and *Colpidium*) were available since very few were carried over during the two-stage transfer of the amebae into the final 98.8% D_2O-Brandwein. Ingestion was observed on the fifth day in 90% D_2O-Brandwein, wherein the number of food organisms was considerably greater (Fig. 1D).

In less heavily deuterated media (30, 50, and 70% D_2O), the form and the velocity of streaming and locomotion of the amebae were not appreciably altered. In fact these specimens seemed quite indistinguishable from control amebae similarly isolated in H_2O-Brandwein. And even when the concentration of D_2O was higher (90 and 98.8%) the alterations in form and activity were rather subtle. There appeared to be a tendency for the pseudopodia to be more slender, more numerous, and more apt to originate near the posterior extremity of the body. Moreover, it consistently appeared that the cylindrical stream of cytoplasm (endoplasm or plasmasol) was unusually broad relative to the surrounding shell of nonflowing cytoplasm (ectoplasm or plasmagel). In other words, a reduced gel/sol ratio (Mast and Prosser, 1932) seemed to be a characteristic of the heavily deuterated amebae. Streaming and locomotion appeared to be accelerated although, in the absence of monopodial forms, measurements of these rates were not obtained.

Pressure Experiments on the Form Stability of Ameba

Pressure-centrifugation studies (Brown and Marsland, 1936; Marsland and Brown, 1936; Landau *et al.*, 1954) have shown that high pressure, in the range up to 8000 psi, imposes a progressive weakening of the gel structure of the peripheral cytoplasm of the ameba. This pressure-induced solation, or weakening, of the plasmagel structure is regularly accompanied by a number of visible changes in the form and activity of the ameba—as studied in the microscope-pressure chamber. At relatively low pressures (2000–4000 psi), the exact level being dependent upon experimental temperature, all pseudopodia stop extending and begin to retract. At low pressures the retraction is incomplete, however, and short, slender, pseudopodial vestiges are maintained in a degree that depends upon the pressure level and the temperature. Higher pressures (4000–7000 psi, again depending on temperature) impose more drastic changes upon the ameboid form (Fig. 2). Forms which are relatively compact initially, slowly round up into inert spheres. More elongate forms, lacking well-extended pseudopodia, may, on the other hand, round into two connected spheres, which finally pinch off separately. But if there are well-extended elongate pseudopodia initially, each breaks into a series of beads or balls, which finally become disconnected (Fig. 3). Usually the bead or ball at the distal extremity of the pseudopodium is slightly larger than the others.

Clearly, the foregoing changes of ameboid form indicate a loss of rigidity (or increase of fluidity) imposed upon the plasmagel system by the pressure conditions. At higher temperatures, however, it requires higher pressures to effect such changes in ameboid form, undoubtedly because the initial rigidity or strength of plasmagel system is greater

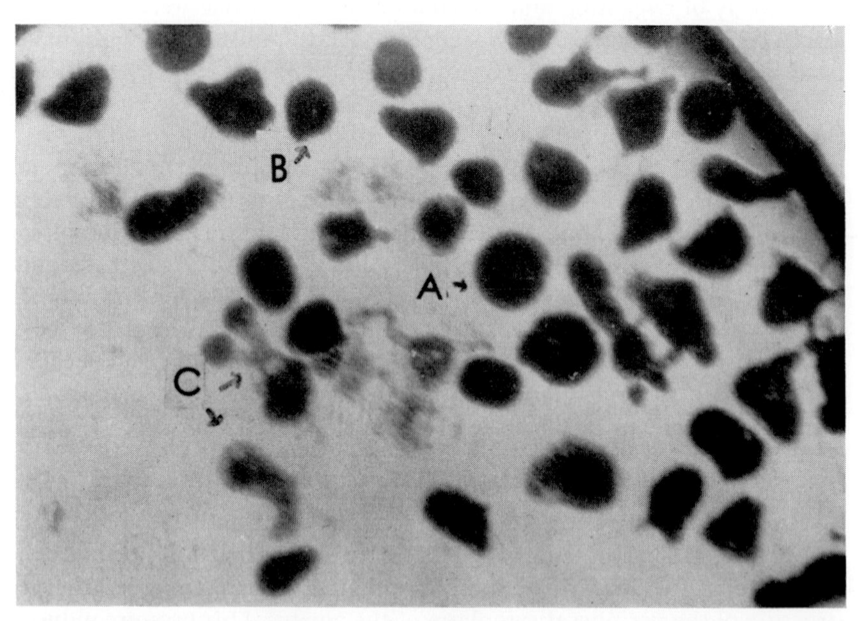

FIG. 2. Criteria for determining the resistance of deuterated amebae to solation by high pressure. These specimens, in 90% D_2O-Brandwein solution, have been in the microscope-pressure chamber for 8 min, sustaining a pressure of 6000 psi. Standard procedure was to make a count at the end of a 20-min compression period, ascertaining the percentage of specimens resistant to the "sphering action" of the pressure. (A) denotes a fully rounded specimen; (B) an intermediate form (excluded from the count); and (C) a lobose, or nonrounded form. The count, at the end of this experiment, was 61% lobose/39% rounded.

(Landau *et al.*, 1954). Accordingly, the magnitude of pressure required to cause a certain loss in the form stability of the ameboid cell may be taken as an index of the gelational state of its plasmagel system and used for an evaluation of the effects of deuteration on the gel system.

STABILIZATION OF AMEBOID FORM BY D_2O

a. Resistance of Deuterated Amebae to Pressure-Induced Rounding. As shown by previous work (Landau *et al.*, 1954), the percentage of specimens in a given population of amebae which is capable of maintaining some degree of lobose form in resistance to the sphering tendency decreases regularly as the pressure increases—other conditions being equal. Thus it has been possible to compare the form stability of specimens immersed in Brandwein solutions containing various concentrations of D_2O with that of amebae immersed in ordinary nondeuterated Brandwein.

In each of these experiments about 100 amebae were immersed in a deuterated (30, 50, 70, and 90% D_2O) or nondeuterated medium, transferred to the microscope-pressure chamber, and allowed to equilibrate

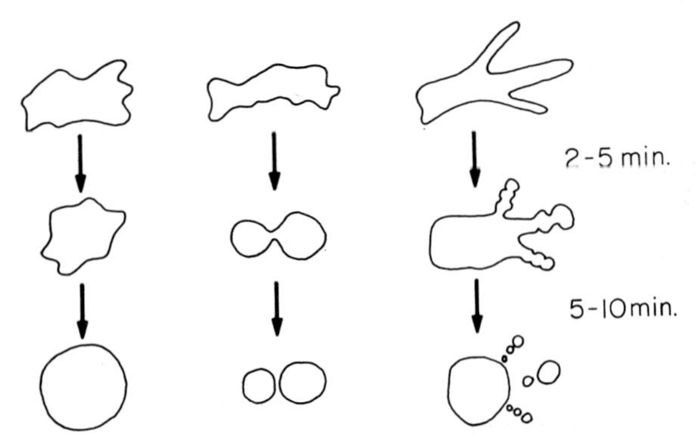

FIG. 3. Instability of ameboid form displayed by amebae exposed to relatively high, sustained, hydrostatic pressure. The critical pressure, at which a beading of elongate pseudopodia (lower, right drawing) occurs, varies according to the D_2O content of the medium (Table I) and according to the experimental temperature. Individuals in a population of amebae differ considerably as to their susceptibility to pressure-induced rounding, but the population as a whole, if from a vigorously growing culture, displays a statistically predictable behavior (cf. Fig. 4).

for 20 min. Then the pressure was built up to a particular level and maintained at this level for 20 min. The amebae were kept under continuous observation at × 30 magnification. At the end of the standard 20-min compression period, a count was made to determine the percentage of specimens that were resistant to the sphering action of the pressure, i.e., specimens which still maintained a plainly lobose form.

The results of these experiments are summarized graphically in Fig. 4. For sake of comparison, this figure also includes a graph in which the previous measurements of Brown and Marsland (1936) on the gelational strength of the plasmagel system of *Amoeba proteus* are plotted as a function of pressure in the same experimental range.

The criteria used to differentiate between lobose and rounded forms, are indicated in Fig. 2. In each count, three categories of form [(1) definitely lobose, (2) intermediate, and (3) definitely rounded] were distinguished, but only the first and third were considered in calculating the percentages. It was assumed that specimens in the intermediate category, which seldom represented more than 15% of a total count, probably would have divided themselves between the other two categories in proportion to the respective percentages.

Several of the points recorded among the 30% D_2O data represent average values derived from only two experiments. All other values rep-

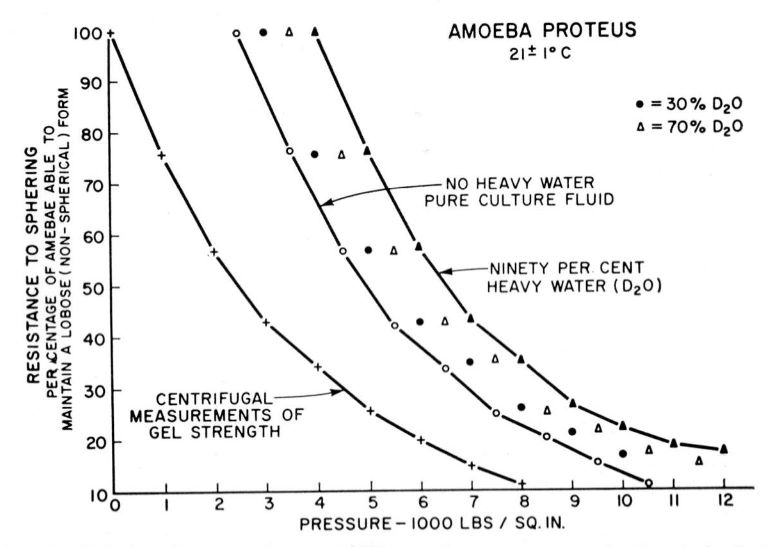

FIG. 4. Relation between form stability and pressure magnitude, derived from amebae immersed in media of increasing D_2O concentration. Form stability appears to be related to the structural state of the plasmagel system. This may be inferred from a comparison with the gel strength–pressure curve (Brown and Marsland, 1936) which is also plotted.

resent averages of at least three experiments involving not less than 250 amebae; and a majority of the points involved four or five experiments. The maximum variation of values obtained with reference to any particular point was ±6%, except in three experiments performed on amebae from an aging culture.

A few experiments utilizing 50% D_2O-Brandwein were also carried out, but only in the 3000–6000 psi range. At each pressure the 50% D_2O values fell between the 30 and 70% D_2O values. These data were incomplete, however, and they are not included on the graph.

Figure 4 shows very clearly that the resistance of the deuterated ameba to the sphering action of high pressure is relatively great, as compared to the resistance of nondeuterated specimens. Moreover, the stabilizing effect is increased with each increase in the concentration of D_2O, as is also shown by plotting the data in a different manner (Fig. 5). And last, the loss of form stability in all the specimens, both deuterated and nondeuterated, appears to be related to a weakening of the gel structure of the plasmagel system, since all the curves are very nearly parallel to each other and to the gel strength–pressure curve.

b. Effects of D_2O on Pseudopodial Beading. At relatively low pressures, elongate extended pseudopodia merely shrink and equilibrium appears to be reached when the pseudopodial length and diameter have

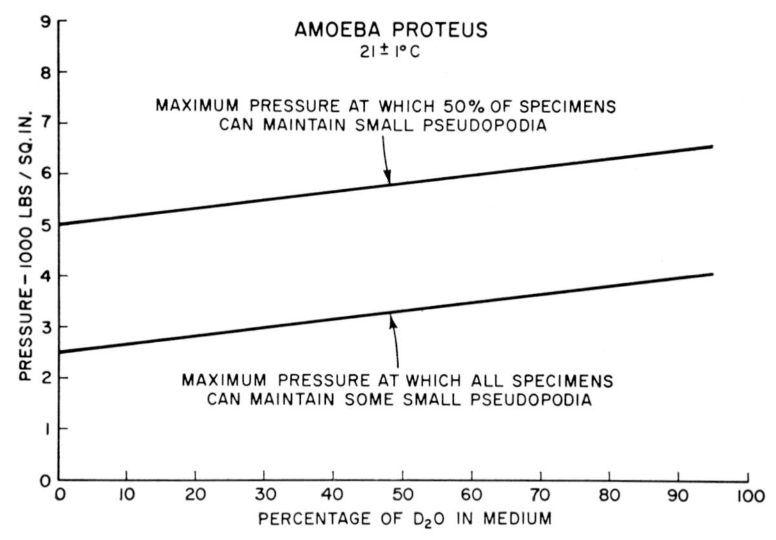

FIG. 5. Relation of D_2O concentration, pressure magnitude, and the capacity of the ameba to maintain its pseudopodia.

been reduced to a greater or lesser degree depending on the pressure magnitude. At higher pressures, on the other hand, elongate pseudopodia become unstable. They begin to bead or ball and within some 2–3 min each has broken up into a linear series of discrete spherules, as is shown in Fig. 3. Apparently the plasmagel wall of the tubular pseudopodium suffers a drastic loss of gel structure. Now the pseudopodium behaves as a fluid cylinder and breaks, under the agency of surface forces, into a series of separated beads. Thus the pressure magnitude required to elicit the beading reaction may be used as an index of the effects of deuteration upon pseudopodial gel structure and stability.

A summary of the observation on pseudopodial beading at varying magnitudes of pressure and in media of increasing concentrations of D_2O is given in Table I. Here again a stabilizing effect of D_2O upon plasmagel structure is plainly discernible. At higher concentrations of D_2O, significantly higher pressures are required before a critical loss of gel structure occurs and the pseudopodium becomes unstable.

TABLE I

EFFECTS OF D_2O ON PSEUDOPODIAL STABILITY IN *Amoeba proteus*[a]

Pressure (psi)	Concentration of D_2O in medium				
	No D_2O	30%	50%	70%	90%
3000	—	—	—	—	—
3500	+—	—	—	—	—
4000	+	+—	—	—	—
4500	++	+	—	—	—
5000	++	++	+	—	—
5500	++	++	++	—	—
6000	++	++	++	+	+—
6500	++	++	++	++	+
7000	++	++	++	++	++

[a] Symbols:—, No pseudopodial beading; +—, slight beading in some experiments, not in others; +, definite beading, all experiments; ++, rapid beading of all elongate pseudopodia.

POSTPRESSURE MASS CONTRACTION PHENOMENON

This reaction, which has been described previously (Landau *et al.*, 1954), was observed in all experiments in which the pressure magnitude was great enough to cause a significant percentage (10–15%) of the specimens to become rounded and more or less spherical during the 20-min compression period. Apparently the reaction represents a generalized contraction of a peripheral plasmagel layer, which appears to be reconstructed very quickly after the pressure has been released. In any event, starting about 20 sec after decompression, 100% of specimens develop a hyaline halo, as the granular cytoplasm quickly shrinks inward away from the cell membrane. At first the hyaline fluid, filling the space between the membrane and the contracting granular cytoplasm is perfectly clear, being devoid of mitochondria and other granular elements (Landau and Thibodeau, 1962), and usually the "elevated" cell membrane displays a smoothly spherical contour. Within another 30 sec, however, bulbous protuberances begin to appear in the contour and vigorous streams of granular cytoplasm begin to pour out from the central mass into the protruding lobes. Thus a number of pseudopodia are formed, extending haphazardly in various direction. Soon, however, one or two of the pseudopodia become dominant and a definitely oriented type of locomotion ensues. Generally speaking, the mass contraction is quicker and more vigorous, forming a broader hyaline zone, when higher pressures (above 6000 psi) are used, but it is plainly discernible at lower pressures, provided these are great enough and maintained long enough to induce a definite rounding of the specimens. The reaction occurred in all media, deuterated as well as nondeuterated; although higher pressures were required to induce a rounding of the deuterated cells.

Postpressure Recovery

Pressures of 5000 psi and less, applied for the standard 20-min period, did not impose any plainly discernible irreversible effects upon either the deuterated or nondeuterated amebae. As to form and locomotion, at least, the pressurized specimen appeared to be normal when examined 24 hr after treatment. With higher pressures, particularly in the range above 8000 psi, the recovery was incomplete. All the specimens showed some pseudopodial activity and locomotion soon after decompression, but 24 hr later quite a few were rounded and motionless.

Discussion

It has been proposed (Landau *et al.*, 1954) that furrowing, or cleavage, in animal cells and movement in ameboid cells generally are achieved by basically similar mechanisms. According to this hypothesis, both furrowing and ameboid movement are dependent upon the contraction of a strongly gelled peripheral layer of the cytoplasm and in both cases metabolic energy is transformed into mechanical energy by a continuing cycle of sol $\leftrightarrows$ gel reactions.

Generally speaking, the effects of deuteration upon the two systems tends to support this hypothesis. Neither system is inhibited by very high concentrations of D_2O. In fact, some evidence (Marsland *et al.*, 1962) indicates that furrowing may be enhanced by heavy deuteration and some of our current observations indicate that ameboid movement may also be strengthened to some degree by such treatment. Other contractile processes, on the other hand, appear to be inhibited completely by heavy deuteration, as in the case of the movement of chromosomes by the mitotic apparatus (Gross and Spindel, 1960; Marsland and Zimmerman, 1963).

The increased resistance of pressure-induced rounding, which becomes more and more evident as the concentration of D_2O is raised, undoubtedly represents an effect upon the gel structure of the plasmagel system. Resistance to surface forces which tend to round the cell falls off with increasing pressure along a curve that closely parallels the gel strength–pressure curve as determined by Brown and Marsland (1936) and by Landau *et al.* (1954). This is true for both deuterated and nondeuterated specimens, but with each increase of D_2O concentration, the initial resistance to rounding is greater and the residual resistance at each higher level is proportionately higher. Also the strengthening action of D_2O upon the gel system of dividing egg cells has been evaluated by the pressure-centrifuge measurements of Marsland and Zimmerman (1963).

Two other kinds of evidence likewise point to a stabilizing action of D_2O upon the structure of the plasmagel system of the ameba. First is the increased resistance to pseudopodial beading (Table I) which is observed with each increment in the concentration of D_2O; and second, is the lower gel/sol ratio observed in the more highly deuterated media. The beading of a cylindrical mass of liquid certainly indicates a certain degree of fluidity and, conversely, a resistance to beading must indicate some sort of structural impediment, in this case, probably, a residuum of structural integrity in the plasmagel part of the pseudopodium.

The reduction in the gel/sol ratio by D_2O appears to be similar to the reduction that occurs with increasing temperature (Mast and Prosser, 1932). Temperature is known to augment the structural strength of the plasmagel system (Marsland and Landau, 1954; Landau *et al.*, 1954), and it may be presumed that D_2O has a similar action. As a sort of physiological adaptation, an ameba with a fortified gel structure does not require as thick a layer of plasmagel, either to stabilize the form of its pseudopodia or to generate mechanical energy for movement.

Deuteration, even when heavy, appears not to interfere very drastically with such aspects of metabolism in the ameba as must provide the energy for movement. If, as has been proposed (Kriszat 1949; Zimmerman *et al.*, 1958), a significant part of this energy is derived from the hydrolysis of adenosine triphosphate (ATP), it must be granted that the deuterolysis of this high-energy phosphate compound provides an effective substitution. This is especially apparent when it is considered that the movement continues with undiminished vigor even when 98.8% of the H_2O in the medium and in the ameba has been replaced by D_2O.

Regardless of whether more, or less, energy is available for forming and sustaining the gel structure, the fact remains that the plasmagel system, as a result of deuteration, is firmer, stronger, and more resistant to the solating effects of high pressure. It seems highly probable that part of this stabilization results from the substitution of relatively stronger D-bonds for H-bonds, as has been proposed by Gross and Spindel (1960). However, little or nothing is known about how the substitution of D_2O for H_2O would affect the stability of the "water shells," or hydration spheres, which, in the sol state, are presumed to surround and protect the bonding sites of a prospective gel structure. Consequently, it is not yet possible to predict how other types of bonding, e.g., salt bridges and disulfide linkages, would be affected by the deuteration of a sol-gel system.

Summary

Specimens of *Amoeba proteus* immersed in deuterated Brandwein solutions (in which 30 to 98.8% of the H_2O content was replaced by D_2O) continued to sustain pseudopodial activity and locomotion for more than 8 days. At lower concentrations (up to 70% D_2O) the deuterated amebae could not be distinguished, on the basis of form or activity, from control specimens. At higher concentrations (90–98.8% D_2O) there was a tendency for the pseudopodia to be unusually slender, elongate, and tortuous; and very frequently they took origin from the posterior extremity of the cell. If anything, the rate of streaming and locomotion was enhanced; and the gel/sol ratio was definitely reduced by heavy deuteration.

Deuteration was found to have a stabilizing effect upon the plasmagel structure of the pseudopodia. This effect was enhanced as the concentration of D_2O was increased. Higher pressures were required to solate the plasmagel to the point of instability, at which the pseudopodium behaves as a cylindrical mass of fluid and breaks into a number of separated beads. In 90% D_2O-Brandwein, it required 6500 psi to evoke a definite beading of the pseudopodia, as compared with 4000 psi for specimens in H_2O-Brandwein solutions.

A stabilization of the plasmagel system was also evident when resistance to the sphering action of high pressure (percentage of specimens capable of maintaining a definitely lobose form) was plotted as a function of pressure, utilizing amebae immersed in media containing 0, 30, 70, and 90% D_2O, respectively. Generally speaking, the resistance–pressure curves were parallel to one another and to the gel strength–pressure curve (Brown and Marsland, 1936). Each increment of D_2O concentration, however, shifted the pressure values upward by approximately 500 psi.

An interpretation of the stabilizing effects of D_2O is discussed, with particular emphasis on the possibility that a substitution of relatively stronger D-bonds for ordinary H-bonds at the intermolecular linkage points of the gel structure may account, at least in part, for the observed phenomena.

References

Brandwein, P. F. (1935). *Am. Naturalist* **69**, 628.

Brown, D. E. S., and Marsland, D. (1936). *J. Cellular Comp. Physiol.* **8**, 159.

Gross, P. R., and Spindel, W. (1960). *Ann. N. Y. Acad. Sci.* **84**, 573.

Kriszat, G. (1949). *Arkiv. Zool.* **1**, 81.

Landau, J. V., and Thibodeau, L. (1962). *Exptl. Cell Res.* **27**, 591.

Landau, J. V., Zimmerman, A. M., and Marsland, D. (1954). *J. Cellular Comp. Physiol.* **44**, 211.

Lewis, G. N. (1934). *Science* **79**, 151.
Lucké, B., and Harvey, E. N. (1935). *J. Cellular Comp. Physiol.* **44**, 211.
Marsland, D. (1950). *J. Cellular Comp. Physiol.* **36**, 205.
Marsland, D. (1956). *Intern. Rev. Cytol.* **5**, 199.
Marsland, D. A., and Brown, D. E. S. (1936). *J. Cellular Comp. Physiol.* **8**, 167.
Marsland, D., and Landau, J. V. (1954). *J. Exptl. Zool.* **125**, 507.
Marsland, D., and Zimmerman, A. M. (1963). *Exptl. Cell Res.* **30**, 23.
Marsland, D., Zimmerman, A. M., and Asterita, H. (1962). *Biol. Bull.* **123**, 484.
Mast, S. O. (1926). *J. Morphol. Physiol.* **41**, 347.
Mast, S. O., and Prosser, C. L. (1932). *J. Cellular Comp. Physiol.* **1**, 333.
Urey, H. C. (1933). *Science* **78**, 566.
Ussing, H. H. (1935). *Skand. Arch. Physiol.* **72**, 192.
Zimmerman, A. M., Landau, J. V., and Marsland, D. (1958). *Exptl. Cell Res.* **15**, 484.

Discussion

Dr. Hoffmann-Berling: I was interested in what you call "mass contraction," which fits very well into the scheme that the contractile material is distributed everywhere but that contractility is limited by some unidentified means to certain portions.

You attribute the transformation from gel to sol to the higher stability of the deuterium bond compared to the hydrogen bond, and you locate the site of the alteration on the structural proteins of the plasmagel itself. Perhaps one should keep in mind that the effects could as well be of an indirect origin; for instance, deuteration and/or high pressure may affect membranes which either keep in or keep out certain ions, let us say, calcium; the ions may be the agent on which the physical state of the cytoplasm depends.

Dr. Marsland: I certainly would grant you this possibility. I think my data are merely suggestive of the possible mechanism.

Dr. Allen: I am not familiar with the work on muscle contraction concerning D_2O. Could somebody tell me what D_2O does to contractile systems in general, let us say, muscle? Does it, for example, tend to favor or inhibit contraction?

Dr. Andrew G. Szent-Györgyi: It has an inhibitory effect.

Dr. Allen: It struck me that if there were an inhibiting action, these data would fit in very well also with the front-contraction idea, according to which inhibition of contraction would reduce the gel/sol or A_t/A_s ratio. That is, if the normal contraction involved a Δl of 30%, let us say, and dropped to 15%, then the "sol," as you call it (or endoplasm), would occupy a greater portion of the cross-sectional area of the pseudopod.

I think it should also be said that all the pressure data that you have obtained can be interpreted in at least two ways. If one considers the effect of pressure only on sol $\rightleftarrows$ gel equilibria, then one could consider that pressure acts only through this equilibrium.

On the other hand, if the ameba is considered as a contractile system, then it might be worthwhile to consider as a model what pressure does to muscle. There it promotes contraction, or at least it clearly increases the contractile tension. According to the front-contraction idea, if pressure were to cause an increase in the extent of contraction, then blunt pseudopodia would form owing to an increase in the gel/sol or A_t/A_s ratio. This is exactly what happens under hydrostatic pressure. On the other hand, if contraction were inhibited by D_2O, then you would expect this agent to antagonize the action of pressure, so that its effects on the deuterated cells would be reduced. This is also what you found. Therefore, I think your data cannot discriminate between the front- and tail-contraction theories.

DR. MARSLAND: Not all of my observations came from the pressure work. If you increase the temperature, the gel/sol ratio goes down. My interpretation would be based on the fact that we know that increasing temperature does strengthen the structural characteristics of the gel. In order to maintain a stable pseudopodium, therefore, you do not need as thick a layer if you have a stronger gel. In line with that, I would consider the D_2O effect on the gel/sol ratio as being comparable to the temperature effect where we had measured the definite increase in gel structure.

DR. ALLEN: Is it not an amazing coincidence that *all* of the endoplasmic or "sol" material that passes down the length of an ameba to the front turns into a gel tube and that the ratio of the cross-sectional areas of a pseudopod can vary over such a tremendous range (from just over 1 to 6 or more)? How do you account for the fact that *all* this material becomes incorporated into the gel tube?

DR. MARSLAND: The thicker the gel wall, I would say, the slower would be the extension of the tip, because more material is going into the wall.

DR. ALLEN: Why do pseudopodia maintain a characteristic shape? Why doesn't the tip of a pseudopodium simply spread out in all directions in all cases?

DR. MARSLAND: It is guided by the gel structure.

DR. ALLEN: What makes it solidify in exactly the amount required to form, say, a cylindrical tube?

DR. MARSLAND: Goodness knows: If I could answer that, I would have the secret of ameboid movement.

DR. ALLEN: This is an important matter, and illustrates how many assumptions of a basic nature are required to explain such a "simple" process as ameboid movement.

DR. MARSLAND: Let me reverse it. How do you explain it?

DR. ALLEN: If all the endoplasmic material becomes contracted in the process of forming the ectoplasmic gel, having a greater stiffness than the endoplasmic material, then I think one needs no further assumptions to explain why tube formation keeps pace with the rate of endoplasmic flow.

DR. MARSLAND: All the material is converted into gel by either theory.

DR. ALLEN: Not necessarily.

DR. MARSLAND: I cannot see why it should not be.

DR. ALLEN: It seems to me you are tacitly assuming that the rate of formation of gel is exactly equal to the rate at which the endoplasm moves forward. There must be a delicate control of one of these processes to prevent endoplasm from forming ectoplasmless blobs at the front end.

Morphological Differences among Pseudopodia of Various Small Amebae and Their Functional Significance[1]

EUGENE C. BOVEE

Department of Zoology, University of California, Los Angeles, California

The many kinds of smaller amebae have pseudopodial morphologies and ameboid movements which are as varied as the amebae themselves. I am much intrigued by these variations and also by the probable fact that these are facets of a basic plan of protoplasmic movement.

Most theories of ameboid movement are based on the very large amebae—*Chaos carolinensis, Pelomyxa palustris,* and *Amoeba proteus* and its relatives *Amoeba discoides* and *Amoeba dubia* (Allen, 1961a, b; Bingley and Thompson, 1962; Buchsbaum *et al.*, 1944; Dellinger, 1906; Goldacre, 1961a; Hyman, 1917; Kavanau, 1963; Mast, 1926; Rhumbler, 1898; Rinaldi and Jahn, 1962, 1963; Schaeffer, 1920; Schulze, 1875; Wallich, 1863). Their movements are somewhat alike and readily observable, and these amebae have long been cultivated in the laboratory.

Other studies have been made of movement in smaller amebae (Abé, 1961, 1962; Bovee, 1956, 1960a, b; Pantin, 1923; Penard, 1905; Ray and Hayes, 1954; Schaeffer, 1920, 1926a, 1931), but these are scarcely noted in the battles about the movements of the giant species.

Among smaller amebae, however, great variety of speed, degree, and finesse of protoplasmic flow and of functional form may be seen. Compared to them the movements of the giant species seem stodgy.

I shall discuss some smaller amebae, particularly from Schaeffer's (1926a) family Mayorellidae. There are some interesting locomotive and pseudopodial formations shown by his genera *Mayorella, Vexillifera,* and *Flabellula,* and by the genera I have added to the family: *Subulamoeba* (Bovee, 1953a), *Oscillosignum* (Bovee, 1953b), *Flagellipodium* (Bovee, 1951, 1953c), and *Vannella* (Bovee, 1963b).

Subulamoeba saphirina (Penard, 1902) forms the clear conical pseudopods of the Mayorellidae (Fig. 1C and E). These serve as piers over and between which the body advances at moderate speed. In rapid

[1] These studies have been supported by Research Grants USPHS #6462 and NIH-AI-01158-06.

advance, however, it throws forward a tubular extension which has a clear conical tip (Fig. 1F and G). This tube becomes the main channel of flow and drives the conical tip forward.

Another ameba, *Oscillosignum proboscidium,* also uses such a pseudopod (Fig. 2), but it bends it and anchors the tip to the substrate (Fig. 2A and C). It then expands that pseudopod and flows into it, riding

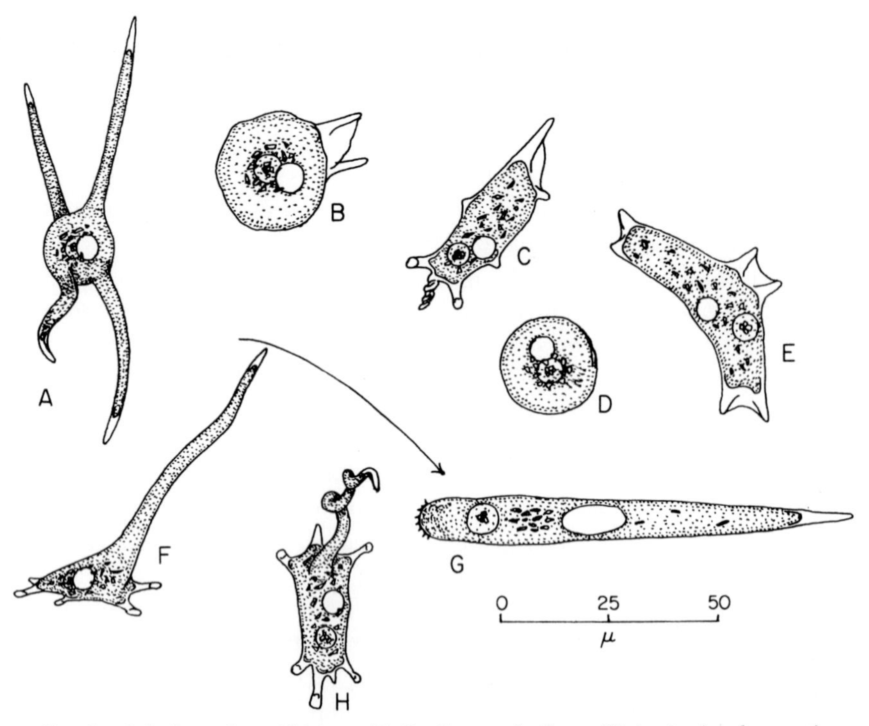

FIG. 1. *Subulamoeba saphirina.* (A) Radiate and afloat; (B) beginning locomotion; (C) in moderate progress, showing conical, mayorellid, determinate pseudopods; (D) at rest, spherical; (E) changing direction, with a new pair of pseudopods formed near the nucleus indicative of the new route; (F) with a long tubular pseudopod thrown forward, to become (G) the rapidly locomotive organism; (H) dorsal view of the organism, with the elongate pseudopod. (After Bovee, 1953a.)

forward over the anchored tip while extending a new pseudopod (Fig. 2B). It also forms shorter pseudopods rapidly, appearing to walk on them (Fig. 2B). The tubular pseudopod is of larger diameter at the rear than at the front end in both *Subulamoeba* and *Oscillosignum.*

Sometimes a mayorellid ameba extends one or more pseudopods ahead of the body and into the water above it (Fig. 3A–E). If the nucleus moves up into the pseudopod, the ameba erects itself, supported on its rear (Fig. 3C–E). It stays erect for some time, and an *Oscillosignum* may extend and wave pseudopods (Fig. 3D). The erect ameba does not resume locomotion *until the nucleus descends to the base of the body,* from where a new pseudopod then extends along a new route of locomotion, followed by the nucleus and the body (Fig. 3B). The site of the nucleus thus seems related to the site of a new pseudopod. (See p. 194.)

If a mayorellid ameba stops, a new pair of pseudopods often forms on one side or the other of the body *near the nucleus* (Fig. 1E). New advance is then in the direction of the new pseudopods.

The conical waving pseudopods of *Vexillifera* sp. (Bovee, 1951, 1956) are of clear protoplasm (Fig. 4). These wave slowly just prior to retraction (Fig. 4B and C). When speeded up by cinematographic projection, these appear impressively active. (See p. 195.)

Amebae of the genus *Flagellipodium* flip out, in 0.5 sec or less, a flagellum-like pseudopod which is vibratile, but does not beat in sine wave patterns seen in typical flagella (Jahn and Harmon, 1963; Jahn *et al.,* 1962). The base of this "flagellipodium" may move along the surface of the organism more or less continuously while it is motile (Fig. 5B). It may be yanked in as fast as it is cast out; it, or one like it, may be flipped out again from any point of the body surface (Fig. 5B and C). The flagellipod may shoot out from the ventral surface of the organism as well as from the dorsal surface, or from the margin. In rapid advance the flagellipod is held rigidly ahead from a clear conical base, like a bowsprit (Fig. 5A). When the ameba halts, the pseudopod may again become motile; may be jerked in again; and may be flipped out again promptly or later. (See p. 196.)

Mayorellid amebae also form radiate floating stages with more or less, long, clear, conical pseudopods (Figs. 6 and 7). These are rigid in some species (Fig. 6A, B, D, and E) but may bend and oscillate in others (Figs. 6C and 7). Sometimes the tips of such pseudopods may be coiled; and this coil moves, seemingly independently of the rest of the pseudopod (Fig. 7A–D). These pelagic stages are the "radiosa" amebae of the literature. (See pp. 197–99.)

Amebae of the genera *Flabellula* and *Vannella* form clear conical pseudopods in the radiate state; and Schaeffer (1926a) placed them in the family Mayorellidae. They are fan-shaped in locomotion with a clear anterior border (Fig. 8). The pattern of protoplasmic flow is active, but difficult to follow in that clear wave. (See p. 200.)

The flabellulid ameba, *Vannella miroides* (Bovee, 1963b) has protoplasmic currents which force their way, seemingly under pressure, through the semisolid parts of the margin, adding to and advancing the

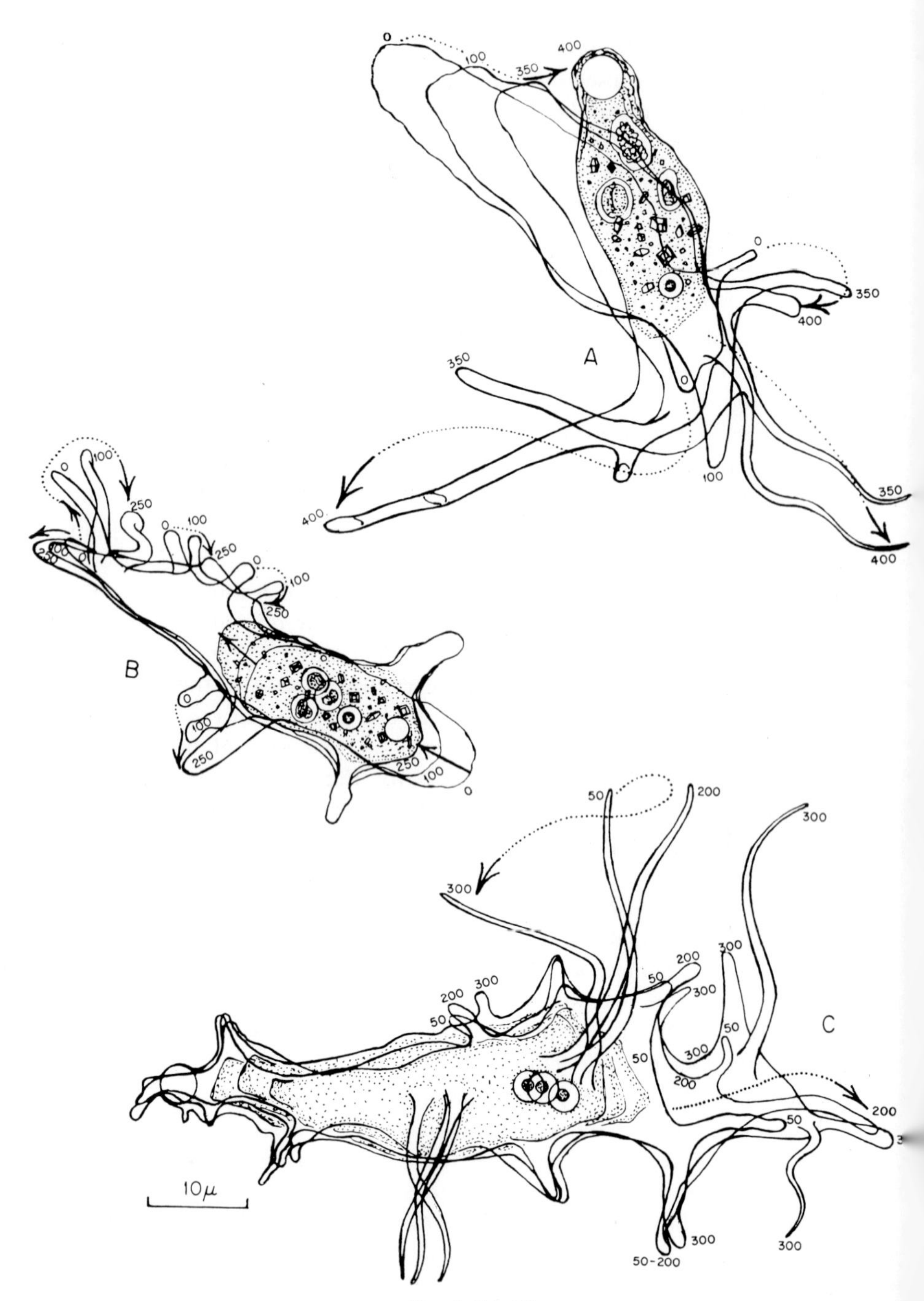

Fig. 2 (A)–(C).

marginal wave (Fig. 9E). The lateral margins of the clear wave appear to contract and retract into the sides of the main body (Fig. 9E). The body trails as a more or less ovate mass (Fig. 9D). (See pp. 201–204.)

On descending from afloat, *Vannella miroides* trails and waves some of the old radiate pseudopods for a time (Fig. 10); and it may form new ones from the upper surface of the body (Fig. 9A–C). Each new pseu-

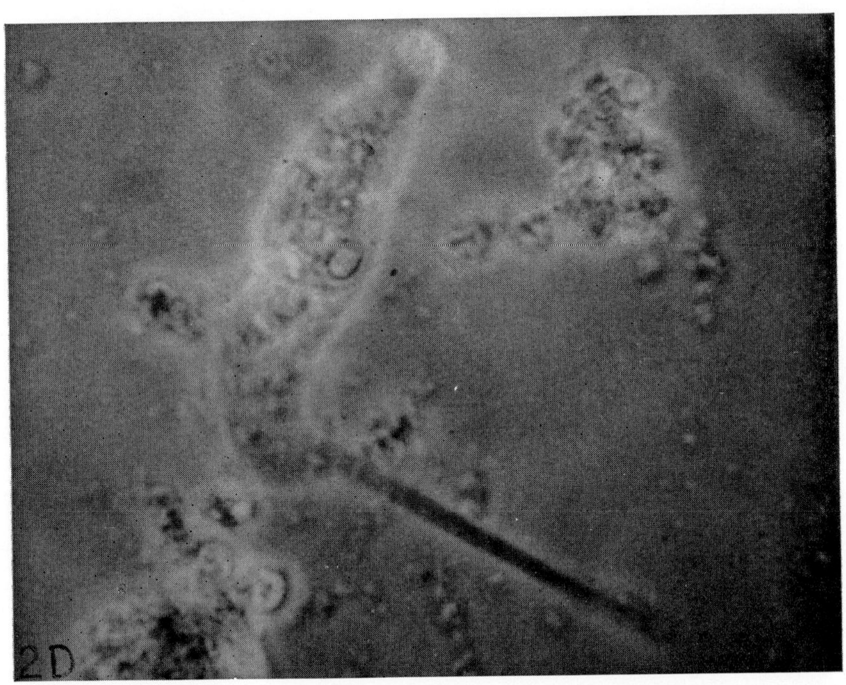

FIG. 2. *Oscillosignum* sp. showing movements plotted from 400 frames of motion picture film (see p. 192). (A) Pseudopod denoting new locomotive path at left, a waving pseudopod at lower right, a supporting pseudopod at upper right, and body movement above; speed of photography, 5 frames/sec. (B) "Walking" movements of pseudopods and forward movements; (C) with formation of a directional pseudopod, lower right; waving pseudopods shown extending above the body. Dotted lines and arrows indicate direction of movement of pseudopodial tips. (D) Photograph of the organism showing waving and directional pseudopods.

dopod starts as a clear half-round bulge, which then extends, fingerlike, 4 to 5 times longer than its diameter, with a clear tip (Fig. 9A). It lengthens spirally until 4 to 6 times that first length, becoming tapering (Fig. 9B). Its distal two-thirds is clear and conical; and the proximal one-third also tapers, but may be granular (Fig. 9C). When extended it may be rigid; or may oscillate by basal movements; or may undulate slowly from base to tip. (See p. 202.)

Other small amebae have other types of locomotion. The Hartmannellidae resemble the flabellulids in gross appearance. They also have a clear advancing margin and a trailing body; but the anterior wave is not so flat as in the flabellulids, nor so fan-shaped (Fig. 11). As a hart-

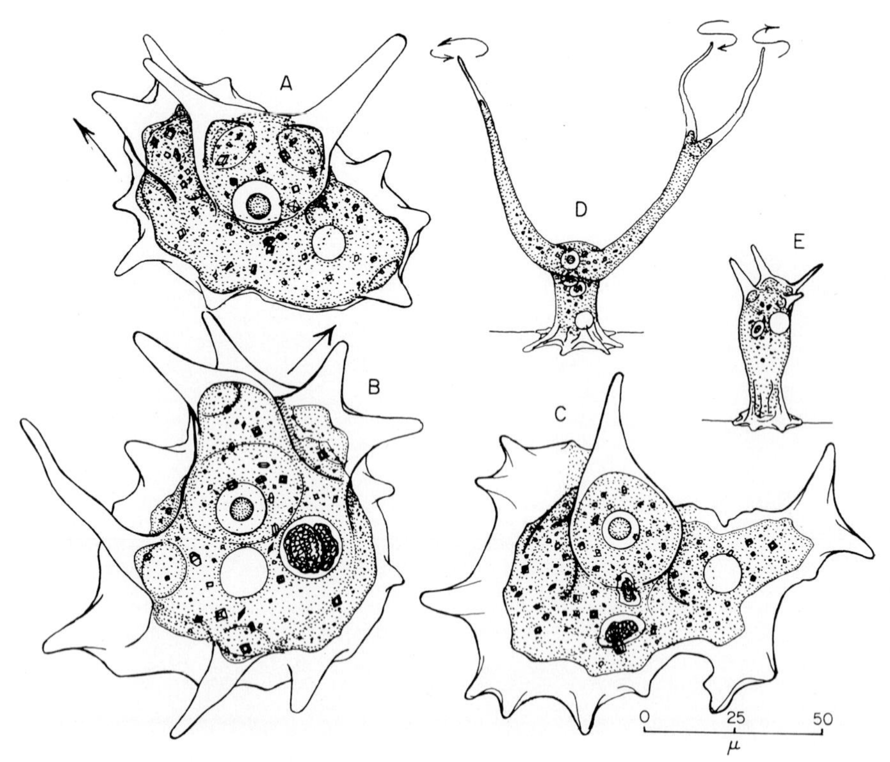

FIG. 3. Erect states of mayorellids—(A)–(C) *Mayorella bigemma*. (A) Nucleus has descended from upright pseudopods; new pseudopods at upper left in direction of arrow; (B) similar, with new pseudopods at upper right; (C) with nucleus in the upright pseudopod; no new pseudopods being formed for locomotion. (D) *Oscillosignum proboscidium* in standing form, with extended pseudopods waving as indicated by the arrows; nucleus in the upright portion; no locomotion. (E) A static, polyplike, erect state of *Mayorella clavabellans*; nucleus in upright body; no locomotion. [(D) After Bovee, 1953a.]

mannellid ameba advances eruptive waves of clear protoplasm burst forward from the granular body mass into the clear semirigid anterior border, spreading it and adding their substance to the anterior margin of it as they, in turn, appear to gelate (Fig. 11C, D, G, and H).

Hartmannellid amebae may form filose pseudopods in groups of 2 to 4 from the upper surface of the body (Ray and Hayes, 1954; Bovee,

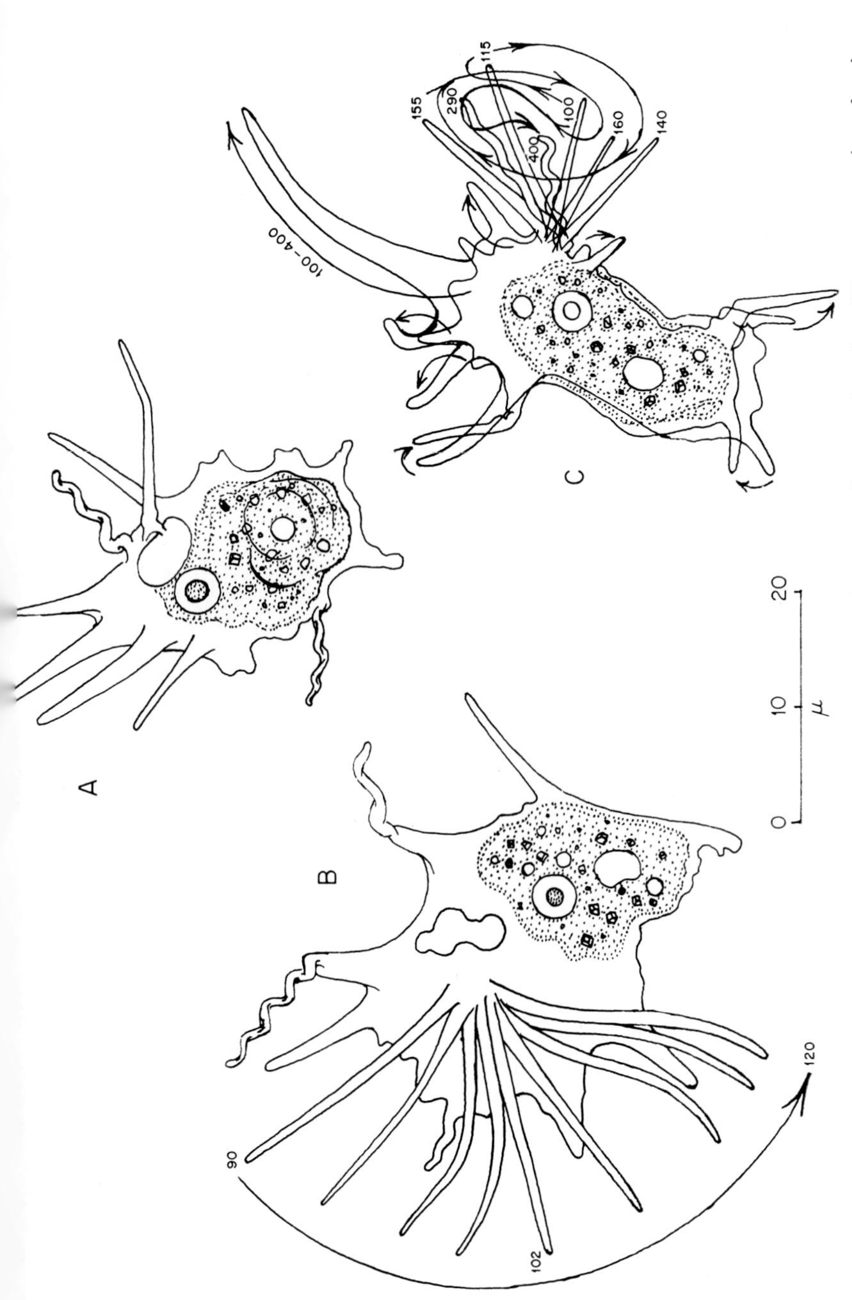

FIG. 4. *Vexillifera* sp. (A) Locomotive individual with three waving, retracting pseudopods; (B) same organism showing movement of a pseudopod swung, in 10 sec, from base just prior to retraction; plotted from 30 frames of motion picture film, photographed at 7 frames/sec; (C) movements plotted from 300 frames of motion picture film of a new pseudopod being extended and an old one being waved and retracted.

1963a). These aid in feeding. Bacteria are dragged down and engulfed (Fig. 11B). (See p. 205.)

In soft agar, hartmannellid amebae burrow (Bovee, 1963a). *Acanthamoeba castellani* then becomes elongated. It pushes new eruptive waves ahead into the agar, adding to the anterior length of the body and to the length of the tunnel (Fig. 12C–F). The body is pressed against and adheres to the wall of the tunnel (Fig. 12C), except along the terminal

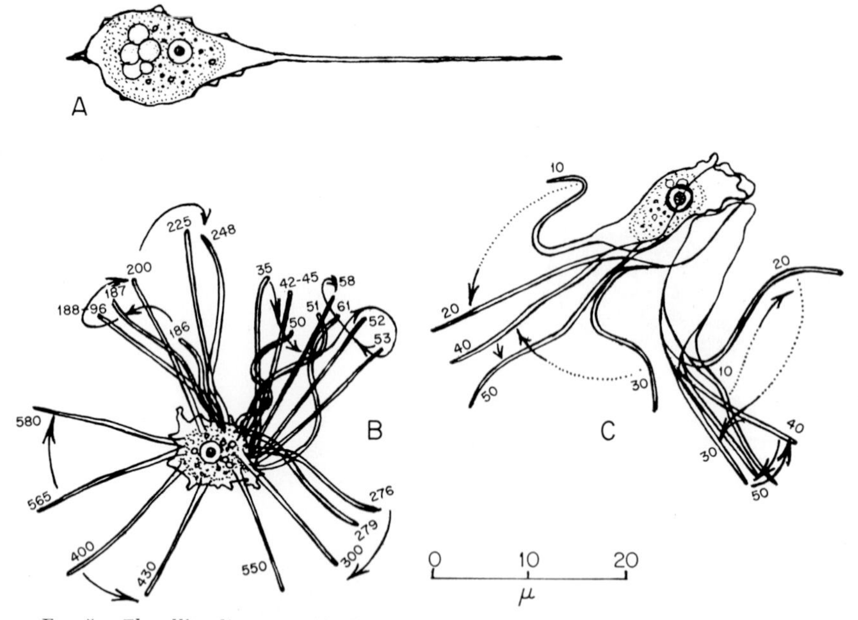

FIG. 5. *Flagellipodium* sp. (A) Locomotive stage with the "flagellipod" rigidly anterior in the direction of locomotion; (B) nonlocomotive organism, showing various positions and movements of the active flagellipod, plotted from 600 frames of motion picture film photographed at 17 frames/sec; (C) "casting" movements of the flagellipod plotted from two separate sequences of 50 frames of motion picture film.

one-third of the body, which is wrinkled and seems to be forcibly contractile (Fig. 12C). Presumably they burrow this way in moist soil where they live. This burrowing form of *A. castellani* resembles the form taken by *Amoeba proteus* in a capillary tube (Goldacre, 1961a) (Fig. 12G). (See p. 206.)

Ameboid stages of some mastigamebae are like the hartmannellid amebae, but they also form pear-shaped biflagellate to tetraflagellate stages (Bovee, 1959; Willmer, 1961) (Fig. 13A–D). Two examples are *Naegleria gruberi* and *Trimastigamoeba philippinensis*. (See p. 207.)

Thick-pellicled thecamebae have a dense, clear ectoplasm. Semiparallel ridges appear on the dorsal surface of the moving ameba (Fig. 14B

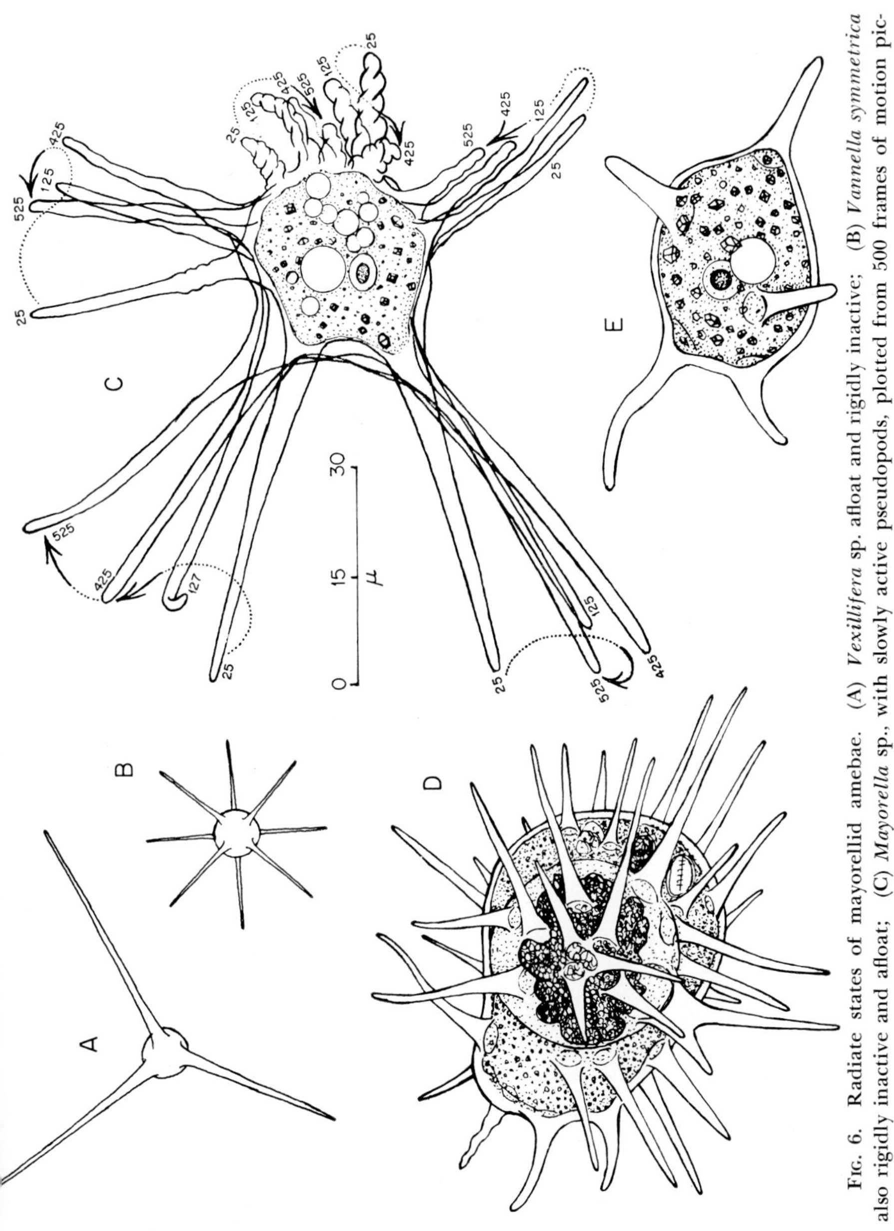

FIG. 6. Radiate states of mayorellid amebae. (A) *Vexillifera* sp. afloat and rigidly inactive; (B) *Vannella symmetrica* also rigidly inactive and afloat; (C) *Mayorella* sp., with slowly active pseudopods, plotted from 500 frames of motion picture film photographed at 14 frames/sec; arrows indicate movements of pseudopodial tips; (D) *Dinamoeba mirabilis*, afloat, pseudopods *very* slowly active (note *large* food vacuole); (E) *Mayorella* sp. rigidly afloat.

and C). Abé (1961, 1962) reports that the ridges are piers of semisolid protoplasm. The moving protoplasm between these carries along inclusions. The contractile vacuole is pressed against these ridges and deformed during locomotion as is also the nucleus (Abé, 1961, 1962; Leidy, 1879; Penard, 1902). Similar ridges are formed by the larger species *Thecamoeba papyracea* and *Thecamoeba sphaeronucleolus,* as well as by *Thecamoeba striata* (Fig. 14A–C). The rolling movement described by

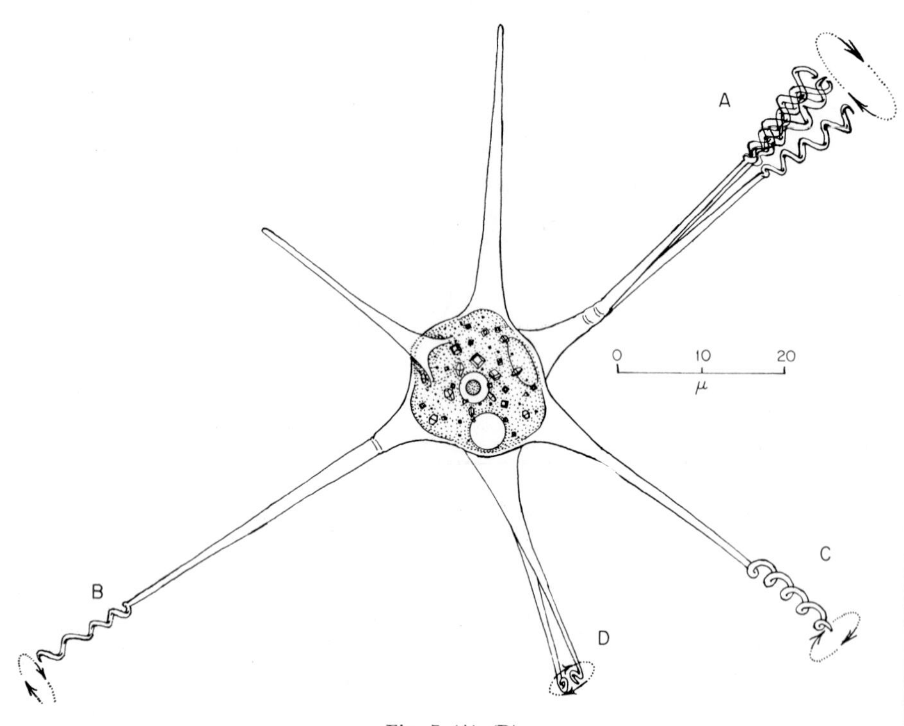

Fig. 7 (A)–(D).

Jennings (1904) does not seem to occur (Penard, 1905; Schaeffer, 1920; Goldacre, 1961a). (See p. 208.)

The ameboid stages of some mastigamebae resemble those of mayorellid amebae (Fig. 15B); that of *Mastigamoeba aspera* is somewhat like *Dinamoeba mirabilis* (Fig. 15C); but they are clearly distinct species. The likeness, however, led the Swiss protozoologist, Penard (1936), to believe that they were the same organism. As an ameboflagellate, *M. aspera* crawls with the flagellum ahead of it, snapping the tip in a hook-like stroke similar to that of the euglenid flagellate *Peranema trichophorum* (Fig. 15B), meanwhile forming and retracting pseudopods. (See p. 209.)

Mastigamoeba setosa is enclosed in a thick plasmalemma studded with chitinoid spines (Fig. 16B). Within this jacket the ameboid movement is like that of pelomyxid amebae. Often, at the same time, its long anterior flagellum is extended and beats in a pattern similar to that of *Mastigamoeba aspera* (Fig. 16A–C). (See p. 210.)

Some of the mastigamebae also beat their flagella in sine waves (Bovee *et al.*, 1963) like those of true flagellates (Jahn *et al.*, 1963; Jahn *et al.*, 1962); and the sine waves may run from base to tip during the swimming

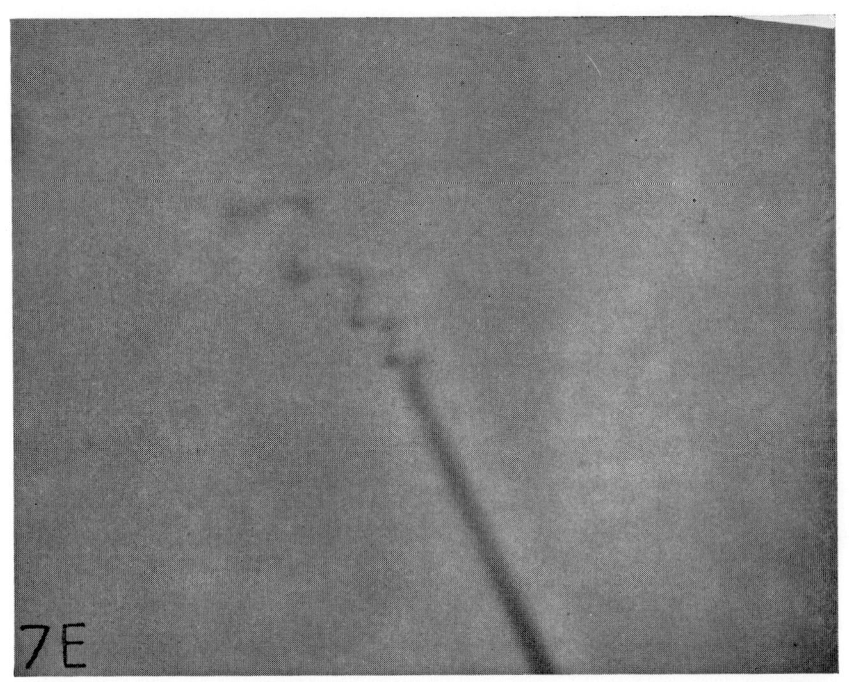

FIG. 7. A radiate, floating mayorellid with six pseudopods (see p. 198). (A)–(D) Oscillating pseudopods with varying degrees of terminal coiling; each pseudopod described a circle at the tip, as indicated, counterclockwise, in 25 frames of motion picture film photographed at 12 frames/sec. (E) Photograph of coiled pseudopodial tip (above).

beat (Bovee *et al.*, 1963) (Fig. 16D–K). *Mastigamoeba steinii* also uses its flagellum in this way (Fig. 17). (See pp. 210–212.)

The mastigamebae may produce a flagellum in 0.5 sec or less—and retract it as swiftly; and by their movements these appear to be true flagella. I do not know of any studies which have been made by electron microscopy of their fine structure.

To assume a different *basic* mechanism for each bizarre protoplasmic movement by these amebae is unnecessary. There is much more reason to consider the similarities, and then try to account for the differences.

It was popular a century ago to consider protoplasmic flow to be basically similar in character wherever found—in plants, ameba, or muscle. Dujardin (1835, 1838), Ecker (1849), Engelmann (1879), and others held such views. Later, as chemistry and physics developed, this postulation became naive. Surface-tension theories bloomed late in the nineteenth century (Berthold, 1886; Bütschli, 1882, 1892; Rhumbler, 1898), waxed, and waned in the first quarter of the twentieth (Furth,

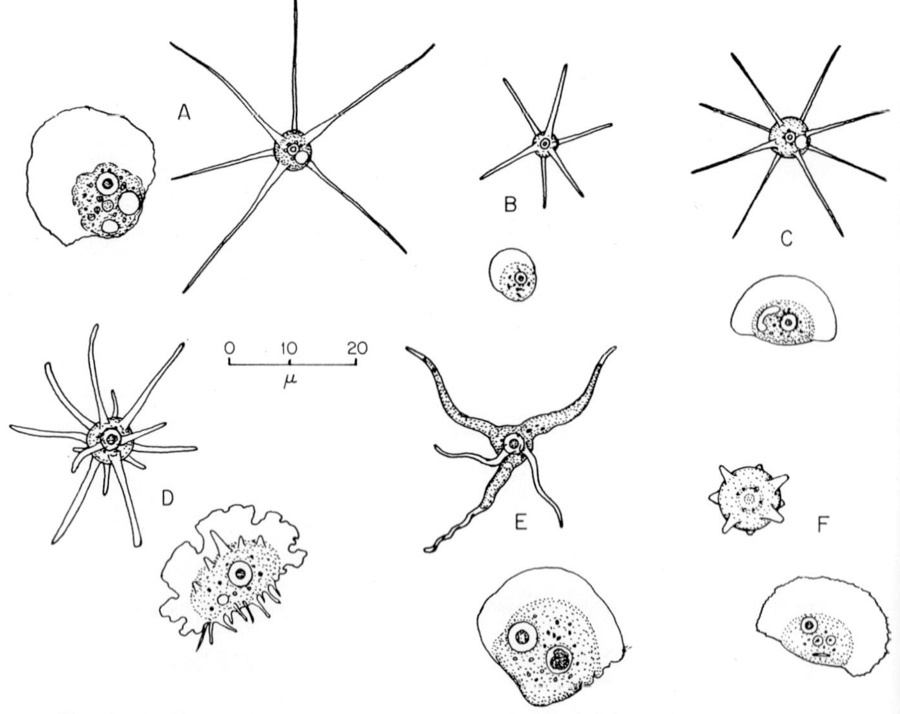

Fig. 8. Radiate and locomotive stages of flabellulid amebae. (A) *Vannella miroides*; (B) *Rugipes vivax*; (C) *Vannella symmetrica*; (D) *Flabellula citata*; (E) *Vannella mira*; (F) *Vannella sensilis*. [(B) and (D) After Schaeffer, 1926a; (C) after Bovee, 1953c.]

1922; Schafer, 1910; Tiegs, 1928a, b). An early denial of the surface-tension concepts was Jennings' (1904) "rolling sac" theory, itself promptly denied (Penard, 1905) and countered with a "walking" concept (Dellinger, 1906).

Studies of colloids and their properties led to syneretic gel-sol concepts of ameboid movement before 1925 (Hyman, 1917; Pantin, 1923). These were promptly developed into a contractile-hydraulic gel-sol explanation of the events of ameboid movement by Mast (1926), one which is still accepted, with modification, by many. Meanwhile, and since, many physical and chemical factors have been employed experimentally to test and postulate bases for the visible events of ameboid locomotion (reviews De Bruyn, 1947; Heilbrunn, 1958; Allen, 1961b).

Szent-Györgyi (1949, 1953), and others especially studying muscle (review, Weber, 1958), and many others studying contractility of cells in the past 25 years (Child, 1961; Hoffmann-Berling, 1954, 1960; Kriszat, 1949, 1954; Loewy, 1952; Nakajima, 1956; Ts'o *et al.*, 1956, 1957) have shown the presence of and importance of actomyosin-like proteins and phosphate energy reservoirs in protoplasmic movements. Still, as late as 1952, it may have been premature to explain ameboid movement, as I then did (Bovee, 1952) via an actomyosin–adenosine triphosphate (ATP) system. At least my paper was generally ignored, but it was cited by Noland (1957) as an attractive hypothesis, still then only an assumption.

More recent information shows plainly that it is no longer premature to assume the generality of such a system; others have already done so (Kamiya, 1959, 1960; Landau, 1959; Oosawa, 1963).

Landau (1959), dealing with ameboid movement, thoroughly reviewed and discussed the literature to that date, and Kamiya has done likewise for protoplasmic movements of mycetozoans and plant cells. Landau gives a sound assumption for an actomyosin-like complex energized by ATP in amebae, having a spatial gradient, and responsible for ameboid movement, and assuming a dynamic equilibrium within the ameba. Kamiya (1959) says this in his review:

"It seems after all that the predominant feature of protoplasmic streaming is based upon the mechanochemical system represented by myosin-like proteins which have by themselves an ATP-ase activity. It looks as if nature had a predilection for using actomyosin-like proteins and ATP as a mechanochemical system for carrying out mechanical work in many cells beside muscle . . . it may be tentatively assumed without denying other possibilities in special cases, that this is the basis of the mechanism of movement widely used in protoplasm in general. It is through it that there are formed different dynamic structures which determine the various types of protoplasmic movement and streaming."

Abé (1961, 1962) also notes the dynamic interaction of morphology and physiology of amebae. Others have stressed the dynamic morphology of amebae in locomotion in relation to their taxonomy and the identity of their species (Bovee, 1953c, 1960a, b; Bovee and Jahn, 1960; Jepps, 1956; Leidy, 1879; Penard, 1902; Schaeffer, 1920, 1926a). Many who have studied ameboid movement also remark on the dynamic interdependence of ameboid form and function (Bovee, 1960b; De Bruyn, 1947; Goldacre,

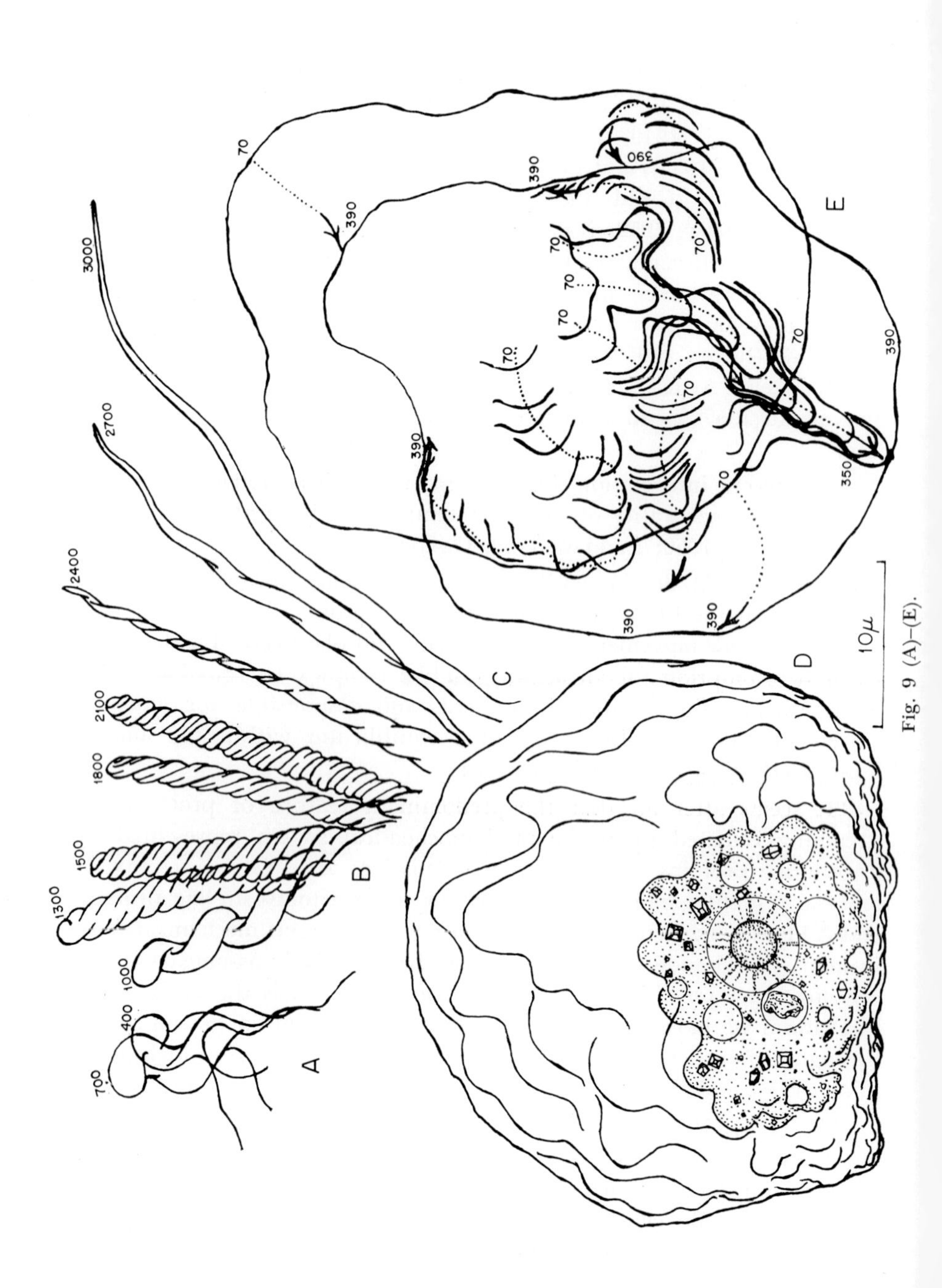

A
700
400
B
1000
1300
1500
1800
2100
C
2400
2700
3000
D
E
70
390
350
10μ

Fig. 9 (A)–(E).

1961a; Griffin, 1961, 1962; Jahn *et al.*, 1960; Jahn and Rinaldi, 1959; Rinaldi and Jahn, 1963; Kamiya, 1959; Noland, 1957).

This dynamicity is a challenge for study. The basic actomyosin–ATP system seems simple enough, but as Kamiya (1959) indicates, there is likelihood that it is not a single stereotyped machine. It seems, instead,

FIG. 9. *Vannella miroides* (see p. 202). (A)–(C) The development of a radiate, waving pseudopod by an actively locomotive individual, plotted from 3000 frames of motion picture film, photographed at 14 frames/sec. (D) The fully locomotive fan-shaped individual. (E) The movements of protoplasmic currents in the anterior border of the locomotive individual; tonguelike flow directly to the anterior border indicates the path of flow of sol from the body to the anterior border; other arcs in sequence as indicated by arrows mark successive positions (each 35 frames) of gel islands formed previously, showing their migrations laterally and back to the body mass; plotted from a sequence of 400 frames of motion picture film photographed at 8 frames/sec. (F) A photograph showing the central sol flow and laterally moving gel islands in the fan-shaped border.

to have many variations. Szent-Györgyi (1960) contends that the machinery of life is based on varied and statistically improbable reactions, because they are the more easily controlled. Self-regulation (i.e., the adaptation of naturalistic terms) is of course the cornerstone of survival.

One deficiency of the theories concerning ameboid movement is that they do not agree on the site for contraction (Weiss, 1961; Rinaldi and

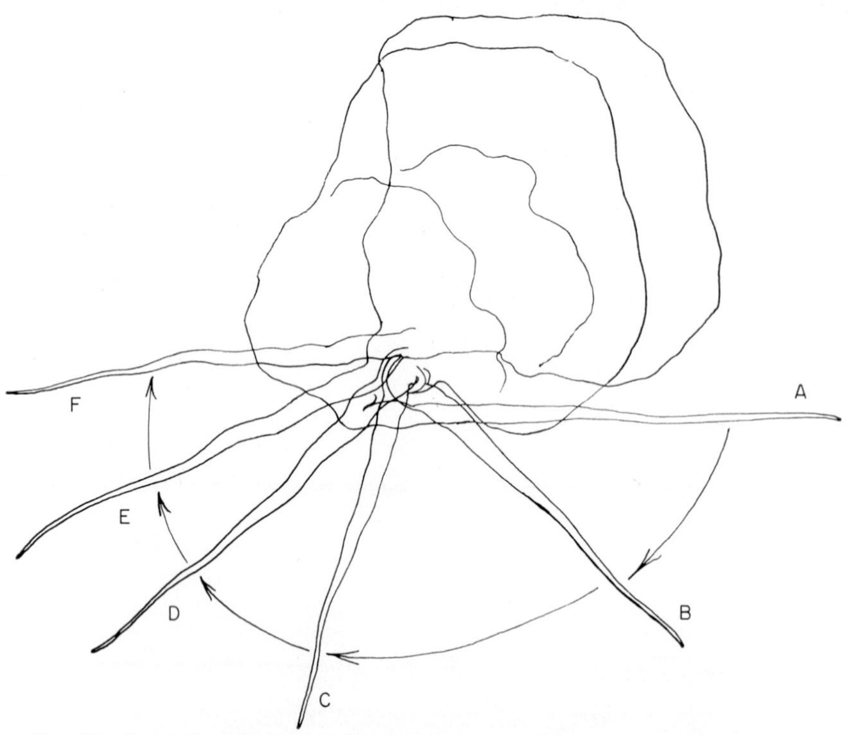

Fig. 10. *Vannella miroides.* Swinging movement of a radiating pseudopod just before its retraction. Movement from positions (A) through (F) occurred in 20 sec. Plotted from 175 frames of motion picture film photographed at 8 frames/sec.

Jahn, 1963). Another difficulty is that they do not take into consideration, and therefore question, whether polymers of actomyosin simply contract, fold, or coil (Goldacre, 1952; Goldacre and Lorch, 1950) or whether as they do that, they shear and slide past one another as well (Allen 1961a; Huxley and Hanson, 1955; Jahn and Rinaldi, 1959), perhaps with a ratchet action (Hayashi, 1961; Noland, 1957). According to one current theory muscle now slides and shears instead of contracting (Huxley and Hanson, 1955, 1960; Weber, 1958); and another recent theory asserts that foraminiferans cast forth self-sliding filaments that glide out from and back to the body along themselves (Jahn and Rinaldi, 1959).

I do not disparage these ideas. I cite them to note some of the intricacy of the problems the organisms have solved. The biomechanochemical systems involved are considerably mediated by the amebae, however, a concept that I and others have stressed elsewhere (Bovee, 1959, 1960a; Kepner and Taliaferro, 1913; Schaeffer, 1917, 1920, 1926a,b; Seifriz,

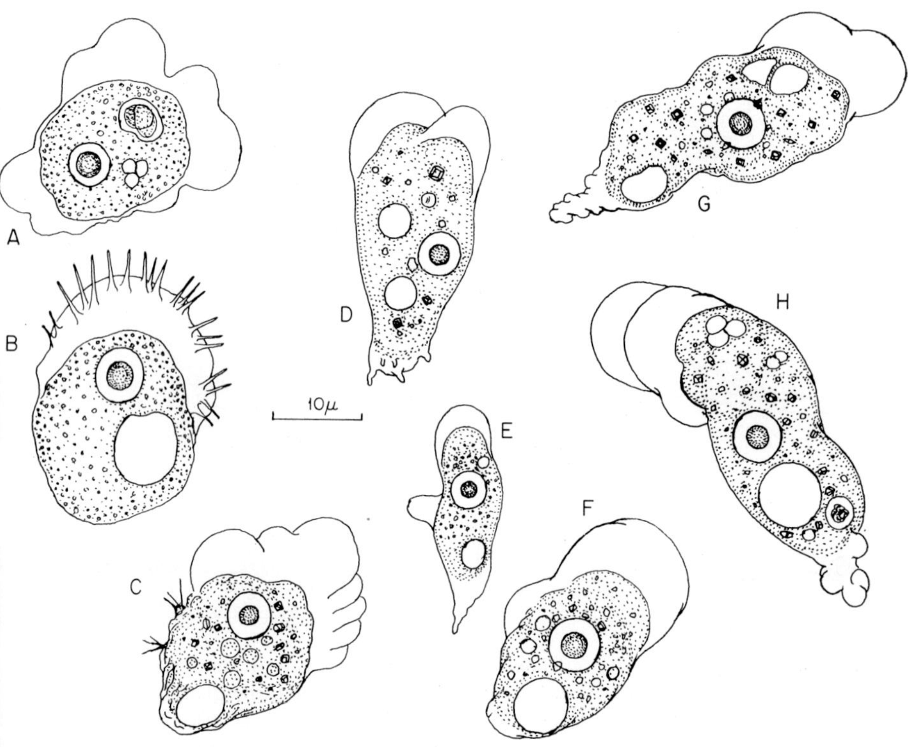

FIG. 11. Various hartmannellid amebae. (A) *Acanthamoeba castellani* (after Volkonsky, 1931); (B) the same, with filose pseudopods; (C) *Acanthamoeba* sp., from original motion picture film; (D) *Hartmannella* sp. (after Chatton and Brodsky, 1909); (E) *Vahlkampfia* sp. (after Wherry, 1913); (F) *Hartmannella* sp., from an original motion picture film; (G) and (H) vahlkampfian hartmannellids from an original motion picture film.

1939, 1942); and the ways in which an ameba uses its machinery, in Jepps' words (1956) "have to fit the circumstances."

However, any ameba adapts and controls its mechanisms only within the limits of its capabilities. Even if its physical mechanisms are highly improbable according to hydrodynamics and thermodynamics, they still must conform to the rules thereof. The difficult, but more durable, types of machinery are perhaps the more often statistically and evolutionarily available. The spontaneously functional, physically probable types are fewer and less controllable.

To the point of triteness, textbooks of general biology state that the nucleus of an ameba or any other cell is vital, and directs its growth, form, and function. Yet only rarely is the nucleus mentioned in theories of ameboid movement, or in discussions of pseudopodial form and function. It appears that the machine is assumed to operate efficiently with-

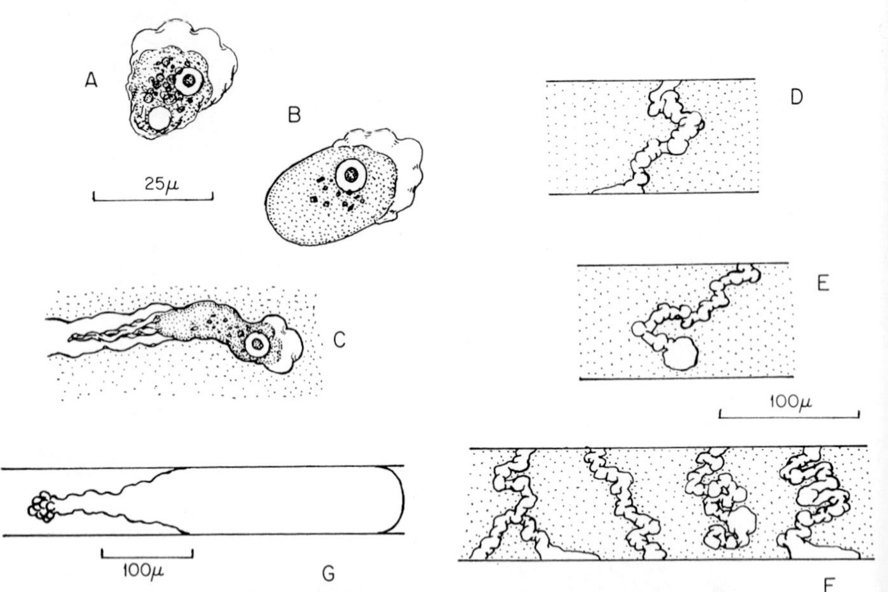

FIG. 12. *Acanthamoeba castellani.* (A) Active trophozoite in fluid overlay on agar; (B) active trophozoite, flattened, in motion between agar and the glass surface of petri dish; (C) active trophozoite burrowing in agar, body pressed against the tunnel wall except for contracting terminal "tail," eruptive waves pushed forward into agar to enlarge the tunnel. (D)–(F) Various tunnels constructed by acanthamebae in agar. The incomplete tunnels were formed by amebae which encysted at the enlarged terminus of the tunnel within the agar. The tunnels running from top to bottom of the agar were made by amebae which moved off between the agar and the glass of the petri dish after completing the tunnel. (After Bovee, 1963a.)

out guidance. I believe that this is a sin of omission. Even though ameboid cytoplasm, as has been shown (Allen, 1959, 1961c; Allen *et al.*, 1960), will continue to run like an "engine" when disengaged and dissociated from the ameba, I am inclined to regard such a circumstance as analogous to the idling motor of the automobile, which provides energy and force to drive the machine only when it is engaged with the other functional parts of the mechanism, at the direction of the operator.

There is, of course, abundant evidence of nuclear involvement in ameboid cytoplasmic form and function (review, Brachet, 1959, 1961; Brachet, 1955; Hirschfield, 1959). All genetics is based on the idea of nuclear control of cytoplasmic function, and there is no need to belabor the genetic point of view here.

Grafting experiments on irradiated and nonirradiated amebae (Daniels, 1952, 1955, 1959) and nuclear transplantations are especially en-

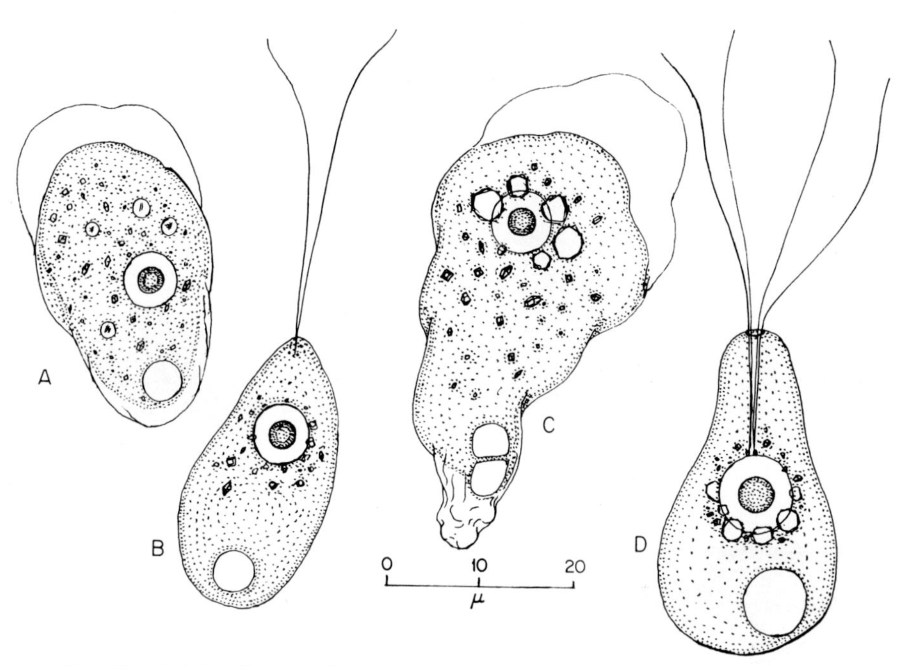

FIG. 13. Polyflagellate amebae. (A) Ameboid state of *Naegleria gruberi*; (B) biflagellate stage of *N. gruberi*; (C) ameboid stage of *Trimastigamoeba philippinensis*; (D) tetraflagellate stage of *T. philippinensis*. [(A) and (B) After Singh, 1952; (C) and (D) after Bovee, 1959.]

lightening as to the closeness of relationship between cytoplasmic and nuclear forms and functions (reviews, Danielli, 1959; Brachet, 1959, 1961).

Daniels shows (1952, 1955, 1959; Daniels and Vogel, 1959) that the organization of ameboid cytoplasm, once severely damaged by radiation, cannot be repaired by a normal nucleus, *unless undamaged cytoplasm is also present* to provide, presumably, some undamaged cytoplasmic templates to be copied. He has also shown that interspecific infusion of cytoplasm is quickly lethal to the recipient, because of probable specific differences in protein polymers (Daniels, 1962).

Danielli and his co-workers show that the cytoplasmic morphology of *Amoeba proteus* may be affected by the nucleus of *Amoeba discoides*. The hybrid is morphologically and functionally intermediate. However, the cytoplasm of the recipient *A. proteus* never becomes entirely like that of the *A. discoides* from which the donor nucleus came, and the *A. discoides* nucleus, in size at least, becomes more nearly that which it should be were it normally an *A. proteus* nucleus (Danielli, 1955; Lorch and Danielli, 1950, 1953). Also it has been observed that there

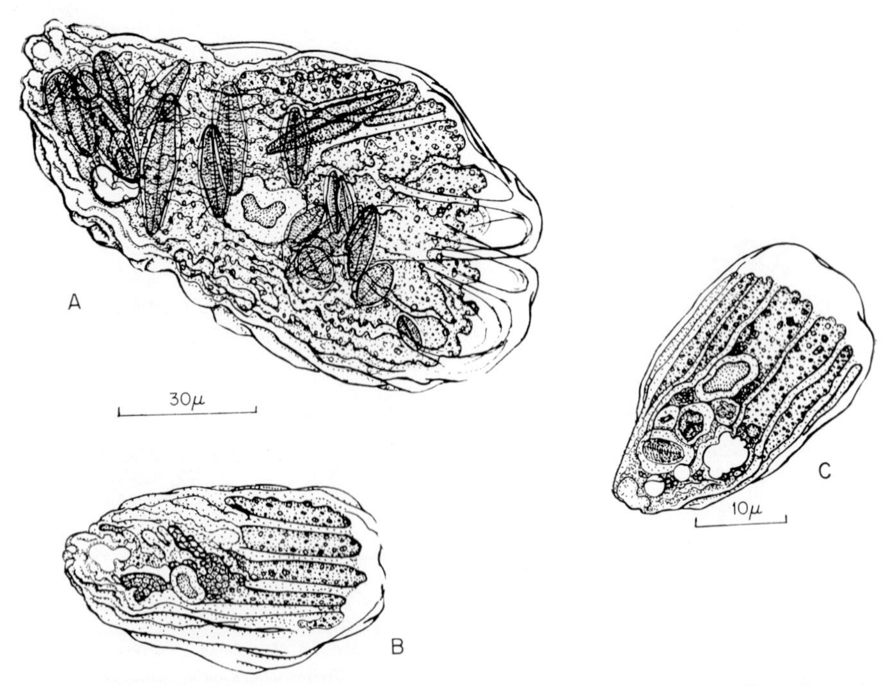

FIG. 14. Thecamebid amebae. (A) *Thecamoeba papyracea*—note distortion of the nucleus under pressure against gel islands; (B) *Thecamoeba sphaeronucleolus*—note parallel gel ridges in anterior portion of the body and the distortion of both nucleus and contractile vacuole by pressure against the gel and the packed inclusions; (C) *Thecamoeba striata,* nucleus and contractile vacuole distorted, and parallel gel ridges prominent in the anterior portion.

is a critical-phase relationship between growth of the nucleus and cytoplasm as evidenced in cross-nuclear transplants from old and young amebae (Comandon and deFonbrun, 1939, 1942), and that certain biochemical capabilities are lost by starved cytoplasm and nucleus (Muggleton and Danielli, 1958). Cross-nuclear transplants have failed completely in *A. proteus* and *A. dubia,* which are presumably close taxonomic relatives (Comandon and deFonbrun, 1939; Rudzinska and Chambers, 1951; Lorch and Danielli, 1953). There must be in ameboid cytoplasm some durable molecules which only a compatible nucleus can modify, and which no nucleus alone can replace.

A long-ignored paper by Heathman (1932) shows that amebae which develop morphologically distinct pseudopods and patterns of ameboid movement, are also antigenically distinct. In an order of arrangement showing both similarities and differences, her studies show that, anti-

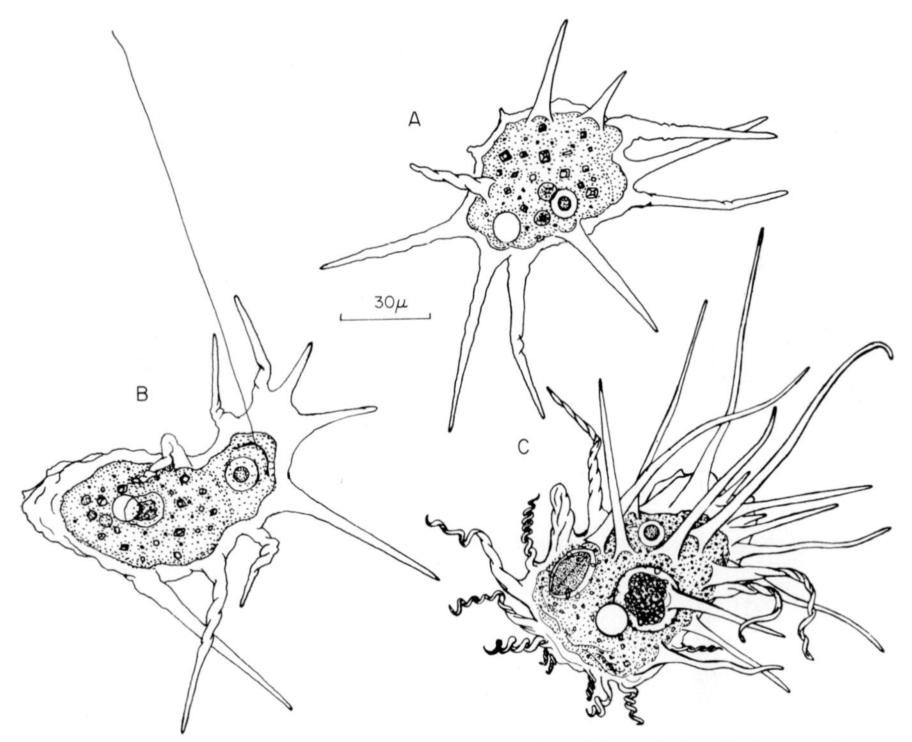

FIG. 15. (A) Ameboid stages of *Mastigamoeba aspera*; (B) ameboflagellate stage of the same organism; (C) *Dinamoeba mirabilis,* showing resemblance to, but also difference from *M. aspera*.

genically: (1) *Amoeba proteus* and *Amoeba dubia* are distinct species, but closely related; (2) *Mayorella conipes* and *Mayorella bigemma* are distinct amebae in a second group intermediate between the *Amoeba* spp. and a third group (3) composed of two more separate species, *Flabellula citata* and *Flabellula mira*. The third group is clearly distinct from, but closer than groups (1) and (2) are, to group (4) composed of varieties of the parasitic *Entamoeba histolytica* (Table I). These relations exactly parallel the morphological and taxonomic distinctions proposed for those amebae by Schaeffer (1926a). Heathman (1932) suggested that there are in ame-

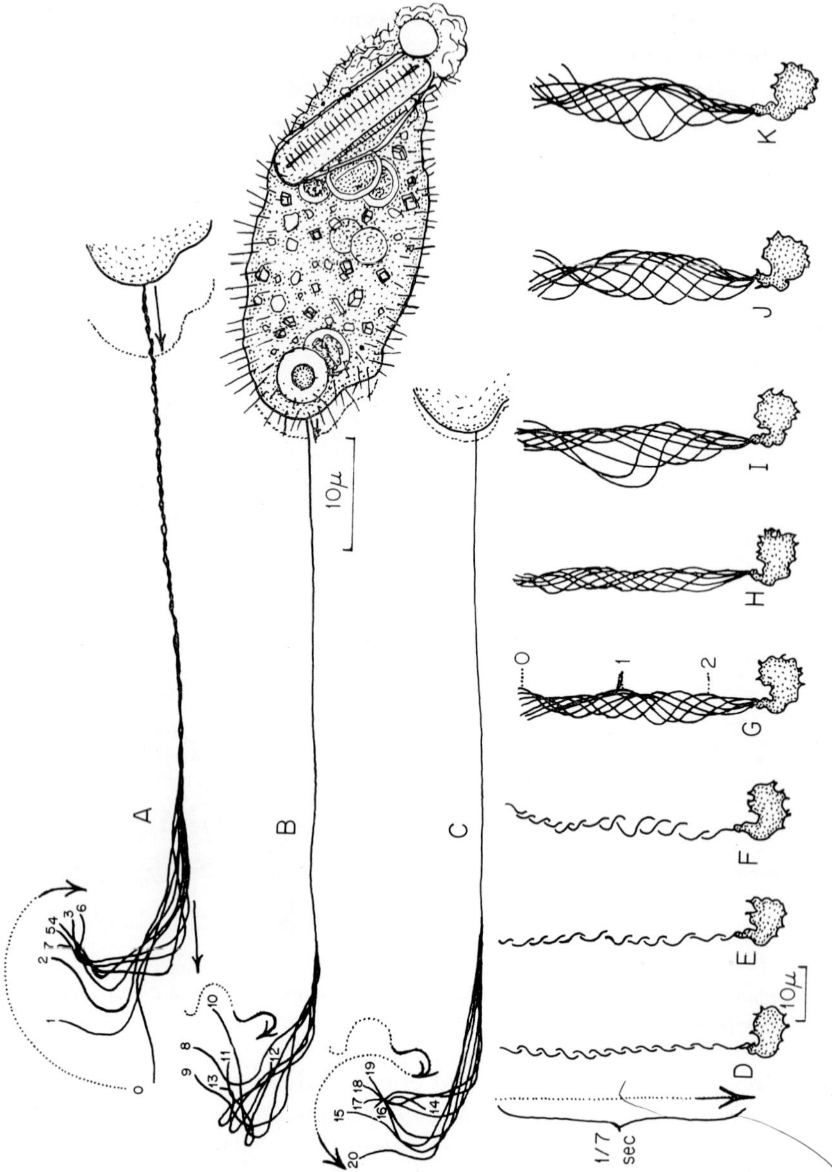

FIG. 16. (A)–(C) *Mastigamoeba setosa.* (A) Shows the hooking power stroke and anterior-posterior wave of the flagellum, in 6 successive frames of motion picture film; (B) the whole organism, showing pelomyxid body form and the pellicular spicules, with part of the recovery stroke of the flagellum plotted from the next 6 frames of motion picture film; (C) the remainder of the recovery stroke in the succeeding 8 frames of film; photographed at 8 frames/sec. (D)–(K) Flagellar movements of *Mastigamoeba* sp.; (D)–(F) three successive sine waves which progressed from tip to base

bae one or more common antigenic properties, perhaps lipoid in nature; and she further suggested that the specific differences are due to specific differences in the protein–carbohydrate complexes which vary among the amebae.

Apparently, then, even *visibly slight differences* between amebae may reflect significant physiological differences in organization which establish the differing organisms as distinct species. Therefore, even slight, but constant visible differences in the morphological patterns of move-

TABLE I

ANTIGENTIC SIMILARITIES AND DIFFERENCES BETWEEN SPECIES[a,b,c]

Immune serum (0.05 ml)	*A. proteus*	*A. dubia*	*M. bigemma*	*M. conipes*	*F. mira*	*F. citata*	*E. histolytica*
A. proteus	4	3	2	2	2	0	0
F. mira	2	0	1	1	4	2	Trace
F. citata	1	0	3	3	3	4	2
E. histolytica	0	X	X	X	Trace	2	4

[a] After Heathman, 1932.

[b] Amount of material used: 0.1 ml antigen of ameba plus 0.5 ml guinea pig complement.

[c] Key: 4, very strong reaction; 3, strong reaction; 2, moderate reaction; 1, weak reaction; trace, very slight reaction; X, not tested.

ment and locomotion are not only valid, but are the most valuable means by which amebae may be quickly and accurately identified. As Schaeffer (1926a) succinctly puts it: "In amebas the shape is dynamic; that is, movement is a function of shape." The converse is also true; and both are ultimately the products of inherent cytoplasmic-nuclear organizations and interactions.

Landau (1959) states that in the presumed actomyosin–ATP system for ameboid motion "the role of the nucleus is multifold." Goldacre (1961a) postulates a self-regulatory feedback system in the cytoplasm of the locomoting *Amoeba proteus.* Thus, he says, "The tail has a means of 'knowing' whether there are other contracting regions in the cell or not." He states elsewhere (Goldacre, 1958a) that the nucleus is intricately involved in the locomotive feedback system, and in other systems which function in the cell, and that the surface membranes assist the feedback operation (Goldacre, 1958b, 1961b).

The feedback systems not only exist from front to rear of the ameba

---

of the flagellum, as flagellar movement began. Plotted from 75 frames (each second frame) of motion picture film photographed at 300 frames/sec with a missile-tracking camera; (H)–(K) sine-wave paths in progress along the flagellum (each fourth frame) from a succeeding sequence of 150 frames of the same film.

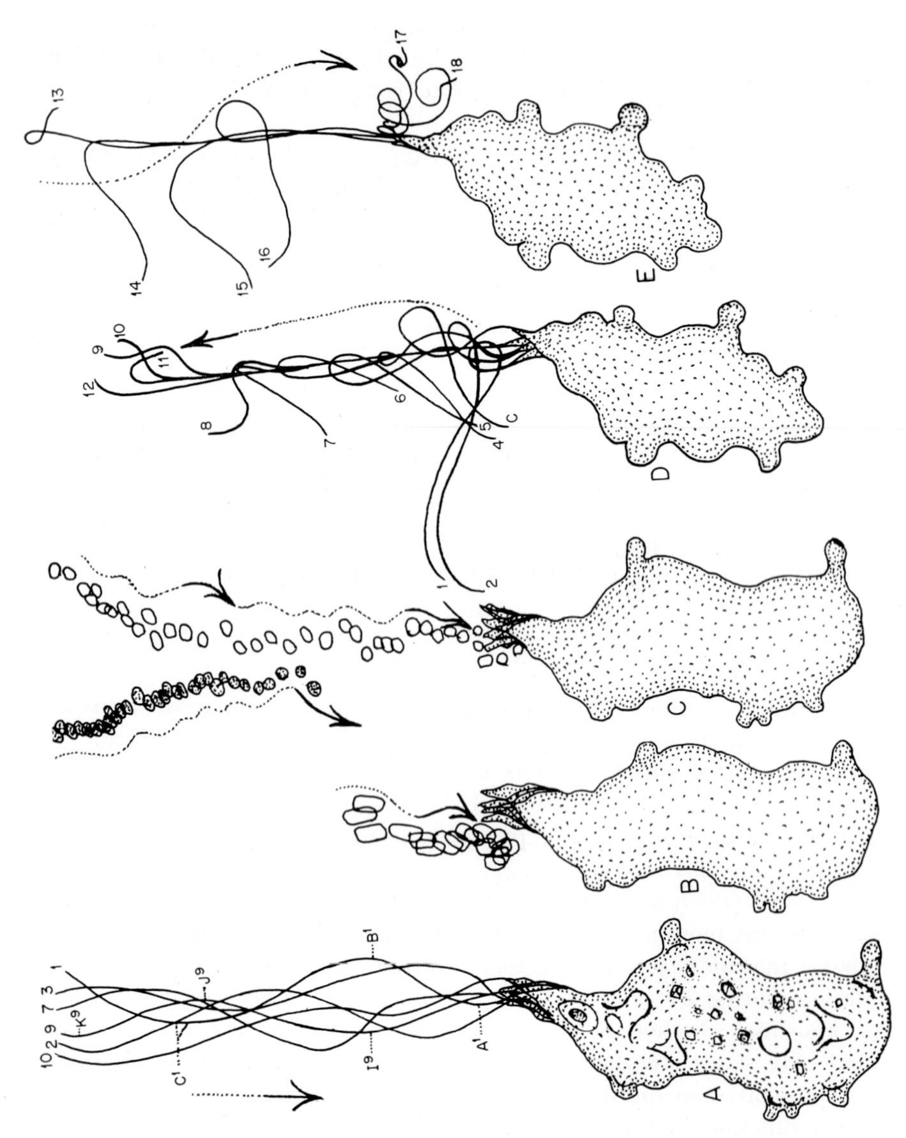

FIG. 17. *Mastigamoeba steinii.* (A) Sine waves in progress from tip to base of the flagellum from 10 successive frames of motion picture film photographed at 17 frames/sec; (B) and (C) paths of granules which moved along the beating flagellum from tip to base, 80 frames, positions of granules shown (each second frame). Arrows indicate sine-wave paths of granules. (D) and (E) Casting and recovery strokes of the flagellum employed in orienting the anterior end of the body in a new direction; eighteen successive frames.

but oscillate from side-to-side and probably up and down as well. The cross-feed oscillation is evident in the formation of pseudopods during continuous locomotion. These arise in many amebae in an alternating pattern at left and then right anterolateral positions (Fig. 18), resulting in a sinusoidal pathway on a flat surface, as repeatedly mentioned by Schaeffer (1916, 1917, 1926a, b, 1931). The dorsoventral feedback relationship is apparent from the fact that new pseudopods form dorsally and old ones are diverted to ventral positions or posterior ones.

These feedbacks may involve even the characteristic structure of the specific carbohydrate–protein complexes of the internal machinery. According to Katchalsky (1963), it may be demonstrated both mathematically and experimentally that certain irreversible, but cyclic, routes for energy exist via these large bio-organic polymers, amounting to rudimentary "memory" patterns within the cell. Also, Chance (1963) suggests some such differences in the structures concerned with energy transfers of the cell, particularly the mitochondrial surfaces and the aggregates attached to them; he indicates that they, too, participate in the regulatory feedback mechanisms.

We might, then, presume that the large molecules, probably proteins which are responsible for the specific antigenic differences among amebae, are also among the molecules that operate differentially and specifically in the feedback and information systems of the various amebae. They and others specific to the locomotor mechanism interact and develop the locomotor morphology of the ameba. Therefore, we should expect each kind of ameba to have some proteins common to all cells, with perhaps minor variation in them, and others more specifically its own. These should interact to produce certain visible features of locomotion and morphology common to all amebae, and still others characteristic only of a certain limited group of amebae, or of even a particular species. These visible features do exist, and they are of both taxonomic and evolutionary import.

An ameba is plainly in control of its activities, and is a self-competent organism. It exhibits an awareness of and responds to its environment with normality and regularity. This has led some writers to use anthropomorphic terms in describing ameboid activities such that they suggest an incipient wisdom on the part of the ameba (Jepps, 1956; Kepner and Taliaferro, 1913; Leidy, 1879; Schaeffer, 1917).

An ameba's responses are not "frozen" into invariable and irrevocable sequences of events. It is keenly attuned, through its informational feedback mechanisms, to the nuances of its environment, and it responds in patterns which are normally and distinctly its own (Bovee, 1960a; Schaeffer, 1926a). These responses not only identify the ameba; but they

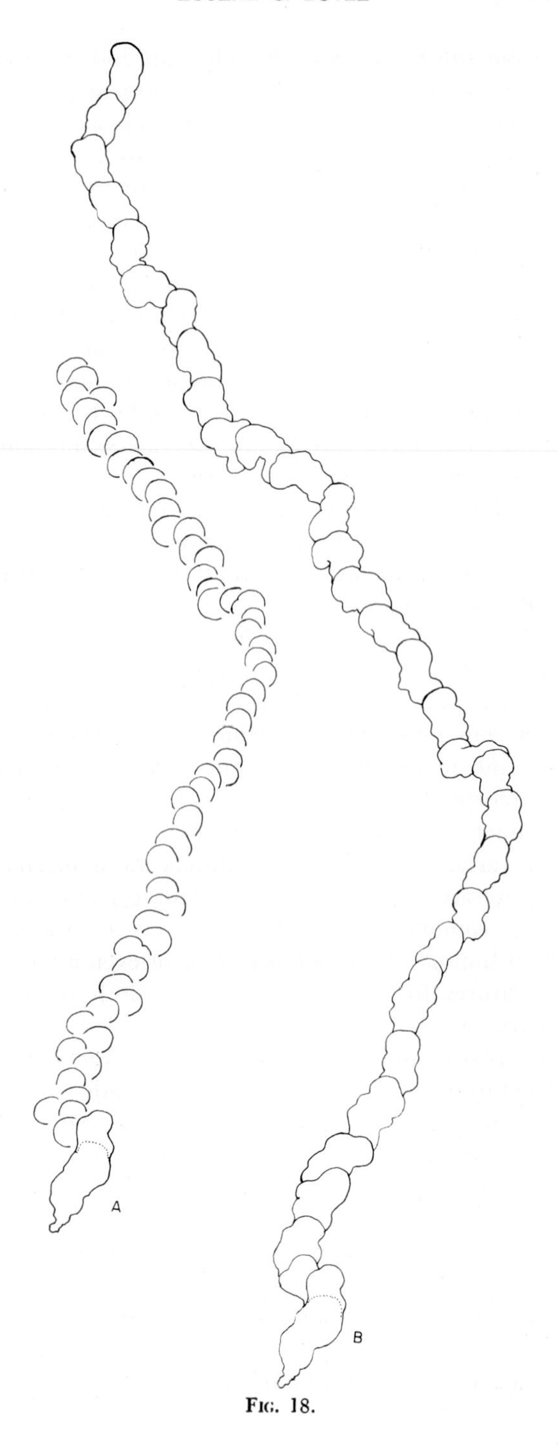

FIG. 18.

are exactly those criteria by which the experimenter determines whether or not, and how, the ameba responds to and is affected by experiment. Abnormality rarely appears in sublethal situations (Bovee, 1960a). To those the self-regulatory mechanisms of the ameba are able to adapt (Bovee, 1960b; Seifriz, 1939, 1942; Schaeffer, 1926a).

The formation of a single pseudopod is often adequate to demonstrate the individuality and normality of an ameba. Although a new pseudopod forms initially in all amebae, so far as I can tell, as a hemispherical bulge at some point on the surface, it always assumes the morphology characteristic of a pseudopod of its species, as it elongates. There is plainly an orderly flow of and incorporation of materials into it which are specifically and sequentially organized into the pseudopodial form characteristic of the species. Otherwise no two pseudopods formed by an ameba should resemble one another; but they always do.

Somehow the notion still generally persists that amebae are shapeless, ever and erratically altering their contours, and move randomly. Such a concept is obviously false. Goldacre (1961a) has called this "the sickly child of literary inbreeding" through a succession of textbooks.

I believe it is a useful business to debate the mechanisms of ameboid movement. However, it often seems to me that some theorists try assiduously to force amebae to conform to their preconceptions of mathematical and physical principles and to formulas (Buchsbaum *et al.*, 1944). Perhaps they are still bemused by this embalmed error of presumed ameboid randomness. I find it singular, if provocative, for example, to contemplate the role of surface films in ameboid movement by considering the random water currents in a flushing toilet, as one theorist suggests (Kavanau, 1962). This neither substantiates nor denies the possible role of surface films in ameboid movement. They may be, and probably are involved. I do not belittle such exercises. Good ideas often come from observations of common events. Theorizing and debate attract attention to the problem, whether or not the debated theories offer a solution. I find both the problem and the debate exciting, and I find the amebae most fascinatingly complex.

We should not, however, lose sight of the organism in the excitement of the debate. As Mazia (1956) suggests, if we are to find out what the cell does and how it does it, we must devise answerable questions, and

---

FIG. 18. Paths of locomotion of a vahlkampfian ameba. (A) Positions of new pseudopods, showing alternating and oscillating sequence resulting in a sinuous pathway of progress. Plotted from a sequence of 1200 frames of motion picture film (each twentieth frame), photographed at 8 frames/sec; (B) the same (each hundredth frame) from a sequence of 3500 frames, showing the tighter sinuous pathway superimposed on the greater sinuous track.

pose them most politely to the cell. Perhaps it will then answer. I believe we will find thereby that we are still describing protoplasmic movement as the blind men described the elephant—in portions with partial truth from each, depending on how and where we have touched the problem. We must grope further, by stages, as Noland (1957) says.

What, then, is the significance of the ameboid movements in smaller amebae? I believe it is that of theme and variation, demonstrating the great variability of a fundamental mechanism of protoplasmic movement, and the specific evolutionary segregation and survival of some of the variants. It is clear that the various amebae are distinct animals. It is evident that they may use their locomotive mechanism in a variety of ways. The elegant and seminal simplicity of the basic parts of the machine and the various bizarre applications of it by amebae are marvelous. They far exceed in inventiveness the theories purporting to explain them.

Material pertinent to this paper will be discussed in the Free Discussions of Parts III and IV.

REFERENCES

Abé, T. H. (1961). *Cytologia* **26**, 378.
Abé, T. H. (1962). *Cytologia* **27**, 111.
Allen, R. D. (1959). *Biol. Bull.* **109**, 339.
Allen, R. D. (1961a). *Exptl. Cell Res. Suppl.* **8**, 17.
Allen, R. D. (1961b). *In* "The Cell" (J. Brachet and A. E. Mirsky, eds.), Vol. II, pp. 135-216. Academic Press, New York.
Allen, R. D. (1961c). *In* "Biological Structure and Function" (T. W. Goodwin and O. Lindberg, eds.), Vol. II, pp. 549-556. Academic Press, New York.
Allen, R. D., Cooledge, J. W., and Hall, P. J. (1960). *Nature* **187**, 896.
Berthold, G. (1866). "Studien über Protoplasmamechanik." A. Felix, Leipzig, Germany.
Bingley, M. S., and Thompson, C. M. (1962). *J. Theoret. Biol.* **2**, 16.
Bovee, E. C. (1951). *Proc. Am. Soc. Protozool.* **2**, 4.
Bovee, E. C. (1952). *Proc. Iowa Acad. Sci.* **59**, 428.
Bovee, E. C. (1953a). *Trans. Am. Microscop. Soc.* **72**, 17.
Bovee, E. C. (1953b). *Trans. Am. Microscop. Soc.* **73**, 328.
Bovee, E. C. (1953c). *Proc. Iowa Acad. Sci.* **60**, 599.
Bovee, E. C. (1956). *J. Protozool.* **3**, 155.
Bovee, E. C. (1959). *J. Protozool.* **6**, 69.
Bovee, E. C. (1960a). *Am. Midland Naturalist* **63**, 257.
Bovee, E. C. (1960b). *J. Protozool.* **7**, 55.
Bovee, E. C. (1963a). *Am. Midland Naturalist* **66**, 173.
Bovee, E. C. (1963b). *J. Protozool. Suppl.* **10**, 10.
Bovee, E. C., and Jahn, T. L. (1960). *J. Protozool. Suppl.* **7**, 8.
Bovee, E. C., Jahn, T. L., Fonseca, J., and Landman, M. (1963). *Biophys. Soc. Abstr.*, p. MD2.
Brachet, J. (1955). *Biochem. Biophys. Acta.* **18**, 247.
Brachet, J. (1959). *Ann. N.Y. Acad. Sci.* **78**, 688.

Brachet, J. (1961). *In* "The Cell" (J. Brachet and A. E. Mirsky, eds.), Vol. II, pp. 771-841. Academic Press, New York.
Buchsbaum, R., Rashevsky, N., and Stanton, H. E. (1944). *Bull. Math. Biophys.* **6**, 61.
Bütschli, O. (1882). *In* "Klassung und Ordnung des Thier-Reichs" (H. G. Bronn, ed.), Vol. I(1), pp. 1-1097. Winter, Leipzig, Germany.
Bütschli, O. (1892). "Untersuchungen über mikroskopische Schaume und das Protoplasma." Leipzig, Germany.
Chance, B. (1963). In "Control Mechanisms in Biological Systems," *Symp. Biophys. Soc.* In preparation.
Chatton, E., and Brodsky, A. (1909). *Arch. Protistenk.* **17**, 1.
Child, F. M. (1961). *Exptl. Cell Res. Suppl.* **8**, 47.
Comandon, J., and deFonbrun, P. (1939). *Compt. Rend. Soc. Biol.* **130**, 740.
Comandon, J., and deFonbrun, P. (1942). *Compt. Rend. Soc. Biol.* **136**, 763.
Danielli, J. F. (1955). *Exptl. Cell Res. Suppl.* **3**, 98.
Danielli, J. F. (1959). *Ann. N.Y. Acad. Sci.* **78**, 675.
Daniels, E. W. (1952). *J. Exptl. Zool.* **120**, 525.
Daniels, E. W. (1955). *J. Exptl. Zool.* **130**, 183.
Daniels, E. W. (1959). *Ann. N.Y. Acad. Sci.* **78**, 662.
Daniels, E. W. (1962). *J. Protozool.* **9**, 183.
Daniels, E. W., and Vogel, H. H., Jr. (1959). *Radiation Res.* **10**, 584.
De Bruyn, P. P. H. (1947). *Quart. Rev. Biol.* **22**, 1.
Dellinger, O. P. (1906). *J. Exptl. Zool.* **3**, 337.
Dujardin, F. (1835). *Ann. Sci. Nat. Zool.* **4**, 343.
Dujardin, F. (1838). *Ann. Sci. Nat. Zool.* **10**, 230.
Ecker, A. (1849). *Z. Wiss. Zool.* **1**, 218.
Engelmann, T. W. (1879). *In* "Hermann's Handbuch der Physiologie," Vol. I, p. 343. Vogel, Leipzig.
Furth, O. (1922). *Arch. Néerl. Physiol.* **7**, 39.
Goldacre, R. J. (1952). *Intern. Rev. Cytol.* **1**, 135.
Goldacre, R. J. (1958a). *Proc. Intern. Congr. Cybernetics, 1st, Namur, 1956,* **1**, 715.
Goldacre, R. J. (1958b). *In* "Surface Films in Chemistry and Biology" (J. F. Danielli, K. G. A. Pankhurst, and A. C. Riddiford, eds.), pp. 278. Pergamon Press, London.
Goldacre, R. J. (1961a). *Exptl. Cell Res. Suppl.* **8**, 1.
Goldacre, R. J. (1961b). *In* "Biological Structure and Function" (T. W. Goodwin and O. Lindberg, eds.), Vol. II, pp. 633-643. Academic Press, New York.
Goldacre, R. J., and Lorch, I. J. (1950). *Nature* **166**, 497.
Griffin, J. L. (1961). *Proc. Am. Soc. Cell. Biol.* **1**, 77.
Griffin, J. L. (1962). *Proc. Am. Soc. Cell. Biol.* **2**, 61.
Hayashi, T. (1961). *Sci. Am.* **205**, 184.
Heathman, L. (1932). *Am. J. Hyg.* **16**, 97.
Heilbrunn, L. V. (1958). *Protoplasmatologia* **2**, 1.
Hirschfield, H. I. (1959). *Ann. N.Y. Acad. Sci.* **78**, 64.
Hoffmann-Berling, H. (1954). *Biochim. Biophys. Acta* **14**, 182.
Hoffmann-Berling, H. (1960). *Comp. Biochem.* **2**, 341.
Huxley, H. E., and Hanson, J. (1955). *Symp. Soc. Exptl. Biol.* **9**, 228.
Huxley, H. E., and Hanson, J. (1960). *In* "Structure and Function of Muscle" (G. H. Bourne, ed.), Vol. I, pp. 183-225. Academic Press, New York.
Hyman, L. H. (1917). *J. Exptl. Zool.* **24**, 55.
Jahn, T. L., and Rinaldi, R. A. (1959). *Biol. Bull.* **117**, 100.
Jahn, T. L., Bovee, E. C., and Small, E. B. (1960). *J. Protozool. Suppl.* **7**, 8.

Jahn, T. L., Fonseca, J., and Landman, M. (1962). *Proc. Am. Soc. Cell. Biol.* **2**, 79.
Jahn, T. L., Harmon, W., and Landman, M. (1963). *J. Protozool.* **10**, 358.
Jennings, H. S. (1904). *Carnegie Inst. Wash. Publ.* **16**, 129.
Jepps, M. (1956). "The Protozoa. Sarcodina." Oliver and Boyd, Edinburgh.
Kamiya, N. (1959). *Protoplasmatologia* **8**, 1.
Kamiya, N. (1960). *Ann. Rept. Sci. Works, Fac. Sci. Osaka Univ.* **8**, 13.
Katchalsky, A. (1963). *In* "Non-muscular Contractions in Biological Systems" *Symp. Biophys. Soc.* (in press).
Kavanau, J. L. (1962). *Science* **136**, 652.
Kavanau, J. L. (1963). *J. Theoret. Biol.* **4**, 124.
Kepner, W. A., and Taliaferro, W. (1913). *Biol. Bull.* **24**, 411.
Kriszat, G. (1949). *Arkiv. Zool.* **1**, 81.
Kriszat, G. (1954). *Arkiv. Zool.* **6**, 195.
Landau, J. V. (1959). *Ann. N.Y. Acad. Sci.* **78**, 487.
Leidy, J. (1879). *U. S. Geol. Surv. Territories* **12**, 1.
Loewy, A. (1952). *J. Cellular Comp. Physiol.* **40**, 127.
Lorch, I. J., and Danielli, J. F. (1950). *Nature* **166**, 329.
Lorch, I. J., and Danielli, J. F. (1953). *Quart. J. Microscop. Sci.* **94**, 445; *ibid.* **94**, 461.
Mast, S. O. (1926). *J. Morphol. and Physiol.* **41**, 347.
Mazia, D. (1956). *Am. Scientist* **44**, 1.
Muggleton, A., and Danielli, J. F. (1958). *Science* **181**, 1738.
Nakajima, H. (1956). *Seitai No Kagaku* **7**, 256.
Noland, L. E. (1957). *J. Protozool.* **4**, 1.
Oosawa, F. (1963). *In* "Non-muscular Contractions in Biological Systems." *Symp. Biophys. Soc.* (in press).
Pantin, C. F. A. (1923). *J. Marine Biol. Assoc. (U.K.)* **13**, 24.
Penard, E. (1902). "Faune rhizopodique du Bassin du Leman." Kundig, Geneva.
Penard, E. (1905). *Arch. Protistenk.* **6**, 175.
Penard, E. (1936). *Bull. Soc. Franc. Microscop.* **5**, 136.
Ray, D. L., and Hayes, R. E. (1954). *J. Morphol.* **95**, 159.
Rhumbler, L. (1898). *Arch. Entwicklungsmech. Organ.* **7**, 103.
Rinaldi, R. A., and Jahn, T. L. (1962). *J. Protozool. Suppl.* **9**, 16.
Rinaldi, R. A., and Jahn, T. L. (1963). *J. Protozool.* **10**, 344.
Rudzinska, M., and Chambers, R. (1951). *Proc. Am. Soc. Protozool.* **1**, 13.
Schaeffer, A. A. (1916). *Arch. Protistenk.* **37**, 204.
Schaeffer, A. A. (1917). *J. Animal Behavior* **7**, 220.
Schaeffer, A. A. (1920). "Amoeboid Movement." Princeton Univ. Press, Princeton, New Jersey.
Schaeffer, A. A. (1926a). *Carnegie Inst. Wash. Publ.* **345**, 1.
Schaeffer, A. A. (1926b). *Quart. Rev. Biol.* **1**, 95.
Schaeffer, A. A. (1931). *Science* **74**, 47.
Schafer, E. A. (1910). *Quart. J. Exptl. Physiol.* **3**, 285.
Schulze, F. E. (1875). *Arch. Mikroskop. Anat. Entwicklungsmech.* **11**, 329.
Seifriz, W. (1939). *Protoplasma* **32**, 538.
Seifriz, W. (1942). "The Structure of Protoplasm." Iowa State College Press, Ames, Iowa.
Singh, B. N. (1952). *Phil. Trans. Roy. Soc. London* **B236**, 405.
Szent-Györgyi, A. (1949). *Science* **110**, 411.
Szent-Györgyi, A. (1953). "Chemical Physiology of Body and Heart Muscle." Academic Press, New York.

Szent-Györgyi, A. (1960). "Introduction to Submolecular Biology." Academic Press, New York.

Tiegs, O. W. (1928a). *Australian J. Exptl. Biol. Med. Sci.* **3**, 1.

Tiegs, O. W. (1928b). *Protoplasma* **4**, 88.

Ts'o, P. O. P., Eggman, L., and Vinograd, J. (1956). *J. Gen. Physiol.* **39**, 801.

Ts'o, P. O. P., Eggman, L., and Vinograd, J. (1957). *Biochim. Biophys. Acta* **25**, 532.

Volkonsky, M. (1931). *Arch. Zool. Exptl. Gen.* **72**, 317.

Wallich, G. C. (1863). *Ann. Mag. Nat. Hist.* **11**, 365.

Weber, H. H. (1958). "The Motility of Muscle and Cells." Harvard Univ. Press, Cambridge, Massachusetts.

Weiss, P. (1961). *Exptl. Cell Res. Suppl.* **8**, 260.

Wherry, W. B. (1913). *Arch. Protistenk.* **31**, 77.

Willmer, E. W. (1961). *Exptl. Cell Res. Suppl.* **8**, 32.

[illegible]

Mechanisms of Ameboid Movement Based on Dynamic Organization: Morphophysiological Study of Ameboid Movement, IV

TOHRU H. ABÉ

Laboratory of Biology, Hosei University, Tokyo, Japan

Since the first observations of ameboid movement and of protoplasmic streaming, an extensive literature dealing with either or both of these subjects has grown, offering us many different unestablished hypotheses on which variously detailed theories have been based. The methods of modern biology have tended to oversimplify the rather formidable complexities of the phenomena observed which take place in living organisms. Analytical studies of these phenomena have had to follow physical or chemical lines. In highly differentiated organisms, the application of physical and chemical methods is often well justified; but the ameba, lacking obvious structural differentiation, is often treated in analytical studies as if it contained no differentiation whatever. Bodies of amebae, however, are actually differentiated, but their structural organization, being dynamic rather than static, has often been overlooked. Earlier research workers who failed to recognize this dynamic organization tended to oversimplify the phenomena of ameboid movement in order to attempt to explain it in terms of seemingly analogous phenomena of physics and chemistry. This trend still prevails in biology, and serves, on the one hand, to advance the science and, on the other, to introduce almost insoluble problems, especially in the study of ameboid movement.

It is of greatest importance to question the reason why most, or perhaps all, hypotheses and theories of ameboid movement and protoplasmic streaming are controversial and conflicting. This situation is apparently due to the fact that, first, there are many different types of ameboid movement and protoplasmic streaming, seemingly quite different from one another, which have in most cases been lumped together and discussed as a single phenomenon, second, cytoplasmic structure and its changes during cell deformation are not discernible, and third, there is a lack of suitable techniques to determine whether the mechanism of streaming is located within the stream itself or in some other part of the cell and whether the distal extension of pseudopods is active or passive.

Because of these difficulties, we have had either to explain the mechanism in purely symbolic terms, to resort to the unanalytical process of "labeling" phenomena as "chemotaxis," "thigmotaxis," etc., or to find seemingly analogous phenomena in physics, chemistry, or physiology, such as contractility, extensibility, viscosity, attachment, spreading, expansion, and viscoelasticity. It is often difficult to justify the use of these terms rigorously in biology.

There is a further fact which needs to be considered. There seems to be a tradition in science to accept experimental results at their face value as valid. Sometimes, however, it appears that we are not sufficiently critical in our interpretations. In such irritable cells as amebae, changes in body contour and protoplasmic streaming in response to a stimulus are rapid and distinct, but well regulated according to the dynamic organization of the body (Abé, 1961, 1962). If based on these considerations, critical reviews of the literature (De Bruyn, 1947; Allen, 1961a) should force us continually to re-examine old and new theories in order to determine what can be accepted as valid.

We should also bear in mind that the variation in types of cell deformation and patterns of cytoplasmic streaming is so enormous that there may well be hidden cause-and-effect relationships between events occurring in different parts of an ameba.

Another problem confronting us is physiological, or at least related to our sensory physiology. One tends to become trapped in a pattern of thought which revolves about some hypothesis or theory, sometimes to an extent involving important aspects of phenomena observed under the microscope. The detection and study of motion falls also in this category. It is difficult or, more probably, utterly impossible to follow accurately, and with the same reliable exactness, by the methods hitherto used, two or more events occurring simultaneously in different parts of a body. This is important because measurements of particle velocities in different regions of the cytoplasm and at different instants are without significance unless combined with the knowledge of events in the posterior and other regions, particularly in the anterior frontal end (Abé, 1963).

Mechanism of ameboid movement and organization of amebae have been discussed in general ways by Pantin (1923), Mast (1926), Goldacre (1956, 1961), and Allen (1961a, b). I have also investigated these problems to some extent using striata and verrucosa groups of amebae, which showed the existence of what is called "dynamic organization" (Abé, 1961, 1962, 1963). At least the first three of the four authors just cited presented the hypothesis that the probable site from which the motive force is generated to induce distal extension of pseudopods can be found in the posterior portion of an ameba body. Contrary to them, Allen presented a new theory in which he suggested that the site is located in the distal end of extending pseudopods. In this respect, my conclusions (Abé, 1961, 1962, 1963) agree with Allen's suggestion, though they differ from it in some points. In this paper I should like to mention principally the ejecting process of the unusually large and solid diatom shell, which naturally tells us something about the dynamic organization and mechanism of ameboid movement in *Amoeba proteus*.

Materials and Methods

The material used in this investigation was a large ameba of proteus type, 200–300 μ in body length, which had been collected from a suburban region of Tokyo and kept in culture for more than 10 years in our laboratory. Usually they are polypodial, but can be induced to take a monopodial form; in this shape the problem of analyzing and describing the inner structures of the cell are considerably simplified. Photographic exposures of 1 sec duration were taken at appropriate intervals (10–30 sec) during the ejection of a digested diatom. Care was taken to include all portions of an ameba's body in every photograph except Fig. 4, so that one could study, correlatively, all the events occurring in different portions of the body. The longitudinal median portion of the body has a considerable thickness. Thus, granules clinging to the dorsal and ventral walls and their inner structures are out of focus in this thickened portion, although the author used a lens of low magnifying power (× 20, Olympus) to take the photographs. However, this is not true of the much thinner peripheral portions of the body. The structural differentiations presented in the figures are limited to those which can be seen within the plane of focus; this can be confirmed in many other photographs, some of which will be published in my forthcoming works. The events which occurred during ejection of this food residue were so complicated that their analysis required several years, during which time so many supporting observations were made that it seemed best to document the observations with the actual photographs, which may then be allowed to tell their own story.

There are only a few reports in the literature of the ejection of food residues by amebae (e.g., Grosse-Allermann, 1909; Neresheimer, 1905), and none of these have dealt with large particles. I have often observed the ingestion of large diatoms, but rarely their ejection. One such event had been recorded photographically and was analyzed in a manner which may shed some light on the relationships between cell deformation and cytoplasmic streaming. The slow-shutter streak-photograph method immediately shows where cytoplasmic streaming or, more probably, movement of granules has occurred; cell deformation is also shown, less clearly in single photographs, as blurred regions of body contour, but, more clearly, by comparing the cell outlines in successively taken photographs. The method must be used intelligently, for different lengths of exposure give quite different pictures. As discussed in my previous paper (Abé, 1963), all particles suspended in the flowing endoplasm move irregularly and at irregular points. An excessively long exposure invariably results in a figure full of granular streaks, whereas too rapid exposure results in a figure with entirely motionless particles. It is also essential to use a phase-contrast or other optical system so that the particles to be registered on film will usually be brighter than their background.

Observations

The sequence of photographs beginning with Fig. 1 shows the entire ejection process of a large and nearly emptied diatom shell lodged in the caudal end of a proteus type of amebae. This cell body had been floating motionless in an attached state owing to mechanical disturbance of the water, and had come to rest finally upon the substratum, to which it adhered. This indicates that the ameba is not flattened in any way, at any instant, from above the body. In the anterior end there is some slow cytoplasmic streaming in the forward direction, as evidenced by the slightly rod-shaped blurred images of particles in the subfrontal and anteriolateral regions. In contrast, no movement was detected in the posterior half of the body. In this stage, the pointed posterior end of the diatom was seen still covered with a thin layer of substance, presumably the plasmalemma.

In Fig. 2 a new surface protuberance is clearly visible (compare Fig. 1) around the posterior end of the diatom, the distal end of which is exposed. This suggests fusion of the plasmalemma and the membranous wall of the food vacuole and their subsequent rupture to form a pore. In its vicinity weak, yet distinct, granular motion can be observed (compare Fig. 5); in spite of this, of the five photographs reproduced here, Fig. 2 shows the least trace of granular movement. At the next moment (Fig. 3), the posterior bulge is rapidly extending, ejecting the diatom from the ameba. At the same time a vigorous cytoplasmic streaming is seen along the axial portion in the direction of the ejected diatom. The stream itself meanders, steering clear of the stationary nucleus and contractile vacuole. The streaks of various lengths throughout the body give a fairly accurate impression of the speeds and directions of particles in motion. Judging from these streaks, motion is most rapid within the narrow channel near the contractile vacuole, and becomes progressively weaker toward the old anterior end (top of the photograph). Stationary particles are seen in various regions of the body.

They are crowded more densely or thickly anteriorly than posteriorly and toward peripheral than toward the median regions. In Fig. 3, vigorous cytoplasmic streaming can be seen along the posterior major portion covering about 96% of the body length and 20 to 60% of the body width. Slow, but distinct, particle movement can be seen even in the so-called ectoplasmic region, apparently contributing to the axial endoplasmic flow. This fact demonstrates the heterogeneity and spongy texture of the so-called ectoplasm (Abé, 1961, 1962, 1963; Goldacre, 1956). More specifically, the so-called ectoplasmic gel layer is so constituted, at least in Fig. 3, that it is penetrated by canaliculi of fluid material in which small particles are suspended and can move.

By the time Fig. 4 was recorded, the ejection process was nearly complete. The caudal bulge is about to cease its growth in length and in transverse dimensions judging from its sharply definable outer contour and very slow motions of particles lying within it.

Another striking feature of the period represented by Figs. 3 and 4 is the formation and growth in number and size of tiny knoblike surface projections along either side of the posterior half of the body.

Figure 5 was reconstructed by tracing Figs. 1 to 4 on the same sheet of paper, taking the positions of the nucleus and contractile vacuole as points of reference. From this reconstruction several changes can be seen which were not obvious from the photographs themselves:

1. The strongest deformation of the body contour occurred in the interval between Figs. 2 and 3 in the original posterior portion just anterior to the bulge: in all probability this reflects a compensation for formation of the bulge.

2. The linearly arranged projections from the lateral edges of the body are not caused by contraction or by wrinkly shrinkage of the body.

3. In various parts of the body there were extensions and retractions. It is particularly noteworthy that the anterior-most portion of the body (top right of figure) continues to advance in spite of the fact that the large particles A and B move posteriorly. This would suggest that fluid matrix in the subfrontal region moved backward in accordance with the vigorous backward flow of the endoplasm in spite of the extension of the very surface of the pseudopod.

4. The stream lines shown in the posterior portion (Fig. 5) were taken partly from Fig. 2 (left side) and partly from Fig. 3 (right side). The former represents the flow preliminary to expulsion of the diatom shell, and the latter is that accompanied by expulsion. The flows represented on the right and left sides of the diagram are correlated clearly with the deformations occurring on the same side. Measurements showed that the direction of the stream on the right-hand side turned at an

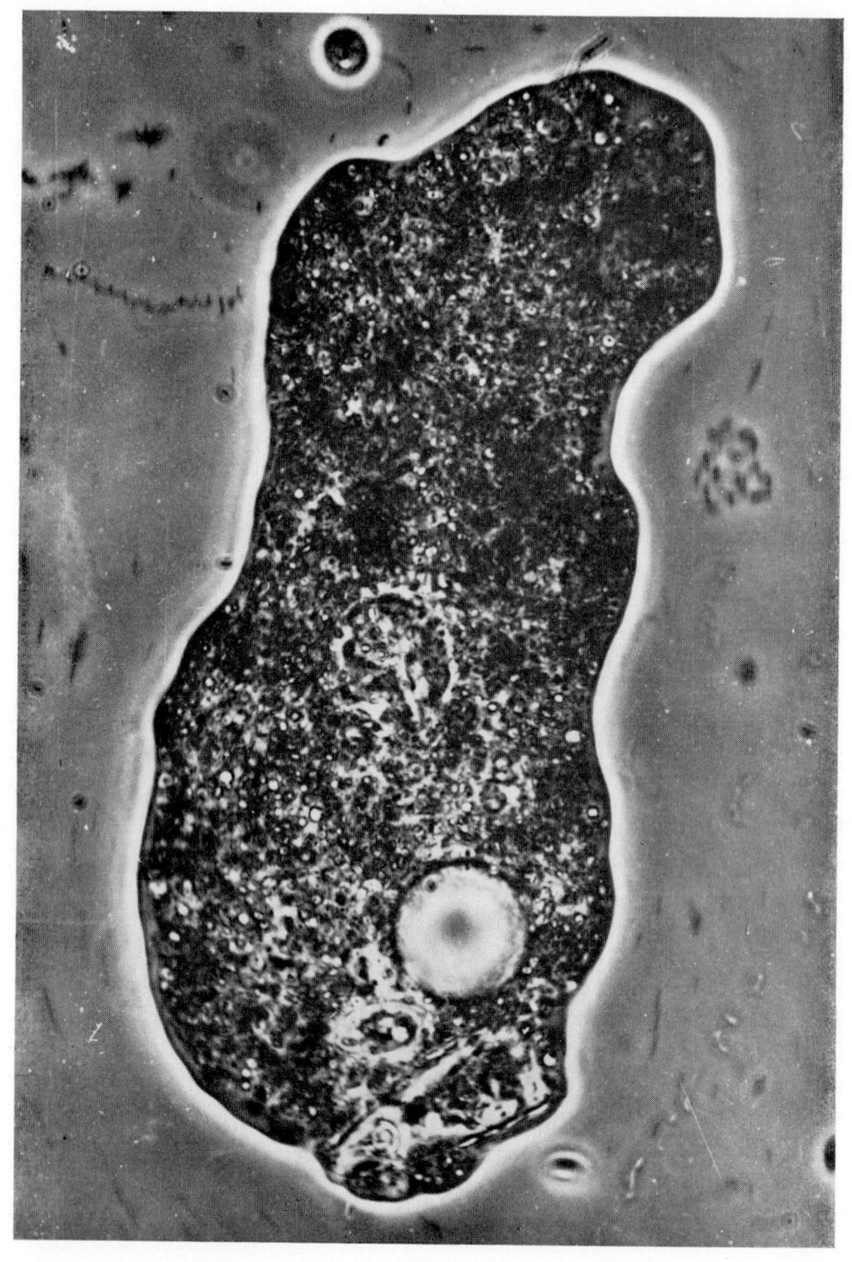

FIG. 1. Diatom lodged in caudal end of proteus type of amebae. In anterior region some slow cytoplasmic streaming in forward direction is seen (rod-shaped blurred images). Posterior end of diatom covered with thin layer of substance.

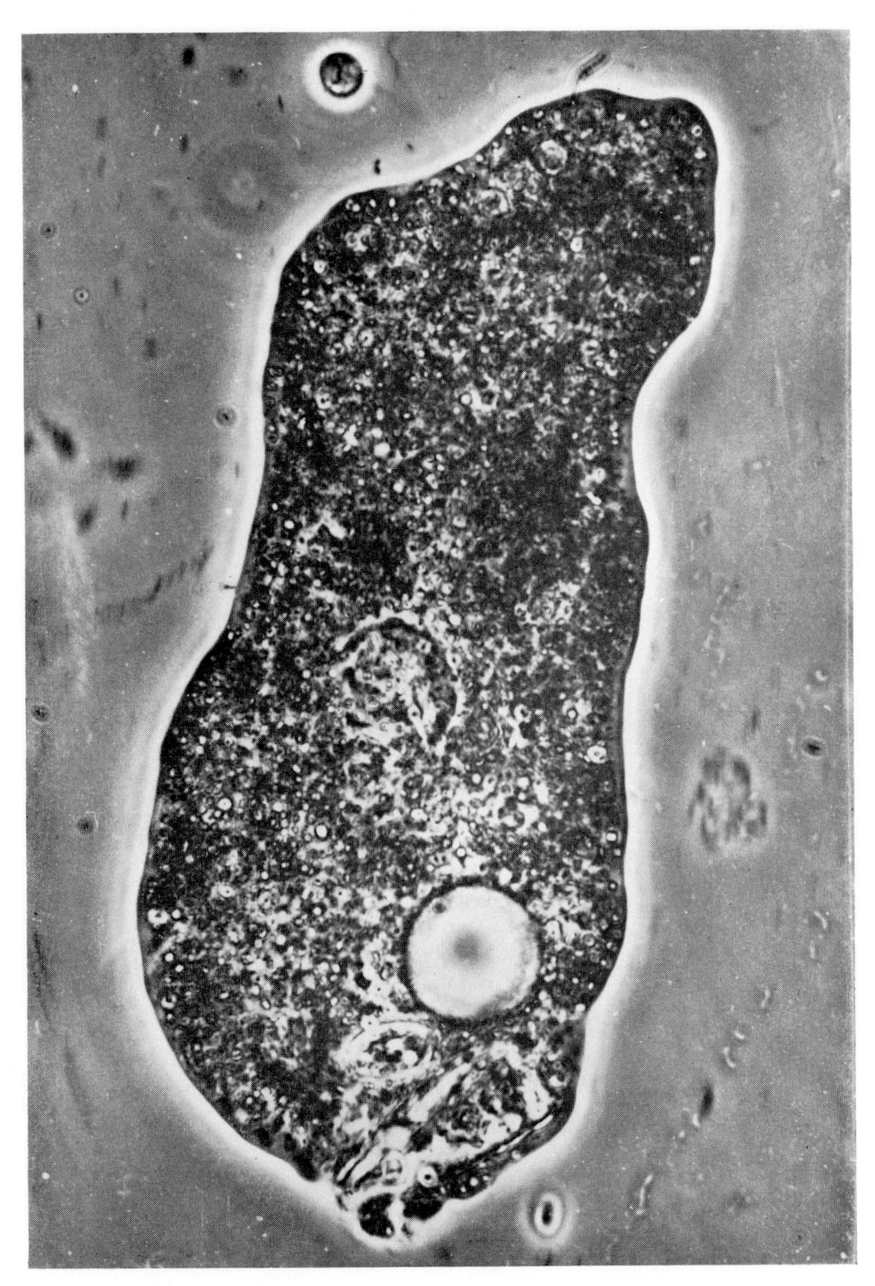

FIG. 2. Posterior end of diatom has new surface protuberance. Weak granular motion is visible in its vicinity.

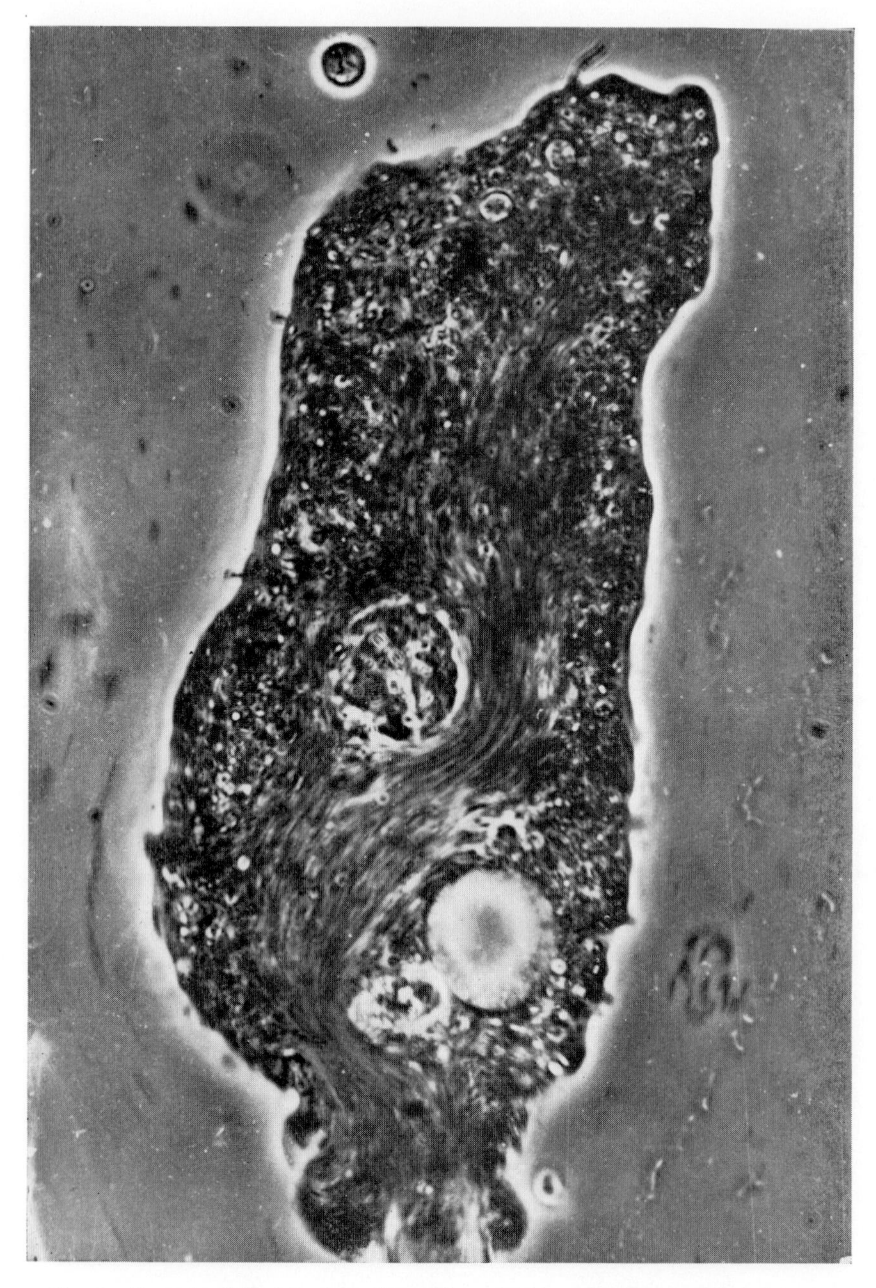

FIG. 3. Posterior bulge ejecting diatom from ameba. Vigorous cytoplasmic streaming in direction of diatom.

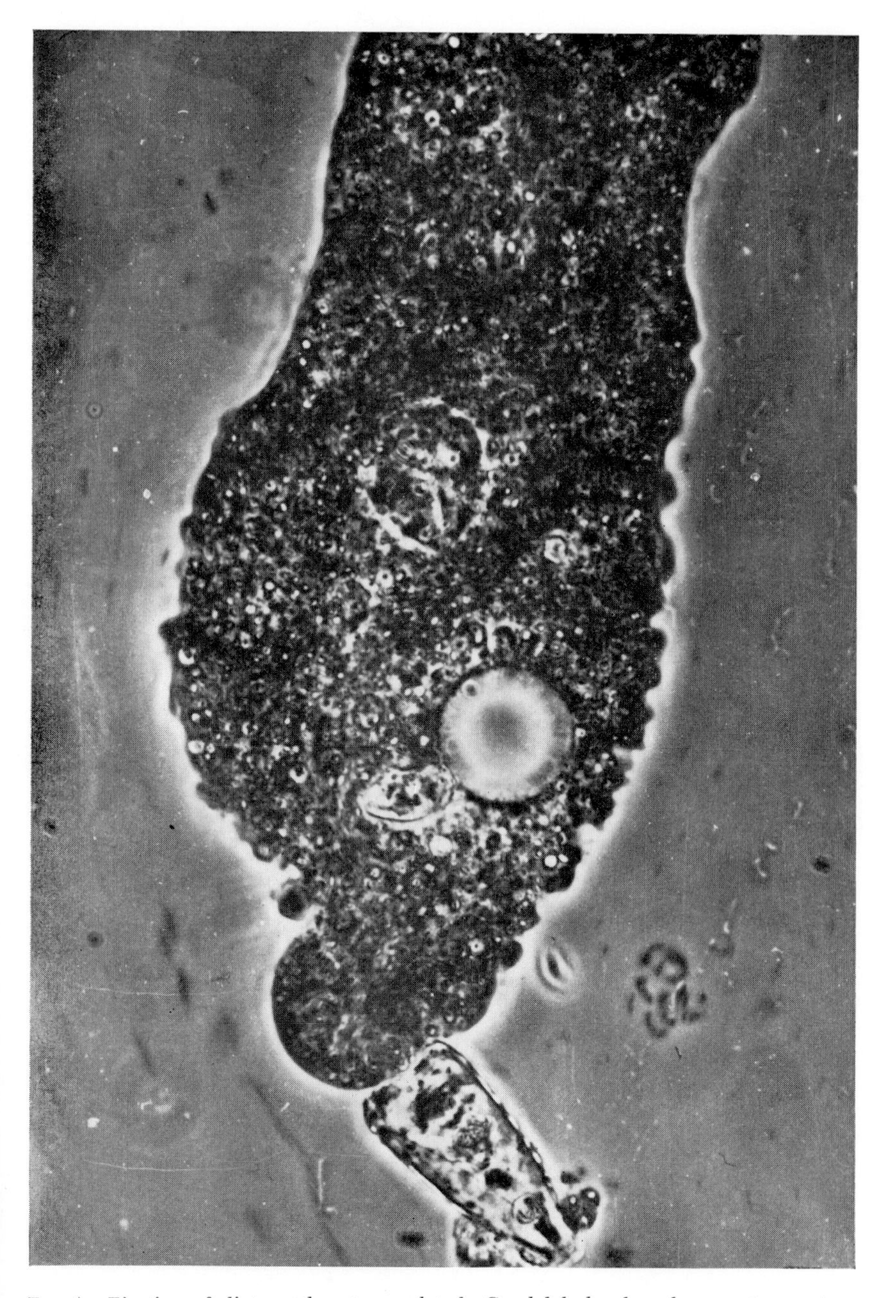

FIG. 4. Ejection of diatom almost completed. Caudal bulge has sharp outer contour.

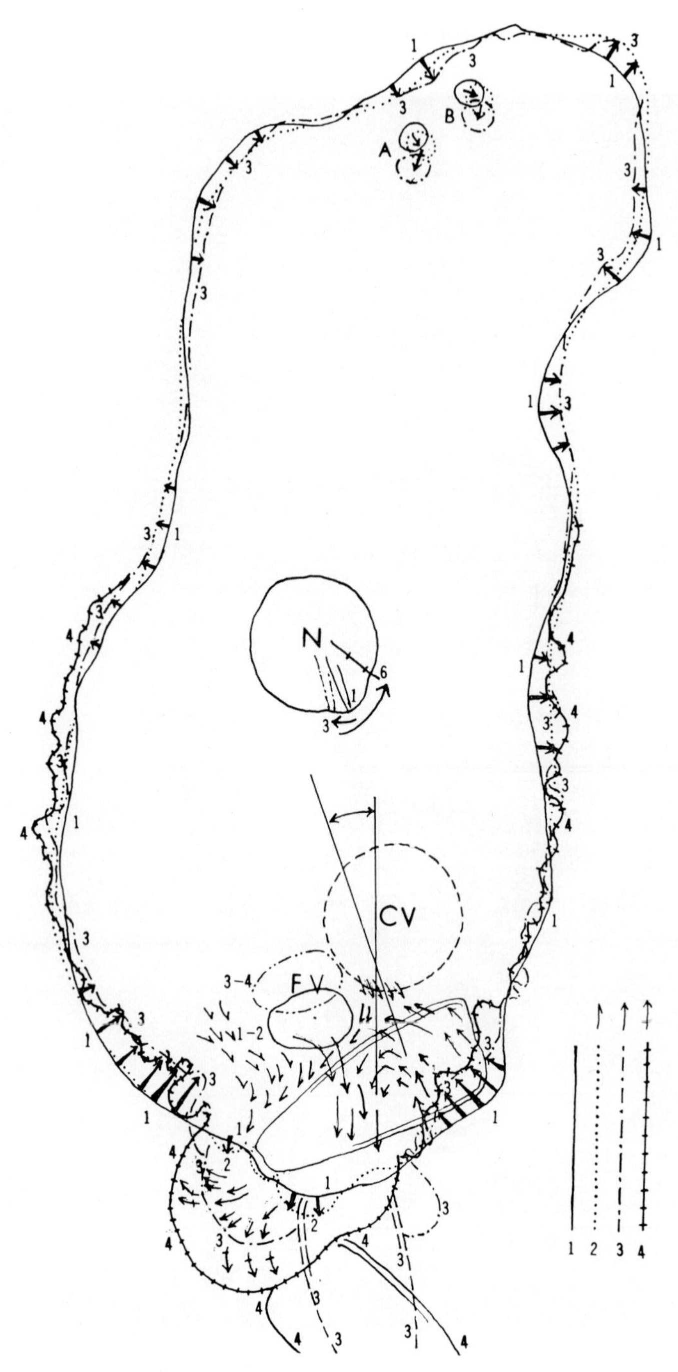

FIG. 5. Tracings of Figs. 1 to 4. N, nucleus; CV, contractile vacuole; FV, food vacuole; A, B, large particles. Four types of solid and broken lines (1–4) represent, respectively, the body contours of Figs. 1–4.

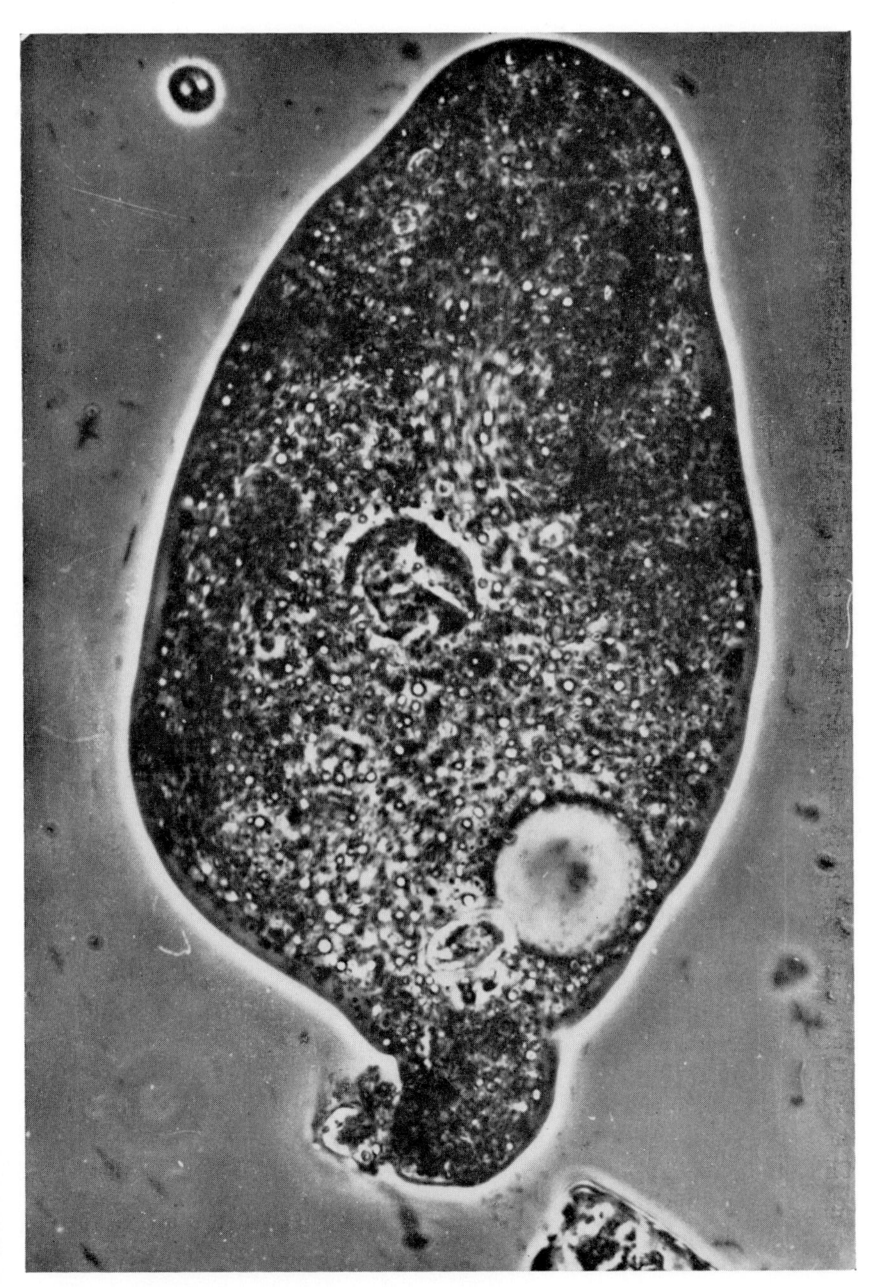

FIG. 6. Ameba is now longer and wider (especially in posterior region) and apparently endoplasmic flow has resumed. A narrow hyaline zone is visible around the body.

angle of approximately 340 degrees from its original anteromedial direction toward the posterior end. This acute change of direction and accompanying phenomena are very unusual and cannot be explained by the assumption of active movement of the fluid matrix or of the granules.

5. The ameba seemed to exert some degree of coordination in ejecting the diatom. The exact manner in which the caudal bulge and bilateral retractions of the body surface were brought forth is suited for rotation of the diatom to minimize its frictional resistance. Rotation was enhanced in part by the initial axial flow (Fig. 2) which pushed the distal end of the diatom in the posterior direction. The retraction of the body surface on the right side then completed the rotation process and partly performed the ejection.

6. The anterior displacement of the arrested smaller food residue (FV) seems paradoxical unless one takes into consideration the fact that the retracted part of the body wall, which happened to push the flared end of the diatom inward, undoubtedly forced, in turn, the food residue (FV) through a gelated region which otherwise would have served to keep it stationary.

7. First there was a slight clockwise, then a much more pronounced counterclockwise, rotation of the nucleus. The former appears to be caused by the vigorous but unusual rearward flow of the endoplasm in contact with the nuclear surface; the latter might have been caused by the normal forward flow of endoplasm in the interval between Figs. 4 and 6, which can be seen, in part, in Fig. 6.

8. The distance between the nucleus and contractile vacuole increased in the interval between Figs. 4 and 6 (longer than the intervals between the other pictures). This movement is apparently not due to contraction of the body, but to the reopening of endoplasmic flow which results in elongation of the body, particularly in its anterior half, and by dint of which the nucleus is carried forward.

By the time Fig. 6 was recorded, the organism had become even flatter than before, and therefore was longer and wider (especially in its posterior region); yet its original organization remains, as suggested by the reopened, forward, endoplasmic flow in the anterior portions of the body. A narrow hyaline zone is formed almost all around the body. This may in all probability be due not to inward retractions of peripheral granules but to extension or formation of hyaline structures.

Discussion and Conclusions

The foregoing appears to be the first documented example of the ejection of a large food residue from an ameba. The details of the process are such that my original conclusions about the importance of dynamic organization in amebae (Abé, 1961, 1962, 1963) appear to be generally applicable, regardless of differences in taxonomic situations of amebae and in size, shape, or types of deformation and locomotion of the cell body.

Perhaps the most noteworthy feature of the ejection process is that the so-called strongest "contraction" of the body occurred not in the region of the body where the rearward endoplasmic streaming originated (i.e., the original front end), but rather in the temporarily formed subfrontal (i.e., old posterior) region of the flow, close to the bulge. Thus, the site of body contraction remained the same even at or after the time when the direction of vigorous streaming reversed temporarily, and the ordinary forward flow reappears invariably in the old anterior end of the body. This seems difficult to account for on the basis of the tail contraction theory (Goldacre, 1956, 1961; Mast, 1926; Pantin, 1923) and perhaps others (Allen, 1961a).

The morphological peculiarities generally accepted as occurring in the portions of the body in which the ordinary endoplasmic stream originates cannot be seen in the old anterior end in which the temporarily reversed vigorous stream starts in this case. The distinct diminution of the body dimensions or the contraction of the body could be seen in the subfrontal portions of the reversed endoplasmic flow (e.g., the old posterior most portions of the body surrounding the unusually formed caudal bulge). In other words, continued solation and carrying away of fluid matrix, which under normal conditions induce so-called "contraction," could not be seen in the old anterior region. Furthermore, the vigorous backward stream did not prevent the continued (although slight) advance of the original front end, nor did it cause continued locomotion of the body in reversed direction. These observations seem particularly difficult to explain in terms of ectoplasmic contraction or of the simplified sol-gel transformation hypothesis. At the same time, they suggest some degree of independence of the pseudopodial extension from the seemingly distinct endoplasmic streaming within. It is to be recalled in this connection that in Fig. 3 the stationary granules are seen along the entire breadth of the original front end, suggesting the existence there of gelled structures; however, not a single arrested particle can be seen along the new frontal surface and within the extending bulge. It will be recalled that in striata amebae, there is a fenestral opening in the posterior end of the body.

There has been a traditional tendency to treat the endoplasmic stream as an entity. But Fig. 3 and subsequent ones, it seems to me, may suggest some differences in physiological properties between the temporarily induced rearward flow and the normal forward flow of endoplasm, because of the fact that the latter is invariably associated with solation in its streamhead regions, whereas the former is not, but instead seems to be supplied with newly evolved sol in its downstream portion. The author's previous writings concerning, e.g., the anteroposterior differences in physiological activities and some other properties of the normally flowing endoplasm of striata ameba (Abé, 1961, 1962) may serve to some degree to explain these peculiarities. It may be permissible for me to suggest that endoplasm lying in the posterior end of the body is newly evolved from gel structures by their solation, and has the least capability for gelation (Abé, 1961, 1962) and the strongest capability for solation acting on gel structures regardless of what it encounters. This may explain why vigorous gelation does not occur along the surface of the bulge into which the newly evolved sol might have been poured, and, at the same time, why the nucleus was least influenced by the rearward stream but was strongly carried forward, showing distinct rotation in accordance with the normal forward flow (Abé, 1961, 1962). According to the views I have already expressed elsewhere (Abé, 1961, 1962), gelation occurs wherever active extension of a pseudopodial tip is carried on. Based on these, one may be allowed to conclude that the formation of the caudal bulge in this case is apparently passive and never active.

The position or line along which the knoblike surface projections are formed corresponds to that of the so-called "lateral fin" in striata amebae. The same kind of structure is seen fairly frequently along the laterals of locomoting, large amebae and also partly or entirely around small species of Amoebidae. The flattening of the body seen in Fig. 6 is not due to passive flattening of the body because no pressure was applied to the body from above. It is due to growth not only in length of the body but also in breadth of the bilateral projections forming a conjoined broad membranous pseudopod, by which the ameba can adhere firmly to and also extend along the substratum. Adherence of the ameba body at the time of ejection of an unusually large food residue is of frequent occurrence and may be, it seems to me, reasonable. Based on the frequent splitting and reunion of the peripheral membranous pseudopods seen quite often in the smaller amebae, the body-contour change occurring in the interval between Figs. 4 and 6 corresponds exactly to that described for tissue culture cells under the heading of "spreading" or "expansion" (Abercrombie, 1961; Ambrose, 1961; Taylor, 1961; Weiss, 1961). However, the use of these terms should not mislead us into believing that the process has more than passing similarity to the spreading or expansion of a liquid drop.

Under normal conditions of locomotion, the posterior end of the ameba body is the ordinary site of the most vigorous solating activity

(Abé, 1961, 1962, 1963). In the present case, it is certain that the diatom surrounded by its vacuolar wall lies in this part of the body filled with freshly evolved sol. At the time of ejection of the diatom, it is apparent that there is formed, through the gelled body wall, a large pore, the diameter of which is a little smaller or larger than that of the diameter of the diatom. Taking into consideration the fact that this region normally has the least gelating activity, there cannot occur rapid gelation all over the pore so as to cover it completely, just as is the case with the discharging pore of the contractile vacuole in striata ameba (Abé, 1961, 1962). Then the membranous wall of the food vacuole might naturally cover the pore and subsequently be pushed out so as to cover at least a major portion of the surface of the caudal bulge, if the fluid endoplasm moves passively under the influence of pressure from within the body.

The explanation for these phenomena is not based on any of the old or current hypotheses, but solely on this author's interpretation of the dynamic organization of the ameba's body. This entails organized gel structures with a considerably higher order of physiological and morphological complexity than has hitherto been recognized. The observations and discussions given in this paper demonstrate that this organization persisted throughout the period in which the polarity of streaming temporarily reversed.

Though the tail contraction hypothesis is not supported, in my opinion, by the facts presented here, it is, of course, possible that the vacuolar wall surrounding the diatom might serve to eject the diatom by dint either of local contractility or turgidity of the body as a whole. But this assumption is insufficient to explain the continued distal extension of the pseudopodial tip. However, it may be a matter of question whether or not the ejection process proper is related to the normal mechanism of ameboid movement. In the case of small food residues, no sign of disturbance in ameboid movement can be detected. The disturbance was shown in such an exaggerated degree in the present case presumably because of the considerably larger size and rigid structure of the food residue, both bringing forth unusually strengthened and causally induced activities in different portions of the body. Not rarely, a much larger diatom is captured and completely digested by *Amoeba proteus* before its ejection. In such cases, locomotion usually stops just as in this case. Therefore, the processes described in this paper are by no means abnormal or unusual.

By analyzing the ejection process, the author has been led to conclude, in agreement with Allen (1961a, b), that the site of generating force or mechanism to bring forth the distal extension of pseudopods and protoplasmic streaming must be sought for in the anterior-most portion of a body or of a pseudopod.

Acknowledgment

The author wishes to express his sincere thanks to Assistant Professor K. Ishii of Hosei University for the extreme care he exercised in taking the photographs in conformity to the author's ideas and suggestions.

References

Abé, T. H. (1961). *Cytologia (Tokyo)* **26**, 378.
Abé, T. H. (1962). *Cytologia (Tokyo)* **27**, 111.
Abé, T. H. (1963). *J. Protozool.* **10**, 94.
Abercrombie, M. (1961). *Exptl. Cell Res. Suppl.* **8**, 188.
Allen, R. D. (1961a). *In* "The Cell" (J. Brachet and A. E. Mirsky, eds.), Vol. II, pp. 135-216. Academic Press, New York.
Allen, R. D. (1961b). *Exptl. Cell Res. Suppl.* **8**, 17.
Ambrose, E. J. (1961). *Exptl. Cell Res. Suppl.* **8**, 154.
De Bruyn, P. P. H. (1947). *Quart. Rev. Biol.* **22**, 1.
Goldacre, R. (1956). *Proc. Intern. Congr. Cybernetics, 1st Namur, 1956,* p. 715. Gauthier-Villars, Paris.
Goldacre, R. (1961). *Exptl. Cell Res. Suppl.* **8**, 1.
Grosse-Allermann, W. (1909). *Arch. Protistenk.* **17**, 203.
Mast, S. M. (1926). *J. Morphol. Physiol.* **51**, 347.
Neresheimer, E. (1905). *Arch. Protistenk.* **6**, 147.
Pantin, C. F. A. (1923). *J. Marine Biol. Assoc. U.K.* **13**, 24.
Taylor, A. C. (1961). *Exptl. Cell Res. Suppl.* **8**, 154.
Weiss, P. (1961). *Exptl. Cell Res. Suppl.* **8**, 260.

Discussion

Dr. Wolpert: I would just like to ask whether the diatom which was egested had a vacuole around it.

Dr. Abé: Apparently not.

Dr. Inoué: In your photograph (Fig. 3), it appeared that there was streaming in the cytoplasm of the ameba near the top of the picture after the diatom was ejected.

Dr. Abé: Streaming began there, not in the posterior.

Dr. Inoué: Then, the first streaming always appears near the surface in the direction in which the ameba moves. Is that correct?

Dr. Abé: Yes.

On the Mechanism and Control of Ameboid Movement

R. J. Goldacre

Chester Beatty Research Institute, London, England

A. Introduction

Ameboid movement is perhaps one of the most striking visible manifestations of life in the cell, and attempts to explain this movement have been made for about 130 years (see review by De Bruyn, 1947). In this century, most investigators have agreed that locomotion is a result of the contraction, at one end, of a tube of plasmagel, which encloses plasmasol. This is expressed in the theories of Pantin (1923) and Mast (1926). It was not possible to say much about the mechanism of the contraction until the isolation of muscle proteins and their reaction with adenosine triphosphate (ATP) (Szent-Györgyi, 1947) suggested that something similar to this reaction was going on in the contracting part of the ameba (Goldacre and Lorch, 1950). Considerable support for the involvement of ATP came from the experiments of Loewy (1952), Ts'o and colleagues (1956), Hoffmann-Berling (1960), Simard-Duquesne and Couillard (1962), and others. The theories of Pantin (1923) and Mast (1926) have been elaborated by Goldacre (1952a, b, 1953, 1954, 1958a, b, 1961) to account for the behavior of amebae in a variety of activities and for the control of locomotion by a feedback process.

In the last few years there has been a considerable increase in interest in the mechanism of ameboid movement, and several new theories have been proposed (Allen, 1961a,b,c; Ambrose, 1961; Bell, 1962; Bingley and Thompson, 1962; Kavanau, 1963). I propose to discuss some of these in the light of experimental evidence presented here.

In this paper I shall consider the location of the motor mechanism of ameboid movement, its nature, and how it regulates itself.

Unless otherwise stated, the species of ameba used in the new experiments reported is *Amoeba proteus.*

B. Is Ameboid Movement a Result of a Pull from the Front or a Push from the Back?

The "fountain-zone" theory of Allen (1961a,b,c) requires a pull from the front as opposed to the theories of Pantin (1923), Mast (1926), and

Goldacre and Lorch (1950), which postulate an active contraction at the back. The following points are opposed to a pull from the front.

1. In a polypodial ameba, a persisting pull from the front of each pseudopod advancing in different directions would pull the cell into pieces. Cytoplasmic movements are strong enough to do this, as, for example, in cell division, and in the tearing of captured prey into pieces by pseudopods (Mast and Root, 1916; Beers, 1924). On the other hand, the existence of more than one rear contracting region (such as

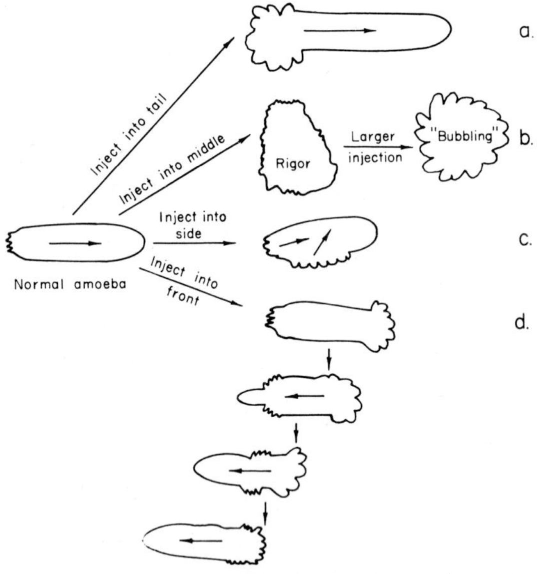

FIG. 1. Effect of microinjection of ATP (2%) into various parts of the ameba. In each, a vigorous contraction of plasmagel occurred at the site of injection immediately, resulting in the wrinkling of the cell membrane over it. These secondary "tails" soon fused with the original tail material. Note reversal of streaming after injection into front end, which then became a tail region.

a tail region and a retracting pseudopod) merely leads to the ultimate fusion of the two contracting regions through the contraction of the part of the cell between them (see, for example, Fig. 1c and d). As cell fragmentation does not occur in ordinary locomotion, the advancing tips must be moving passively.

2. A pull from the front requires the "pumping-out" of the tail, resulting in the wrinkling of the membrane in it; if the membrane at the tail region is broken, the external medium should be sucked into the cell. This does not happen—cytoplasm always escapes (de Bary, 1864; Goldacre, 1961).

3. A glass spring microbalance, made in the shape of a two-pronged fork on the de Fonbrune microforge, showed, by the squeezing together of the prongs when it was pushed into the tail region of amebae, that there was an active contraction there (Goldacre, 1961). There was no effect on the prongs when it was placed in the "fountain zone," indicating an absence both of a contraction [which should occur according to Allen's (1961a,b,c) theory], and of an expansion.

4. While the shape of the ameba is continuously changing, and the cytoplasm circulates and occupies each different part of the cell in turn, the tail is the only permanent surface feature. The leading pseudopod in a polypodial ameba may retract from time to time, but the tail remains as a permanent contracting region of the cell (Goldacre, 1961). If the motive force were located in the anterior end of the cell, it would be extinguished when the leading pseudopod retracted. Analysis of cine film of the polypodial *Amoeba discoides* showed that this happened about every 3 min.

Now follow a number of experiments which indicate that the tail is a special, unique region of the cell, which is compatible with its being a motor region and incompatible with the location of the motor in the front.

5. The results of the microinjection of ATP into the ameba were briefly described by Goldacre and Lorch (1950) and more extensively by Goldacre (1952c). Figure 1 summarizes the results of microinjection into several hundred amebae of small amounts of 2% ATP solution (neutralized with NaOH) into various sites of the cell. In each case, a vigorous contraction began immediately, and the membrane near the site of injection wrinkled as if from the contraction of the plasmagel underlying it. This wrinkling was similar in appearance to a large "tail" or uroid, and suggests that in the normal ameba's tail there is a continuous release of ATP by some enzyme reaction.

The effect of the injected ATP lasted about 1 min, and the solution could be injected repeatedly to produce the same effect.

It is noteworthy that when two wrinkled tail regions were simultaneously present in the cell (Fig. 1c and d), they soon fused into one, owing to the eventual contraction of the region between them, for each produced a propagated contraction, the contracted material liquefying to flow away along the central channel of the cell.

Control injections were done with other substances, none of which caused a contraction or wrinkling of the membrane. These were: Chalkley's medium, Ringer's solution, $MgCl_2$ equimolar with the ATP injected, distilled water, 0.1% neutral red, and 2% lissamine green.

That the contraction observed with 2% ATP was not produced by a mere mechanical stimulation of the cell was also shown by injection of a series of progressively more dilute solutions of ATP, which produced progressively less effect. Below 0.4% ATP, there was hardly any response. It is to be noted that the actual effective concentration was much lower than these figures, owing to diffusion beyond the point of injection.

An interesting effect owing to the reversal of the direction of streaming by the injection of ATP into the advancing tip of the ameba (Fig. 1d) was the appearance of the nucleus and contractile vacuole in the new advancing tip of the ameba (which had previously been the tail end). Since the nucleus normally maintains a fairly central position in the cell, and is not carried along in the stream like the small granules, this suggests that the nucleus is held back or "filtered out" by a network of plasmagel strands in front of it which are absent behind it, so that when the stream is artificially reversed, the nucleus can enter the tail. Further evidence for this plasmagel network is given below (Figs. 5 to 7).

6. The microinjection of heparin (1:1000 solution, in amounts of about 3 times the volume of nucleus) produced results which were the opposite of those made by ATP (Goldacre, 1952c). Heparin caused a spherical pseudopod to be blown out at the site of injection (Fig. 2). This was in contrast to the cylindrical one formed normally when the walls were stiffened by the plasmagel tube. Often the whole of the cell would blow out into the pseudopod, as if that were the weakest part, giving way under the pressure within the cell (Fig. 2a). Sometimes the spherical pseudopod was pinched off. Often the nucleus moved into the spherical pseudopod as if all plasmagel network resistance to its passage along with the plasmasol stream were dissolved.

A large quantity of heparin liquefied the cell completely. The heparin influenced a volume of cytoplasm many tens of times greater than the volume injected. It appears that heparin both liquefies existing gel and prevents the formation of fresh gel.

The local absence of plasmagel caused by heparin in the advancing end of the cell indicates that no active contractile process can go on there and that events there are passive.

7. When an ameba is placed in neutral red solution for a few seconds, the plasmagel tube is stained red. If the ameba is then replaced in its natural medium, the dye is progressively released into the tail as the tail overtakes each portion of the red plasmagel tube in turn, and the plasmasol which streams forward from the tail is colorless (Goldacre, 1952a), so that the tail contains all the dye by the time the ameba has streamed through its own length. Nothing happens at the anterior end. It is interesting in this connection that stretched and unstretched

muscle have a different adsorptive power for dyes. Margaria (1932, 1934) found an apparent change of pH of 1.4 units in muscle stretched in the presence of phenol red. This was reversible. The release of neutral red thus suggests an active process in the ameba's tail, i.e., a contraction producing a decrease in dye adsorption.

It has been suggested by Bingley and Thompson (1962) that the neutral red accumulates in the tail because of an electrical potential gradient which they reported in the cell. This cannot be so, because when

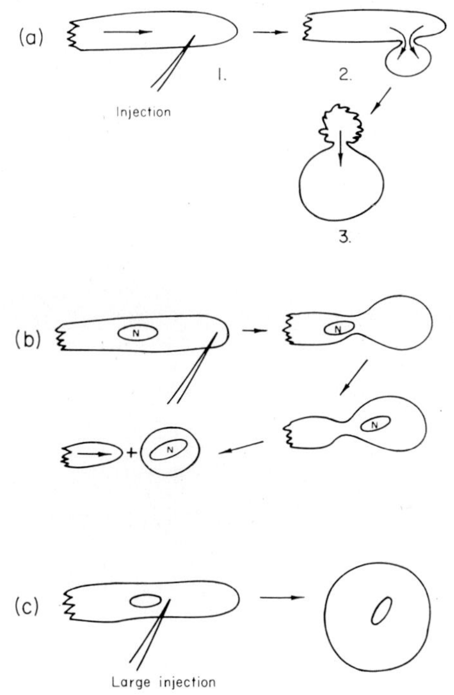

FIG. 2. Microinjection of heparin (1:1000 solution). Note liquefaction of plasmagel at site of injection resulting in the blowing out of a spherical rather than a cylindrical pseudopod.

oppositely charged dyes (such as eosin and the common anionic indicators) were injected into the cell, they spread evenly throughout the cytoplasm (Clark, 1943) and did not accumulate at the anterior end of the cell. This experiment also disproves Kavanau's (1963) suggestion that neutral red goes to the tail because water is driven in the opposite direction to the stream of particles which, he says, are jet-propelled. Kavanau's mechanism should apply to all dyes, and not exclude the anionic ones; moreover, it would merely cause circulation, not accumulation of the dye solution.

8. When amebae are exposed to about one-quarter saturated, aqueous solutions of volatile fat-soluble substances, including anesthetics (such as chloroform, ether, and benzene), the plasma membrane becomes raised away from the granular cytoplasm owing to an increase in thickness of the hyaline layer, except at the tail, where the membrane and plasmagel remain attached (Goldacre, 1952b). This again shows that the tail is a special region.

9. If ATP is being used up continuously in the ameba's tail, one might expect to find phosphate ions from its breakdown there. A rather long-

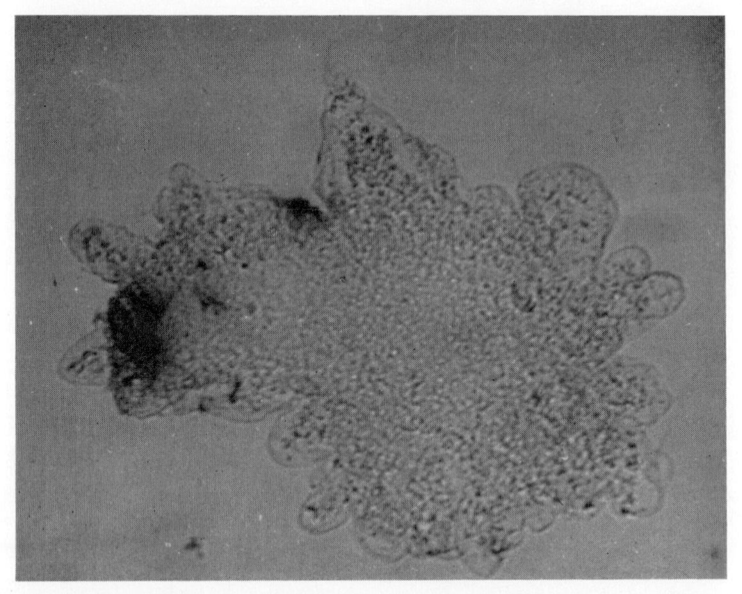

FIG. 3. Ameba showing, as blackening, the intense blue color produced in the tail when a reagent producing a blue color with phosphate ions is dropped over it when streaming actively (see text).

shot experiment was carried out in the hope of testing this. The medium was sucked away from amebae streaming actively in a drop on a slide and replaced by a drop of a solution which turns an intense blue with phosphate ions (ammonium molybdate, 0.1%, and stannous chloride, 0.1%, in *N* sulfuric acid). This solution, though toxic, permitted ameboid movement to continue for 2 to 3 min, but the tails began to turn blue within a few seconds, so that the experiment was over before the amebae had stopped streaming. It was also found that prodding the ameba with a blunt needle [which would normally produce a local contraction (Goldacre, 1952b)] produced a blue color at the place touched (Fig. 3). Control amebae previously fixed in osmic acid gave no blue color, as would be expected, for phosphate ions would soon diffuse away in the fixative. These results are consistent with the production and breakdown of ATP at the tail and site of stimulation, respectively. There might be other interpretations of the blue color, however, owing to the extreme sensitivity of the reagent and to the fact that the amebae were not alive (though fixed by the reagent) at the end of the experiment.

These nine experiments indicate that the active process, which drives the ameba along, is located in the rear of the cell and cannot be in the front.

C. The Nature of the Ameba's Tail

The experiments in Section B suggest that the ameba is driven along by the continuous production, in the tail region, of ATP, which would cause contraction and then liquefaction of the plasmagel in the tail. The resulting plasmasol would flow up the central liquid channel of the ameba to gel once more on the walls of the advancing pseudopods.

The question arises, how would the continuous production of ATP in the tail come about? The continuous production of anything at any fixed part of the ameba seems at first sight impossible, since the cytoplasm circulates every 2 min and occupies each part of the ameba in turn, so that, for example, any enzyme which might be producing the ATP in the tail would be moving continually out of place; moreover, it is possible to cut off the tail (or the front end) of the ameba without interfering, except momentarily, with ameboid movement (Radir, 1931); a new tail (or front end, respectively) forms again to replace the old one immediately.

A clue was given by the fact previously mentioned that anesthetics raise the membrane except at the tail. Inspection of the ameba under high-power magnification showed that in the normal ameba, also, the membrane was in contact with the granular plasmagel in the tail, but elsewhere was insulated from it by a hyaline layer which, however, was much thinner than in anesthetized amebae (Goldacre, 1952b, 1961). This is in contrast to Mast's (1926) much-reproduced diagram which shows hyaline layer over the whole surface of the ameba including the tail.

It seemed likely that this membrane-plasmagel contact in the tail was the cause of the enzyme reaction there, and this was tested by causing the membrane to contact the plasmagel by five different methods:

1. Prodding with a blunt needle. Whenever the membrane was pressed right across the hyaline layer, a response to touch occurred (Goldacre, 1952b).

2. Passing an electric current (Fig. 4). Membrane and cytoplasm moved

in opposite directions, and cytoplasm pressed against the membrane on the anodal side of the cell. These then contracted and became the tail (Goldacre, 1958a), causing the well-known streaming toward the cathode (Mast, 1931).

3. Sucking sharply at one end of a capillary tube in which an ameba has been squeezed. Hyaline fluid could be sucked from one end of the cell to the other (like water through a sponge), raising the membrane at one end of the cell and pressing it against the plasmagel at the other. The latter end became the tail. The end which was the tail could be reversed at will by this method (Goldacre, 1958a).

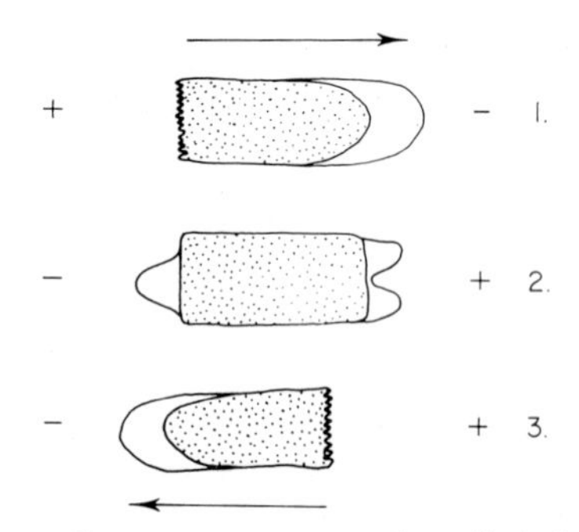

FIG. 4. Effect of passing a direct current across the cell. Cell membrane and granular cytoplasm move in opposite directions to press together on the anodal side (+) of the cell. This contact causes a contracting tail region there, and the ameba then streams actively toward the cathode (—).

4. Pressing on the cover slip over an ameba, to flatten it to several times its original area. This caused the membrane to contact the plasmagel over most of the surface of the ameba, which contracted all at once when the cover slip was raised, pulling the ameba into a spherical shape. A similar effect was produced by sucking an ameba repeatedly in and out of a narrow tube so as to squeeze its sides. On release it rounded up. This is a well-known technique for producing spheres for volume measurement.

5. Replacing the external medium with oil. A striking phenomenon occurred. The amebae were washed in distilled water, the water was sucked away as much as possible, and about half a minute allowed for the evaporation of the last trace, shown when gentle blowing caused the surface of the ameba (as seen in the low-power microscope) to wrinkle. The ameba was immediately covered with oil, whereupon pulsations began. Under high power, it was seen that the granular cytoplasm slowly (1–3 μ/sec) moved outward until it touched the cell membrane; immediately it pulled away rapidly (10–20 μ/sec) by about 20 μ. This plasmagel

expansion-contraction cycle of absorption of hyaline fluid followed by contraction and syneresis to produce insulating hyaline fluid between membrane and plasmagel again occurred about every 10 sec. A somewhat similar phenomenon has been reported by Kavanau (1963).

These five experiments indicate that it is the membrane-plasmagel contact which causes contraction in the tail. This explains why cutting off the tail has no effect—new membrane becomes pressed against the plasmagel at once.

One supposes that there is an enzyme on the membrane which has a nondiffusible substrate in the plasmagel, from which it is insulated over the surface of the ameba except in the tail [and also retracting pseudopods (Goldacre, 1961)]. Contact in the tail would produce ATP continuously, and this would cause continuous contraction in the tail. How this is controlled by feedback is discussed in Section E,1.

The presence of an occasional clear blister or occasional trapped food vacuole in the tail does not invalidate the argument above, since each contact between plasmagel and membrane causes contraction over a considerable area beyond it (as does an injection of ATP), and any pockets of passive material in between contacts would merely be pulled along until they were eventually disposed of.

D. Experimental Evidence for Plasmagel Network

The heterogeneous structure and consistency of the cytoplasm is of interest in relation to the maintenance of the position of the larger organelles, such as the nucleus and contractile vacuole. Information about this was obtained as follows:

Actively streaming uncompressed specimens of *Amoeba proteus* were photographed with a high-power objective (×95, N.A. 1.3) with an exposure of 3 sec. This gave an optical section in which the particles in the moving cytoplasm were streaked in the direction of their movement, whereas those in the stationary cytoplasm gave sharp images (Figs. 5 and 6). The apparent plasmagel network structure is correlated with the observation that the finest cytoplasmic granules flow smoothly forward, larger ones start and stop and may move sideways before proceeding forward (as if moving around an invisible obstruction), and the largest particles, such as nucleus and large food vacuoles, are held back completely (Goldacre, 1952c; Marsland, 1950; and, in a striata ameba, Abé, 1961). Similar observations have been reported for the movement of these cytoplasmic particles in a centrifugal force field (Allen, 1960; Harvey and Marsland, 1932).

Figure 7 shows diagrammatically the distribution of plasmagel and

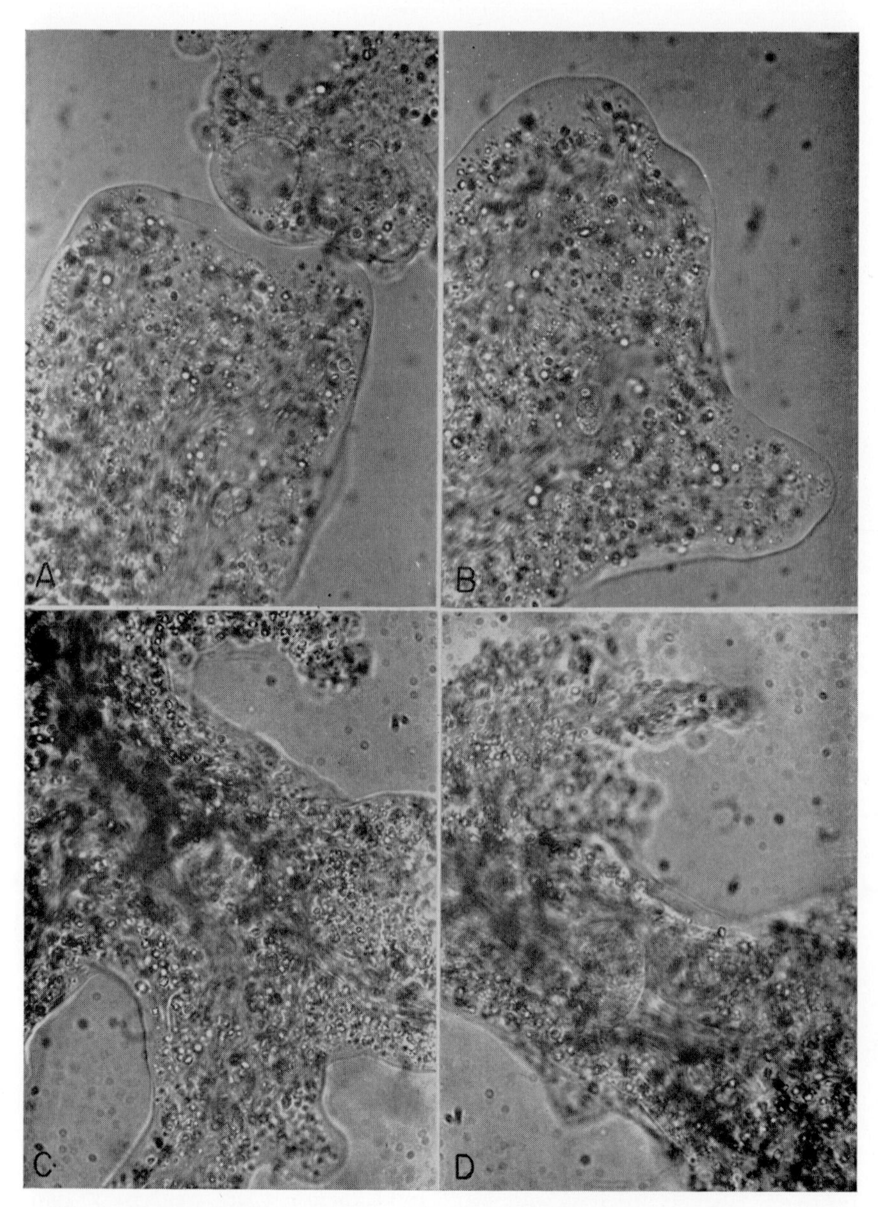

FIG. 5. Photos showing existence of interlocking plasmasol-plasmagel network in *Amoeba proteus.* High-power optical section with long exposure (3 sec) causing streaking of moving particles along direction of movement. Note how nucleus is held back in C and D by the narrower channels in front of them. For corresponding drawings showing position of plasmasol-stream network in black and plasmagel network in white, see Fig. 6A, B, F, G.

plasmasol in the ameba (drawn in the monopodial form for simplicity), based on evidence from many photos of the type shown in Fig. 5, from direct microscopical observation, and from time-lapse microcinematography. The hyaline layer and most of the finer plasmasol channels are omitted for clarity.

The diagram shows at once why the nucleus is not carried forward with the other particles surrounding it in the axial stream, but main-

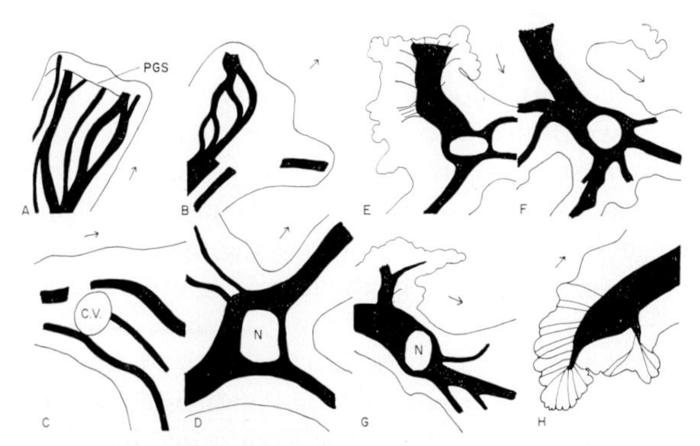

FIG. 6. Drawings showing position of plasmasol-stream network (black) and plasmagel network (white); taken from many photos similar to those in Fig. 5, at various parts of the ameba. Note how nucleus is held back in D, E, F, and G, and the contractile vacuole in C, by the narrower channels in front.

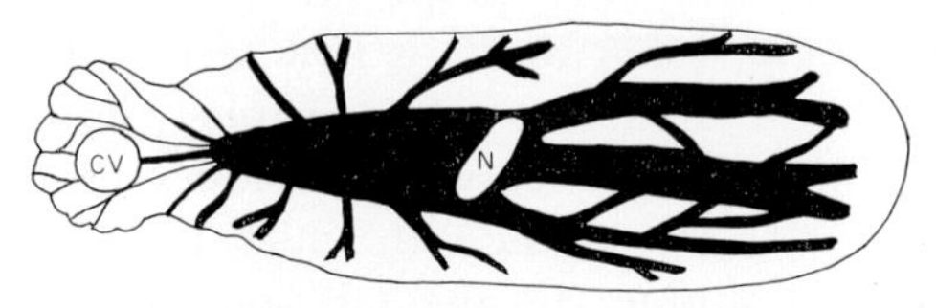

Fig. 7. Composite picture showing distribution of plasmagel (white) and plasmasol stream (black), based on many photographs similar to those in Fig. 5, on cine films, and on direct observation. The smallest plasmasol streams are omitted for clarity. This distribution of plasmagel explains why the nucleus and contractile vacuole maintain a central and posterior position, respectively.

tains a position in the cell a little behind the center. The plasmasol stream behind the nucleus is wide enough to allow its passage, but the plasmasol streams in front of it are too narrow (Figs. 5 and 6D, E, F, G). The nucleus is thus "sieved out." Thus, when ATP was injected into the front end of the ameba so as to cause reversal of flow (Section B,5), it was possible for the nucleus to move up to the anterior end (the former tail) because there was a wide channel with no plasmagel network to obstruct it in this region (Fig. 7). Other means of liquefying the resisting plasmagel network in the anterior half of the cell also allowed the nucleus to move to the front, for example, microinjection of heparin (Section B,6) and heating to just below the point (about 40°C) when the plasmagel completely liquefied and the cell rounded up; this was preceded by very fast streaming, and the nucleus appeared in the front end of the cell. The contractile vacuole appears to be trapped in the thick, soft plasmagel in the tail formed by the pulling inward of the walls by the tail contraction, so that there is no free wide channel in front of it. Large food vacuoles sometimes became trapped in the tail region, also, being pulled around for long periods. They sometimes appeared on cine films, especially in amebae slightly flattened by the cover slip for interference microscopy. The nucleus is rarely trapped in this position because it seems to have the power to liquefy slowly the plasmagel network in front of it, and slowly ploughs a channel through it, as Figs. 5C, and D show. The power of the nucleus to liquefy plasmagel very near it is suggested also by the following observation: Small pieces were cut with a micromanipulator from amebae; nucleated fragments of volume several times that of the nucleus soon became quite spherical and still, as if entirely composed of plasmasol, whereas anucleate fragments, of the same size and at the same time, had irregular shapes, as if containing some plasmagel. Larger nucleated fragments, which had some cytoplasm further away from the nucleus, behaved quite differently and streamed actively.

The tail has a relatively high proportion of "ectoplasm" (Mast and Prosser, 1932; Marsland, 1956) which is now seen to be plasmagel; but this is penetrated by numerous very fine channels leading to the membrane (Fig. 7). These are probably the residues of the much wider channels at the anterior end of the cell, which have been overtaken by the tail and squeezed thinner by the tail contraction. Through these channels can be seen, in some amebae, trains of cytoplasmic granules moving in single file toward the axial streamlike red cells in fine capillaries. It is possible that finer ones exist which are difficult to detect. Only the broader ones could be photographed by the technique used in Fig. 5, owing to blurring by the advance of the tail region during the intentionally long exposure given. However, by detaching amebae from their grip on the glass so that they fell to the bottom of a hanging drop, and streamed without advancing, this difficulty could be partly overcome.

When an injection of ATP was made just below the cell membrane, sometimes a large, clear bubble or blister was formed under the membrane; this always then contracted vigorously, driving the fluid inward and away from the site of the injection (Goldacre, 1952c). This shows that the production of ATP by membrane-plasmagel contact in the tail region (Section C) would tend to drive the contracted and ATP-liquefied plasmagel inward into the posterior axial stream, and the fine channels in the tail plasmagel would provide for this. This observation is thus consistent with the continuous production of ATP in the tail, at the membrane-plasmagel interface.

E. Self-Regulating Mechanism

Now follows one of the most interesting consequences of this system—its power to regulate itself. No other theories of ameboid movement deal with self-regulation in the cell.

1. Speed Regulation

It is noticeable that in fast amebae the hyaline cap is thick, about 20 μ thick in *A. proteus,* whereas in sluggish amebae, such as those in an old culture, it is only 1–2 μ thick or even less. The hyaline layer is only visible at the anterior end of sluggish amebae, whereas in fast-moving amebae it extends down the sides back almost to the tail. This hyaline layer volume is a function of the speed of the ameba and appears to be the physical basis of a speed governor based on the feedback principle (Goldacre, 1954, 1958a, 1961). Briefly, the more vigorous the tail contraction, the greater the amount of hyaline fluid squeezed out from the tail's spongelike plasmagel. This flows forward like a tide (Allen and Roslansky, 1958) and raises the membrane off the surface of the plasmagel, to reduce the plasmagel-membrane contact and hence the rate of ATP production in the tail, resulting in a reduced contraction there. This is a negative feedback process in which an increase in tail contraction increases the hyaline fluid which reduces the membrane-plasmagel contact which reduces the tail contraction. This tends to keep the speed of streaming constant; the speed is, in fact, constant in monopodial amebae (Pantin, 1923; Pitts, 1933; Hahnert, 1932; Goldacre, 1958a, 1961). The speed decreases as the number of pseudopods increases (Pitts, 1933) and an explanation for this, in terms of the above feedback mechanism, has been given by Goldacre (1961).

Negative feedback regulators can be induced to generate oscillations under suitable conditions of time lag, etc. (Wiener, 1948), and a spectacular demonstration of the existence of the feedback speed regulator in the ameba was provided by the cytoplasmic oscillations that occurred when the ameba was flattened with a cell compressor to a critical thickness (Goldacre, 1954, 1958a, 1960, 1961). These had a frequency of about 1/sec—a rate about 50 times as fast as pseudopod formation and retraction.

All living things are goal-seeking organizations, and their apparent purposiveness is a result of built-in negative feedback governors of various kinds (Ashby, 1960; Potter and Auerbach, 1959). The first mechanochemical feedback regulator in the cell was described by Goldacre (1954, 1958a), and the first biochemical self-regulating device (feedback inhibition) was described by Umbarger (1956) and extended by Magasanik (1961). The mechanism described previously for regulating speed in amebae seems also to be involved in regulating where they go, as indicated below.

2. Trophic Mechanisms in the Ameba

An ameba is a goal-seeking organism and, for example, discovers where food is and stays there; it also avoids noxious substances. Although amebae congregate densely and remain almost stationary near a local supply of food organisms, when unfed they move rapidly in an exploring manner and do not appear to be influenced in their direction of movement by individual food organisms unless almost touching them. Observations and cine films revealed that pseudopods of an exploring and unfed ameba frequently retracted from a food organism when a further 10% protrusion would have engulfed the organism, whose suitability as food was demonstrated shortly afterward when it was engulfed by another pseudopod.

The explanation appears to be as follows. Unfed amebae move with only one or a few pseudopods in a direction unrelated to that in which food lies and at relatively high speed until by chance they contact food organisms. These induce pseudopods in greatly increased numbers over the surface of the amebae, causing a reduction in speed, according to the inverse relation between speed and number of pseudopods observed empirically by Pitts (1933) and explained in terms of a feedback mechanism by Goldacre (1961). As a result, the amebae would congregate where their speed is reduced, i.e., in places of food, like molecules of liquid distilling from a warm to a cold spot.

Thus it seems that the feedback mechanism regulating speed of locomotion in terms of number of pseudopods would also serve to lock the ameba on to its goal, i.e., food. What happens when amebae get within very close range of food is described by Jeon and Bell (1962) and Bingley *et al.* (1962), who found that certain tissue extracts provoked pseudopod formation. It is interesting that one of the active substances was heparin, which was also found by Goldacre (1952c) to cause a pseudopod to be pushed out at the site where microinjected (in strong contrast to ATP, which caused the ameba to move away from the site of injection). These opposing effects are illustrated in Figs. 1 and 2.

It is a curious fact that whereas amebae will avoid a drop of toxic substance in a wide dish by veering away from it, they are unable to avoid it if they are placed in a narrow capillary tube in which they cannot turn around and are orientated so as to move toward the toxic substance at the far end. Being unable to reverse, they are moved forward by the inexorable contraction of their tail region until they meet the poison and are killed (Goldacre, unpublished).

A clue to the mechanism of negative chemotaxis was provided by the observation that many toxic substances induce the monopodial form in *A. proteus* in concentrations well below the rapidly lethal concentration; for example, neutral red, which was extensively used in this work because of its visibility, induced the monopodial form at about 0.002% (Goldacre and Lorch, 1950) and killed rapidly at about 0.05%, and a wide variety of fat-solvent type anesthetics made the cell monopodial below the concentration at which the cells were immobilized (Goldacre, 1952b).

In the induced monopodial state, the contraction of the tail and the advance of the leading pseudopod were unaffected. In a naturally polypodial ameba, this must mean that the reagent inhibited selectively the formation of lateral pseudopods, of which there are usually several on each side of the main axis. Analysis of cine films of normal ameboid movement in *Amoeba discoides* revealed that about every 3 min the leading pseudopod retracted and one of the lateral pseudopods became the anterior end of the cell, so that the cell moved off at about 50 degrees to its previous direction. In this way the cell followed a wavy, exploring path. Photos of an ameba rendered monopodial by neutral red and of a normal polypodial ameba in water are given by Goldacre and Lorch (1950).

It seems clear from the observations above that negative chemotaxis would operate as follows: the ameba would move about in the concentration gradient until one side of the ameba arrived at a concentration capable of causing the monopodial form; pseudopod formation on that side would be inhibited and motion would only be possible in other directions.

F. Summary

1. Reasons are given for locating the motor mechanism of ameboid movement in the rear end or "tail" of *Amoeba proteus* rather than in the front.

2. The tail is a distinct physiological, as well as morphological, region of the cell, and behaved like a region of the cytoplasm into which ATP had just been injected with a micropipette.

3. When cell membrane was brought into contact with the plasmagel across the intervening hyaline layer by each of five different methods, contraction occurred in the region of contact.

4. A theory of ameboid movement was elaborated which explains how a continuous supply of ATP could be provided at the site of cytoplasmic contraction in the tail by membrane-plasmagel contact, and how the movements are sustained and regulated by a negative feedback process involving the variable volume of the hyaline layer. This theory was extended to explain the details of the processes whereby amebae congregate where food is and avoid harmful substances.

5. Reasons for the central and posterior positions of nucleus and contractile vacuole were given in terms of the distribution of a plasmagel network extending throughout the cell, which was demonstrated by photographs taken by time exposures.

6. Microinjection of heparin prevented formation of plasmagel and caused local solation of existing plasmagel. This enabled the nucleus to enter the anterior end of the cell, which is normally forbidden to it by the plasmagel net.

Acknowledgments

This investigation has been supported by grants to the Chester Beatty Research Institute (Institute of Cancer Research: Royal Cancer Hospital) from the Medical Research Council, the British Empire Cancer Campaign, the Anna Fuller Fund, and the National Cancer Institute of the National Institutes of Health, U.S. Public Health Service.

References

Abé, T. H. (1961). *Cytologia* **26**, 378.
Allen, R. D. (1960). *J. Biophys. Biochem. Cytol.* **8**, 379.
Allen, R. D. (1961a). *In* "Biological Structure and Function" (T. W. Goodwin and O. Lindberg, eds.), Vol. 2, p. 549. Academic Press, New York.
Allen, R. D. (1961b). *Exptl. Cell Res. Suppl.* **8**, 17.
Allen, R. D. (1961c). *In* "The Cell" (J. Brachet and A. E. Mirsky, eds.), Vol. 2, p. 135. Academic Press, New York.
Allen, R. D., and Roslansky, J. D. (1958). *J. Biophys. Biochem. Cytol.* **4**, 517.
Ambrose, E. J. (1961). *Exptl. Cell Res. Suppl.* **8**, 54.
Ashby, W. R. (1960). "Design for a Brain," 2nd ed., p. 55. Chapman and Hall, London; see also *ibid. Electron. Eng.* **20**, 379 (1948).
Beers, C. D. (1924). *Brit. J. Exptl. Pathol.* **1**, 335.
Bell, L. G. E. (1962). *J. Theoret. Biol.* **3**, 132.
Bingley, M. S., and Thompson, C. M. (1962). *J. Theoret. Biol.* **2**, 16.
Bingley, M. S., Bell, G. E., and Jeon, K. W. (1962). *Exptl. Cell Res.* **28**, 208.
Clark, A. M. (1943). *Australian J. Exptl. Biol. Med. Sci.* **21**, 215.
de Bary, A. (1864). "Die Mycetozoen," W. Engelman, Leipzig, Germany.
De Bruyn, P. P. H. (1947). *Quart. Rev. Biol.* **22**, 1.
Goldacre, R. J. (1952a). *Intern. Rev. Cytol.* **1**, 135.
Goldacre, R. J. (1952b). *Symp. Soc. Exptl. Biol.* **6**, 128.
Goldacre, R. J. (1952c). Ph.D. Thesis, London University, pp. 109-115.
Goldacre, R. J. (1953). *Nature* **172**, 593.
Goldacre, R. J. (1954). *Excerpta Med.* **8**, 408.

Goldacre, R. J. (1958a). *Proc. Intern. Congr. Cybernetics, 1st, Namur, 1956,* pp. 715 and 726. Gauthier Villars, Paris.
Goldacre, R. J. (1958b). *In* "Surface Phenomena in Chemistry and Biology" (J. F. Danielli, K. G. A. Pankhurst, and A. C. Riddiford, eds.), p. 278. Pergamon, New York.
Goldacre, R. J. (1960). *Cybernetica* **2**, 117.
Goldacre, R. J. (1961). *Exptl. Cell Res. Suppl.* **8**, 1.
Goldacre, R. J., and Lorch, I. J. (1950). *Nature* **166**, 497.
Hahnert, W. F. (1932). *Physiol. Zool.* **5**, 491.
Harvey, E. N., and Marsland, D. A. (1932). *J. Cellular Comp. Physiol.* **2**, 75.
Hoffmann-Berling, H. (1960). *In* "Comparative Biochemistry" (M. Florkin and H. S. Mason, eds.), Vol. 2, p. 341. Academic Press, New York; see also *ibid. Biochem. Biophys. Acta* **14**, 188 (1954).
Jeon, K. W., and Bell, L. G. E. (1962). *Exptl. Cell Res.* **27**, 350.
Kavanau, J. L. (1963). *J. Theoret. Biol.* **4**, 124.
Loewy, A. G. (1952). *J. Cellular Comp. Physiol.* **40**, 127.
Magasanik, B. (1961). *In* "Biological Approaches to Cancer Chemotherapy" (R. J. C. Harris, ed.), p. 35. Academic Press, New York.
Margaria, R. (1932). *Boll. Soc. Ital. Biol. Sper.* **7**, 557.
Margaria, R. (1934). *J. Physiol. (London)* **82**, 496.
Marsland, D. A. (1950). *J. Cellular Comp. Physiol.* **36**, 205.
Marsland, D. A. (1956). *Pubbl. Staz. Zool. Napoli* **28**, 182.
Mast, S. O. (1926). *J. Morphol.* **41**, 347.
Mast, S. O. (1931). *Z. Vergleich. Physiol.* **15**, 309.
Mast, S. O., and Prosser, C. L. (1932). *J. Cellular Comp. Physiol.* **1**, 333.
Mast, S. O., and Root, F. M. (1916). *J. Exptl. Zool.* **21**, 33.
Pantin, C. F. A. (1923). *J. Marine Biol. Assoc. U. K.* **13**, 24.
Pitts, R. F. (1933). *Biol. Bull.* **64**, 418.
Potter, V. R., and Auerbach, V. H. (1959). *Lab. Invest.* **8**, 495.
Radir, P. (1931). *Protoplasma* **12**, 42.
Simard-Duquesne, N., and Couillard, P. (1962). *Exptl. Cell Res.* **28**, 85, 92.
Szent-Györgyi, A. (1947). "Muscular Contraction." Academic Press, New York.
Ts'o, P. O. P., Bonner, J., Eggman, L., and Vinograd, J. (1956). *J. Gen. Physiol.* **39**, 325.
Umbarger, H. E. (1956). *Science* **123**, 848.
Wiener, N. (1948). "Cybernetics," Wiley, New York.

Discussion

Dr. Bovee: I hesitate to cast myself in the role of a partial peacemaker in the wonderful argument that seems to have been engendered here between these two theories. I think it is generally agreed there is some sort of fibrous network formed at the anterior end. Whether it actually begins contraction there seems unresolved.

In my short paper presented in 1952, I suggested that in a pressure system the development of the gel at the anterior end might be in the form of fibrous network which, because of the pressure of the flowing endoplasm, might enter into a contraction that would be, in the older terminology, isometric and show no appreciable shortening at first. But as this ectoplasm reached the rear end, under different conditions favoring isotonic contraction there, a visible shortening would result.

Presumably this shortening might be related to the membrane contact Dr. Goldacre has proposed. This would produce pressure from the rear and permit both theories to have at least some relationship to one another.

DR. MARSLAND: I would prefer to think in terms of a tube-wall contraction rather than tail contraction; very likely contraction can occur in any part of the tube wall.

Is it not more likely that the locus of action of the ATPase enzyme is the contractile protein itself, as in the case of muscle? It might then be the ATP that is being carried on the membrane rather than the enzyme.

DR. GOLDACRE: That is a possibility, if enough ATP were carried on the membrane.

DR. GRIFFIN: Your film of the twitching movements demonstrates very clearly the oscillation in the cytoplasm, but I could see no movement of the membrane and no change in thickness of the hyaline layer. I wondered if you could see such a change in the films you have made?

DR. GOLDACRE: I looked for that, but the amplitude of the oscillation is only a few microns, and if you work out what it would be in the membrane itself, it comes to very much less than that.

DR. GRIFFIN: Do you think that a change in hyaline-layer thickness might be occurring above or below the plane of focus?

DR. GOLDACRE: No. I think it would be too small to see.

DR. GRIFFIN: Your postulated control mechanism invokes a movement of the membrane and the making and breaking of cytoplasm-membrane contacts. I had somehow had the impression that your observations supported this hypothesis.

DR. GOLDACRE: There is a diagram to show the feedback mechanism of a speed governor controlling the constant speed.

DR. GRIFFIN: Clearly, an oscillation occurs, but there is in your film no indication that the membrane is involved. Also, I know of no other clear evidence that membrane contact does induce cytoplasmic contraction.

DR. GOLDACRE: There are the 5 experiments reported in my communication. In a rapidly moving ameba, the hyaline layer is thick, perhaps 20 μ; in sluggish amebae, such as in old cultures, it may be only a micron thick. This suggests that the volume of the hyaline fluid is related to the speed of the ameba. I think I have shown how the variable volume of the hyaline layer could be the physical basis of a speed governor.

DR. GRIFFIN: It is an interesting hypothesis. I just don't see any evidence for it. That some sort of feedback occurs during ameboid movement, I have no doubt.

DR. GOLDACRE: I think the hyaline fluid flows in the way Allen and Roslansky showed with their earlier interference microscopy with flattened ameba. That study was done under optical conditions which were better than those used for later work with roughly cylindrical amebae. There they postulate the syneretic fluid flows forward as a tide from the contracting tail.

DR. KITCHING: I should like to ask Dr. Goldacre how this self-regulating mechanism of propulsion is modified in response to external conditions; in other words, what is the basis of behavior in the ameba?

DR. GOLDACRE: I think you can use this feedback mechanism to show two things: First it explains Pitts' observation on polypodial amebae. They move more slowly than monopodial specimens because the total contraction rate in the cell is constant and shared over all the retracting pseudopods, leaving less for the tail. It also explains why they congregate near food. Food induces many pseudopods by some unknown mechanism which slows them down and, therefore, they congregate where food is. It is like molecules going from a hot place to a cold place. They distill over on to the cold place because their movement is slower there.

DR. KITCHING: That does not really explain the question. It is why the pseudopods grow where there is food.

DR. GOLDACRE: I am not attempting to explain that, but am using that fact to explain why they congregate. Recently Bell and Jeon at Kings College have shown that heparin and extracts of hydra and various ciliates, when placed in a capillary tube near an ameba, will cause pseudopods to form. I have shown that injected heparin does the same thing. ATP and heparin have the opposite effect on locomotion, for ATP causes the cell to move away from and heparin toward the site of injection.

DR. KITCHING: Schaeffer in 1917 demonstrated responses toward food material. But the problem is: How does the food material affect the moving ameba?

DR. GOLDACRE: Heparin will do it, and heparin-like extracts of hydra. A most beautiful film Bell and Jeon have made at Kings College shows amebae "playing football" with small round pieces of hydra tissue put into a petri dish. The rolling ball, pushed forward by any ameba nearby, activates other amebae to move toward it whenever it comes within range of them.

DR. INOUÉ: Could you tell us what forces you found with your elegant microspring balance in different portions of the ameba?

DR. GOLDACRE: The force in the tail or the contracting pseudopod—it was of the order of a hundredth of a milligram. Elsewhere there was no pulling whatever.

DR. INOUÉ: No pulling or expanding?

DR. GOLDACRE: No.

DR. INOUÉ: Only in the tail?

DR. GOLDACRE: Only in the tail and contracting pseudopods.

DR. MARSLAND: I would like to know whether the mucoprotein is considered to be equivalent to what Mast called plasmalemma, or is it possible that we have a mucoprotein layer, a protein layer (plasmalemma), and then the true cytoplasmic membrane?

DR. GOLDACRE: I think they are two quite separate things, and the electron microscope suggests the mucoprotein is the fringe which is seen on the outside of the Danielli double-layered plasma membrane, elements about 1000 A long and 100 A in diameter.

DR. WOLPERT: Really, it is much more protein than polysaccharide. The membrane analysis is about 32% lipid, 12% polysaccharide, and about 24% protein; so far we cannot account for 100% of the mass in terms of these substances.

The Motive Force of Endoplasmic Streaming in the Ameba[1]

NOBURÔ KAMIYA

Department of Biology, Faculty of Science, Osaka University, Osaka, Japan; Department of Biology, Princeton University, Princeton, New Jersey; and The Marine Biological Laboratory, Woods Hole, Massachusetts

The subject matter of the present study is concerned with motive force measurement of endoplasmic streaming in a free-living ameba, and motive force changes under normal and experimental conditions. For most of our experiments, *Chaos chaos* was used, but sometimes *Amoeba proteus* was used for comparison. They were fed with *Tetrahymena* and cultured in a medium of low ionic strength (Prescott and James, 1955).

The motive force responsible for protoplasmic streaming was first measured in the slime mold *Physarum polycephalum* by what we called the "double-chamber method" (Kamiya, 1940, 1942, 1953, 1959). The measurement was based on the null method; that is, the motive force was found from the counter pressure that just stopped endoplasmic streaming. In order to conduct such an experiment, a chamber divided into two compartments was constructed and the dumbbell-shaped myxomycete plasmodium was so placed that two plasmodial masses were connected with each other by a plasmodial strand penetrating the central partition.

The method used for slime mold is applicable, at least in principle, to other cells showing ameboid movement. However, amebae, even in the case of *Chaos chaos,* are much smaller than the plasmodium of *Physarum,* and we are inevitably confronted with the technical problem of how to construct a double chamber fitted to them.

Double Chambers Applicable for Studies on the Ameba

In making such a chamber, I worked out several methods of which two are most convenient for many experimental purposes.

(*a*) *Agar capillary method.* One method is to construct an agar chamber. We first prepared a pair of glass capillaries, each of which had

1 This study was supported in part by a Research Grant (RG8691) from the Institute of General Medical Science, U.S. Public Health Service, administered by Dr. R. D. Allen of the Department of Biology, Princeton University.

an orifice constructed at the end by fire-polishing with a microflame (Fig. 1A). The tip was then ground off slightly into a flat end (Fig. 1B). Two such capillaries were placed so that they faced one another in a pair of glass sleeves mounted in a Lucite frame. Then, a glass fiber just

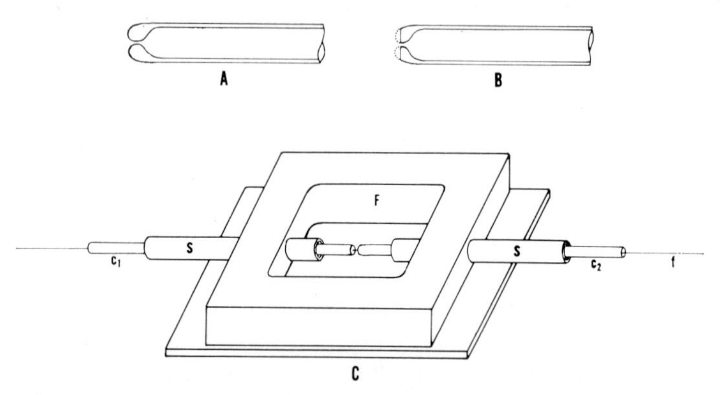

FIG. 1. (A) A fire-polished end of a glass capillary. (B) The same capillary after the tip (the dotted parts) was ground off. (C) The chamber in perspective before agar is poured. F—Lucite frame; S—glass sleeves to hold the capillaries (C_1 and C_2); f—the central glass fiber penetrating the capillaries through their terminal constrictions.

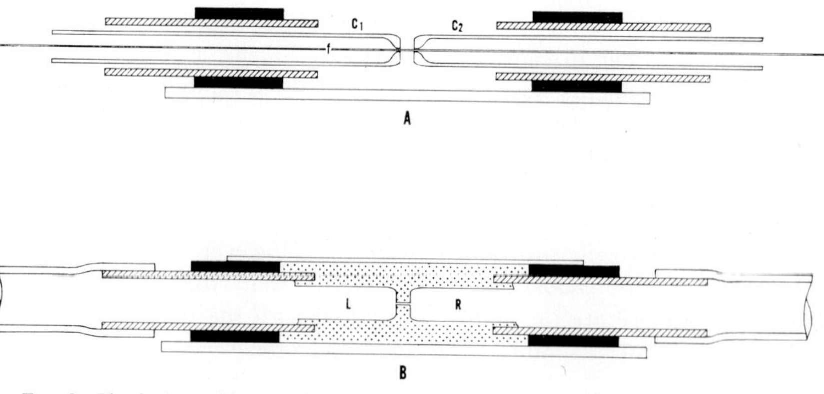

FIG. 2. Vertical sections at the median plane of the chamber. (A) Before agar is poured. Black parts—Lucite frame; shadowed parts—glass sleeves to hold the glass capillaries (C_1, C_2); f—central glass fiber. (B) After molten agar was poured and the central glass fiber and a pair of capillaries were removed. Dotted parts—agar gel; L, R—left and right chambers, respectively.

small enough to fit through the holes was introduced through both capillaries (Fig. 1C, Fig. 2A). When the pair of capillaries had been brought in a proper position, molten agar (4% in Prescott-James solution) was poured into the frame, and a cover slip was placed on top.

After the agar had set into a stiff gel, the central glass fiber was pulled out first, and then the two glass capillaries. Two cylindrical hollow spaces (L and R) connected by a narrow channel were left behind in the agar gel (Fig. 2B); these spaces were then filled with Prescott-James solution. By this method, we can now make a narrow tunnel with uniform thickness and of any desired length between the two chambers.

An ameba is introduced from one end by applying an appropriate hydrostatic pressure difference (usually in the range of 20–80 mm water column) between the two sides, L and R. What is a suitable bore for the

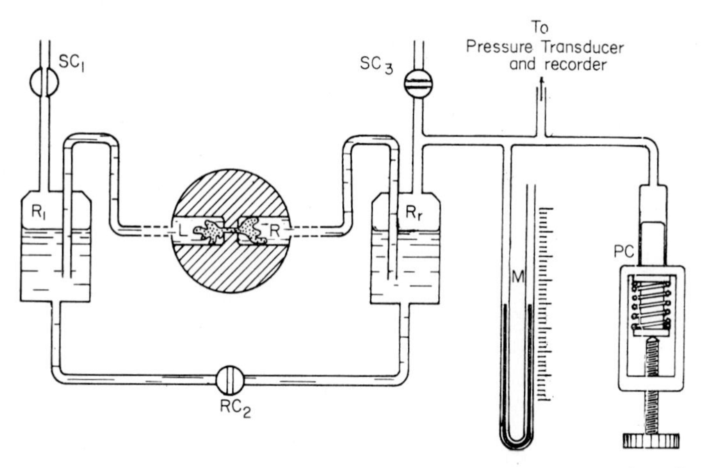

FIG. 3. A diagram of an arrangement for controlling and measuring the hydrostatic pressure difference between the two chambers into which an ameba is separated through a hole. The encircled, shadowed part is enlarged. L, R—left and right chambers, respectively; R_l, R_r—left and right reservoirs, respectively; PC—pressure controller; M—manometer; SC_1, SC_2, SC_3—stopcocks.

channel depends naturally on the material used. For *Chaos chaos,* a channel with a bore of 50–130 μ is appropriate; for *Amoeba proteus* a suitable diameter is in the range of 30–60 μ.

In order to control and measure the hydrostatic pressure difference across the channel penetrating the central partition, chambers L and R are connected to reservoirs R_l and R_r whose water level is kept exactly the same before the experiment by opening the stopcocks SC_1, SC_2, and SC_3 (Fig. 3). As shown in the figure, R_r is connected to a vertical water manometer and pressure controller composed of a 10-ml injection syringe and a screw. Water reservoir R_l is left open to the atmosphere. Pressure difference between the chambers L and R induced by the pressure controller manually is read from the manometer or recorded electrically by means of a pressure transducer on a recorder.

(*b*) *Constricted orifice method.* The other method which is particularly convenient for certain physiological experiments is to use a glass capillary with a fire-polished constriction at its end (cf. Fig. 1A). First, a piece of Lucite block having a central pool (P) in the form of an inverted cone and a pair of built-in stopcocks (SC_1, SC_2) is prepared (Fig. 4A and B). The construction is such that a pair of glass capillaries (C_1 and C_2) connect the pool P and small spaces under stopcocks SC_1 and SC_2 respectively, the narrowed orifice of C_2 being at the center near

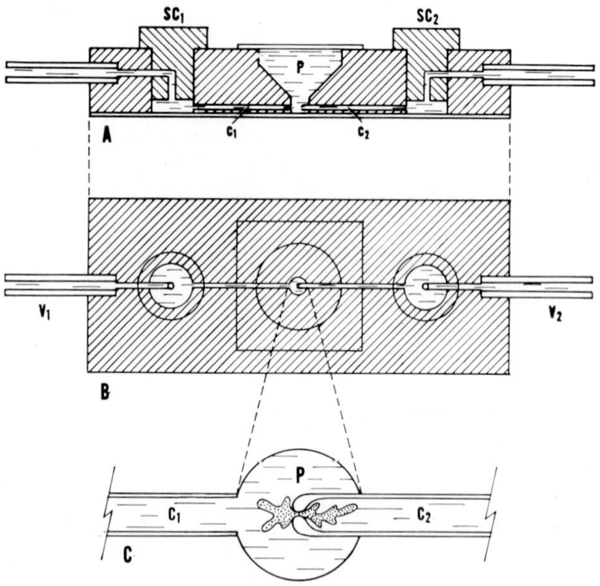

FIG. 4. A Lucite chamber arranged for the constricted orifice method. (A) Side view. (B) Top view (about the actual size). (C) The bottom area of the pool with the constricted orifice of a capillary loaded with an ameba (enlarged). P—pool; C_1, C_2—capillaries; SC_1, SC_2—stopcocks; V_1, V_2—vents.

the bottom of the pool; the bottom of the Lucite block is closed with a cover slip and sealed with epoxy resin. The entire volumes of the pool P, capillaries C_1 and C_2, and stopcocks (SC_1 and SC_2) are filled with Prescott-James solution. The vents of the chamber, V_1 and V_2, are connected to reservoirs R_l and R_r just as before (cf. Fig. 3).

After several amebae are introduced from the top, into this pool, the top of the pool is covered with a piece of cover slip. The amebae gradually sink to the bottom. If negative hydrostatic pressure is applied to capillary C_2, the water in the pool P starts flowing into C_2. As soon as one of the amebae is carried with the water current toward the hole and blocks it, what appears to be a pseudopodium is induced to form into the hole by the pressure gradient. Later, a considerable amount of endoplasm is transported into the capillary C_2. Thus, we have divided an ameba into two parts—one in the pool P and the other in the capillary C_2—the two being connected by a narrow neck of cytoplasm at the hole (Fig. 4C). While looking at the endoplasmic streaming at the neck by an inverted microscope from below, the hydrostatic pressure in the capillary (C_2) is controlled and measured with the arrangement shown in Fig. 3.

This method is convenient for use in experiments which test the effect of various chemical reagents, since we can easily replace the greater part of a solution in the pool P with another by using a pipette and closing the stopcocks SC_1 and SC_2 to keep the ameba in position. It is also possible by this method with a pair of nonpolarizable electrodes to measure electrical potential differences between the two media to which two different parts of ameba are exposed.

Behavior of an Ameba under an External Pressure Gradient

First, we shall discuss the behavior of an ameba in the double chamber under an external pressure gradient. For such observations the agar double chamber is more advantageous than the glass, since in the former, the connecting capillary is evenly thick and no shadow is formed because of the small difference between the refractive indices of the 4% agar gel and water. As soon as an ameba is carried by the current of water so that it closes the entrance of the agar capillary, a hyaline cap is formed toward the interior of the capillary. The endoplasm then "pours" into the capillary to form an induced pseudopodium [Fig. 5(1)]. This induced pseudopodium formed in the agar capillary lengthens rapidly at first, but less rapidly later under the same external pressure gradient (for instance, 30 mm of hydrostatic head across 0.5-mm agar wall). When the pseudopodium has proceeded some distance into the agar capillary, it often ceases to advance or may even retract against the external pressure gradient as if the ameba were actually resisting the pressure imposed upon it from the outside. In such a case we need to apply greater pressure differences in order for further pseudopod formation to occur.

When the hemispherical advancing front of the pseudopodium reaches the exit of the narrow channel, a more-or-less spherical pseudopod is formed at the advancing front in the other chamber [Fig. 5(2) and (3); Fig. 6]. At this time the speed of endoplasmic inflow through the hole into that pseudopodium suddenly increases and is accompanied by an increase of its surface area. It is at this stage that the ameba is most vulnerable to bursting at its front end. The external pressure gradient must then be made smaller so that a too rapid increase in the surface area of the front end is avoided. A series of pictures of *Chaos chaos* forced to move under pressure gradient from one chamber to the

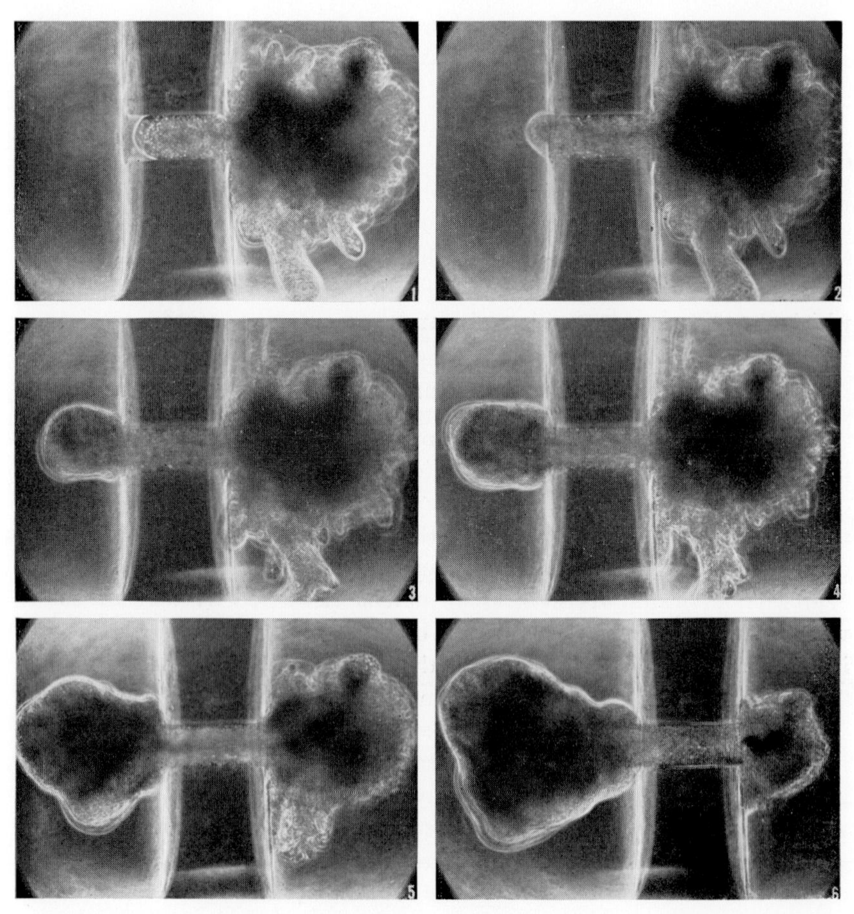

FIG. 5. An ameba (*Chaos chaos*) being introduced into one chamber from the other through agar capillary under pressure gradient. Six photographs were taken at unequal intervals in 3 min. Hydrostatic pressure in the right chamber was higher than in the left, at first by 60 mm and later by 30 mm of water column. Diameter of the capillary—115 μ; wall thickness—280 μ.

other through a channel penetrating the central agar wall are shown in Fig. 5.

The pressure necessary to suck an ameba into a capillary varies greatly depending upon the morphological site at which the suction is applied and also on the physiological condition of the animal. Amebae starved for several days are more resistant to bursting than those soon after feeding. Although an ameba can be sucked in from any part of the body, the advancing tip of the naturally formed pseudopodium can be sucked most easily with a minimum of pressure difference. If the side wall of a pseudopod or the tail region of the animal is brought into the capillary by pressure, this part is now converted into the advancing front which is characterized by a hyaline cap formation and a smooth

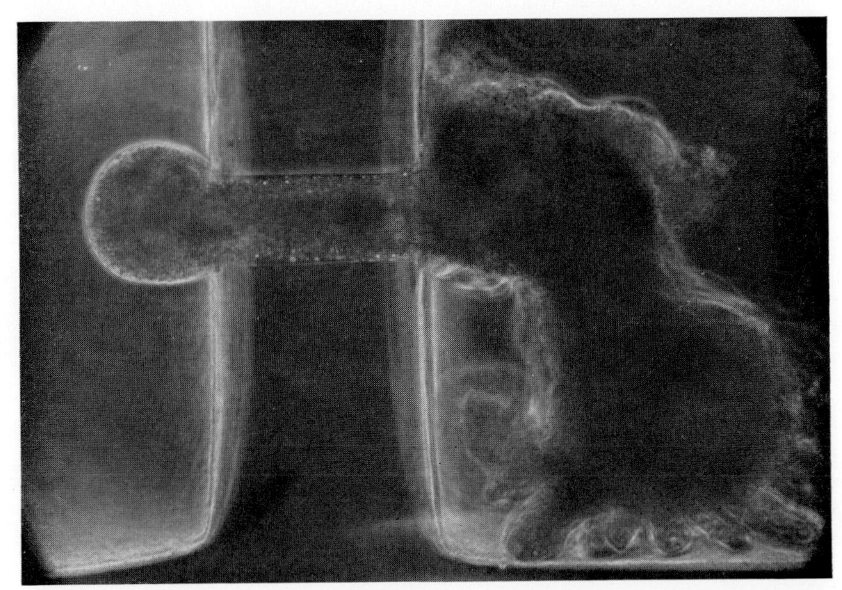

FIG. 6. Formation of spherical pseudopod (*Chaos chaos*) at the orifice of the capillary. The chamber used is the same as that in Fig. 5. Pressure difference—25 mm of water.

surface. Irrespective of what part of the ameba is first sucked from one end, or pushed from the other end, into the capillary, the artificially induced, endoplasmic streaming is usually accompanied by the formation of a hyaline cap at the advancing front as was pointed out by Mast (1931a). Often this is also true of an ameba which is not confined in a capillary but has much greater free surface.

When such an ameba is divided nearly equally into two chambers with a narrow streaming channel in-between, one-half of the body of the ameba, toward which the endoplasm is made to stream artificially for an extended period, always takes the form of an advancing front, whereas the other half of the body, from which the endoplasm is removed, takes the form of the tail. If a considerable amount of endoplasm is forced to move from one end to the other by external pressure, then, when the pressure is released, the endoplasm tends to "spring back" sometimes from the hyaline cap side, or the "head," toward the "tail" characterized by a wrinkled surface. In this case, therefore, morphological characteristics and the direction of the motive force behind streaming are just the opposite to the normal ameboid movement. At any rate, an ameba always behaves in the experimental situation as if its morphological characteristics of the head and tail are a consequence of endoplasmic flow rather than its cause, since the front end of the endoplasmic streaming, no matter whether it is spontaneous or induced by pressure gradient, invariably assumes the form of head and the rear end, the form of tail. Although the artificially induced head and tail were not identical in their organization with those formed under normal conditions, this fact may be significant in considering the cause and effect in the events involved in ameboid movement.

Measurement of the Motive Force

When a single ameba extends into both chambers across the central septum (Fig. 7) and when there is no pressure difference between the two, the endoplasm in the connecting channel moves toward one direc-

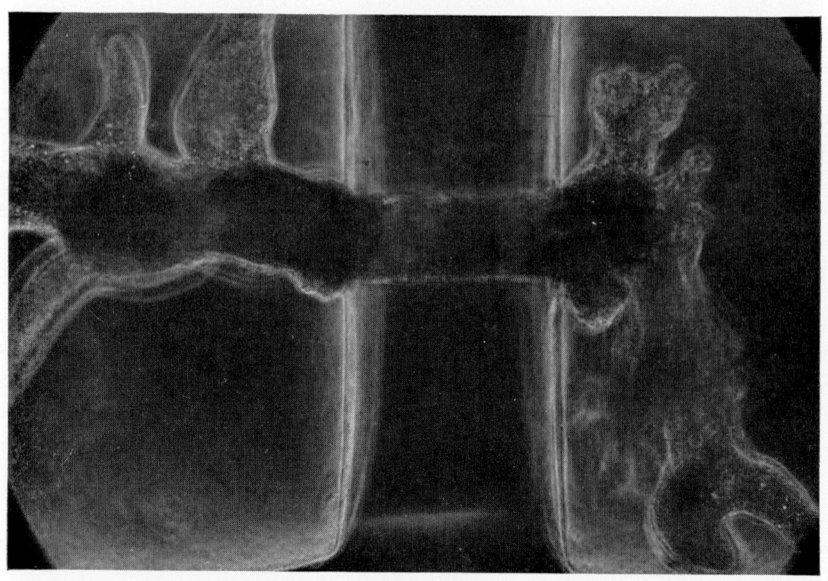

FIG. 7. A single ameba (*Chaos chaos*) divided into two chambers across the capillary in the central wall. The photograph was made after the endoplasm in the connecting strand had been kept at a standstill for several minutes. Note the formation of pseudopodia on both sides of the central wall.

tion or the other by the motive force generated in the living ameba. Under such circumstances the endoplasm can be influenced extremely sensitively by the hydrostatic pressure difference established artificially between the two chambers. By controlling the direction and magnitude of the pressure gradient, it is possible to accelerate, retard, or reverse the natural endoplasmic flow along the connecting capillary. By this means, it is not difficult to find just how much pressure is necessary to keep the endoplasm in the connecting capillary at a standstill.

Keeping the endoplasm stationary in the connecting capillary is so delicate an operation that endoplasm starts flowing in either direction if the counter pressure is out of balance by as little as 0.5 mm of water or less. There is no perceptible "play" between the movement of the endoplasm and external pressure difference. Therefore the balance pressure can be determined extremely accurately. The counter pressure to keep the endoplasm at a standstill is usually less than 15 mm of water. Allen and Roslansky (1959) introduced a pseudopodium of *Chaos carolinensis* into a glass capillary and applied a hydrostatic pressure inside the capillary against the advancing pseudopodium for a few seconds to stop the endoplasmic streaming. They found that the minimum value to stop the streaming was 1.10 cm which is very close to the value obtained by the double-chamber method.

Dynamoplasmograms

In order to keep the endoplasm in a channel between the two parts of the ameboid body, it is necessary to modify the counter pressure continuously. This is because the motive force generated in the ameba changes spontaneously under constant external conditions. By plotting a series of successive values of the balance pressure against time, or by recording the pressure change electrically using a pressure transducer, we obtained some rather complicated curves which denote exactly how the motive force behind the endoplasmic streaming in the connecting channel has changed.

The curve representing the motive force of the protoplasmic streaming has been referred to by myself as a dynamoplasmogram (DPG) (Kamiya, 1942, 1953, 1959). Figure 8 shows an example of the dynamoplasmogram of *Chaos chaos.* Circles plotted represent instantaneous values read by another observer every 10 sec from the manometer. It is shown here that the motive force has changed in the range of ± 15 mm. The points where the curve intersects the baseline represent the moments when the motive force was zero and its direction had reversed spontaneously. Generally, motive-force curves of the ameba do not manifest themselves as regular waves, such as those in myxomycete plasmodium, but it is clear from this figure that the magnitude and direction of the force are constantly changing.

It is noted in this case that even though the endoplasm is kept quiet by this means in the connecting capillary, the two ends of the ameba, which are outside of the capillary, show normal and vigorous ameboid movement accompanied by endoplasmic streaming, and new pseudopodia or tails are being constantly formed. Two parts of the ameba on the opposite sides of the central septum behave in this case as though

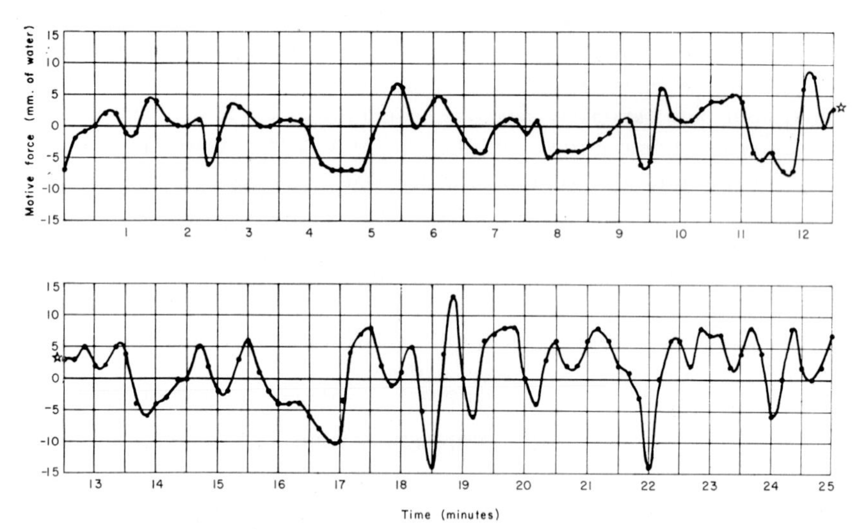

FIG. 8. A dynamoplasmogram of *Chaos chaos* in a normal state (temperature, 23°C).

they are two independent amebae so long as the endoplasmic flow in the connecting capillary is kept immobile.

Motive Force at Different Sites of an Ameba

The motive force responsible for endoplasmic streaming in an ameba is generally not the same in different parts of the body.

In order to measure the motive force at different sites of one and the same ameba simultaneously, we placed a pair of capillaries with fire-polished constrictions at their ends in such a way that the constricted ends closely faced each other at the bottom of the pool similar to that shown in Fig. 4. By this means, it is not difficult to load an ameba in a manner shown in the inset of Fig. 9. As is shown there, an ameba body is separated into three parts, capillary C_1, pool P, and capillary C_2

through two necks at the capillary ends. The pool P (cf. Fig. 4) is uncovered in this case and its free water surface is well balanced with those of reservoirs R_1 and R_r (cf. Fig. 3). Each of the capillaries, C_1 and C_2, is connected to an independent manometer and pressure controller. With this twin system, it is possible to keep the endoplasm at both necks immobile and thus to measure the motive force at the two different parts of the same ameba simultaneously.

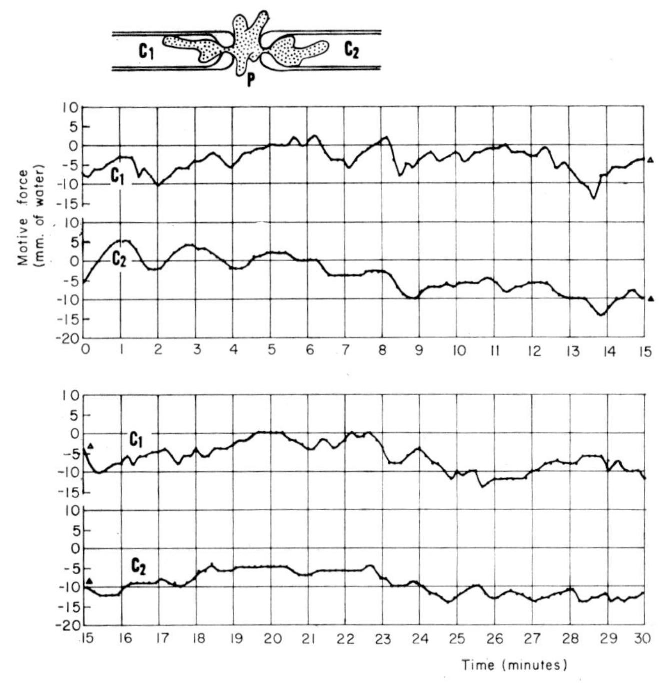

FIG. 9. Dynamoplasmograms at two loci of one and the same ameba (*Chaos chaos*). The upper curve shows the balance pressure in C_1, and the lower curve that in C_2.

Figure 9 shows an example of this kind of measurement. Balance pressures in C_1 and C_2 were read alternately every 5 sec from the two manometers, i.e., successive values of the balance pressure in one of the two capillaries, C_1 and C_2, were determined at 10-sec intervals. The upper curve in Fig. 9 shows the balance pressure in C_1 in reference to P, and the lower curve that in C_2. From this figure it can be seen that there is a general parallelism between the upper and lower curves. Since the upper and lower curves represent the hydrostatic pressures in the capillaries C_1 and C_2 which were just sufficient to keep the endoplasm at a standstill at the necks of the capillaries, the parallelism of the two curves means that changes in motive-force generation occurred both distally or proximally relative to the central portion. This in turn may imply that the motive forces are governed in this case primarily by the central part of the ameba, that is, the part formed in the pool P.

The parallelism between the two curves, however, is true only in their general trend. If we compare the two in detail it is clear that they are by no means identical, showing that the force behind the endoplasmic streaming is different in different portions of the cell.

Effect of ATP, AMP, and Pyrophosphate[2]

The ameba double-chamber method gives us a unique opportunity to study quantitatively the effect of various chemical and physical agents

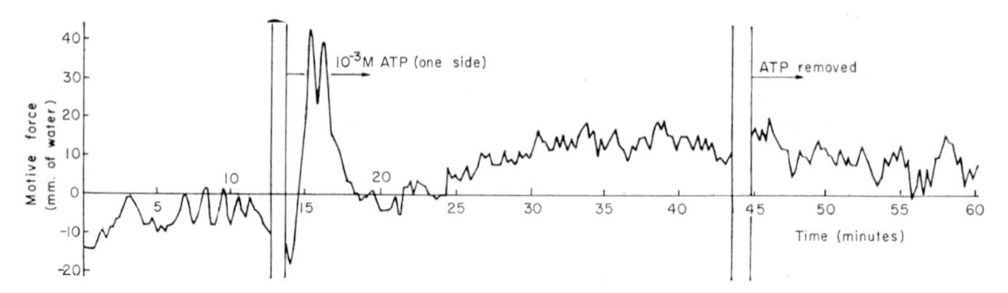

FIG. 10. The effect of 10^{-3} *M* ATP on motive-force generation in *Chaos chaos*. ATP dissolved in Prescott-James solution (adjusted to the pH of the original Prescott-James solution with NaOH) was administered to only one-half of the ameba (P in Fig. 4) whereas the other half (C_2 in Fig. 4) remained in the plain Prescott-James solution.

on the motive force. Now I would like to give some examples of the results obtained by this method.

In view of the paramount physiological importance of adenosine triphosphate (ATP) and related substances in mechanochemical systems involved in various cellular motions, we shall first consider how endoplasmic streaming reacts when these substances are applied extracellularly. Figure 10 shows a dynamoplasmogram of an ameba when 10^{-3} *M* ATP was applied to one side (pool side in Fig. 4) of the animal. The initial part, up to 13 min, represents motive-force generation under the normal condition of an ameba. By admitting ATP thereafter, the motive force increased conspicuously for a few minutes. Since the pressure was controlled on the side (C_2 in Fig. 4) where there was no ATP, the fact that more pressure was needed to stop the inflow of the endoplasm showed that endoplasm tended to flow predominantly from the ATP

---

[2] I am indebted to Dr. H. Nakajima for his cooperation in these experiments.

side toward the opposite side where no ATP was present. Although this does not necessarily mean that ATP would act in this case as an energy donor, this result confirms the observation of Goldacre and Lorch (1950) on microinjected ATP. Although the effect of ATP is transient, such a reaction is not observable with adenosine monophosphate (AMP) at the same concentration, as shown in Fig. 11. But it does occur as conspic-

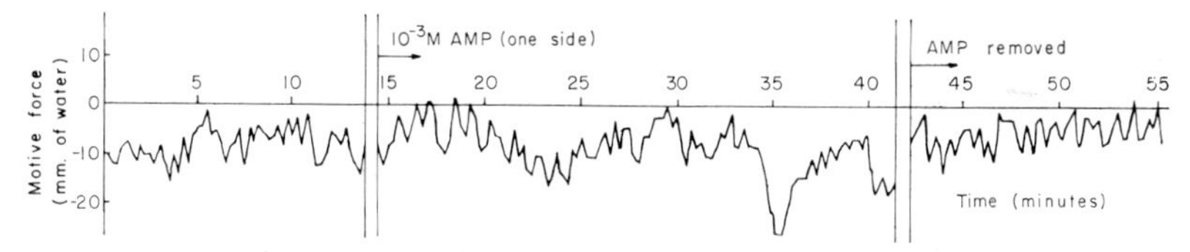

FIG. 11. The effect of 10^{-3} *M* AMP. No appreciable change in motive-force production is observable.

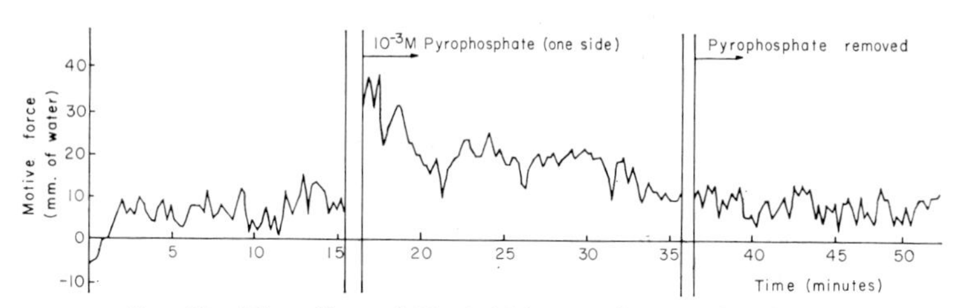

FIG. 12. The effect of 10^{-3} *M* inorganic pyrophosphate.

uously with inorganic pyrophosphate as with ATP (Fig. 12). Orthophosphate has no effect.

Motive-Force Generation in Relation to Bioelectrical Phenomena

Now I would like to discuss the relation of motive-force generation to some bioelectrical activities of the ameba. A series of experiments along this line have been performed in collaboration with Dr. Ichiji Tasaki.

Figure 13 shows the effect of electric current on motive-force production. An ameba is placed in a glass capillary chamber so that two parts of the body are separated by the constriction at the capillary end, and a direct current of about 5×10^{-8} amp was applied through nonpolarizable electrodes across the body of the ameba as is indicated by the inset. The dynamoplasmogram in the initial 4 min shows the motive-force generation under normal conditions. Contrary to the previous figures, the motive force is designated here (and also in Figs. 14 and 15) as plus when the balance pressure inside the capillary is negative, i.e., when the endoplasm tends to flow from the left to the right (cf. the inset of

Fig. 13), and as minus in the opposite case. The direction of the current is taken as positive when the outside of the capillary is anodal and negative when it is cathodal. From Fig. 13, we see that as soon as the current is applied with the cathode on the right the ameba tends to move toward the cathode, since a positive pressure head of nearly 20 mm water is needed on that side to stop the flow. On breaking the current, the motive force shows a "rebound" before it returns to the original level. When the

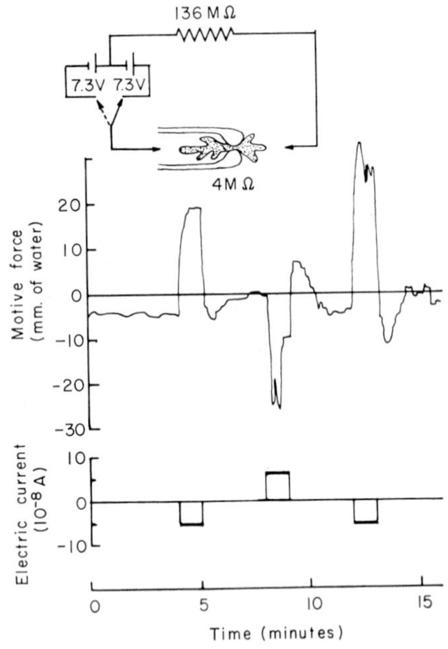

FIG. 13. The effect of electric current through ameba (for details see text).

current is applied in the opposite direction, the motive force is immediately reversed as shown by the downward trend of the motive-force curve. The same process can be repeated many times. It must be noted in this case that the hyaline cap is invariably formed at the cathodal end (Mast, 1931b; Hahnert, 1932).

In the experiment shown in Fig. 14 the direction of the current was reversed every 2 min. The direction of the motive force was so changed without exception so that the endoplasm tended to move toward the cathode. Thus the cathodal galvanotaxis of ameba has been now firmly established on a quantitative and dynamic basis.

Instead of applying electric current through the ameba, it is also possible to measure and record the normal bioelectric potential difference between the two ends of an ameba from the outside. In order to do this the two external media are connected to amplifying and recording circuits through a pair of nonpolarizable electrodes (Ag-AgCl-agar).

By this means it was made clear, first of all, that the pressure-induced acceleration or reversal of the endoplasmic streaming does not alter the potential difference between the two ends of an ameba appreciably. In other words, there is no evidence to show the occurrence of a streaming potential in the ameba just as in the case of the slime mold (Kamiya and Abe, 1950).

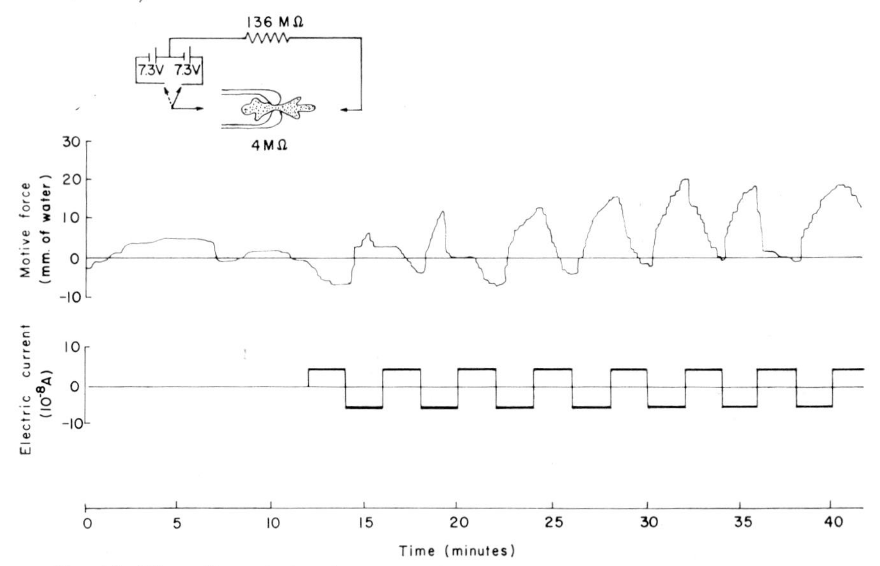

FIG. 14. The effect of electric current in the form of alternating square waves.

The extracellular potential difference between the two parts of one and the same ameba changes spontaneously even though the endoplasm in the connecting capillary is kept at a standstill. The upper curve in Fig. 15 is an example of the record of the extracellular potential change of *Chaos chaos* when the streaming was kept quiet at the connecting capillary. A characteristic feature of this curve is that there are numerous rapid and conspicuous potential changes, or "spike potentials," of variable amplitude, but the potential comes back quickly to the original level. Without going into the details of the physiological nature of these spike potentials, it suffices to say that they are accompanied by a temporary drop in membrane impedance and, second, that they are reversibly eliminated by anesthetics, such as cocaine and ether, and by the absence of bivalent cations.[3] All these characteristics are in common with those of action potentials in nerve or muscle fibers, except that the

---

[3] Details of the work will be dealt with elsewhere by I. Tasaki and N. Kamiya.

spike potential in ameba does not obey the "all-or-none" law. There seemed to be no clear correlation between production of spike potentials and pseudopod formation or other visible morphological changes.

Beside the rapid spike potentials it was noticed here that the resting potential level fluctuated spontaneously within the range of several millivolts. The lower wave train shows the spontaneous changes in the motive force as recorded simultaneously with a pressure transducer. Here we see that there is some correlation between the two. In this graph, the upper curve represents the potential of pool P (cf. Fig. 4) in reference to the inside of the capillary C_2, and the lower curve shows the balance pressure in the pool in reference to the pressure inside the capillary.

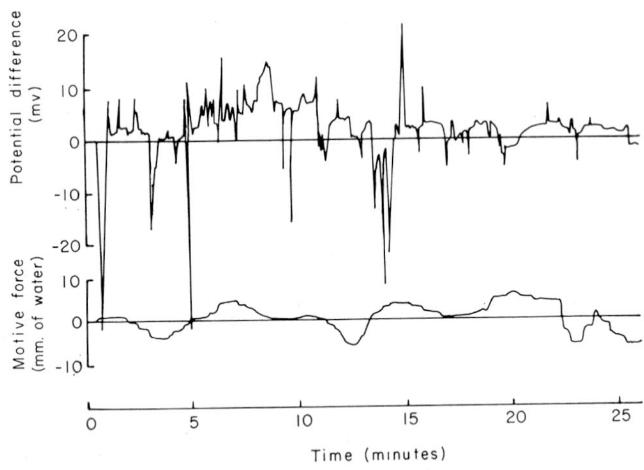

FIG. 15. Simultaneous records of extracellular potential difference and the motive force responsible for endoplasmic streaming in *Chaos chaos* (for details see text).

In the case of plasmodia, it was demonstrated some time ago (Kamiya and Abe, 1950) that the generation of the motive force is accompanied by changes in outside potential difference, and that the phase of the potential wave lags behind that of the dynamic wave. The relation between the electrical and dynamic activities in ameba seems to be similar to but not so distinct as that in the slime mold. The parallelism between the electrical and dynamic activities often ceases when the ameba is treated with monovalent and bivalent cations, anesthetics, etc. The causal relations of the two activities remain to be analyzed.

Concluding Remarks

In the present work I have described the methods worked out for measuring the motive force of the endoplasmic streaming in free living amebae and presented some facts revealed by taking advantage of the methods. The ameba double chamber should be applicable with some technical modifications to the study of motile systems of much smaller cells exhibiting ameboid movement and also for measuring extracellular potential differences of these cells.

Although the mechanism of ameboid movement has not been discussed previously, it is my hope that experimental facts to be obtained by the previously described method will help one to understand the nature of the ameboid movement of a variety of cells from a new angle and to analyze this complicated phenomenon on a quantitative basis.

Summary

1. Methods were introduced for measuring the motive force responsible for endoplasmic streaming in free-living amebae.

2. When a single ameba is loaded in an agar capillary double chamber, or in a glass capillary through its constricted end, an external hydrostatic pressure gradient can counterbalance the endoplasmic streaming.

3. The balancing of the motive force with an external pressure gradient is so delicate that the endoplasm starts flowing in one direction or the other when the counter pressure deviates from the balancing points by less than 0.5 mm of water.

4. A new "pseudopod" is formed artificially in the hole by the externally applied pressure gradient.

5. When the endoplasm is forced to flow by the external pressure gradient, the advancing front always assumes the morphology of the head characterized by smooth surface and often by hyaline cap formation. The rear region, on the contrary, shows the morphology of the tail characterized by a wrinkled surface.

6. Spontaneous changes in the motive force behind endoplasmic streaming are represented in an undulating curve (dynamoplasmogram) with variable periods and amplitudes. The motive force in *Chaos chaos* changes usually within the range of ± 15 mm of water under normal physiological conditions.

7. The motive forces can be measured simultaneously at different loci of the same ameba. They are correlated with each other, but not identical in pattern.

8. ATP (10^{-3} *M*) and inorganic pyrophosphate (10^{-3} *M*) augment the motive force conspicuously, whereas AMP and orthophosphate have no effect at the same concentration.

9. Motive-force changes during cathodal galvanotaxis were studied.

10. Pressure-induced artificial acceleration, retardation, or reversal

of the streaming is not accompanied by the perceptible potential change, i.e., there is no evidence of a streaming potential.

11. The extracellular potential difference between two regions on the ameba surface changes spontaneously. There are slowly fluctuating potential changes and rapid potential changes, or "spike potentials," of variable amplitudes. The causal relationships of the dynamic and electrical activities are still obscure.

Acknowledgments

The present work was started in the summer of 1962 at The Marine Biological Laboratory at Woods Hole where I enjoyed the privileges of Senior Lalor Fellow through the courtesy of the Lalor Foundation. The major part of this work was developed and elaborated during my sojourn at the Department of Biology, Princeton University.

It is a pleasant duty for me to express my sincere gratitude to Dr. C. Lalor Burdick, the director of Lalor Foundation, Dr. Arthur K. Parpart, the Chairman of the Department of Biology, Princeton University, and especially to Dr. Robert D. Allen for all the support and encouragement which made the present work possible.

References

Allen, R. D., and Roslansky, J. D. (1959). *J. Biophys. Biochem. Cytol.* **6**, 437.

Goldacre, R. J., and Lorch, I. J. (1950). *Nature* **166**, 497.

Hahnert, W. F. (1932). *Physiol. Zool.* **5**, 491.

Kamiya, N. (1940). *Science* **92**, 462.

Kamiya, N. (1942). *In* "The Structure of Protoplasm" (W. Seifriz, ed.), pp. 199-244. Monogr. Am. Soc. Plant Physiol., Ames, Iowa.

Kamiya, N. (1953). *Ann. Rept. Sci. Works, Fac. Sci. Osaka Univ.* **1**, 53.

Kamiya, N. (1959). *Protoplasmatologia* **8**, 3a.

Kamiya, N., and Abe, S. (1950). *Colloid Sci.* **5**, 149.

Mast, S. O. (1931a). *Protoplasma* **14**, 321.

Mast, S. O. (1931b). *Z. Vergleich. Physiol.* **15**, 309.

Prescott, D. M., and James, T. W. (1955). *Exptl. Cell Res.* **8**, 256.

Discussion

Dr. Wolpert: I just wondered to what extent you feel your work fits in with the work of Bingley and Thompson.

Dr. Kamiya: They measured potential difference between the two loci of the interior of the ameba, using one or two microelectrodes. Recently, Dr. Tasaki and I studied some electrical characteristics of *Chaos chaos*. In some of our work we used microelectrodes. Using two microelectrodes, we could also detect fluctuations of the potential difference within a single ameba, but we are not certain what kind of potential difference it is we are measuring, since the tip of the microelectrode is so easily covered with a coagulum in the ameba cell. In such a case, a potential difference measured between the exterior and interior, which is at first as high as 100 mv in Prescott-James solution, is unstable and tends to become smaller. Furthermore, when the electrode is pulled out of the cell, there remains some potential difference. Bingley and Thompson did not say anything in their paper about this point.

According to our work (Tasaki and Kamiya, unpublished), the ohmic resistance of the membrane is about the same order as that of the neuronal or muscle fiber membrane. The resistance of the ameba is of the order of one-tenth of that of Prescott-James solution. If there is an ohmic current within the ameba body, as Bingley and Thompson insist, there must be a rather high external potential difference when measured with the constricted orifice method which I have just described. But actually the external potential difference between two electrically insulated halves of the ameba is only of the order of a few millivolts. This is decisive evidence showing that the occurrence of the internal current postulated by Bingley and Thompson would not actually be the case. The microelectrode method is very advantageous if it is adequately used for proper materials, but at the same time we must be cautious in interpreting what we have measured with microelectrodes, especially in an organism such as ameba or slime mold.

CHAIRMAN DE BRUYN: Does your microelectrode penetrate the mucopolysaccharide layer?

DR. KAMIYA: Yes.

DR. ZIMMERMAN: Some time ago, Drs. Marsland and Landau and I reported that ATP increases pseudopodial stability in amebae. Your data with ATP seem to confirm this very well. The question I should like to ask you is: Have you tried using low concentrations of ATP? We found that when higher concentrations of ATP were applied to cells, there was a generalized contraction of the entire cell. When we used lower concentrations of ATP, in the range of 5×10^{-4} *M*, the cells moved in apparently normal fashion, but when tested with pressure, they had an increased pseudopodial stability.

DR. KAMIYA: We tried, so far, 1×10^{-4} *M* and 2×10^{-4} *M* ATP in the low-concentration range. At 2×10^{-4} *M* ATP, temporary augmentation of the motive force is clearly observed. Even at 1×10^{-4} *M* the moderate effect of ATP is still perceptible.

DR. ZIMMERMAN: We suggested at that time that ATP could be the energy source for ameboid locomotion.

DR. HIRSHFIELD: I noticed from the films that, as the cytoplasm poured through the bridge, the side that emerged was rounded and then pseudopods formed later. How long did it take before the reappearance of the normal pseudopods was noted?

DR. KAMIYA: Only a couple of minutes.

DR. HIRSHFIELD: Do you think there is any particular significance to the initial rounding as it emerged through the port?

DR. KAMIYA: I think it is a noteworthy fact. The ameba invariably forms a spherical or oval head when it emerges through the pore under pressure. I suppose this is caused by the stretching of the surface layer or membrane which has a certain amount of tensile strength, which in turn resists mechanically the increase in the surface area brought forth by the inflow of the endoplasm. This is probably why the advancing front takes a spherical form. As a matter of fact, the advancing front is most vulnerable to bursting when the rate of increase in surface area is at its maximum, i.e., at the moment when it reaches the orifice of the capillary and starts increasing in surface area.

DR. HIRSHFIELD: When it becomes the tail, does it assume the pseudopodial characteristics as it emerges on the other side? Is it rounded again? Will it do this at will and continue to do so?

DR. KAMIYA: Yes.

DR. ANDREW G. SZENT-GYÖRGYI: What is the concentration of ATP in the ameba?

DR. KAMIYA: I don't know the concentration of ATP in *Chaos*. In the slime mold (*Physarum*), a normal average content of ATP is 0.4 μ M/gm, according to Hatano and Takeuchi.

DR. ANDREW G. SZENT-GYÖRGYI: I would be very careful to interpret the experiments where addition of ATP to the intact cell leads to contraction. Such an observation does not necessarily mean that ATP serves as the energy source or even that the contraction mechanism is similar to that of muscle. Let us take the analogy of muscle. ATP has been tried on intact muscle fiber and caused contraction; this experiment was cited as proof for a direct role of ATP in muscle contraction. In fact, contraction was accompanied by an action potential and action of ATP was, therefore, not a direct one on the contractile proteins. When ATP was microinjected, the muscle did not respond with contraction, and this is what is expected since the concentration of ATP in muscle is high, about 6 μM/gm. Its action in resting muscle is prevented by the relaxing factor system.

I do not know how to interpret those systems where addition or injection of ATP leads to contraction. It may mean that these systems work quite differently from muscle, or that the regulatory systems are quite different, or that there is no ATP present at rest. The system is still much too complicated, and the various possibilities should be sorted out experimentally. If ATP is a part of the contractile mechanism and serves as direct source of energy, then there must be some type of mechanism which prevents its action when the system is not contracting. These control mechanisms should be destroyed or inactivated for ATP to act if its function is similar to that of muscle. ATP should act on killed cells but not on an intact system.

DR. KAMIYA: At present I am not in a position to insist on anything definite about the mechanism of ATP effect on the motive-force production in ameba except to say that ATP and inorganic pyrophosphate are specifically effective and AMP and orthophosphate have no effect at all. Some experiments conducted on the slime mold, however, are instructive in this connection. If ATP is applied externally to the slime mold at the concentration of 5×10^{-3} M, the motive force gets gradually stronger in the first 10 min or so until it reaches a definite level which is much higher than the normal. On the other hand, when 10^{-3} M ATP of one-tenth the volume of the plasmodium is microinjected into a slime mold, augmentation of the motive force is immediate and conspicuous. According to an unpublished result of Takata, it is likely that the motive force of streaming becomes greatest when the ATP content of the mold is made higher than the normal level by about 10^{-4} M. When the ATP level becomes still higher, the motive force of the streaming begins to decrease. These experiments are in conformity with the view that the ATP is the direct energy source of the streaming and the normal level of ATP available for the movement in the slime mold is suboptimal. Whether or not this is true of ameba is still to be dealt with.

DR. ALLEN: As I understand it, when you cause an induced pseudopod to go through one of these channels, when it comes out the other side it is spherical. Is that right?

DR. KAMIYA: Always.

DR. ALLEN: This is the point I was trying to bring up this morning. I don't see why a pseudopod should always take on a cylindrical shape if the endoplasm flowed by hydrostatic pressure. One would then have to postulate some additional mechanism for laying down the gel wall in a cylindrical pattern.

Perhaps it should be pointed out in connection with Dr. Kamiya's experiments that an increase in motive force on treatment with ATP may not necessarily mean that chemical energy of ATP has been utilized. The cell normally moves against an internal resistance. Lowering that resistance without lowering the contractile tension would also cause an increase in the net motive force as he measures it.

DR. GOLDACRE: I think the experiment you are referring to is this: first, the pseudopod was spherical, and when it reached its normal diameter it continued on as a cylinder.

DR. KAMIYA: Cylindrical pseudopodia are formed when artificially induced rapid inflow of endoplasm is stopped. As long as the artificial inflow continues, an ameba rarely forms cylindrical pseudopodia.

DR. ZIMMERMAN: To consider one point, I have recently shown that the surface of amebae has ATPase activity; furthermore the data strongly suggest that ATP can also enter the cell. Independently, Sells, Six, and Brachet (*Exptl. Cell Res.* **22**, 246, 1961) have also reported that amebae can split ATP presumably by means of enzyme localized at the cell surface. Their methods of analysis were entirely different from those which I employed. Thus in two different laboratories evidence indicates that amebae can hydrolyze exogenous ATP.

DR. ANDREW G. SZENT-GYÖRGYI: I am not for or against the particular proposals discussed here. I think, though, that one ought to measure how much ATP an ameba contains and what is the reason that this ATP, if there is any, does not act in the resting condition. The ATP either has to be compartmentalized and effectively separated from the contractile component, or present in an "inactive form," which I doubt a great deal. There must be some reason that injected ATP acts differently from ATP present originally. What is the control mechanism which prevents its action? Of course it is also possible that there is no ATP in the ameba, though such a result would be rather surprising.

DR. BOVEE: I just wanted to make one short comment to the effect that in many kinds of amebae the initial pseudopodial form is always a hemisphere. It then may become extremely long and tenuous, sometimes twisted, and sometimes with a conical tip. This seems to be generally true, but I have no explanation for it.

Relative Motion in *Amoeba proteus*[1]

THEODORE L. JAHN

Department of Zoology, University of California, Los Angeles, California

The movement of protoplasm usually can be detected by the movement of granules and/or crystals that are bound or enmeshed in the protein framework of the cell. This is especially true of *Amoeba proteus* and other amebae, in which the cytoplasm abounds in granules and crystals. These objects are very prominent under dark-field illumination and the larger ones are easily photographed by a variety of methods, using either cinematographic or single-frame photographs of either brief or long exposures which produce streak pictures of the tracks of the granules. During the past 2 years granular movement has been photographed in my laboratory by Dr. Robert A. Rinaldi and Mr. James R. Fonseca, and movement of carmine particles attached to the pellicle has been studied by Mrs. Jane Jorgensen Russell.

I plan to discuss some of these results in relation to the contraction-hydraulic and frontal-contraction theories of movement of *Amoeba*. My conclusions, as already expressed (Jahn and Rinaldi, 1963; Rinaldi and Jahn, 1962), are that our records of granule paths support the contraction-hydraulic theory and cannot be explained on the basis of frontal-contraction. I also plan to present some old data of other investigators, which essentially invalidate the frontal-contraction theory, even though some of these data have been interpreted previously (Allen, 1961a, b) in favor of this theory.

Contraction-Hydraulic versus Frontal-Contraction Mechanisms

Since the term "fountain streaming" is a misnomer and therefore is likely to be confusing, and since the contraction-hydraulic theory is sometimes erroneously called the "sol-gel" theory, it seems wise to point out the essential similarities and differences of these theories. In both theories it is assumed that both the sol and the gel are composed of a spongy contractile protein network in which the various granular inclusions are enmeshed. This gives both the sol and the gel the well-recognized characteristics of non-Newtonian flow. Also it is assumed in both theories that the difference between sol and gel is the denseness of the

[1] Aided by National Institutes of Health, grant #6462.

protein framework, i.e., in the number of protein fibers per unit volume of the framework. Many of the enmeshed granules are free to undergo Brownian movement in both the sol and the gel (Mast, 1926), but the range of movement of any given granule in gel is more restricted as would be expected on the basis of a greater number of fibers (or of alveolar walls) per unit of volume.

In brief, the contraction-hydraulic and the frontal-contraction theories, as now expressed, differ primarily on two basic points: (1) The site of the contractile force, i.e., front versus rear; and (2) The effect of the force on the plasmasol, i.e., pull versus push.

Relative Motion of Parts of the Ameba

Movement of the Pellicle

Movements of the pellicle seem to be complex, especially in polypodial amebae, and to a certain extent undetermined. There are two principal schools of thought, namely, that (1) new pellicle is formed at the anterior end and old pellicle is incorporated into the gel at the posterior end, and (2) the pellicle moves forward at the same speed as the net movement of the protoplasm so that particles attached to the pellicle at a given distance from the tip remain in the same location relative to the ameba as it moves forward. The first idea is strongly supported by the observations of Goldacre (1952a, b, 1961) and the second by the work of Griffin and Allen (1960), who studied the movements of particles attached to the surface of the ameba. Two years ago we made serial photographs of carmine particles near the advancing tip of *A. proteus* and *Chaos carolinensis* which could support the contention of Griffin and Allen (1960) to the extent that new plasmalemma did not seem to be forming *at or near* the advancing tip. However, our results also indicate that new plasmalemma may be formed slightly behind the advancing tip, i.e., more than one diameter back from the tip.

For example, in Fig. 1, granule A, attached to the particle, moved through positions 1 to 8 in successive photographs taken at 10-sec intervals. One might assume that, if new membrane were formed at the tip, this granule would have moved to one side of the pseudopodium. The statement can be made for the successive positions of granules C, D, and the later positions of B, since all of them seem to have remained in more or less the same position in relation to the tip of the pseudopod. Furthermore, granules B, C, and D are almost in the same relationship to each other in position 8 as in positions 6 and 7. However, it is obvious that granules A and B are much farther apart in position 8 than they were in position 1. Therefore, new membrane must have been added someplace between granules A and B as they moved from position 1 to position 8. We have a considerable amount of similar data which show the same thing.

From these data it is not possible to say exactly where this new membrane was formed, and, unfortunately, we have not been able to continue these observations. However, it is obvious that the whole pellicle cannot move forward without the addition of new areas. This is contradictory to the suggestion of Griffin and Allen (1960). New pellicle must be formed, and, in the organism shown in Fig. 1, this presumably occurred within a short distance, possibly one or two diameters behind the advancing tip.

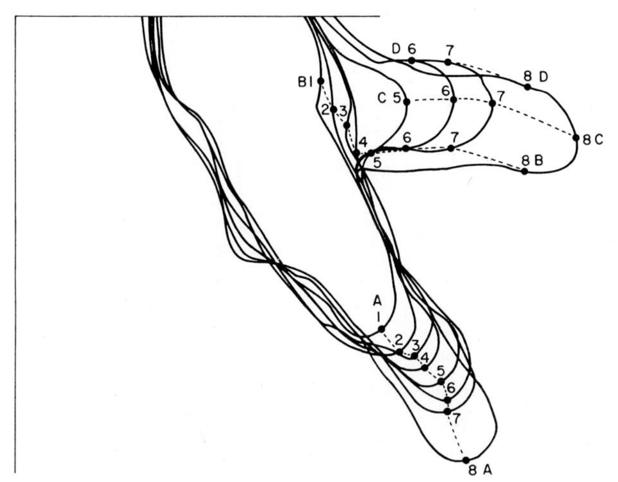

Fig. 1. Successive positions of carmine particles on pseudopodia of *A. proteus*.

Goldacre (1961) has offered an alternative electrical interpretation for experiments involving the use of carmine granules, basing his conclusions on the use of oil droplets. In his published oil-droplet experiments the drop at the anterior tip tended to remain at the tip, just as the carmine granules in our own experiments. However, regardless of the relative merits of carmine and oil, the results of Goldacre (1961) with the use of several techniques (oil droplets, glass filaments, wrinkle contours) convincingly demonstrate that new plasmalemma must be formed someplace in the advancing pseudopodium. In a polypodial ameba, formation of new plasmalemma seems necessary even if a moderately unreasonable amount of plasmalemma deformation is assumed.

The same must be true of *Mayorella,* because it seems inconceivable that the whole membrane would move *forward* in relation to the almost vertical, determinate pseudopodia, which are formed in the second quarter of the advancing fan-shaped cell, and in relation to the cell certainly move *posteriorly* through the third and fourth quarters to be retracted and absorbed at the posterior end. If the idea of Griffin and Allen (1960) were applied to the whole membrane of *Mayorella* we would have to visualize a movement of particles of membrane anteriorly over

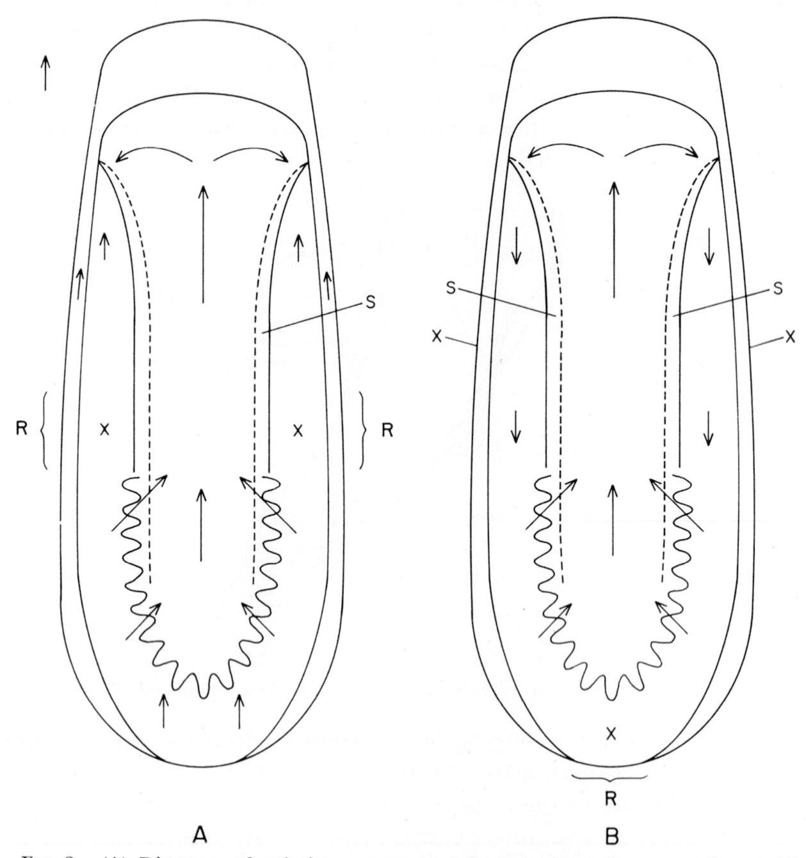

FIG. 2. (A) Diagram of relative movement of protoplasm in an ameba attached near middle of body. Wavy line denotes region of solation. (B) Same for ameba attached at rear: X, Nonmoving protoplasm; R, region of attachment; S, shear zone.

the peak of each of the determinate pseudopodia as the pseudopodia move backward. This would require more stretching and contraction of the surface of the membrane than seems possible with the assumption of anything like a surface area in which the various sections are fixed in position in relation to each other.

One possibility in polypodial *Amoeba* and *Chaos,* and also in *Mayorella* is that (1) new membrane is formed whenever and wherever the existing membrane is placed under sufficient tension, and (2) old membrane is absorbed whenever the tension is sufficiently decreased, especially in the posterior end. This possibility does not seem to be contradictory to the data of Goldacre (1952a, b, 1961, 1962).

Movement of the Gel

The direction of protoplasmic flow that might be expected in a monopodial ameba on the basis of the contraction-hydraulic theory is shown in Figs. 2A and B. In Fig. 2A it is assumed that the ameba is attached someplace in its mid-region and in Fig. 2B it is assumed to be attached at the rear end.

A polypodial *A. proteus* or *C. carolinensis* really moves in three dimensions so that locomotion is more of a "walk" or a roll, as described by Wilber (1946). Mr. Fonseca and I have taken cinematographs from the side of *C. carolinensis* and have confirmed most of Wilber's observations. Whenever an ameba (*Amoeba* or *Chaos*) is observed moving monopodially the location of the attachment usually is not determined, and furthermore it often cannot be determined by simple observation.

In Fig. 2A the gel is shown as moving forward in the anterior and posterior regions but not in the mid-region. This can be seen in streak (8-sec, dark-field) photographs. An example is shown in Fig. 3 in which there is a short mid-lateral pseudopodium in and near which the granules are not moving in the main direction of locomotion. However, in the posterior and anterior regions of this ameba, the granules of the gel are moving forward. Forward movement also is noticeable in the insert of Figs. 3 and 4. These streak pictures alone do not tell us the direction of movement, but the forward flow can easily be seen in some of the cinematographs of *A. proteus* and also of *Mayorella* in which the speed of motion is increased by a factor of 8. The relative immobility of the granules of the mid-gel region also can be seen in Fig. 5 in which the granules of the major posterior end (and also of a small withdrawing pseudopodium) are moving rapidly.

Figure 6 shows one type of movement in a retracting pseudopod of a polypodial ameba as the pseudopod undergoes a rapid decrease in diameter, presumably by elastic as well as active contraction. An interpretation of lateral movement adequate to explain the lines at right angles to the major axis is precluded by the fact that the center of the lower part (flowing sol) shows no evidence and the upper end shows relatively little evidence of lateral movement. The numerous short lines at right angles to the major axis are interpretable as evidence of radial contraction of the gel tube caused by a sudden drop of pressure of the sol such as that resulting from formation of a new lateral pseudopodium. This type of movement can occur easily at the beginning of a pseudopodial retraction and is rapidly converted into the more characteristic type shown in Figs. 3 and 5.

If the sol were actually pulling the closed posterior end of the gel tube forward, as postulated by the frontal-contraction eversion hypoth-

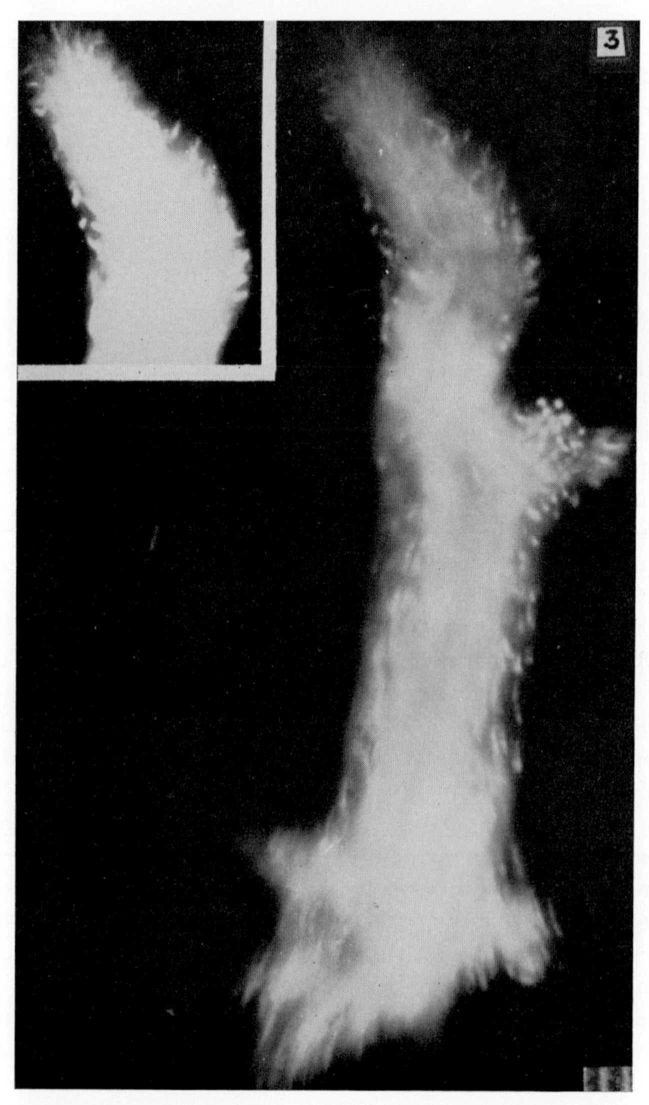

FIG. 3. Streak photograph of moving *A. proteus*. Note that movement is more rapid at anterior and posterior ends than in the mid-body region. Exposure: 7 sec. Scale: 10 μ. (After Rinaldi and Jahn, 1963.)

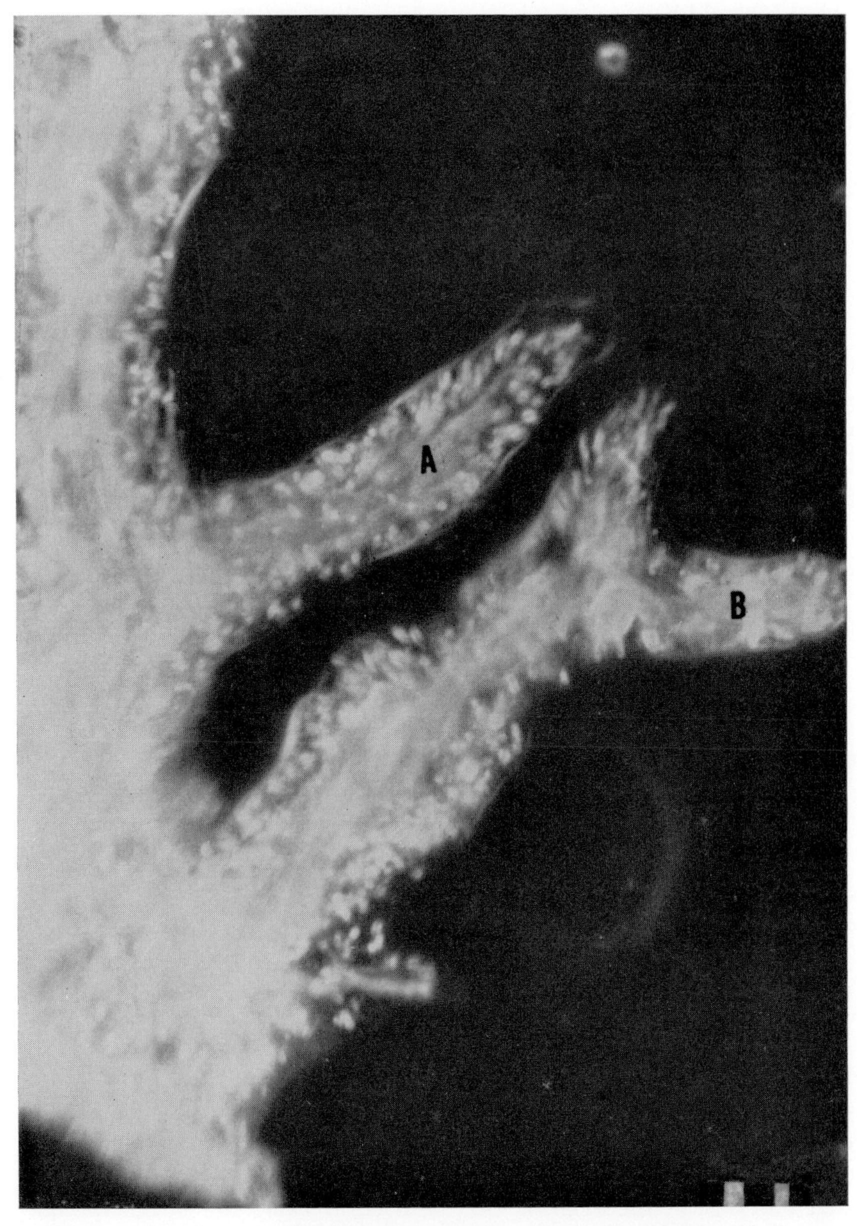

FIG. 4. Advancing pseudopodia of *A. proteus*. In A the granules of the new gel are advancing and also moving peripherally. Exposure: 8 sec. Scale: 10 μ. (After Rinaldi and Jahn, 1963.)

esis, we would have to postulate attachment only at the extreme posterior end in Fig. 6, but attachment over a much longer area in order to explain the data of Figs. 3 and 5. For reasons to be discussed later, it is highly improbable that any adequate attachment exists.

It is possible, under certain conditions, for the movement of protoplasm to be as shown in Fig. 2B. This type of flow looks exactly like a

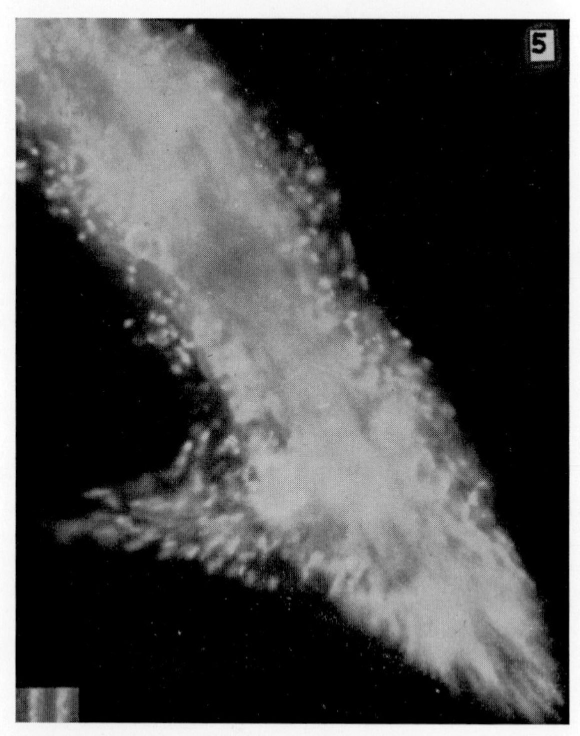

FIG. 5. Central and tail regions of *A. proteus*. Note that gel movement is rapid in the posterior region but almost zero in the mid-body region. (Rinaldi and Jahn, 1963.)

fountain, and we have an excellent moving picture showing this type of flow in *C. carolinensis*. However, organisms with this type of flow definitely are *not* locomoting, and are obviously anchored at some point. In Fig. 2B the anchor point is shown at the posterior, but it is possible that blockage at the anterior end, or friction along the sides, especially in a tight-fitting glass tube, could produce the same effect. It is interesting that this fountain-type flow can easily be explained on the basis of the contraction-hydraulic theory, with the driving force being exerted at the base of the fountain—a position where it occurs in real fountains.

This has been observed by several investigators and is described for *A. proteus* by Mast (1926, p. 400) as follows:

"Occasionally, however, monopodal specimens are found in which all but the tip of the posterior end is free and in which this end is so attached that locomotion is prevented. In such specimens the movements of the plasmasol and the plasmagel are precisely the same as they are in moving specimens; but in reference to points in space, the plasmagel

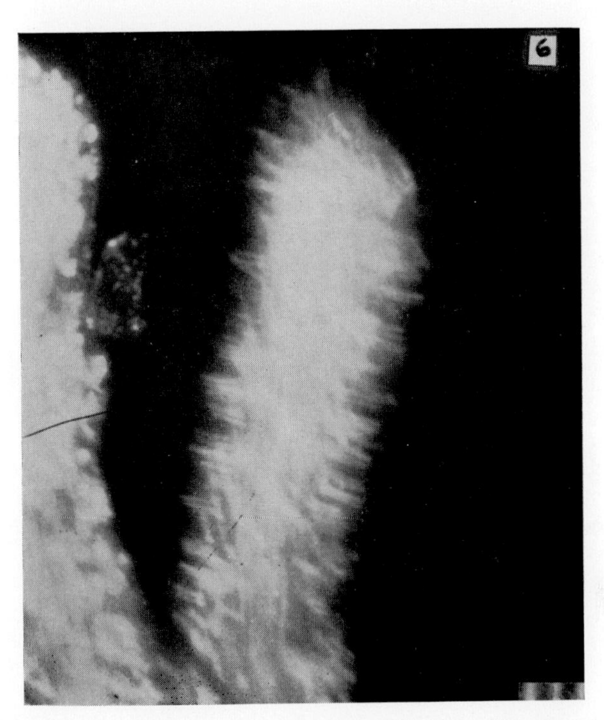

FIG. 6. Pseudopod which has just begun to withdraw. Showing inward radial movement of granules. Exposure: 5 sec. Scale: 10 μ. (After Rinaldi and Jahn, 1963.)

actively moves backward instead of being at rest. This gives the appearance of typical fountain streaming, so much discussed in the literature on *Amoeba*."

Movement of Very Small Granules through the Gel and the Peripheral Hyaloplasm

In Fig. 7 a section is outlined in which there are small granules that move forward along the edge of a gel region, between the plasmagel and plasmalemma, in the hyaline ectoplasm. Forward flow is not restricted to granules of this very small size, as frequently a few isolated granules 3–4 μ in diameter can be seen flowing in the plasmagel. The smallest granules can be observed only by means of dark-field microscopy. They are difficult to observe, but they can be seen clearly by careful focusing, especially with a cardioid condenser. In streak pictures

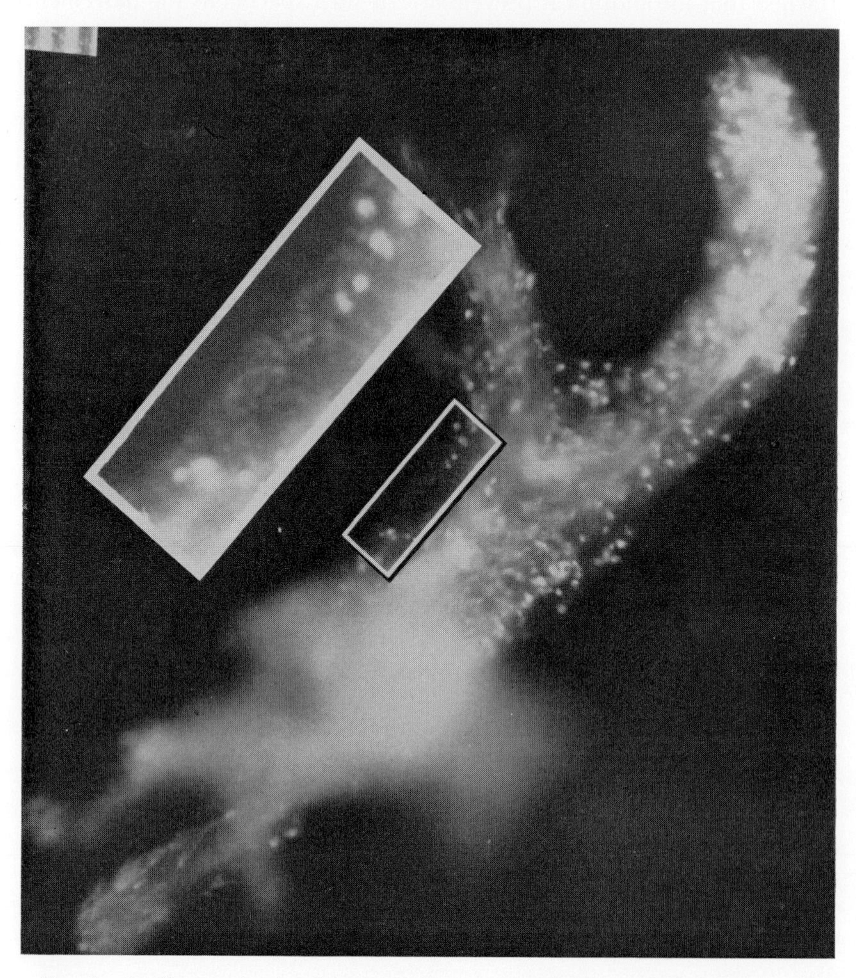

FIG. 7. Small granules in the gel region of *A. proteus*. Exposure: 8 sec. Scale: 10 μ. (After Rinaldi and Jahn, 1963.)

(Fig. 7) one can observe a number of the small granules. They move forward and may leave faint streaks. They move much slower than granules of the central region and in general faster than the larger granules of the peripheral region.

The important thing about the movement of these small granules is that *they move in the direction of the main stream of flow of the plasmasol.* Furthermore, they move forward, not only between the plasmagel and the plasmalemma, but also along zigzag paths *through the interstices of the plasmagel.* Such small moving granules can be seen to move past and around many of the larger fixed granules of the gel, as if they were bumping into invisible fibers of the spongy structure of the gel. This type of movement of granules in *A. proteus* has not been described previously, but Abé (1961) has described somewhat similar movement of granules in *A. striata.*

Movement of the Sol

The fact that the sol moves forward is, of course, easily observed and can be seen in streak photographs (e.g., Fig. 4). When the sol approaches the tip of the pseudopodium it spreads laterally as shown in Fig. 3, insert, and sometimes there is a backward movement near the periphery. These movements also can be demonstrated by means of shadowgraphs made from cinematographs.

A backward movement might occur on the basis of either the hydraulic-contraction or the frontal-contraction theories, but on the basis of the frontal-contraction theory a backward movement is an essential part of the locomotor mechanism. I have not found any statement to this effect in Dr. Allen's publications, but a backward movement of the contracting sol is definitely shown in a figure (Allen, 1961b, Fig. 26) demonstrating the postulated contraction of the gelating sol.

In order for such a frontal-contraction eversion system to work, the contraction of fibers must occur *during* the eversion. Each fiber is assumed to turn 180 degrees and to be shortened and fastened in position in the gel at the end of the turn. This means that the shortening must continue until the fiber is finally fastened into position. Since the everting material has a definite thickness (the radius of the pseudopod) the material from the center of the sol is everted to the periphery and that from the periphery to the inner portion of the gel tube. The material that travels from the center of the sol to the periphery of the gel must move in a greater arc than the material from the periphery of the gel. Consequently this central material must be pulled backward at the periphery in order to complete the arc.

However, it also is possible to explain the backward flow on the basis of hydrodynamics, and furthermore to explain why it does not always occur. Figure 8 is modified from a summary of fluid mechanics by Kenyon (1960, p. 183) in a section concerned specifically with the pattern of flow due to a sudden expansion by fluid flowing in pipes. As illustrated by the diagram in Fig. 8a, there is a striking and noticeable backward flow upon an expansion of a laminar flowing fluid at the orifice where the fluid begins its expansion from the diameter of pipe A to pipe B.

Figure 8b depicts protoplasm flowing as a fluid. When the protoplasm leaves the plasmagel tube the cross-sectional area of flowing increases because it is no longer restrained at the tip by the tube. Consequently, this would result in a backward flow at this region similar to that of any laminar flow through an orifice. The orifice in the case of an ameba is the simultaneously formed (plasmagel) tube. In addition, the plasmagel tube through which the stream is flowing forward in an

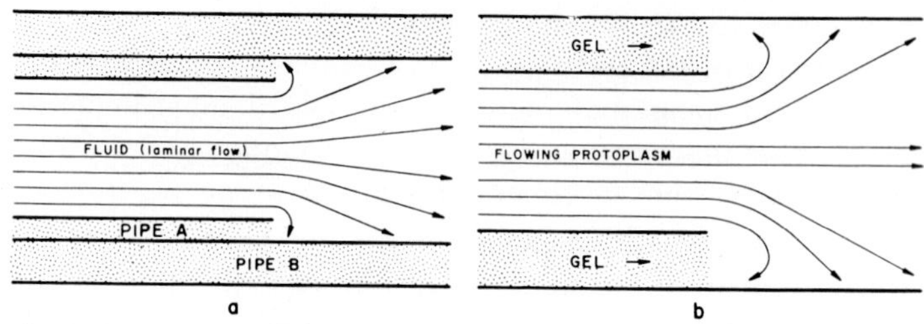

FIG. 8. Behavior of a fluid upon expansion from a smaller to a larger tube, illustrating the turbulence (backward flow) produced at the orifice. (a) Flow of fluid in a pipe. (After Kenyon, 1960.) (b) Flow of protoplasm. (After Rinaldi and Jahn, 1963.)

ameba can itself move slightly forward by stretching under hydraulic pressure (indicated by the arrows) and also is being extended by gelation at the end of the tube.

If the profile of new gel being formed (Fig. 8b) is sloping forward and outward toward the plasmalemma rather than perpendicular as shown, there will be no necessity for a backward movement. Some of the shadowgraphs of Rinaldi and Jahn (1963) show forward movement only. Radial movement, of course, also occurs but cannot be shown in this type of shadowgraph used.

Allen (1962) states that the "endoplasm must shorten as it thickens just before or during its turning at the front of the cell to form the ectoplasmic tube." If we assume tentatively that a frontal-contraction eversion system were to exist, the contraction would have to continue during the eversion, regardless of whether it had started previously. Therefore, we can disregard the possibility that it shortens "just before . . . its turning" as inconsequential.

General Direction of Movement of Granules in Anterior Third of an Advancing Pseudopodium

The results described previously demonstrate that there is a flow of small granules forward, not only in the central plasmasol, but also in the hyaline ectoplasm on the outside of the plasmagel tube, and furthermore, through the interstices of the spongy structure of the gel. In brief, there is a forward movement of tiny granules throughout the entire cross section in the anterior third of an advancing pseudopodium but it is most obvious in newly formed, lateral pseudopodia when the total length of pseudopodium is only 1–2 times its diameter. This forward flow is exactly the type that can be expected on the basis of the Mast (1926) theory; it is merely a flow of syneretic material from the contracting region. It is not predictable from the frontal-eversion assumption (Allen, 1961a), which requires that there be a posteriorly directed flow of syneretic material. In a later paper (Allen *et al.*, 1962, p. 116) it stated that the hyaline cap is formed from syneretic fluid from the eversion zone and that: "It seems likely that hyaline cap fluid is recirculated toward the tail in a channel beneath the plasmalemma." This channel would be the lateral hyaline ectoplasm. These same authors further aver: "The plasmalemma slides over the ectoplasmic tube, providing a channel through which the hyaline cap fluid may (in fact must) circulate toward the tail, where it somehow finds its way through the weakened ectoplasmic tube to rejoin the ectoplasmic stream." A later figure (Allen, 1962, pp. 116–117) shows arrows indicating a reverse flow in the hyaline ectoplasmic layer, but there has never been presented any evidence that such a reverse flow exists. Our observations show that *the small granules of the hyaline ectoplasm actually move forward rather than backward.* Therefore no reverse flow of syneretic fluid is present, and no amount of supposition can produce it.

The general forward flow of material which occurs throughout the entire cross section of the ameba (plasmasol, plasmagel, hyaline ectoplasm) can be predicted from contraction-hydraulic postulates. It cannot be predicted nor accounted for by the present version of the frontal eversion idea which requires forward flow in the plasmasol only, no forward flow in the gel, and a reverse flow of the hyaline ectoplasm along a postulated but unproved route. For these reasons it is concluded that the posterior contraction-hydraulic pressure explanation agrees best with the facts.

Analysis of Other Data Pertaining to Mechanism of Movement

Transmission of Tension to the Posterior End

According to the frontal-contraction eversion concept, the contracting plasmasol at the anterior end exerts a tension which is assumed to pull axial plasmasol forward. In order for a force so positioned to pull the posterior end of the ameba forward it must be transmitted to the posterior plasmagel. However, evidence has been given (Allen, 1960) that the area between the posterior end of the axial plasmagel and the posterior plasmagel (i.e., the solation or so-called "recruitment zone") is one of very low consistency. To pull the tail the tension must, therefore, be transmitted across this area of solation, which, together with the shear zone, has the lowest consistency (i.e., viscosity) of the entire ameba (Allen, 1960). Since no bridge is therefore provided between the posterior gel and sol, then either the gel must move forward by contraction, or else be sucked forward by axial plasmasol. If the posterior gel contracts, then a frontal contraction and eversion pulling from the anterior end is superfluous. If the posterior gel is drawn forward by suction, then the same suction should pull the posterior portion of the shear zone backward, and there is no evidence available that this occurs. Therefore until *evidence* is presented that the area in question is actually bridged by fibrous structures of sufficient strength to maintain a frontal eversion while causing and accompanied by a posterior inversion, the mere conjecture of their presence (Allen, 1961a) against evidence to the contrary (Allen, 1960) renders the scheme unacceptable.

Furthermore, if the back end of the ameba is being pulled forward by suction, cutting a small hole in the plasmagel in this region should result in a flow of external medium into the hole rather than a flow of protoplasm into the medium as ordinarily seen by micrurgists, and which has been demonstrated very elegantly by Goldacre (1961).

Another interesting consideration is that the unlimited posterior narrowing and elongation of the tail, described by Goldacre (1958), would be impossible if the tail were being pulled forward by the sol instead of actually contracting as described by Goldacre.

Formation of Lateral Ridges

It is generally recognized (Leidy, 1879; Mast, 1926; Schaeffer, 1920) that the lateral ridges on the pseudopodia of *A. proteus* are formed by the flow of granular plasmasol through large interstices of the gel to the outside of the plasmagel tube and thence forward between the plasmalemma and the gel tube, forming a ridge, which later consists of gel with a core of flowing plasmasol, and which may beome a secondary pseudopodium. The fact that during formation these ridges progress forward, with little or no circumferential spreading, is a strong indication that the driving force must be pressure from the posterior end. This is the same pressure that drives the small granules forward throughout the entire cross section of the ameba.

ORIGIN OF SYNERETIC FLUID

It has been calculated on the basis of measurements or refractive index (Allen and Roslansky, 1958) that the protein content of a moving ameba is about 3.95% at the posterior end, 3.33% at the anterior part of the sol, and about 1.00% in the hyaloplasm.

It is generally assumed that during the folding of an elongated protein molecule some of the intermolecular bonds become intramolecular, and consequently there are fewer exposed groups to which water and other small molecules could be attached. Therefore, the amount of water and other smaller molecules attached to the protein is reduced. This idea was applied to *Amoeba* by Goldacre and Lorch (1950) and Goldacre (1952b). Conversely, it is assumed that water would become attached to the protein during the unfolding process.

In the contraction-hydraulic theory it is assumed that part of the framework folds and becomes detached, thereby going into solution, at the time and also at the site of contraction, in the mid-posterior region. It is also assumed on the basis of the bond conversion mentioned previously that water is squeezed out of the molecular framework during contraction. If the contraction is at the posterior end and is a gradual process involving slow, molecule by molecule, dissolution of part of the protein framework, some water would be released from partly folded molecules which would still be attached to the framework. This would provide surplus water which would be pushed forward ahead of the protein molecules which would be released only after they are completely folded and detached from the protein framework. Water carried by this forward flow would accumulate at the front end of the ameba until it is used during gelation and in the hardening process of the gel which follows. The amount of water absorbed during gelation would include the excess water released during the solation process and forced forward as part of the flowing sol, and that which filters forward through the gel, and also that which flows forward in the peripheral hyaloplasm. In this way a gradient of water would be maintained throughout the length of an ameba.

In the fountain-streaming or frontal-contraction eversion theory the framework is assumed to become more compact by folding of either or both molecules and fibers at the anterior "fountain" zone, thereby squeezing water of syneresis into the hyaline cap. It is also assumed that this syneretic fluid flows *backward* in the lateral hyaline layer between the gel and the plasmalemma.

If one assumes that contraction is at the anterior end and that the water is released during contraction, the surplus water at the anterior end is easily explained. However, the next question is how it is transported to the posterior end so that it can be available during solation. Since the actual movement of granules in the peripheral hyaline zone is

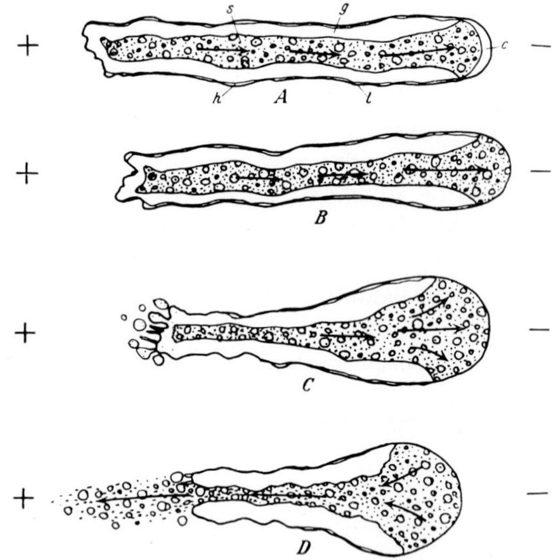

FIG. 9. The effect of galvanic current on a monopodial ameba moving toward the cathode. —, Cathode; +, anode; arrows, direction of streaming. *A*, very weak current; *B*, *C*, *D*, progressively stronger current. Note that in a current of moderate density the hyaline cap disappears and the plasmasol extends to the plasmalemma, and that in stronger current the cathodal end expands and the anodal end contracts and finally breaks after which the granules in the plasmasol flow toward the anode, indicating that they are negatively charged. (After Mast, 1931c.)

the reverse of that postulated in the frontal-contraction eversion hypothesis; it cannot move backward as proposed by Allen (1961, 1962). Furthermore, even if it were transported posteriorly in the peripheral hyaline zone it would have to cross the wall of the gel tube in order to reach the solation zone, where according to the fountain-zone theory the folded molecules of the gel presumably become unfolded. The only other possibility would be a reverse flow through the gel itself, and this also has been disproved.

For these reasons, it seems as if the data of Allen and Roslansky (1958) as far as these data pertain to the origin and the cycling of syneretic fluid, are best explained on the basis of the contraction-hydraulic theory as originally explained by Allen and Roslansky (1958) rather than on the basis of the frontal-contraction eversion hypothesis as suggested by Allen (1961).

Effect of Electric Current

The effect of direct electric current on locomotion of ameba has been studied by Mast (1931c) and Hahnert (1932). These investigators have

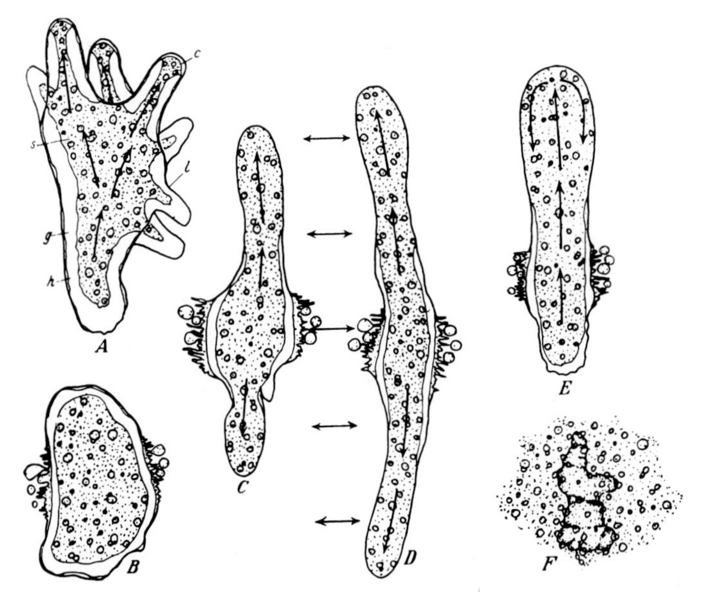

Fig. 10. The effect of alternating current. *A*, before circuit was closed; *B–F*, successive stages after closure; arrows, direction of streaming; double-ended arrows, direction of the current. Note that the ameba orients perpendicularly to the direction of the current, that the plasmagel in the pseudopods at this time is very thin or absent, that the plasmagel contracts violently at the surface directed toward the poles, that blisters are formed on these surfaces, and that the ameba eventually breaks here and then disintegrates. (After Mast, 1931c.)

demonstrated that in an organism moving toward the cathode the effect of current is to inhibit gelation at the cathodal (i.e., anterior) end. However, this inhibition of gelation does not stop the ameba (Fig. 9) which continues to move forward *in spite of the fact that the gel tube is not extended, and the anterior advancing portion is all sol.* Since there is no new gel tube to confine the sol it spreads out to form a large low-viscosity structure which may have more than twice the diameter of the ameba. On the basis of the fountain-zone contraction theory this would be impossible; nevertheless, it occurs. A possible alternative suggestion

(Heilbrunn and Daugherty, 1939) that cathodal movement is caused by a positive charge on the particles is disproven by Mast's (1931c) demonstration that the charge on the granules is negative.

Mast (1931c) also demonstrated that an alternating current causes a rounding of the ameba and then a contraction of the gel. This contraction causes the formation of long pseudopodia at right angles to the electric field (Fig. 10). These pseudopodia contain no gel and, therefore, could not possibly be extended by means of a frontal-contraction eversion mechanism.

Effect of Illumination

It is well known that the effect of illumination of the advancing tip of a pseudopodium, i.e., of the fountain-zone region, causes immediate

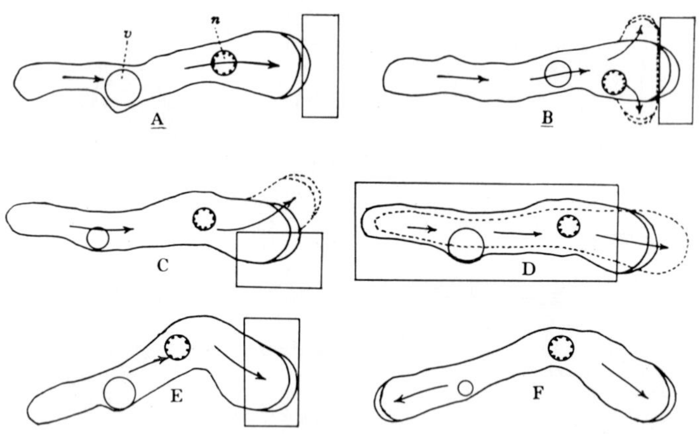

Fig. 11. Response of *Amoeba* to localized illumination. Only the rectangular areas were brightly illuminated. A–D, same specimen; F, same specimen as E, a few moments later. (After Mast, 1932.)

gelation of the anterior end (Mast, 1932). However, this gelation does not inhibit forward flow of the main stream of sol which continues forward, unimpeded, and branches out laterally, as shown in Fig. 11, through weak places in the end of the gel tube. If the sol were being pulled forward, in accordance with the fountain-zone contraction theory, a gelation of the anterior end should block the forward movement, at least temporarily, or until new contracting forward areas could be formed. However, this does not occur.

Furthermore, if the whole body of an ameba locomoting in very dim light is brightly illuminated, the sol of the anterior tip is gelated, exactly as described previously (Mast, 1931b). However, when the anterior end gelates, the closure of the gel tube is complete; there are no weak places, as mentioned previously, and the newly formed closed end of the tube begins to contract (Fig. 12). This causes a reverse flow and this reversed flow progresses backward exactly like the upstream waves of reversal so characteristic of *Physarum,* described by Kamiya (1959; and earlier) and Stewart and Stewart (1959) and by other investigators. It seems to me as if this single observation of Mast completely invalidates the frontal-contraction eversion theory, and it is especially intriguing because the only experimental manipulation necessary is opening of the condenser diaphragm of the microscope. We often have observed such reversals

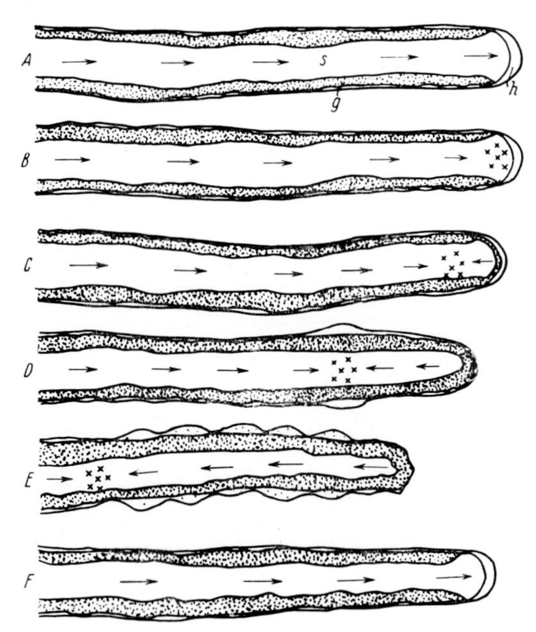

FIG. 12. Processes observed in a pseudopod in the response to rapid increase in illumination. Arrows, direction of flow in the plasmasol; x, regions at rest. Note that the response begins by cessation in streaming at the tip of the pseudopod immediately back of the hyaline cap and continues by cessation farther back accompanied by streaming in the reverse direction at the tip. (After Mast, 1931b.)

when we switch by means of a mirror from tungsten filament lamps used for visual observations to carbon arcs for cinephotomicrography. A comparable diagram showing flow from both ends toward the middle was published by Mast in another paper (Mast, 1931a) with the comment that it was "repeatedly observed."

Still another figure (Mast, 1931b, Fig. 2) shows essentially the same phenomenon except that when the two streams meet they cause a ballooning of the body because of the increased hydraulic pressure in that region (Fig. 13). Mast (1931b, p. 141) described this as follows:

"A few moments after the increase in illumination streaming stopped, first immediately back of the hyaline cap at the tip of both pseudopods, then successively farther back, but before streaming stopped at the posterior end it had begun again at the anterior end. Here, however, it now continued in the opposite direction. Soon after this it also began again at the posterior end, but in the original direction. There was thus at this time backward streaming in the anterior portion and forward streaming in the posterior portion of the amoeba (Fig. 2,b). This resulted in the formation of a marked enlargement near the middle. A few moments later there was a sudden break in the plasmagel farther forward (Fig. 2,b,x) and the plasmasol immediately rushed to-

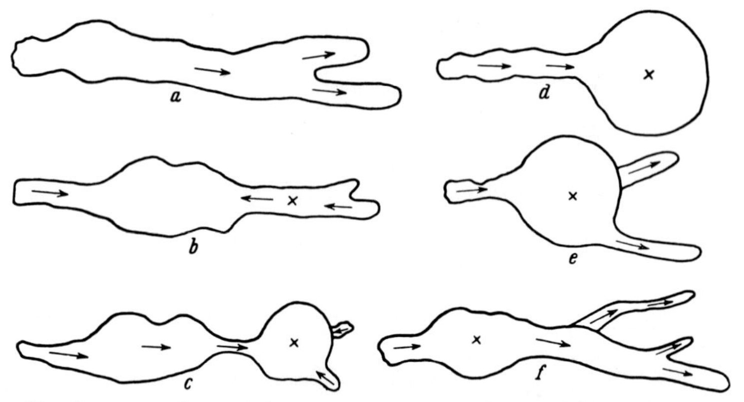

FIG. 13. Processes observed in the response of an ameba to rapid increase in illumination. Arrows, direction of flow in the plasmasol; x, regions at rest. Note that the response consisted of a complex series of phenomena and that in this series there was at no time complete cessation of flow in the organism. (After Mast, 1931b.)

ward this break, resulting in the formation of another enlargement which eventually contained nearly all the substance in the amoeba (Fig. 2,d). From this, new pseudopods developed, after which the amoeba soon returned to its original form (Fig. 2,f)."

It also seems quite significant that illumination of all of an ameba except the fountain-zone area and the advancing tip increases the *rate* of streaming and causes a decrease in diameter of the organism, as well as increases in the thickness of the gel tube and in the gel-sol ratio (Mast, 1932). These facts are easily explainable on the basis that light converts the layer of sol next to the gel into gel, as explained by Mast (1932). However, if we assume the contractile force is in the unilluminated anterior end, it is difficult to see how illumination of the posterior 90% or more could increase the rate of streaming, especially since the diameter of the flowing stream is decreased.

Some of the above experiments of Mast (1931a, b) have been misinterpreted by Allen (1961a) who states, "There are many aspects of the ameba's responsiveness to stimuli which lend excellent support to the

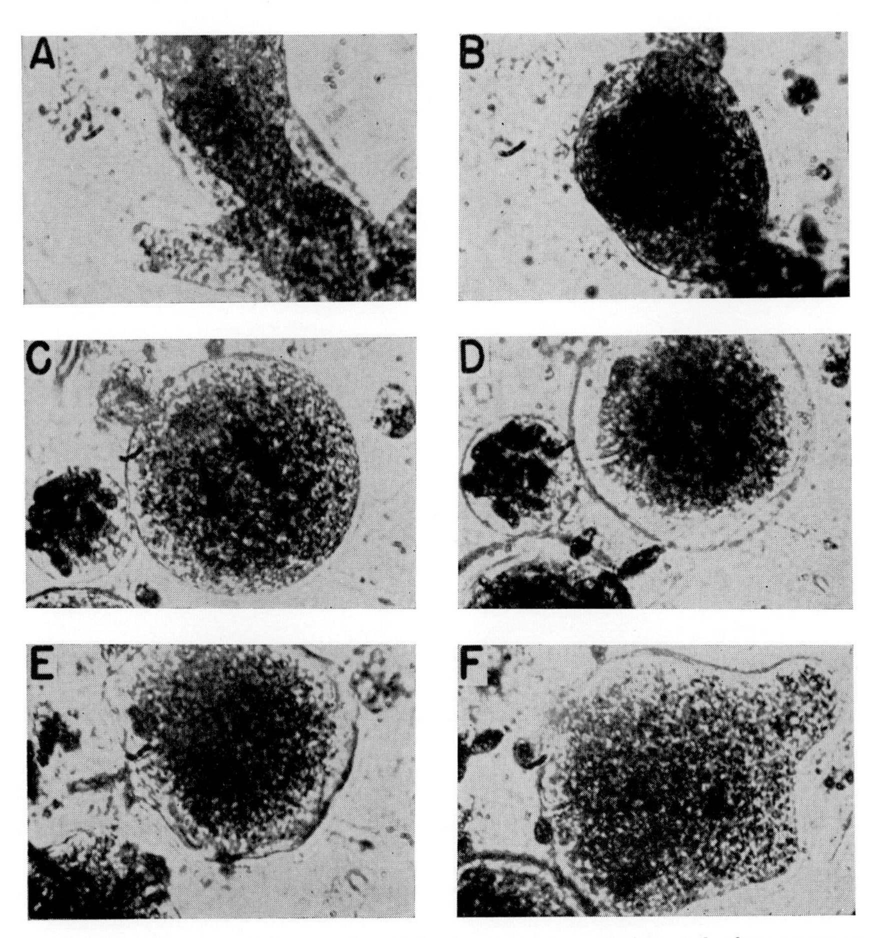

Fig. 14. Changes of form and activity in *A. proteus* during and after exposure to a pressure of 6000 psi at 25°C. (A) Normal form at atmospheric pressure. (B) Same specimen at 6000 psi, maintained for 5 min. Note the rounded form of the main cytoplasmic mass and the pinched-off part of the large pseudopodium (above). (C) Fifteen minutes later pressure still maintained. Specimen has rotated about 90°. Note complete rounding of the main mass and of the pinched-off part of the cytoplasm which now lies to the left. (D) Fifteen seconds after pressure was reduced to the atmospheric level. Note the marked contraction of the granular cytoplasm (plasmagel) and the development of a broad hyaline zone between the granular cytoplasm and the cell membrane. (E) Ninety seconds after decompression. Note first signs of ameboid activity. (F) Seven minutes after decompression. Ameboid activity now quite vigorous. (From Landau *et al.,* 1954.)

fountain zone contraction theory. First, both mechanical stimuli and light are effective only if applied locally to the fountain zone." Mast's observations, corroborated by our own, demonstrate that light also is very effective when applied to the whole organism. Furthermore, these observations do not "lend excellent support to the fountain zone contraction theory" but are quite contradictory to the fountain-zone theory.

Effect of High Hydrostatic Pressure

One type of evidence which supports the contraction-hydraulic theory is the definite correlation between contraction and solation. This is easily observable in the posterior end of an ameba if one believes in the theory, but to nonbelievers the observation may seem equivocal. An explanation on a molecular basis was given by Goldacre and Lorch (1950).

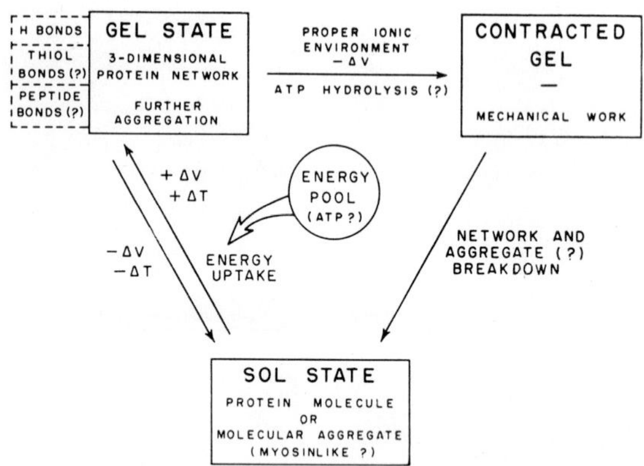

FIG. 15. A schematic representation of the sol-gel phenomenon. (After Landau, 1959.)

This explanation is still completely valid in principle, although I would tend to apply it to some rather than to all the framework of the gel, leaving a loose spongy network in the sol as assumed by Mast (1926). Furthermore, the correlation of solation and contraction is strongly supported by studies on the effect of high hydrostatic pressures on ameba protoplasm (Landau *et al.*, 1954), especially during conditions under which the ameba is not moving. If an ameba is subjected to pressures of 6000 psi (Fig. 14) all the protoplasm becomes a sol of low viscosity and the ameba becomes spherical. If the pressure is suddenly released the sol is converted into gel and the gel contracts violently with one massive contraction, leaving a layer of hyaline material around the periphery. Within seconds after this contraction occurs the granules move into the hyaline zone, indicating that the granular region that was gel is now at least partly solated. Shortly afterward it is obvious that the ameba has normal sol-gel ratios. On the basis of this and other evidence the correlation between contraction and solation was incorporated into the diagram of sol-gel relationships (Fig. 15) devised by Landau (1959). The current uncertainties that concern this diagram seem to involve the role of adenosine triphosphate (ATP) and the mechanics of contraction, as pointed out by Landau (1959), rather than the relationship between contraction and solation, which seem to have a very solid foundation.

If we were to assume that the anterior contraction theory is true, we would have to discard this diagram (Fig. 15) almost completely and invent a new one in which gelation and contraction would be correlated with each other, as would solation and molecular extension. The assumption that these proposed correlations are true would then necessitate a re-evaluation of the theory of high-pressure effects on protoplasm. This is a task which I gladly leave for consideration by the experts on high pressure who, of course, have the alternative of retaining their present interpretations which strongly support the contraction-hydraulic theory.

Summary

It has been demonstrated by studies of the movement of carmine granules attached to the membrane of *A. proteus* and *C. carolinensis* that new membrane is not formed at the advancing tip of a pseudopod but that it may be formed a few diameters behind the tip.

It is demonstrated that, in the cross-sectional area of the anterior third of large advancing pseudopodia and especially in newly formed, lateral pseudopodia, all the protoplasm is moving forward, including the sol, gel, peripheral hyaloplasm, pellicle, and syneretic fluid through the gel.

A reanalysis is presented of data of other investigators concerning (1) the low viscosity of the posterior solation zone, (2) the formation of lateral ridges, (3) the origin and cycling of the syneretic fluid, (4) the effect of direct and of alternating electric current, (5) the effect of illumination, and (6) the effect of changes in hydrostatic pressure.

It is pointed out that all of these data can be explained on the basis of the contraction-hydraulic theory of posterior contraction of the gel tube and that most of the data discussed are contradictory to the fountain-zone or frontal-contraction eversion theory. It is concluded, therefore, that the fountain-zone contraction theory is invalid.

References

Abé, T. H. (1961). *Cytologia (Tokyo)* **26**, 378.

Allen, R. D. (1960). *J. Biophys. Biochem. Cytol.* **8**, 379.

Allen, R. D. (1961a). *Exptl. Cell Res. Suppl.* **8**, 17.

Allen, R. D. (1961b). *In* "The Cell" (J. Brachet and A. E. Mirsky, eds.), Vol. II, pp. 135-216. Academic Press, New York.

Allen, R. D. (1962). *Sci. Am.* **206**, 112.
Allen, R. D., and Roslansky, J. D. (1958). *J. Biophys. Biochem. Cytol.* **4**, 517.
Allen, R. D., Cooledge, R. R., and Hall, P. J. (1962). *J. Cell. Biol.* **12**, 185.
Goldacre, R. J. (1952a). *Symp. Soc. Exptl. Biol.* **6**, 128.
Goldacre, R. J. (1952b). *Intern. Rev. Cytol.* **52**, 135.
Goldacre, R. J. (1958). *Intern. Congr. Cybernetics, 1st,* Namur, 1956, p. 715. Gauthier-Villars, Paris.
Goldacre, R. J. (1961). *Exptl. Cell Res. Suppl.* **8**, 1.
Goldacre, R. J. (1962). *In* "Biological Structure and Function" (T. W. Goodman and O. Lindberg, eds.), Vol. II, pp. 633-643. Academic Press, New York.
Goldacre, R. J., and Lorch, I. J. (1950). *Nature* **166**, 497.
Hahnert, W. F. (1932). *Physiol. Zool.* **5**, 491.
Heilbrunn, L. V., and Daugherty, K. (1939). *Physiol. Zool.* **12**, 1.
Jahn, T. L., and Rinaldi, R. (1963). *Ann. Meeting Biophys. Soc., 7th, New York, 1963,* Abst. MD1.
Kamiya, N. (1959). *Protoplasmatologia,* **8**, 1.
Kenyon, R. A. (1960). "Principles of Fluid Mechanics." Ronald Press, New York.
Landau, J. V. (1959). *Ann. N. Y. Acad. Sci.* **78**, 487.
Landau, J. V., Zimmerman, A. M., and Marsland, D. A. (1954). *J. Cellular Comp. Physiol.* **44**, 211.
Leidy, J. (1879). *U. S. Geol. Surv. Territories* **12**, 1.
Mast, S. O. (1926). *J. Morphol. Physiol.* **41**, 347.
Mast, S. O. (1931a). *Protoplasma* **14**, 321.
Mast, S. O. (1931b). *Z. Vergleich. Physiol.* **15**, 139.
Mast, S. O. (1931c). *Z. Vergleich. Physiol.* **15**, 309.
Mast, S. O. (1932). *Physiol. Zool.* **5**, 1.
Rinaldi, R. A., and Jahn, T. L. (1962). *J. Protozool.* **9**, (Suppl.), 16.
Rinaldi, R. A., and Jahn, T. L. (1963). *J. Protozool.* **10**, 344.
Schaeffer, A. A. (1920). "Amoeboid Movement." Princeton Univ. Press, Princeton, New Jersey.
Stewart, P. A., and Stewart, B. T. (1959). *Exptl. Cell Res.* **17**, 44.
Wilber, C. S. (1946). *Trans. Am. Microscop. Soc.* **65**(4), 318.

The Comparative Physiology of Movement in the Giant, Multinucleate Amebae[1]

JOE L. GRIFFIN

Department of Anatomy, Harvard Medical School, Boston, Massachusetts

The movement of the large carnivorous amebae, particularly *Amoeba proteus* and *Chaos carolinensis* (*Chaos chaos* L., *Pelomyxa carolinensis* Wilson), has been investigated extensively (cf. Allen, 1960b, 1961a, b; De Bruyn, 1947; Noland, 1957). Recent work on these organisms by Allen and co-workers (Allen, 1960a; Allen and Cowden, 1962; Allen and Roslansky, 1958, 1959; Allen *et al.*, 1960) led Allen to propose the frontal-contraction theory of ameboid movement, which suggests that (1) at the front of each pseudopod, the endoplasm contracts as it becomes everted, and simultaneously shortens, and stiffens to form the advancing edge of the ectoplasmic tube and (2) this contraction pulls the axial endoplasm forward and thus furnishes the motive force for movement. Prior to the work of Allen, the most widely accepted theory of ameboid movement was the tail contraction or pressure gradient theory, derived from the ideas of Ecker (1849), Schulze (1875), Hyman (1917), and Pantin (1923) and elaborated by Mast (1926, 1932); see Goldacre and Lorch (1950), Bovee (1952), Allen and Roslansky (1958), and Landau (1959) for current concepts of tail contraction. According to the tail-contraction theory, the posterior part of the ectoplasmic tube contracts, forcing forward the endoplasm, which forms by solation of the tail ectoplasm during or after contraction and which gelates at the front to form ectoplasm again. Mechanisms other than these two have been proposed (cf. Allen, 1960b; De Bruyn, 1947; Noland, 1957; Bingley and Thompson, 1962; Bell, 1961; Kavanau, 1963) but have usually been considered to be inadequate to explain the observed phenomena or have failed to suggest experimental approaches.

Pelomyxa palustris is a giant, herbivorous ameba that differs from the carnivorous amebae in its habitat, food, cytoplasmic inclusions, external morphology, and other characteristics (Allen, 1960b; Kudo, 1946,

[1] Research supported by grants from the National Institute of Allergy and Infectious Diseases of the National Institutes of Health; AI-03410 at Brown University and AI-05395 at Harvard Medical School.

1957; Griffin, in preparation). In the 1870's movement in *P. palustris* was described by Greeff (1874), who discovered the organism, and by Schulze (1875). The most complete investigation to follow was that of Mast (1934).

From the earlier descriptions and from my initial observations it was clear that *P. palustris* differed from the carnivorous amebae in several aspects of ameboid movement and would furnish interesting material for a comparative study.

Methods and Materials

Chaos carolinensis (strain CH 1, see Griffin, 1960) and *A. proteus* (PROT 1) were grown in mass cultures fed on *Tetrahymena* (Griffin, 1960, 1961). *Chaos illinoisensis* (*Pelomyxa illinoisensis,* Kudo, 1951) was obtained from E. W. Daniels and was maintained according to Kudo (1951). This organism is clearly not closely related to *Pelomyxa palustris* (Griffin, in preparation) and can not remain in the genus *Pelomyxa.*

Pelomyxa palustris was collected from four locations: Falmouth, Massachusetts, in a ditch across the street from the previous site (Griffin and Allen, 1959; Mast, 1934); Congers, New York, from the northeast corner of Congers Lake (Kudo, 1957); Salisbury Cove, Maine, from Beaver Pond (W. H. Lewis, personal communication); and Cheshire, Connecticut, from a swampy area west of Waterbury Road, just north of the intersection with Byam Road (site discovered by Robert Iorio). Samples of bottom material and water were taken in wide-mouth glass or plastic bottles or large plastic dishes and kept below 23°C in the laboratory. Some organisms from Falmouth survived up to 9 months in gallon jars while some organisms from Salisbury Cove survived more than 12 months in 350-ml screw-cap plastic bottles.

Quartz capillaries were pulled as described by Allen *et al.* (1960) but were cut so that an ameba was first pulled into a relatively large opening and was gradually compressed as it moved into the narrow portion. This technique was easier and succeeded more often than did pulling an organism directly into a narrow capillary.

Reproducible compression of organisms beneath cover slips was achieved very simply by placing two narrow strips of lens paper[2] beneath the cover slip of a standard wet mount. As the aqueous medium was slowly removed from the preparation, the cover slip was finally supported by the lens paper.

---

[2] Ross Optical Lens Tissue, manufactured by A. Rosmarin, New York, was used for large amebae.

Results and Observations

General Characteristics of Movement in *Chaos carolinensis*

Most biologists are familiar with the appearance of *C. carolinensis* during normal locomotion, as in Fig. 1c (see Allen, 1960b; Kudo, 1946, for references). On repeated scanning of culture dishes containing 5000–10,000 amebae, no normal specimens of *C. carolinensis* were observed to exhibit continuing monopodial progression, although smaller organisms sometimes remained monopodial for as long as 4–5 min. Instead, alternate extension of pseudopods seems characteristic of the carnivorous amebae, and the retracting pseudopods usually contribute the supply of endoplasm for extending pseudopods. Advancing pseudopods (Fig. 1c, to right) were preceded by hyaline caps and had smooth contours whereas retracting pseudopods (Fig. 1c, to left) were more pointed and irregular in outline. The length–width ratios for single pseudopods of *C. carolinensis* varied widely and depended on the physiological state of the organism, but such ratios were often between 5 and 10 during locomotion of extended organisms and rarely exceeded 20 (cf. Kudo, 1946).

Figures 1a–c show the recovery of one specimen of *C. carolinensis* following physical agitation. In general, inactive forms of the carnivorous amebae formed smooth spheres only when moribund or under severe or unusual conditions such as when warmed or under high hydrostatic pressure.

It has been observed frequently that, prior to retraction, reversal of streaming direction in a pseudopod of *C. carolinensis* is usually accomplished by a wave of reversal which passes from the base of the pseudopod toward the still advancing tip (cf. Allen, 1961a). During the fairly rapid passage of such a wave of reversal there is no obvious decrease in pseudopod diameter that might indicate a wave of contraction of the ectoplasmic tube.

There is, in the carnivorous amebae, a general tendency for large inclusions to be carried in the rear of the cell. Figure 5 illustrates a particularly obvious example of this tendency, a specimen of *C. carolinensis* that had just fed heavily on *Tetrahymena*.

General Characteristics of Movement in *Pelomyxa palustris*

Pelomyxa palustris was rather limited in the range of shapes assumed. Figures 2a, 3d (the middle organism), and 4 illustrate the usual morphology during active locomotion. Continuing monopodial progression is characteristic of *P. palustris*. During movement, the anterior end lacks a hyaline cap but a hyaline space can usually be seen beneath the tail membrane, as in Fig. 8. In some organisms hyaline blebs appeared and

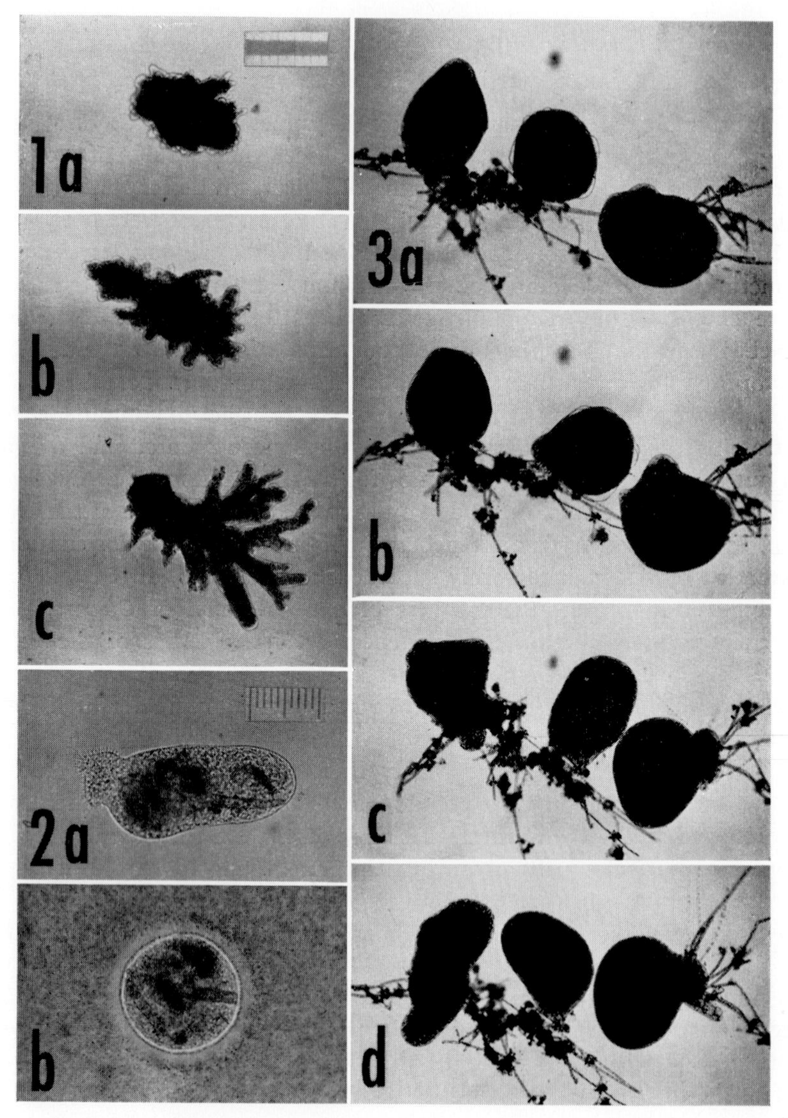

FIG. 1a–c. Photographs of a specimen of *C. carolinensis* to show the resumption of locomotion after agitation in the bore of a pipette. See text for explanation. Scale: 500 μ in 10 μ units.

FIG. 2a–b. Photographs of a small specimen of *P. palustris* from Maine. (a) Taken during normal locomotion, and (b) after the agitation caused by addition of a suspension of carmine particles. An extracellular coat, 15 μ thick, is revealed in (b). Scale: same as in Fig. 8.

FIG. 3a–d. Three specimens of *P. palustris* from Falmouth photographed during resumption of movement after agitation in the bore of a pipette. See text for explanation. Scale: same as in Fig. 1a.

disappeared on or near the tail during movement. Such a bleb is visible on the right side of the organism in Fig. 4. In large specimens loaded with glycogen spheres, as in Figs. 4 and 7a, a widening of the hyaline space was often observed at the surface just behind the fountain zone.

Figure 6 (a 3-sec photographic exposure) reveals the pattern of streaming of the specimen of *P. palustris* shown in Fig. 4. Recruitment of endoplasm occurred over the posterior two-thirds of this cell. This photograph shows the backward movement of cytoplasm at the anterior end of the ectoplasmic tube and the narrow girdle of cytoplasm that is stationary in relation to the substratum ("Gürtel" of Schulze, 1875). Mast (1934) did not agree on the existence of this "Gürtel."

Specimens of *P. palustris* were approximately circular in cross section and did not form ectoplasmic ridges, as *C. carolinensis, C. illinoisensis,* and *A. proteus* do. During normal movement, most organisms had length–width ratios of about 3 with extremes of 4 and 1.5. Large inclusions were carried in the front of the cell, as in Fig. 4.

Mast (1934) reported a clear difference between the endoplasm and ectoplasm of *P. palustris* in the density of packing of the characteristic cytoplasmic vesicles (see vesicles in Fig. 8). I was unable to confirm this observation. Those vesicles forming the outside border of the ectoplasmic tube usually appeared firmly bound and angular in outline but, in organisms having many vesicles, those in the endoplasm also appeared tightly packed and angular. Also, in organisms having fewer vesicles, even those vesicles at the periphery seemed spherical.

Waves of reversal, such as were seen in *C. carolinensis,* were never observed in *P. palustris.* Instead, the first sign of reversal was usually cessation of movement at the very front with simultaneous formation of an ectoplasmic barrier across the front of the tube and elevation of the membrane by hyaline fluid. Figure 7a shows an advancing front and Fig. 7b shows the same area just after reversal. The ectoplasmic border so obvious in Fig. 7b may be what Mast (1934) considered to be the "plasmagel sheet" which was supposed to close the front of the ectoplasmic tube during forward movement. This border and the hyaline layer appeared on reversal and not during forward movement, as Mast (1934) apparently thought. I observed, as did Schulze (1875), that the endoplasm apparently comes into direct contact with the membrane at the front. Another sort of reversal, which occurred during normal locomotion, was the occasional streaming from the tail region into and out of the tail piece or urosphere. In Fig. 4 the urosphere was somewhat larger than usual because such a stream of cytoplasm had just entered it.

The response of *P. palustris* to agitation depended on the physiological state of the organism and the severity of the agitation, but the fol-

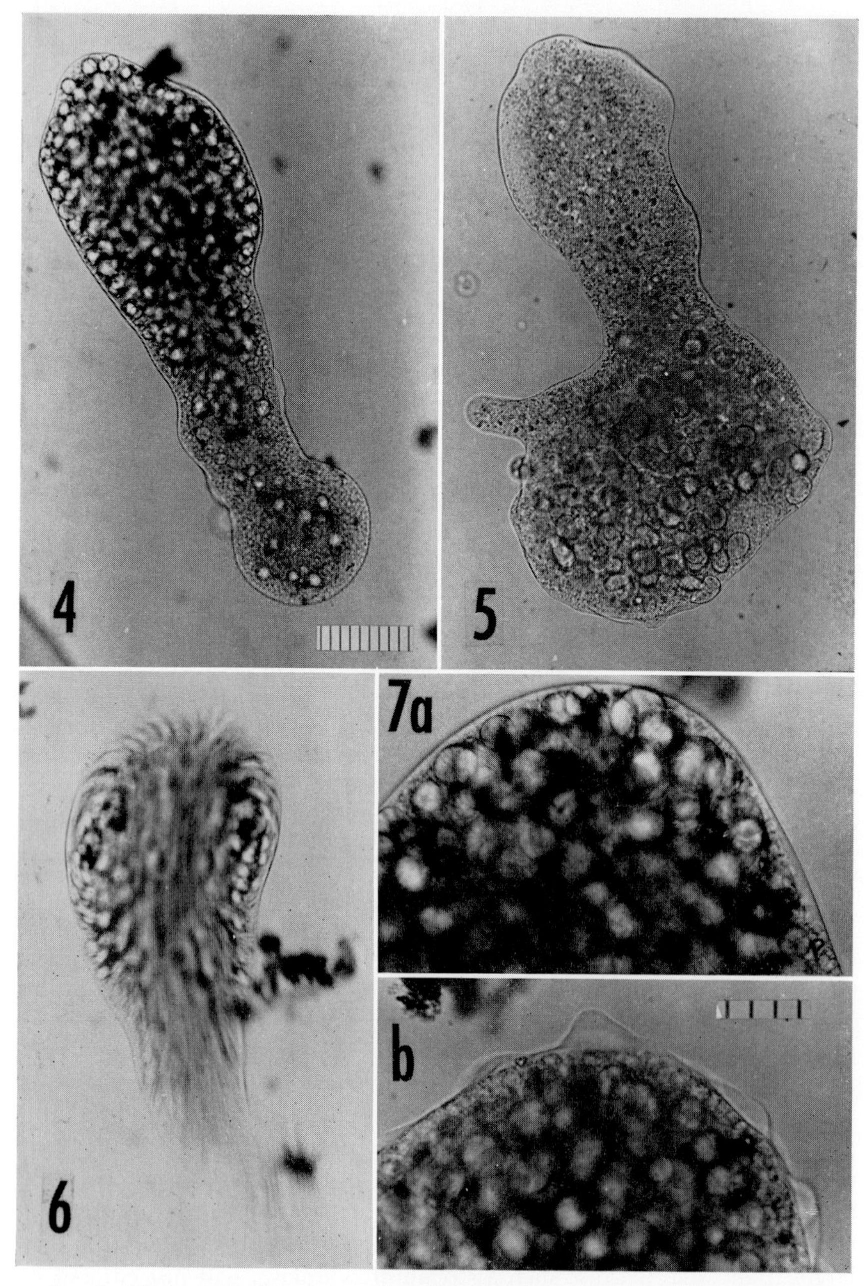

FIG. 4. Photograph of an unusually clear, well-fed specimen of *P. palustris* from Falmouth. Scale: 90 μ in 10 μ units.

FIG. 5. Photograph of a specimen of *C. carolinensis* that had just resumed move-

lowing sequence of events was usually seen in organisms in which agitation caused a complete cessation of streaming. (1) The organism formed a smooth sphere. (2) The membrane was elevated by hyaline material (Fig. 3a, middle). Some organisms formed villi over the surface at this stage (Fig. 2b). (3) A single broad pseudopod was extended, lacking a hyaline cap (Fig. 3b, middle). (4) Most of the time, this pseudopod became the body of the organism, the hyaline material was resorbed and the normal morphology was resumed (Figs. 3c and d, middle, and 2a, actually photographed before agitation). The organisms to the left and right in Figs. 3a–d did not stop streaming completely when agitated and did not show a hyaline layer during recovery. In Fig. 3d the organism to the left was streaming in two directions and the cell on the right extended a narrow pseudopod near the tail, while forward streaming continued.

In natural populations, there are, of course, some exceptions to the usual phenomena. One interesting small *Pelomyxa* from Maine exhibited an anterior hyaline cap continuous with a thick hyaline layer extending back to the tail. In this specimen small particles, which moved freely by Brownian motion, passed into the hyaline fluid from the tail ectoplasm and moved forward to the hyaline cap at a rate somewhat slower than the rate of endoplasmic streaming. This observation suggests strongly that the hyaline fluid may be produced at the rear of the cell and resorbed at the front or that the particles might have moved under the influence of a potential gradient of the sort found in *A. proteus* by Bingley and Thompson (1962). However, according to Gicklhorn and Dedjar (1931), the membrane potential of *P. palustris* is similar in magnitude but opposite in sign from those reported for the carnivorous amebae. Therefore, it would be dangerous to generalize Bingley and Thompson's results to *P. palustris*.

Ingestion of Food and Response to Environmental Stimuli

Chaos carolinensis responds to tactile stimulation or contact with motile food by forming food cups which can surround and entrap motile protozooans (Allen, 1960b, 1961a; Kudo, 1946). In Fig. 9, a small food cup had been initiated in response to the presence of two *Tetrahymena.*

---

ment after feeding heavily on *Tetrahymena*. Scale: same as in Fig. 4.

FIG. 6. Photograph, exposed for 3 sec, to show the streaming pattern of the specimen of *P. palustris* in Fig. 4. Scale: same as in Fig. 4.

FIG. 7a–b. Photographs of part of a specimen of *P. palustris* similar to that in Fig. 4. (a) The front of the organism during active locomotion and (b) the same area just after the cessation and reversal of streaming. Scale unit: 10 μ.

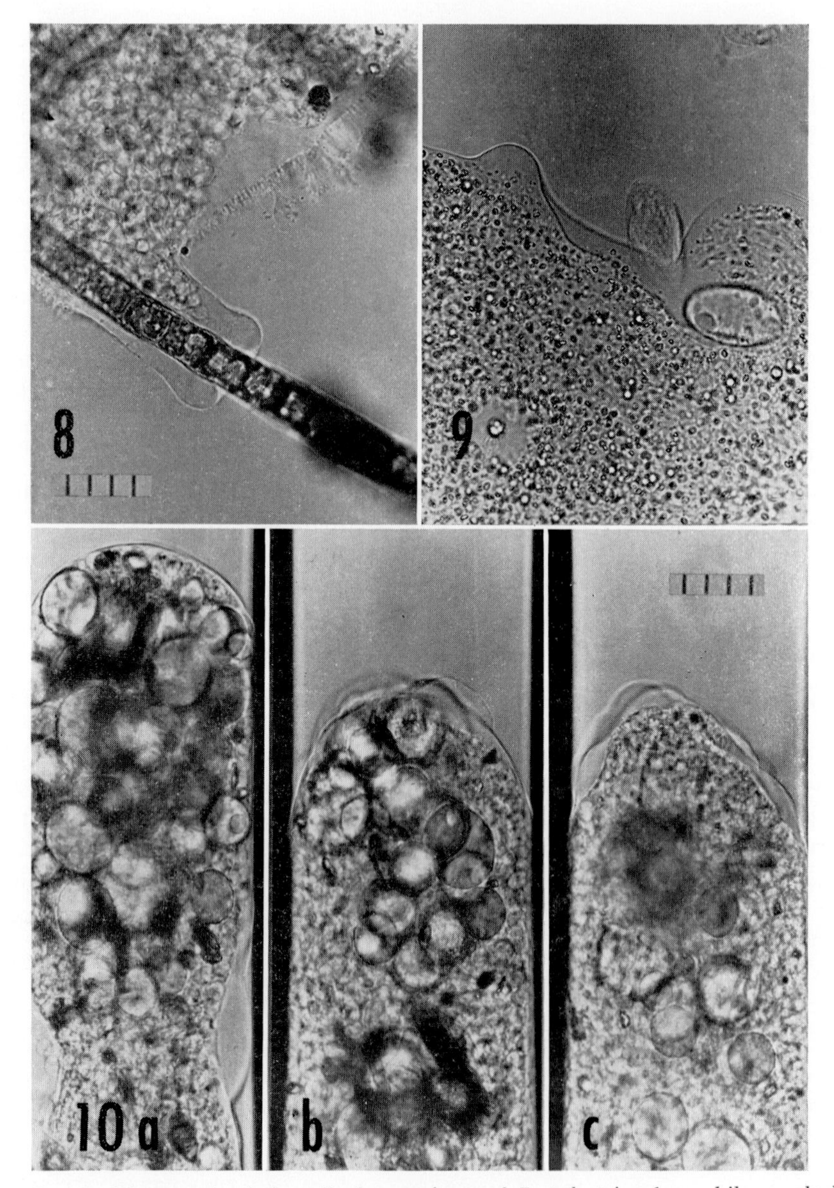

FIG. 8. Photograph of the tail of a specimen of *P. palustris* taken while an algal filament was being ingested. Scale unit: 10 μ.

FIG. 9. Photograph of an early stage of food cup formation in *C. carolinensis*. Scale: same as in Fig. 8.

FIG. 10a–c. Photographs of a specimen of *P. palustris* streaming in a capillary. In (a) streaming proceeded in both directions from the constricted region at the bottom of the picture. In (b) and (c) the capillary was moved down slightly to show the end advancing in (a), just after reversal (b) and about 15 sec later (c).

The extensions to the left and right and the wave just starting over the top exhibit the typical hyaline layer which precedes the moving cytoplasm (cf. Allen, 1961a). [See Bovee (1960) for references to feeding in amebae.]

Pelomyxa palustris formed no food cups or pseudopodial extensions for the capture of food and did not respond to motile organisms or to contact with micromanipulating tools. No ingestion of motile organisms has been reported for *P. palustris.* Instead, filamentous algae, sand grains, decaying vegetable matter, and other debris adhere to the villi at the tail and are pulled into the cell from the rear, as reported by Kudo (1957). The tail of the feeding organism in Fig. 8 illustrates the appearance of the villi, the hyaline layer beneath the villi and surrounding the algal filament, the alveolar appearance of the cytoplasm, and the bacterial inclusions.

In *C. carolinensis* and other carnivorous amebae the tip of advancing pseudopods (the fountain zone) is sensitive to light, and sudden exposure to a moderately intense light beam causes an immediate cessation, diversion, or reversal of streaming in the exposed pseudopod (cf. Allen, 1960b, 1961a; Mast, 1932).

Specimens of *P. palustris* did not respond to light intense enough to stop streaming in *C. carolinensis* even when the whole organism was included in the spot of light. During some purely qualitative experiments, *P. palustris* slowed and finally stopped streaming when exposed to very intense light for a minute or more. The possibility that heat rather than light caused the response was not excluded. After the intense light was removed, streaming resumed only after several minutes.

A_t/A_s Ratios

Allen (1960b, 1961a) discussed the fact that, in the carnivorous amebae, the cross-sectional area of the ectoplasmic tube (A_t) always exceeds the cross-sectional area of the endoplasm (A_s). The ratio A_t/A_s was considered by Allen to be related to the degree of contraction (shortening and thickening) undergone by the endoplasm as it everted to form the ectoplasm. For the calculation it is necessary to select a boundary between endoplasm and ectoplasm and to assume that the relative thicknesses of endoplasm and ectoplasm seen in the optical section apply around the perimeter of the cylindrical pseudopod.

From eight streak photographs of *C. carolinensis,* such as Fig. 11a, A_t/A_s ratios between 1.6 (Fig. 11a) and 2.6 were calculated, with most values grouped around 2.

The first A_t/A_s ratios for *P. palustris* were determined from motion pictures of early summer organisms from Falmouth. The ratios from

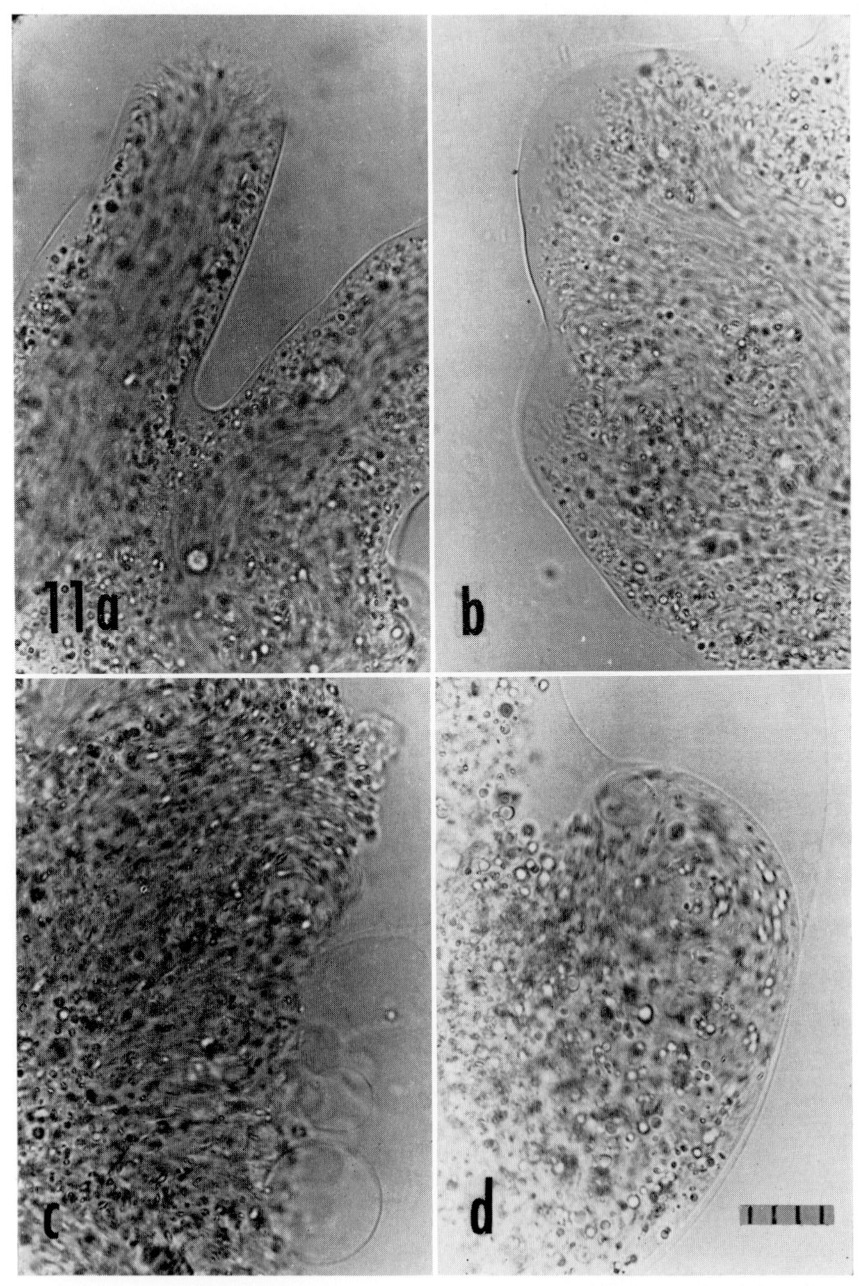

FIGS. 11a–d. Photographs of *C. carolinensis,* exposed for 1 sec to show the pattern of streaming. Scale unit: 10 μ. (a) A noncompressed cell exhibiting the usual streaming through an ectoplasmic tube. (b) Streaming in a specimen compressed beneath

eleven determinations varied between 0.6 and 1.3 with the median at about 0.9. These organisms were not observed from the side and it was possible that a thicker layer of ectoplasm could have been present along the bottom of the cell. The next spring determinations from streak photographs were made on 19 large organisms from Falmouth and Congers, viewed from the top and from the side, and A_t/A_s ratios between 1.0 and 1.9 were found. From a series of six streak photographs of a single, small, fall organism from Falmouth, viewed from the side, A_t/A_s values between 0.8 and 1.3 were determined.

Because there was variability within this single organism and others observed and because an assymetry was visible in several cells, one can conclude only that it seems probable that specimens of *P. palustris* can have a true A_t/A_s of less than 1.

Streaming in Capillaries

In quartz capillaries, specimens of *C. carolinensis* were forced to assume a monopodial type of streaming. Amebae that fit snugly, particularly those that exhibited fountain streaming in place, often developed extensive hyaline caps at the front, and elongated, wrinkled tails at the rear.

Specimens of *P. palustris* in capillaries usually streamed in a manner quite like the normal monopodial movement. However, in contrast to *C. carolinensis,* the front had no hyaline cap, and it was the tail that accumulated hyaline fluid. In Fig. 10a the front lacks a hyaline cap, and hyaline fluid accumulated at a constricting region from which streaming proceeded in both directions. Figures 10b and c show the same organism just after streaming reversed, and a few seconds later. Note, as mentioned previously, the tendency for large, inert inclusions to collect in the front and to leave the tail. Specimens in tight capillaries reversed the direction of streaming relatively often and did not continue streaming in a single direction, as was common in normal movement. This reversal seemed to be associated with the accumulation of hyaline fluid at the tail. It may be necessary that this fluid reach the front for the normal continuous change of endoplasm to ectoplasm to take place.

Streaming in Broken Cells

Observations were made on the streaming in naked cytoplasm from *C. carolinensis* reported by Allen *et al.* (1960), using some modifications

---

a cover slip to a thickness of about 55 μ. (c) The streaming pattern of naked cytoplasm isolated from a compressed specimen. (d) An apparent propagated contraction of naked cytoplasm at the edge of a syneretic vesicle. This contraction seemed to pull the cytoplasm beneath the vesicle toward the edge.

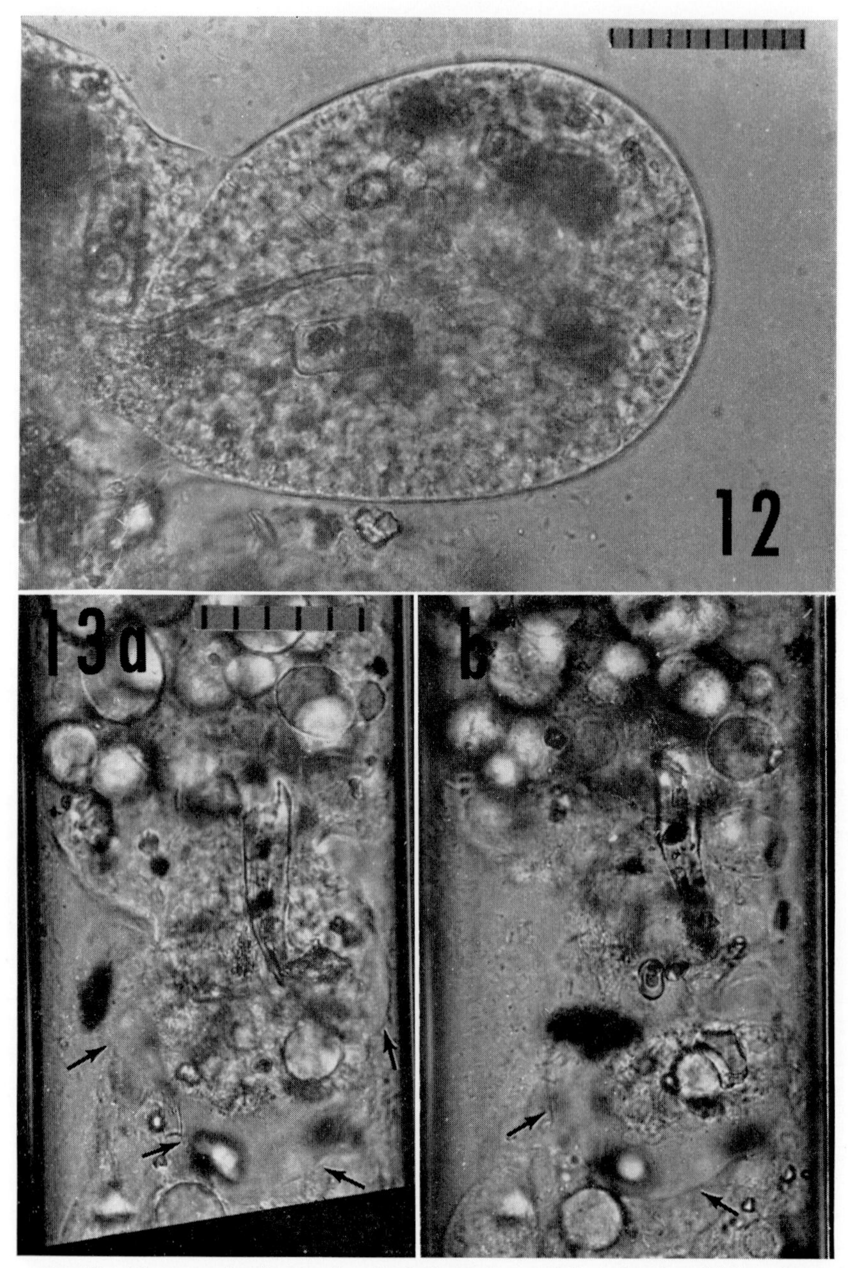

FIG. 12. Photograph of part of a crushed specimen of *P. palustris,* which re-formed an ectoplasmic tube beneath the original membrane and continued streaming for about 15 min. Scale: 100 μ in 10 μ units.

FIG. 13a–b. Photographs of the contraction of an isolated tail region from a spec-

of technique. In cells broken in capillaries by shattering the capillary under oil (Allen *et al.*, 1960) or by pulling an organism into a capillary rapidly enough to break the anterior membrane the following four stages were seen to develop from the normal pattern of streaming: (1) Streaming occurred in an intact ectoplasmic tube that lacked a membrane. The endoplasm did not disperse into the medium. (2) Streaming of broad, coordinated loops having an endoplasmic and an ectoplasmic arm followed the loss of the characteristic tubular organization. (3) The movement of individual cytoplasmic inclusions revealed streaming of thin independent loops, some of which extended some distance out into the medium (as in movies shown at the symposium). (4) After the movement stopped, the absence of the cell membrane in the areas previously streaming could be confirmed and the membrane could be identified as a wrinkled mass at the original rear of the cell. This continuing movement, which is arbitrarily divided into four stages above, developed directly from the pattern of streaming in the intact cell, and, during all the stages of streaming, apparently less rigid endoplasmic material approached the front or loop rapidly and formed a more structured ectoplasmic material that receded more slowly from the front.

Specimens of *C. carolinensis* were also broken, after compression beneath a cover slip, by tapping the cover slip with a pencil. Figures 11a–d are 1-sec streak photographs which show the pattern of streaming in the uncompressed cell (Fig. 11a), after compression (Fig. 11b), and in naked cytoplasm (Figs. 11c and d). The streaming patterns in Figs. 11b and c seem similar to the movement reported by Abé (1961, 1962, 1963) in *Thecamoeba*. Note, in Fig. 11c, the integrity of the naked cytoplasm, the streaks or blurs which show the movement of particles, and the formation of precipitation membranes enclosing what seems to be syneretic fluid. While the cytoplasm in Fig. 11d was streaming, a large vesicle, presumed to contain syneretic fluid, formed and covered a part of the surface. That part of the cytoplasm beneath this membrane did not contract, but apparently was pulled toward the edge of the vesicle and contracted as soon as it was exposed to the medium. A new vesicle, slightly out of focus, can be seen forming at the site of contraction. This inhibition of contraction beneath such membranes and propagated contraction at the edges of the membrane was observed repeatedly and seems to be a general phenomenon. Streaming usually continued for less than 10 min in the inorganic culture medium (Griffin, 1960) which

---

imen of *P. palustris* broken in a capillary. (a) An intermediate stage of the contraction, and (b) the terminal stage. The tail membrane, elevated by hyaline fluid, has been indicated by arrows. Scale: 50 μ in 10 μ units.

contained calcium. A similar sort of streaming was observed in naked cytoplasm isolated from *C. illinoisensis* and from *A. proteus,* although in *A. proteus* streaming usually continued for less than 30 sec. Kavanau (1963) may have seen such streaming in *A. proteus.*

Four different results were observed when specimens of *P. palustris* were broken in capillaries or under cover slips: (1) The whole organism or smaller fragments could re-form membranes and resume movement as in the intact cell. (2) Streaming units, with complete ectoplasmic tubes, could form inside the original membrane, as in Fig. 12. (3) Lysis of the cell could occur, during which the cell membrane and the membranes of the cytoplasmic vesicles would break down. The cell membrane was not preserved after the integrity of the cell was lost, as was the membrane of the carnivorous amebae. (4) In capillaries, isolated tails, with the posterior membrane intact, were observed to empty the cytoplasm within them until almost nothing was left. Early stages of such a contraction were similar in appearance to Fig. 11c, and Figs. 13a and b show an intermediate and a terminal stage of such a contraction. There was an elevation of the posterior membrane by hyaline fluid during such an emptying. The material extruded by or anterior to such an area did not move and the vesicles underwent lysis. It did not seem, therefore, that these tails could have been emptied by tension from the front.

Discussion and Conclusions

For the purpose of discussion, the following basic assumptions have been made. First, the motive force for movement is probably generated by protoplasmic contraction. This view has been held by most recent workers in the field (cf. Allen, 1960b, 1961a; De Bruyn, 1947; Bovee, 1952; Goldacre, 1961; Landau, 1959), including most of the participants in this section of this symposium. Recently, this view has been strengthened by the results of Simard-Duquesne and Couillard (1962a, b) who isolated an actomyosin-like adenosine triphosphatase from *A. proteus* and observed a nondirected contraction of glycerinated amebae on addition of adenosine triphosphate (ATP). Second, the clear correlation between the appearance of hyaline fluid and various phases of movement (this study; Allen and Cowden, 1962; Goldacre, 1952, 1961; Landau *et al.,* 1954) suggests that the separation of hyaline fluid from the granular phase can be taken as an indication of contraction accompanied by syneresis (expulsion of interstitial fluid), occurring somewhere in the cell.

Mechanisms not invoking contractility have been proposed (cf. Allen, 1960b; De Bruyn, 1947) including recent suggestions involving electrophoresis (Bingley and Thompson, 1962), "surface extension" (Bell, 1961), and "jet propulsion" (Kavanau, 1963), but direct evidence supporting any of these theories seems to be lacking at present.

It is obvious that *P. palustris* differs from *C. carolinensis* in almost all of the details of movement, just as it differs in other physiological characteristics. What seem to be significant differences are summarized in Table I.

TABLE I

SUMMARY OF OBSFRVED DIFFERENCES RELATED TO MOVEMENT

Property	*C. carolinensis*, *C. illinoisensis*, *A. proteus*	*Pelomyxa palustris*
1. Characteristic form in locomotion	Polypodial (alternate pseudopods), ectoplasmic ridges, wrinkled tail	Monopodial, cylindrical body
2. Initiation of movement	Many pseudopods, gradual decrease in number	Extrusion of hyaline fluid over surface, single pseudopod
3. Hyaline cap or layer	Forms at front	Forms at rear
4. Reversal of direction	Frequent, wave of reversal from pseudopod base to tip	Ectoplasmic barrier forms and hyaline material is extruded at old front
5. Large inclusions	Carried in tail	Carried at front
6. Ingestion of food	Anterior food cups, motile food	Algae pulled in at tail
7. Light sensitivity	In anterior fountain zone, rapid response	Relatively insensitive, slow response
8. A_t/A_s ratio	Greater than one	Apparently less than one at times
9. Behavior of naked cytoplasm	Continues streaming in fountain and loop patterns	Contraction of tail of broken organisms, no streaming in naked cytoplasm

Of the theories of ameboid movement currently available, only the frontal-contraction theory proposed by Allen (1960b, 1961a), seems, in my opinion, adequate to account for the observed details of movement in *C. carolinensis*. This theory seems to offer a reasonable and consistent explanation for the properties of *C. carolinensis* listed Table I.

One of the difficulties in deciding on the correctness of one or another of the theories of ameboid movement is that much of the evidence we have is compatible—or at least not incompatible—with several theories. This point was discussed by Allen (1961a) at the Leiden symposium.

The streaming of naked cytoplasm is, however, an observation that cannot be reconciled with the tail contraction theory. In fact, the tail-contraction theory predicts the very outcome of the cell rupture experiment observed with *P. palustris,* where streaming occurs only if a contracting tail remains. The fact that streaming in *C. carolinensis* continues after the membrane and the ectoplasmic tube are disrupted must be taken into account in the evaluation of theories of ameboid motion. It is not sufficient to explain this away as "abnormal." If this were an acceptable explanation, then 20 years of research on the biochemical activities of isolated mitochondria and chloroplasts could be dismissed in the same manner.

Although it seems clear that tail contraction to create a pressure gradient cannot account for all the phenomena observed in the carnivorous amebae, it is possible that a tail contraction does perform work in the cell. As is well known, the posterior part of the ectoplasmic tube does shorten and particles move closer together (cf. Allen, 1960b; De Bruyn, 1947; Goldacre, 1952, 1961). From the evidence presently available, this shortening could be passive, caused by the endoplasm being pumped out from the front, but it is possible that it could be active, actually contributing to the work done at the front.

It is clear that *P. palustris* differs in almost all details of movement from *C. carolinensis.* Is it possible to decide whether the differences are due to different mechanisms of movement or whether they might reflect ecological adaptations based on a common mechanism of movement? The following facts concerning movement in *P. palustris* bear on this question:

1. Allen (1961a) listed, for the carnivorous amebae, a number of characteristics that were compatible with or tended to support the frontal-contraction theory. *Pelomyxa palustris* exhibited none of those characteristics and none of those listed for *C. carolinensis.*

2. If the appearance of a hyaline layer can be assumed to indicate a contraction with syneresis, no indication of frontal contraction was observed whereas several observations, in particular, movement in capillaries and isolated tail contraction, indicated the local appearance at the tail of syneretic fluid.

3. The observation, in one cell, that small particles moved from the rear to the front in the lateral hyaline layer also suggests that contraction with syneresis occurs at the tail, although the possibility that this movement resulted from a potential gradient cannot be excluded.

4. If the apparent A_t/A_s ratios for *P. palustris* represent true ratios below 1, the frontal-contraction theory could not apply to these organisms, since the ratio must exceed 1 if the endoplasm shortens and thickens

as it forms ectoplasm, as the frontal-contraction theory proposes (Allen, 1960b, 1961a).

None of the evidence supports the presence of frontal contraction in *P. palustris,* and several observations suggest that contraction of the tail does occur. Some of the observations, particularly negative results, such as the failure to obtain fountain streaming in naked cytoplasm, must be taken with some caution, since the integrity of the cytoplasm in *P. palustris* apparently depends on the presence of the membrane, just as it seems that the integrity of the membrane is lost as the cell lyses.

The distinct backward movement of inclusions at the front of the ectoplasmic tube revealed in Fig. 6 and the slight thickening of the hyaline layer beneath the membrane in the same region, just behind the front (Figs. 4 and 7a), at one time suggested that a frontal contraction, occurring a bit later during eversion of the endoplasm than in the carnivorous amebae, could occur in *P. palustris.* However, it now seems that this is not the case and that the hyaline fluid is extruded at the tail and accumulates just behind the fountain zone prior to its resorption at the front.

Although the evidence supporting tail contraction in *P. palustris* is not as complete as the evidence supporting frontal contraction in *C. carolinensis* and the other carnivorous amebae, it seems reasonable to conclude that these two types of organisms differ in the site of application of the motive force, just as they differ in all other physiological characteristics.

Allen (1961a) said of the tail-contraction theory of ameboid movement that it "offers a reasonably satisfactory and consistent explanation of continuous progression of a monopodial ameba, but little more." Since *P. palustris* does, in fact, exhibit "little more" than a monopodial progression, it would seem that this opinion has been supported.

Mast (1934) concluded: "The essential factors involved in locomotion in *Pelomyxa palustris* as in *Amoeba proteus* are (1) maintenance of turgidity of the system; (2) localized attachment of the plasmalemma to the substratum and the plasmagel; (3) transformation of plasmasol into plasmagel at the anterior end and vice versa at the posterior end; and (4) continuous difference in the elastic strength of the plasmagel in different regions." Points 2 and 3 of Mast would still seem to be correct. However, the conclusion that *P. palustris* is similar to *A. proteus* in the mechanism of movement seems almost certainly to be incorrect. Mast (1926, 1932, 1934) apparently followed Hyman (1917) in adopting the idea that a gel exerted tension or tended to contract on gelation. Current concepts of tail or ectoplasmic tube contraction (cf. Allen and Roslansky, 1958; Goldacre and Lorch, 1950; Goldacre, 1952, 1961; Bovee, 1952;

Landau, 1959) postulate that the processes of gelation and contraction are separated in time and space. Also, there is no indication that an intact *P. palustris* is a turgid system as Mast thought. Although the cytoplasm does seem to expand on lysis, this probably reflects only the breakdown of the vesicles and a change in the physical nature of the cytoplasm because, in lysing immobile cells, there was no tendency for the cytoplasm to pour out of the unlysed portion in the direction of the wave of lysis.

In conclusion, the movement of *C. carolinensis* (and other carnivorous amebae), on the basis of available evidence, seems to be compatible with the frontal-contraction theory (Allen, 1960b, 1961a), whereas other theories available at present do not seem able to account for all the observable details. Indications that the frontal-contraction theory applies to *P. palustris* are lacking, and evidence available supports rather strongly the concept that tail or ectoplasmic tube contraction produces the motive force.

Material pertinent to this paper will be discussed in the Free Discussion of Part II.

References

Abé, T. H. (1961). *Cytologia* **26**, 378.

Abé, T. H. (1962). *Cytologia* **27**, 111.

Abé, T. H. (1963). *J. Protozool.* **10**, 94.

Allen, R. D. (1960a). *J. Biophys. Biochem. Cytol.* **8**, 379.

Allen, R. D. (1960b). *In* "The Cell" (J. Brachet and A. E. Mirsky, eds.), Vol. II, pp. 135-216. Academic Press, New York.

Allen, R. D. (1961a). *Exptl. Cell Res. Suppl.* **8**, 17.

Allen, R. D. (1961b). *In* "Biological Structures and Function" (T. W. Goodwin and O. Lindberg, eds.), Vol. II, pp. 249-56. Academic Press, New York.

Allen, R. D., and Cowden, R. R. (1962). *J. Cell Biol.* **12**, 185.

Allen, R. D., and Roslansky, J. D. (1958). *J. Biophys. Biochem. Cytol.* **4**, 517.

Allen, R. D., and Roslansky, J. D. (1959). *J. Biophys. Biochem. Cytol.* **6**, 437.

Allen, R. D., Cooledge, J., and Hall, P. J. (1960). *Nature* **187**, 896.

Bell, L. G. E. (1961). *J. Theoret. Biol.* **1**, 104.

Bingley, M. S., and Thompson, C. M. (1962). *J. Theoret. Biol.* **2**, 16.

Bovee, E. C. (1952). *Proc. Iowa Acad. Sci.* **59**, 428.

Bovee, E. C. (1960). *J. Protozool.* **7**, 55.

De Bruyn, P. P. H. (1947). *Quart. Rev. Biol.* **22**, 1.

Ecker, A. (1849). *Z. Wiss. Zool. Abt.* **A1**, 218.

Gicklhorn, J., and Dedjar, E. (1931). *Protoplasma* **13**, 450.

Goldacre, R. J. (1952). *Intern. Rev. Cytol.* **1**, 135.

Goldacre, R. J. (1961). *Exptl. Cell Res. Suppl.* **8**, 1.

Goldacre, R. J., and Lorch, I. J. (1950). *Nature* **166**, 497.

Greeff, R. (1874). *Arch. Mikroskop. Anat. Entwicklungsmech.* **10**, 51.

Griffin, J. L. (1960). *Exptl. Cell Res.* **21**, 170.

Griffin, J. L. (1961). *Biochim. Biophys. Acta* **47**, 433.

Griffin, J. L., and Allen, R. D. (1959). *Biol. Bull.* **117**, 413
Hyman, L. H. (1917). *J. Exptl. Zool.* **24**, 55.
Kavanau, J. L. (1963). *Developmental Biol.* **7**, 22.
Kudo, R. R. (1946). *J. Morphol.* **78**, 317.
Kudo, R. R. (1951). *J. Morphol.* **88**, 145.
Kudo, R. R. (1957). *J. Protozool.* **4**, 154.
Landau, J. V. (1959). *Ann. N. Y. Acad. Sci.* **78**, 487.
Landau, J. V., Zimmerman, A. M., and Marsland, D. A. (1954). *J. Cellular Comp. Physiol.* **44**, 211.
Mast, S. O. (1926). *J. Morphol. Physiol.* **41**, 347.
Mast, S. O. (1932). *Physiol. Zool.* **5**, 1.
Mast, S. O. (1934). *Physiol. Zool.* **7**, 470.
Noland, L. E. (1957). *J. Protozool.* **4**, 1.
Pantin, C. F. A. (1923). *J. Marine Biol. Assoc. U.K.* **13**, 24.
Schulze, F. E. (1875). *Arch. Mikroskop. Anat. Entwicklungsmech.* **11**, 329.
Simard-Duquesne, N., and Couillard, P. (1962a). *Exptl. Cell Research* **28**, 85.
Simard-Duquesne, N., and Couillard, P. (1962b). *Exptl. Cell Research* **28**, 92.

FREE DISCUSSION

Dr. Allen: A number of statements have been made of a conservative nature regarding theories of ameboid movement from Drs. Marsland, Jahn, and Goldacre. I appreciate conservatism, but for an unusual reason. I would like to see a theory developed which is conservative in that it requires the least number of untestable assumptions. In other words, I like to use Occam's razor. I think one valuable approach, which is too little followed, is to list the additional assumptions required to explain ameboid movement by the various theories. The result can be a list of surprising length.

The ameba is a physically closed system, at least as far as forces and deformations are concerned. I would go so far as to say that it is impossible to tell by simple microscopic observations on an intact cell whether streaming endoplasm is pushed by pressure or is pulled by a contraction. To cite an example of the difficulties involved in deciding such matters on the basis of observation, suppose that we have an ameba in a capillary. I shall show a film of a *Chaos chaos* specimen enclosed in a capillary made of carrageenan by a modification of Dr. Kamiya's very beautiful ameba double-chamber technique which has been devised by my student, Vincent Reale.

If one places an ameba in such a capillary and allows it to stream of its own accord, there are several possible outcomes, depending on the size of the capillary and the physiological state of the ameba. But in my film the occurrence of continued streaming in the fountain pattern will be seen. The cell surface slips in this tube, so there is no locomotion. After only a few minutes of streaming (which goes on for a period of several hours), the tail becomes enormously attenuated at the expense of material that had been in the cylindrical body. At the same time, the hyaline cap becomes considerably enlarged due to accumulation of hyaline fluid, which Dr. Cowden and I have shown is produced by syneresis. This is actually an interpretation from refractive index measurements, but I think it can be justified.

Our interpretation of the swollen hyaline cap is that the fluid cannot circulate back into the tail; therefore, the tail elongates. This phenomenon was described briefly by Dr. Goldacre a number of years ago, but in terms of his theory, the observation produces an amusing paradox; by contracting, the tail *elongates!*

Let me describe another situation that can happen under only slightly different circumstances. If the capillary is a little bit larger so that there is no restriction of the flow of hyaline cap fluid back into the tail, the cytoplasm often streams sporadically in bursts. The cytoplasm rushes ahead, but it begins to do so first at the very front, then a wave of acceleration passes from the front back toward the tail, decreasing in amplitude as it passes.

One interpretation of this sporadic streaming which makes sense in terms of the front-contraction theory is that tension is developed between the rim of the ectoplasmic tube and elastic elements which we believe are probably present in the stream. The waves of acceleration occur as the cytoplasm yields to the tension. The result is elastic waves which are progressively damped by the viscous elements of the cytoplasm. In the rear of the cell, streaming is virtually constant in velocity.

Now, this observation can be turned to exactly opposite purposes, and it can be assumed that there is tension being developed in the tail ectoplasmic tube, resulting in the formation of a pressure gradient which builds up, causing rupture at the front. Of course, one has to assume also some special properties for the ectoplasmic tube to explain why the first acceleration occurs at the front and not at the tail.

So far, you have two interpretations which are exact mirror images of one another, each requiring additional assumptions.

Now, I think the test between the two hypotheses, front-contraction and tail-contraction, may possibly be solved in a situation like this. It has already been pointed out that it is possible to break an ameba and have the cytoplasm continue to stream. In this case, the membrane, as Cooledge, Hall, and I showed a number of years ago, is usually swept toward the old tail region and is no longer in intimate contact with the streaming cytoplasm, contrary to what Dr. Goldacre has claimed. We are certain about this, and the experiments are easily repeated. *Immediately* after breaking the cell, the cytoplasm continues to stream in its normal fountain pattern. Then after a few minutes the fountain pattern breaks down radially into loops, and the very same rushes of cytoplasm are observed, with the same acceleration waves passing away from the bends of the loops. It should be noted that this occurs now in broken cytoplasm preparations showing two-way flow where there is no longer any chance that pressure might be the motive force.

I think this is at present the only *crucial evidence* with which we can decide between these two hypotheses. If it could be shown that movements in broken cells were caused by some completely extraneous force, I would perhaps concede the possibility of explaining some aspects of ameboid movement by pressure. But until some other adequate explanation can be found for this phenomenon, I am still forced to believe that the front contraction idea is the best explanation for continuous ameboid movement by cylindrical pseudopodia. This does not exclude the possibility that in other species of amebae, such as *Pelomyxa palustrus,* the situation might be quite different. I also do not exclude the possibility that hemispherical or bulbous "pseudopods" induced e.g., by heparin, might form in the manner Drs. Goldacre, Jahn, and Marsland would prefer to believe.

I think Dr. Marsland's point that the cytoplasm of an ameba can exist in two states was well taken. He refers to them as "sol" and "gel." I prefer to call them "contracted" and "relaxed" states. We must recognize that these are both interpretations of the movements we observe; the consistencies we infer from the movements. It seems equally possible, on the one hand, that solation and gelation could take place at random points within the cell as Dr. Marsland stated, and, on the other hand, that contractions and relaxations could take place at random points within the cell.

As long as we face the problem that every observation on *intact* cells is susceptible to alternative interpretations, I do not really see how it is possible to distinguish by observation alone between the font- and tail-contraction theories. What we have been trying to do most recently is make a decision between these theories on physical grounds, using polarized light. I am sorry to report that the necessity of making arrangements for the conference made it impossible for me to finish the experiments, but we are looking for evidence of photoelasticity in the endoplasm as evidence for the existence of elastic elements under fluctuating contractile tension.

Dr. Marsland: Your crucial experiment, I think, can also be interpreted according to the sol-gel hypothesis. I do not go so far as Dr. Goldacre in saying that membrane contact between the gel and the membrane is essential. I do not know; possibly it is. But assuming that it is not, you do not know where the gel is; you do not know where the sol is. You cannot differentiate with your eye unless there is movement.

Dr. Allen: I assume you are referring to the broken cell experiment. When streaming occurs, there is no difficulty discriminating between what you call sol and gel. The latter is displaced as a block, whereas the former shows some shear.

Dr. Marsland: It seems to me that when the "fountain pattern" breaks down, there is a degeneration of the organization, and what you may be seeing is just contraction.

DR. ALLEN: This is exactly what I have been saying.

DR. MARSLAND: You cannot specify which end of this membraneless mass is anterior or posterior or even external or internal because it is a disorganized mass of protoplasm, and these reactions may be occurring randomly.

DR. ALLEN: It is only rarely that cytoplasm in these preparations could be called a "disorganized mass." It is perhaps unfortunate that Dr. Griffin's pictures were shown instead of some others we have that show this is more typically a very highly organized movement.

DR. MARSLAND: Can you differentiate the regions in terms of the intact animals?

DR. ALLEN: Yes, indeed. The bend of the loop toward which the most rapid flow occurs always corresponds to the front of the organism.

DR. MARSLAND: You have loops going in the other direction, don't you, when you have the direct streaming?

DR. ALLEN: No, the bends of the loops always correspond to the previous front end of the organism.

DR. MARSLAND: How do you get two-directional streaming in the same place?

DR. ALLEN: Our theory explains it as a contraction at the bend of the loop which applies tension to one arm and compression to the other. Thus the bend remains stationary and equal masses of cytoplasm are propelled in different directions. The different velocities reflect the deformation (shortening) on contraction.

DR. MARSLAND: Still I declare it is not possible to say what the original orientation is.

DR. ALLEN: We certainly know where the front of the cell was, and where the bends of the loops are.

DR. MARSLAND: You know where it was, but do you know where it is? There is, perhaps, no such thing any more.

DR. BURGERS: May I ask you, Dr. Allen, to tell us once more how you envisage this contraction? Do you have fibers that are fixed on the front side?

DR. ALLEN: I am afraid I do not have a complete mechanical analysis of this, but the best way to illustrate my feeling for how it works is this: Suppose that you have, as a model of a cytoplasmic loop, a muscle fiber which is bent into the shape of a loop with one end in a clamp. Let us assume only reasonable properties for a muscle fiber, namely, that it gets stiff when it contracts. Since it is bent, it is stretched on the outer side. Under these circumstances, if the contraction is somehow propagated away from the clamped region, this muscle will straighten progressively as the wave of stiffening passes along it.

I think in the case of the ameba the cytoplasm of a pseudopod can be thought of as if it were made of a fused circle of clamped muscles, the clamped side representing the ectoplasm and the free side representing the endoplasm. In this way, when a contraction is propagated in the direction of the endoplasmic arms of these fused loops, the contraction stays at the bend. The result is a building forward of the ectoplasmic tube (or its displacement backward in uattached cells) and continuous endoplasmic advance.

DR. BURGERS: So you must suppose that there are fibers in all of these loops and that there is some mechanism which straightens them out?

DR. ALLEN: That is right, but I would settle for an anisotropic gel not necessarily differentiated into fibers.

DR. BURGERS: And contraction plays a part. The main point is that the structure will push forward.

DR. ALLEN: Yes.

DR. BURGERS: After a certain time when the front has moved a certain distance forward, do you suppose that these fibers can still go further?

DR. ALLEN: We do not really have much evidence about "fibers" in living ameba cytoplasm. There have been electron micrographs that have shown fibers in fixed cells; these are very likely "vital artifacts." I think one can conclude from the fact that pseudopodia have a good deal of stiffness, that there must be considerable mechanical structure present. If some kind of anisotropic structure (not necessarily fibrillar) can propagate a contraction either steadily or at a varying velocity, then the pseudopod would be expected to advance steadily or sporadically.

DR. BURGERS: I would be prepared to accept the presence of many types of structures. What bothers me is that while you can use this structure for a certain distance, you must rebuild it at the new spot to go still further.

DR. ALLEN: What I think is that we have a continuous loop of contractile material which is somehow continually organized in the tail region. We don't see very clearly what happens at the tail.

DR. BURGERS: Is it a closed loop?

DR. ALLEN: Movements in the tail region are too complicated for the structure to be this permanent.

DR. BURGERS: It is easy to assume that the loops attach themselves to the substratum like a caterpillar tractor tread.

DR. ALLEN: Exactly.

DR. SHAFFER: Why does the bottom arm move backward?

DR. ALLEN: It moves backward only *relative to* the loop or the cell. If the cell remains stationary because it lacks traction, then any particle will follow an elliptical path through the cell, as was shown years ago.

DR. WOLPERT: What I cannot understand is that if you have contraction only at the front, how can you get the movement going in both directions, especially in isolated cytoplasm where there is no barrier or fixed point?

DR. ALLEN: According to one of Newton's laws, when a body is displaced by a force, the same force also displaces its surroundings in the opposite direction.

DR. WOLPERT: I would expect the arms to move in the same direction and accumulate at the front point. If I have a spring in the form of a U and contract my spring at the bend, then all regions tend to move toward the point of contraction.

DR. HAYES: This is a point of confusion which often came up in Dr. Jahn's talk. The point of confusion is what the reference set of coordinates are. Are they a reference frame fixed with respect to the substrate or a reference frame fixed with respect to the ameba, and will one of these reference frames move with respect to the other, with respect to the substrate? Then it is possible for all the particles to be moving forward, even though some of them must move forward slower than others.

With respect to the frame of reference in the animal, then, across any section, the principle of conservation of matter requires that you have as much material going backward as you do forward, and it is possible then for particles in the gel to be moving forward in one frame of reference system. But so long as they are moving forward slower than the ameba moves, they will be moving backward in the frame of reference in which the ameba is fixed. I think this is the point of confusion here.

DR. ALLEN: Very well put.

DR. SHAFFER: If the endoplasm can transmit tension, obviously it can be drawn forward to the front of the cell. But the key requirement of the fountain-zone theory is that the tension should carry the endoplasm round the bend and advance the front of the cell. I do not think any exposition of the theory has ever explained precisely how this could occur. Dr. Allen has suggested a bent muscle fiber as an analogy. But doesn't the muscle straighten on contraction only because of the fixed mechanical relationship between opposite sides of the fiber? Is the transverse organization of the endoplasm sufficient for a similar relation to hold in the ameba? Besides, if a bundle of radially arranged, U-shaped, muscle fibers were closely packed inside a tube, could the inner arms, in fact, be everted by contraction at the bends? Presumably, if it were feasible, some animals would use the contraction of bent longitudinal muscles, arranged in this way, to evert a tubular structure; but I am not aware of any that do so.

DR. ALLEN: In reply to the comment made by Dr. Wolpert, the contraction occurs only on one side of the bend; the other (contracted) side has stiffened. With a model like this, contractile material will, I believe, "flow" through the bent region regardless of whether this portion is anchored or free to move.

In answer to Dr. Shaffer, I would agree that we have not explained *precisely* how the "contraction at a bend" principle works, but I think we have suggested a plausible type of mechanism which explains the motions seen in dissociated cytoplasm and in some types of amebae. We have been looking for a model system such as a tubular structure capable of everting by this mechanism.

PART III

Cytoplasmic Streaming, Locomotion, and Behavior of Specialized Ameboid Cells

Introduction

Douglas Marsland

Washington Square College, New York University, New York, New York

Currently there are two main types of contraction hypotheses in regard to the mechanisms of ameboid movement, and each has received support from a number of investigations. The first, which I should like to designate as the *tube-wall contraction hypothesis,* had very early origins (Ecker, 1849 and Schulze, 1875), although more specific modern formulations awaited the work of Mast (1926), Marsland and Brown (1936), Goldacre and Lorch (1950), and Zimmerman, Landau, and Marsland (1958). The second, called the *fountain-zone contraction hypothesis,* was formulated very recently by R. D. Allen (1961a) on the basis of excellent work by this investigator and a number of co-workers (Allen, 1955, 1960, and 1961b; Allen and Roslansky, 1958; and Allen, Cooledge, and Hall, 1960).

In proposing the fountain-zone hypothesis, Allen (1961a) has listed a number of reasons which have led him to believe that tube-wall contraction, as conceived by Mast, must be eliminated as a possible mechanism in ameboid movement. It is my belief, however, that many of these criticisms are based upon an unduly restricted concept of the tube-wall contraction theory. Therefore, I would like to present a broader concept, and in this presentation I shall retain the terminology of Mast.

Broad Concept of the Tube-Wall Contraction Hypothesis

In my opinion, the tubular plasmagel system of ameba cannot be regarded as a system of inert pipes into which an inert fluid is forced by contractile processes localized strictly in the posterior, or "tail" region, of the ameba. Every part of the system, all parts of both plasmagel and plasmasol, are capable, I think, of active metabolic change from moment to moment. In the complex interplay of stimulation and response, any part of the plasmagel may initiate contraction or undergo solation and any part of the plasmasol may convert itself into plasmagel. On the average, the plasmagel system tends to maintain a tubular form, either branched or unbranched. It seems probable, however, that any of the main or side branches may display—as a result of local changes in the physiological situation—a variety of structural reorganizations. The lumen of the tube may be occluded by localized gelation at any point along its length; the lumen of the tube may increase or decrease by a changing gel/sol ratio; contractile tension in the plasmagel wall may increase or decrease either generally or locally; or the wall of the plasmagel tube may give way, as the result of localized solation, allowing another branch, or pseudopodium, to extend out laterally. Generally, perhaps, contractile tension in the plasmagel system may tend to be maximal in the posterior parts, but the active ameba must be granted a full latitude of variation in fulfilling the complex requirements of its locomotion.

The foregoing concept as to the nature of the tubular plasmagel system of the ameba may be broader than was originally intended by Mast. In this case it may be called the Marsland concept of the Mast hypothesis; or more simply, the Mast-Marsland tube-wall contraction hypothesis of ameboid movement. In any event this broader concept provides, I think, a sounder basis for comparing the fountain-zone and tube-wall contraction mechanisms, and when this is done I think it will be found that the latter mechanism cannot be eliminated on the basis of our present knowledge.

References

Allen, R. D. (1955). *Biol. Bull.* **109**, 339.
Allen, R. D. (1960). *J. Biophys. Biochem. Cytol.* **8**, 379.
Allen, R. D. (1961a). *Exptl. Cell Research Suppl.* **8**, 17.
Allen, R. D. (1961b). *In* "Biological Structure and Function" (J. W. Goodwin and O. Lindberg, eds.), Vol. II, p. 549. Academic Press, New York.
Allen, R. D., and Roslansky, J. D. (1958). *J. Biophys. Biochem. Cytol.* **4**, 517.
Allen, R. D., Cooledge, J. W., and Hall, P. J. (1960). *Nature* **187**, 896.
Ecker, A. A. (1849). *Wiss. Zool. Abt.* **A1**, 218.
Goldacre, R. J., and Lorch, I. J. (1950). *Nature* **166**, 497.
Marsland, D. A., and Brown, D. E. S. (1936). *J. Cellular Comp. Physiol.* **8**, 167.
Mast, S. O. (1926). *J. Morphol. Physiol.* **41**, 347.
Schulze, F. E. (1875). *Arch. mikroskop. Anat. u. Entwicklungsmech.* **11**, 329.
Zimmerman, A. M., Landau, J. V., and Marsland, D. (1958). *Exptl. Cell Research* **15**, 484.

The Role and Activities of Pseudopodia during Morphogenesis of the Sea Urchin Larva

TRYGGVE GUSTAFSON

The Wenner-Gren Institute, University of Stockholm, Sweden

We have quite sophisticated ideas about how deoxyribonucleic acid (DNA) codes ribonucleic acid (RNA), how RNA codes protein synthesis, how enzymes bring about metabolic reactions, and how metabolites regulate these events, but we are still ignorant about the final steps in the causal chain between the genes and the shape they control. In order to bridge the gap between molecular biology and morphology it seems logical to try to define the cellular forces that bring about changes in shape and how the forces become properly directed. If we could show that a morphogenetic process can be reduced, for example, to pseudopod activity, we could later on try to define the molecular basis for this activity, and how it becomes established at the right time and place. The sea urchin embryo is very useful for such studies: it has relatively few cells, there is little growth or cell division during the main morphogenetic events, it is transparent, and its walls are only one cell layer thick. It rapidly develops into a shape which is sufficiently complex to pose important and puzzling morphogenetic problems, but is not so complex as to overwhelm the observer. Our main experimental approach has been to study the development by means of time-lapse cinematography. A simple technique was worked out for holding the larvae in a constant position during filming. For bibliography of our papers see our review (Gustafson and Wolpert, 1963) where also the work of our colleagues from Japan, particularly Dan and Okazaki, is discussed.

The main morphological changes of the larva are summarized in Fig. 1 and comprise changes in thickness and curvature of the wall, strong, directed extension of some regions, release of cells from the wall which migrate and form a more-or-less regular mesenchyme pattern, the laying down of a complicated skeletal system, and the perforation of cell sheets and the surrounding hyaline membrane to give rise to a mouth. As to the forces that bring about these changes, moderate deformations of the wall are explicable in terms of forces arising within the cell sheet themselves, "intrinsic forces." Changes in cell contact due to changes in adhesion and/or tension in the cell membrane and changes in adhesion to the hyaline membrane appear to be responsible for variation in thickness and curvature of the wall and the release of cells from it. Strong deformations of the wall, on the other hand, require "extrinsic forces," i.e., forces taking their origin outside the cell sheet. The main mechanism for these "directed long-range translocations of a cell sheet," and also for the migration of free cells, is thus related to pseudopodial activities.

A map of the prospective significance of the different zones of the egg is shown in Fig. 2. Zone 1 gives rise to the primary mesenchyme cells

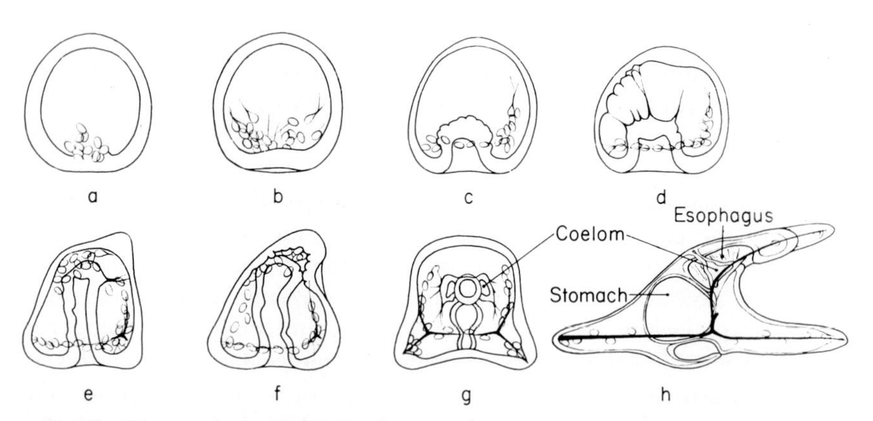

FIG. 1. Diagrammatic drawings of some stages of a developing sea urchin larva to show the main morphological changes. (a) An early mesenchyme blastula; the cells at the vegetal pole lose adhesion for each other and enter the blastocoele to form the primary mesenchyme. (b) The primary mesenchyme cells migrate by means of filopods. (c) An early gastrula in side-view, the primary mesenchyme cells have arranged themselves into a characteristic pattern, and the archenteron rudiment has formed a rounded invagination due to a decrease in contact and hence a change in packing of its cells. (d) The rounded invaginated region extends by means of pseudopodial activity at its tip; note the "cones of attachment" in the ectoderm formed where pseudopods attach. (e) The pseudopod-forming cells have detached from the archenteron tip which begins to attach to the mouth region by means of pseudopods. (f) The archenteron tip is in direct contact with the mouth region where the ectoderm is slightly invaginated. The uppermost region of the archenteron tip will later on become delineated by a constriction to form a sac. (g) (Ventral view) the sac has become extended and subdivided into two sacs by means of pseudopodial activity. A skeleton has also become laid down by the primary mesenchyme, and the "arm buds" begin to protrude and will later on become extended due to the tension produced by the growing skeleton. (h) (A pluteus larva) the mouth is completely formed, the archenteron is subdivided by constrictions, and the skeleton has extended the wall into characteristic processes, the four arms and the dorsal *Scheitel*. The coelom has become pulled out and subdivided into two sacs. Pseudopod-forming cells from the coelom pull out themselves and become free mesenchymal strands that line up around the esophagus and form a contractile network around it. The pseudopods have also extended the coelom (normally the left one) into a tube, the primary pore-canal.

which will lay down the skeleton. In the late blastula they begin to lose adhesion for each other and for the hyaline membrane and begin to round up and show a pulsatory activity. Thus they tend to force each other apart and enter the blastocoele (Fig. 1a). They pile up at the vegetal pole and then spread along the blastula wall to form a regular pattern comprising a ring with two clusters from each of which a branch of mesenchyme is formed (Fig. 1c). The formation of this pattern of mesenchyme appears to require directed cell migration, and the sea ur-

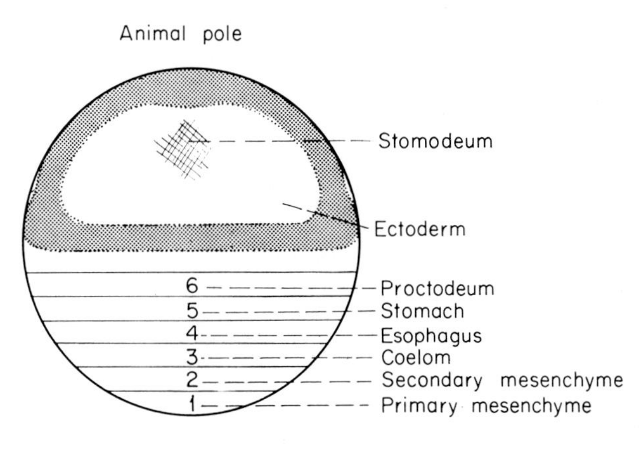

FIG. 2. A map of the presumptive regions of the egg. All the zones indicated have not yet been mapped by vital staining technique, cf. Hörstadius (1935), and the diagram should be considered as an extrapolation from available results. The relative proportions of zones 1–6 is overscaled for convenience. The stippled zones correspond to thickened regions of the ectoderm, but must be considered entirely schematic.

chin larva, therefore, provides an almost unique system for studying such a process *in vivo*.

The piled up cells gradually begin to "shoot out" numerous filopods into the blastocoele (Fig. 1b). The filopods are about 0.5 μ in diameter, but may be thinner. They may reach a length of 30–40 μ and can extend as quickly as 30 μ in 3 min. There is a considerable variability in their arrangement. The first ones project separately from the cell surface, but later on they may be confined to a small protrusion from the advancing end. Although quite thin, they may be stiff enough to be moved bodily across the blastocoele like long, thin bristles. They may also bend and show wavelike motions at their base, which contribute to their ability to explore the blastocoele and its wall. In the thickest filopods one may see a traffic of granules up and down. The filopods can attach to the wall, but only with their tips. Upon attachment, the extension of the filopods appears to stop.

The filopods provide the basis of cell movement throughout the migration of the primary mesenchyme. When a filopod has attached to the wall it contracts and brings about a movement of the cell body. Filopods that do not make contact or make very unstable contacts may continue to probe the wall but, more often, rapidly collapse (e.g. 25 μ shortening in 8 min) and are withdrawn into the cell body, or form an irregular mass, apparently incapable of reorganizing into a new filopod at once.

The filopodial exploration, attachment, and contraction does not only account for the cell movement, but also for its apparent directedness and thus for the development of the regular mesenchyme pattern. The filopods shoot across the blastocoele, and, since this is apparently liquid, they cannot be guided either by it or the wall. The random manner in which they shoot out is a further indication of this lack of guidance. Their subsequent exploration also lacks any obvious directedness and appears to be quite random. The cells also show no tendency to continuous polarized movements along the wall, and after moving in one direction they may move in any other. They may, therefore, transiently show a considerable dispersal and deviation from the final ringlike arrangement. Sometimes they migrate halfway to the animal pole, but sooner or later return to the ring level if they have surpassed it. The direction of their movements is, in fact, determined by the contraction of those filopods that make sufficiently stable contacts with the wall. If only one filopod is formed and brings about a movement, filopods formed later on may move the cell in another direction. If there are many filopods there is, in a sense, a competition between them, each trying to move the cell in its direction and the final movement is determined by the relative strength of the contact between the filopods and the wall. The cell bodies will, therefore, finally line up along the zones of the ectoderm where the adhesiveness is highest and so form the characteristic pattern. The increase in order goes hand in hand with a progressive fusion of the filopods into a syncytium that attains the shape of a cable from which innumerable thin short filopods extend, exploring the wall together, and, therefore, with a high probability of finding the most adhesive zone.

The early dispersal of the mesenchyme cells past the ring level and their subsequent return is most difficult to account for in terms of mechanisms based on contact guidance, but is easily explained in terms of random filopodial exploration and the gradual development of the pattern of selective adhesiveness of the ectoderm in relation to the mesenchyme cells, or in the adhesive properties of the filopods themselves. The variability in the manner in which the final pattern is reached probably reflects the lack of precision in the timing of various events, e.g., the release of the mesenchyme cells from the blastula wall, their formation of filopods, their development of adhesiveness, their fusion into syncytia, and the changes in the adhesiveness of the ectoderm. In such a mechanism as that postulated here, the same final state will be reached regardless of variation in the timing of the individual events.

It would be interesting to know the nature of the adhesiveness of the ectoderm. Studies on slightly compressed larvae, where the ectoderm is extended and its cells partially separated, show that the cells are mutually connected only at their inner and outer ends, but where the contact is low there are only outer connections (Gustafson, 1963). A strong packing

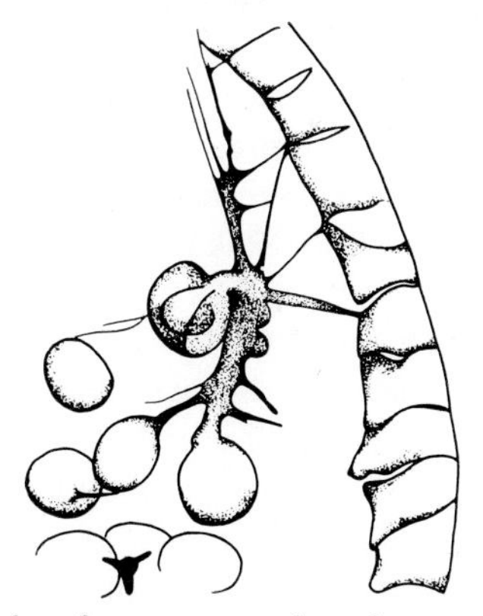

Fig. 3. Filopodial branches from a ventrolateral syncytial cable of the primary mesenchyme, and the filopodial attachments to the ectoderm at the inner points of contact between the ectoderm cells.

of the cells and hence a thickening of the ectoderm is evidently due to the presence of inner contact points. We have recently found that it is to these contact points that the filopods attach, particularly the inner ones (Fig. 3). The cell surface as a whole is not adhesive. This explains the important fact that the mesenchyme cells tend to line up along the thickened zones of the ectoderm. But why do the filopods attach so badly to the animal plate which is the thickest region? To be accessible for the filopods, one may visualize that the adhesive surfaces must be partially uncovered by a tension in the sheet that tends to separate the cells or, that the tension is involved in some other way. In the animal plate the strong contact along the whole length of the cells may prevent changes. Some justification for this hypothesis will be given later.

The ectoderm can evidently be regarded as a "template" for the mesenchyme pattern in the sense that it presents a pattern of variation in adhesiveness which is reflected in the distribution of the mesenchyme cells, and the pseudopodial mechanism thus contributes to the coordination of the embryo, quite as well as inductive stimuli do in other cases.

The mechanism for directed movement outlined here is not new in that it was suggested in general terms by Weiss (1947) as one of the three possible mechanisms by which cells could be guided, i.e., selective fixation, selective conduction, and selective elimination. The observations we have described are perhaps the first clear-cut example of selective fixation, however, cf. Twitty and Niu (1958) and Twitty (1949). One of the novel features of a mechanism such as we have outlined is the use of long filopods to explore the substrate. The great length and number of the filopods and their slenderness make them a powerful and economic tool for this purpose. Another advantage is that they both localize the target and bring about the movement of the cells toward it. And, finally, they secure a constant end result in spite of an imperfect timing of the events in the embryo. It is, therefore, not surprising to find that pseudopodial mechanisms also play a role in other events where directed long-range translocation is needed, cf. Nakai and Kawasaki (1959) and Nakai (1960).

Even when the mesenchyme pattern concerned is attained, pseudopods are continually making and breaking contacts with the ectoderm, there is a "turnover" of the contacts. The cells are now, however, joined by the cablelike structures and no longer move independently, but considerable rearrangements are still possible, and as soon as the adhesiveness of the ectoderm changes, a corresponding change takes place in the distribution of the cable-linked mesenchyme cells. These changes underly the development of the progressively complicating skeletal pattern. The skeleton is formed as a triradiate crystal in the syncytial mass in the two mesenchyme clusters, and the three radii continue to grow within the syncytial cables and later on bend and branch in a characteristic way (Fig. 1f–h). A crystallographic mechanism is responsible for the formation of its initial triradiate shape and for the tendency of its radii and branches to grow straight. The skeleton is, however, laid down with random orientation, but within less than half an hour it attains a correct orientation so that two of its radii point along the ringlike cable and the third along a cable branch extending upward. This reorientation is brought about by the tension in the cables which meet at the point where the triradiate structure is laid down. The tension is produced by the filopods which are "guided" by the adhesion of the ectoderm. Filopods attaching to adhesive regions are also responsible for all bends and branches of the skeleton including spine formation, since they deform the cable and bring about rearrangements of the mesenchyme.

It is interesting to find that experimental interference with the "viscosity" of the filopods brings about a more complicated skeletal pattern since the cells can form more complicated filopodial connections (cf. von Ubisch, 1957). Furthermore, mesenchyme cells from a species with a complicated skeleton transplanted to a species with a simpler skeletal pattern can, probably due to the characteristic "viscosity" of their cytoplasm, respond to adhesive regions which the host's own mesenchyme cells normally do not detect, and a skeleton with some characteristics of the "complicated" donor forms in the "simple" host (cf. von Ubisch, 1957).

Skeletal mesenchyme collects at some thickened regions of the ectoderm, the "arm buds," and forms skeletal branches (Fig. 1g), and the pressure of the growing branches with the mesenchyme at their tips extend the "arm buds" into long slender arms. Filopodial activity is thus, indirectly, responsible for the arm extension.

The archenteron rudiment corresponds to zones 2–6 in the diagram (Fig. 2). The first step in its invagination involves a decrease in contact between the cells (owing to changes in adhesion and possibly in tension) and their tendency for rounding up thus leads to a change in cell packing (Fig. 1b and c). This change is particularly strong in zone 2 and is similar to that in zone 1, but the decrease in contact is not complete. Pulsations also occur, probably due to the reduced contact, and these may help the invagination by breaking bonds between cells and by assisting in the transfer of cytoplasm from the outer to the inner edge. A further rounding up or change into pear-shape of the cells would only make the rudiment round up further and bring about no further elongation, and invagination, therefore, tends to stop. The pulsatory activity of the cells, however, persists and may become extremely vigorous. Gradually the pulsatory cells begin to shoot out pseudopods (filopods), often directly from the pulsatory "lobopodia." They therefore mainly shoot upward, though their direction is variable and they "explore" the blastocoele and the wall (Fig. 1d). Many of the pseudopods attach to the wall. When they fail to do so, they tend to "collapse" and are withdrawn. The pseudopods are of quite different dimensions and are often branched at their tips. The branches operate independently from each other in the sense that some of them may detach and collapse or continue to probe the wall, whereas other branches are still attached, holding the pseudopod extended. Sometimes the branches seem to fuse into a sheetlike perforated structure that seems to slide along the wall. The attached pseudopods shorten and exert a tension on the ectoderm and on the archenteron. This tension is made evident by the formation of cones of attachment in the ectoderm, i.e., cells where the pseudopods attach tend to be pulled inward (Fig. 1d), and sometimes the whole wall is deformed. The pseudopod-forming cells may also pull themselves out from the archenteron tip.

The correlation between the onset of pseudopodial activity and the onset of continued invagination is clear (Fig. 4) and, if the attachments of the pseudopods is prevented or the pseudopod-forming cells detach too early from the archenteron tip, the invagination stops. It is thus evident that it is the pseudopods that extend the archenteron. Calculations of the force required to bring about invagination have shown that they need only be very small, 10^{-2} dyne, and this is quite consistent with our knowledge of cellular forces. Work in progress with time-lapse cinematography in my laboratory by Dr. Shigeru Hamano, University of Hokkaido,

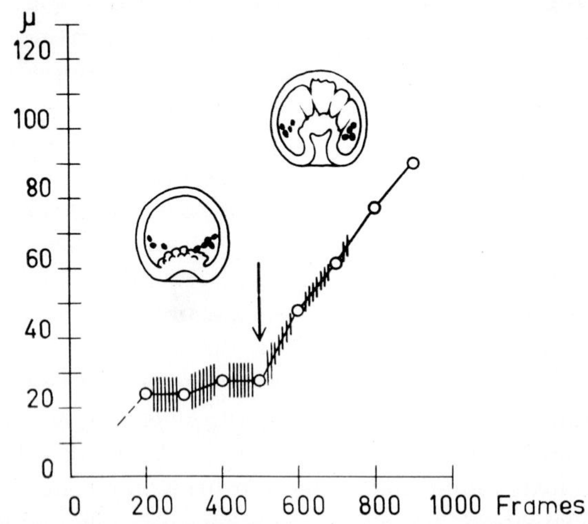

FIG. 4. Diagram of the invagination of the archenteron. The curve shows the length of the archenteron cavity plotted against time. The extent of the pulsatory activity is proportional to the length of the cross lines on the curve. The appearance of the first pseudopods at the archenteron tip is indicated by the arrow. The diagram shows the two steps in the process of invagination and the correlation between the onset of the second step and the appearance of pseudopods. A thousand frames correspond to 5 hr, 33 min.

has shown that pseudopods also play a role in gastrulation of a fish embryo (*Danio rerio*) which is transparent enough to allow the observations of the cells in the mesendodermal region. This suggests that pseudopods play a more or less general role for long-range translocations during morphogenesis.

Gastrulation also involves the problem how movement becomes directed, and one can, in principle, account for this problem in the same terms as the directed movement of the primary mesenchyme cells, i.e., in terms of random exploration of pseudopods for adhesive surfaces. In fact, both kinds of pseudopods attach to the inner contact points between the ectoderm cells (Fig. 5a) (Gustafson, 1963). That the archenteron extends along the dorsal and not along the ventral wall is probably partly a matter of relative distance (Fig. 1c and d). The archenteron tip is brought closest to the dorsal wall owing to early changes in curvature of the ectoderm, and the probability for contacts is hence greatest there. Furthermore, because of the extension of this wall, its cells tend to separate, and the contact points would, therefore, become partly uncovered and accessible for the pseudopods. This interpretation is supported by the observation that a rupture in the animal plate makes it possible for pseudopods to attach there (Fig. 5b). When the invagination has proceeded almost to the animal pole, the pseudopod-forming cells

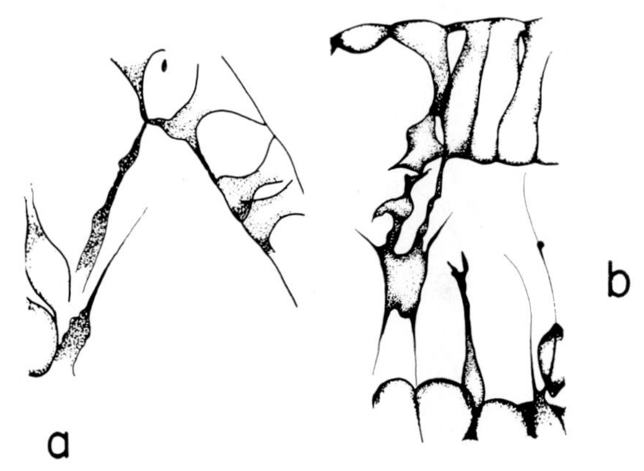

FIG. 5. (a) The attachment of an archenteron tip pseudopod at the contact points between adjacent ectoderm cells. (b) Relation between the animal plate (above) and pseudopods from the archenteron tip (below). The ectoderm cells in the animal plate are normally in close contact along the whole of their length, but owing to a rupture in the plate followed by a retraction of the cells to the left, an edge of a cell in the plate becomes available for pseudopod attachment.

detach and the archenteron tip thus becomes free (Fig. 1e). The remaining pseudopods attach to the mouth region which, for reasons still unknown, is very adhesive. The tension in these pseudopods is high and may pull the oral ectoderm inward, but also bends the archenteron to the mouth region where it becomes anchored (Fig. 1e and f). Gastrulation is a process that is quite variable, but the end result appears unaffected by this variability, and this is again explicable in terms of pseudopodial activity and adhesiveness of the wall.

The secondary mesenchyme cells also migrate by means of pseudopods attaching to the points of junction of ectoderm cells (Fig. 6) (Gustafson, 1963), even to the same point as a primary mesenchyme filopod. That the two mesenchyme patterns are still not identical is probably because of the fact that the secondary mesenchyme cells are multipolar and their pseudopods may attach at widely distant points of the wall making the pattern more irregular than that of the primary mesenchyme. In spite of this one may, with some imagination, see a similarity between the pattern of the two types of mesenchyme.

One class of the secondary mesenchyme cells are characterized by their content of pigment granules. These cells move by means of short pseudopods and long filopods. The granules can be used as markers for the cytoplasmic movements. We have seen how granules, thicker than the average diameter of a filopod, moved up through it to its tip, but the traffic was one way and the granulum remained at the tip when the filopod collapsed (Gustafson and Wolpert, unpublished). The pigment

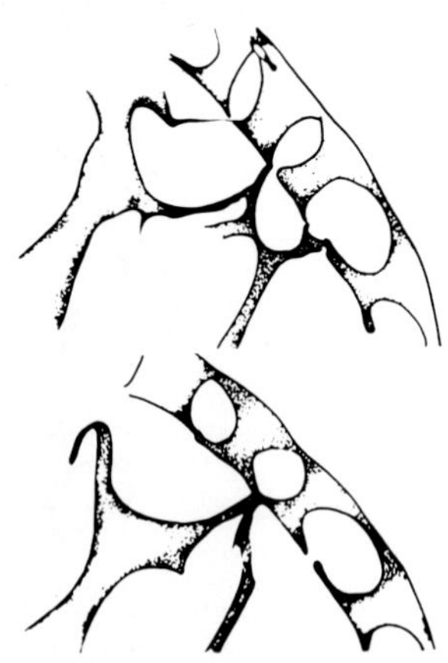

FIG. 6. The attachment of secondary mesenchyme pseudopods at the contact points between adjacent ectoderm cells.

cells also illustrate how similar end results may be reached in different ways. In *Psammechinus* the pigment cells detach from the archenteron tip at the end of invagination, and later on some of them always collect at the arm tips. In *Echinocardium* the cells detach already before invagination has started and migrate in the spaces between the ectoderm cells, but their final distribution is similar to that in *Psammechinus*. This would again be the consequence of random filopodial exploration and attachment to adhesive regions within the ectoderm.

To what extent do the principles discussed also account for apparently complicated processes later on in development? It seems that a dominant role is also played here by changes in cell contact, pulsatory and pseudopodial activities. Much of the apparent complexity of these processes appears to arise from spatial relationships which make the study difficult.

The invagination of the archenteron was brought about mainly by activities in zone 2, the future secondary mesenchyme cells (Fig. 2). When these have detached, the archenteron tip corresponds to zone 3, the coelom region, which begins to show a pulsatory and pseudopodial activity as in the preceeding zones (Fig. 1g). The pseudopods extend the sheet in a similar way as the archenteron is extended. Cones of attachment in the ectoderm indicate that they exert a considerable tension. The role of pseudopods in the coelom extension is illustrated by observations in vegetalized larvae where an everted coelomic rudiment often occurs at the vegetal pole. Pseudopods that may here shoot out into the sea water have no surface to attach to, and no extension occurs. On the other hand, if they attach to the slide, coelom sheet may become pulled out. The subdivision of the coelom into two sacs seems to be accounted for by the fact that the anterior sheet of the unpaired rudiment is attached to the oral ectoderm and by the inability of the pseudopods to attach to the animal plate (Fig. 8).

When the coelom sacs are pulled backward (Fig. 1h), some of the pseudopod-forming cells pull themselves out of the coelom and form mesenchymal strands, a "coelom mesenchyme," attached to the esophagus wall (Fig. 7a and b). These strands and also the coelom pseudopods begin to show an intense, periodic contractility. The development of such a contractility in slowly contractile elements suggests a close relation between the two types of contractility. Anyhow, the contractions apparently help the lining up of the strands around the esophagus. The situation may be likened to the tightening of elastic bands around a cylinder. Later on the strands form thin side-processes which fuse so that a muscular cage is formed around the esophagus, but the details of this process and how the contractility becomes peristaltic remains to be investigated. The great ability of the coelom mesenchyme strands to form side-links may reflect their origin from coelomic cells which normally attach to each other along the whole of their edges and form a sheet.

The backward extension of the coelom sacs along the esophagus by means of pseudopods (Fig. 7) is similar to gastrulation (Fig. 1d). Since the left sac is larger than the right one it can be further extended and forms a tube, the direction of which would be determined in the same way as the direction of the free mesenchymal strands (Fig. 7). Finally, the pseudopods at the tip of the tube attach to the adjacent dorsal ectoderm and extend the tube right up to it. After that, these pseudopods have only the "choice" to direct more or less radially (Fig. 7e), and their tension probably contributes to break up the blind end of the tube as well as the ectoderm to which it attaches, leading to a free passage through the tube into the coelomic sac, and so the primary pore-canal has formed. The underlying processes appear to be simple and involve no novelty.

The double cell layer in the mouth region, the ectoderm, and the coelom-esophagus sheet are thinned out and the cells are torn from each other by the rapid contractility in the coelomic pseudopods and the peristaltic activity of the esophagus muscles; and this, probably together with an enzymatic activity of the esophagus that softens the hyaline layer, finally causes a rupture of the double cell sheet, and thus the mouth perforation occurs (Fig. 8c–h). The invention of this important step does not appear to require basically new mechanisms, only a proper adjust-

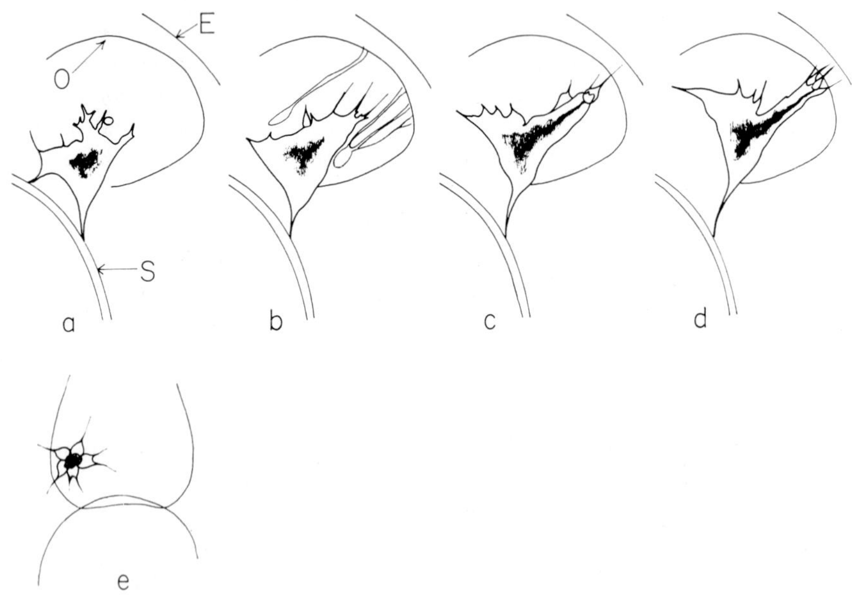

FIG. 7. Semischematic drawings of a larva in profile, ventral side to the left, showing the rounded esophagus, O, and the left coelom sac which extends backward along the esophagus by means of pseudopods and finally forms a tube, the primary pore-canal that attaches to the dorsal ectoderm, E. In (e) the end of the tube is seen from the dorsal side; note the radiating pseudopods that tend to break up the end of the tube and the ectoderm to which the tube has attached. In (b) some of the mesenchymal elements pulled out from the coelom are indicated.

ment of the time and intensity relations between the underlying mechanisms.

The following zone, 4, gives rise to the esophagus and zone 5 to the stomach (Fig. 2). The further morphogenesis of these zones also involves some pulsatory activity, but this is not followed by pseudopod formation. However, their final shape is influenced by the contractile elements around the esophagus. For instance, the peristaltic activity pumps water into the thick-walled stomach rudiment which thereby becomes extended into a thin-walled vesicle (Fig. 1h). The development of the sphincter systems at the border between the compartments of the gut and the anal sphincter remains to be elucidated.

From our studies we got the general impression that the molding of the shape of the embryo is determined by the time-space distribution of a few cellular activities. Both the time when these activities become overt

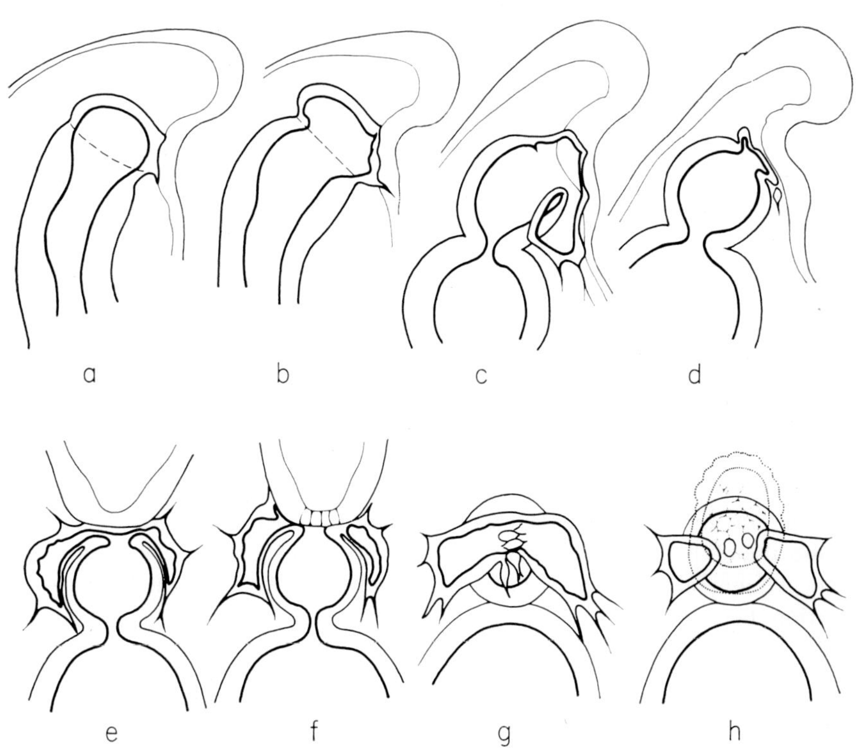

Fig. 8. Semischematic drawings to illustrate the development of the coelom and the mouth. (a–d) Larvae in optical median section [(c) somewhat to the right of the median plane]; (e) and (f) dorsoanimal view; (g) and (h) ventral view. The coelom rudiment becomes delineated and increases its contact with the ectoderm and becomes pulled out by pseudopods into two sacs. As a result of the rapid periodic contractions in the pseudopods and the coelomic strands around the esophagus, the anterior coelom sheet and the ectoderm to which it attaches becomes torn and the cells separate leading to the formation of a mouth. The edges of the oral invagination of the ectoderm is indicated by dotted contours in (h).

and their nature appear to be related to the position of the regions along the animal vegetal axis (Fig. 2). The changes are, in fact, time-graded in that decrease in contact, pulsatory activity, and pseudopod formation all start in zone 1 and progressively spread upward into the consecutive zones. The "intensities" of the activities are also graded and fall off along the axis although to a different extent. The decrease in contact is complete in zone 1, and the cells are released before pseudopods appear, whereas in zone 2 they have more or less to pull themselves out of the cell sheet, and in zone 3 only few cells can pull themselves out, most of them remaining together forming a thin sheet, and in the following zones the cells keep together still more strongly. The pulsatory activity occurs in zones 1, 2, and 3, and also tends to occur in zone 4 and 5 at least. The pseudopod activity only occurs in zones 1, 2, and 3, occasionally in zone 4, and not in the following ones. It seems that the different activities are related, since a decrease in contact is always correlated with a pulsatory activity, and a strong pulsatory activity is followed by a pseudopod formation. One may, therefore, suggest that these activities reflect quantitative variations of only one or two cellular parameters along the egg axis, perhaps in some properties of the cell membrane. It is possible that the cellular activities within the ectoderm can be added to this "spectrum" of activities of the mesendoderm, since, also, the main morphogenetic differences between the ectodermal regions involve differences in cell contact. The primary mensenchyme and the animal plate cells would thus represent the opposite ends of this "spectrum," and the high adhesion between the ectoderm cells may also allow the mesenchyme pseudopods to adhere there, provided the contact points are accessible.

The justification for these extrapolations is that they urge us to think about pseudopodial activity in relation to other cellular phenomena. And, we may also begin to visualize vaguely that there might be a close relation between the spectrum of morphogenetic activities of the cells and the biochemical and physiological gradients, so famous among embryologists. To establish some kind of relation between the molecular and the morphological level was at least our point of departure and is still our goal. Some old results of ours, which seem to have stirred up some controversy through the years now appear to be closer to general acceptance. I refer to the role of mitochondria in differentiation and to the effect of the lower blastomeres upon the formation and distribution of mitochondria (Gustafson and Lenicque, 1952). Although electron micrographs only show some vague structural differences and no difference in the total number of mitochondria along the egg axis, the differences in Janus green staining (Czihak, 1962) indicate that the number of actively respiring mitochondria is greatest near the animal pole. (This is in line with our conclusions drawn in 1952 with regard to different stages in mitochondrial development.) The decrease in the proportion of ectoderm cell and the parallel decrease in "active" mitochondria brought about by subjecting the cleaving egg to lithium (Gustafson and Lenicque, 1952) perhaps follows from a block in the formation of functional RNA and hence synthesis of new proteins, since chloramphenicol, which has this chemical effect, interferes with morphogenesis just as lithium does (cf. Lallier, 1962; Hörstadius, 1963), and lithium retards the onset of formation of new proteins as chloramphenicol does in many systems. If we return to adhesion, it seems that the high adhesiveness of the ectoderm cells corresponds to the formation of new proteins, perhaps related to the formation of actively respiring mitochondria, whereas cells with low adhesiveness—and also the ability to form pseudopods, perhaps a more primitive type of behavior—appear to have fewer respiring mitochondria. The relation among active respiration (or, more specifically, oxidative phosphorylation), adhesion, and pseudopod formation will in any case be investigated. The ideas themselves only intend to symbolize a way of thinking about a link between molecular events and morphology.

One may ask how a system operating with such a restricted number of mechanisms may still give rise to such a sophisticated form as the pluteus larva. In the preceding discussion we have tried to outline the manner in which this is possible. A formal and perhaps amusing way to illustrate this problem more directly is to consider an irregular polygon for which one has only one rule: connect all junctions between lines. A progressively complicated pattern arises. This example also illustrates our basic approach which is essentially to view development as a historical process. One can only view, for example, the normal development of the coelomic sacs and the primary pore-canal as the results of a series of simple and partially interlinked steps, the results of the later ones being determined by what has occurred before and therefore more complicated and more difficult to predict. A correlate of this is that an early strong disturbance that is not corrected by the steady readjustments of the pseudopodial-ectoderm relations bring about the development of a shape that is hardly recognizable as a sea urchin larva, although all the instructions are well considered by the cells. There is thus in one sense a certain "all or none" character of normal development, the choice being determined by the extent of the early disturbances. It is remarkable that the cells of dissociated early larvae can aggregate and later develop into a fairly normal larva (Giudice, 1962), but once a sorting out of cells and blastula formation have occurred there are no restrictions in the system that prevent the instructions from bringing about the whole series of consecutive processes of normal shape formation.

A special aspect of the self-complication is the possible inductive stimulus that the primary mesenchyme may emit to the ectoderm cells even after the blastula stage. The ectoderm acquires a pulsatory and pseudopodial activity in some regions of the ventral side, i.e., in the curvature between the arm rudiments which are brought about by these activities. One may be permitted to suggest that the ectoderm cells have acquired certain mesenchymal properties as a result of their close earlier contact with the primary mesenchyme clusters.

But let us leave all these speculations and return to the central topic of this paper. I hope that this review has shown that pseudopods play an important role in morphogenesis, and that their advantage is connected with their ability to produce a force as well as to localize the target, even a distant one, and so to give the movement a proper direction. And since they may be extremely thin and require little material they can be formed in large amounts. The pseudopods are evidently well-suited tools for integration of structures formed by regions far apart. Their attachment does not require highly specific types of adhesiveness, and this makes the process of integration "cheap." Their continued exploration, attachment, and detachment, furthermore, allow a rather extensive variation and lack of precision of the underlying events in different regions and secure a rather constant end result. And, although my contribution to the understanding of central mechanisms of pseudopod activity has been more meager than it was intended to be, one may visualize that the pseudopod activity is closely related to other basic cellular activities.

References

Czihak, G. (1962). *Arch. Entwicklungsmech. Organ.* **154**, 29.
Giudice, G. (1962). *Develop. Biol.* **5**, 402.
Gustafson, T. (1963). *Zool. Bidrag Uppsala* **35**, 425.
Gustafson, T., and Lenicque, P. (1952). *Exptl. Cell Res.* **3**, 251.
Gustafson, T., and Wolpert, L. (1963). *Intern. Rev. Cytol.* **15**, 139.
Hörstadius, S. (1935). *Pubbl. Staz. Zool. Napoli* **14**, 251.
Hörstadius, S. (1963). *Develop. Biol.* **7**, 144.
Lallier, R. (1962). *Experientia* **18**, 141.
Nakai, J. (1960). *Z. Zellforsch. Mikroskop. Anat.* **52**, 427.
Nakai, J., and Kawasaki, Y. (1959). *J. Embryol. Exptl. Morphol.* **7**, 146.
Twitty, V. C. (1949). *Growth* **13**, Suppl. 133.
Twitty, V. C., and Niu, M. C. (1958). *J. Exptl. Zool.* **108**, 405.
von Ubisch, L. (1957). *Pubbl. Staz. Zool. Napoli* **30**, 279.
Weiss, P. (1947). *Yale J. Biol. Med.* **19**, 235.

Discussion

Dr. Jaffee: I wonder how you fit exogastrulation into your view of how the archenteron extends?

Dr. Gustafson: The mechanism of exogastrulation is quite different from that of normal gastrulation. If we plot the length of the archenteron rudiment during normal gastrulation we get the curve in Fig. 4. If we plot exogastrulation we get a curve which starts off in the same way, but after the first phase of invagination the extension stops owing to a failure of the pseudopodial mechanism. Later on, when there is an increase in the osmotic pressure within the blastocoele, the archenteron rudiment becomes reverted and extended outward. The elongation of the archenteron rudiment during exogastrulation is thus brought about by an external force quite as the elongation during normal gastrulation, but it is an osmotic and not a pseudopodial force in normal larvae.

CHAIRMAN MARSLAND: Then it is an absence of pseudopodial activity at the critical time.

DR. GUSTAFSON: Yes; the archenteron rudiment does not become attached to the inside of the ectoderm, because of a premature release of the pseudopod-forming cells from its tip. The osmotic pressure that gradually increases in the blastocoele can, therefore, cause a reversion and outward extension of the archenteron rudiment. We have many films that show this process.

DR. REBHUN: Has any electron microscopy been done on these pseudopods to see whether or not they do have internal structure of some sort?

DR. GUSTAFSON: I think Dr. Wolpert can answer this question. I have also some preliminary electron-microscopic observations of pseudopods. They look quite empty. However, I know that Dr. Sylvan Nass has observed some structures inside the pseudopods. Another point of interest is that there are no signs of terminal bars or any other structure at the pseudopodial tips which may be involved in their attachments to the ectoderm.

DR. REBHUN: I wondered. Usually in electron micrographs, the microvilli appear empty. By several techniques, the freeze-thawing technique, and by use of acrolein fixation, it is possible to demonstrate rodlike structure within the center of almost all microvilli. Dr. Dougherty and I have looked at structures which I would tentatively suggest are similar to the microspikes of Dr. A. C. Taylor. They are not seen in formalin or ordinary osmium fixation.

DR. WOLPERT: We haven't looked at pseudopodia in great detail. Most pseudopodia are seen in cross section. You sometimes see a central vesicle and begin to think it represents a central core. Dr. Mercer, with whom I have worked on this, prefers to believe it represents endoplasmic reticulum.

DR. REBHUN: In Dr. Dougherty's acrolein preparations, the cores look much more like Dr. Roth's "spindle tubules" than like endoplasmic reticulum.

DR. WOLPERT: I think Dr. Gustafson pointed out that you often find a granule moving rapidly foward in a pseudopod, and in certain cases moving back again. This argues against any fixed internal structure.

DR. GUSTAFSON: Yes, and the granules may be much thicker than the filopod causing a local extension. So I think there is no room for any thick solid rods within the filopods.

DR. NAKAI: I studied some electron micrographs of pseudopods from the nerve fiber in tissue culture, and the diameter was about from 0.25 to 0.3 μ. I couldn't find any structures in filopodia under the phase microscope, but electron micrographs show small vesicles which are probably ER.

Can you see the movement directly under the microscope? I mean is it moving fast?

DR. GUSTAFSON: Yes; the movements of the pseudopods may be particularly fast during bending. If the pseudopods bend at the base, their movements may be quite easy to follow in the microscope.

DR. NAKAI: What was the diameter of the pseudopodia?

DR. GUSTAFSON: On the average about 0.5 μ, but at the base they may exceed a micron; at the tips they may also approach the resolution limit of the microscope.

Movements of Cells Involved in Inflammation and Immunity

ROGER ROBINEAUX

Hôpital Saint-Antoine, Paris, France

Cells encountered in the phenomena of inflammation and immunity are not very good material for the study of cellular movement from a fundamental point of view. Many facts can, however, be gathered from observing them.

For several years, we have devoted a good deal of work to the study of the movement of cells involved in inflammation. This work has been resumed recently (Robineaux and Pinet, 1960). More recently still, we have studied in tissue culture the movement of cells involved in immunity.

The technique of observation used in all these cases has been time-lapse microcinematography in phase or interference contrast, subsequently analyzed frame-by-frame. We shall present here the results obtained in a study of cells of the adult guinea pig spleen. The cells were cultured under a dialysis membrane.

Material and Methods

The spleen cultures came from adult guinea pigs. Before splenectomy, the animals were bled by cardiac puncture and the spleens removed by lobotomy and placed in a sterile petri dish containing 10 ml of Hanks' solution. After removal of the capsule, the spleen was cut into fragments of less than 1 mm^3 in culture medium.

NUTRIENT MEDIUM

The medium consisted of 70% Parker 199, adjusted to pH 7.2 with sodium bicarbonate; 5% chicken embryo extract; 25% guinea pig serum, either autologous serum or a homologous serum from a pool of normal guinea pigs stocked at —20°C; solution of antibiotics and antifungal agents—penicillin (200 units/ml), streptomycin (50 μg/ml), mycostatin (100 units/ml).

CULTURE TECHNIQUE

We have chosen to use the cellophane strip technique described by Rose *et al.* (1958). We used a standard size (2-ml) multipurpose chamber; cover slips less than 0.15 mm in thickness were required for use with high-power and oil-immersion objectives.

Four tissue fragments were placed on a cover slip in a small drop of nutrient medium. The cover slip was then completely covered with a strip 4 cm wide of Visking dialysis membrane previously sterilized in alcohol and washed in Hanks' solution. The silicone rubber gasket which forms the walls of the culture chamber was then put in place, followed by the second cover slip which closed the chamber. The whole chamber was maintained between two metal retaining plates by four screws. Two needles were placed in the gasket; one permitted the filling of the chamber above the cellophane strip with about 2 ml of nutrient medium, the other served as an air vent; both could be removed after filling.

Results

General Aspects

Beginning as early as the first hours of cultivation and continuing up to the third day, many cells could be observed migrating out from the explant toward the periphery in the space delimited by the explant, the cover slip, and the dialysis membrane. The most peripheral cells became flat and stationary when peripheral movement was limited by the dialysis membrane. In this zone are found neutrophilic and eosinophilic granulocytes, lymphoid cells of all ages, and many undifferentiated cells, with and without macrophagic activity. Cells with macrophagic activity typically clean the intermediary zone rather rapidly.

Between the third and the ninth days, the intermediary zone becomes progressively organized: at about the tenth day, when many cell divisions appear, it assumes its most interesting aspect. The evolution of this zone will continue for 1 or 2 weeks.

It is easy to remove the entire culture on the cellophane strip for fixation and staining. The study of a fixed and stained preparation on about the fifteenth day shows all the intermediate types between reticular cells and mature plasmacytes. This method of culture then seems to represent a good system for obtaining *in vitro* all the stages of plasmacyte differentiation. We shall now outline what can be found in the study of cells in this series.

Cell Movements

Two kinds of movement can be seen in this culture:

1. The movements of single cells: reticular cells, free or in syncytium, middle-size and small lymphocytes, plasmablasts (hemocytoblasts of the immunologists), proplasmacytes and plasmacytes, histioblasts and mono-

cytes, and polymorphonuclear leucocytes ("polymorphs") in the third day of cultivation. One can also study movements during mitosis in these different types of cells.

2. Cell interactions, of which the following aspects are the most characteristic: (*a*) Movements of cells in reticulolymphoplasmacytic islets; (*b*) emperipolesis observable between small and middle-size lymphocytes and reticular cells or certain cells in mitosis.

Movement of Single Cells

1. *Free Reticular Cells.* The two types of cells encountered in the cultures can be distinguished clearly by the activities of their membranes. In cells functioning as macrophages, these membranes are of the undulating type, pinocytosis is intense, and the cytoplasm is filled with phagocytized material. In cells without macrophagic activity, there is no undulating membrane, no pinocytosis, and movement is restricted to "microbubbling." These two types of cells are not polarized, and their locomotion is generally weak.

2. *Syncytia.* Numerous reticular cells organize themselves into syncytia containing numerous nuclei with nucleoli and cytoplasm containing large amounts of phagocytized material. The movements of the membrane are of the undulating type. The pinocytotic vacuoles which form at the periphery of the cell enter the cytoplasm in waves forming lines parallel to a tangent to the cell surface.

3. *Middle-Size and Small Lymphocytes.* These movements conform to the classic descriptions. In media free of plasma, they advance highly polarized, rapidly emitting many short hyaloplasmic veils at the anterior pole; these are plainly visible in phase contrast. In the small lymphocyte, these veils are smaller than in the middle-size lymphocyte, but they are just as active as the middle-size ones; the speed of migration is the same for the two types of cells. In the small, spherical lymphocytes, these veils can resemble by their length and fineness the undulating membranes of histiocytes.

In the course of their locomotion, they may have the so-called "hand-mirror" shape, but many remain oval or somewhat rectangular. The posterior part of the cell, dense and contracted, adheres very little to the substratum. We have never observed a lymphocyte change its polarity; it always moves straight forward, striking other cells, going around them, or engaging in emperipolesis. It can stop, and while stationary can lose its polarity, sending out only temporary ameboid pseudopodia. At this time, none of the characteristic production of hyaloplasmic veils occurs, as in the moving cell.

4. Plasmacytic Series. The type of culture utilized has permitted us to follow the dynamic behavior of plasmacytes of different ages.

a. Plasmablast. This cell possesses much cytoplasm and an anteriorly located nucleus; it forms hyaloplasmic veils quite comparable to those of polymorphs; in addition, one can observe in the front of the cell during locomotion a moderate pinocytic activity. The hand-mirror form described for lymphocytes is clearly visible and a "tail" region of the cell appears then with clarity. The cell leaves behind, on the substratum, some trailing cytoplasmic fibrils (Fig. 1). These cells can change their polarity and one can then observe between the new anterior pole and the remains of the old one, the formation of a cortical gel (Fig. 2).

b. Proplasmacyte. At this stage the cell has reduced its volume and no longer shows hyaloplasmic veils, but only temporary cytoplasmic protrusions which then become the front of the cell in locomotion. The nucleus is generally situated in a median position, with the centrosome behind it still visible; the "tail" of the proplasmacyte hardly adheres at all to the substratum (Fig. 3).

c. Plasmacyte. These cells, which are very easily identified in our cultures, can locomote easily but slowly; while the speed of migration of polymorphs and of lymphocytes is comparable (20–35 μ/min), the plasmocytes move much more slowly (20–25% as rapidly); as for the lymphocytes, the locomotion is polarized, and the cells do not change their polarization. When the cell is moving, it takes on a trapezoidal form, the longest side located at the front of the cell. The nucleus is median in location; the cell then resembles a cylindrical epithelial cell, with the mitochondria occupying principally the anterior part of the cytoplasm. The movement occurs without intervention of the hyaloplasmic veils by a mass cytoplasmic streaming relative to the substratum. It is clear that that maturation is accompanied by a radical modification of dynamic behavior in speed as well as morphology (Fig. 4).

5. Histiocytes. a. Histioblasts. This type of cell, quite rarely encountered in our cultures, moves very little; when it does move, it resembles all histiocytic cells: large undulating membranes and intense pinocytic activity. One can recognize in this type a bipolarity on the same axis.

b. Monocytes. These are histiocytic elements with feeble locomotion and hyaloplasmic veils of the undulating membrane type. This cell can also change polarity. In Figs. 5 to 7 it is to be noted that the change in orientation of the cell is accompanied by the formation of a new membrane. The gelated zone easily observable behind the hyaloplasmic veils at the anterior pole disappears before the veils themselves disappear, as the poles change. The gel zone then reappears after the hyaloplasmic veils

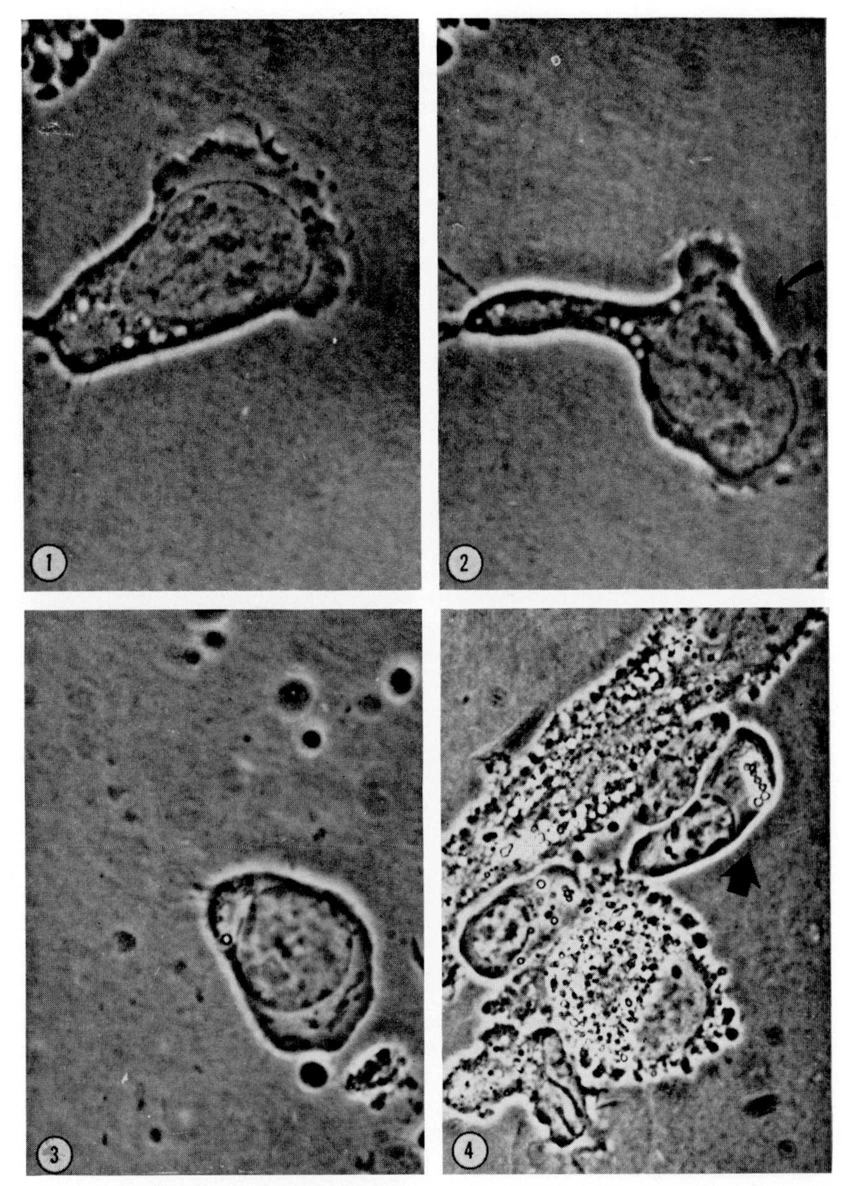

FIG. 1. Plasmablast exhibiting numerous hyaloplasmic veils at the anterior end. Note the pinocytic activity. Magnification: × 1368.

FIG. 2. The same plasmablast showing a gelated zone between the old and the new veil formation. Magnification: × 1368.

FIG. 3. Proplasmacyte. No hyaloplasmic veils are visible at the anterior pole and the cell has an ameboid movement. Magnification: × 1368.

FIG. 4. Mature plasmacyte resembling a cylindrical epithelial cell. No hyaloplasmic veils are visible. The cell flows in mass on the substratum. Magnification: × 900.

appear at the front of the cell. The change of polarity is thus accompanied by a parallel change in the distribution of gel. It should be added that the tail of the monocyte adheres to the substratum and leaves behind easily visible cytoplasmic fibrils.

6. Mitoses. Of all the cells encountered in our cultures three types can be distinguished: reticular cells coming directly from the reticulum,

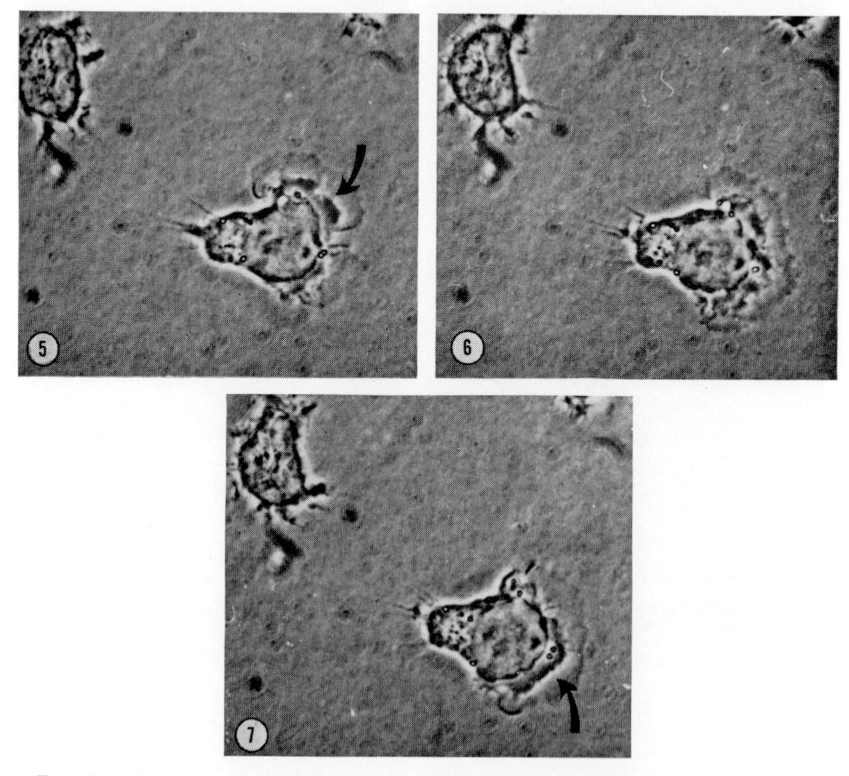

FIGS. 5, 6, 7. Monocyte. The polarity of the cell changes and at the same time the gelated zone is reorganized behind the veils of the anterior pole (10 sec between each picture). Magnification: $\times$ 1254.

plasmablasts derived from elements of the reticulum, and middle-size lymphocytes which we think issue from germinative centers of the explant. A point of some interest was the fact that movements of the cells disappeared during division, only to return after division and separation into daughter cells was completed. It is, for example, very striking to see the middle-size lymphocytes mobilize themselves rapidly after their division and recoup their original mode of interphase movement.

7. Clasmatosis. This is the segregation of one part of the cytoplasm lacking a nucleus; it is easily seen in the cells of the plasmacytic series. Clasmatosis can occur also in reticular cells. This segregation occurs after the production of a large ameboid process, which becomes somehow separated, apparently by "strangulation." During this process, cytoplasmic streaming is clearly visible, especially in dividing plasmablasts.

Cellular Interactions

1. Emperipolesis. Rapid cellular contacts between small and middle-size lymphocytes, on the one hand, and reticular cells, on the other (with

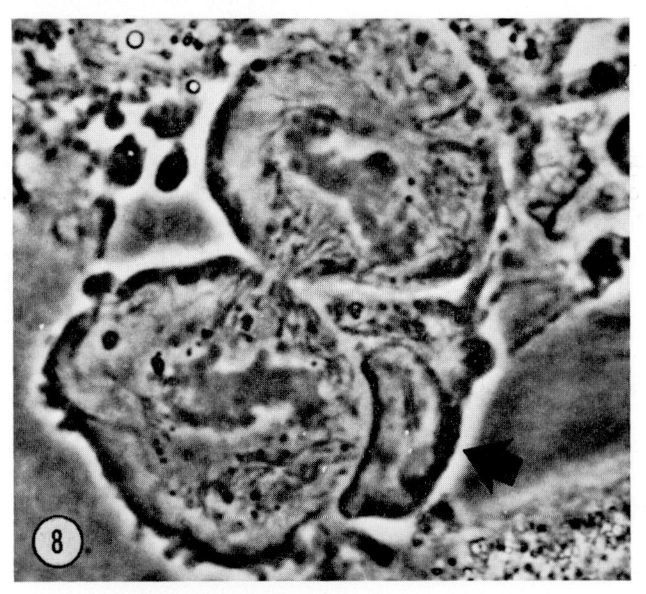

FIG. 8. Mitosis of a plasmablast. Emperipolesis by a lymphocyte inhibits bubbling of the two daughter cells. Magnification: × 1900.

or without macrophagic function) are very frequent. They are similar to those which one can observe between lymphocytes and cancer cells in culture. This emperipolesis represents, probably, a surface phenomenon; it must be distinguished from certain situations where lymphocytes, plasmacytes, and monocytes are included and moving into the syncytium.

A particular type of emperipolesis is to be noted in middle-size lymphocytes and plasmacytes in mitosis. Cell contact is generally established at the moment of cytokinesis, and on the cytoplasmic bridge; it is to be noted that at the moment of contact, the bubbling which is so violent at this stage of mitosis stops in both of the daughter cells (Fig. 8). The movement continues once the lymphocyte has broken off contact. We have observed this phenomenon with plasmablasts, but not with the middle-size lymphocytes or with the reticular cells.

2. *Lymphoplasmacytic Reticular Islets.* We were able to observe such an islet in culture for 6 days. The center is formed by reticular cells becoming progressively flat. The reticular cell exhibits weak membrane activity on the part of the microvilli, but without any microscopically visible pinocytosis. The population which surrounds this cell changes with time. At the beginning one can observe essentially lymphocytes; 48 hr later, plasma cells are in the majority; after 72 hr, large mononucleate cells appear; then by the fifth to sixth day, the plasmacytes are present once again. The movements of the cells which surround the reticular cell are the same as those described earlier for single cells.

In addition, there are two other types of cell movements to be mentioned:

a. Emperipolesis of lymphoid cells bordering on the reticular cell, or crossing on top of, or beneath the reticular cell.

b. Close relations between reticular cells, on the one hand, and plasmacytes and mononucleate cells (proplasmacytes?), on the other. On accelerated projection of a film, it is almost impossible to distinguish the closely associated cytoplasms of the two cells.

It should be added that during this entire period of observation of 6 days, the reticular cells were not observed to divide.

Discussion and Conclusions

We can thus review the observed facts seen in our cultures concerning cell movements:

1. All the migrating cells, except mature plasmacytes, exhibit hyaloplasmic veils at their anterior pole. These veils take on, in the monocytes, the typical character of the undulating membrane. In general, hyaloplasmic veils vary from one type of cell to another in number, size, and speed. It seems to be true that one can observe some intermediate types of veils between those described in the polymorph and those in cells with histiocytic function.

2. Stationary histiocytic cells, syncytia, macrophagic reticular cells, and histioblasts produce undulating membranes which are large and numerous. These cells are not polarized, except for the histioblasts.

3. Nonmacrophagic reticular cells produce only microvilli.

4. The tail of locomoting cells is clearly visible at the opposite end of the cell from the hyaloplasmic veils, except in the mature plasmacytes. In the tail one can observe contractions, particularly in histiocytes.

5. The histioblasts which are rich in undulating membranes, and

polymorphs which are rich in hyaloplasmic veils, can have a double polarity and extend in two opposite directions. If one of the two poles disappears, the hyaloplasmic expansions disappear first, the adjacent gel layer persists, becomes denser, and organizes itself as the tail for the cell during movement.

6. If the cell has two "anterior" poles, then these are separated by a cortical gel.

7. Cytoplasmic streaming is visible only in polymorphs; streaming is not continuous even though its direction is the same as the direction of locomotion. It is doubtful whether the streaming plays a role in the locomotion of the cell.

8. The emperipolesis of the lymphocytes characterized by fast contacts is observed in nonmacrophagic reticular cells and plasmablasts at the time of cytokinesis.

9. Some prolonged contacts can be established, on the one hand between middle-size lymphocytes or plasmacytes and a macrophagic reticular cell rich in phagocytized material, or, on the other hand, between young or old plasmacytes and nonmacrophagic reticular cells. These relations have completely different aspects in the two types of reticular cells.

How should we interpret all these observations?

Cell Movements

Lewis (1931, 1939, 1942) and De Bruyn (1944, 1945, 1946, 1947) believe that the movements of the ameba and of leucocytes are essentially of the same nature. Their conclusions are based on: (1) the similarity of form of these cells during movement; and (2) the behavior of lateral protuberances and of granules, showing that in the leucocytes as in the ameba, the lateral parts of the cytoplasm are stationary. There exists at the surface a superficial gel layer where processes of contraction take place. These contractions would explain a forward displacement of the plasmasol. A solation must occur in the body, since plasmasol is constantly present at the anterior pole.

In regard to the observations, it seems that insufficient attention has been paid to the production of hyaloplasmic veils in the leucocyte (with the exception of mature plasmacytes).

These numerous and active veils are quite visible with phase and interference contrast microscopy. Their existence suggests a mechanism of migration analogous to that described by Ambrose (1961) for moving fibroblasts on a solid substratum. The migration was postulated to be due to the undulations of the cell membrane apposed to the substrate. These undulations would themselves be produced by contractile fibrils embedded in the cytoplasm, parallel and close to the contact surface. Such a mechanism evidently could be invoked to explain movement in leucocytes.

It is, however, evident that some contractile phenomena are observable in leucocytes in motion, in particular at the posterior pole. We have previously presented a series of interference contrast pictures of this (Robineaux and Pinet, 1960). This progressive and continued contraction was marked by the migration of two vacuoles toward the Golgi zone, and the anteriolateral production of a hyaloplasmic expansion.

It is, besides, easy to observe the stationary gelated lateral zones of a cell in movement, just as they were described by De Bruyn (1947).

The formation itself of undulating membrane in a stationary cell shows the production of gelated zones which are linear and radial, and between which form pinocytosis vacuoles; their orientation indicates the direction of the retraction of the membrane. The coexistence within the leucocytes of gel formations and of movement is certain; the variation of movement probably bears some strict relationship with the variation of gel structure.

If one compares these observations with the observations of Goldacre (Goldacre and Lorch, 1950a,b; Goldacre, 1961) on the ameba, one finds agreement concerning the formation *de novo* of membrane in the front of leucocytes (at least during pinocytosis) and concerning tail contraction.

But, the existence of gelated zones in other parts of the cell in the anterior pole of a leucocyte in movement or in the undulating membranes of stationary cells of macrophagic function would support the presence of local zones of contraction wherever the movement is produced.

A possible relationship between the zone of contraction and cytoplasmic streaming, such as has been suggested by Allen (1961a,b; Allen *et al.*, 1960) in the frontal contraction theory, is at present impossible to determine in the case of leucocytes.

In the particular case of movement in plasmacytes, we can affirm that movement is in strict agreement with the production of hyaloplasmic veils. Numerous in plasmablasts, these structures are correlated with a high velocity of migration. Less numerous in proplasmacytes, they are correlated with a less rapid locomotion. There are no veils in mature plasmacytes where the movements have become slower and where mass cytoplasmic streaming is observable. The progressive disappearance of veils in the course of plasmacyte maturation is perhaps in agreement with the development of endoplasmic reticulum in these cells. The extreme development of this reticulum might confer on the mature plasmacyte some aspects of its characteristic movement.

Thus, besides the phenomena of contraction, the movements of the hyaloplasm appear to play a role of great importance in the movement of leucocytes. The movement of the leucocyte would then seem to resemble, in part, both the ameba and the fibroblast.

The production of these veils raises the question of the formation of new membranes, or perhaps of the reutilization of material from resorbed membranes. As far as the ameba is concerned, the formation of new membranes during locomotion is contested by Griffin and Allen (1960) and more recently by Wolpert (1963). This appears true for the ameba, but would not seem applicable to leucocytes, because of the importance of providing a new supply of membrane for the formation of pinocytotic vesicles. The relation between movement and the production of new surface as described by Bell (1961) appears to us to be especially applicable to leucocytes, with the exception of mature plasmacytes.

Cellular Interactions

We have discussed this question elsewhere (Robineaux *et al.*, 1962a,b) in detail, but will now summarize some of the principal cellular interactions.

1. Emperipolesis. We have observed the same phenomena in tissue cultures of spleen under dialysis membrane as described by Humble *et al.* (1956) in agar cultures, between lymphocytes and neoplastic cells; between lymphocytes and megakaryocytes; and between lymphocytes and megaloblasts. In our experiments, these surface relations become established between lymphocytes and reticular cells, and between lymphocytes and plasmablasts in mitosis.

No interpretation can, as yet, be given to this phenomenon. The suggestion of Humble *et al.* (1956), according to which the lymphocytes are the mobile sources of enzymes or of metabolites necessary to the cell for growth or mitosis still remains only a hypothesis. It is possible that the speed of their movement, the passage of one reticular cell to another, can suggest a transport function, but one cannot say more. However, a recent study by Loutit (1962) marshals many arguments in favor of a trophic role for lymphocytes.[1]

2. Lymphoplasmacytic Reticular Islets. These islets have been described by Undritz (1950) in fixed and stained lymph node smears. They were found again by Thiéry (1960) in fresh preparations of dissociated lymph node, and in thin sections studied by electron microscopy. We have just described them in differentiated cultures of spleen.

These islets have both an anatomical and functional significance: The

[1] Two varieties of lymphocytes exist. One type incorporates tritiated thymidine very slowly; it plays a role in immune phenomena. The other variety with a short tritiated thymidine incorporation time has no known function (Everett *et al.*, 1963).

close relation between proplasmacytes and reticular cells suggests a transfer system.

It is important to recall at this time that the experiments conducted by Sharp and Burwell (1960) and by Fishman (1961) have advanced the notion of interrelationships between macrophages and antibody-forming cells in sensitized systems in tissue culture.

Finally the passage, by micropinocytosis, of macromolecules from a reticular cell into a plasmacyte (Thiéry, 1962) is in support of some kind of messenger system.

A possible immunological significance of strict relationships between reticular cells and plasmacytes might be suggested: some information (in the form of ribonucleic acid?) might be transmitted from the reticular cell reservoir containing antigenic structure to the plasmacyte which forms specific protein antibodies.

Summary

The culture of adult guinea pig spleen under dialysis membrane has made it possible to obtain a population of growing, differentiating cells. The cells encountered are involved in the phenomena of inflammation and immunity. Their movements, their divisions, and their interactions with other cell types have been studied. Particular attention has been given to the plasmacytic series, of which all stages of development of the various movement patterns could be found. The anatomical and physiological relations between plasmacytes and reticular cells have been studied from a dynamic point of view.

References

Allen, R. D. (1961a). *In* "The Cell" (J. Brachet and A. E. Mirsky, eds.), pp. 135-216. Academic Press, New York.

Allen, R. D. (1961b). *Exptl. Cell Res. Suppl.* **8**, 17-31.

Allen, R. D., Cooledge, J. W., and Hall, P. J. (1960). *Nature* **187**, 896-899.

Ambrose, E. J. (1961). *Exptl. Cell Res. Suppl.* **8**, 54-73.

Bell, L. G. E. (1961). *J. Theoret. Biol.* **1**, 104-105.

De Bruyn, P. P. H. (1944). *Anat. Record* **89**, 43-63.

De Bruyn, P. P. H. (1945). *Anat. Record* **93**, 295-315.

De Bruyn, P. P. H. (1946). *Anat. Record* **95**, 177-192.

De Bruyn, P. P. H. (1947). *Quart. Rev. Biol.* **22**, 1-24.

Everett, N. B., Caffrey, R. W., and Rieke, W. O. (1963). *In* "Leukopoiesis in Health and Disease," N.Y. Acad. Sci. In preparation.

Fishman, M. (1961). *J. Exptl. Med.* **114**, 837–856.

Goldacre, R. J. (1961). *Exptl. Cell Res. Suppl.* **8**, 1-16.

Goldacre, R. J., and Lorch, I. J. (1950a). *Nature* **166**, 497-500.

Goldacre, R. J., and Lorch, I. J. (1950b). *Intern. Rev. Cytol.* **1**, 135-164.

Griffin, J. L., and Allen, R. D. (1960). *Exptl. Cell Res.* **20**, 619-622.

Humble, J. G., Jayne, W. H. W., and Pulvertaft, R. J. V. (1956). *Brit. J. Haematol.* **2**, 283.
Lewis, W. H. (1931). *Bull. Johns Hopkins Hosp.* **49**, 29-36.
Lewis, W. H. (1939). *Arch. Exptl. Zellforsch. Gewebezücht.* **23**, 1-7.
Lewis, W. H. (1942). *In* "Structure of Protoplasm" (W. Seifritz, ed.). Iowa State College Press, Ames, Iowa.
Loutit, J. F. (1962). *Lancet* **24**, 1106-1108.
Robineaux, R., and Pinet, J. (1960). *Ciba Found. Symp. Cellular Aspects Immunity*, pp. 5-40.
Robineaux, R., Pinet, J., and Kourilsky, R. (1962a). *Compt. Rend. Soc. Biol.* **156**, 1025.
Robineaux, R., Pinet, J., and Kourilsky, R. (1962b). *Nouvelle Rev. Franc. Hematol.* **2**, 797-811.
Rose, G. G., Pomerat, C. M., Shindler, T. O., and Trunnel, J. B. (1958). *J. Biophys. Biochem. Cytol.* **4**, 761-764.
Sharp, J. S., and Burwell, R. G. (1960). *Nature* **188**, 474.
Thiéry, J. P. (1960). *Ciba Found. Symp. Cellular Aspects Immunity*, pp. 59-91.
Thiéry, J. P. (1962). *J. Microscopie* **1**, 275-286.
Undritz, E. (1950). *Folia Haematol.* **70**, 32-42.
Wolpert, L. (1963). Personal communication.

Discussion

Dr. Rhea: Do you have any evidence of active movement on the part of the nucleus, or was it all passive?

Dr. Robineaux: This is an important problem. In the cells of these cultures the nuclei are very deformable, especially in polymorphs and macrophages, and to a lesser degree in lymphoid cells. However, these are probably passive movements.

Dr. Rebhun: I noticed two very interesting things in some of your films. In one, surface bubbling appeared only at telophase, while the cell was smooth during the rest of mitosis.

In one of your other films, an elongated cell with pseudopods at the polar regions divided, but the mitosis occurred perpendicularly to the axis of the cell. The furrow cut in, in a direction parallel to the long axis. Do these two things occur often in these cells?

Dr. Robineaux: Concerning the mitosis of reticular cells, it is usually seen that: (1) the cells are bipolar, and (2) the spindle is perpendicular to the polar axis. I am not sure whether this is always true.

The bubbling is ordinarily seen at anaphase and telophase, but may also be seen earlier, for example, in metaphase. In the case that you referred to, two lymphocytes established cell contact with the reticular cell (emperipolesis); the bubbling appeared after the departure of the lymphocytes in telophase.

Dr. De Bruyn: I would like to comment on the nomenclature of these cells, even though it was stated in your film that the problem of nomenclature is not as yet settled. All of the cells that you called "plasmablasts," I would call "medium-sized or large lymphocytes." Similarly, I believe that your plasmacytes are cells in the erythrocytic series.

If I understood you correctly, cells (lymphocytes, or whatever they were) were believed to penetrate the macrophage or reticular cell. I have seen things like this, too, but I always assumed that these cells moved between the cover glass and the larger cell. If really true, I think penetration of one cell by another would be an extremely interesting phenomenon.

DR. ROBINEAUX: To answer your first question, I do not agree that our plasmacytes are erythroblasts. The size, shape, and color characteristics of these cells are those of plasmacytes. In fact, there are no erythroblasts in our cultures. Certainly an ultrastructural study could settle the matter.

About the relations between lymphocytes, plasmacytes, monocytes, and reticular cells in syncytia. First, lymphocytes in emperipolesis: I think this is a surface phenomenon. I asked Pulvertaft what he thought, but he was not certain whether the phenomenon was intracytoplasmic or at the surface. Second, lymphocytes, plasmacytes, monocytes inside syncytia: this phenomenon is similar to what is seen in cancer cells and megakaryocytes. These cells are included in an empty round surface in the cytoplasm, and the cell membranes are quite distinct. It could not be a superposition, because the depth of field is very thin and the details are just as distinct in the syncytia as in the reticular cells.

DR. BISHOP: Have you studied antibody production or release?

DR. ROBINEAUX: I think this kind of culture will permit many studies of this kind to be carried out, but I have had no experience in this field. The animals from which the cultures came were not sensitized, but the media were heterologous, containing chicken embryo extract.

Relaxation of Fibroblast Cells

H. Hoffmann-Berling

Max-Planck-Institut für Physiologie, Heidelberg, West Germany

Introduction

The term "primitive motility," which is part of the title of this Symposium, may be understood to cover two subjects: (1) primitive mechanisms of motility and (2) motility of primitive organisms. This article deals with some aspects of the second subject and is concerned with the motile system of fibroblasts.

To summarize some facts known from earlier experiments: fibroblasts and other nonmuscular cells (Hoffmann-Berling, 1954a), thrombocytes (Bettex-Galland and Lüscher, 1961), amebae (Hoffmann-Berling, 1956), and probably slime mold plasmodia (Nakajima, 1960) make use of the same general mechanochemical principle involved in muscular contraction for the purpose of generating movement. Evidence in support of this view has been gathered by applying the techniques of muscle physiology to nonmuscular objects. This includes: (1) extracting the cellular contractile proteins and comparing their reactions to those of actomyosin, isolated from muscle; and (2) cytolyzing the cells without further disintegrating the contractile structures. A subtle means of doing this is to immerse the cells into cold glycerol. The contractile structures remain insoluble; after removing the glycerol their reactions may be compared to those of glycerinated muscle fibers or myofibrils.

Both lines of investigation have rendered corroborating results:

1. Superprecipitation, dephosphorylation of adenosine triphosphate (ATP), solubility reactions, and viscosity changes, i.e., the reactions evoked by ATP and characteristic of muscular actomyosin, are exhibited by cellular contractile proteins as well. Deviations are quantitative and may be explained by a lower specific rate of ATP-splitting of the nonmuscular contractile proteins (Table I) (Bettex-Galland and Lüscher, 1961; Hoffmann-Berling, 1956; Nakajima, 1960).

2. Whole cells or cell layers, which by the glycerination procedure have been depleted of their endogenous ATP and have been rendered permeable to ATP added from outside, shorten or develop tension if immersed into an ATP-containing bath. On a unit cross sectional area basis, the tensile strength exhibited by a contracting fibroblast amounts to one-hundredth that of skeletal muscle (Hoffmann-Berling, 1956). The

difference is due to the different concentrations of contractile protein; that in a fibroblast, is 100 times less than that in muscle (Table I).

The physical nature of muscular actomyosin as a complex protein is prerequisite to its functioning in muscular contraction and relaxation. Evidence that contractile proteins of cellular origin contain an actin-like and a myosin-like component is indirect and has been derived mainly from viscosity measurements. In only one case has the complex nature of a cellular contractile protein been established by separating the com-

TABLE I

CONTRACTILITY AND CONTRACTILE PROTEINS OF MUSCLE AND NONMUSCULAR CELLS

Starting material	Contractile protein (% fresh weight)	Active tension (kg/cm^2, 20°C)	ATPase activity (μM P mg N^{-1} min^{-1}, 20°C)	Shortening velocity (% standard length, 20°C)
Skeletal muscle	12	5	2.4	200
Smooth muscle	2.5	0.3	0.075	12
Fibroblasts, sarcoma cells	0.2[a]	0.02[a]	0.015[a]	0.2[b]
Thrombocytes	3[c]	—	0.028[d]	—
Slime mold plasmodia	—	—	0.1[e]	—

[a] Hoffmann-Berling (1956).
[b] Hoffmann-Berling (1954a).
[c] Bettex-Galland and Lüscher (1962).
[d] Bettex-Galland and Lüscher (1961).
[e] Nakajima (1960).

ponents. The contractile protein of human thrombocytes on fractional extraction gives rise to two kinds of proteins, which re-establish the extremely high viscosity of the native complex and the type of viscosity reactions characteristic of actomyosin, if the thrombocyte components are either recombined or if each of them is allowed to react with its complementary component, taken from muscle (Bettex-Galland *et al.*, 1962). Contractile proteins, cross-reacting in one of the highly specific chemomechanical reactions of contractile proteins, cannot be entirely different in their molecular organization. Taking all the facts into consideration, there is little doubt that the principle of ATP-induced contraction is an elementary, evolutionary acquisition and is not restricted to the highly specialized muscle cell.

Relaxation in Muscle and in Other Cells

Muscle and other biological systems deliver mechanical work by alternating between activity and rest. A contracted muscle relaxes; a fibroblast cell, rounded up in mitosis, returns to its former expanded state.

The indentity of the contractile mechanisms seems to be well established. Is relaxation too—as a process, which originates from inhibition of contraction—common to both muscle and nonmuscular cells? The problem may be posed more precisely: If in muscle and in other cells contraction results from a reaction of the contractile structures with ATP, and if in the living object both components, ATP and the contractile protein, are in close contact, how do muscle and cells avoid a permanent contraction and to what an extent is the inhibitory reaction identical in muscles and in other cells?

The first contribution, bearing on this problem, was made by Marsh (1952). He found that muscle homogenates contain an extractable agent which hinders contraction, if added *in vitro* to a system of washed myofibrils and ATP. According to recent investigation the outstanding properties of the agent are:

1. It is particulate and can be sedimented out of homogenates by high-speed centrifugation (Portzehl, 1957). The particles, which have been termed "relaxing grana" probably are the desintegration products of the sarcoplasmic reticulum, which is known to penetrate from the outer muscle membrane into the interior of the fibers (Weber, 1960).
2. The grana are not relaxing agents *per se*. They become so by a reaction with ATP and magnesium ions (Hasselbach and Weber, 1954). The mechanism of the process which ultimately impedes contraction, is a matter of controversy (Briggs and Fuchs, 1960; Weber, 1959), and will be discussed here only as far as it has some bearing on the experimental behavior of the nonmuscular objects.
3. If calcium ions in excess of 10^{-6} mole/liter are added, relaxation ends and contraction starts, despite the presence of active grana. Addition of calcium is a convenient means of demonstrating that the contractile apparatus under the conditions of physiological relaxation is kept functional.

We first consider the action of relaxing grana, isolated from skeletal muscle and applied to glycerinated fibroblast cells. The cells derived either from chicken sclera or from chicken skeletal muscle, had been cultivated *in vitro*. From Fig. 1 it may be seen that 0.5 mg grana protein/ml are sufficient to reduce the shortening of the extracted cells to an insignificant amount. There is full contraction, however, if the experimental system containing cells, Mg-ATP, and grana is treated with 10^{-4} mole/liter of calcium (Kinoshita *et al.*, 1963).

The inhibitory action, which muscular grana exert on nonmuscular cells, could be a reaction having no physiological significance. However, if the experimental situation corresponds to the reactions *in vivo*, fibroblast cells on homogenization should release relaxing grana with an activity similar to that of muscle grana. In the experiments cited in Fig. 2, grana isolated from sclera fibroblast tissue cultures had been applied to glycerinated cells. Comparison with the curves of Fig. 1 shows that, on the basis of their protein content, the cellular grana nearly equaled the efficiency of the muscle grana. The cellular grana too are rendered ineffective and the cells contract if calcium ions are introduced into the experimental solution (Kinoshita *et al.*, 1963).

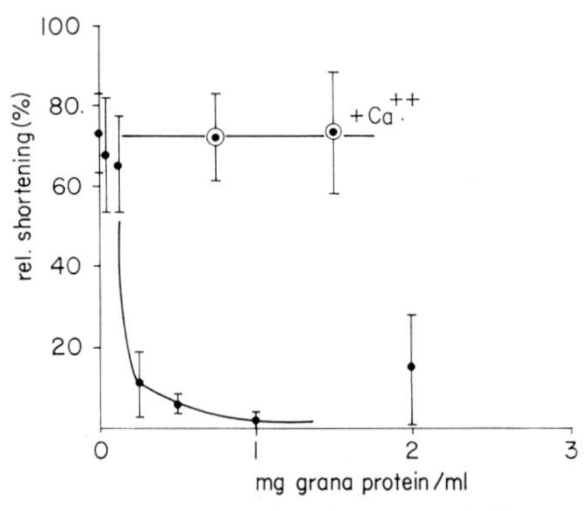

FIG. 1. Inhibition of cellular contraction by grana isolated from rabbit skeletal muscle. Fibroblast cultures, derived from chicken skeletal muscle and stored in glycerol for 69 days, were incubated 30 min at 30°C in a solution, which contained (in moles per liter): KCl 0.08; histidine buffer (pH 7.2), 0.02; $MgCl_2$, 0.005; potassium oxalate, 0.005; ATP, 0.005; and the grana concentrations indicated. Each point of the curves is derived from 15 to 20 cells and gives the change of the main cellular diameter in per cent of the original value. Small, filled-in circles—without calcium ions added; open circles with dotted centers—with 10^{-4} mole/liter $CaCl_2$ added before incubation.

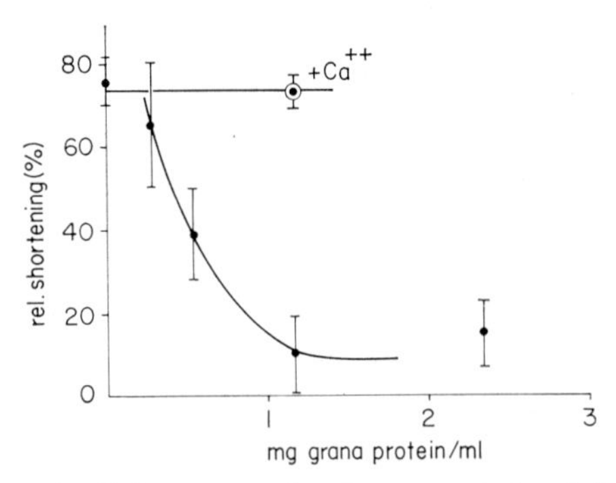

FIG. 2. Inhibition of cellular contraction by grana, isolated from sclera fibroblasts cultivated *in vitro*. The test cells were derived from chicken skeletal muscle, cultivated *in vitro* and were stored in glycerol for 4 days. The other conditions are the same as in Fig. 1. Small, filled-in circles—without calcium ions; open circles with dotted centers—with 10^{-4} mole/liter calcium ions.

It should be pointed out, that unfractionated suspensions of the grana have only a limited effect. Muscle grana in a state of partial purity, which, however, possess full activity on muscle preparations, impede cellular contraction only in a narrow range of grana concentration (Fig. 3). Relaxation has turned out to be hindered by contaminating particles, which probably derive from the muscle or sclera cell nuclei, and which are retained in the grana preparations. Since the interfering particles sediment faster than the relaxing grana themselves, the suspensions can be cleared from contamination by sucrose density-gradient centrifugation. The mode of action of the interfering particles is not known; enzymatic analysis indicates that the particles do not impede

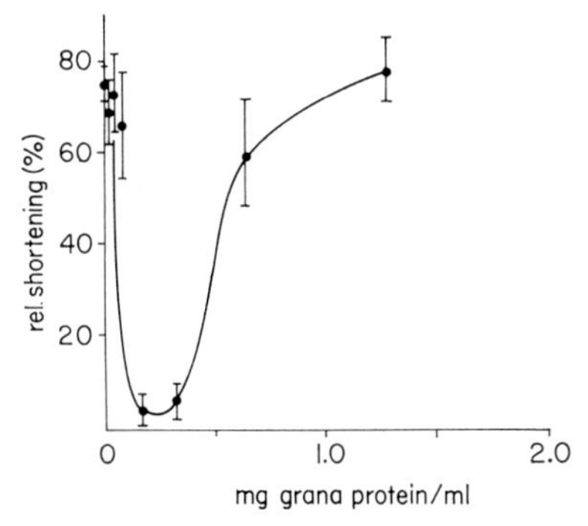

FIG. 3. Inhibition of cellular contraction by unfractionated suspensions of skeletal muscle grana. Conditions are the same as in Fig. 1.

relaxation by producing adenosine diphosphate (ADP) from ATP. This mode of action had been considered, since high concentrations of ADP are known to interfere with the activity of the relaxing grana (Makinose and Hasselbach, unpublished data). The interfering particles are mentioned here because they have precluded cellular relaxation in many of the starting experiments.

Means of Observing the Process of Relaxation

In muscle the process of relaxation may be followed in more than one way: either by recording mechanical activity or by measuring the rate of ATP dephosphorylation. Both reactions are intimately linked, and the relaxing grana will retard or impede both. In glycerinated nonmuscular cells the contractile dephosphorylation of ATP cannot be estimated since the concentration of the contractile structures and their enzymatic activity are extremely low and obscured by ATPases of other chemical nature. To obtain clearer indication that the mechanisms of granar action in muscle and cells correspond, one should compare closely muscular and cellular relaxation and the conditions on which they depend. One means of doing this is by applying poisons that are known to block granar activity. As an example, I will confine myself to the experiments with mersalyl,[1] an organic mercury compound which has turned out to be an extremely valuable tool.

Relaxing grana pretreated with mersalyl and applied to myofibrils prove to be inactive. The procedure of inactivation may be modified by incubating the grana first with ATP and adding the mersalyl later. The results of the two procedures are different: Whereas in the first case the

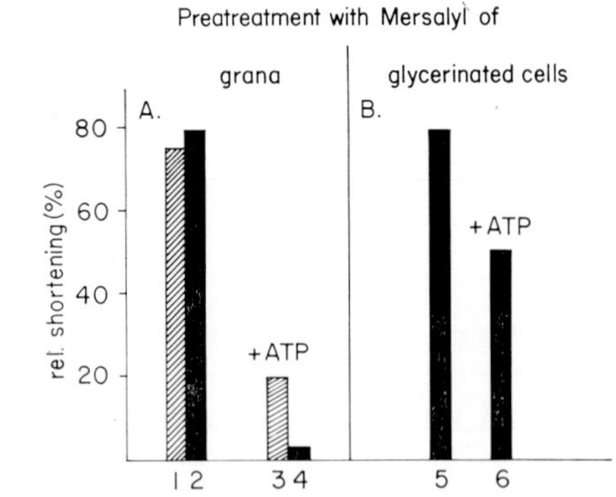

FIG. 4. Inactivation of the isolated relaxing grana (A) and of the relaxing system, retained in glycerinated cells (B), by mersalyl. (A) Grana derived from skeletal muscle (1,3) or from sclera fibroblast cultures (2,4) were incubated 5 min with 2×10^{-4} moles/liter mersalyl in an ice bath. Five minutes later the suspension was made 10^{-2} *M* in cysteine. Preincubation was performed either in the absence (1,2) or in the presence of ATP (3,4). Granar activity was tested afterward on sclera fibroblasts as described in Fig. 1. (B) Whole glycerinated fibroblast cells were pretreated with mersalyl and cysteine either in an ATP-free or in an ATP-containing solution as described under A. After pretreatment the cells were incubated in a suspension of ATP and skeletal muscle grana and the shortening of the cells was recorded as in Fig. 1.

grana are irreversibly deprived of any functional activity and stay ineffective even if mersalyl is subsequently complexed with cysteine, in the second case ATP exerts a protecting effect—the granar function can be restored by adding cysteine (Hasselbach and Makinose, 1963). The effect of ATP may be compared to the protecting action that some substrates exert on the corresponding enzymes.

If grana pretreated in one or the other way are applied to cells, the results completely parallel those obtained with muscle. There is full relaxation of the cells when the preliminary steps of deactivating and

[1] Salicyl-(γ-hydroxymercuri-β-methoxypropyl-)amidoorthoacetate.

reactivating the grana have been performed in an ATP-containing solution; if ATP was absent during the poisoning procedure cysteine failed to restore and the cells contracted. These effects hold, regardless of whether the grana have been derived from skeletal muscle or from tissue culture cells (Fig. 4) (Kinoshita *et al.,* 1963).

Since there is no reasonable explanation for the protecting action of ATP, the effects do not contribute to a better understanding of the mechanism of granar function. However, the phenomena may be taken as further evidence that the mechanism of granar action on cells intimately corresponds to their action on muscle.

Conclusion

Muscle fibers or glycerinated cells are rather dense objects, into which relaxing grana, whose diameter is about 100 mμ, could hardly penetrate. The inhibitory effect of grana, which are kept outside, must be of an indirect nature. Two kinds of mechanisms have been considered: relaxation may result because some activating agent, perhaps calcium ions, indispensable for contraction, are withdrawn from the fibers and stored in the grana (Ebashi and Lipmann, 1962; Hasselbach and Makinose, 1961), or because the grana release some agent, which penetrates into the fibers and interferes with contraction (Briggs and Fuchs, 1960; Parker and Gergely, 1960). Muscle physiologists are inclined to believe, that both processes—the withdrawal of calcium ions and the release of a relaxing substance of low molecular weight—play a role. Anyhow, the inhibitory influence, which originates in the grana, can penetrate only short distances into the contractile structures. Full contractile inhibition is achieved only if the muscle fibers have been disintegrated and evenly dispersed as a suspension of myofibrils (Hasselbach and Makinose, 1963).

Contradiction to this statement seems to come from the well-known fact that relaxing grana will effectively interfere with the contraction of whole glycerinated fibers, i.e., of objects of 50–100 times the diameter of a myofibril. However, relaxation of such large objects results from an indirect effect. The grana (probably by releasing a cofactor, Briggs and Fuchs, 1960; Parker and Gergely, 1960) seem to restore functional activity to the fibers' own relaxing system, which have been rendered ineffective by the glycerination procedure. It is the endogenous relaxing system of the fibers, restored by the grana, which ultimately interferes with contraction (Makinose and Hasselbach, 1960; Hasselbach and Makinose, 1963). Crucial evidence comes from experiments in which the endogenous system has been irreversibly deactivated by mersalyl. I will not give the details here, but stress the fact that glycerinated cells behave much as muscle fibers do. That is to say, glycerinated cells relax to the extent that the cell's own relaxing system is restored. Grana added from outside the cells are devoid of any inhibitory activity if the endogenous system has been destroyed previously, for instance, by presoaking the glycerinated cells in mersalyl and cysteine (Kinoshita *et al.*, 1963).

I would not mention these rather complicated facts, if they did not have experimental consequences. Earlier experiments have indicated that the cytokinesis of a fibroblast is generated (or initiated) by a local contraction, which in a glycerinated cell may be provoked by adding ATP. In the living and in the glycerinated object, the contractile process under the appropriate conditions confines itself to the equatorial parts, whereas the polar parts of the cell do not participate in the movement and stay relaxed. The experimental facts were consistent with the hypothesis that cells contain a relaxing system, which during cytokinesis is confined to the polar parts of the cells, but is lacking or inactive in the equatorial area. Experimental evidence seemed conclusive but was tedious to gather, since storing the cells in glycerol for more than 1 hr deprived the cellular relaxing system of its activity. As a consequence, the local constriction process was extended into a uniform cellular contraction (Hoffmann-Berling, 1954b).

Since grana applied from outside produce an effect by restoring the cell's own equipment, any uneven distribution of the cellular relaxing system in the living cell should produce the original effect if the system is inactivated by glycerination and afterward restored to its former activity. As may be seen from Fig. 5, cells killed during the start of cytokinesis, stored in glycerol for a time sufficient to render cytokinesis impossible with ATP alone, resume their cleavage movement if incubated in a suspension containing ATP and grana. The contraction of the equatorial area is localized and does not lead to an over-all shortening of the cell. On the contrary, the cleavage effect is optimal if the grana are applied in a range of concentration such that shortening of the cells is just prevented. In those experiments, where either no grana or insufficient amounts of them had been applied, the contractile process spread to the poles and gave rise to a uniform contraction, which mechanically interfered with constriction. In those other experiments where excessive amounts of grana had been added, constriction was also prevented; grana under these circumstances interfered with the contraction of the equatorial area (Kinoshita and Hoffmann-Berling, 1963).

It is important to note that any agent which impedes the functional activity of either the added grana or of the intrinsic relaxing system, gives rise to a uniform cellular contraction and renders cleavage impossible. Uniform contraction results if the glycerinated cells have been pretreated with mersalyl and cysteine, which renders the intrinsic system inactive. The same result occurs if calcium is added to the incubation mixture. This treatment renders the endogenous relaxing system ineffective and, in addition, interferes with the activity of the grana outside (Kinoshita and Hoffmann-Berling, 1963).

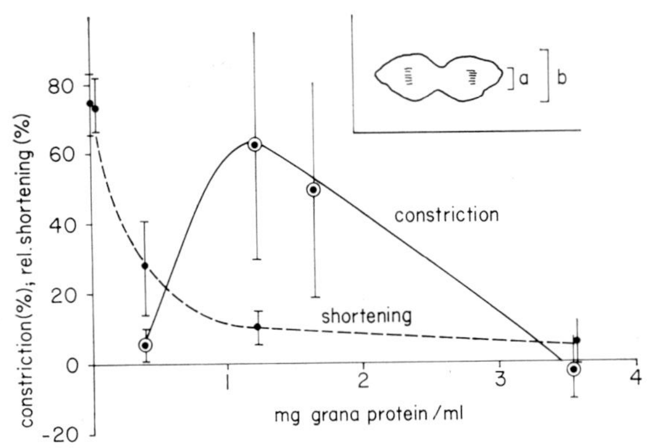

Fig. 5. Constriction of glycerinated fibroblast cells in early telophase under the influence of ATP and of grana from skeletal muscle. Skeletal muscle fibroblasts were stored in glycerol for 4 days and exposed to the same conditions as in Fig. 1. Each point of the curves is derived from measurements on 8–12 cells in early telophase. Relative shortening of the cells is recorded as in Fig. 1. Relative constriction (in per cent of the maximal transverse diameter of the cells) is given by $(a_t/b_t) - (a_0/b_0) \times 100$, where the meaning of a and b is indicated by the diagram; 0 indicates the values recorded before and t after the incubation in ATP plus grana.

Taken as a whole, the experiments then produce the following additional evidence: (1) Inhibition of contraction in fibroblasts and in muscle is due to grana of an identical physical nature. (2) The mechanisms of granar action in muscle and fibroblast cells correspond. (3) Cells do contract locally. This has been tacitly assumed in many of the lectures given earlier in this Symposium. In this respect the behavior of these cells differs from that of muscle. (4) Fibroblast cytokinesis results from a local contraction, which in the appropriate stage of mitosis is set to work by an uneven distribution of the cellular relaxing system between the polar and the equatorial regions of the cell.

References

Bettex-Galland, M., and Lüscher, E. F. (1961). *Biochim. Biophys. Acta* **49**, 536.

Bettex-Galland, M., Protzehl, H., and Lüscher, E. F. (1962). *Nature* **193**, 777.

Briggs, N. F., and Fuchs, F. (1960). *Biochim. Biophys. Acta* **42**, 519.

Ebashi, S., and Lipmann, F. (1962). *J. Cellular Biol.* **14**, 389.

Hasselbach, W., and Makinose, M. (1961). *Biochem. Z.* **333**, 518.
Hasselbach, W., and Makinose, M. (1963). *Conf. Biochem. Muscle Contr., Detham, 1962.* In press.
Hasselbach, W., and Weber, H. H. (1954). *Biochim. Biophys. Acta* **11**, 160.
Hoffmann-Berling, H. (1954a). *Biochim. Biophys. Acta* **14**, 182.
Hoffmann-Berling, H. (1954b). *Biochim. Biophys. Acta* **15**, 332.
Hoffmann-Berling, H. (1956). *Biochim. Biophys. Acta* **19**, 453.
Kinoshita, S., and Hoffmann-Berling, H. (1963). In preparation.
Kinoshita, S., Andoh, B., and Hoffmann-Berling, H. (1963). In preparation.
Makinose, M., and Hasselbach, W. (1960). *Biochim. Biophys. Acta* **43**, 239.
Marsh, B. B. (1952). *Biochim. Biophys. Acta* **9**, 247.
Parker, C. J., and Gergely, J. (1960). *J. Biol. Chem.* **235**, 3449.
Portzehl, H. (1957). *Biochim. Biophys. Acta* **26**, 373.
Nakajima, H. (1960). *Protoplasma* **52**, 413.
Weber, A. (1959). *J. Biol. Chem.* **234**, 2764.
Weber, H. H. (1960). *Arzneimittel-Forsch.* **10**, 404.

Discussion

Chairman Marsland: This paper is exceedingly interesting to me. Perhaps I might comment initially that these results fit in with the experiments that Dr. Zimmerman and I have done over the past few years. We have been inducing cleavage prematurely, regardless of the state of the nucleus or whether the cell is developmentally "ready to divide" or not, by drastic centrifugation under hydrostatic pressure. Certain granules—and we think we may have identified them—are thrown to one pole of the cell where they initiate a solating or relaxing effect, so that the cell begins to divide prematurely at any stage of the normal division cycle.

Dr. Wolpert: I would like to give one piece of evidence that supports, on living cells, Dr. Hoffmann-Berling's observation on the localized contractions in cells. As you probably know, there is some controversy as to the site of the forces. There is one experiment I would like to mention. If you take the sea-urchin egg at the time it elongates before cleavage and use a Swann and Mitchison elastimeter, that is, if you bring to the surface a small micropipette and suck out a bleb, then by measuring deformation against hydrostatic pressure, you get a measure of the resistance to deformation.

If you place the pipette at the polar region, suck out a bleb, and keep the pressure constant, then as cleavage proceeds, the deformation gets larger. However, if you do exactly the same experiment but now place the pipette across the furrow region, then you find that the bleb gets smaller and can pull right out of the pipette. This seems to me to be quite direct evidence that the poles relax and the furrow contracts during cleavage.

Dr. Edwin Taylor: Do I understand correctly, that anaphase is prevented from taking place by the presence of an intercellular relaxing factor?

Dr. Hoffmann-Berling: Are you referring to the contraction of the chromosomal fibers?

Dr. Edwin Taylor: Yes.

Dr. Hoffmann-Berling: If one tries to evoke this kind of movement in a glycerinated cell, taken from anaphase, the addition of adenosine triphosphate (ATP) results in an over-all contraction of the cytoplasm. As a consequence, the spindle is compressed, and any possible action of the chromosomal fibers is obscured for simply mechanical reasons. If the contraction of the cytoplasm can be prevented by the application of relaxing grana, the way may be opened to study the motions in the central parts of the cell.

Dr. Edwin Taylor: Is the contraction of the chromosomal fibers the active step in the process or the relaxing step?

Dr. Hoffmann-Berling: I cannot tell. To elucidate the role of ATP, a comparison with such systems which require ATP for contraction (muscle) and other systems which require ATP for relaxation (*Vorticella*) would be required. This has not been done yet.

Dr. Teru Hayashi: I would like to comment on the idea that somehow the actomyosin or contractile protein of a slowly contracting cell might be different because of its lower ATPase activity. There is another possible explanation. The ATPase activity may be quite the same in the slow-contracting cell and the fast-contracting muscle, but when the ATPase activity is measured on a milligram protein basis, an anomalous low value may be obtained due to the presence of nonenzymatic protein which has not been separated from the contractile protein. Ordinary actomyosin procedures depend on solubility properties of the protein, and another protein having the same solubility properties as actomyosin would be difficult to separate.

In our laboratory, Dr. Margit Nass has extracted actomyosin from different stages of frog embryos and purified them to the utmost according to such procedures. Now, if the ATPase activity is measured as Dr. Hoffmann-Berling has indicated, it is found to increase with increasing age of the embryo. From this, one might say that the different stages of the frog embryo contain different actomyosins.

However, an additional purification procedure may be employed, using specific actomyosin antibody. If each of the extracts from these embryonic stages is treated with anti-actomyosin, an antigen-antibody precipitate forms which can be separated out and tested for ATPase activity. Progressively more precipitate is obtained from later stages, and the enzyme activity of the antigen (actomyosin) is unaffected by the antibody. The ATPase activity, per milligram antibody-precipitable protein, is essentially the same for all stages. It seems possible, therefore, that the actomyosin from a slowly contracting cell may have just as high ATPase activity as actomyosin from muscle, but it is tested under conditions where it is present in low concentration mixed with nonenzymatic protein of similar solubility properties.

The Movement of Neurons in Tissue Culture

J. NAKAI

Department of Anatomy, School of Medicine, University of Tokyo, Hongo, Tokyo, Japan

Introduction

The neuron is an extremely complex system which contains an almost inexhaustible supply of problems to be resolved in the future, not only in morphology and function of the fully developed nervous system, but also in the mechanisms of the remarkable changes which take place in embryonic development.

The activites of neurons to be discussed are those seen in phase-contrast cinemicrographic records of embryonic cells cultivated *in vitro.* Such observations were made also by the pioneers in this field such as Harrison (1910), Lewis and Lewis (1912), Levi (1934), and Weiss (1934) which were thoroughly reviewed by Levi (1934) and Weiss (1955). Lewis' (1950) observations and the analysis of his motion pictures taken of the same material as used in the present report have been found to be very accurate and concise. The present author intends to present a summary of some time-lapse, phase-contrast cinemicrographic records of chick embryo sensory neurons in tissue culture. The motion picture and its analyses may offer some evidence either for or against earlier findings and interpretations. Many of the studies have already been published (Nakai, 1955, 1956, 1960; Nakai and Kawasaki, 1959; Nakai *et al.*, 1961); some represent previously unpublished work. The observations will be presented in the following order: (*1*) migration of neurons, (*2*) extension and retraction of filopodia and fibers, (*3*) peristaltic movement, (*4*) adhesiveness of filopodia and retraction force, (*5*) migration of particles in the neuron and axoplasmic flow, and (*6*) the area responsible for motility. While such a classification is convenient, it is not necessarily the only one which is reasonable, since the various phenomena are clearly interrelated.

Migration of Neurons

The migration of neuroblasts and neurons occurs during establishment of the nervous system, particularly, of the neural tube (Fujita, 1962) and of the peripheral autonomic nervous system.

There is a general tendency in tissue culture for tissues to become flat on the cover slip because of the migration of tissue cells. The be-

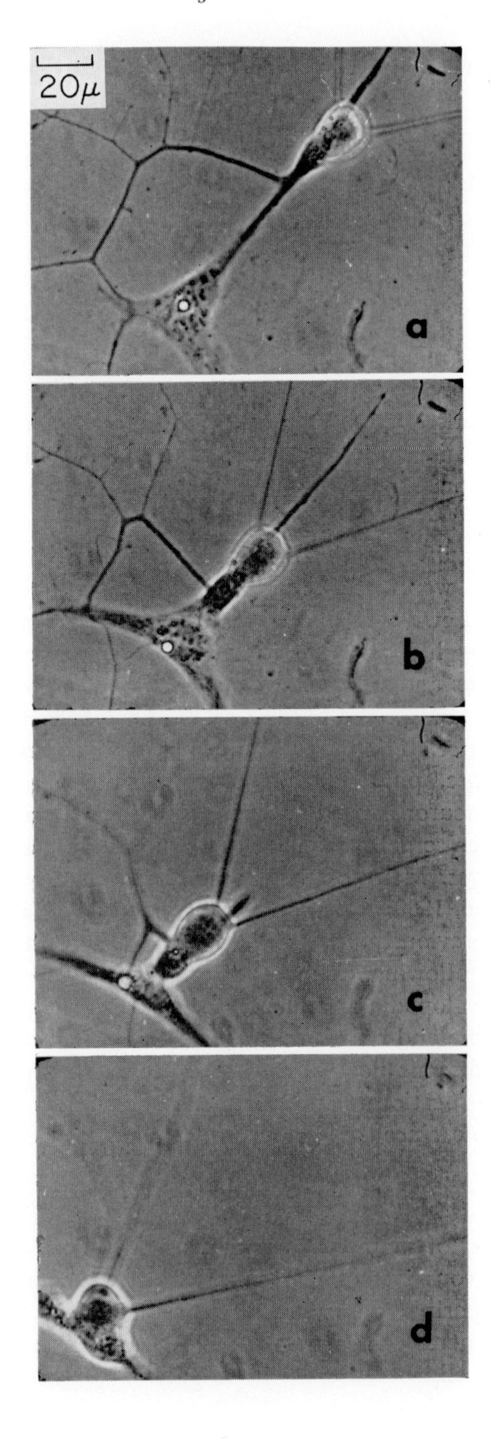
20μ
a
b
c
d

havior of spherical spinal ganglia in the present studies is not exceptional; cells of the capsule and neuronal processes grow out radially on the glass surface, resulting in loosening of the ganglion. After a few days or a week, one can see scattered neurons entangled in their nerve plexus in the explant. Some neurons are observed apart from the original group. Do these neurons migrate, and if so, do they do so actively or passively? Lewis (1950) stated simply that "they are still capable of migrating a little."

A growing nerve fiber roves on the solid surface (Harrison, 1914). The fiber retracts while its tip sticks firmly to some obstacle, then the cell body is drawn toward the tip to some extent, or travels throughout its entire length.

Occasionally the cell body does not move, and a fiber may finish its retraction without any disturbance. In another case, an alternative retraction and extension of fibers or a tripolar neuron may drive the cell body in various directions. A rocking motion of a neuron and its displacement within a small area are usually seen in the initial phase of regeneration of fibers from a spherical neuron (Nakai, 1956, Fig. 1). It is caused by retraction and extension and by peristaltic movement of the cell.

Migrating fibroblasts and other cells are abundant in cultures of spinal ganglia and often collide with neurons and nerve fibers; they may drag the neurons or bend the fibers, causing slight displacements or deformations of neurons. Figure 1 shows a remarkable case of migration of a neuron both actively and passively.

Thus the neurons migrate, apparently by extension and shortening of its fibers, and also by traction exerted by other cellular elements.

Extension and Retraction of Filopodia and Fibers

The axon grows in length at an average rate of 1–2 mm/day. It is obvious that, if a particular cell is followed for several days (Nakai, 1956, Fig. 15), the fiber elongates not only at the tip but also in other parts. Retraction of an axon or a collateral over its entire length or shortening of a part of a fiber (Fig. 1) have also been observed.

Filopodia of the growth cone and of the stem of the axon extend and retract at an average rate of 1 μ/min. When they protrude and swing, they appear to be firm but elastic, and they become bent when the tip

---

FIG. 1. A neuron migrates actively by retracting its fibers and/or passively by traction of a migrating cell. Note changes of a collateral and its branches at the upper left (a, b, and c). Twelve-day chick embryo, spinal ganglia after 3 days in culture. Selected film frames from a cine record; time intervals: 0, 97, 186, and 247 min. Scale: 8 mm = 20 μ.

collides with an obstruction. On retraction, they become flaccid and serpentine.

An axon or a collateral sometimes retracts in this manner, while a new sprout originates from the opposite pole of the cell body.

Similar phenomena have been observed sometimes in filamentous or fibrous processes of nonneuronal cells. Rhizopodia of the Foraminifera show a resemblance (Allen, 1961), although they move about 100–200 times faster than the filopodia of neurons.

It may be presumed that when the filopodia or fibers extend, they are differentiated into ectoplasm and endoplasm; however, as they retract, the endoplasm flows proximally prior to retraction of the ectoplasm.

Additional evidence is required to prove the concept that the filopodia consist of a gelated outer ectoplasm and an inner solated endoplasm (Lewis, 1950; Weiss, 1955). An unusual phenomenon tentatively called "ecdysis" may contribute evidence to this interpretation and also to that of axoplasmic flow mentioned later: A growth cone crossed the stem of another fiber adherent to it at the base and to some object at the tip. The fiber proximal to the cone showed undulation. Then, viscous material in the cone flowed back into the fiber leaving "an empty capsule" connected by a thin tube to the fiber. The capsule jerked and changed its form in a shape of a glove while being drawn by the fiber and decreased in size.

Peristaltic Movement

In young cultures the cell bodies of sensory neurons shows peristaltic movement just when new sprouts are to be protruded. Violent undulation in the motion picture gives an impression that the cell "squeezes out" its contents. An oval and eccentrically located nucleus is thus displaced and sometimes shows a rotatory movement (Nakai, 1956).

Peristaltic movement is observed more or less constantly in the stem of a nerve fiber either locally or on the whole surface.

During the formation of a collateral, a series of rhythmic contractions is observed in the region of a prospective bud for more than an hour. In the course of these events, fine longitudinal parallel stripes in the cytoplasm or on the surface of the fiber become flexed and wrinkled, and the dense granules along the opposite side of the bud are constantly displaced. The opposite side of the fiber does not show any remarkable change and keeps a smooth margin during this process.

When filopodia or collaterals stick to a fixed obstacle which disturbs their retraction and extension, a small vibration is seen along the whole length. Such filopodia do not extend straight, but instead show a winding path between two fixed points. There is a repetition of extension and retraction in a zigzag course which appears to be quite different from that of retraction without any disturbance. When this occurs in a filopodium or a collateral arising from the stem of a fiber which continues to elongate, the base of the filopodium or collateral shifts position at first in the direction of growth of the fiber and then backward. The shifting motion is repeated until the tip escapes from the obstruction.

Some dense bulges which probably correspond to the "local bulges" observed by Lewis (1950) are also seen on the surface of an axon; these are the remnants of retracted collaterals. They appear "darker" than the rest of the axon in phase contrast and move back and forth along the axon as he described.

These observations lead to a concept that the local development and extension of the surface layer of a fiber, ectoplasm, take place repetitively. In electron micrographs of cultivated neurons (Nakai and Yamauchi, unpublished data) the protoplasmic membrane is locally obscure or wrinkled, which suggests "local minute weakness of a gel layer" (Lewis, 1950).

Adhesiveness of Filopodia and Their Traction Force

The filopodia are sticky and adhere to obstructions at the tip or sometimes to each other. There are, however, differences in the duration of adhesion to obstacles or even to the same material. The filopodia often retract after merely touching an object; or they may stick to some obstruction and continue to draw for from several minutes to hours or more. As reported in detail (Nakai, 1960) they may stick to a macrophage and slow its migration; they may pull a small obstacle or adjacent fibers together to make a bundle; or they may tear off a fragment of cell debris.

As these evidences suggest, adhesiveness of the filopodia, their tensile strength, and their retraction exert a "traction force." This force was tentatively estimated in a filopodium to be about 3×10^{-10} dyne.

Migration of Particles in the Neuron with Reference to Axoplasmic Flow

Pinocytotic vacuoles are brought in at the tip of the growth cone not only by the action of undulating membranes but also by the filopodia (Lewis, 1950; Hughes, 1953; Godina, 1955; Nakai, 1955, 1956). These vacuoles travel at an average velocity of 1 μ/min in the cone and in the cell body. After passing the cone, they migrate faster in the fiber stem (at the rate of 2–5 μ/min). In the growth-cone vacuoles, mitochondria and dense granules, including lysosomes, are visible for a longer time and migrate in various directions. In the fiber they travel mostly proximally but sometimes stop or reverse direction for a short distance. Other particles

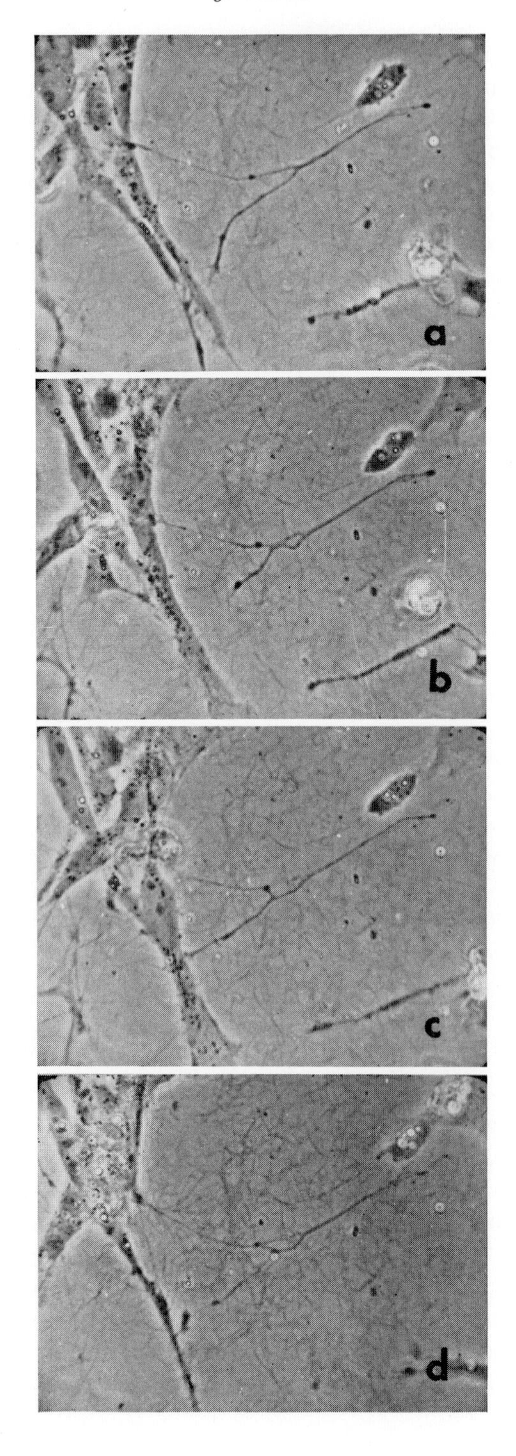
a
b
c
d

migrate in the opposite direction and often collide with the pinocytotic vacuoles. Two sequences of the film shown demonstrate the whole process of pinocytotic vacuoles traveling from the growth cone to the fiber until they disappear in the perikaryon.

Another sequence of film shows a flat, quadrangular cell with double nuclei pressing the neck of a fiber down to the glass surface. The tip of the fiber distal to the cell continues active pinocytosis and undergoes consequent swelling, and the cell does not shift its position. The fiber under the cell escapes to the margin of the cell and gives rise to a new sprout which begins pinocytosis and increases in thickness. The original dichotomized tip becomes a small bulb. During these events, the fiber proximal to the cell decreases in diameter and regains thickness after a new cone emerges. Damming of the fiber is probably due to uptake of pinocytotic vacuoles. Changes in the thickness of a fiber depends upon the material supplied from the growth cone and not from the cell body. The theory of "proximodistal flow" of the axoplasm (Weiss, 1944; Weiss *et al.*, 1962) is not appropriate so far as migration of visible particles (probably including some invisible endoplasmic substance in neurons) in tissue culture is concerned. Lubinska (1963) reports additional evidence of bidirectional flow of axoplasm in adult animals by measuring cholinesterase (CHE) at the cut ends of the peripheral nerves.

Active Sites for Neuron Motility

The previous investigators (Lewis, 1945; Weiss, 1955) have assumed that the active center of neuron motility (such as elongation of fibers and axoplasmic flow) resides to a great extent in the cell body. Although the organizing center must ultimately be in the cell body, the observations of movement characteristic for each kind of particle, local peristalsis of the surface layer, etc., lead the author to propose a concept of multiple active sites within the neuron.

In 1925, Levi (Levi, 1934) observed that a fragment of an axon tip experimentally isolated from the cell body, "trophic center," grew at the normal rate for 12 hr. Hughes (1953), reported a similar observation briefly. A film sequence presented here shows the activity of the peripheral part of a fiber isolated from the cell body (Fig. 2). The activity of filopodia at the growth cone appears to be quite normal, and the fiber shortens or elongates and undulates while the tip moves actively. More-

---

FIG. 2. Activities of a nerve fiber cut and isolated from its cell body. Notice movements of the growth cone and an interaction between dichotomized tips. Eleven-day chick embryo, spinal ganglia after 30 hr in culture. Selected film frames from a cine record; time intervals: 0, 23, 124, and 323 min. Scale: 8 mm = 20 μ.

over, an initial process of fasciculation occurs between dichotomized tips. The cut end never shows any movement during observation for more than 5 hr.

This film shows that extension and retraction of filopodia and of fibers and peristaltic movement of the fiber do not require the presence of the cell body within a limited time. Such evidence tends to support the concept of multiple motility centers to explain neuron movement.

On the other hand, the collision of dense granules with pinocytotic vacuoles in the neuron or that of a vacuole with a filamentous mitochondrion in a cell of the human fibrosarcoma (Gey *et al.*, 1954) is caused by migrations of particles in different directions apparently in the same plane. In the latter case, a pinocytotic vacuole which is larger and moves faster collided with a mitochondrion and broke it into two parts.

The fact that two different kinds of particles move in their own directions, each at a particular velocity and yet in the same plane, suggests the possible autonomy of the particles.

Summary

A series of time-lapse, phase-contrast cinemicrographs of chick embryo neurons in tissue culture and their analyses have been presented. These demonstrate dynamic activities such as migration of neurons, extension and retraction of filopodia and neuronal fibers, the peristaltic movements of the fibers and the cell body, and the movement of particles within the neuron. Axoplasmic flow and the center of movements were briefly discussed in relation to the observations and theories reported by previous authors.

References

Allen, R. D. (1961). Personal communication.
Droz, B., and Leblond, C. P. (1962). *Science* **137** (3535), 1047.
Fujita, S. (1962). *Exptl. Cell Res.* **28**, 52.
Gey, G. O., Shapras, P., and Borysko, E. (1954). *Ann. N.Y. Acad. Sci.* **58**, 1089.
Godina, G. (1955). *Z. Zellforsch.* **42**, 77.
Harrison, R. G. (1910). *J. Exptl. Zool.* **9**, 787.
Harrison, R. G. (1914). *J. Exptl. Zool.* **17**, 521.
Hughes, A. (1953). *J. Anat.* **87**, 150.
Levi, G. (1934). *Ergeb. Anat. Entwicklungsgeschichte* **31**, 125.
Lewis, W. H. (1945). *Anat. Record* **91**, 287.
Lewis, W. H. (1950). *In* "Genetic Neurology" (P. Weiss, ed.), pp. 53-65. Univ. Chicago Press, Chicago, Illinois.
Lewis, W. H., and Lewis, M. R. (1912). *Anat. Record* **6**, 7.
Lubińska, L. (1963). Personal communication.
Nakai, J. (1955). *Anat. Record* **121**, 462.
Nakai, J. (1956). *Am. J. Anat.* **99**, 81.
Nakai, J. (1960). *Z. Zellforsch.* **52**, 427.

Nakai, J., and Kawasaki, Y. (1959). *Z. Zellforsch.* **51**, 108.
Nakai, J., Takata, C., and Kawasaki, Y. (1961). *Acta Anat. Nippon* **36**, 356.
Weiss, P. (1934). *J. Exptl. Zool.* **68**, 393.
Weiss, P. (1944). *Anat. Record* **88** (Suppl. 4), 464.
Weiss, P. (1955). *In* "Analysis of Development" (Willier, Weiss, and Hamburger, eds.), pp. 346-401. Saunders, Philadelphia, Pennsylvania.
Weiss, P., Taylor, A. C., and Pillai, P. A. (1962). *Science* **136** (3513), 330.

Discussion

Dr. Ling: It would be interesting to know the diameter of the filopodia. If you estimate the diameter at 0.1 μ, for example, then the power would perhaps be similar to a contracting frog muscle. On the other hand, if it were 1.0 μ, then the power would be much smaller.

Dr. Nakai: The diameters of the filopodia are between 0.25 and 0.3 μ.

Dr. Ling: In that case, these must be rather powerful structures.

Dr. Abé: I would just like to say that the term "bubbling" when applied to the activities of tissue cell surfaces during mitosis is frequently misleading; bubbling can be seen only in the time-lapse cinemicrography. At normal speeds these projections would be called "pseudopodia."

Dr. Robineaux: I think that is a very important point. With time lapse, it must be realized that only a part of the real time-scale is used for the exposure of the film. A big part of the time is not used, and during this part of time many things can be happening.

Dr. Abé: If you look at a bird in flight with time-lapse films, movement of the wings may be obliterated from the projection screen, and only the body of the bird may appear moving swiftly. One cannot be too critical in interpreting the results of time-lapse films; otherwise confusions in terminology cannot be avoided.

Intracellular Movement and Locomotion of Cellular Slime-Mold Amebae

B. M. Shaffer

Department of Zoology, Cambridge University, Cambridge, England

When feeding, the cells of most cellular slime molds, or Acrasina, are small soil amebae, living a solitary existence and ingesting bacteria. After the food supply has been depleted, the cells, attracted by acrasin, form aggregates, which then behave as unitary organisms, or "grex"; these can move over the ground or partly through the air, and finally differentiate into stalked, fruiting bodies (Bonner, 1959). A great deal has been learned of the factors that orient the individual cells and the aggregates, but until recently, the cells had maintained their "black-box" status virtually intact. The attempt to deprive them of this status was motivated, among other things, by the difficulty of understanding certain features of movement within aggregates. Cells streaming toward aggregation centers stick together end to end in chains. The tip of a cell that has approached such a chain from the side receives no guidance if it makes contact with the middle of a cell in the chain, but is turned in the direction the chain is advancing when the back end of this cell moves past it (Fig. 1; Shaffer, 1962). This behavior, which has been called "contact following," can be readily explained if a cell's surface remains stationary relative to the ground along its sides, and is created at its front end and removed at the back end. If this is so, we can suppose that the tip of the incoming cell is not guided by a surface to which it adheres as long as this is stationary, but follows it if it recedes. Yet the mere existence of cell chains seems incompatible with a stationary surface. How can the cells hold on to one another unless the surface is relatively permanent and moves along with a cell? It seems especially unlikely that the regions of intercellular contact should be areas of maximum surface turnover.

The fact that the adherent cells in an aggregation stream are commonly not in a single file but lie many abreast does not raise any additional mechanical problems, whether the lateral cell surfaces are stationary or carried along with the cells. However, in an old aggregation stream, several layers of cells may lie on top of one another, and a grex may be up to about fifty cells high (Raper, 1940). Electron microscopy shows that the cells in a grex are closely packed together (Mercer and Shaffer, 1960); and we also know that they move forward by their own individual efforts, and that, though they may change position, the majority of them maintain the same velocity relative to the ground (Raper, 1940; Bonner, 1952; Bonner *et al.*, 1953). To account for this it has been commonly assumed —though without any evidence—that the cells are miniature editions of the large solitary amebae, such as *Amoeba proteus* or *Chaos chaos*—which I may perhaps be permitted to refer to as conventional amebae—and that their casings of gelled ectoplasm constitute a sufficiently rigid stationary framework for each cell to crawl forward over the side surfaces of its neighbors just as it would do over the ground. But, of course, this raises yet again the controversy, started by Wallich exactly a century ago and still going strong (Griffin and Allen, 1960; Goldacre, 1961; Wolpert and O'Neill, 1962), as to whether the side surface of a conventional ameba is stationary or advancing. This has already been fully discussed in this Symposium. There is no dispute that, if the surface does travel forward with the cell, it must be temporarily stationary in some regions if in-

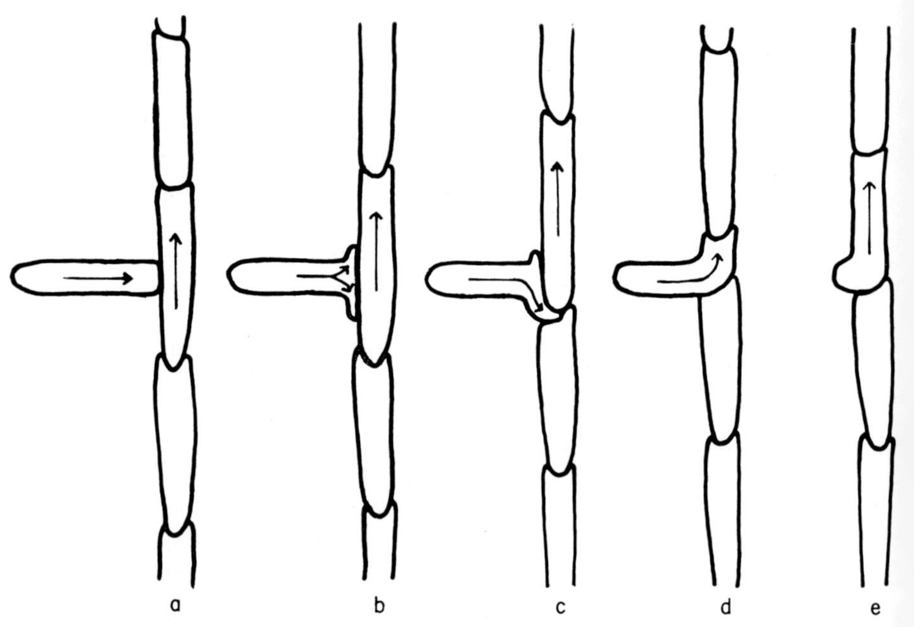

FIG. 1. Contact following. (a) A cell has been attracted to an aggregation stream, perpendicularly to the long axis of the stream, and has just made contact with the middle of the lateral surface of a stream cell. (b) The incoming cell can turn either upstream of downstream, or, as represented here, in both directions. (c–e) It is guided in the right direction toward the aggregation center only when the back end of the stream cell to which it is adhering moves past it. (c) This guidance first affects the most upstream part of the anterior end of the incoming cell.

ternally generated motive power is to be transmitted to the ground. However, it is not too obvious how a three-dimensional network of such stationary regions could provide an adequate mechanical skeleton for an aggregate built of cells that behaved in this way. Even if the cell surface had the properties ascribed to it in the Goldacre (1961) model, it might well not have sufficient rigidity to support the propulsion of an overlying cell, except at the rear, where it adhered to the granuloplasm. And if this were the case within an aggregate, the rear ends of all the cells would presumably have to lie on top of one another. No one who has examined sections of a slime-mold grex has reported this to be the case. These reflections led to an attempt to observe the behavior of the slime-mold cell surface directly.

The surface movement of the conventional ameba has been conventionally examined with particles of various sorts (Bütschli, 1892; Jennings, 1904; Schaeffer, 1920; Pantin, 1923; Mast, 1926; Griffin and Allen, 1960); and despite the multiplicity of hypotheses and corresponding observations, the majority view is that most such particles move forward as fast or faster than the cell, though Goldacre (1961) has claimed that such movements are due to electrophoresis and not to the motion of the surface itself. However, with underwater, temporarily monopodial, slime-mold cells (*Polysphondylium violaceum*), it is quite clear that all particles of all sorts and sizes, after attaching to any part of the cell surface except the extreme anterior tip, do not significantly move either forward or backward relative to the ground until they arrive almost at the rear end. They then begin to travel forward, though more slowly than the cell itself, until they reach the rear extremity, where they collect as a cloud that is towed along behind the cell. This behavior essentially mimics contact following, but as the particles are passive, they provide clearer evidence that the lateral surface of the cell must be stationary.

This conclusion applies to a cell in a monolayer. But would the surface resist deformation sufficiently for other cells to move on top of it as if they were on the ground, if it had the so-called fluid or at least almost unbelievably passively extensible structure often postulated for the conventional ameba, especially as any backward slip of the top surface of a cell relative to the bottom surface could be summated from one cell layer to another? Conceivably an inherently deformable surface could be made rigid enough if it everywhere adhered to an underlying gel. Is this in fact the situation?

The best conditions for phase microscopy obtain in sandwich preparations, in which the cells lie between glass and a thin sheet of agar and move more nearly as they do on an ordinary agar plate than they do under water. In sandwiched cells, unlike conventional amebae, no hyaline layer can usually be seen between the granuloplasm and the side surface. As this remains true even when a new hyaline pseudopod is being produced at the expense of an old one at the other end of the cell, it must be easier for the hyaloplasm to pass *through* the granuloplasm than to penetrate between it and the surface, although limited parts of the surface obviously must become detached from the granuloplasm whenever a new pseudopod is formed. However, even if the surface adheres to the granuloplasm, does this really have a stationary, outer gel layer to provide the necessary rigidity?

The smaller inclusions of the granuloplasm are in constant agitation, and often the internal movements are more noticeable than the locomotion of the whole cell. The most striking behavior is shown by apparently aqueous vacuoles, a dozen or so per cell, ranging in diameter from about 0.4 μ down to the limits of visibility. They are not part of the contractile-vacuole system, nor is there any clear indication that they are produced by pinocytosis. They continuously dart about the granuloplasm with equal facility in all directions, on their longer excursions traveling in fairly straight lines as much as a third the length of the cell or even further, and taking about a second to do so. As the cell's velocity is at least several and often many times less than this, the forward component of vacuolar movement associated with the cell's advance is relatively inappreciable. Small granules of unknown nature and of about the same size as the motile vacuoles also are continuously perturbed, but their average excursions are much shorter; and the mainly rod-shaped mitochondria, 1–2 μ long, move about less than the granules. The free movement of all these inclusions throughout the length and breadth of the granuloplasm—within the limits of resolution, right up to the boundary with the hyaloplasm at the front and up to the cell surface at the sides —shows that the granuloplasm, unlike that of the conventional ameba, neither undergoes a regular circulation nor as an appreciable outer layer of different consistency.

Apart from the light they throw on cell locomotion, these movements have their own intrinsic interest. What is their basis? The motile vacuoles are displaced very much farther than are extracellular particles in Brownian movement in water. Moreover, the contractile vacuoles, even when as small or smaller than the motile ones, are static in a stationary sandwiched cell, and occupy a zone of relatively constant position in the middle of a cell that is continuously moving in a particular direction. Any hypothesis must account for the fact that motile vacuoles traveling in opposite directions may pass one another, without collision, at a distance that can hardly be resolved, and that the granules occasionally suffer as great displacement as the vacuoles. The inclusions might be moved by adhesion to submicroscopic motile fibrils or reaction from them (Jarosch, 1958), by electrical forces, or by asymmetrical exchanges of some kind with their environment.

Since the hyaloplasm freely flows out of the granuloplasm whenever a new pseudopod is formed, and flows in again almost quicker than the eye can follow when a suitable sandwich preparation is jarred, why do not the inclusions in the granuloplasm penetrate into the hyaloplasm, and perhaps even get stranded there if that environment does not support movement? One obvious but not unilluminating observation is that the granuloplasm never fragments into pieces separated by hyaloplasm, nor does a zone of hyaloplasm ever segregate in the middle of the granuloplasm. This implies that the granuloplasm, despite its fluidity, porosity, and ability to change shape, is, nevertheless, a unit, presumably being kept together by a sufficiently strong boundary or by containing some sort of fibrillar felt. However, such a felt can hardly prevent the inclusions from escaping, since they can move through the granuloplasm so readily, unless the felt is itself motile and they move by adhering to it. If it is the boundary that traps them, it must be continuously intact; for there is no cycle, as there is in a conventional ameba, in which the granular material periodically invades the hyaline cap and rushes forward right up to the plasmalemma at the tip. In a very thin or starved cell, the inclusions are markedly more concentrated in the anterior half. But if the boundary is filtering them off, why do they not pile up against it? The answer may be that when a cell is continuously moving in one direction, although material for the new surface must ultimately be supplied to the hyaloplasm from the granuloplasm and must presumably bear some direct or indirect relation to the surface material removed at the rear end, the total flow through the granuloplasm may be volumetrically insignificant and quite unable to overcome the dispersive effect of the inclusions' own movements. However, when a pseudopod is being produced from the opposite end of the cell, the smaller inclusions do move predominantly toward it, rushing past the nucleus and contractile vacuoles on a tide of hyaloplasm and then slowing up when they reach the leading part of the granuloplasm. Unfortunately, perhaps partly because of the lability of the hyaloplasm of separate cells, electron microscopists studying these organisms have not primarily directed their attention to the difference between hyaloplasm and granuloplasm and the nature of the boundary between them. But unpublished photographs of aggregated cells by Mercer and Shaffer do show cytoplasmic areas free from all visible structure except for the smallest unattached granules, and areas that are packed with organelles, without any membrane being apparent in-between. This agrees with Wohlfarth-Bottermann's (1960) description of *Hyalodiscus*. Conceivably the inclusions could be held back not by a membrane, but by an electric field.

Let us return to the question of locomotion. We suggested that the cells within a grex would be able to obtain the requisite traction even if the cell surface were inherently highly deformable, if this were made fairly rigid and inextensible by adhesion to a stationary underlying gel. But we now find that the outer part of the granuloplasm is not gelled, and that instead the whole of it moves forward, with nearly uniform velocity, inside an apparently stationary cell surface. Moreover, although we cannot determine the consistency and behavior of the interior of a hyaline protrusion by the same methods, it is improbable, considering its greater mobility, that much of it is a gel, and it is doubtful whether it could give adequate support for intragrex movement even if it were one; direct evidence that it is not will be presented later.

It is conceivable that the grex cells could move through permanent rigid tunnels of secreted "slime" if this filled the 200 A space between their lateral surfaces; but if they did, it is difficult to see how they could so readily overtake one another, and it would scarcely be expected that the space between adjacent plasma membranes would be the same at the front and back of the cells as along the sides, as in fact it is. The much more probable conclusion is that the sides of the slime-mold cell are encased in a surface layer that is below the limits of resolution of the light microscope, and is not only stationary but is itself highly resistant to deformation.

Is there any other evidence of such rigidity? An underwater cell gently detached from the substratum keeps its shape, floating with its pseudopods held out quite stiffly. Before reattachment, mechanical reaction to pseudopodial growth often makes the cell spin round as a whole, with its processes maintaining their relative positions. As the same rigidity is shown by all parts of the cell, whether containing granuloplasm or only hyaloplasm, it seems reasonable to attribute it to the cell surface. Moreover, a fluid cell surface would hardly be compatible with the cells' ability to survive and to move and even aggregate fairly normally when floating at an air-water interface.

What then is the motive power for cell locomotion? Contraction on a large scale is shown by underwater aggregation streams, especially those encased in slime sheath, that are still flowing into an aggregation center. Immediately after they are severed peripherally and detached from the substratum they start to shorten and thicken, and within a minute or two are a fraction of their original length. This also demonstrates that a stream is prevented from contracting by attachment to the substratum. What part of the cell, if any, contracts actively and how could it produce locomotion? If there were contractile fibrils in the granuloplasm, they could squeeze out hyaloplasm, but how could the resultant hyaline pseudopods progress any further? The granuloplasm could still be important in propulsion if it could drive itself forward by actively shearing on the rigid cell surface. But this could not explain how a hyaline pseudopod could be withdrawn from one end of a cell and protruded at the other, while the granuloplasm hardly changed its position. Moreover, if the granuloplasm did push the cell forward, one would expect it to displace any hyaloplasm ahead of it—considering that this can so readily flow right through it—and lie flush against the advancing tip. Indeed, any satisfactory theory of cell locomotion must explain why the leading part of the cell is always hyaloplasm.

Let us consider the movements of the hyaline pseudopods. Those of a normal unaggregated cell on agar in air advance discontinuously and only over the agar. Not only do pseudopods in different parts of the cell advance more or less independently and in any sequence, so too do limited sections of the leading edge of a single broad pseudopod. A cell that is covered with water is far less constrained and can protrude many pseudopods from all over its surface up into the water; this makes locomotion much more irregular. And as large parts of the surface may be out of contact with the substratum, much of the cell is often rounded up. A single hyaline pseudopod can produce many cylindrical subpseudopods, which may be called "pseudodigits"; these are not confined to one plane but can stick out in any direction from any part of its surface. They all advance discontinuously and in any sequence, some advancing but once, others several times. Successful pseudodigits are followed and enlarged by the main mass of the pseudopod, and eventually, if locomotion continues in the same direction, by the granuloplasm. Unsuccessful digits are withdrawn.

Paradoxically, often one or more very fine filaments are abruptly shot out from the tip of a digit in the first stage of withdrawal (Fig. 2). Their spikiness contrasts with the rounded profiles of all extensions that are capable of further growth. A digit may be completely withdrawn before the granuloplasm reaches it. But if a filament is attached to the substratum at its outer tip, withdrawal may be considerably delayed. Not infrequently it remains quite stationary, sticking out from the side of the cell until the whole of the rest of the cell has moved past it. Then its base is jerked round after the cell, the rest of it follows, and it is dragged along behind the cell much like a particle stuck on the outside, though once free from the substratum it is soon withdrawn. This again shows that the side surface of the cell is stationary, for there is no static internal structure to which the withdrawal thread can be anchored, and the rest of the surface could hardly flow round its base without displacing it. This is shown even more clearly by those withdrawal threads that stick up stiffly into the water. Because they are unattached to the substratum, they are much shorter-lived, but nevertheless do not move forward with the cell.

One pseudodigit may be withdrawn at the same time as another on the same hyaline pseudopod is extending. This proves that a pressure generated by the granuloplasm cannot be the sole propulsive force,

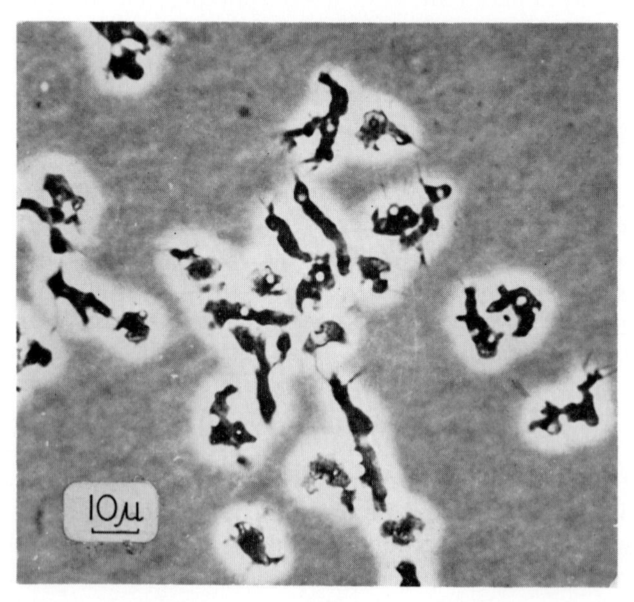

FIG. 2. Polysphondylium violaceum cells in a layer of water. The rigid filose cell processes have all been extruded as the first stage in the withdrawal of pseudodigits. In some cases they extend between cells, but they have also been formed by the separate cells.

unless the extension of a pseudodigit is an elastic deformation of the surface, which can occur at the expense of withdrawal elsewhere. That this is obviously not the case is shown, for example, by the shape of the digits, by the fact that withdrawal is not a simple reversal of extension, and by the fact that it is never partial but always complete (unless a growing digit invades the shrinking one). We must conclude that at least some of the motive force is generated within the pseudodigits themselves.

One could postulate some complicated structure for the hyaline core to account for this. One possibility would be a modification of Kavanau's (1963) model of pumping endoplasmic reticulum. The layer of reticulum would have to be attached to but separable from the actual surface, and the whole cycle of breakdown and repair at the tip would have to occur within the hyaloplasm without the intervention of the granuloplasm; but other objections apart, it seems improbable that much reticulum will be found in the hyaloplasm.

Kamiya (1962) and Jarosch (1958) have presented considerable evidence that the propulsive force both for cytoplasmic streaming and for the movement of certain cell organelles is generated at the boundary of some rigid structure, respectively, a cytoplasmic gel layer and the organelles themselves. It is tempting to suppose that the hyaloplasm and perhaps the granuloplasm too of the slime-mold ameba could be propelled forward by a force generated at its boundary with the cell surface, but considering the known turbulence of the granuloplasm and the probable consistency of the hyaloplasm, it is not obvious why such a force should advance the cell rather than merely circulate its contents. Instead we may postulate that locomotion is produced by one or more changes taking place within the substance of the rigid surface layer rather than at its boundary with the cytoplasm: active *contraction* of area, by rearrangement of material, or by removing it; or active *enlargement,* by the production or liberation of new surface material at the tip, as Bell (1961) has suggested happens in the conventional ameba, and also perhaps by the expansion of material after its entry into the surface.

This reminds one of the long-standing controversy as to whether or not growth of plant cell walls is dependent on turgor pressure (Burstöm, 1961). In any case, both in fungi and in higher plants, it is a question of extending a structure of considerable rigidity, a property emphasized by the word "wall." In slime-mold cells too, the surface must be a sort of wall, if either its contraction or its growth is responsible for locomotion: if the surface were "fluid" or almost indefinitely deformable, any amount of material introduced at the front of the cell could be transported backward wherever the surface was not adherent to the substratum, and then contracted and removed, without its advancing the cell (Fig. 3a). It must, therefore, be impossible to displace the surface on one side of a monopodial cell backward relative to that on the other sides. This is an independent reason for believing that the surface has the mechanical properties we have already ascribed to it.

The shape of pseudodigits that are being withdrawn is obviously not determined by surface tension, as is especially clear from the unbeaded filaments they form, some of which are at least as thin as can be resolved with the light microscope. This shows that the rigidity of the surface is retained during retraction. This process cannot easily be accounted for by the removal of the fluid core as a result of negative hydrostatic pressure generated by the extension of the surface elsewhere. So it is probable that the surface of the shrinking digit actively contracts, forcing out most of the hyaline core. It is even plainer that the initial lengthening of the filaments cannot be due to a negative pressure inside them; and we may suppose that constriction of the digit traps some of the fluid core in its distal end, and this—or at least the surface material in it—is then squeezed out to make new surface. The rigid, contracting, surface layer cannot be more than half as thick as the resultant thread; that is, in the thinnest ones visible, it cannot be more than 500 A thick. This is an upper estimate in that there is no reason to suppose that the whole of the fluid hyaloplasm can be converted into surface, and the withdrawal threads may well still contain a fluid core. During the next

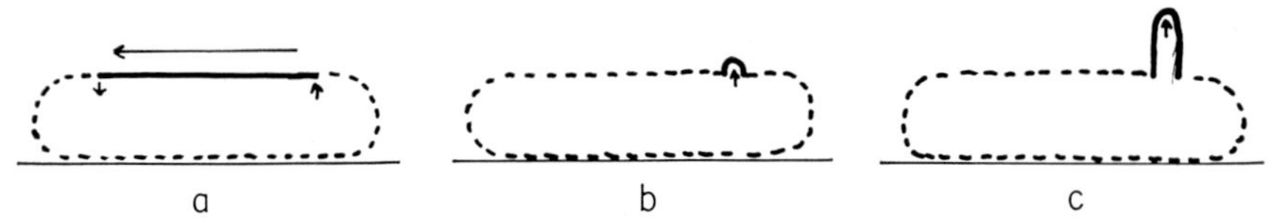

FIG. 3. (a) If the cell surface were fluid or so deformable that any part of the lateral surface that was not attached to the substratum could be displaced backward relative to the rest of it, material (solid line) could be introduced into the surface at the front of the cell and contract or be removed at the rear without moving the cell forward. (b) In contrast, if the surface is rigid, new material introduced at any point must produce a bleb. (c) A rigid cylindrical projection formed by continuing to add new surface in a limited region.

stage of withdrawal a thread shortens while remaining quite stiff. The shorter the thread becomes, the more likely it is to pivot stiffly about its base till it lies flat against the neighboring cell surface. Withdrawal must involve not only contraction of the surface but also the eventual return of its substance into the fluid hyaloplasm in the main pseudopod. It is important to note that all this can occur at what is still the anterior end of the cell.

If the surface actively contracts, motive power would not be generated only at the front. Indeed, the roughly triangular profile of a separate monopodial cell on an agar plate—the apex at the rear of the cell—is difficult to explain if the surface resists passive deformation, unless the surface does contract progressively as it approaches the rear. We may suppose that in normal locomotion, contraction of the surface at the back of the cell increases the internal hydrostatic pressure; that when this is high enough, the surface suddenly starts to grow at its weakest point; and that consequently the pressure falls and growth stops. Thus the pseudopods advance in spurts. It would be slightly more difficult to explain the abrupt, discontinuous nature of locomotion if all the motive power came from extension of the surface. But this is not to say that none of it does. However, the surface may perhaps expand not because work is done on it, but because energy is not expended to keep material contracted or excluded. This is a possible explanation of the very large number of hyaline blebs formed by a cell poisoned with dinitrophenol, which uncouples oxidation from phosphorylation.

The most interesting type of locomotory behavior is seen in sandwich preparations in which there is enough water in the sandwich layer to allow the cells to produce cylindrical pseudodigits in this plane. A member of a chain of aggregating cells frequently produces a lateral digit which it applies to the side of the cell ahead of it (Fig. 4). Initially

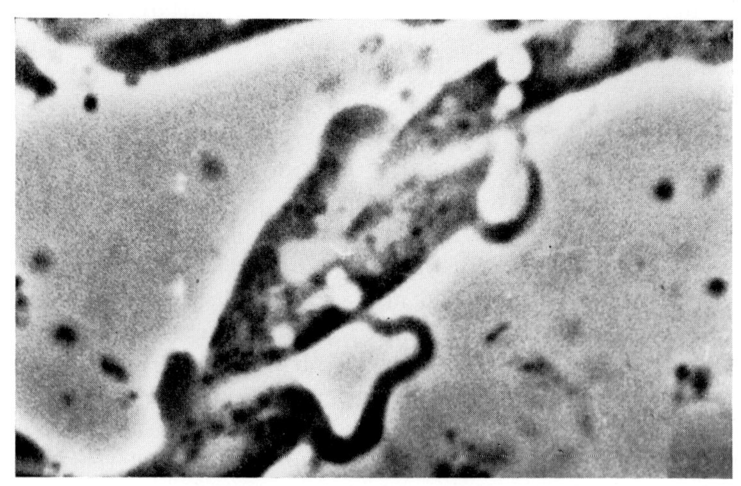

FIG. 4. Pseudodigits applied by aggregating *Polysphondylium violaceum* cells to the sides of adjacent cells. Magnification: × 2500.

this digit is short, but as it lengthens, being free except at its tip and its base, it arches outward. Often one sees two free digits with their tips adhering to one another, continuing to lengthen and thereby growing into loose coils. Bulges and kinks in either of these digits move away from its tip, showing that new rigid surface is being produced there. More unexpectedly, they gradually approach its base, and disappear when they reach it. From the fact that the smallest irregularities share this fate, there can be no doubt that there is an actual translocation of the surface and not merely a propagation of symmetrical or asymmetrical contractions and relaxations. The surface must, therefore, be simultaneously removed at the base. This is confirmed by the behavior of particles adhering to the outside: these are carried to the base and then become stationary. New material for the surface must continuously travel up the interior of the digit, and this proves that the interior is more fluid than the outside —as we have already supposed—for otherwise it would be impossible for kinks to move toward the base. The individual digits look rather similar to those described by Holtfreter (1946) in amphibian cells. They also resemble myelin bodies, which Holtfreter (1948) used as models to illustrate his theory of cell movement, though they are undoubtedly more complex. A digit, which is completely hyaline when observed in the light microscope, behaves very much like an entire conventional ameba that cannot obtain traction on a substratum: some or all the fluid axial material moves forward to the tip, is incorporated into a rigid external tube, moves backward, and is returned to the fluid interior.

Perhaps the most illuminating point is that surface can be removed at an annular site encircling the base of a growing digit. Clearly, variation in the shape and position of the sites where surface material is added and where it is removed, and perhaps where it contracts and where it expands, can give a cell a large repertoire of movements. But just as surface propulsion requires that the surface should have certain mechanical properties, so these in turn impose certain geometrical limitations. For example, surface addition or expansion can occur along any extent of the edge of a flattened cell; but in some small circular area or "point" on the upper surface, it would necessarily produce a small bleb, and in a linear area it would form a ridge (Fig. 3b). If further growth is limited to a small area of the surface of the bleb, a cylindrical structure will result (Fig. 3c). Rapidly changing patterns of high points and ridges are, in fact, seen projecting up into the air from the upper surface of some particularly large and active *Dictyostelium discoideum* cells. If such a cell is covered with water, it presents an astonishing sight as a forest of long papillae leap up all over it.

It is theoretically possible for an annular source or an annular sink to encircle any part of the cell without deforming any other region of the surface (Fig. 5), if we ignore restrictions imposed by the environment. Surface material could be added to or removed from either or both sides of such an annulus, depending on local intra- and extracellular conditions.

Of course, the surface may be altering over much larger areas than this. The separate cells have no permanent rear, unlike conventional ameba, but nearly the whole of the surface covering what is temporarily the back is normally contracting, though even here there may be limited areas of surface growth. Almost the whole surface contracts in sandwiched cells that round up in response to jarring. And when a pseudodigit is "exploded" by the rest of the cell entering it, new material must be introduced over all its surface. All these possibilities mean that mechanical rigidity is coupled with extreme behavioral flexibility.

What determines the sites of all these activities? We have little enough information about this, and even less space to discuss it in. Whereas a conventional ameba changes direction in response to such factors as light (Mast, 1932) and a microdissection needle (Goldacre, 1952) only when they impinge on the granuloplasm, it is easy to see that the hyaline pseudopods and pseudodigits of a slime-mold cell can turn in response to chemotactic and mechanical guidance by local differentials in their rates of advance; and, indeed, given their method of locomotion, it could hardly be otherwise. We may suppose that positive chemotactic agents act directly on the surface, presumably by weakening or plasticiz-

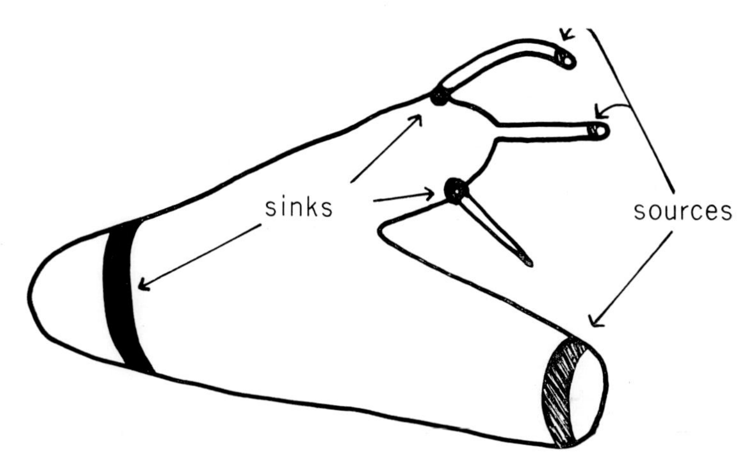

FIG. 5. Annular sources and sinks of surface material could encircle any part of a cell without deforming other regions of the surface.

ing it, and so promote the introduction of new material. Contact following can be accounted for if mechanical tension has the same effect.

We can now return to the problem posed initially of how cells hang together in chains. The lateral surface, if it is stationary, must be made at the front and removed at the back; but this seems to imply that the surface at the front of one cell must be moving in the opposite direction from that at the back of the cell ahead of it, and therefore shearing on it (Fig. 6a). However, we have seen that both with whole aggregation streams and with the withdrawal threads on individual cells, contraction and removal of the cell surface is hindered by attachment to another surface—though obviously this hindrance cannot be complete or movement would be impossible. Attachment may also be expected to strengthen the cell surface and so hinder its expansion. We may, therefore, assume that the adherent end surfaces of two cells will tend to stabilize each other, and so to be relatively permanent and to travel along with the cells. This can be reconciled with the requirement of a stationary lateral surface if the surface sources and sinks are annuli encircling the cells (Figs. 6b and 7). If we made a model of such a cell chain out of a column of cylindrical tin cans, it would be held together by sticking the permanent top of each can to the permanent bottom of the next one. The source of the side surface would be the rim round the top, and the sink, the rim round the bottom. It is easily understandable that the tension generated by movement should affect the source and the sink differen-

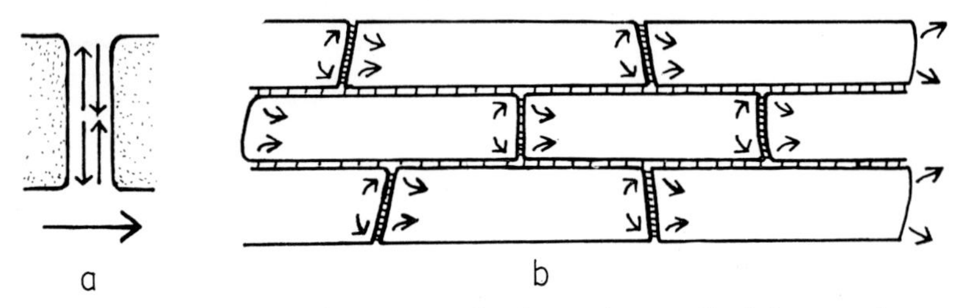

FIG. 6. (a) In an aggregation stream, the front of one cell, if it were the source of the surface, would apparently have to shear on the rear of the cell ahead of it, if this were the site of surface removal. (b) The end surfaces of stream cells could be fairly permanent and easily adhere to one another if surface formation and removal were limited to the lateral regions of these surfaces. The side surface is stationary.

tially, because the surface is in a different state in these areas. The mechanical relations between adherent cells tend to eliminate the spurts characteristic of the locomotion of separate cells; this occurs most completely when a small number of cells are arranged in a closed ring.

What is the relationship between the type of locomotion described here and that of the conventional ameba? We have all long ago abandoned the notion that the conventional ameba has a particularly simple organization, but what has perhaps not been so generally appreciated is just how specialized it is—perhaps, in its own way, as much as are the plasmodial slime molds. By Nature's standards it is unconventional in its surface coat and its nuclear membrane (Pappas, 1959; Mercer, 1959), and I suggest, above all in its locomotor apparatus, if, as is generally believed, this is in the interior. These specializations are all doubtless due to its relatively gigantic size. The slime-mold arrangement with the motor in the surface, where it is subject to immediate environmental control, certainly seems both simple and efficient. If it is, indeed, primitive, what evolutionary path has led to the large solitary ameba? Presumably, the thin surface layer proved inadequate to move so large a mass, and virtually the whole interior was pressed into service for this task, the granular ectoplasm carrying out essentially the same cycle as the slime-mold cell surface. We may note parenthetically that the detachment of this ectoplasm from much of the ameba's surface, possibly because of the production of a large volume of syneresis fluid, raises the problem of how the motive power is to be transmitted to the ground. This is especially acute if it is true that the surface from which the static ectoplasm may

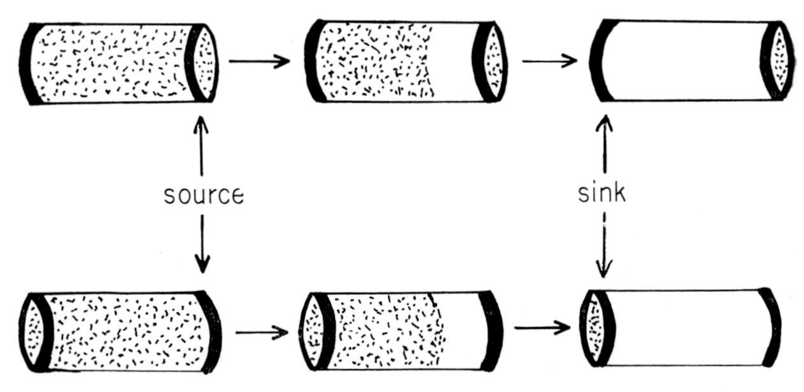

FIG. 7. Hypothetical behavior of the surface while a cell in an aggregation stream advances its own length. Top, the cell is represented as seen from slightly ahead; bottom, from slightly behind. The surface originally present is stippled, new surface is white. The end surfaces are permanent and travel along with the cell. The lateral surfaces are stationary relative to the ground, and continuously made at the annular source and removed at the annular sink.

be separated by a fluid layer is itself fluid or extremely extensible and moving forward. It would seem that the ectoplasm would have to adhere to the plasmalemma in exactly the same regions as this adhered to the ground; this suggests that one of these adhesions locally alters the plasmalemma so as to induce the other adhesion on the opposite side of this membrane. If adhesion to the ground were the primary factor, this would ensure that the ectoplasm attached itself to the plasmalemma on the side of the cell next to the ground.

It seems not improbable that many types of solitary ameboid cell of about the same size as the slime-mold ameba will be found to use essentially the slime-mold method of locomotion. But of more general significance will be to discover to what extent this is used by metazoan cells.

Fauré-Fremiet (1929) observed ruffled membranes on the hyaline parts of invertebrate choanoleucocytes while the granuloplasm was quiescent;

and Holtfreter (1946, 1948) published extensive observations showing that the various pseudopodial movements of all types of cells from early amphibian embryos could occur when the granuloplasm was inactive or even absent, from which he concluded that the cell surface provided the force for locomotion. Holtfreter assumed a fairly permanent surface that carried the cell along with it by its contraction and relaxation. However, it is not too clear how a permanent surface of this kind can be capable of deforming itself into long thin projections and yet be able to transmit the force for cell movement. Ambrose (1961) has supposed that fibroblast movement is due to patterns of contraction and relaxation just under the surface. In his model, this layer does not enter the surface undulations and ruffles. If the surface is relatively rigid, inextensible, and permanent, the proposed contraction could doubtless throw it into lamellate folds that were able to transmit sufficient force to the substratum to advance the cell. But again, such a surface would seem incapable of forming filose projections, although very many types of cells can do so. Sea-urchin mesenchyme cells project and subsequently contract narrow cylindrical pseudopods to move themselves into position during development (Gustafson and Wolpert, 1961). And in the bug *Rhodnius,* the epithelial cells can use very fine pseudopods as much as 150 μ long to pull tracheoles toward themselves (Wigglesworth, 1959), though we have no information about how these processes are extended and withdrawn again. Pseudopods of this and a variety of other kinds could be produced by the slime-mold mechanism.

We may conclude that in slime-mold cells, the locomotive force is generated by one or more of four processes: contraction or relaxation of surface material in situ, or its addition or removal; that the surface is rigid; that addition and removal can take place, with certain geometrical restrictions, in any region of the cell; that the surface can be a patchwork of new areas and permanent ones; and that all these changes can have been taking place in a cell that appears in the electron microscope to be enclosed in a continuous "unit" membrane and to have this membrane everywhere as closely adherent to those of adjacent cells as it would be in an undifferentiated metazoan tissue.

Acknowledgments

I am extremely grateful to Drs. J. T. Bonner, L. E. R. Picken, and M. G. M. Pryor for their critical reading of the manuscript of this paper.

References

Ambrose, E. J. (1961). *Exptl. Cell Res. Suppl.* **8**, 54.
Bell, L. G. E. (1961). *J. Theoret. Biol.* **1**, 104.
Bonner, J. T. (1952). *Am. Naturalist* **86**, 79.

Bonner, J. T. (1959). "The Cellular Slime Molds," 150 pp. Princeton Univ. Press, Princeton, New Jersey.
Bonner, J. T., Koontz, P. G., and Paton, D. (1953). *Mycologia* **45**, 235.
Burström, H. (1961). *In* "Encyclopedia of Plant Physiology" (W. Ruhland, ed.), XIV, pp. 285-310. Springer Verlag, Berlin.
Bütschli, O. (1892). "Untersuchungen über mikroskopische Schäume und das Protoplasma," 234 pp. Engelmann, Leipzig.
Fauré-Fremiet, E. (1929). *Protoplasma* **6**, 521.
Goldacre, R. J. (1952). *Symp. Soc. Exptl. Biol.* **6**, 128.
Goldacre, R. J. (1961). *Exptl. Cell Res. Suppl.* **8**, 1.
Griffin, J. L., and Allen, R. D. (1960). *Exptl. Cell Res.* **20**, 619.
Gustafson, T., and Wolpert, L. (1961). *Exptl. Cell Res.* **24**, 64.
Holtfreter, J. (1946). *J. Morphol.* **79**, 27.
Holtfreter, J. (1948). *Ann. N.Y. Acad. Sci.* **49**, 709.
Jarosch, R. (1958). *Protoplasma* **50**, 93.
Jennings, H. S. (1904). "Contribution to the Study of the Behaviour of the Lower Organisms," 257 pp. Carnegie Inst., Washington, D. C.
Kamiya, N. (1962). *In* "Encyclopedia of Plant Physiology" (W. Ruhland, ed.), XVII/2, pp. 979-1035. Springer, Berlin.
Kavanau, J. L. (1963). *J. Theoret. Biol.* **4**, 124.
Mast, S. O. (1926). *J. Morphol.* **41**, 347.
Mast, S. O. (1932). *Physiol. Zool.* **5**, 1.
Mercer, E. H. (1959). *Proc. Roy. Soc.* **B150**, 216.
Mercer, E. H., and Shaffer, B. M. (1960). *J. Biophys. Biochem. Cytol.* **7**, 353.
Pantin, C. F. A. (1923). *J. Marine Biol. Assoc. U.K.* **13**, 24.
Pappas, G. D. (1959). *Ann. N.Y. Acad. Sci.* **78**, 448.
Raper, K. B. (1940). *J. Elisha Mitchell Sci. Soc.* **56**, 241.
Schaeffer, A. A. (1920). "Amoeboid Movement," 156 pp. Princeton Univ. Press, Princeton, New Jersey.
Shaffer, B. M. (1962). *Advan. Morphogenesis* **2**, 109.
Wigglesworth, V. B. (1959). *J. Exptl. Biol.* **36**, 632.
Wohlfarth-Bottermann, K. E. (1960). *Protoplasma* **52**, 58.
Wolpert, L., and O'Neill, C. H. (1962). *Nature* **196**, 1261.

Discussion

Dr. Allen: I would like to point out a similarity between slime-mold amebae and certain small species of marine and fresh-water amebae which I and several others here have studied.

It seems to be generally true of the smaller ameba with extensive hyaloplasmic regions that these regions have a higher refractive index and a greater electron density than the cytoplasm in which the inclusions are contained. This may mean that this material is not a simple fluid at all, but rather a gel capable of various kinds of contractile movements. Presumably it is more or less homogeneous except for the presence of fibrillar material as observed by Professor Wohlfarth-Bottermann.

In most of these small ameboid cells, this hyaloplasm is characteristically located toward the advancing front of the cell; for example, in *Hyalodiscus,* a semicircular sheet of this material advances in all directions within a 180-degree arc. This is in contrast to the situation with the large amebae, in which the hyoplasm is a low refractive index material, which is probably produced by syneresis. This latter hyaline fluid appears periodically only in a restricted area of the cell.

DR. BOVEE: Dr. Allen has hit upon exactly the thing I wanted to mention. Most of the amebae I showed in my film had hyaline pseudopods, particularly the one having the long structured pseudopod with the coiled tip. I find it hard to believe that all hyaloplasm should necessarily be fluid. In fact, I would suspect the opposite.

DR. SHAFFER: I entirely agree. The pseudopods and pseudodigits of my cells possess considerable rigidity. My point was only that the surface is more rigid than the interior, for which the best evidence is perhaps the fact that a pseudodigit can continue to grow at its tip while bends in it and particles adhering to its surface move toward its base. But the surface material is derived from the hyaline interior.

DR. REBHUN: I noticed a number of phenomena in your cells. First, the vacuoles undergo this jumping or saltatory movement, but the elongated dark particles also sometimes displayed rather long, saltatory jumps. In many of your cells there were rows of particles on the inner part of the surface which did not change position relative to each other, as if they were being held rather rigid in some form of layer which was not visible.

Next, there were areas in your cells in which the particles did not seem to move relative to each other but around which a fair amount of movement was taking place. However, having seen the film just once, I do not know how widespread this is in your material. It would certainly suggest there is considerable structure, although one can't say if it has anything to do with movement of particles or cells.

DR. SHAFFER: Yes, the particles do sometimes undergo as extensive movements as do the motile vacuoles. As for your other points, there is indeed evidence of considerable structure in the granuloplasm. The inclusions do not saltate in perfectly straight lines, and the contractile vacuoles generally maintain their station in the middle of the cell. It is also true that small areas of the granuloplasm may be relatively quiescent, though only temporarily. However, it is not true that the particles nearest the surface commonly are fixed in position. Whatever organization the granuloplasm has, it differs fundamentally from that of the large amebae in that there is no regular circulation of the inclusions, and no consistency difference between the core and the exterior.

DR. WOLPERT: If the cells in the aggregate do move, would they have the front-to-back contact? Would you be able to put the annular regions on the side? What about the side cells? Isn't the contact there just as good and strong?

DR. SHAFFER: The entire surface of a cell within an aggregate is closely adherent to that of its neighbors. The model proposed allows by far the greater part of the cell surface—front, rear, and sides—to be stationary relative to the surface with which it is in contact. Of course, contact has to be broken whenever surface is removed, but at any instant the surface sinks occupy only a small fraction of the total surface area. Withdrawal of the surface is, indeed, hindered by adhesion to a nonliving substratum, as several observations show; but obviously it cannot be entirely prevented by it, or movement would be impossible. Presumably this applies also to adhesion to other cells.

DR. GRIFFIN: The large carnivorous amebae will follow oil drops on their surface, as Dr. Marsland discussed, and can become trapped behind large internal vacuoles. The behavior of one of your amebae might be accounted for by a similar response to the tail of the ameba in front, caused by a contact reaction or a posterior secretion of acrasin.

DR. SHAFFER: Undoubtedly cells show strong chemotaxis toward rear ends of aggregating cells, if these are not masked by other cells. Once cells are in contact in a chain, surface growth may be regulated not only by acrasin but also by nondiffusible factors, including mechanical ones.

CHAIRMAN MARSLAND: I have frequently observed perfectly hyaline pseudopodia coming out, particularly from the back end of *Amoeba proteus*; they appear just as stable to pressure as granular pseudopodia. I think the idea that hyaline material has the capacity to gelate or assume some sort of structural arrangements is perfectly valid.

Cytoplasmic Streaming and Locomotion in Marine Foraminifera[1]

ROBERT D. ALLEN

Department of Biology, Princeton University, Princeton, New Jersey and Marine Biological Laboratory, Woods Hole, Massachusetts

The Foraminifera (or "forams") are a group of highly diversified protists with complex life cycles. They emit long, filamentous pseudopodia ("filopodia"), which usually branch and fuse to form complex networks ("reticulopodia"). Bidirectional cytoplasmic streaming in all parts of this network is one of its most striking characteristics.

There are a number of cogent reasons for studying cytoplasmic streaming in forams. The first is the intrinsic fascination of these organisms: in no other animal cells can such rapid, widespread, and complex cytoplasmic streaming be seen.

Second, it is a matter of considerable theoretical importance to know what taxonomic relationship the Foraminifera (and the Radiolaria, which show similar streaming) bear to the other ameboid organisms. The gross morphologies and patterns of movement differ, but does this mean that the basic underlying mechanisms are different? Could one mechanism have evolved from the other, or both from a common mechanism? The taxonomy of the sarcodines will remain somewhat obscure until we are in a position to answer these questions.

Third, cellular and developmental biologists are becoming increasingly aware of the important role that autonomous tissue cell movements play in embryonic development in organisms as different as slime molds (Shaffer, 1963), sea urchins (Gustafson, 1963), and amphibians (Holtfreter, 1946). The problem of understanding tissue cell movements runs parallel, in many respects, to those in the sarcodines. Pseudopodia are found in both kinds of cells with nearly every conceivable dimensions and shape, from cylindrical "lobopodia" with rounded or pointed tips to flat sheets (referred to as "hyaloplasmic veils" by tissue culture workers) on the one hand, and filopodia or reticulopodia, on the other.

Neurons, in particular, exhibit a movement pattern which is surprisingly similar to that in forams, except that it is perhaps 100–200 times

---

[1] Supported by a Research Grant (RG-8691) from the Institute of General Medical Sciences, National Institutes of Health, U. S. Public Health Service.

slower (Godina, 1955; Nakai, 1956, 1963). This similarity in movement pattern suggests a similar mechanism. The relative ease with which forams can be grown in the laboratory and handled in experimental studies might lead one to expect that they would be the material of choice as a "model" for neuron motility studies which are basic to an understanding of brain development and perhaps to some important aspects of brain function.

Earlier Descriptive Accounts and Theories of Foraminiferan Movement

There have been many fascinating descriptions of streaming and locomotion in living forams, some written in very poetic terms. For example, in one oft-quoted passage from Leidy's (1879) monograph, the motions of the reticulopodial network were compared to the motions a spider's web might exhibit if it were made from streaming cytoplasm under control of the spider. If the fact were added that streaming in every part of the network is bidirectional, then the description would be remarkably complete. Among the other authors who have described streaming and locomotion in forams are Dujardin (1835), Schultze (1854), Doflein (1916), Sandon (1934, 1944), Le Calvez (1938), Jepps (1942), and Jahn and Rinaldi (1959).

Unfortunately, all of the former studies on this subject have been hampered by one or more of the following considerations:

1. All have dealt only with fully developed reticulopodial networks. In some respects, observations on developing, degenerating, or regenerating networks are more revealing.
2. Until the recent study of Wohlfarth-Bottermann (1961), there was virtually no information available on reticulopodial ultrastructure. Even now, much remains to be learned.
3. Up to the present time there have been no published quantitative studies of the dynamic aspects of streaming or pseudopodial movement.
4. Nearly every investigator has approached the problem of streaming and locomotion with certain mistaken or at least preconceived notions about either the nature of flow itself, the rheological properties of the cytoplasm, pseudopodial ultrastructure, or the molecular basis of contractility. Furthermore, there has been a tendency to describe observations in terms of a particular model or theory, instead of allowing the observations first to stand alone, then be examined in the light of one theory or another.

The theories themselves have evolved about as one might expect. Virtually all the fluids with which we are familiar flow only when pressure is applied to them. It is, therefore, quite natural that biologists with mechanismic leanings would assume, often tacitly, that cytoplasmic streaming must be caused by pressure. However, that this generalization need not be true should have been obvious to anyone who has stirred raw egg-white with a fork. This viscoelastic fluid, and others like it can be displaced by tensile forces, whereas purely viscous fluids require pressure to make them move. It seems now to have been proven in the case of the rotatory cytoplasmic streaming in Characeae that pressure is not the motive force; instead a shearing force is generated at a "gel-sol interface" (Kamiya, 1959; see also papers of Hayashi and Kuroda in this Symposium).

In the case of Foraminifera, it was first pointed out by Sandon (1934) that the velocity of streaming is independent of pseudopodial diameter, and that there was no sign of any sol-gel differentiation similar to that in the ameba. Doubtless, it must have occurred to earlier workers that if pressure were the cause of flow, the pseudopod would have to consist of two tubes, one each to accommodate the inward and outward streams. A single-channel pseudopod could not accommodate bidirectional streaming; yet streaming in both directions does occur in all parts of the network, as Jahn and Rinaldi (1959) have stressed, and these and other workers have asserted that particles in opposing streams collide and reverse. In view of the ultrastructure which has been shown for these pseudopodia, these "collisions" may have been misinterpretations of spontaneous reversals of particles belonging to "streams" which terminate and reverse direction at various points along the network.

The objections to pressure as the cause of cytoplasmic flow in reticulopodia became very convincing after publication of the observations of Jahn and Rinaldi (1959). They stressed the presence of bidirectional streaming in all parts of the network, even in parts of it which had been excised from the body. Furthermore, they observed flow in a loop pattern in the very smallest pseudopodial branches. On these bases, they rejected the pressure hypothesis in favor of an "active-shearing" model which assumed a particular ultrastructure for reticulopodia and an "active-shearing mechanism," the exact nature of which was left open. It was assumed that pseudopodial branches contained one or more folded filaments, the arms of which were contiguous over most or all of their length; a shearing force was assumed to be developed between the two arms, displacing them in opposite directions. As well as making a definite break with the then popular theory of ameboid movement, the active-shearing model offered the possibility of explaining foram streaming along the lines in which muscle contraction (Hanson and Huxley, 1955) and *Nitella* streaming (Kamiya and Kuroda, 1956) had been explained.

The active-shearing theory seemed plausible in the absence of detailed evidence to the contrary from studies either of the ultrastructure or of the dynamics of reticulopodia. However, the electron-microscopic study of *Allogromia laticollaris* by Wohlfarth-Bottermann (1961) has left little doubt that the specific ultrastructure demanded by the active-shearing theory is absent. In fact, both light- and electron-microscopic studies now point to the pseudopodia consisting largely of a loose bundle of fibrils[2] of different sizes separated by empty spaces. If this information is correct, it would be difficult to imagine how an active shearing mechanism could operate over the distances represented by these empty spaces.

Wohlfarth-Bottermann also pointed to the existence of filaments within the fibrils. It seems more likely that contraction might occur at the level of the fibril because of interactions at the filament level. So far, none of the ultrastructural data shed any light whatever on the mechanism of streaming or of contraction if it is involved.

Materials and Methods for the Present Study

The organism adopted for the present study is a new species of *Allogromia* (Figs. 1 and 2) discovered and cultivated by Dr. John Lee. It is currently designated *Allogromia* sp. strain NF (Lee, 1963). The organisms are cultivated on 1% agar in sea water on petri plates sprinkled with milligram quantities of bakers' yeast. Within a few days, the *Allogromia* grow to large numbers.

Single organisms are obtained for study by adding a few milliliters of sea water to a plate and brushing the surface of the agar lightly. The sea water containing dislodged specimens is then decanted into a small dish, from which single organisms can be selected before they become attached again.

Movements are recorded on Plus-X film at either 8 or 16 frames/sec with an Arriflex camera driven by a synchronous motor. Of a variety of optical systems which have been tried, Reichert Anoptral appears to yield the best contrast for both the hyaloplasm and inclusions. Interference contrast and polarized light have also been used, but the results of these studies are not ready for publication.

---

[2] The terms "fibril" and "filament" have been used in many different situations in biology and require some redefinition in each situation. *Fibrils* in this case are membrane-bounded protoplasmic units, barely visible under best optical conditions, several of which make up a pseudopod. These "fibrils" are displaced *in toto* and are not tubes through which cytoplasm flows. Each fibril, in turn, has within it submicroscopic *filaments*.

A Vanguard Motion Analyzer[3] and a Perceptoscope[4] projector were used for plotting the motions of particles.

Results

OBSERVATIONS OF CYTOPLASMIC STREAMING IN *Allogromia* SP. STRAIN NF.

Extension and Retraction of Pseudopodia

Advancing pseudopodia almost always extend straight ahead into the medium from the cell body until they make contact with, and adhere to, the substratum. Once attached, pseudopodia seldom advance any great distance in a straight line; instead they tend to bend at intervals at angles of from 5 to 30 degrees usually in the same direction (Figs. 1 and 2).

Attached pseudopodia with definite bends in them tend to straighten out as the most proximal attachment points are lost. Apparently tension between a more distant attachment point and the cell body is responsible for this straightening. If a retracting pseudopod becomes detached, the pseudopod shows some elastic return, indicating that it was under tension, and then begins to wave about, showing that it is considerably less rigid than when originally extended; sometimes pseudopodia bend at angles, wave back and forth, or form a loose coil, which may flail about briefly before disappearing into the cell body. Smaller branch pseudopodia shrivel and disappear within seconds on becoming suddenly detached.

Pseudopodia which retract slowly show what will henceforth be referred to as the "fibril-droplet transition": although originally most of the body of the pseudopodia has been more or less cylindrical in shape with only a few accumulations of cytoplasm at attachment points and the tip (Fig. 2), on slow retraction, cytoplasm originally in the form of the fibrillar pseudopodial mass becomes converted into the substance contained in "droplets" which may appear at certain places in the network, or may be transported along the network and into the cell body. The conversion of this droplet material to fibrils during pseudopodial extension will be discussed later.

The most obvious difference between extending and retracting pseudopodia is their rigidity; the former are quite stiff and unyielding, whereas the latter are flaccid and bend easily when stressed by weak water currents.

---

[3] The Vanguard Motion Analyzer is available from the Vanguard Instrument Co., 184 Casper St., Valley Stream, New York.

[4] Perceptual Development Laboratories, 6767 Southwest Avenue, St. Louis, Missouri.

Proliferation of Pseudopodia by Branching and Splitting

Newly formed pseudopodia almost invariably form branches diverging at acute angles (usually 15–30 degrees) from the direction of the main tip. The most common method of branch formation is as follows:

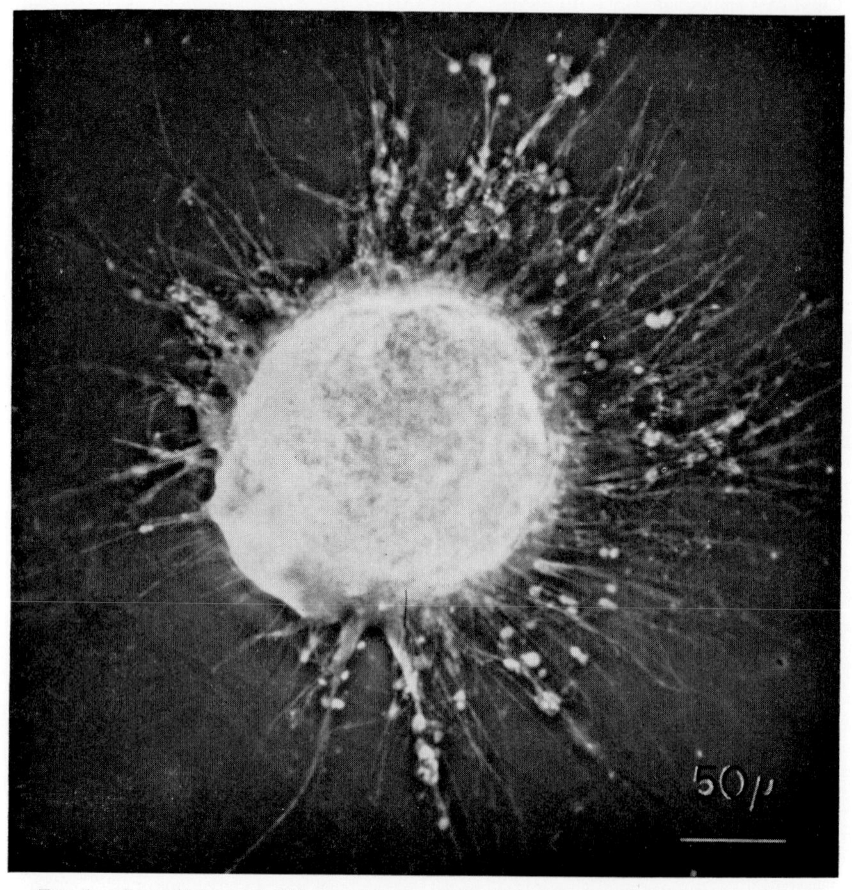

FIG. 1. A specimen of *Allogromia* sp. strain NF. (obtained from Dr. John Lee) feeding on dried bakers' yeast and bacteria. Scale, 50 µ.

after the original "trunk" pseudopod has made one or more slight bends, tension develops in the pseudopod as a whole between the body and the most proximal attachment point. When the pseudopod pulls away from this attachment point, a slender thread of cytoplasm remains (Fig. 2); immediately it begins to grow in length and diameter and serves as a branch pseudopod. We can infer from this either that the branch pseudopod (or a loop of cytoplasm that formed it—see discussion to follow) pre-existed at the attachment point, or that a loop was formed by pulling a straight fibril away from an attachment point. As we shall see later, the former possibility seems more likely.

Branched pseudopodia characteristically split when their branches

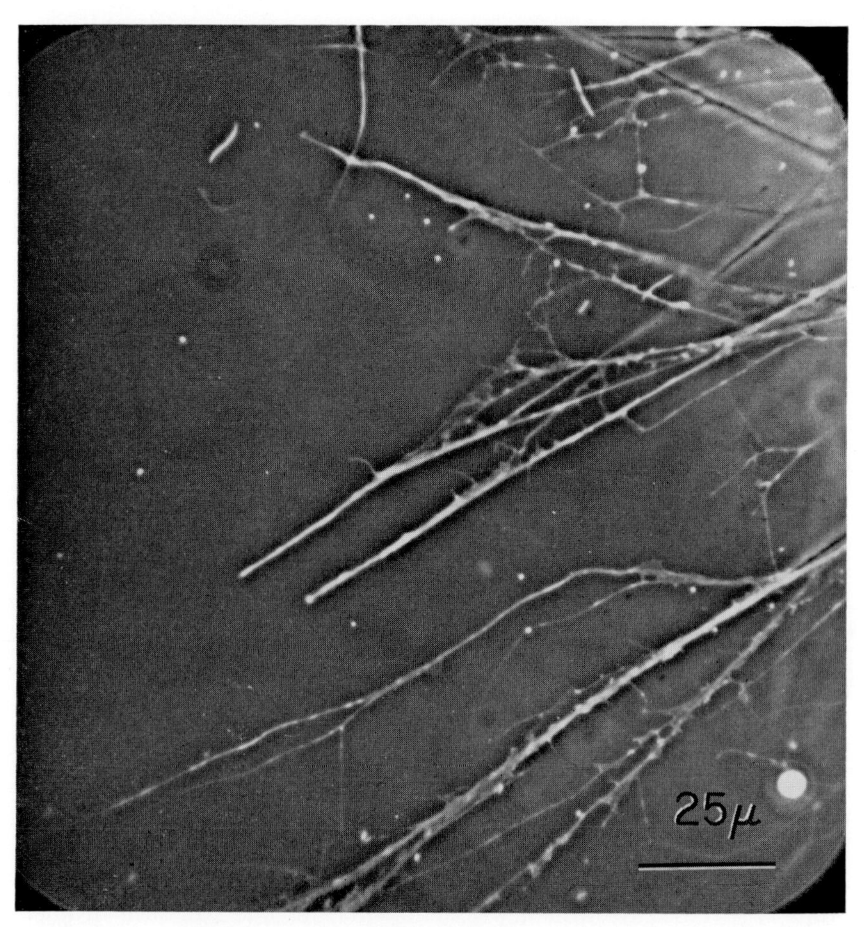

FIG. 2. A portion of the reticulopodial network of a specimen of *Allogromia* sp. strain NF. larger than that shown in Fig. 1. Note the difference in the sizes of various pseudopods, droplets present at tips, junctions, and attachment points; note also small sheetlike areas apparently attached to the substratum. Scale, 25 μ.

diverge at wide angles and develop tension. At the point of splitting, there is usually a thin web of apparently fluid cytoplasm (Fig. 2). Observations on living organisms suggest that the moving fibrils are embedded in some kind of tacky matrix which is apparently not preserved by fixation for electron microscopy.

The splitting process is reversible; tips which have moved apart cause splitting of the trunk pseudopod behind them. If they approach one another, reducing the tension which causes splitting, then the branches fuse again.

Fusion of Branches to Form the Reticulopodial Network

Branches from different trunk pseudopodia may meet at any angle whatever and still fuse. The composite pseudopod then shows particles moving at the velocities characteristic of the two branches that fused; these velocities may be either quite similar or quite different.

A fused branch may show splitting into the original two streams as the result of tension developed within the network. In newly developed networks splitting typically progresses proximally, but in older networks, in which some of the branches point toward the cell body after making several turns in the same direction, splitting may also proceed distally.

The Fibril–Droplet Transition

Filopodia and reticulopodia of forams may appear almost perfectly cylindrical at low magnification, but when examined with high-resolution optics they can be seen to have thickened regions and various other irregularities, particularly at the pseudopodial tips, at attachment points where dropletlike masses are usually present, and at the junction of branches where weblike structures can be seen (Figs. 1 and 2).

For descriptive purposes, I have used the terms "droplet" and "fibril" to denote only the shape of the portions of the cytoplasm that we see in the microscope; their possible functional significance will be discussed later. However, the formation and disappearance of these droplets appears to be an important aspect of cytoplasmic streaming and locomotion in forams. For example, when a pseudopod is diminishing in mass, it does so not so much by decreasing its length as by converting its fibrillar pseudopodia to the form of these droplets which are then transported along the network to the cell body. Presumably the material in these droplets is reused to form new pseudopodia, for one sees such droplets being transported toward sites of pseudopodial extension. In extending pseudopodia, accumulations tend to occur mostly at attachment points and near or at the advancing tip.

There are two experimental situations which demonstrate the fibril droplet transition in an exaggerated form; in one of these its reversibility is particularly well seen.

1. As was pointed out by Sandon (1934), injurious stimuli sometimes cause a pseudopod to break up into a series of short droplets and rods. By chance, we were able to record a case of pseudopodial contact between

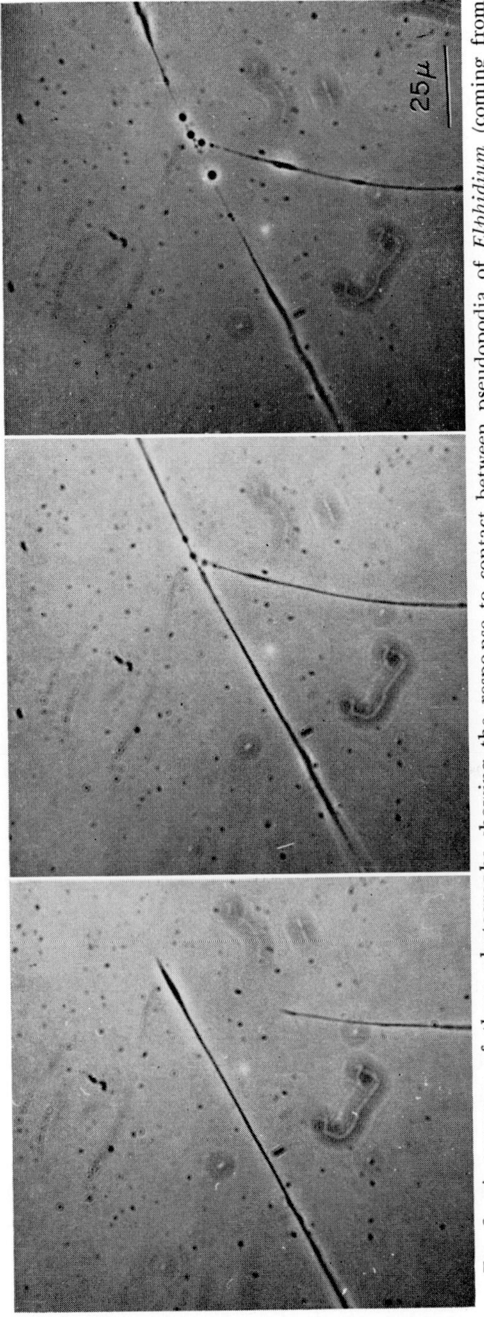

FIG. 3. A sequence of three photographs showing the response to contact between pseudopodia of *Elphidium* (coming from the left) and *Bolivina* (from below) collected at Roscoff, France. Note the formation of droplets in both pseudopodia on contact. Scale, 25 μ.

specimens of the genera *Elphidium* and *Bolivina* at Roscoff, France. In Fig. 3 it can be seen that, although both organisms responded to this contact, in one, the pseudopod rapidly disintegrated into large droplets. Subsequently, one of these droplets was carried away intact, and the material of the others was somehow "spun" into fibrils again.

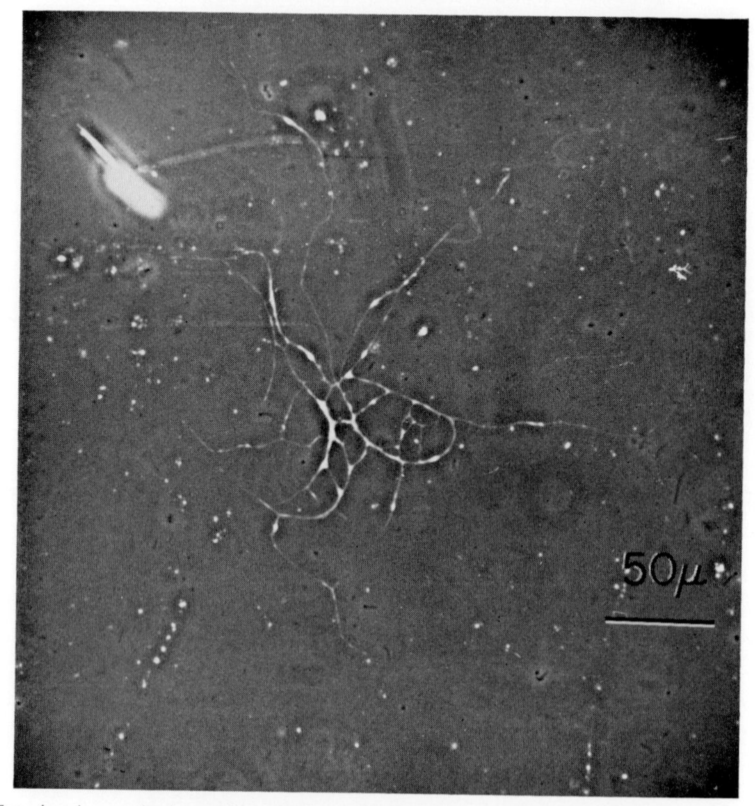

FIG. 4. An excised portion of the reticulopodia network of *Allogromia* sp. strain NF. about 20 min after excision. Note the beginning of droplet formation. Scale, 50 μ.

2. Experimentally excised portions of the reticulopodial network continue to show virtually normal streaming, as Jahn and Rinaldi (1959) showed, but within minutes begin to show the first effects of a gradual degeneration process, the end result of which is the appearance of droplets and thickened regions throughout the network (Fig. 4). During the degeneration process, the fibrillar parts of the network become thinner and show less and less vigorous cytoplasmic streaming. The final stages of this streaming process involves *unidirectional* streaming of a few particles toward a droplet, into which they disappear. If an intact portion

of the network makes a chance contact and fuses with the excised portion, there is an immediate response. The material within the droplets is "rescued" by being rapidly reincorporated into the fibrillar organization of the network, and bidirectional streaming is again established throughout the whole fused network.

A Preliminary Analysis of Particle Motion Velocities in Reticulopodia

Although the initial impression one receives of the streaming in the reticulopodial network of a foram is one of extreme complexity, there

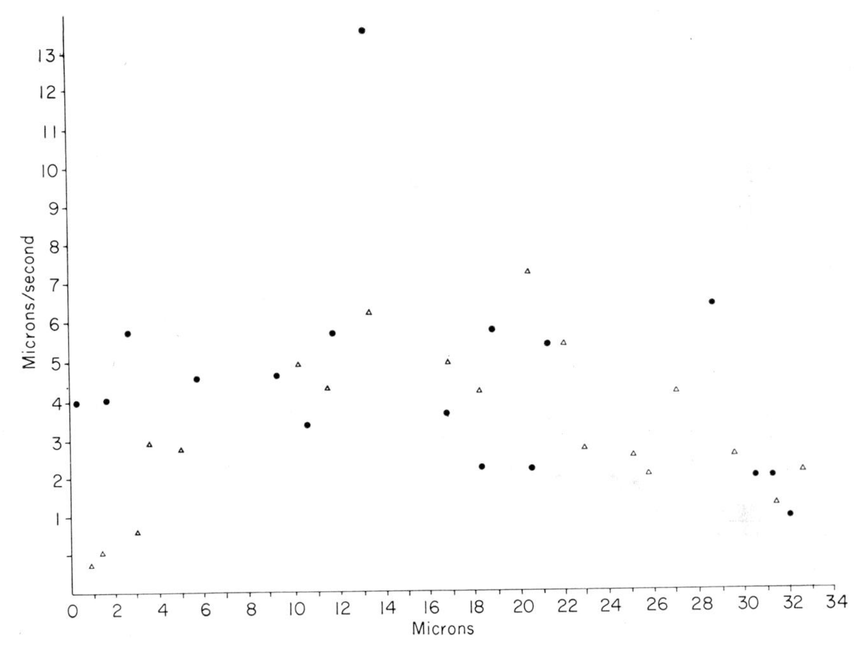

FIG. 5. A record of the velocities exhibited by various particles during a 5-sec interval of time along a section of a trunk pseudopodium between two attachment points 33 μ apart. The one closer to the body is the origin. Note the two distally moving particles which are traveling at a velocity several times that of the slowest particles.

are certain features of this streaming which appear to be universal and, therefore, probably important.

1. The tips of branch pseudopodia or unbranched filopodia may extend, remain stationary, or retract, but in any event bidirectional streaming continues.

2. Branch pseudopodia from the same trunk pseudopod quite often exhibit streaming at different velocities.

3. Large, trunk pseudopodia may show at one time particles traveling at different velocities. The most rapid particles may be moving at several times the velocity of the slowest ones (Fig. 5).

4. Larger pseudopodial branches (over 1 μ in diameter) usually show particles moving at several velocities in both directions (Fig. 6).

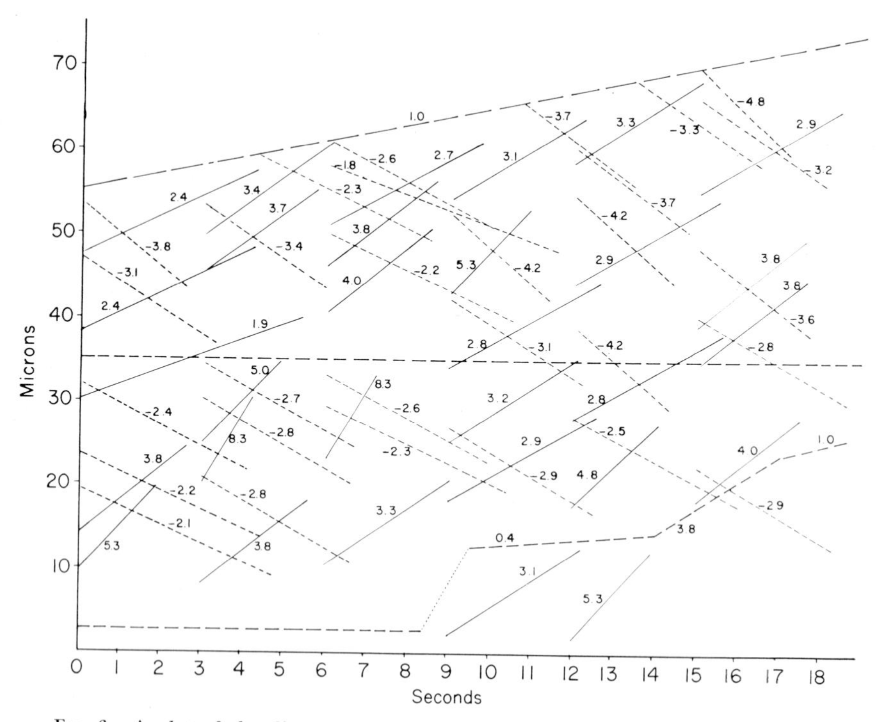

FIG. 6. A plot of the distance traversed in time by the tip of a pseudopod 3 μ in diameter (— —), with various distally (———) and proximally (- - - - - - - -) moving particles. The attachment points and the times at which they yielded to tension transmitted along the pseudopod are shown as broken lines parallel to the x axis. (Figure 8 shows the morphology of the pseudopod during the period in which these measurements were made.)

5. In larger branches, a single particle may be found to change its velocity abruptly in the vicinity of points of attachment to the substratum (Figs. 6 and 7). Frequently these attachment points can be discerned only after the pseudopod pulls away from them, leaving new branch pseudopodia (Fig. 8).

6. In the exceedingly fine pseudopodial branches of young organisms (Fig. 9), the velocity at which the tip advances is not nearly so uniform as in larger pseudopodia (cf. Figs. 7 and 10). The velocity of

particles is much more variable and can be followed accurately except at the very tip where we have never been certain that we have tracked the same particle passing over the tip and returning. Characteristically there is a bulge or a droplet at or near the tip into which the particles disappear. Some appear to come out again after only a little delay, but

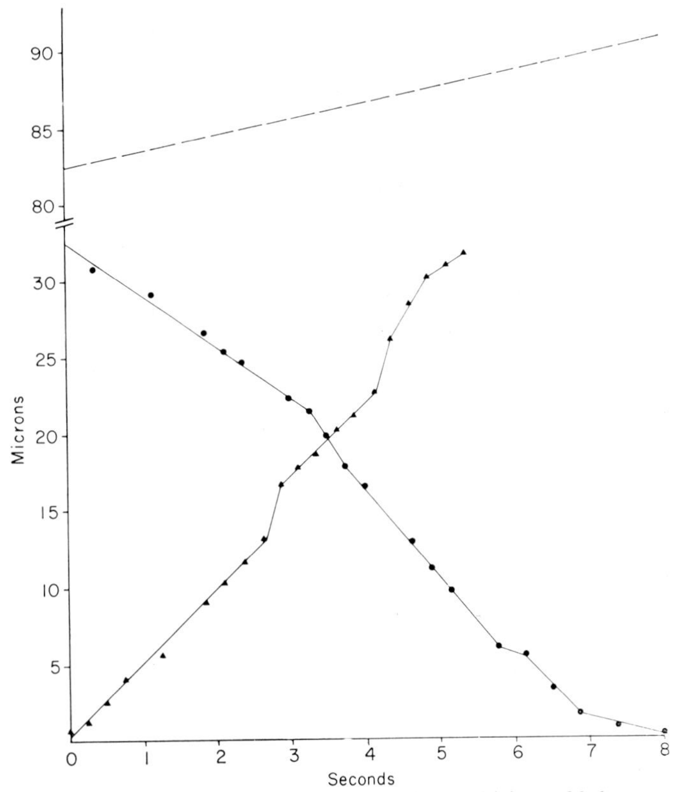

FIG. 7. A record of the motions of two particles which could be traced for a period of 8 and 5 sec, respectively, in relation to the advancement of the tip. This analysis was made on the same pseudopod as that in Fig. 6. (Figure 8 shows the morphology of the pseudopod during the period in which the measurements were made.)

others remain in the droplet for some time. Particle e in Fig. 10 illustrates this problem: it may or may not be identical with particle g. If so, then it returned from the tip much more slowly than it approached it. We thought we could see this quite often in casual observations, but when faced with the problem of recording and analyzing such an event, we could not be certain of having kept the same particle in view while passing over the tip region.

7. It was first noted in the analyses, and subsequently confirmed in casual observations, that a number of peculiar events take place at attachment points. First, there is characteristically a thickening there which does not remain constant in size or shape (Figs. 2 and 8). Second, particles enter the attachment-point region and characteristically speed up and slow down; they may then either (1) pass on (usually at a differ-

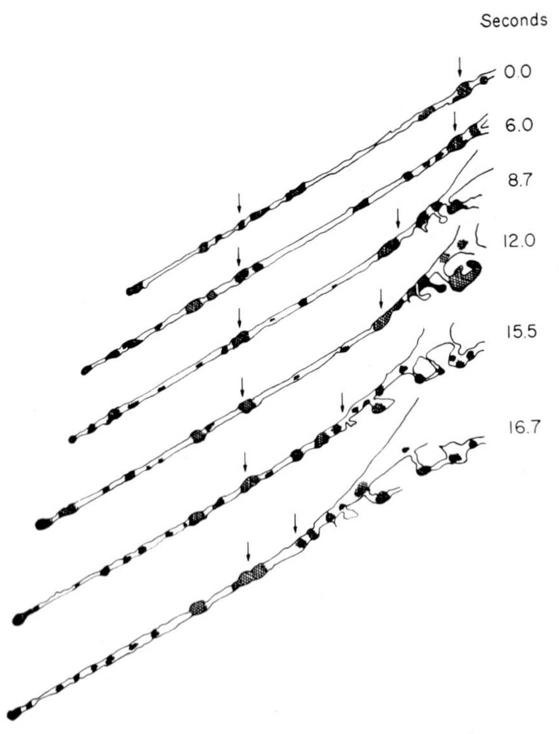

FIG. 8. Tracings from the motion picture records analyzed in Figs. 6 and 7. Arrows show positions of known attachment points revealed by the presence of a branch pseudopod as the main pseudopod was pulled away.

ent velocity as in Figs. 6, 7, and 10), (2) disappear in the thickened cytoplasm at the attachment point, or (3) reverse their direction.

Each organism is somewhat different from the last; therefore, it cannot be stated that these analyses will apply to all organisms similarly analyzed in the future. However, the features pointed out above have been seen enough times to instill some confidence that they are widespread, and, therefore, of importance. Data of this type are the raw material of which intelligent models to explain this and other kinds of motility must be made.

Cytoplasmic Streaming in, and Movement of, the Cell Body

The cell body of *Allogromia* sp. strain NF. is so transparent that cytoplasmic streaming can be seen at the entire periphery by changing the level of focus of the microscope. The pattern of streaming within the body is very complex. Streaming seems to be most active in the

FIG. 9. A portion of the network of a young organism (*Allogromia* sp. strain NF.) showing pseudopodia close to the resolution limit of the microscope emanating from a thin sheet of hyaline cytoplasm traversed at points by barely visible fibrils. Particles streamed along these fibrils (not elsewhere on the hyaline sheets) and out into the pseudopodia. Some remained stationary while others moved. (The analysis shown in Fig. 10 was made of one of the pseudopodia in the group shown, but at a later time when the morphology had changed somewhat.) Scale, 10 μ.

cortical region. Often the cytoplasm contains many clear vacuoles, around which thin streams of particles stream in a tortuous path, as if the whole cortical region were a mass of reeled-in pseudopodial fibrils. The velocity of streaming in the cell body is not significantly different from that in the pseudopodia.

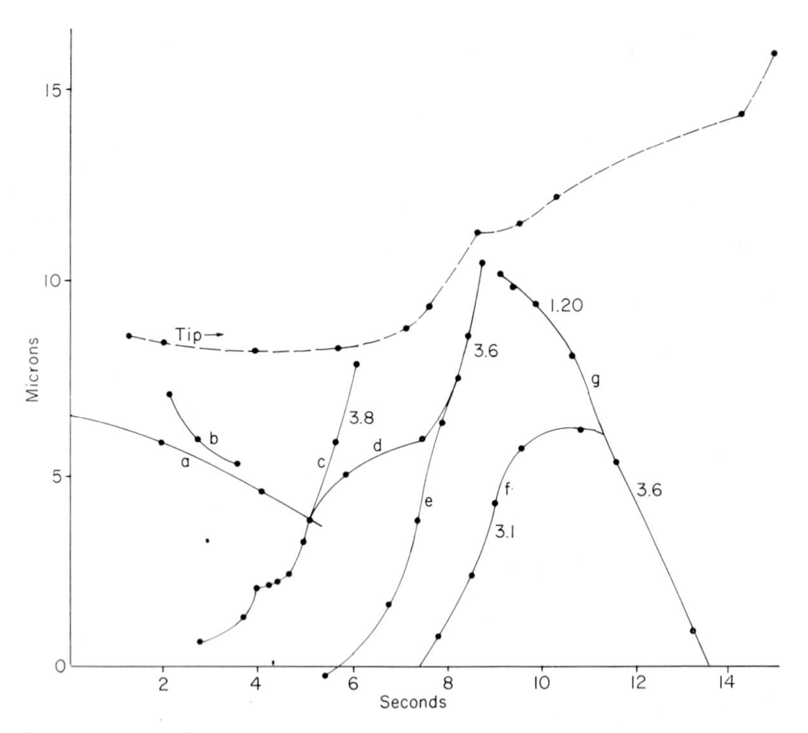

FIG. 10. An analysis of the advance exhibited by the tip of a reticulopod such as shown in Fig. 9, and the motions exhibited by particles in or on the pseudopodium. Note the irregular tip velocity (cf. Figs. 7 and 8) and the changes in the velocities of particles advancing to the same region of the pseudopod (particles d and f) while particles c and e went through the same region unimpeded. It is not certain whether or not particles e and g are identical because e disappeared momentarily into a droplet at the tip of the pseudopod.

The cell body exhibits independent "ameboid" movements which are apparently not related to events in the reticulopodial network. In fact, some cells have been observed to move slowly in this manner without any pseudopodia. The cell becomes deformed, but not continuously as is the case with amebae. Instead, the cell surface moves in "lurches," more or less as it would be expected to if some kind of strain were applied to the cell surface until a yield point was reached.

Discussion

Reticulopodial Ultrastructure

There are at least seven separate lines of evidence that bear on reticulopodial ultrastructure.

1. The obvious mechanical rigidity of extending pseudopodia is particularly remarkable in the light of their fineness. This must indicate the existence of structural anisotropy, perhaps fibrillar or filamentous structures.

2. The optical anistropy expected of structurally anisotropic material is present. Schmidt (1937) reported positive, axial birefringence in *Miliola* filopodia. Using Dr. Shinya Inoué's rectified, oil-immersion system, I was able to see the birefringence of the network and noticed that certain sections of it increased gradually in birefringence over a few seconds, but then suddenly became isotropic again. This could indicate the development of temporary strain in elastic structures which was relieved by flow.

3. It is even possible to see with the light microscope some of the fibrillar structure in trunk pseudopodia at places where they spread out. As Jahn and Rinaldi (1959) reported, the particles in a particular region follow tracks at the same velocity; in addition, under good optical conditions one can see the fibrils themselves.

4. The dynamic behavior of reticulopodia, especially their branching, splitting, and fusion, all point to a fibrillar organization of the network.

5. The recent electron-microscopic study of Wohlfarth-Bottermann (1961) has confirmed the suspicions of earlier authors that the reticulopodium does, in fact, consist of a loose bundle of fibrils of different sizes surrounded largely by empty space (empty, at least, after fixation and exposure to the electron beam). Each fibril is a protoplasmic unit containing such structures as mitochondria and surrounded by a unit membrane. The electron microscope also has revealed a filamentous ultrastructure for the fibrils.

6. The behavior of the pseudopodia in fusion, splitting, and branching suggests that there may be a tacky substance or "glue" or some sort which keeps the membrane-bounded fibrils, of which the pseudopod is composed, together. If this is true, this material does not survive the preparatory procedures for electron microscopy.

7. The fiber-droplet transition is a feature which tells us little, so far, about reticulopodial ultrastructure other than that the fibrils are unstable under certain local physiological or environmental conditions.

We need a great deal of further information about the structure of the droplets when forming and when disappearing.

Although there remain many fundamental uncertainties regarding the details of reticulopodial ultrastructure, a concept is beginning to emerge which offers an explanation for several previously incomprehensible aspects of foram movement and behavior. Jahn and Rinaldi's (1959) concept of the smallest branches being made up of one or more folded fiber loops would seem to be probably correct except that each fiber (and not the whole filopod) is surrounded by a membrane. The only reservation concerning this concept springs from the fact that the existence of a true loop at the filopodial tips remains to be proven, either morphologically or by a clear demonstration of a single particle following an uninterrupted U-shaped path at the tip.

One of the most puzzling observations which this new concept explains is the flow of membrane in different directions in different parts of the pseudopodial surface. As long as the pseudopod was considered as a single protoplasmic unit, the alternative explanations of this involved either high-velocity gradients within the membrane, or the absence of a membrane—an interpretation which Jahn and Rinaldi (1959) correctly pointed out was "physiological heresy."

The manner in which branches form at attachment points suggests that not all the folded fibrils terminate at the tips of branches. The termination of some of these loops at points along the trunk pseudopods and branches would explain the gradual (and apparently stepwise) diminution of diameter toward the periphery and the instantaneous appearance of branches when a trunk or branch pseudopod is pulled away from an attachment point.

The Fibril-Droplet Transition

Although the appearance of droplets due to stimulation or injury had been noted in passing by earlier workers (esp. Sandon, 1934), the normal occurrence of a similar phenomenon throughout the network during establishment and destruction of the network has not received the attention it deserves. The observations presented here would suggest that it is a phenomenon of central importance. Unfortunately, we do not know in which of several ways it should be interpreted.

It must be borne in mind that what we have described as "droplet formation" is merely deformation of a body which could be either an active (i.e., contractile) or passive process. If passive, it could represent either deformation due to "solation," that is, to decreased internal resistance to tension at the surface of the fibrillar unit (i.e., surface tension and/or elastic forces); alternatively the fibrillar unit might be "reeled in" intact at attachment points and other places and stored in droplets to be later "paid out" again.

In interpreting the fibril droplet transition, it should be emphasized that in the extreme case the whole filopod breaks up into droplets, usually connected by a fine fibril (Fig. 3). This situation is quite reminiscent of the breakup of *Amoeba proteus* pseudopodia into droplets on sudden application of very high hydrostatic pressures (Marsland and Brown, 1936). In the latter case, the breakup appears to be due to surface tension acting on a cylindrical mass of fluid that has suddenly lost its gel structure. However, we cannot exclude the possibility that the foram fibrils contract actively into the little droplets which we see. In normal reticulopodial activity, only some of the fibrils in any part of the network transform into droplets. Under these circumstances, it is difficult to see why solated fibrils should respond to surface tension when the normal tackiness of fibril units in a pseudopodium would suggest that adhesive forces should tend to keep them elongated.

It is probably premature to draw any serious analogies between the fibril droplet transition and simple, abstract models of complex rheological events in ameba such as the Landau (1959) scheme for "sol-gel reactions." It should be pointed out, however, that the application of this model would place the following restrictions on events in the organism: (1) Since the model assumes that contracted gel must solate before it can contract again, droplet formation would have to occur only in contracted material; (2) solated material in the droplet would have to be formed into fibrils again without the aid of contractile forces; and (3) contraction could play no role in the droplet→fibril direction of the transition.

A second analogy that could be made is to actomyosin sols and gels on the addition of adenosine triphosphate (ATP). Actomyosin sols show a viscosity drop on the addition of ATP but cannot perform mechanical work or develop tension because they lack elastic structure, being in the sol state. Actomyosin gels, on the other hand, show contraction and syneresis on the addition of ATP, but do not solate between contraction cycles. These analogies, then, are also not very helpful.

A third possible analogy might be to the behavior of intact muscle, in which the primary contractile event is an elasticity change resulting in shortening and stiffening of elastic components, but which also involves an increase in the viscous resistance against which the muscle must work (Gasser and Hill, 1924; Pryor, 1952). The elasticity change of intact muscle is imitated by muscle models, and by actomyosin gels and fibers, but the contractility of these model systems does not necessarily involve the same changes in viscous resistance.

While not wishing to propose a formal model to explain the consistency changes associated with cell movement here, I would like to point out that it may be unnecessary to assume, as it is assumed in the Landau model (and others) that a change of viscosity is a necessary and integral part in the recovery from contractile processes. This concept originally grew out of the Mast (1926) theory of ameboid movement, which rests on rheological interpretations which have been questioned (Allen, 1960, 1961a). It may be more generally true that contractile events involve primarily only the elastic components of cytoplasmic structure; what has been termed solation may involve changes in the viscous properties, stiffness, and yield point of the material, and may be either secondary to or unrelated to the contractile event.

The Mechanism of Pseudopodial Extension and Cytoplasmic Streaming

The readily observed fact that the reticulopodia exert tension and perform work would suggest strongly that they are contractile. The absence of close association between fibrils in a pseudopod seems to rule out an active-shearing force between fibrils as an explanation of contractility at the pseudopod level. Therefore, it must be the fibrils themselves that are contractile, probably through events occurring at the level of the filaments, observed by Wohlfarth-Bottermann (1961), or below.

Leaving aside for the present all problems of how contractility occurs at the submicroscopic level because of lack of evidence, let us see how pseudopodial extension and cytoplasmic streaming could be explained in terms of fibrillar contractility.

The growth of a rigid filopodium indicates that some kind of gel structure must be built up in it. This could be done either by addition and stiffening of less rigid material (not necessarily sol!) on either the proximal or the distal side. The first possibility can be rejected as it would result in only forward cytoplasmic streaming in the advancing pseudopod. The addition of gelated material at the tip in the same manner as in the ameba would account for bidirectional streaming unless the gelled part of the fibril were attached to the substratum.

The model by which, in principle, a folded contractile fibril can displace its two ends in opposite directions by a propagated contraction anchored at the bent portion has been described in detail elsewhere (Allen, 1961a,b).[5] In the case of pseudopod extension in forams, it is

[5] A further discussion of the mechanics of the "contraction anchored at a bend principle" is to be found in the Free Discussion following Parts III and IV of this Symposium.

only necessary to assume that a contraction is anchored at the bend of the folded fibril (in this case at the tip of a pseudopod) so that the material passing away from the tip is contracted, and that passing toward the tip is relaxed and about to contract. If we assume that the stiffness of the "springs" in contracted fibrils increases as in muscle, then we have a satisfactory explanation for the existence of a stiffened gel rod in newly formed pseudopodia.[6] A drop in the stiffness of this rod on relaxation could account for the relative flaccidity of pseudopodia in retraction.

The "contraction anchored at a bend principle" described previously and elsewhere also serves as a convenient explanation of bidirectional streaming throughout the network. The force of contraction is applied as tension to one side of the loop and compression to the other side, forcing the two sides to move in opposite directions. This model has the same advantage as the active-shearing hypothesis in that it invokes directed forces (in this case tension and compression) rather than the nondirectional force of pressure which would be clearly inapplicable here. The contraction model has a distinct advantage over the active-shearing model in that it assumes no ultrastructure which cannot be demonstrated, and it invokes no new and mysterious processes or forces. This is not, however, to deny the possible existence of such forces at the filament level or below; at present we have no information about what occurs at that level.

An apparent disadvantage of the contraction hypotheses is that one would expect the resistance to movement to increase with the length of the pseudopod if all the force were applied at the tips of branches. Instead of slower motion, long pseudopodia seem to show the most rapid motion (cf. Fig. 5). To account for this it seems probable that long pseudopodia (and short ones as well) must have "booster engines" at intervals along them to keep the fibrils moving. A series of folded filaments extending different lengths along the pseudopodium would, in effect, deploy just such a series of booster engines (in the form of the bent regions of folded fibrils). When such a long pseudopod is pulled laterally by another pseudopod, it leaves behind tiny branch pseudopodia

[6] Neither the present author, nor Jahn and Rinaldi (1959) were able to confirm reports in the early literature concerning the existence of a "stereoplasmic rod" in the reticulopodia of the marine foraminifera examined. It seems likely that earlier authors (e.g., Doflein, 1916 and others) were misled by diffraction artifacts and strong convictions about the necessity for a rigid rod to provide the observed stiffness. At present all that need be assumed is that at least one of the fibrils is rigid; since it is not discernible in phase contrast, it very likely does not differ significantly from other fibrils in refractive index.

at intervals of from 10–30 μ apart, showing the previous location of these boosters.

It is quite clear that there occur along the network regions in which particles move in both directions at a wide variety of velocities, some several times as rapid as the slowest movement. There are also some particles which are stationary. The slower particles can be explained as belonging to folded filament loop units extending for different distance out into the network; some doubtless extend out to the tips, while others terminate at various attachment points. The fastest particles (some over 13 μ/sec) are difficult to account for unless we assume some kind of "pick-a-back" mechanism in which one folded fibril unit functions while itself in motion riding on another one, so that the maximum streaming velocity observed is the sum of the velocities gained from the two (or perhaps more) units combined. Such a situation would not be hard to imagine in a loose bundle of fibrils and could easily give rise to the variety of velocities often observed.

Using the convenient contraction at a bend principle we could almost construct a tentative model of reticulopodial dynamics, except for the fact that we know so little about the fibril droplet transition. In general, these droplets must represent "reservoirs" of "fibril-forming materials"—or, perhaps, folded fibrils themselves which have been reeled in for storage. In the latter case it would be easy to postulate a way in which they could be paid out to form the rigid structures present in pseudopodia. However, if the fibrillar material solates completely in forming droplets, it would be difficult to imagine how this could be spun out again by tension unless there were enough elasticity left to orient filaments into a form that could later contract. Similar problems are encountered in the organization of the mitotic spindle (Inoué, 1959; see also paper in this Symposium).

The maintenance of bidirectional streaming into and out of expanding and retracting parts of the network would seem to require some activity on the part of the cell body, at least to the extent of reeling in contracted arms of folded filaments, which as we have suggested may be "pushed" by the compressive force of the contraction at a bend mechanism toward the cell body. The presence of a complicated pattern of cytoplasmic streaming within the cell body would seem to indicate the possibility that the cell body participates to this extent. However, it must be recalled that streaming can continue for one to several hours in the absence of the cell body, but sooner or later fails.

As stated previously, the author's tentative model based on the contraction at a bend principle is incomplete in many respects. However, it has the advantage of offering a similar model to that previously suggested for the *Chaos-Amoeba proteus* group of amebae. We still lack sufficient data on the mechanical properties of the cytoplasm in either organism to construct a detailed mechanical model to illustrate exactly how the principle would work. The mechanical conditions are somewhat different in the cases of amebae and forams; in amebae one would expect the lateral components of the contractile forces to balance out because of the radial symmetry of the pseudopod. In the case of forams we have no reason to expect that such a symmetry exists; therefore it seems likely that pseudopodia should be found to bend when the contractile force exerted at the tip exceeds the stiffness of the pseudopod. This may explain the pronounced tendency of foram pseudopodia to bend at intervals, typically at attachment points. A close match between the lateral component of the force of contraction and the stiffness might easily result in the activities which have been reported for unattached pseudopodia, such as bending, waving, flailing about, and in extreme cases coiling.

Summary

The cytoplasmic streaming, locomotion, and general behavior of marine Foraminifera with reticulopodial networks has been discussed in general and with particular reference to observations carried out on *Allogromia* sp. strain NF.

The principal processes in the establishment of the network are pseudopodial extension, attachment, branching, splitting, fusion, and retraction.

Cytoplasmic streaming is bidirectional in all parts of the intact network as asserted by earlier workers, but unidirectional streaming may occur in excised portions. Data are presented to show how the velocity of particles varies along different parts of the network. Changes in velocity are particularly obvious at or near attachment points.

An important aspect of reticulopodial dynamics is the transition between fibrillar material of the pseudopod and droplets which appear at attachment and branching points, at or near the tip, and at various other points along the network. The possible nature of this transition is discussed.

Two experimental situations were observed in which fibrils are readily converted into droplets. One case involved an intergeneric pseudopodial contact; the other occurred in excised portions of a network. In one case the recovery of material in droplets formed in an excised portion of network was effected when a portion of the original network fused with it and thus "rescued" it from further degeneration.

The possible mechanisms of pseudopodial extension and cytoplasmic streaming are discussed in the light of the present work and of a recent electron-microscopic study indicating that the pseudopodial structure is basically a loose bundle of fibrils of different sizes containing filamentous material. On the basis of this and other information, the active-shearing model is rejected in favor of an incomplete, but more attractive model based on fibrillar contractility.

Pseudopodial extension and bidirectional cytoplasmic streaming can both be explained in terms of the contraction anchored at a bend principle applied to a folded contractile fibril. Long pseudopodia consist of many folded fibrils, many of which terminate along the pseudopodia. These may serve as booster engines and help overcome the added friction involved in streaming in long pseudopodia. Multiple velocities observed in trunk pseudopodia may require a "pick-a-back" deployment of folded fibrillar units, in which some function while themselves riding on top of others.

The model which is proposed rests on the same mechanical principle previously proposed as a partial explanation of ameboid movement in the *Chaos-Amoeba proteus* group.

Acknowledgments

We wish to acknowledge the assistance of Mr. W. Reid Pitts, Jr., Mrs. Eleanor Benson Carver, and Mrs. Prudence Jones Hall in analyzing the motion picture films.

References

Allen, R. D. (1960). *J. Biophys. Biochem. Cytol.* **8**, 379.
Allen, R. D. (1961a). *In* "The Cell" (J. Brachet and A. E. Mirsky, eds.), Vol. II, pp. 135-216. Academic Press, New York.
Allen, R. D. (1961b). *Exptl. Cell Res. Suppl.* **8**, 17.
Doflein, F. (1916). *Zool. Jahrb. Abt. Anat. Ontog. Tiere* **39**, 335.
Dujardin, F. (1835). *Ann. Sci. Nat. Zool.* **14**, 343.
Gasser, H. S., and Hill, A. V. (1924). *Proc. Roy. Soc.* **B96**, 308.
Godina, G. (1955). *Atti Accad. Naz. Lincei Rend. Sci. Fis. Mat. Nat.* **18**, 104.
Gustafson, T. (1963). This Symposium.
Hanson, J., and Huxley, H. E. (1955). *Symp. Soc. Exptl. Biol.* **9**, 228.
Holtfreter, J. (1946). *J. Morphol.* **79**, 27.
Inoué, S. (1959). *In* "Biophysical Science" (J. L. Oncley *et al.*, eds.), pp. 402-408. Wiley, New York.
Jahn, T. L., and Rinaldi, R. A. (1959). *Biol. Bull.* **117**, 100.
Jepps, M. W. (1942). *J. Marine Biol. Assoc. U.K.* **25**, 607.
Kamiya, N. (1959). *Protoplasmatologia* **8**, 1.
Kamiya, N., and Kuroda, K. (1956). *Bot. Mag. (Tokyo)* **69**, 546.
Landau, J. V. (1959). *Ann. N. Y. Acad. Sci.* **78**, 487.
Le Calvez, J. (1938). *Arch. Zool. Exptl. Gen.* **80**, 163.
Lee, J. J., and Pierce, S. (1963). *J. Protozool.* **10**, 404.
Leidy, J. (1879). *U.S. Geol. Surv. Rept.* **12**, 324 pp.
Marsland, D. A., and Brown, D. E. S. (1936). *J. Cellular Comp. Physiol.* **8**, 167.
Mast, S. O. (1926). *J. Morphol.* **41**, 347.

Nakai, J. (1956). *Am. J. Anat.* **99**, 81.
Nakai, J. (1963). This Symposium.
Pryor, M. G. M. (1952). *In* "Deformation and Flow in Biological Systems" (A. Frey Wyssling, ed.). Amsterdam.
Sandon, H. (1934). *Nature* **133**, 761.
Sandon, H. (1944). *Nature* **154**, 830.
Schmidt, W. J. (1937). *Protoplasma* **27**, 587.
Schultze, M. (1854). "Uber den Organismus der Polythalamien." Leipzig, Germany.
Shaffer, B. (1963). This Symposium.
Wohlfarth-Bottermann, K. E. (1961). *Protoplasma* **54**, 1.

Discussion

Dr. Jahn: I would like to commend Dr. Allen on these very excellent motion pictures. This is the second time I have seen them, and I have been unable to detect any differences between the two species of *Allogromia*. So far as I know, anything that can be said about one can be said about the other. I would confirm the statement of mine that Dr. Allen quoted to the effect that particles go out to the tip, turn around, and come back.

I would also like to underscore the point that when the cell body is being moved by the pseudopodia, there is two-way movement in all of the pseudopodia. In a large pseudopodium, perhaps 80% of the granules may be going in one direction; but if the microscope is focussed carefully, the other 20% of the granules will be found going in the opposite direction. There is always two-way streaming.

One other point I would like to make is that I am not disturbed by the fact the electron microscopists have not yet discovered the "legs of the millipede," because they have also not discovered whatever causes active sliding in *Chara, Nitella,* in the stamen hair cells of *Tradescantia,* and elsewhere where it has been described. Possibly a single mechanism will turn up sooner or later that will explain all of these phenomena. It is also possible Dr. Allen has a good idea on this subject.

Dr. Allen: Since you brought up the active-shearing idea, I should perhaps point out that the fibers which make up a pseudopod, and which are moving in opposite directions, consist apparently of cytoplasmic elements, each of which is surrounded by a membrane. It is rather difficult to imagine how your "millipede legs" could be acting between these membranes, which are separated (at least in fixed material) by several millimicrons.

Chairman Marsland: It is apparent from the film that hydrostatic pressure could not explain protoplasmic streaming in Foraminifera.

Dr. Lee: I have been looking at these Foraminifera for some time, and am interested in their mechanism of movement. Polyoral organisms are even more interesting than these monoral ones; there is always a dominant "head" or "chief" region which advances in one direction and all the other "heads" follow behind. It looks to me like a pulling action in which one chief region becomes stronger than the others.

In these polyoral individuals, occasionally the head end will bud off; about a half hour before this head end is budded off, a new chief takes over and the old head end then begins to trail behind until it is budded off.

The feeding process is also very interesting. These form "pools," as I like to call them, at the end of pseudopodia which are effective bacterial traps. One sequence you saw in the film was really very beautiful. The trap spreads over the bacterial surfaces and all the bacteria are somehow caused to adhere and get carried back toward the body.

CHAIRMAN MARSLAND: According to this hairpin loop contraction idea that you have proposed, wouldn't it necessitate some backward movement in the contracting part of the loop as well as the forward movement?

DR. ALLEN: The mechanism I proposed requires relative movement of the two arms of the loop in opposite directions. Movement relative to the substratum depends on what part is attached. If, for example, the contracted arm were attached, then the bend would be expected to advance. The contracted arm would be "built forward" by new material becoming incorporated and stiffened at the bend.

CHAIRMAN MARSLAND: Yes; but wouldn't there have to be just an approximation of the particles as a result of the contraction which would be toward the point of attachment?

DR. ALLEN: They should come closer together in approaching the bend, or at any other place where contraction is postulated. Of course, approximation of particles only *suggests* contraction; physical evidence is required to demonstrate a contraction.

Filopodial Movement in *Cyphoderia ampulla* (Ehr.)

ROBERT E. BERREND[1]

Department of Zoology, University of Wisconsin, Madison, Wisconsin[2]

Filipodia of *Cyphoderia ampulla* (Ehr.)

Cyphoderia ampulla (Ehr.) (Fig. 1) is one of the Testaceafilosa De Saedeleer, 1934. Together with the genus *Campascus* Leidy, it forms the family Cyphoderiidae Deflandre, 1953. The test of the animal is colorless to yellowish, and is composed of small round or elliptical scales. It is flask-shaped in outline with a tapering neck curved in such a way that

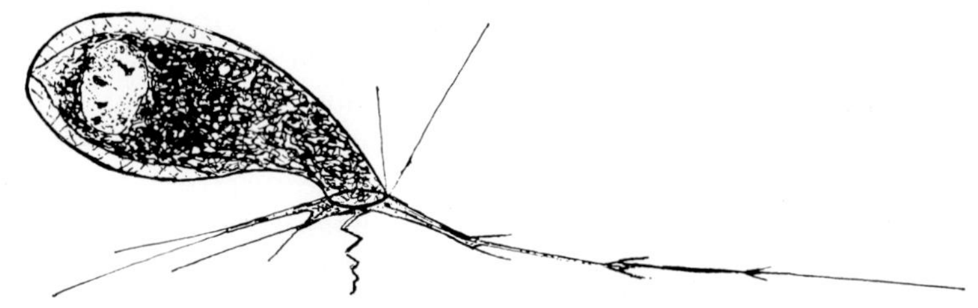

FIG. 1. *Cyphoderia ampulla* (Ehr.).

the aperture is oblique to the main axis of the test. The cytoplasm, densely filled with reserve scales and glycogenic grains does not completely fill the test. The large nucleus is located in the posterior portion of the cell body. Slender ectoplasmic filopodia originate from a mass of cytoplasm—the podostyle—which protrudes from the mouth of the test.

It is convenient on the basis of locomotion and the manner of deployment to divide the filopodia of *Cyphoderia* into three types: directive, nondirective, and secondary. The directive filopodium determines the direction in which the animal moves. It is of interest to note that if two directives should be present, the animal moves along a line which bisects the angle between the two. It is obvious that this situation cannot exist for long unless the two directives are nearly parallel; the larger the angle of divergence the sooner the two directives become opposed to one

---

1 *Editor's note:* In the absence of Dr. Berrend, his paper was read at the Symposium by Dr. Lowell E. Noland of the University of Wisconsin, under whose supervision the study was carried out.

2 Present address: San Francisco State College, San Francisco, California.

another and forward movement ceases. Movement begins again when dominance is gained by one of the two, or a new directive originates. Nondirective pseudopodia vary in number from many to, rarely, none. They are more abundant when the animal is moving slowly, is on the underside of a surface, or when the surrounding water is in motion. Secondary pseudopodia spread out laterally from either of the two primary types of filopodia and appear to serve as holdfast devices.

Directive Filopodia

There is a tendency among nonplanktonic ameboid Protozoa to utilize a monopodial or some functional equivalent of a monopodial form of movement. The more pronounced this monopodiality, the greater the distance covered and the greater the area over which the animal may browse. If we postulate an ameboid form with a radial distribution and random dominance of pseudopodia, it is evident that, for the organism to move any distance, some degree of dominance must be maintained for a period of time. The longer the time through which the dominance acts the further the excursion of the organism in that direction. In the absence of trophic responses a random origination of dominant directive pseudopodia will tend to restrain the animal within an area inversely proportional to the rate of replacement of one directive by another. Actually most ameboid forms established an anterior end by various means (e.g., uroid, bilateral tests). In undisturbed naked amebae the "limax" pattern of progression is, therefore, the rule. The behavioral patterns of the Testacealobosa range from a limax pattern which is particularly well developed in *Hyalosphenia minuta* to a "crawl" pattern such as is found in *Lecquereusia* sp. In the latter case along an axis which represents an extension of the median sagittal axis of the test the replacement pseudopodium is formed between the anterior lip of the test and the base of the actively advancing pseudopodium, which in its turn is in a similar relationship, above and in front of its predecessor now retracting. The result is the persistence of a relatively permanent anterior end in the animal, which accordingly can be expected to cover a large territory in its movements.

Cyphoderia ampulla, as a representative of the Testaceafilosa also demonstrates monopodiality, but here it is combined with frequent random changes in direction. The net result is that the range of the animal is small compared with its rate of movement, covering an area of only a few square millimeters in 24 hr; the range is not appreciably larger in 96 hr.

In normal progression there is one directive filopodium along which the animal moves. This directive passes through three stages.

Stage I. The filopodium first originates as a nondirective. It then extends rapidly and has at this time a basal width of 1.5 μ when measured from 10 to 15 μ from the test. At intervals along its length the filopodium widens on contact with the substratum producing a series of thickened areas or nodes connected by narrower internodal sections. These nodes are 1.5 to 2 times the width of the adjacent internodal sections and are slightly elliptical, with the main axis lying in the direction of the directive pseudopodium. Stage I lasts slightly more than a minute.

Stage II. The filopodium becomes wider at the base, 3 μ or more, and very long, commonly from 130 to 160 μ in animals with a test length of from 120 to 140 μ. Lengths of 300 μ may occur. During this stage the animal is moving forward, therefore shortening the filopodium proximally. Concurrently the distal portion continues to extend at least as rapidly as the test is moved ahead. Very delicate secondaries originate at the nodes. Stage II lasts from 3 to 4 min.

Stage III. When the filopodium no longer extends as rapidly as the test advances, the senile stage commences. The base may now be 9 μ or more in width. The widening tendency progresses to the tip of the pseudopodium which becomes less sharply pointed. The number of secondaries increases at the nodes and they are longer and thicker than in stage II. During the last portion of this stage the new directive may be passing through stage I. Once this new directive is established, the old directive detaches from the substratum and the remnant slowly contracts (candles) into the cell body at the podostyle.

Nondirective Filopodia

Nondirective filopodia are variable in number and length. In the beginning they are not distinguishable from stage I directives. Their principal function appears to be that of adhesion to the substratum. They originate at any point on the podostyle and in a stationary individual radiate in all directions around the mouth of the test. In a moving animal locomotion produces a constant forward displacement of their origins causing those filopodia that are in contact with the substrate to trail behind the test. These filopodia appear to function as do the strands of mucoid material trailing behind the tests of the Testacealobosa, i.e., they adhere to the substratum and stabilize the test by elastically opposing the directive pseudopodium. The nondirective filopodia are commonly retracted *en baionnette* within 2 min. Should the animal be swept free of the substratum, the nondirectives become more rigid. They form angular bends and tend to send out secondary filopodia—a behavioral pattern which would seem to facilitate their snagging in detritus and thus reduce the possibility of the animal being swept away by currents.

Secondary Filopodia

Secondaries form at the attachment nodes of directives and occasionally of nondirective pseudopodia. Their function is apparently that of increasing the contact area of the parent pseudopodium with the substratum. They form an acute angle to the axis of the parent pseudopodium, the apex of the angle being toward the test. The impression they give is that of guy wires or anchors which resist slippage of the main pseudopodium. This conjecture is strengthened by the observation that secondaries may also be formed along the proximal internode when the animal's direction of movement is such as to pull a nondirective or a codirective in a lateral direction.

Structure and Function of Filopodia

Cyphoderia ampulla is one of the largest of those Rhizopoda that are classed as Filosa. It is a very active animal, and when moving by means of a stage II directive filopodium, often travels through its own length (140 μ) within a minute. The size of the pseudopodia and their persistence makes *C. ampulla* particularly useful for observations of the structure of filopodia.

De Saedeleer's (1932) definition of filopodia differs from those of his predecessors in several important details. He notes that the frequency of fusion between separate pseudopodia is common rather than rare or absent as previously stated by many authors. In addition, he stresses the importance of branching and finally the presence of the sheets of protoplasm which form basally between filopodia or laterally along a filopodium (Fig. 2).

This latter observation had been previously made by Bělař (1921) on *Pamphagus hyalinus*. He interpreted this lateral spreading with its resulting accentuation of the original central core, as evidence of an axial filament similar to that of the Foraminifera. De Saedeleer, on the other hand, considered that no axial filament was present, either in the small or large pseudopodia. The appearance of a filament being an illusion formed by a transformation of the cortical cytoplasm of a pseudopodium to a more fluid state. The solation of the cortex results in lamellar spreading, the untransformed medullary portion then appearing as an axial filament.

Filamentous structures that Bělař describes as extending into the tubular portion of the filopodium were considered by De Saedeleer to be illusory. He states, ". . . pour nous, ces axes, vestiges de pseudopodes en voie de transformation, ne s'observent que dans les lamelles et leur présence ne peut en aucune façon faire interpréter ces pseudopodes comme des pseudopodes réticulés, car il leur manque le caractère essentiel de ces derniers: les courantes granulaires." The present observations made on *C. ampulla* tend to support the concept of Bělař rather than that of De Saedeleer.

Using the phase microscope, bright contrast 45 ×, a dark line 0.5 μ in width can be seen to extend through the length of the pseudopodium except near the distal end where the sides of the pseudopodium are barely wider than this line. The dimension of this core does not vary and when seen as it passes through the distal nodes it is the same as in

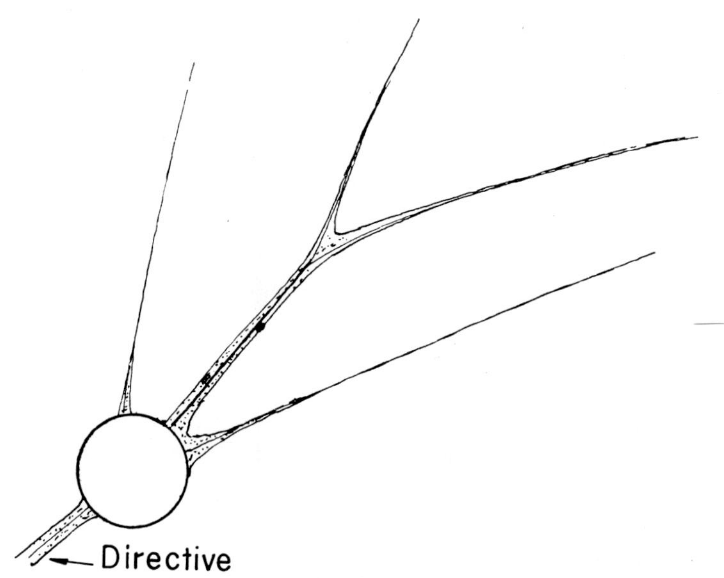

FIG. 2. Basal confluence of nondirective pseudopodia showing axial core.

the expanded base of the pseudopodium. This core is seen to branch into the secondary filopodium if these become large enough to show it. In the consolidation of two or more parallel pseudopodia, the cores approach one another as the degree of fusion progresses until they in turn fuse (Fig. 2).

Additional evidence for the presence of a stereoplasmic core is found in the observations of function and misfunction of the pseudopodia. The usual means of retracting the pseudopodium is *en baionnette,* that is, the pseudopodium bends in a series of sharp angles and eventually folds into the cytoplasmic mass at the podostyle.

It has been observed that under some conditions, treatment with potassium salts in particular, that the *en baionnette* method of pseudopodium return is replaced almost completely by candling, a process in which the rheoplasm runs basipetally to the cell body. Although this type of retraction usually includes the stereoplasmic axis to such a degree that only a small portion of it protrudes distally, occasionally a considerable length of the axis may be exposed. In the latter case, the naked stereoplasmic core can again be seen to be of a constant diameter. This axis may then be recovered either by folding *en baionnette* or by a progressive basal incorporation into the cell body.

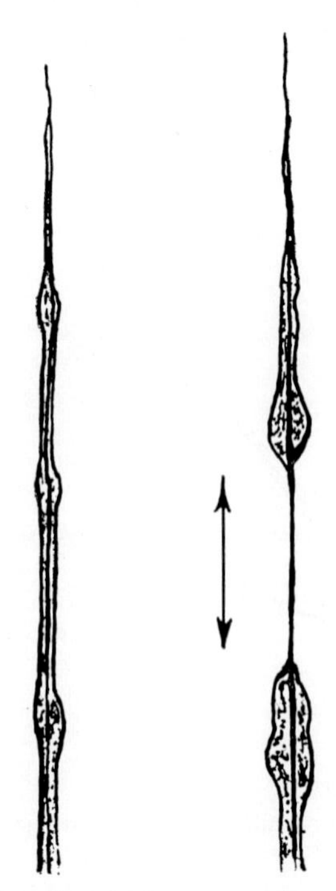

FIG. 3. A break in the rheoplasmic layer exposing the apparently naked stereoplasmic axis.

On several occasions treatment with sodium salts (0.006 N) produced a proximal break in the rheoplasm (Fig. 3). The rheoplasm contracted on either side of the break, the proximal portion moving into the cell body while the distal segments concentrated around the most proximal nodal attachment. This break and withdrawal of the rheoplasm exposed the axis which remained connecting the two masses. The distal protoplasm was then either recovered by *en baionnette* bending of the axial filament or the exposed section of the axis broke and was left behind along with the distal portion of the filopodium.

Rarely in the culture solution and with varying frequency in the single salt solutions used, the rheoplasm on a filopodium fragmented centripetally and each fragment rounded into beads of cytoplasm strung along the core (Fig. 4). These beads form a graded series, the size of the beads being directly proportional to the width of the pseudopodium at that locus. The largest beads are, therefore, proximal and nodal. In a 0.006 *N* $CaCl_2$ solution beading is only a temporary phenomenon, the protoplasm quickly flowing back together and reconstituting the complete filopodium. In 0.006 *N* K_2HPO_4 or $(NH_4)_2HPO_4$ solutions beading is always accompanied by loss of the protoplasm involved. There is evi-

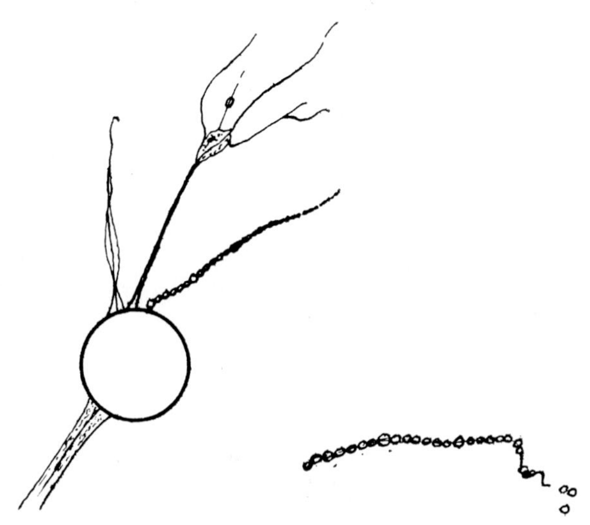

FIG. 4. A drawing to show the effects of sodium ions (0.006 *N*). Note beading of a nondirective pseudopodium. Detail: a single filopodium treated with 0.006 *N* potassium ions showing beaded rheoplasm slipping from the stereoplasmic core.

dently not only a change in state in the rheoplasm in these latter solutions but also a change in the stereoplasm. In potassium and to a lesser extent in sodium and ammonium solutions the beaded segment of the pseudopodium shortly shows evidence of Brownian bombardment which produces a trembling movement in the internodal portions. Eventually the stereoplasmic core itself fragments, and clusters of beads are moved away by the flow of the solution. In several instances beads were observed to slip one by one from a broken strand of stereoplasm (Fig. 4, detail).

The preceding observations cannot, I believe, be explained by any means other than the presence of an organized core in the pseudopodium. Moreover, there are indications that the mechanism for the *en baionnette* return typical of filopodia is dependent on the presence of such a core.

De Saedeleer (1932) states that filopodia do not contain granular currents of rheoplasm. He asserts that the apparent granules in a drawing by Bělař (in Hartmann, 1927) of the filopodia of *Pseudodifflugia gracilis* are in reality of external origin. Observations on *C. ampulla* made in the present study indicate that these pseudopodia often do contain granules. The granules in *C. ampulla* are small and reniform. They are highly refractile under a light microscope, are brighter than the cytoplasm in bright-contrast phase, and darker that the cytoplasm in dark-contrast phase. They appear in both directive and nondirective pseudopodia and are usually confined to the more proximal portions of these structures. Exceptions to this latter statement occur when treatment by single salt solutions of low pH (Na in particular at pH 4.4) causes the nondirective pseudopodia to be drawn out into extremely long strands. Here the granules are present at considerable distance from the body of the animal. In small filopodia, such as nondirectives and secondaries, these granules form an outward bulge in the cell membrane as they are moved along and around the pseudopodium. Since this does not occur in the larger basal portion of the directive or in the larger nodes it is probable that this bulging of the membrane is caused by the presence of the more resistant stereoplasmic axis of the pseudopodium. The granules move independently of each other. It is not uncommon to have two granules moving in opposite directions along the axis of the pseudopodium collide, or nearly collide, then swing around each other as they continue along their original course. To a lesser extent the granules are independent of the major flow in the filopodium; granules may move proximally in a rapidly elongating pseudopodium or distally in one which is flowing into the cell body. Generally, however, the majority of the granules move with the prevailing direction of the flow in the pseudopodium.

The structure of a filopodium can be deduced from the preceding. There is present a core or axis of gelled protoplasm of approximately 0.5 μ in diameter. This core is surrounded by a sheath of fluid protoplasm in turn bounded by a thin cell membrane. Initially the cross section of a filopodium is circular and in the smaller portions may remain so but as the structure becomes larger and older there is a lateral spreading. Whereas the stereoplasmic core seems to be of constant diameter, the amount of rheoplasm increases as the filopodium ages. Some of this increase may be apparent, rather than real, caused by a lateral spreading of the rheoplasm. There is a temptation to think of the cortical cytoplasm becoming more solated—De Saedeleer so considers it—but there is no direct evidence that this is the case. A change probably occurs in the membrane at this time as is indicated by the enhancement of adhesiveness in the more proximal nodal and to a lesser extent in the internodal areas. Probably, however, the major proportion of the increase in width is caused by protoplasm flowing from the cell body into the filopodium. Since filopodia are ectoplasmic and the protoplasm is extremely hyaline, this flow is easily overlooked. In *C. ampulla* the appearance of granules basally and their movement outward along the pseudopodium establishes that such a flow is present and considerable.

Discussion

In many rhizopods, pseudopodium formation and movement should be considered separately. In most laboratory amebae this distinction is normally academic, since such a large volume of the animal is involved in the formation of a pseudopodium that spatial displacement of the organism is automatically involved. However, in *C. ampulla* the difference between the volume of the pseudopodium and the volume of the cell body is very great, and it is evident that the two (formation and movement) were best considered as separate, though related, phenomena.

The formation of a filopodium is rapid. As previously reported in this paper the rate of formation for a directive pseudopodium, exceeds, then equals, and finally falls behind the rate of movement of the organism. If the filopodium is produced distally, it is evident that the production of axis from cortex requires that the rheoplasm streams at least as rapidly as the animal moves forward. The rate of protoplasmic streaming in *C. ampulla* has not been satisfactorily determined although it is evident that the velocity meets this requirement. The groundplasm of the filopodium is nongranular. Those large granules that are present are few, proximal, and their positions and movement indicate a considerable degree of autonomy from the principal streaming in the structure; certainly, they are not satisfactory indicators of general protoplasmic flow. As they do, however, give some insight into possible velocity, their rate of flow was determined. This was found to be about 250 to 300 μ/min. Rates of streaming indicate that velocities of the required order are easily attained by protoplasm. Measurements of 0.5 μ granules moving along the reticulopodia of *Diplogromia* showed that 500 μ/min is not uncommon, whereas in the slime mold, *Didymium nigripes,* the velocity of streaming may reach 1250 μ/min (Vouk, 1913).

As mentioned above the volume of the filopodia of *Cyphoderia* is small compared with the volume in the body of the animal. Close observation indicates that there is no displacement of the cell body into the pseudopodium in the manner that is characteristic of the naked Lobosa. In *Cyphoderia* it appears that the animal streams along the axial filament. The filopodium thus prepares a pathway along which the animal moves.

We may postulate that the axis of the directive filopodium forms a reaction surface. As such it provides the necessary surface for the pseudopodial sol to stream distally to where, by sol–gel transformation, it extends the reaction surface. The rheoplasm of the base of the pseudopodium and probably that of the podostyle streams against the axial filament. As already noted, at points along its length the filopodium is in close and adhesive contact with the substratum. These nodes in addition to adhering to the substratum are centers of secondary filopodium formation, thus increasing the contact area. Each nodal locus functions as an anchor for the axial filament. Streaming along such a reaction surface which in its turn is fixed in its position external to the cell body must result in movement of the organism. Conversely, if the reaction surface were not fixed and were of smaller mass than the organism, it would be ingested by the organism, cf. the feeding of *Amoeba vespertilio* on filamentous algae or *Histomonas* on bacteria. In *Cyphoderia* the retrieving of a pseudopodium by the candling process suggests just this sort of process.

Several observations support the contention that the axial filament must serve as the locomotor reaction surface, just as the cytoplasm of the cortex or of the podostyle must serve as the actual locomotor force. Both the Frey-Wyssling (1947) and Loewy (1949) theories require that the reaction surface be a comparatively rigid structure. This condition is only satisfied by the axial gel. The cortex, although slightly elastic, does not have enough rigidity to serve in itself as a reaction surface. If the filopodium were in contact with the substratum throughout its length, this would not be a valid objection since the substratum itself could serve as a traction surface, as it apparently does in some slime molds (Loewy, 1949). It has been observed, however, that only rarely do the internodal sections of the filopodium adhere to the surface. The most proximal internode is often lifted at a considerable angle above the surface. The animal occasionally moves along its pseudopodium at a rate faster than the streaming of the peripheral rheoplasm. When this occurs the rheoplasm is seen to wrinkle in advance of the oncoming mass of the organism. If the cortex were serving as the reaction surface, such wavelike folding could not occur as it would be pulled taut in opposition to the movement. Since the axis is the only portion of the pseudopodium with sufficient rigidity to be free from the substratum, except at anchor points, and to serve as a reaction surface, this function must reside in the gelled axis.

To summarize, a filopodium is an organelle consisting of a delicate cell membrane over an actively streaming cortex, the rheoplasm, which in turn surrounds a gelled core. A filopodium is formed by the streaming of the rheoplasm along a reaction surface, which in this case, is the stereoplasmic axis. Stereoplasm is produced by a distal sol–gel transformation of the rheoplasm. If a filopodium functions as a directive simultaneously with its distal elongation, it is found that the cell body advances along the proximal section. It is suggested that this advance of the cell body is what would be expected from a streaming action against a reaction surface firmly fixed to the substratum, as is the axial rod of the filopodium.

References

Bělař, K. (1921). *Arch. Protistenk.* **94**, 93-160.
De Saedeleer, H. (1932). *Arch. Zool. Exptl. Gen.* **74**, 597-626.
De Saedeleer, H. (1934). *Mem. Museé R. Hist. Nat. Belg.* **60**, 112 pp.
Frey-Wyssling, A. (1947). *Exptl. Cell Res. Suppl.* **1**, 33-42.
Hartmann, M. (1927). "Allgemeine Biologie," 756 pp. Jena, Germany.
Loewy, A. G. (1949). *Proc. Am. Phil. Soc.* **93**, 326-329.
Vouk, V. (1913). *Denkschr. Akad. Wiss. Wien. Math-Naturw. Kl.* **88**, 652-672.

Discussion

Dr. Bovee: I shall show a film which will demonstrate many of the points in Dr. Berrend's paper. However, I would like to disagree on one point: I don't think the axial rod is formed at the time the pseudopod is extended from the body. These pseudopods almost always extend upward into the water and then down. The head forms a node of attachment. There seems to be a gelation of this strip between the point of origin at the shell and the point of attachment. The next part of the pseudopod then may drop down and form a second point of attachment and the rod seems to appear in that area.

Dr. Noland: The first ones that fall down are quite narrow, and it would be very hard to see the rod. However, they are extremely straight, suggesting they must have some stiffening structure.

Dr. Bovee: Where they have formed a node, you can see the gelation between the node and body. The pseudopod may go forward in sections and there may be one, two, and possibly a third node by the time it is extended. As each node makes contact, there is apparently a local change sufficient to show contrast and suggest that this axial rod is in formation.

Dr. Allen: I would be a little bit doubtful about the existence of this rod on the basis of the fact that a number of excellent older investigators, such as W. J. Schmidt, F. Doflein, and others, described such rods and drew them in pictures of Foraminifera, but in later studies with phase contrast and polarized light, no one has succeeded in demonstrating these rods. They certainly may occur in some organisms, but I think that the burden of proof of their existence should lie with the observer. It seems to be assumed that such a rod must exist to explain the rigidity. Actually, all that is required to explain the shape and rigidity of these pseudopodia is the presence of gel structure.

Dr. Noland: It is hard to interpret the straight, rigid strand between the droplets of solating cytoplasm in any other way.

Dr. Bovee: Also the bending and folding in segments.

Chairman Marsland: Did Dr. Berrend use phase contrast?

Dr. Noland: Yes.

Dr. Kitching: The axopods of Heliozoans can be made to show similar bending effects and they certainly do have an axial rod.

The Axopods of the Sun Animalcule *Actinophrys sol* (Heliozoa)

J. A. KITCHING

Department of Zoology, University of Bristol, Bristol, England

Introduction

The "sun animalcule" *Actinophrys sol* Ehrenberg (Heliozoa) has a spherical body usually about 30 μ in diameter, from which project a considerable number of "axopods." These axopods are usually regarded as the equivalent of pseudopodia, although in fact pseudopodial activity is shown by lobes and processes which project from their bases rather than by the axopods themselves. There is a single central nucleus in *Actinophrys,* but there are many nuclei in the related form, *Actinosphaerium.* In both genera the ectoplasm is vacuolated, and the endoplasm dense. After fixation and staining, each axopod is seen to contain an axial rod (or skeleton) which is rooted deep within the body of the organism. In *Actinophrys* the axopods rest upon the nuclear membrane (Bělař, 1923), and a similar relation has been reported for *Actinosphaerium nucleofilum* (Barrett, 1953) unlike other species of that genus. The axial rods are known to be composed of parallel fibers (Roskin, 1925; Rumjantzew and Wermel, 1925; Wohlfarth-Bottermann and Krüger, 1954; Anderson and Beams, 1960).

Actinophrys moves very slowly. If you are lucky, watching *Actinophrys* down the microscope is about as exciting as watching the minute hand of a clock. Although *Actinosphaerium* has for long been reported to roll itself along, Kuhl (1951), using time-lapse photography, has described a gliding movement in which the organism appears to row with its axopods. When I have seen two *Actinophrys* separate, as at the close of binary fission, the protoplasmic bridge between them has become more and more attenuated, as though they were pulling it apart. There was no obvious activity of the axopods.

If small Protozoa collide with the axopods of *Actinophrys,* they are liable to stick (Looper, 1928; Kitching, 1960). Often by their own struggling they come to make contact with the lobose axopod bases or body surface, and this contact appears to provoke the quite rapid outgrowth of a membranous funnel, which usually comes to surround the prey and enclose it in a food vacuole. Occasionally, the prey may appear to

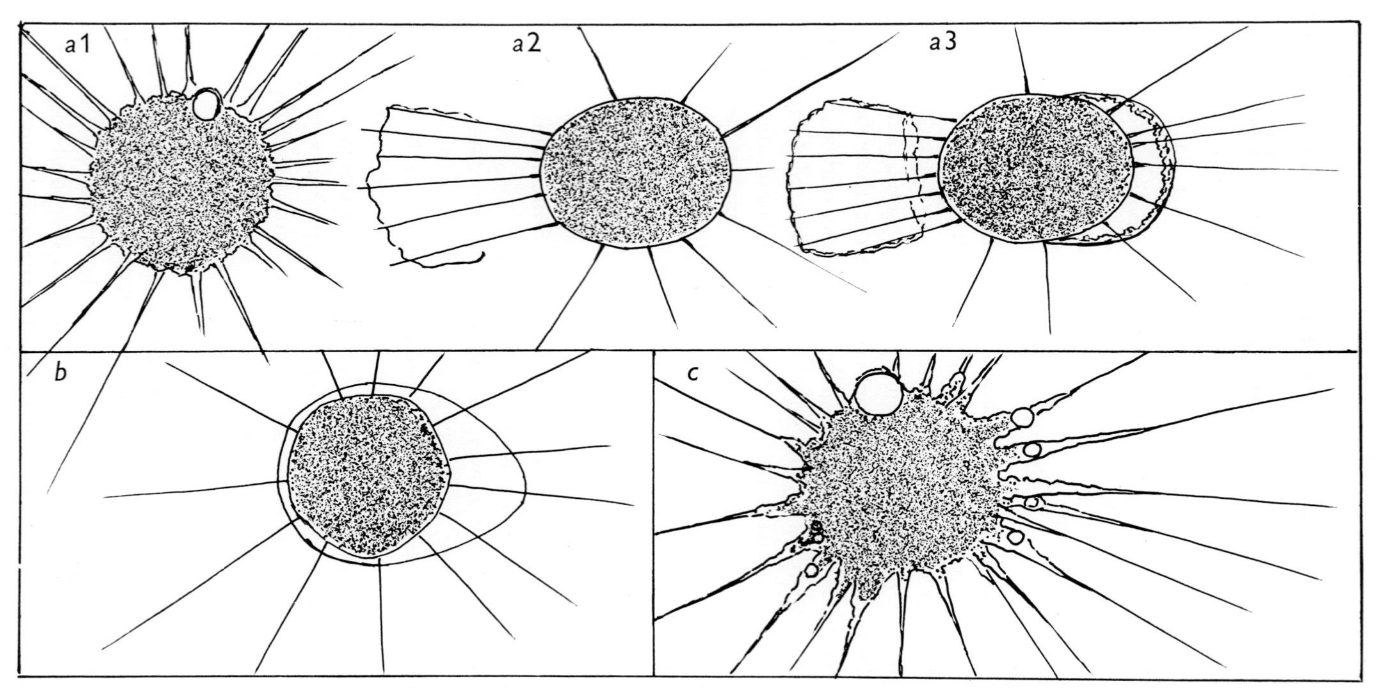

FIG. 1. (a) *Actinophrys sol* treated in a hanging drop with 0.1% bovine plasma albumin at 0 min, and with 1% bovine plasma albumin at 12 min; a1 before treatment; a2 (with an incomplete thin skin) at 5 min; a3 (with an additional thick skin on one side) at 25 min. (b) A complete skin of moderate caliber of the type normally formed in response to 0.1% bovine plasma albumin. (c) Lobulation of axopod bases in response to 0.1% bovine plasma albumin in $M/64$ $MgSO_4$. (From Kitching, 1962.)

move along the axopods toward the body of the *Actinophrys* as though conveyed, but normally the irregular jerks of the prey itself obscure any such axopodial conveyor activity as may exist.

On treatment of *Actinophrys* with a solution of plasma albumin or certain other proteins, a "skin" lifts off from the body surface (Kitching, 1960, 1962). If the solution has been strong (e.g., 1%), the skin is thick and normally remains for some hours in position, separated by a small thickness of water from the body of the *Actinophrys*. If the solution has been rather weak (e.g., 0.1%), the skin breaks and is cast off on one side of the *Actinophrys*. It progresses quite rapidly outward along the axopods, and in this process draws the axopods together so that the *Actinophrys* looks as though it were suspended from the cords of a parachute (Fig. 1). The remains of the skin are cast off from the tips of the axopods. After desquamation, the body surface of the *Actinophrys* is unusually smooth and rounded.

On treatment with hydrostatic pressures of the order of 4000 psi or more (at 15°C), the axopods of *Actinophrys* collapse and disintegrate. Within a few minutes of release of the pressure, new axopods grow out again (Kitching, 1957).

Fine Structure

Thin sections were prepared from *Actinophrys sol* fixed in cold, buffered osmium tetroxide solution, 5–10 days after feeding, concentrated to a pellet, and embedded in Araldite. They were stained with uranyl acetate.

The grosser features of the cytoplasmic structure are illustrated in Figs. 2 and 3. (Figures 3 and 4 also illustrate the skin formed in response to plasma albumins.) The dense endoplasm contains much tubular reticulum. The ectoplasm is highly vacuolated. Both contain mitochondria and also some electron-dense bodies which are no doubt to be identified with Bělař's "lipoidal" bodies. These latter lie principally in a zone about halfway between the nucleus and the body surface. Smaller and even denser granules form, together with small more or less empty vacuoles, a single layer at the surface of the body; and similar granules and vacuoles extend irregularly over the axopods. Some of these vacuoles contain peripherally disposed pegs of dense material.

The axial rods traversing the cytoplasm of *Actinophrys sol* are seen in transverse section to consist of a number of parallel fibers disposed in a very regular way. The profiles of the fibers form a double spiral (Figs. 5 and 6), and there are indications of radial links between some corresponding fibers in the two spirals. The number of fibers ranges at least from 68 to 129, and they are 15–20 mμ thick.

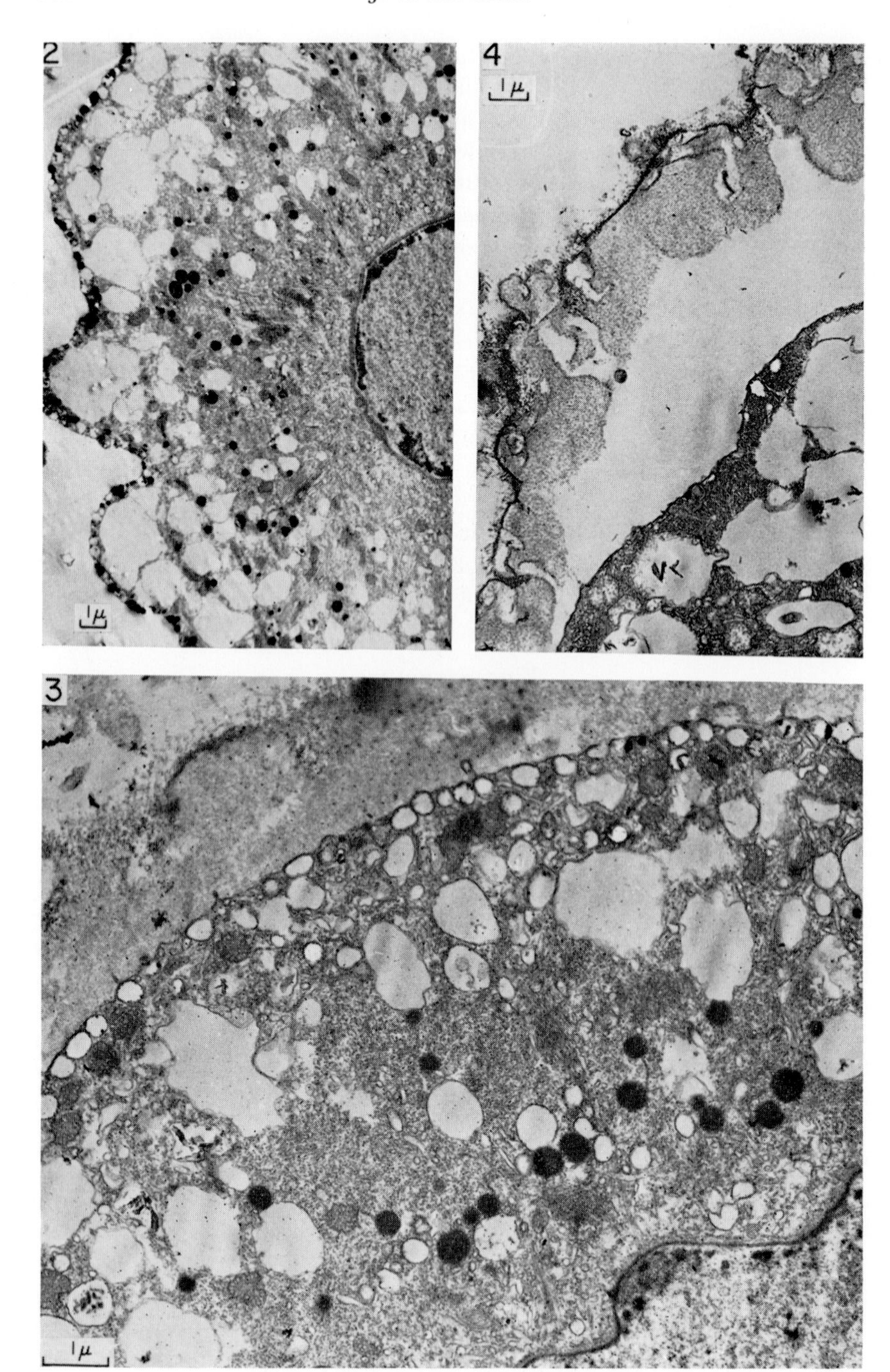
2
1μ
4
1μ
3
1μ

In oblique section the fibers of the axial rods form the characteristic pattern shown in Fig. 7. In its middle region the configuration simulates a median longitudinal section of an axial rod, and the fibers are well spaced; at each end it grades into a tangential longitudinal section, and as it does so the fibers become more densely packed.

The inner ends of the axial "rods" (or fibrillar bundles) rest upon the outer nuclear membrane, but the photographs obtained so far do not reveal any further details of this relationship. Tubular elements of ectoplasmic reticulum partly surround the axial rods and run parallel with them.

The axial rods have been traced continuously from the nuclear membrane into axopods (Figs. 9 and 14). In those axopods having the largest diameter in the sections, and, therefore, presumably cut at no great distance from the general body surface, the double spiral arrangement of constituent fibrils is well maintained (Figs. 12 and 13). In smaller axopodial sections the fibers are less regularly arranged and fewer in number, and in the smallest sections—presumably representing distal regions of the axopods—the fibers are only faintly discernible, few in number, and disarranged (Fig. 11). I do not know to what extent the disarrangement may be caused by fixation, but I see no positive reason at present to discredit it. In some sections the fibers cut within the body appear optically hollow (Figs. 5 and 6). This feature is even more strongly marked in some sections through axopods (Fig. 10). The hollow appearance is maintained at all levels of focus.

Movements

Attempts to study the conveyance of food material from the site of capture along axopods to the body surface have so far proved rather unsatisfactory. Active kinds of prey thrash about and make contact with the body surface of their own accord, and so provoke the outgrowth of a membranous funnel, while small inactive species remain stuck in the same place for long periods. Accordingly I have studied the more vig-

---

FIG. 2. Low-power electron micrograph of untreated *Actinophrys sol* showing nucleus, dense endoplasm, vacuolated ectoplasm, and superficial layer containing many small vacuoles and dense surface granules.

FIG. 3. *Actinophrys sol* fixed after treatment for 5 hr with 1% bovine plasma albumin Fraction V, showing nucleus (bottom right), endoplasmic reticulum, dense bodies, mitochondria, vacuoles, empty vacuoles but no surface granules in superficial layer, and diffuse skin material.

FIG. 4. *Actinophrys sol* fixed after treatment for 5 hr with 1% bovine plasma albumin Fraction V. There are very few superficial vacuoles, but a thick layer of skin lies outside the body surface.

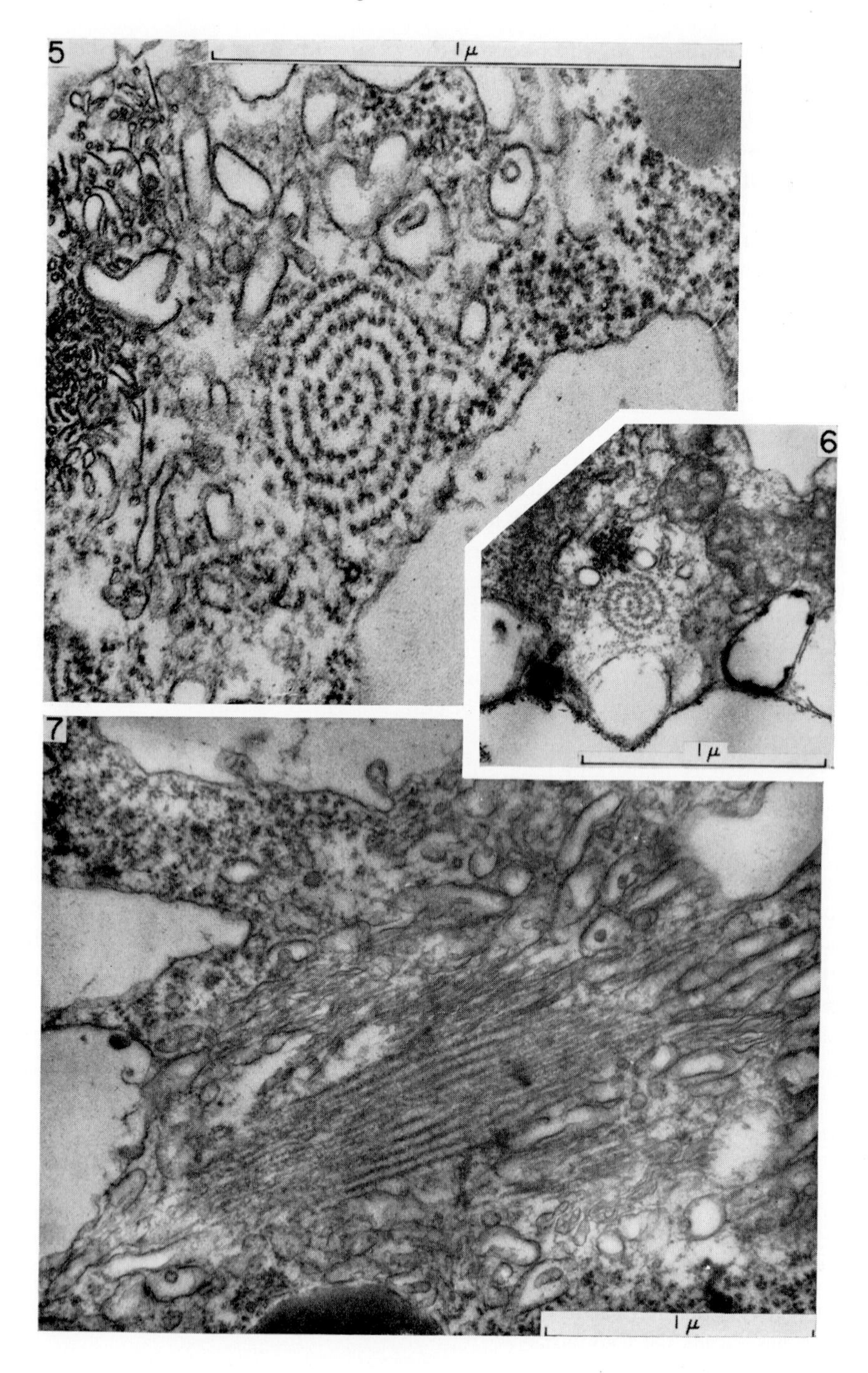
5
1 μ
6
1 μ
7
1 μ

orous process of the removal of skins from the body surface to the axopod tips.

Whatever the process by which skins are first lifted up off the body surface in response to a solution of plasma albumin, progression of the skin to the axopod tips cannot be explained as a consequence of osmosis. Complete skins do not progress, no doubt because they are held in place by the body of the organism within them. Only incomplete or broken skins move out, and it cannot be supposed that any effective osmotic pressure would build up within these. The outward progression of skins and debris must be attributed to a conveyance by the surface of the axopod itself. Once the skin is free, it may move outward quite rapidly, for instance at a rate of 20 μ/min. It therefore seems likely that skins are conveyed outward by the movement of the axopod surface itself.

When the axopods of a living *Actinophrys* are examined with an objective of sufficient resolution, minute granules are to be seen scattered irregularly within them. They appear to correspond with the surface granules seen by electron microscopy. It was convenient to observe the movements of these granules with a Baker interference microscope, using the $\times$ 100 water-immersion objective, which shows these up well without the inconvenience of the viscous drag on the coverglass caused by immersion oil.

In an untreated *Actinophrys* the granules in the axopods are mostly stationary, but make occasional excursions centrifugally (outward) or centripetally along the axopods. During the movement of the skin outward along the axopods there was no sign whatsoever of any movement of granules outward. On the contrary there was a well-marked countermovement of granules to the bases of the axopods, at a rate of up to 30 μ/min. It seems likely that this downward movement affects the protoplasm surrounding the axial rods or bundles, and that it accounts for the observation (Kitching, 1962, p. 362) that on elevation of a skin the axopods become thin and needlelike.

From these observations we are led to suggest very tentatively that a shearing force can be exerted between a very thin layer of superficial protoplasm and a slightly deeper layer containing the surface granules. The precise depth of these granules within the axopod is hard to de-

---

FIG. 5. Section through outer endoplasm. The axopodial skeleton is cut in transverse section, showing the double-spiral arrangement of the parallel fibers. Elements of endoplasmic reticulum surround the axopodial skeleton and run parallel with it.

FIG. 6. Section through outer ectoplasm and body surface showing superficial vacuoles and axopodial skeleton cut transversely.

FIG. 7. Oblique section through axopodial skeleton.

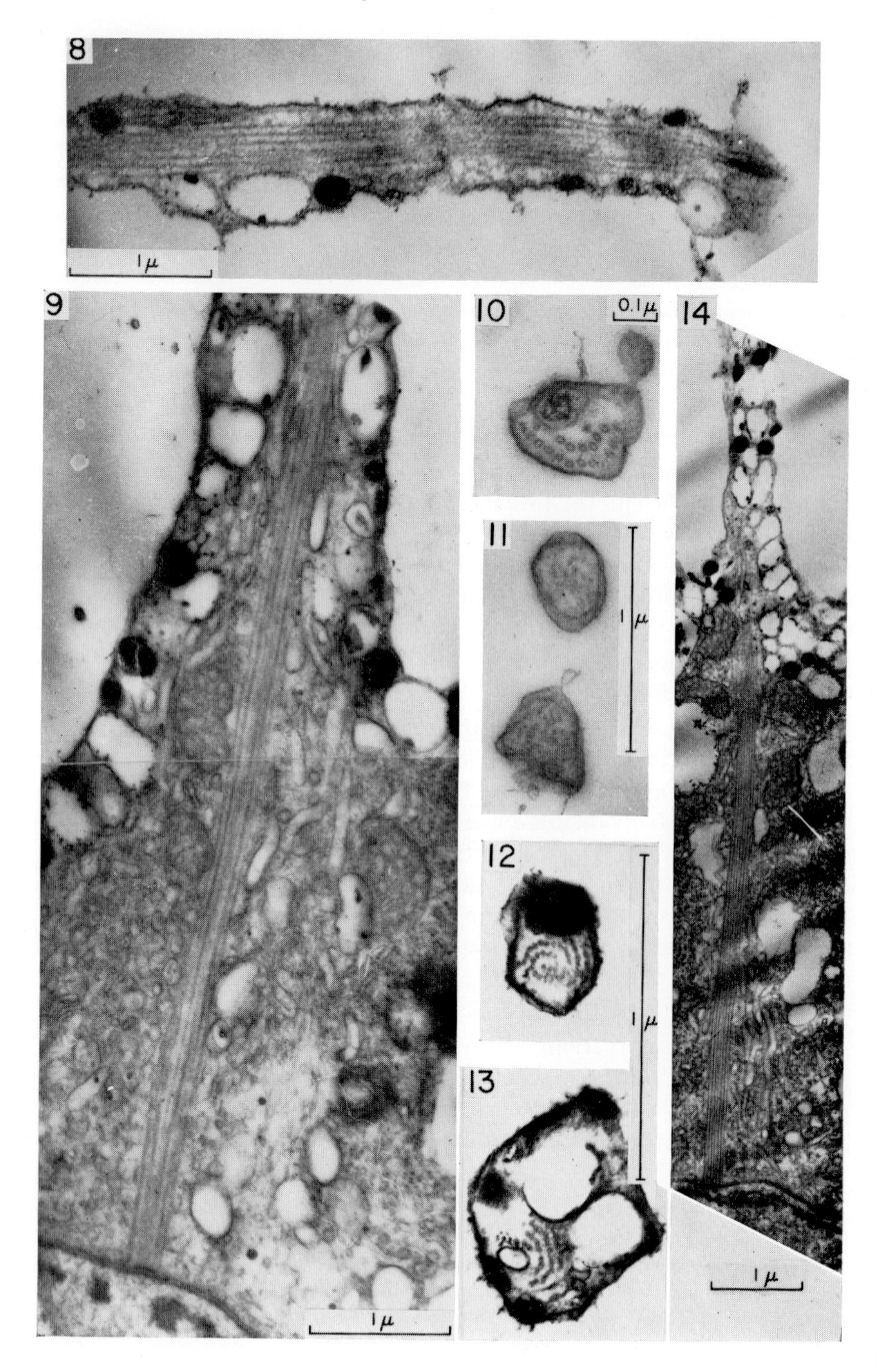
8
1 μ
9
1 μ
10
0.1 μ
11
1 μ
12
1 μ
13
14
1 μ

termine, as it might be disturbed by the process of fixation. Movement of ectoplasm upon endoplasm is well established in other materials, as reviewed by Kamiya (1959), and the countermovement of ectoplasmic horns and of endoplasm in isolated drops of protoplasm from *Nitella* or *Chara* seem to offer an appropriate parallel.

Nevertheless, it seems possible or even likely that the movement of cytoplasm centripetally from the axopods to the body proper is merely one aspect of the striking changes that are induced by a dilute solution of certain proteins. I do not know for certain where the skin-forming material comes from. It seems too much to be explained by the combination of the skin-provoking protein with an extraneous coat of body surface. Thus it is presumably exuded. After prolonged treatment (e.g., 5 hr) with a solution of plasma albumin, the number of the superficial dense granules is reduced (Fig. 3), and in many cases the surface vacuoles are also reduced in number or even almost completely missing (Fig. 4). Granular cytoplasm now makes direct contact with the plasma membrane, and the underlying ectoplasmic vacuoles are larger, as though vacuolar confluence had taken place. It seems possible that superficial granules and vacuoles have been discharged and that they may have contributed the skin-forming substance. The source of the superficial granules will require further investigation; the dense peripheral pegs within the superficial vacuoles may throw some light on this. In any case, it is likely that a drastic change takes place in the physical condition of the superficial ectoplasm, which may well be associated with the centripetal movement of axopodial protoplasm.

The mechanism of progression of *Actinophrys* remains unknown, and further investigations by time-lapse photography are desirable, as carried out by Kuhl (1951) on *Actinosphaerium*. Kuhl's suggestion that *Actinosphaerium* uses its axopods as oars, pivoted in the body surface and actuated by movements of the alveolar ectoplasm, does not appeal to me. As a speculation, a mechanism for the centrifugal conduction of

---

FIG. 8. Longitudinal section of an axopod.

FIG. 9. Axopodial skeleton running from nuclear membrane into an axopod.

FIG. 10. Transverse section of an axopod, showing optically hollow fibers of axopodial skeleton, diminished in number and in partial disarray.

FIG. 11. Transverse section of two axopods, showing fibers in disarray.

FIG. 12. Transverse section of an axopod with spiral of fibers and superficial granule.

FIG. 13. Transverse section of axopod with incomplete spiral of fibers and two vacuoles.

FIG. 14. Axopodial skeleton running from nuclear membrane into axopod. Tubes of endoplasmic reticulum are seen running parallel with it.

rejecta along the axopods could act as a means of repelling the whole organism from an unfavorable environmental stimulus locally exerted; centrifugal surface movements of axopods in contact with the substrate would convey the *Actinophrys* away from the site of stimulation. In the same way, but perhaps less probably, centripetal movements of the axopod surface, such as might help to convey prey to the body proper, could draw the whole organism toward a favorable area of the environment. This mechanism would not require central coordination. However, these suggestions are purely fanciful.

Function of the Axopodial Skeleton

The constituent fibers of the axopodial skeleton have about the same diameter as the subfibers of the nine peripheral fibers of a cilium, and like these are optically hollow. I do not know how they grow, but it seems likely that they grow outward from the nuclear membrane. This presumably provides an anchorage for them, and perhaps also the structural and biochemical organization which leads to the formation of fibers. The aggregation of electron-dense material—also recognizable in life—at the inner surface of the nuclear membrane may have significance in this connection. The composition of the fibers of the axopodial skeleton is unknown, but they are likely to consist of a protein. The reconstitution of the axopods on release from high hydrostatic pressure presumably involves the outgrowth of axopodial rods, so that in respect of rather rapid outgrowth there is also a resemblance to the fibrillar components of cilia. Although in my *Actinophrys* the axopods normally showed little or no contractile activity, those of *Actinosphaerium* shorten and draw food down into the body (Mackinnon and Hawes, 1961). The nature and extent of movements carried out by an organelle such as a cilium must depend largely on the disposition of contractile structures and on the ability to propagate a contraction, rather than on any outstanding contractility of the fibers concerned. Thus the fibers of the axopodial skeleton could turn out to be rather like those of cilia. Their disposition in *Actinosphaerium* and in other Heliozoa should prove interesting.

The very long thin form of the axopods seems to demand a supporting skeleton. Given that fibers of the kind found in *Actinophrys* are the basic component out of which such a skeleton must be constructed, it seems to be of some interest to consider the structural arrangement of these fibers from an engineering point of view. I am greatly indebted to Professor Sir Alfred Pugsley, F.R.S. of the Department of Civil Engineering, University of Bristol, for his comments on this subject. It is assumed that the fibers are bound together laterally, so that they act as a sheet. A closed tube would have the best resistance in flexure and torsion, and the fact that a closed tube is not used suggests to me that there are good biological objections to it. Perhaps, for instance, the double-spiral structure permits a change of the number of the constituent fibers, either during growth or in relation to anatomical position, without disturbance of the general structure or symmetry. A single spiral would have good flexural resistance but would tend to twist about a shear center situated outside it and on the side away from the opening. In a double spiral, with openings on opposite sides, provided that the two spirals were linked by radial rods, there would be excellent resistance to flexure and torsion.

Acknowledgments

This work has been made possible through the cooperation of Mr. Edward Livingstone, who grew many *Actinophrys* for me, of Mrs. Joyce Abblett, who fixed and embedded the material, of Mr. Maurice Gillett, who cut sections, of Mr. Alan Bassett, who instructed me in the use of the Philips EM200 electron microscope, and of Mrs. W. M. Dyer, who also helped me with it on many occasions. I am indebted to the Company of Biologists Limited for permission to reproduce Fig. 1 from *The Journal of Experimental Biology*.

References

Anderson, E., and Beams, H. W. (1960). *J. Protozool.* **7**, 190.
Barrett, J. M. (1953). *J. Protozool.* **5**, 205.
Bělař, K. (1923). *Arch. Protistenk.* **46**, 1.
Kamiya, N. (1959). *Protoplasmatologia* **8**, 3a, 1-199.
Kitching, J. A. (1957). *J. Exptl. Biol.* **34**, 511.
Kitching, J. A. (1960). *J. Exptl. Biol.* **37**, 407.
Kitching, J. A. (1962). *J. Exptl. Biol.* **39**, 359.
Kuhl, W. (1951). *Protoplasma* **40**, 555.
Looper, J. B. (1928). *Biol. Bull.* **54**, 485.
Mackinnon, D. L., and Hawes, R. S. J. (1961). "An Introduction to the Study of Protozoa." Clarendon Press, Oxford, England.
Roskin, C. (1925). *Arch. Protistenk.* **52**, 207.
Rumjantzew, A., and Wermel, E. (1925). *Arch. Protistenk.* **52**, 217.
Wohlfarth-Bottermann, K. E., and Krüger, F. (1954). *Protoplasma* **43**, 177.

Discussion

Dr. Bovee: I would like to thank Dr. Kitching for providing me with a partial explanation for the folding phenomenon exhibited by helioflagellates in the transitional state preparatory to the forming of flagella. As nearly as I could estimate, these individuals had 32 original axopods which folded (first in groups of four). Then all simultaneously shortened to about half of their original lengths. The double-spiral arrangement you have suggested could be a good mechanical device allowing not only a telescoping motion to provide for this shortening but also a twisting motion as well.

DR. KITCHING: I think it is important to study your particular animal and see what the fibrillar arrangement is in that.

DR. COWDEN: Were you ever fortunate enough to catch any of these in the act of replication?

DR. KITCHING: Not yet.

DR. LING: I am most fascinated by what you have found and would like to make a few comments. First, there are some well-known differences between egg albumin and bovine serum albumin. One striking difference is that serum albumin has very strong binding capacity for negative charges, which the egg albumin does not share. Dr. Gustafson has told me that he and also Dr. Ponder have observed specific effects of serum albumin on other types of cells. This being the case, then perhaps serum albumin reacts with negative charges on the surface of the cells in a way which is more intense than egg albumin does. This may lead to the difference of response.

Second, I should like to recall what Dr. Nakai reported about the bubbling action at muscle cell surfaces in response to contact with the end of a nerve fiber and also Dr. Abé's comment that this may bear some resemblance to the formation of pseudopodia. Well, let us make a big jump and assume that there are some basic similarities in this phenomenon and the phenomenon of film formation. Is it possible that the nerve in the culture is already secreting some cationic component similar, let us say, to acetylcholine, and, like bovine serum albumin, will have effect on the surface and produce a similar kind of action as that brought about by bovine serum albumin on Heliozoan filopods?

Finally, I would like to suggest the use of two compounds which are now readily available and which we have studied to some extent. One is polylysine and the other is polyaspartic acid. We have found polylysine to have enormous effects on the cell surface.

DR. KITCHING: I have already published a paper using this very interpretation for differences between the effects of egg albumin and plasma albumin. It may interest you to know that the lifting of the skins can be prevented by the presence of certain concentrations of certain salts, of which magnesium sulfate is the most active. I suggest that the sulfate is adhering to the albumin and the magnesium to the body surface, thus separating the two.

DR. ALLEN: Is there any structure at the base of these axopods that might resemble a centriole?

DR. KITCHING: The fibers rest upon the outer nuclear membrane. They do not appear to cross between the two membranes. I have not been able to resolve any further structural information. It is possible that with better technique something might be found.

DR. NOLAND: I think I recall some forms that have the flagellum and axopodium coming off the same basal granule.

When these organisms are flattened between a cover glass and slide, have you noticed that there is locomotion? This occurs even though the cell is much flattened.

DR. KITCHING: I have never tried this. In fact, I take every precaution to prevent any such happening!

DR. BOVEE: Dr. Allen asked, I believe, if there was any evidence of centrioles connected with the axopods. I don't have electron-microscope evidence at all, but the helioflagellate that Dr. Noland mentioned has a basal granule. Whatever this granule is, it seems to be functioning as a centriole, both in the origin of an axopod and as the origin of the flagella.

FREE DISCUSSION

Dr. Allen: I have been very much interested by Dr. Abé's observations, not only those he presented today on diatom egestion, but his earlier careful descriptions of *Amoeba striata* organization and movement. In this species, he pointed out the presence of a "fenestra" or window in the posterior ectoplasm where one would expect endoplasm to escape if it were under high hydrostatic pressure. Dr. Abé has been very careful in drawing conclusions from these observations, but I infer from his papers that he regards the most essential aspect of movement as not the generation of flow by pressure, but rather the "building forward" of the gel layer (ectoplasm or plasmagel). Dr. Abé and I speak different languages, both nationally and scientifically; as a morphologist, he speaks in terms of structure, while I prefer to think in terms of forces and deformation. However, our respective attentions are focussed on the same process, which he refers to as the building of new gel at the front, and which I believe may be a contractile process at the front. I think our views are rather similar.

Dr. Abé: Dr. Kanno in my laboratory has performed an interesting experiment. A proteus-like ameba was sucked into a capillary with a bore of about 50–70 μ and allowed to establish vigorous unidirectional locomotion. Then a smaller capillary with a bore of about 12 μ was thrust about 10 μ into its tail region and cytoplasm sucked out. Movement in a forward direction continued even though over half of the cytoplasm was withdrawn from the cell. How do you think this should be interpreted?

Dr. Marsland: I think that the terminology should be changed. Tail contraction, I think, is too limited a concept. I think that contraction can occur, and perhaps does occur, predominantly in the tail, but I think we must recognize that any restricted part of the tube or any considerable part of the tube wall may participate in the contraction process.

I should also like to make certain my comments this morning are not misunderstood. I do not think that these loops and ring configurations in isolated cytoplasm are without meaning; I think perhaps they are significant. But I think that it is possible that they are derivatives of a cytoplasmic network structure. I think it is impossible at the moment to eliminate the possibility of a front contraction, possibly in the fountain zone, but I also think, in the light of everything that I have observed and all the observations that I have heard presented here, that the tube wall cannot be eliminated, either, as a site of contraction in *Amoeba proteus*.

Dr. Lee: I think it is obvious that we have been dealing with a whole lot of different creatures today. I would like to ask either Dr. Jahn or Dr. Allen if they would agree with me that there are two kinds of pseudopodial movement in the Foraminifera. I think there are thicker pseudopods which are the larger, principal force-generating pseudopods and smaller feeding pseudopods. I am very much attracted to Dr. Jahn's hypothesis of active-shearing. It seems to me that this mechanism is quite applicable to this particular phenomenon.

I would like to know what Dr. Allen thinks about it.

Dr. Allen: One of the most striking aspects of streaming in "large" pseudopodia of Foraminifera is that files of particles move at separate and distinct velocities which are more or less constant. What we have shown that is new, I believe, is that the velocity changes significantly in the vicinity of points of attachment to the substratum.

The active-shearing hypothesis was an excellent formal model as long as it was compatible with all the information about pseudopodial dynamics and ultrastructure. However, now that Dr. Wohlfarth-Bottermann has shown that the individual "streams" in a large pseudopod are separate, membrane-bounded protoplasmic bodies, there appears to be no ultrastructural basis for the "millipede legs" Drs. Jahn and Rinaldi proposed. As far as pseudopodial dynamics are concerned, the active-shearing model so far does not account for the velocity relationships we have domonstrated, or for the "fiber-droplet transition." I think we are still in the fact-gathering stage in the study of most of these phenomena; it is desirable to have as many testable models as possible, in order to suggest decisive experiments.

DR. JAHN: I think that the granules move very much the way Dr. Allen described today. We have made similar measurements. We used a simpler shadowgraph technique, very similar to what Dr. Kamiya used for slime-mold streaming. I do not see so far how this gives us the key to the mechanism.

DR. INOUÉ: I find that I am among several others who cannot understand the simple mechanics of Dr. Allen's frontal-contraction theory, and I wondered if we could have the opportunity of having this clarified a bit.

DR. ALLEN: Could you be more specific?

DR. INOUÉ: Well, how do you get a consistent mechanical picture of an ameba moving forward with a frontal-contraction theory?

DR. ALLEN: In view of the extent to which I am now finding this simple idea is misunderstood, I think I should try to clarify it. I think the best short cut to understanding this frontal-contraction model is to think about how to construct an ameba (in the mind's eye) from the simplest units of function.

As I described in the Free Discussion of Part II, and as I will show by means of a film, the cytoplasm of broken amebae continues to stream. The simplest "units of streaming" are hairpin-shaped loops of cytoplasm from a few microns to a few tens of microns wide, and roughly as long as the cell from which they came. I think we "understand" the streaming which these cytoplasmic loops perform in terms of a propagated contraction occurring in cytoplasm entering the bend of this loop. This statement is based on the fact that detailed motion analysis of cytoplasmic streaming in such loops agree in every way with a simple contraction model of the following kind.

As a more tangible substitute for a cytoplasmic loop, imagine a long muscle capable of slow conduction of a long-lasting contraction. Imagine further that one end of such a muscle were attached upright in a clamp of suitable size so that its other end hung down because of the relaxed muscle's flaccidity. Now if stimulation were applied near the clamp, the already upright portion of the muscle would shorten and stiffen until the contraction reached the bent region. When the contraction reached the bend, the greater tension on the stretched outer portion would cause the muscle to straighten; the accompanying stiffening would "freeze" the straightened portion and thus serve to anchor the contraction always at the bend. The end result would be the straightening out of the muscle by continuous displacement of the relaxed arm of the loop. If the same experiment were done in a free-floating (i.e., not clamped) muscle, then the bend would remain stationary while the relaxed and contracted arms were displaced toward and away from the bend, respectively. So far, I think the mechanics poses no serious problems.

Now, how can such hairpin loops be put together into a pseudopod showing continuous locomotion? Imagine arranging a group of hairpin loops in a radially symmetrical pattern such that all of the uncontracted arms (which are thinner) are gathered in a solid bundle, and the (thicker) contracted arms are arranged in a circular pattern around the uncontracted arms. Now, if you just release a wave of contraction from the bend of each loop inward toward the massed uncontracted arms, then these will advance toward the bend, and their mass will pass through the bend and become incorporated into the tubular structure composed of stiffened contracted arms.

Relative to the bends of these fused loops, the uncontracted material (we can now call it endoplasm to bring us back to real, rather than imaginary, amebae) is displaced forward, and the contracted material (ectoplasm) is displaced backward. Ordinarily, the ectoplasm is anchored at some point through the plasmalemma to the substratum, so that only the endoplasm is displaced forward.

In the film you will see the reverse of this imaginary construction of an ameba pseudopod. In broken cells, the cytoplasm gradually tears itself apart into the "units of streaming" on which this idea is in part based.

CHAIRMAN DE BRUYN: I can understand how this might work in a tubular structure, but what about the initiation of pseudopod formation in spherical cells?

DR. ALLEN: When it comes to this situation, we have very few facts to go on. In *Amoeba* and *Chaos*, the first evidence of impending pseudopod formation is the strictly localized formation of a hyaline cap at the point where the pseudopod will appear. Dr. Griffin has seen just the opposite behavior in *Pelomyxa*, which leads him to believe in a somewhat different mechanism in that organism.

There are two ways to account for pseudopod formation in an initially spherical cell. The first is to assume it is initiated by pressure from generalized contraction, and that the pseudopod "pops" out of a weaker spot in the ectoplasmic layer. I do not deny this is a possibility.

The other possibility is to explain it in terms of the hairpin loop structure that I propose to maintain steady-state streaming in cylindrical pseudopods. But in this case one must assume that prior to pseudopod formation there occurs a localized radial orientation of the mechanochemical system. This oriented material, properly anchored in the ectoplasmic layer of a spherical ameba, could serve as the "clamp" anchoring a ring of fused hairpin loops (which I conceive pseudopodial structure to be) and permit them to push out the plasmalemma. What we need is more data with which to decide among these and perhaps other possibilities.

DR. MARSLAND: Could you describe what happens to the membrane in your model?

DR. ALLEN: The model has no specific predictions about membrane dynamics as Dr. Goldacre's model does. Nevertheless, there are plenty of published data by Jennings, Mast, Griffin, and Allen, and now by Wolpert to show that the surface layer of the ameba (and it seems probable that this includes the mucus coat and plasmalemma) acts as a semipermanent sack which slides about and gets deformed in response to form changes in the cell. I feel quite certain that Dr. Goldacre's formation-dissolution cycle is incorrect for the large amebae, but it might apply to species we haven't studied.

DR. ANDREW G. SZENT-GYÖRGYI: I am rather ignorant about the systems discussed here and I am completely confused. I have the feeling that we were presented with an extremely large number of phenomenological observations, some of them contradictory, and attempts are made to describe these observations as if the mechanisms controlling contractility would not exist or were absolutely stable and unchanging.

I would like to recall the beautiful demonstration by Dr. Hoffmann-Berling that different parts of a fibroblast cell may be active or inactive depending on the life cycle of the cell. I wonder whether or not the various phenomena and rather contradictory observations could not be better correlated by assuming the presence of a control system, let us say similar to the relaxing factor system in muscle. The activity of such a system may depend on the state of the ameba and on the various experimental conditions and may decide which part of the cell will be contractile at a given time. Since at least certain kinds of glycerinated amebae contract on addition of adenosine triphosphate, the techniques described by Dr. Hoffmann-Berling seem to be applicable, and the contractility of the ameba and of the pseudopods and the presence of control systems at various states perhaps may be tested. It is very difficult for me to see how one could arrive at a clear-cut decision of ameboid movement on the basis of the information which has been presented here.

DR. ZIMMERMAN: Dr. Allen, in view of the fact that a hyaline cap may appear in the middle of an ameba where a new pseudopod arises, would that suggest to you that this portion of the ectoplasm has the ability to contract, or do you believe the endoplasm underneath this area is causing it to be pushed out?

DR. ALLEN: I do not see any reason why cytoplasm in any part of the cell could not respond to local conditions by either contracting or relaxing. If it contracted, syneresis could occur. Still, it would be difficult to prove where the hyaline cap fluid originated.

DR. ZIMMERMAN: Do you agree that the ectoplasm can contract?

DR. ALLEN: Yes, provided we agree on terms. I take "contraction" to mean a change in elastic modulus resulting in production of tension under isometric conditions and performance of work under isotonic conditions. What one *sees* in amebae as evidence for or against contraction is *shortening*, which may be due either to active contraction or to passive deformation. If we see shortening in the tail or fountain-zone regions (it occurs in both places), we can only conclude that *perhaps* contraction is the cause. On the other hand, if no shortening is observed, there could not be any performance of work.

The ectoplasmic tube may shorten somewhat in its anterior most portion, depending on conditions, and it certainly shortens in the tail. For this reason, I would favor Dr. Goldacre's version of the tail contraction theory over Dr. Marsland's, for in *Amoeba proteus* and *Chaos* there is excellent evidence to show that most of the tube wall does *not* shorten and, therefore, could not be contracting.

However, there are certain ameboid cells (e.g., *Difflugia*) in which various parts of the ectoplasm clearly do "contract" and show clear localized syneresis. The fact that syneresis occurs is the basis of the measurements of endoplasmic and ectoplasmic refractive indices that Dr. Cowden and I recently made. Those early experiments had some shortcomings, as Dr. Goldacre pointed out. Now we have succeeded in making our pseudopodia cylindrical in carrageenin capillaries.

DR. ZIMMERMAN: Are your newer results with carrageenin capillaries in agreement with your earlier studies, and will they be published?

DR. ALLEN: Yes, there is good agreement to the extent that the refractive index of the endoplasm is lower. The results will be written up later this year.

DR. ZIMMERMAN: As an observer without any axe to grind, I think this is the first place where the two contraction theories appear to be in close agreement, especially insofar as the ectoplasm having the ability to contract.

DR. ALLEN: We certainly agree on the major point, that contractility is the basis for streaming. I think this follows from a vast number of observations and experiments. For this reason, I tend to reject models such as Kavanau's, which break with this body of evidence and depend on hypothetical ultrastructure which no one has been able to confirm.

My personal feeling regarding the opposing contraction theories is that a definite decision between them cannot now be made, but may not be long in coming. I believe that the frontal-contraction theory is more promising in that it explains with fewer assumptions a larger number of phenomena. These phenomena of intact cells can also be explained for the most part by tail contraction, but requiring many more qualifying assumptions. Where the tail contraction or pressure theory runs into serious trouble is in explaining streaming in naked or dissociated cytoplasm. Here the pressure theory runs into a fundamental law of hydrodynamics, and the only tenable assumption which saves the theory is that the observations themselves are invalid because they are made on something less organized than an intact cell. This assumption, if widely accepted, could wipe out two decades of advances in cellular and molecular biology.

Dr. Griffin: I have been asked some rather pointed questions about the streaming loops of naked cytoplasm from *Chaos carolinensis,* with particular reference to the events occurring at the tail end of the loops during the movement of particles from the endoplasmic arm to the ectoplasmic arm. Neither the films shown during my talk nor those Dr. Allen showed at Leiden happened to include the posterior part of the loops, but we can say that particles simply reverse direction and start to move forward, just as occurs during the recruitment of endoplasm in the intact cell. Dr. Allen's film shown on Thursday evening did include the posterior part of an intact ectoplasmic tube lacking a membrane.

A model system has occurred to me that might clarify the pattern of events that could be occurring. Suppose that we have a muscle fiber in which the more rigid contracted part that is associated with a moving action potential occupies a length of 10 units and in which contraction causes a shortening of 50% and a doubling of cross-sectional area. Now, imagine what would happen if we fused the ends of a relaxed fiber 30 units long to form a circular loop. We then stimulate this fiber and damp the potential on one side of the stimulus so that the action potential passes in one direction and continues around the circle until fatigue occurs. At any given instant, a contracted and a relaxed arm of the loop, both 10 units long, would be present. Because of the rigidity of the contracted portion, both parts would be straight and would lie side by side, connected at both ends. If this system were constrained in some way to prevent a rotatory motion, as by enclosing in a capillary, the following events would occur. The action potential would be localized at the bend at the front between the relaxed and the contracted arms. Material in the relaxed arm (corresponding to endoplasm) would move into this bend and contract, and the contracted arm (ectoplasm) would move away at half the forward speed of the relaxed arm (since the cross-sectional area is doubled). The contracted material would approach the opposite end (rear), relax, and proceed forward in the relaxed arm under tension propagated through the relaxed arm.

This model would account for recruitment at the rear, the relative velocities of streaming, the more rigid ectoplasmic and the less rigid endoplasmic arms, and the cross-sectional area differences. Certainly, the organization of streaming units in ameba cytoplasm would be expected to be more tenuous than in muscle, but it seems that loop streaming could be explained by such a model. By integrating a number of such units in a three-dimensional pattern, the streaming of the intact cell could also, in principle, be accounted for.

Dr. Shaffer: One would expect rotation of the zone of contraction rather than streaming to occur under certain conditions. You have not mentioned this happening in the ameba system.

Dr. Griffin: In a free loop one might have a rotational movement or perhaps a combination of rotation and streaming. However, in the broken cells that I have observed, the loops have been inside capillaries or the posterior part of the loop has had other cytoplasmic material around it. I have not seen free loops so far.

Dr. Burgers: I have been impressed very much by the ideas presented on the movement of amebae. They raise several questions in which mathematicians can be interested and which can be used as starting points for the construction of certain models. The properties of these models can be analyzed, and biologists can decide whether the models look appropriate or not.

Some scientists have suggested that the seat of the locomotion of an ameba may be found in the surface or skin which surrounds it. This idea leads to the following mathematical problem: Suppose we have a closed surface which is deformable and which is filled with an incompressible fluid in order to keep the interior volume constant. Assume that it is possible to produce certain tensions in fibers embedded in this surface, or to release these tensions, and that the structure is of such a nature that tension produces a certain curvature of the surface. I shall suppose that at each point of the surface a state of tension with a certain curvature, or a state of relaxation, can be produced arbitrarily, by effects connected with the life of the ameba. The mathematician will then be able to work out the forms which a surface can assume, if it is endowed with such a mechanism. For instance, a certain distribution of curvature may lead to the appearance of a protruberance on the surface, which even can grow out into something with the shape of a tube (closed at its end) or of a finger.

A further possibility is obtained when we also give to the surface elements the property that they can attach themselves at will to some surface upon which the ameba happens to be resting. It is then possible, by the exertion of appropriate contractions producing curvature, combined with absence or release of tension in other parts of the surface, that the surface can roll over in a certain direction. This could represent the beginning of a displacement in that direction. If next some elements situated more forwardly with reference to the direction of displacement will attach themselves to the substratum, while the elements which previously had attached themselves would release their hold, another forward movement could be made, and so the process might continue, somewhat in the manner of a vehicle moving on caterpillar bands.

The types of movement and deformation discussed by Dr. Allen can be produced in a similar way, if there are sufficient *internal* structural elements, which can contract themselves to produce a certain curvature, or can release contraction.

In this scheme, localized tensions and release of tension either in the surface or in certain internal structural elements take the place of "pressures" in the interior substances, and I would believe that a much greater variety of shapes and motions can be obtained in this way.

In all cases mentioned, the mathematician can give a general description of possible forms of motion and can work out how contraction and release of tension must be distributed in order to produce various definite results.

From the mathematical analysis the subject then returns to the biologist, to whom it puts two definite questions. The first one is: Is it admissible to assume that various elements of the surface, or internal structural elements, can be affected by the processes of life in such a way that they contract here, and release tension there, while at the next moment this happens at some other place, and so on? Can one assume that the structural elements themselves have a power to decide when to contract or to release? In view of the fact that effects are resulting which we must consider to be definite features of the life of the ameba, we must assume that there is some form of cooperation of the various elements. They cannot behave completely capriciously or anarchically. This introduces the question whether information is transmitted between them, and whether the transmission is assured by some material system, or whether there are nonmaterial forms, unknown as yet, of inducing schemes of cooperation?

The second question refers to internal motions and to the problem: How much structure does exist in the interior substance of the cell, which in so many respects seems to be rather fluid? (This same question will also refer to the surface, if the outer surface is not a definite structure but is something that can be made up from material which at other times is in the interior.) We face the question whether structural elements (and these even might be large molecules, or complexes of molecules) can make and unmake connections between themselves, so that for a certain period of time they are able to transmit forces and to exert pulls or tensions, etc., while at another moment they may release each other, so that arbitrary displacements can take place, with the possibility that still later new contacts are made and are held again for some time. This question first of all involves problems of physical (or chemical) nature concerning contacts and bonds between molecules. But also here the problem arises: How does it come about that patterns appear to be followed? Does this need transmission of information between the cooperating parts? Theoretical studies concerning self-organizing systems have shown that certain patterns of connection can establish themselves more or less automatically through a process of repeated trial, starting from connections which originally were more or less haphazard. However, there must be some preliminary structure before anything can happen, and there must even be something of the nature of a conceptual activity, leading to the idea of cooperation and of pattern, as otherwise the term "trial" is completely hanging in the air.

With the last remarks, I am touching upon the mystery of life. My intention was to point out that mathematical discussion of forms of motion, and of schemes of contraction and release of tension which can produce them, may be helpful as a means for analyzing patterns of cooperative activity, so that biologists will have more definite structures to consider and can discuss them from their points of view.

DR. THIMANN: I have been listening to discussion on the ameba and trying to bring the movement of amebae into line with movement in other cells and in some plants. Perhaps the plant cell offers some insights, since the cytoplasm moves and the cell remains stationary, so you do not have the added complication of the movement of the cell itself.

If we think about the movement of cytoplasm in the plant cell, it seems to me our picture is relatively clear, because in order to propel liquid cytoplasm, you have to have two things. One is a spring that pushes it and the other is a solid anchor for at least one end of the spring. We used to think the solid anchor was on the cell wall, until it was shown that fully plasmolyzed cells, in which all the material was out of contact with the cell wall, could still continue streaming at a normal rate. So we know the solid anchor is not the wall, but is at least semisolid gel-like cytoplasm, about which Dr. Kamiya has spoken so eloquently. So here you have a relatively simple system: a gel with springs attached to it. The springs are all oriented in the same direction, so that every time any one of them pings off it drives a little liquid in a certain direction. The orientation may be clockwise or anticlockwise but is uniform around the cell, so at a continually macromolecular level you have a driving system which maintains a rotation.

Now, if we try to apply that to an ameba, you see at once we need our solid anchor which presumably may be the gel tube, but that the driving force is at the molecular level and does not involve any contraction; it may be located at any point all over the surface of whatever gel material there is. This, I think, makes it easier to visualize how the ameba behaves when it is in the narrow capillary into which Dr. Kamiya drives it, because it is in there that presumably the tube structure is distorted or something happens to it, so that streaming can take place from one end to the other and can be readily reversed by slight pressure. I think this could only occur if the whole driving force was at the molecular level and, therefore, located virtually all over the cell.

Dr. Wolpert: Another point which we should consider is the control of ameboid movement. So far this subject has been discussed only in terms of Dr. Goldacre's hypothesis. There are data by Dr. Bell and Mr. Jeon at Kings College which I think must be considered. This concerns the chemical induction of pseudopods. They have shown that, as Dr. Goldacre already mentioned, you can cause a pseudopod to form anywhere on the ameba, even at the tail, by the application of such substances as an extract from *Hydra*.

I think, considering the mechanism of ameboid movement, we have to understand how something applied to the outside surface of the cell, without any mechanical deformation, can somehow direct the movement. The one clue which may be relevant to this is their extension of Bingley and Thompson's ideas on the electrical potential gradient. They find that these substances that direct movement lower the electrical potential. Whether or not this is the real control mechanism, I think it is terribly important to consider such matters of control as to whether and where a pseudopod will form.

Dr. Hoffmann-Berling: Dr. Allen, if I understood correctly, you explained the moving forward of a pseudopod in terms of contraction of fused hairpin loops.

Dr. Allen: That is right.

Dr. Hoffmann-Berling: If I understood your idea correctly, what you require is not actually a contraction but a stiffening, which is lengthwise contraction. I would say that under these circumstances if you are inclined to get a model from muscle, the paramyosin system of certain catch muscles would serve your purpose much better. This system has the ability to "freeze in" at a given length without shortening.

Dr. Allen: Let me make this point clear. As I now envisage this model, both shortening and stiffening are required. Both are normal accompaniments of generalized contractile processes. The shortening is required to do the work of moving the cytoplasm and, therefore, the cell. The stiffening is required to maintain the contraction at the bends of the hairpins, i.e., between the fountain zone and the advancing rim of the ectoplasmic tube.

Dr. Hoffmann-Berling: There must be a compensatory elongation. Where does this occur?

Dr. Allen: In the tail endoplasm region, the distance between particles increases, showing that the substance in which they are embedded elongates. Then at the front of the cell they come together again where we believe contraction occurs.

Dr. Marsland: The particle in front is about to become stationary. I would think it would be almost inevitable that the other particle would catch up to it.

Dr. Allen: Yes, this alternative interpretation follows from your model. The other interpretation is shortening. This is another nondiscriminating observation, since such motions are compatible with both contraction models. I believe that probably one cannot decide what is a contraction and what is a passive deformation just by looking at particle motions.

Dr. Marsland: In fact, I think all the data can be interpreted both ways.

Dr. Allen: When it comes to particle motion data, I agree with you, and I, therefore, call attention to a large gap between the observations made and the conclusions drawn by Dr. Goldacre and Dr. Jahn.

Dr. Zimmerman: It was generally agreed that it would be desirable to have evidence of the existence of fibrillar organization in the hairpin loops on which you base much of your model. The loops are at present inferred from streaming patterns, but no organization has been demonstrated. I think your model would be strengthened by a demonstration of the presence of such an organization.

Dr. Allen: I certainly agree that any structure postulated to contract anisodiametrically must have some kind of anisotropic structure. I doubt whether this has to be fibrillar or not. As fixation methods improve, we can hope for electron micrographs of such structure if it is present.

In the meantime we do have very strong evidence that the endoplasm does have optical anisotropy or birefringence, indicating at least a diffuse orientation at the molecular level, and perhaps fibrils. Work on ameba birefringence is still in progress, and a more complete report than has hitherto appeared will soon be submitted for publication.

PART IV

Non-Brownian and Saltatory Motion of Subcellular Particles, and Mitotic Movements

Introduction

David W. Bishop

Carnegie Institution of Washington, Baltimore, Maryland

To a fascinated nonparticipant in this vigorous field of ameboid and protoplasmic streaming biology, the material presented at this Symposium suggests that heroic efforts are being made to bring unity to a chaotic situation which, in reality, may better be viewed as a packet of processes. That an underlying basis for movement in many cells may be attributed to musclelike proteins can hardly be doubted. That a *single* operational process, on the other hand—functional at all times, throughout the cell, regardless of the intrinsic and experimental state of the cytoplasm—must account for what one sees or measures during the span of observation, is not very clear, and is perhaps not really to be expected. Indeed, the ameba, at least, knows something about what is going on at both ends and, what is more, also in the middle even if there remain diametrically opposed theories as to where contraction occurs. I would emphasize the comments of Dr. Hoffmann-Berling, Dr. Andrew G. Szent-Györgyi, and others that control and regulation mechanisms may, in various fashions, act continuously throughout the cell. Perhaps, more complete analytical descriptions of relatively simplified parts of the system might aid in an elucidation of the problems; certainly, the ingenuity and elegance of some of the experimentation reported here would indicate that the time for such illumination may not be far off.

If I might comment further on the concept of "musclelike" proteins as the possible, if not probable, physicochemical basis for cytoplasmic movement, it might be well to remind ourselves that these manifest themselves in a surprising and increasing number of places. Not only are such molecular components found in muscle and certain other cells singled out here, but they have presumably also been localized in, for example, the mitotic apparatus, liver mitochondria, sea urchin egg cortex, cilia, and sperm tails. Surely we cannot expect all of these to function precisely alike. On the other hand the more pressing question here is: If we are to regard the streaming or cyclotic movements in intact or fractured amebae as a function of contractile protein, it seems surprising that this has not been adequately demonstrated in plant cells like *Chara*, *Nitella*, and *Nicotiana* which show comparable movement. It would be sheer madness, I suppose, but perhaps provocative, to suggest that other unexplored bases of motility might also operate—to propose, as Seymour Cohen has recently reminded us, that faithfully following the currently fashionable party line may obscure interesting and significant alternatives, the exploitation of which might pay out healthy dividends.

Since our discussion will move into the area of fibrillar systems of nonameboid material may I, first, record an objection to what seems like an unnecessary search for fibrillar systems in ameboid cells. If fibrils are present and accounted for, so much the better; but I believe Dr. Thimann made a good point when he suggested that we accept a molecular, that is, dispersed, basis and not feel driven to a visible fibrillar organization for motility. I should, in fact, like to point to two misleading parables apropos of fibrillar contraction which have developed in the study of my pet laboratory affection, namely, spermatozoa.

Thanks to the long-standing emphasis on fibrillar structure for motile systems and the ubiquity of the famous nine-plus-two arrangement of longitudinal filaments in the axial bundle of cilia and flagella, these elements have quite naturally been regarded as the physical basis for motility in these organelles. They may so prove, but the evidence thus far seems largely indirect. Remember that from sperm flagella contractile proteins have been extracted which, chemically and serologically, are somewhat similar to muscle protein, and which, under some circumstances of gel formation, show periodic oscillatory movements. Moreover, glycerol-extracted cell corpses can be prepared from sperm which respond to adenosine triphosphate (ATP) and the relaxing factor derived from various sources.

The ultrastructural correlation persists that the filaments are responsible for movement of an organelle which must be, simultaneously, in a state of shortening and elongation at various elements along its length. But for those who would wish to involve the entire filament in a state of contraction, I would remind them that in some spermatozoa like those of the insect, *Notonecta,* the flagella are 10–12 mm long and the functional ratio of the filaments is on the embarrassing order of about one-half million to one. A different type of discrepant evidence involving the fibrils as *the* basis for motility comes from various attempts to tag the musclelike proteins with antibodies formed against myosin or actomyosin. The primary fibrils themselves do not react, but the relatively clear matrix adjacent to them does. So, perhaps, a preoccupation with formed fibrils as the only site of reactive motile molecules may prove more misleading than helpful.

Echinochrome Granule Motion in the Egg of *Arbacia punctulata*

ARTHUR K. PARPART[1]

Department of Biology, Princeton University, Princeton, New Jersey and Marine Biological Laboratory, Woods Hole, Massachusetts

General

The movement of cellular inclusions independent of Brownian motion and of cytoplasmic streaming has frequently been observed in a large variety of cells. In his recent comprehensive review of protoplasmic streaming, Kamiya (1959) summarized our present knowledge and concepts concerning the independent motion of protoplasmic particles. The present paper presents some data and ideas about the independent motion of the echinochrome granules in the cytoplasm of the unfertilized egg of *Arbacia punctulata* (Parpart, 1953).

The echinochrome granules averaged 1.3 μ in diameter with a spread of 0.9 to 1.6 μ. Their distribution over an area of ca. 1100 μ^2 in the unfertilized egg is illustrated in Fig. 1, and their distribution over the same area for the same egg, after fertilization, is presented in Fig. 2. In the former case the granules are in active independent motion, whereas in the latter figure the granules have ceased moving and have increased about threefold in number.

The independent motion of the echinochrome granules in unfertilized *Arbacia* eggs was examined by TV microscopy (Parpart, 1951; Parpart and Hoffman, 1956). The image on the monitor was recorded on movie film at a rate of 1 frame/sec. The eggs were mounted on a microscope slide so that their jelly coat became lodged between the slide and a tilted, fixed cover slip. Thus a desired solution could be rapidly flushed around the eggs and the action of a specified agent studied. Eggs were never compressed between slide and cover slip.

The majority of the echinochrome granules are moving about in a 5–10 μ cytoplasmic layer near the surface. The granules are unaffected by the normal cytoplasmic streaming, whether they are at rest or in motion. This motion is entirely random and is of a start-stop nature. This is illustrated by the typical motion of the granule plotted in Fig. 3.

[1] This research was supported in part by the Whitehall Foundation.

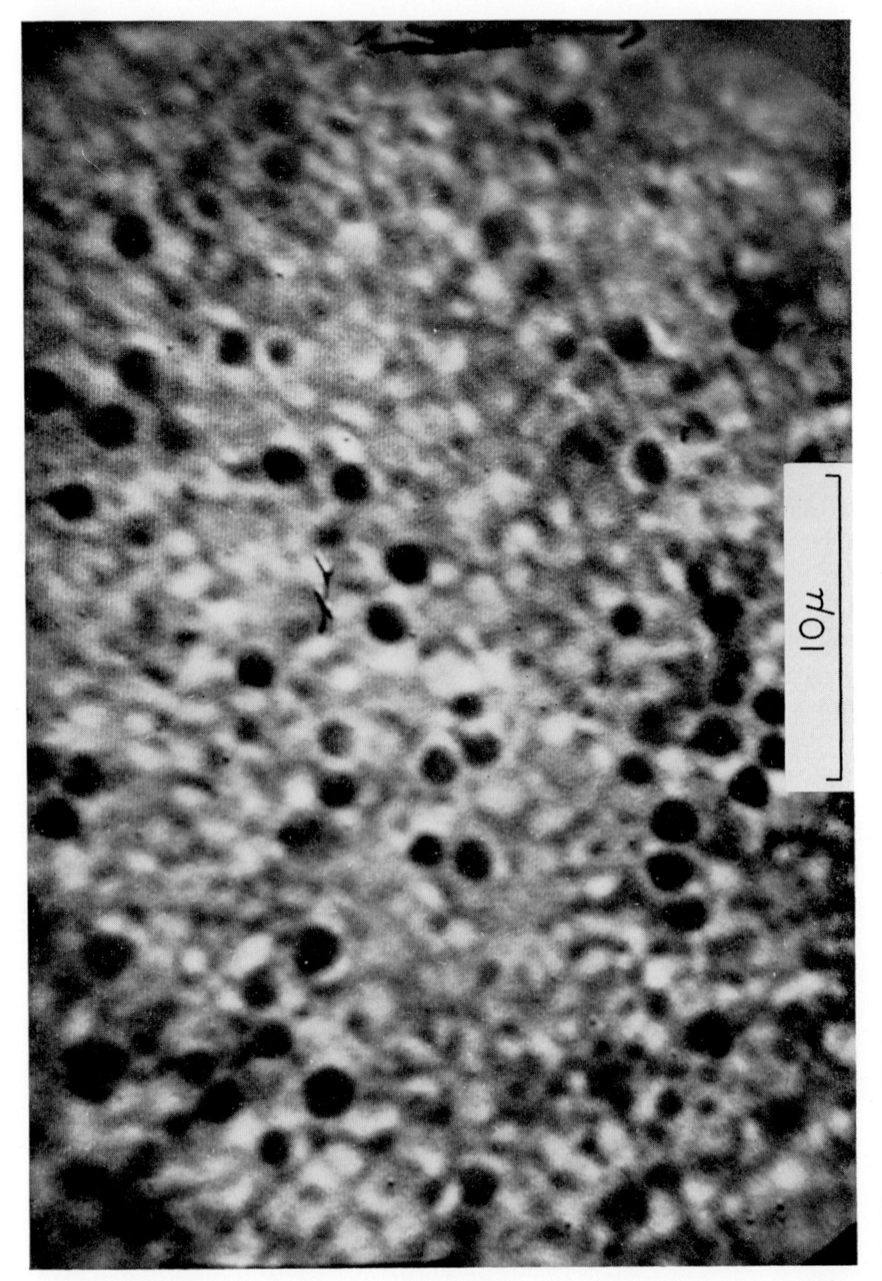

FIG. 1. Photograph of television monitor image of the upper surface, ca. 2 μ below the plasma membrane of the unfertilized egg of *Arbacia punctulata*. The optically dense spherical bodies are the echinochrome granules. The magnification of a 10 μ stage micrometer is included. The sweep lines of the monitor are visible.

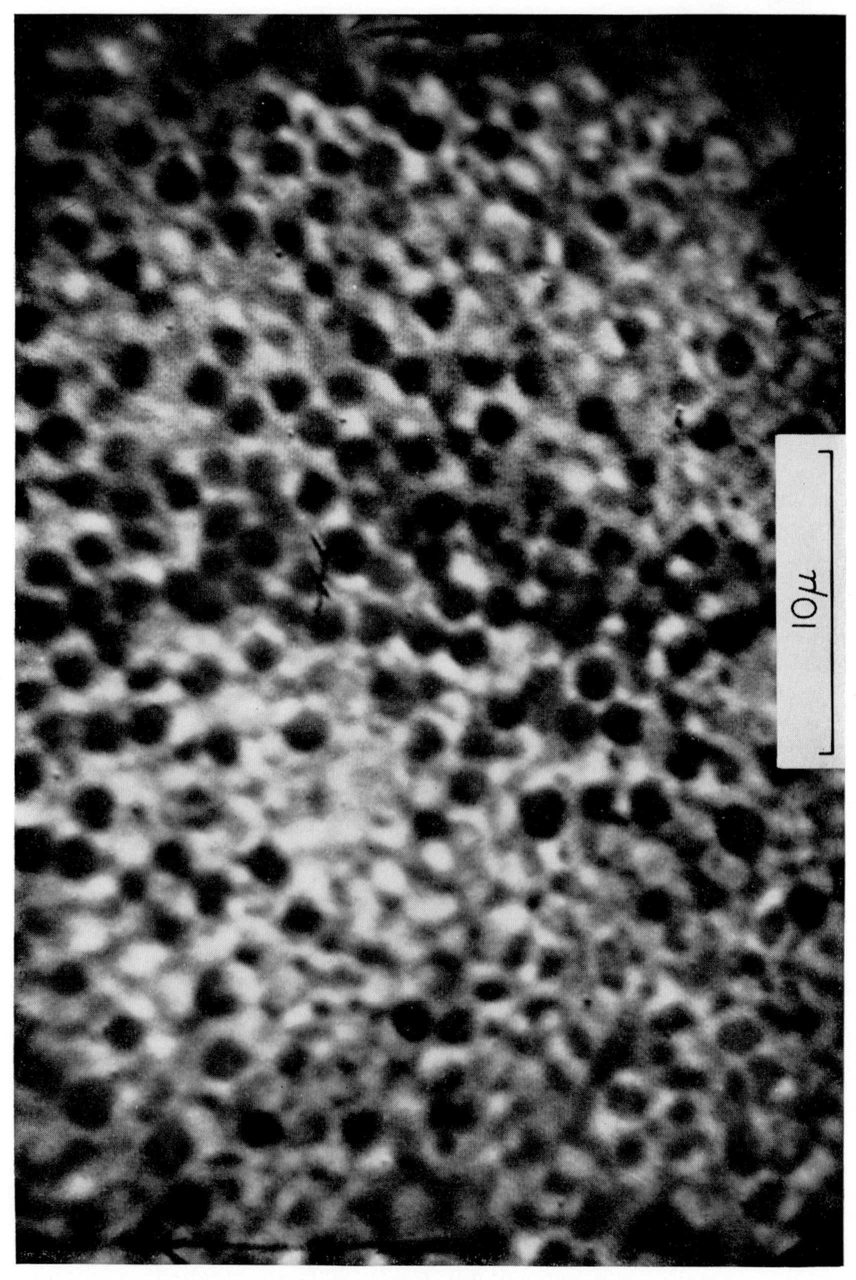

Fig. 2. Photograph of the same region of the same egg as in Fig. 1, 10 min after fertilization. All other conditions are the same as for Fig. 1.

Studies on several hundred granules of thirty egg cells at 20°C gave an average rate of movement of 0.6 μ/sec with a spread of 0.3–0.8 μ/sec. These values are from one-third to one-fifth the rate of ameboid movement of marine amebae (Pantin, 1924); and from one-quarter to one-two-hundredth of cyclosis in plant cells (Kamiya, 1959). A striking feature of the motion, observable in Fig. 3, is the fact that a granule will vibrate,

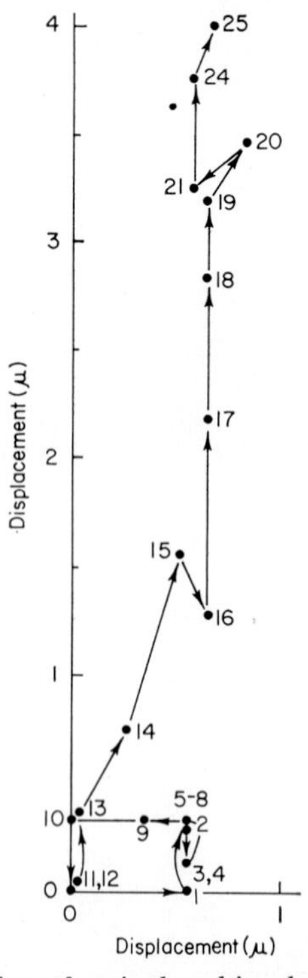

Fig. 3. Analysis of the motion of a single echinochrome granule in an unfertilized *Arbacia* egg. Determined by a motion analyzer from movie recording, at 1 frame/sec, of the monitor of a TV microscope. The numbers at each point represent successive times in seconds. At 25 sec the granule plunged out of the focal plane of the microscope. Temperature, 20°C.

Brownian motion, for a period of time (2–3 sec) at one point and then suddenly move 1–5 μ in a given plane in space.

Advantage of this latter characteristic has been taken in analyzing the action of various factors on the motion of these echinochrome granules. Frame-by-frame analysis of the motion of a minimum of one hundred granules was carried out for each factor studied. The time required for each granule to go from a state of rest to motion through 3–5 μ was determined. Each such motion was recorded as 1 displacement.

Each displacement required a certain time, and the data were averaged for each 10 displacements as to equivalent time spans. The "control" curve in Figs. 4 and 5 illustrates the spread and mean of these time

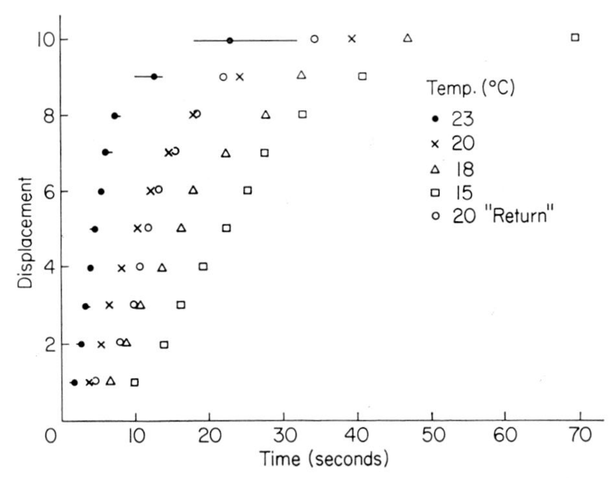

FIG. 4. *Arbacia* eggs were equilibrated at the temperatures given in the figure. The curve for "20°C Return" was for the same egg cells returned from 15° to 20°C and brought to equilibrium (25 min). Method is outlined in text. Displacement—100 echinochrome granules at each temperature.

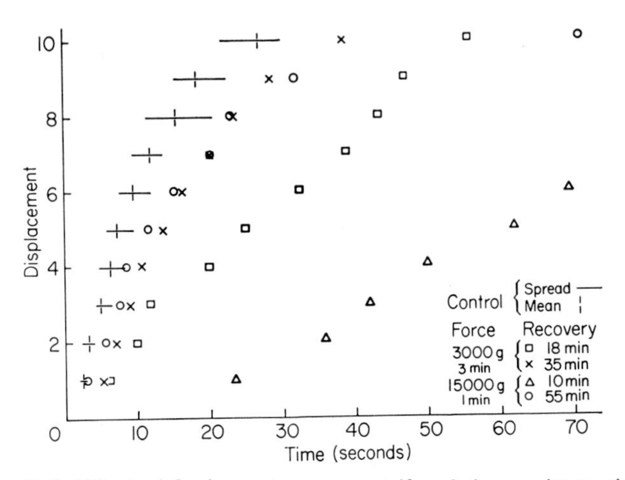

FIG. 5. Unfertilized *Arbacia* eggs were centrifuged in an isosmotic sucrose-sea water density gradient at forces 3000 and 15,000 *g* for 3 and 1 min respectively, according to the method of Harvey (1932). Then eggs were quickly washed in sea water and echinochrome granule motion studied at the "Recovery" times noted on the graph. Temperature 22°C. Displacement—100 echinochrome granules. Method is outlined in text.

measurements. Thus the behavior of a random population of granules under the influence of various factors was determined.

The action of various factors on the motion of echinochrome granules is recorded in Figs. 4 to 8. These figures record the displacement of 10 time-groups of 100 echinochrome granules against the time required

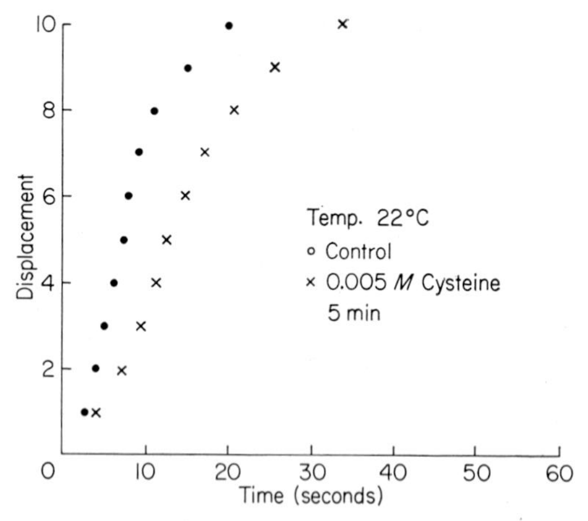

FIG. 6. Unfertilized *Arbacia* eggs exposed to 0.005 *M* cysteine dissolved in sea water for the time noted (15 min). Displacement—100 echinochrome granules. Method is outlined in text.

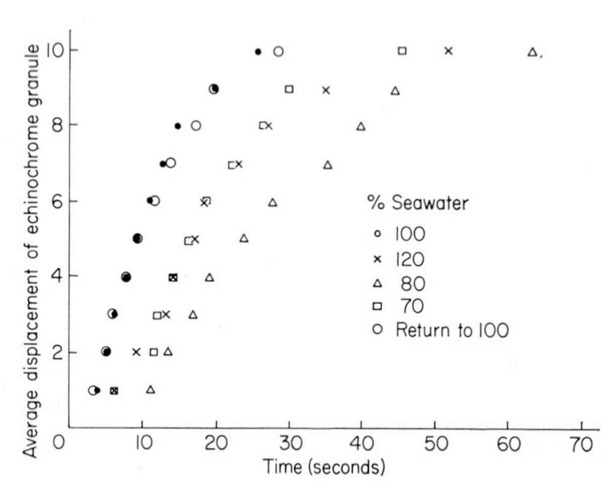

FIG. 7. Unfertilized *Arbacia* eggs exposed to the hypertonic and hypotonic concentrations of sea water noted on the graph. The curve "Return to 100%" was on the cells that had previously been equilibrated in 70% sea water. All equilibrations were for 30 min. Temperature 20°C. Displacement—100 echinochrome granules.

for such displacement. The factors studied were: temperature; centrifugal force; cysteine; tonicity; sodium iodoacetate and sodium fluoride. In all cases the action, when properly controlled, was found to be reversible.

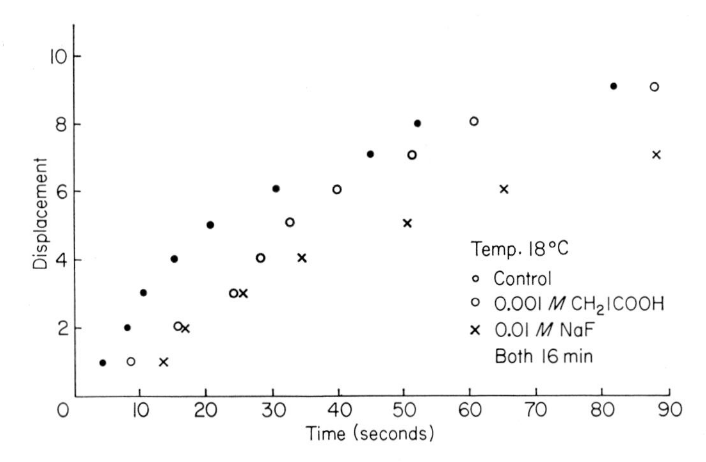

FIG. 8. Unfertilized *Arbacia* eggs exposed for a period of 16 min to (1) 0.001 *M* $NaCH_2ICOOH$ dissolved in sea water and (2) 0.01 *M* NaF dissolved in sea water. Displacement—100 echinochrome granules. Method is outlined in text.

Discussion of Factor Action

Temperature

The decrease in the rate of motion of the granules with decrease in temperature is notable in Fig. 4. The temperature coefficient (Q_{10}) for the motion of the granules varies between 4.2 and 5.7. This is a surprising action if one examines it from the point of view which Marsland (Marsland and Landau, 1954) has shown, namely, decreasing temperature over the same range leads to an increase in the sol structural character of the layer of cytoplasm ("cortical") in which the majority of the echinochrome granules are located. However, if one views the effect of decreasing temperature as a decrease of the fibrous structure of this cytoplasm, then it might appear that these fibers may play some role in the motion of the granules.

However, measurements on the distance that the individual granules will move have shown that it is not affected by temperature; even though its rate is. This is in accord with numerous data in the literature which established that the heat/tension ratios of muscle fibers undergoing isometric contraction are the same over a wide range of temperatures.

It would appear that the action of temperature on the motion of the echinochrome granule lends support to the concept that these granules are moved about the cytoplasm by fibers attached to them.

Centrifugal Force

It has long been known that the echinochrome granules have the highest density of any of the granules in the cytoplasm of the *Arbacia* egg. Thus, in the unfertilized egg, under centrifugal force they go to the centrifugal pole of the egg. Eggs centrifuged by the method of Harvey (1932) were examined for the alteration in motion of these granules. The results are given in Fig. 5. Immediately after centrifugation, whether at low (3000 *g*) or high (15,000 *g*) force, these granules have ceased all motion. Recovery of motion occurs gradually, somewhat faster at lower than higher force, until it is completely reversed. Recovery of the normal rate and distance of motion of the echinochrome granules even at the higher force, occurs long before the reorientation of the yolk and other cytoplasmic granules. Recovery of the normal rate and distance is also evident long before the re-establishment of cytoplasmic streaming.

The data on centrifugal force suggest that during its application the fibers normally attached to the echinochrome granules are torn loose and that one of the first acts of recovery is the re-formation of these fibers. This recovery occurs in the peripheral 5–10 μ of the cytoplasm. It is this latter region that is partially gelled, at 20°C, in the unfertilized egg. It is conceivable that some of the fibers formed in this gel also attach to the echinochrome granules. Such fibers, if contractile, could account for their motion.

Cysteine

Unfertilized *Arbacia* eggs exposed to low concentrations (0.001–0.005 *M*) of cysteine dissolved in sea water show a gradual diminution in the rate and distance of motion of their echinochrome granules. After from 20 to 50 min exposure all such motion is stopped. It has been demonstrated that cysteine induces a gelation of the egg cytoplasm such that no cytoplasmic granules of any sort can be moved on exposure to centrifugal forces of 15,000 *g* (Parpart and Ballantine, 1962). It has also been observed that if these eggs are returned to sea water within 15 to 20 min the effect of cysteine is completely reversible. A typical slowing of echinochrome granule motion after exposure for 5 min is presented in Fig. 6. It was also observed that at this 5-min exposure time all cytoplasmic streaming had ceased even though echinochrome granule motion had been only partly slowed.

Similar to the data for the effect of temperature, those with exposure to cysteine and the recovery suggest that there is an optimal degree of gelation of the "cortical" cytoplasm for maximal granule motion. As cysteine continues its action and the fibers of the cytoplasmic gel become more numerous the motion is impeded and stopped. This stoppage is similar to that which occurs on fertilization except for its reversibility if the exposure is not continued too long.

TONICITY

Another factor that will alter the gel-sol relations of the cytoplasm is the amount of water in it. The variables involved in shifting water into or out of the egg cell are numerous. However, an experimental evaluation of such shifts is presented in Fig. 7.

Alteration of the cytoplasmic water content whether toward hypertonicity or hypotonicity always produces a decrease in the rate but not the distance of motion of the echinochrome granules. Slight shifts toward the hypotonic (80% sea water) have a more pronounced effect than a larger (70% sea water), but still physiologically reversible, shift. All these effects are reversible if the egg is returned to sea water within an hour of the change of tonicity. Eggs equilibrated with 60% or 150–180% sea water have echinochrome granules that have ceased moving. This reaction is also reversible on return to sea water.

Unfortunately there are no data in the literature, by the present methods of study of *in vivo* gel-sol transformations, on the influence of the amount of cell water on the state of the gel-sol ratio. The data displayed in Fig. 7 would indicate that the condition of the hypothetical fibers which move the echinochrome granules is optimal in sea water. Any alteration of the water content of the cytoplasm lowers the contractile rate of such fibers. It is also possible that the contractile rate of the fibers may depend on a critical concentration of the Ca^{++} and/or Mg^{++} in the cytoplasm, which would be changed by the water shifts. This line of reasoning has some support from the studies of Bozler (1954). He observed that minute changes in the Ca^{++} or Mg^{++} concentration markedly changes the contractile properties of the isolated and reconstituted muscle fibrils which he studied. Another suggestive bit of evidence is the fact that if unfertilized *Arbacia* eggs are exposed to 0.005 *M* ethylenediaminetetraacetate dissolved in sea water, the motion of the echinochrome granules is stopped within 3 to 5 min. This is reversible if the eggs are returned to sea water.

IODOACETATE AND FLUORIDE

The influence of sodium iodoacetate (0.001 *M*) and of sodium fluoride (0.01 *M*) is recorded in Fig. 8. These metabolic inhibitors act primarily on the "recovery" mechanisms of muscle fibrils. They might, therefore, be expected to slow down gradually the rate of motion of the fibrils that are here hypothesized as responsible for the motion of the granules. This was found to be the case, since not only did it take a long time, ca. 2 hr, for these inhibitors to stop the motion of the echinochrome granules, but it was observed that the distance of travel of the granules decreased too. Addition of adenosine triphosphate (ATP) to the sea water environment of the egg cell had no effect upon the echinochrome granule motion. This is probably due to the inability of ATP to enter the cytoplasm.

Summary and Discussion

It is obvious from the foregoing that the weight of experimental evidence at present available favors the concept that the random motion of the echinochrome granules of the unfertilized *Arbacia* egg may be associated with contractile fibrils in the cytoplasm. The data strongly suggest that each echinochrome granule has a number of such contractile fibrils attached at many points on its spherical surface. The motion of these granules occurs in unpredictable directions in the three planes of cytoplasmic space. The three-planar motion has not been stressed before since all measurements were made in a two-planar layer of cytoplasm. However, when one watches the motion of these granules under the microscope or on movie film the echinochrome granules are continuously moving into and out of the focal plane of the microscope field (see legend of Fig. 3). The most dramatic motion of this sort occurs within 1 to 2 min of fertilization. If the upper 2 μ of the cytoplasm are under TV microscope observation this two-dimensional field rapidly fills up with echinochrome granules which quickly travel from deeper in the cytoplasm directly to this surface region (cf. Figs. 1 and 2). The majority of these granules, but by no means all, thus come to rest in the outermost cytoplasm of the egg.

The arrangement of the contractile fibrils is, of course, purely speculative. They may be thought of as attached at many points on the surface of the echinochrome granule. The other ends of these fibrils may be attached 10–20 μ away from the granule or to the extensive villar structures of the plasma membrane of the egg. The latter point of view seems justified from the rapid manner with which an echinochrome granule 20–30 μ deep in the cytoplasm is "pulled" to the subsurface layer. The possibility of attachment of these fibrils to fixed gelled regions of the cytoplasm has some confirmation from the fact that as the temperature of the cell is decreased the motion of these granules decreases markedly, and stops completely at about 8°C, even though the cytoplasm is going over to a more sol state as the temperature decreases. The reported actions of cysteine (Parpart and Ballantine, 1962), hypertonicity and hypotonicity, and of the metabolic inhibitors also tend to confirm such a picture.

In addition, it would appear that these contractile fibrils, if they exist, must be continuously formed and broken down. The action of centrifugal force forces many of these granules from the centripetal region of the egg through 75 μ of cytoplasm to the centrifugal end of the egg. Yet within 5 to 10 min after removal from the centrifuge these granules are again in motion and after 1 hr, even when 15,000 *g* is used they are very nearly back to the original rate and distance of motion. This is well before any cytoplasmic streaming can be observed.

The next logical step would be an attempt to demonstrate such fibrils by electron microscopy. All such attempts have failed since we have never been able to find echinochrome granules in electron micrographs of sections of *Arbacia* eggs. This led to an examination of the influence of a great variety of cytological fixing agents on the echinochrome granules of eggs under observation by TV microscopy. It was observed that none of the standard fixatives would preserve the echinochrome granules. In fact it was found that the fixatives lead to an explosive decomposition of the echinochrome granules (Parpart *et al.,* 1961).

The foregoing concepts are by no means new, except in so far as they apply to the *Arbacia* egg. The recent excellent reviews of Abé (1962) and Kamiya (1959) have stressed somewhat similar ideas for plant cells and amebae. Their articles give the pertinent literature.

There remains one further aspect of these observations on echinochrome granule motion. It would appear that this method of study of this motion opens up the possibility of study of gel-sol transformations in any cell where such granule motions can be recorded. Such studies could be carried out in all planes of the cytoplasm, and thus make possible not only analysis of region-by-region gel-sol conditions but also of the influence of a variety of agents on such transformations.

References

Abé, T. H. (1962). *Cytologia* **27**, 111.
Bozler, E. (1954). *J. Gen. Physiol.* **38**, 735.
Harvey, E. B. (1932). *Biol. Bull.* **62**, 155.
Kamiya, N. (1959). *Protoplasmatologia* **8**, 1.
Marsland, D., and Landau, J. V. (1954). *J. Exptl. Zool.* **125**, 507.
Pantin, C. F. A. (1924). *Brit. J. Exptl. Biol.* **1**, 519.
Parpart, A. K. (1951). *Science* **113**, 483.
Parpart, A. K. (1953). *Biol. Bull.* **105**, 368.
Parpart, A. K., and Ballantine, T. V. N. (1962). *Biol. Bull.* **123**, 485, and 508.

Parpart, A. K., and Hoffman, J. F. (1956). *J. Cellular Comp. Physiol.* **47**, 295-303.
Parpart, A. K., VanNorman, E., and Bernhardt, J. C. (1961). *Biol. Bull.* **121**, 403.

Discussion

Dr. Wolpert: It is well known that the eggs of various species of sea urchins have different properties. The egg of *Psammechinus miliaris* also shows cytoplasmic movements. Dr. Gustafson and I have made time-lapse movies of the unfertilized egg, and the cytoplasm shows very clear twitching movements and large oscillations that are inhibited by carbon dioxide. In this behavior it is similar to some of the observations made by Dr. Parpart on *Arbacia*. However, there is an important difference with respect to the presence of a gel-like cortex—a matter that I would like to discuss with Dr. Parpart and the people concerned with hydrostatic pressure. Dr. Mercer and I have examined centrifuged eggs, both fertilized and unfertilized, in the electron microscope and could find no region in which the larger particles, mitochondria and yolk granules, were prevented from moving by a gel-like region. We have not been able to detect, by any means, a typical cortex in *Psammechinus*.

Dr. Parpart: The entire cytoplasm of the *Arbacia* egg is in active streaming motion prior to fertilization. After fertilization all such motion is stopped in the outermost layer of cytoplasm, 5–10 μ thick. As you focus further into the cytoplasm you begin to pick up slower motion, especially as the mitotic apparatus goes into full action. However, the internal cytoplasmic streaming continues, whereas near the surface there is no streaming at all.

Dr. Wolpert: Then the two eggs must be quite different.

Dr. Zimmerman: I would be surprised if there were not a cortex in *Psammechinus* just as in *Arbacia*. In *Arbacia*, there is a region near the surface of the egg about 5–7 μ thick which has physical characteristics different from those found in the interior of the cell. As Dr. Parpart has said, if you centrifuge a fertilized *Arbacia* egg, even at very high forces, the echinochrome pigment vacuoles remain in the cortical plasmagel layer. This plasmagel layer becomes very rigid prior to division in the *Arbacia* egg. I do not know, exactly, what experiments you conducted, but very likely the plasmagel layer in *Psammechinus* becomes rigid just prior to division, as it does in *Arbacia*. Your results would depend on the stage of development at which your studies were conducted, since we have reported that there are fluctuations in the strength of the plasmagel in *Arbacia* during its division cycle. If you tested the strength of the cortex of *Psammechinus* at a stage when the cortex was relatively weak, you would record a corresponding low value for the gel. Therefore, I would be surprised if there are any major discrepancies between the different eggs.

Dr. Rebhun: It appeared to me in your pictures of centrifuged eggs that some of the yolk granules were participating in this motion. I have never seen this in a normal egg and have never seen it in your films before.

Dr. Parpart: I avoided mentioning this. It is much harder to detect. One Vidicon tube in the TV system had a high sensitivity. In fact, in the near ultraviolet it permitted exclusive delineation of the echinochrome granules. On the other hand, the Vidicon tubes used for the present studies picked up yolk and other granules. These yolk and other granules certainly are also in motion of similar type. It is not rapid.

Dr. Gustafson: May I take the opportunity to mention an observation I have made on a cytoplasmic change under the influence of cysteine and other SH compounds in the egg of *Echinocardium*. In the same range of concentrations of SH compounds as you used, a marked local contraction of the egg occurs, leading to the formation of a large concavity, presumably corresponding to the future ventral side of the embryo. The effect was irreversible, i.e., persisted after transfer of the egg into pure sea water. The same effect can be obtained with histidine. Have you studied the effect of histidine on the cytoplasmic movements?

DR. PARPART: No; I have not tried histidine, but glutathione has somewhat the same effect as cysteine. Incidentally, if you expose unfertilized eggs to cysteine for certain periods of time and then return them to sea water, there is a reversal of this whole change and actual cleavage occurs. We have gotten up to eight-cell stages with perfectly good cortex around the egg, that is, no breakdown of cortical granules but good cleavage of the egg.

DR. ZIMMERMAN: I have a question to ask Dr. Parpart about the gelating effect. How do you establish the gelation in *Arbacia*?

DR. PARPART: I am not certain of this. I only say it is gelation because the yolk and other granules cease to move about in streaming, whereas in the interior similar particles are actively moving about in streaming. This is a normal fertilized egg. With the cysteine-treated egg, the gelation or cessation of motion of yolk granules goes deeper and deeper with time of exposure.

CHAIRMAN BISHOP: May I ask you if you envisage the fibers as permanent or temporary?

DR. PARPART: They are thought to be broken and re-formed continuously.

CHAIRMAN BISHOP: Is there evidence from electron micrographs of fibrillar connections?

DR. PARPART: We cannot get electron micrographs of these granules. If you attempt to observe them under the microscope and put osmium under the cover slip, the granules explode.

Characteristic Movements of Organelles in Streaming Cytoplasm of Plant Cells

S. I. Honda, T. Hongladarom, and S. G. Wildman

Department of Botany and Plant Biochemistry, University of California, Los Angeles, California

Introduction

Observations on organelle movements in hair cells of tomato and tobacco, and mesophyll cells of spinach are recorded here by phase-contrast cine photomicrographs in color at 24 frames/sec. Previously Guilliermond (1941) and Jones *et al.* (1960) filmed organelle behavior in plant cells by time-lapse photomicrographs. Honda *et al.* (1961, 1962) without time-lapse photography recorded novel features of the interplay between chloroplasts and mitochondria as well as illustrating some rapid changes in mitochondrial forms.

In general, organelle movement is passive, being the result of Brownian motion or kinoplasm (streaming cytoplasm) movement. Only in case of mitochondria, chloroplast jackets, protuberances of the chloroplast jackets and the cytoplasmic network does it appear that some independent motion occurs. These motions are relatively localized and do not result in any substantial swimming movement or locomotion. Still, the movements shown here are remarkably dynamic and illustrate that time-lapse photography is unnecessary and even detrimental for the recording of most organelle motion in the uncultured, mature, higher plant cell.

Some systems considered here may not be classified as organelles. However, since we are concerned with organelle movement in streaming cells it is relevant to show the behavior, for example, of kinoplasm as it carries different organelles about the cell.

Mitochondria

The particles identified here as mitochondria are highly pleomorphic bodies encompassing all previously reported variations in shape and size. We consider these forms to be normal since the cells can stream for days and, as evidence of normal metabolic behavior, will produce starch in chloroplasts when illuminated. The mitochondria appear the most labile organelles in the plant cell. The slightest injury or other irritation causes them rapidly to assume spherical shapes before conspicuous changes occur in other organelles. Indeed, the appearance of appreciable numbers of rounded mitochondria is a sign that the cell is about to expire.

Although all forms of mitochondria may be seen easily in cells of spinach leaves, potato tubers, and sweet potato roots, the hair cells of tobacco and tomato are especially suited for photography at one site or of a particular mitochondrion because virtually all variations of form may occur in a short time. For example, the following sequence of events was recorded on cine film in 105 sec. An elongated mitochondrion formed a loop by end-to-end fusion (cf. Gey *et al.*, 1956; Tobioka and Biesele, 1956). Two arms were extended from the loop (cf. Frédéric, 1958) and the loop opened near the junction with one arm. A second, short, rodlike mitochondrion joined with the opened loop by end-to-end fusion (cf., sticky mitochondria reported by Gey *et al.*, 1956). A thread connecting the newly added mitochondrion with the opened loop traversed the length of the added portion in a way appearing to turn the arm inside out. Later, the remaining branch, formed before the loop opened, was retracted. Then a new arm was extended from a location near the site of the previous retraction. The arm formed by fusion with the second mitochondrion was retracted and the result was a very elongated mitochondrion with no branches. A sphaerosome collided several times with the elongated mitochondrion without showing any tendency to fuse or stick. Finally, a third mitochondrion fused end-to-end with the elongated mitochondrion.

In addition, we have observed loop formation by longitudinal splitting of a portion of a mitochondrion (cf. Dangeard, 1958), but we have not observed complete longitudinal splitting. Fusion of spherical mitochondrial granules occurs (cf. Sorokin, 1941; Tobioka and Biesele, 1956; Jones *et al.*, 1960), but we have never observed full length, side-to-side fusion of rodlike mitochondria (cf. Tobioka and Biesele, 1956). Syncytial complexes of several mitochondria occur (cf. Dangeard, 1958; Palade, as reported by Novikoff, 1961). Common mitochondrial forms, showing thick portions tapering to thin threads, resemble those forms that Robertson (1960) interprets as occurring in the formation of mitochondria from the endoplasmic reticulum. Rarely, the process of mitochondria becoming trapped in pulsating loops of kinoplasm may be seen. Mitochondria at times enter between the lobes or platelets of Golgi bodies and may or may not emerge. Although not recorded on cine film, we have observed the coalescence of mitochondria with the cytoplasmic network upon application of pressure on cells and their reappearance from the network after the release of pressure.

Apart from arm extensions and retractions, which have been described as ameboid (cf. Sorokin, 1941), and local contractions and relaxations (pulsations), mitochondria show only passive motion. They often move on paths defined by the cytoplasmic network. Sometimes the mitochondria in following invisible tracks suddenly move out of focus. These deviations may result from a twisting motion of strands of kinoplasm (cf. Kamiya and Seifriz, 1954) which carry along the mitochondria, sphaerosomes, Golgi bodies, chloroplasts, and nuclei.

We have recorded a transformation between chloroplast structures and mitochondria (Honda *et al.*, 1961, 1962; Spencer and Wildman, 1962; Wildman *et al.*, 1962). Chloroplasts *in situ* are surrounded by jackets of material not containing chlorophyll (Figs. 1 to 3). The jackets, integral parts of chloroplasts, are in continuous motion and at times extend long protuberances even from chloroplasts without any obvious amount of jacket. These protuberances may segment into particles resembling mitochondria.[1] In addition, mitochondria may coalesce with the chloroplast protuberances and lose their individual identities or only partially fuse together. It is stressed here that the segmentation of chloroplast protuberances does not appear to be a very widespread activity in terms of numbers.

Many plant physiologists and biochemists are reluctant to accept these segmented particles from chloroplast protuberances as mitochondria, in the sense that they contain respiratory enzymes, even though the particles appear and behave *in situ* as mitochondria do in plant cells without chloroplasts. The biochemical properties of the particles from protuberances remain to be characterized. However, there is no rigorous evidence with purified plant preparations that respiratory systems with dehydrogenases, electron transport enzymes, phosphorylating enzymes, and cytochrome oxidase are limited, in fact, to one type of plant particle with cristae. Two possibilities, among several, to account for mitochondrial behavior of particles derived from chloroplast protuberances are: (*a*) Mitochondria may be incorporated into chloroplast jackets by a fusion or melting of their limiting membranes and become invisible and later by segmentation of chloroplast protuberances the mitochondria may regain their individuality; and (*b*) particles segmented from

[1] Some of the confusion about the pleomorphism of chondrioconts, plastids, and mitochondria (cf. Guilliermond, 1941; Sorokin, 1941, 1955) may be related to transformations between mitochondria and jackets of nongreen plastids in epidermal cells (Figs. 2 and 3). These plastids are much smaller than chloroplasts found in mesophyll cells and, not appearing green, they may not be recognized as chloroplasts. However, the presence of chlorophyll in these plastids is readily revealed by fluorescence microscopy.

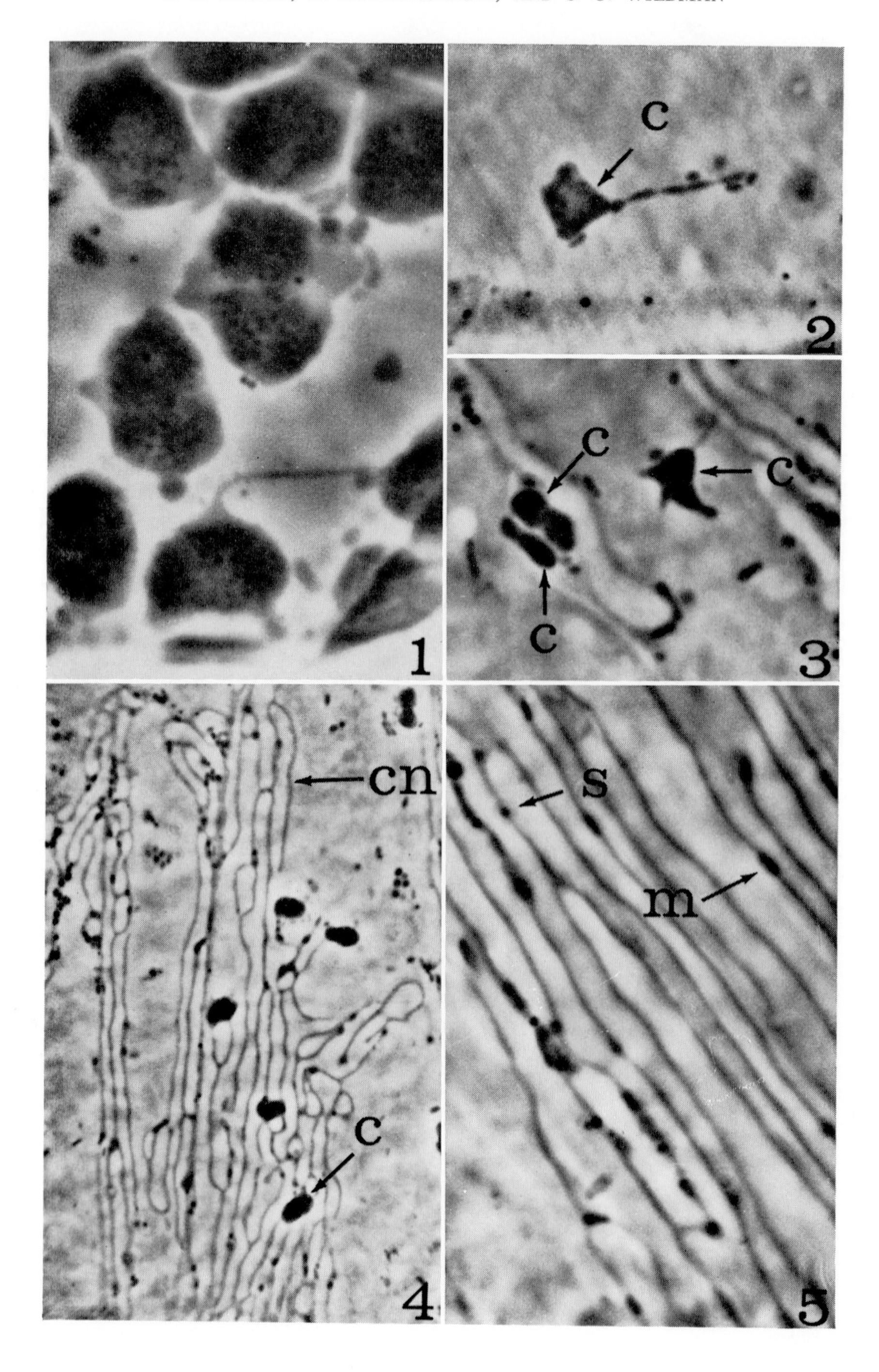
c
2
c
c
c
c
1
3
cn
s
m
c
4
5

chloroplast protuberances may develop new biochemical properties upon their removal from the internal milieu of chloroplasts. The concept of mitochondria developing enzymatic activities is not a long step from mitochondria being biochemically heterogeneous or pleotropic (cf. Mercer, 1960; Avers, 1961).

The reluctance to accept segmented particles from chloroplast protuberances as mitochondria seems partly related to definition. We use the classical sense of the term "mitochondrion" (Benda, 1902); that is, threadlike body, because in the living cell we can only identify an observed particle by morphological traits and behavior. It should be pointed out that all those particles called mitochondria by us appear as threadlike bodies at one time or another even though they assume other shapes during intervening periods. Indeed, dynamic pleomorphism is characteristic of mitochondria, and this property sharply distinguishes them from the behavior of sphaerosomes. A contemporary definition of plant mitochondria (Mercer, 1960) requires the presence of cristae mitochondriales and the ability to oxidize fatty acids and to exhibit oxidative phosphorylation with Krebs cycle acids.

It is often tacitly assumed that cristae and the ability to exhibit oxidative phosphorylation are coincident properties of mitochondria. There is little doubt that respiratory enzymes reside in animal mitochondria with cristae; rather, the concern is the localization of enzymes within the mitochondria. There are only a few cases, limited to rat liver and beef heart mitochondria, which attempt to establish the local-

---

FIGS. 1 to 5. Positive phase-contrast photomicrographs were obtained with a Zeiss GFL microscope equipped with a V Z phase-contrast condenser, Neofluar phase objective lenses, Komplan ocular lenses, basic body II with photocell and meter, and Zeiss camera back. Eastman Double-X film, cine type 5222, was exposed with electronic flash and printed on No. 6 grade Agfa Brovira paper except for Fig. 1, which was printed on F-5 Kodabromide paper. All subjects were living cells mounted in distilled water. Figure 1—grana are visible as dark gray spots in oval-shaped regions within the irregularly shaped, light gray jackets of spinach mesophyll chloroplasts. One extended chloroplast protuberance is visible. Magnification: $\times$ 4000. Figure 2—jacketed chloroplast with beaded, extended protuberance (c) photographed in a living tomato hair cell. Chloroplasts of this type do not appear green by phase microscopy but reveal the presence of chlorophyll by fluorescence. Magnification: $\times$ 2000. Figure 3—jacketed chloroplasts (c) with and without chloroplast protuberances photographed in a living tomato hair cell. These structures do not appear green but contain chlorophyll. Magnification: $\times$ 2400. Figure 4—cytoplasmic network (cn), formed by strands and sheets of kinoplasm which may completely surround small vesicles or vacuoles, with chloroplasts (c) photographed in a living tomato hair cell. Magnification: $\times$ 1200. Figure 5—cytoplasmic network with out-of-focus sphaerosome (s) and mitochondrion (m) photographed in a living tomato hair cell. Magnification: $\times$ 2400.

ization of respiratory enzymes on cristae or other fragments of mitochondrial membranes (Siekevitz and Watson, 1956, 1957; Watson and Siekevitz, 1956; Ziegler *et al.*, 1958; Fernandez-Moran, 1962). Even in these cases, the correlations can be questioned. For studies of this type Novikoff (1961) points out the necessity for morphological evidence showing the origin of fragments and the establishment of satisfactory sampling procedures. Sjöstrand (1956) mentions the possibility of an association of enzymes on artificially formed structures from lipids and proteins in mitochondrial extracts.

The presence of cristae is not in itself sufficient to show that a particle has the conventional biochemical properties of a mitochondrion. Yotsuyanagi (1959) found particles with cristae in a mutant yeast which had no cytochrome oxidase, succinic dehydrogenase, and other respiratory enzymes (Ephrussi and Slonimski, 1955; Yotsunyanagi, 1955). Avers (1962) found about nine times more plant particles with cristae per cell than particles stained with Janus green B or the Nadi reaction (Avers, 1961; Avers and King, 1960) which are among the most reliable agents for localizing electron transport enzymes and cytochrome oxidase (cf. Novikoff, 1961).

It is not excluded that aerobic respiratory enzymes can reside elsewhere than in mitochondria. The evidence from staining with Janus green B and the Nadi reaction indicate the presence of electron transport enzymes and cytochrome oxidase in bacterial cells without mitochondria (for evidence and discussion see Novikoff, 1961). Furthermore, Perner (1953) found that the sphaerosomes rather than the mitochondria showed a positive Nadi reaction. He suggested that contaminations of plant mitochondrial preparations with sphaerosomes have led to conclusions that plant mitochondria possess cytochrome oxidase.

Regardless of definitions, do the particles segmenting from chloroplast protuberances possess cristae? Electron micrographs bearing on this point are obtained fortuitously. There are very few published electron micrographs which even show the chloroplast jackets as obvious jackets. In general, most electron micrographs show no cristae in the protuberances or the critical regions are insufficiently defined. These protuberances, however, are never in a visibly different, extremely extended configuration seen in the living cell during segmentation. Thus, these regions in most electron micrographs may not represent the state of protuberances about to segment.

We have seen suggestions of cristae in some of our micrographs (Hongladarom, unpublished electron micrographs), and M. Nittim and F. V. Mercer (private communication) have found a case of well-defined cristae in a protuberance of a bean chloroplast. Unfortunately, as in other cases, electron microscopy cannot permit a decision on direction of movement. Thus, cristae may not occur in a segmenting particle but rather in a fusing mitochondrion. Our attempts to fix cells at the moment of segmentation of chloroplast protuberances during observation by phase microscopy have not been successful.

With regard to plant mitochondria in general, we feel that a complete extension of homology between animals and plants to the organelle level is not warranted or desirable. Since plant "mitochondrial" preparations are grossly heterogeneous and because of a peculiarity of plant respiration they are insensitive to cyanide (cf. Chance and Hackett, 1959; Hackett, 1959; Bonner, 1961), it is particularly pertinent to determine rigorously if there are characteristic particles bearing characteristic respiratory systems. The unorthodox conclusion of Perner (1953) that cytochrome oxidase is in the sphaerosomes rather than in the mitochondria should not be dismissed cavalierly.

The Cytoplasmic Network

The cytoplasmic network is kinoplasm (cf. the streaming cytoplasm, Kamiya, 1960) in anastomosing strands and sheets which surround small vesicles or vacuoles in a pattern resembling a network (Figs. 4 and 5). Portions of the network may be a visible manifestation of the endoplasmic reticulum (ER). Porter (1961) summarized the reasons why the endoplasmic reticulum may be recognized by light microscopy. Others have interpreted similar networks in different cells as being endoplasmic reticulum (Rose and Pomerat, 1960). Fawcett and Ito (1958) have shown by electron microscopy that their network visible by light microscopy was, indeed, the endoplasmic reticulum.

The network may be very extensive or it may be difficult to find in plant cells. It is localized in a layer adjacent to the central main vacuole. Strands of kinoplasm (the streaming cytoplasm) traverse the central vacuole from wall to wall. The diameter of such strands can be smaller than the diameters of sphaerosomes or mitochondria which distort such strands during their passage across the vacuole. Sheets of kinoplasm also may extend through the main vacuole and completely partition the vacuole. Both the strands and sheets of kinoplasm contain network vesicles. Neighboring strands of the network may stream in opposing directions. However, it is rare that such network strands are exactly in the same plane of focus although inclusions may pass from one strand to another. Kinoplasm is not always visible but its presence may be inferred from the streaming of cell inclusions in ordered arrays without deviations from invisible tracks. As with the visible network, neighboring streams of invisible kinoplasm may move in opposing directions. At times mitochondria and sphaerosomes swirl about in local vortices of kinoplasm and then continue in a more orderly procession.

The network may stream vigorously or it may display little or no motion. Kinoplasm moves the various cell inclusions about the cell and, indeed, the Golgi bodies and the nucleus are always surrounded by kinoplasm in the form of the cytoplasmic network. This is consistent with the notions that the Golgi bodies are intimately connected with the ER system and that the ER system forms an envelope around the nucleus (cf. Porter, 1961). Where network movement is extremely slow or nil, the mitochondria tend not to show ameboid motion. They remain in a given configuration without change. When streaming of the net recommences the mitochondria begin to show pulsations and pleomorphic changes.

At times small portions of the network strands separate from the rest of the network under normal conditions. These drops of kinoplasm appear like isolated drops of protoplasm described by Yotsuyanagi (1953). We have not observed protrusions of small arms from the drops although the surface of the drops appears to pulsate. These drops of kinoplasm resemble particles called "leucoplasts" by Perner (1958). Also during streaming, the network sometimes forms a collection of vesicles which becomes separated from the main body of kinoplasm and floats free in the central vacuole. These collections of vesicles assume a spherical shape. We have never observed the balls of free floating network to return to the cytoplasm.

Individual strands of kinoplasm in the network sometimes behave as mitochondrial threads except that they are still connected to the general network. Contortions of such strands are especially well shown on the surface of hexagonal inclusions of U_1 tobacco mosaic virus in a previous cine film (Hongladarom *et al.*, 1961; Honda *et al.*, 1962). As with mitochondria, the motion of network strands may result from the movement of surrounding cytoplasm but swellings and contractions and segmentations are less easily explained.

Certain treatments, such as pressure, stop kinoplasm motion and parts separate from the network in forms sometimes resembling extremely thick, elongated mitochondria. These particles or closed tubes of the network may be mistaken for pathological forms of mitochondria (cf. Dangeard, 1958). These particles can be distinguished from mitochondria by their great size and lack of pleomorphism after release of pressure. They usually show local contractions and relaxations to give the effect of pulsations moving about the particle surface. Smaller spherical network pieces may become trapped in network vesicles together with sphaerosomes and mitochondria. The mitochondria under these conditions usually become spherical. Cells usually recover from pressure treatments that result in the above described alterations.

At certain degrees of pressure application, mitochondria can be made to coalesce with the cytoplasmic network. The mitochondria lose their individual identity in this process. Upon release of pressure, the mitochondria re-form from the network. In untreated cells, mitochondria are often connected to the network by sticky, very fine threads. During streaming the mitochondria may be pulled by the threads connected to the network and the threads may distort the mitochondria and finally snap apart.

The similarities in behavior and appearance of mitochondria and strands of network and the transformations between them suggest to us a close relationship. Robertson (1960) has suggested that mitochondria originate from the endoplasmic reticulum. Rudzinska and Trager (1959) and Glauert and Hopwood (1960) suggest that the bacterial membranes, which resemble endoplasmic reticulum in other organisms, may assume the functions of both the mitochondria and the endoplasmic reticulum of other organisms.

It is our experience that the mitochondria, the cytoplasmic network, and the chloroplast jackets and protuberances may be transformed reversibly into each other under certain conditions. We have noted that in untreated cells when there is an extensive amount of network, there are fewer visible mitochondria. When there are many mitochondria, there is little jacket material around the chloroplast.

A classic representation of a mature plant cell is that of a sheet of cytoplasm bounded on the cell wall side by the plasmalemma and bounded on the vacuolar side by the tonoplast. However, the tonoplast area must be many times greater than the area of the plasmalemma. Virtually the entire cytoplasmic network may be exposed to the central vacuole. Thus, the constantly changing, vastly convoluted surface of the cytoplasmic network exposed to the vacuole represents the tonoplast. A useful oversimplified analogy is to imagine a thin layer of cytoplasm pressed against the cell wall from which hangs an enormously complicated anastomosing system of tubules, some of which extend through the central vacuole from one cell wall to another. The outer surface of each tubule constitutes part of the tonoplast. It can also be imagined that the small vacuoles entirely surrounded by the network and also some portions of the central vacuole are in almost direct contact with the cell wall. These regions of contact would undergo a continual change in position as the cytoplasmic network changes its position during streaming.

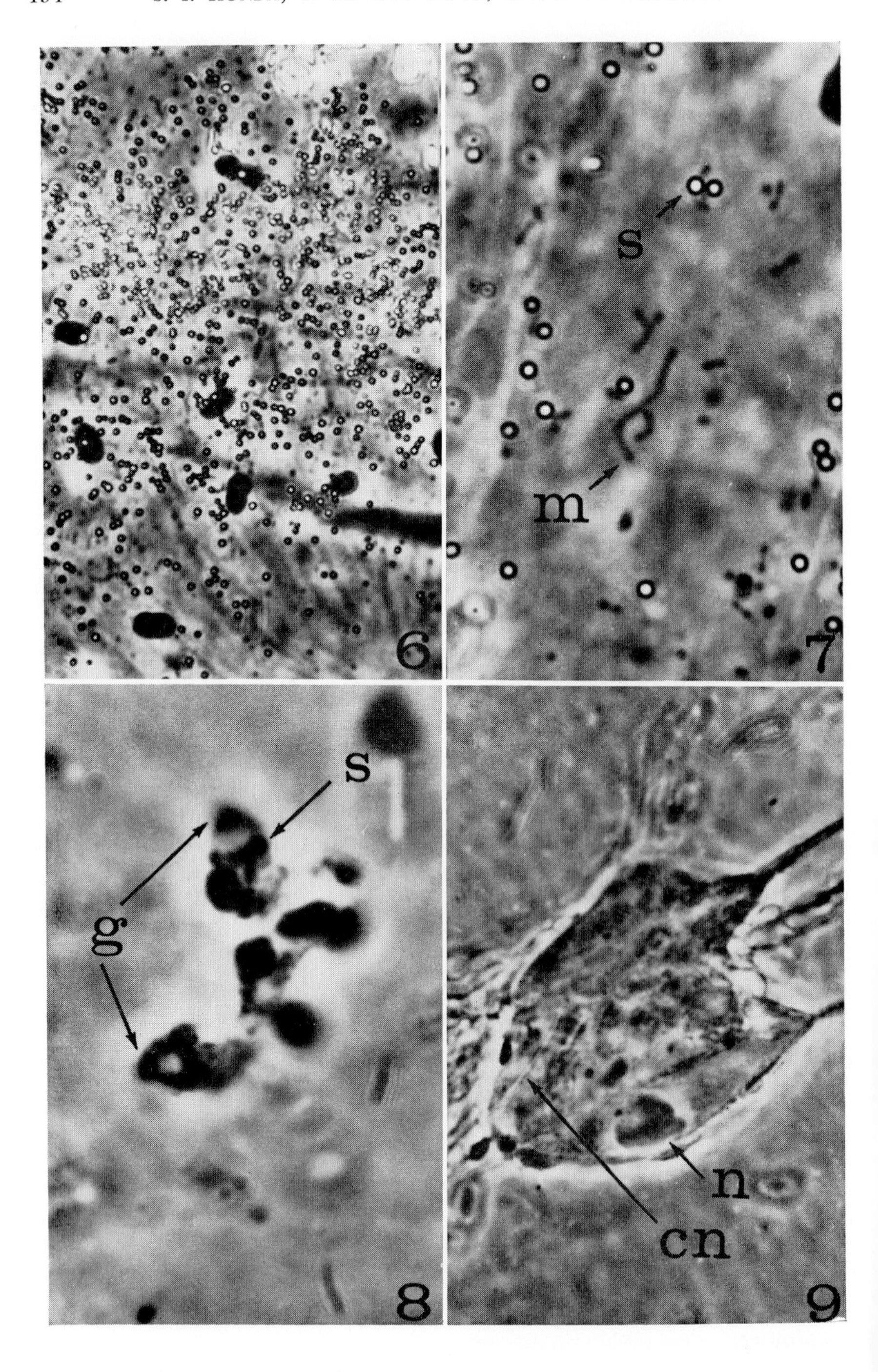
6
s
m
7
s
g
8
n
cn
9

Sphaerosomes

Sphaerosomes, spherical refractile bodies about 0.2 to 0.5 μ in diameter, exist in large numbers in plant cells (Figs. 6 and 7). Formerly, these particles were called "microsomes" by botanists before biochemists (cf. Claude, 1943) appropriated the term in referring to artifacts derived from the endoplasmic reticulum of animal cells (cf. Palade and Siekevitz, 1955). Steffen (1955) and Perner (1958) have reviewed the known properties of the sphaerosomes and the means for distinguishing them from proplastids, lipid droplets, and mitochondria. Virtually nothing is known about their biochemical properties since plant biochemists have ignored the existence of these numerous bodies in plant cells and in "mitochondrial" preparations. In higher plants from five different families, the sphaerosomes are the most numerous particles of one type visible in a cell by light microscopy (cf. Wildman and Cohen, 1955).

Although the sphaerosomes occur throughout the cytoplasm they appear chiefly in the faster moving layer of the kinoplasm near the main vacuole. Therefore, the bulk of the sphaerosomes will not be visible when chloroplasts imbedded in the cortical gel are being observed. However, both sphaerosomes and mitochondria are sometimes observed to pass between the stationary chloroplasts and the cell wall. The sphaerosomes may be easily distinguished from all other cell inclusions by their bright, spherical appearance in dark-field illumination and their rapid translocation by the kinoplasm. Sphaerosomes display no independent motion and show only Brownian motion while being transported by kinoplasm. As with the mitochondria, the sphaerosomes usually move in paths defined by the cytoplasmic network but they are not restricted to the paths. The sphaerosomes may be observed to enter between the lobes or plates of the Golgi bodies.

Golgi Bodies

Although the Golgi bodies here have not been rigorously identified by comparison of electron micrographs with light photomicrographs of

FIGS. 6 to 9. Same equipment and conditions as for Figs. 1–5. See page 489. Figure 6—large number of sphaerosomes photographed in a living tobacco hair cell. Chloroplasts and mitochondria are out of focus. Magnification: × 960. Figure 7—sphaerosome (s) and branched mitochondrion (m) photographed in a living tobacco hair cell. Several sphaerosomes and mitochondria are visible. Magnification: × 1200. Figure 8—complex of Golgi bodies (g) with closely associated sphaerosomes (s) photographed in a living tomato hair cell. Magnification: × 2400 Figure 9—lobed nucleus with nucleolus (n) photographed as it was suspended by the cytoplasmic network in the central vacuole of a tomato hair cell. A strand of the cytoplasmic network (cn) is closely appressed to the nuclear surface. Magnification: × 1200.

identical particles, these structures resemble the Golgi bodies, apparatus, complexes, or dictyosomes in other cells (Fig. 8). These structures satisfy the criteria of Pollister and Pollister (1957) who present an especially useful summary for the interpretation of dark and light images in the recognition of Golgi bodies by light microscopy. Also there are present in the cells we observe, complex structures which possess the fine structures (Hongladarom, unpublished electron micrographs) now accepted as characterizing the Golgi bodies (Dalton and Felix, 1954). These structures show the typical associations of small vesicles with the edges of compressed or flat sacs in close proximity with endoplasmic reticulum (cf. Dalton, 1961).

The Golgi bodies occur only in the cytoplasmic network and are moved slowly about the cell by kinoplasm. Spilling of network vesicles over the surface of the Golgi bodies may contribute to the vesicular aspects of regions in and around the Golgi bodies as revealed by electron microscopy.

We have observed that both mitochondria and sphaerosomes, traveling along the same strands of cytoplasmic network that carry the Golgi bodies, may enter between the lobes or plates of the Golgi bodies (Fig. 8). When trapped between the lobes of the Golgi bodies, the mitochondria and sphaerosomes show Brownian motion. Occasionally, on larger Golgi bodies there appear to be remnants of chloroplasts. Rarely have trapped particles been observed to leave the Golgi bodies. The trapped particles around the Golgi bodies may also correspond to the granules and vesicles associated with the Golgi bodies revealed by electron micrography. Observations of the Golgi bodies do not impart the notion that in living cells the vesicles or granules are produced in the Golgi bodies, as is often inferred from electron micrographs of dead cells. Rather, we feel that cell detritus accumulates around these structures.

The Golgi bodies display an unusual stiffness or rigidity to their forms. This stiffness is extremely characteristic. In contrast, other cell inclusions except crystals appear to possess a certain amount of plasticity.

The Nucleus

In the living, uninjured cell the saucer-shaped nucleus rarely, if ever, appears homogeneous and rounded. It is irregularly lobed and furrowed (Fig. 9). Although surrounded by the cytoplasmic network the nucleus moves only slowly even while suspended in the interior of the vacuole away from the cortical gel. Since the cells we observe are mature and will not divide, the nuclear observations represent only one aspect of the resting stage. We see, for example, no chromosomes or nuclear fission. However, the nucleoli often present a heterogeneous appearance, in some cases extending protuberances into the nuclear body (Fig. 9). It is rare to see more than one nucleolus per nucleus in these cells.

Some aspects of a close apposition of the cytoplasmic network on the nuclear surface may be mistaken for a furrowed surface (Fig. 9). However, the typical network strands may be traced out into the cytoplasm. Superficially, it appears that the cytoplasmic network streams into and out of the nucleus. It is clear, however, that the cytoplasmic network only covers the nuclear surface. The whole complex appearance is consistent with portions of the network being endoplasmic reticulum (cf. Fawcett and Ito, 1958) which envelopes the nucleus, as shown by electron microscopy (cf. Porter, 1961).

Immobile mitochondria and sphaerosomes may be seen on the nuclear surface. If there is little streaming in a network strand, the mitochondria and sphaerosomes may stay in prolonged proximity to the nucleus. Chloroplasts also may be carried to the nuclear surface and stay or be carried away again.

In common with mitochondria, the irregularly shaped nucleus becomes rounded when the cell is injured. This may account for the rounded appearance of nuclei in many fixed cells. Nuclei in cell extracts made with most types of media also appear rounded, swollen, fragmented, or even nonexistent.

Chloroplasts

Chloroplasts generally are immobile, being imbedded in the cortical gel which is interposed between the cell wall and the vacuole. Some kinoplasm may pass between the immobile chloroplasts and the cortical gel because mitochondria and sphaerosomes sometimes stream between the cell wall and the chloroplasts. Usually the mitochondria and sphaerosomes stream around the chloroplasts rather than across the chloroplast face on either the vacuolar or cell wall side.

When displaced from the cortical gel and caught in the kinoplasm the chloroplasts move about the cell. They readily may become immobile again. Sometimes the chloroplasts rotate in the strands of kinoplasm. Presumably the chloroplast rotation reflects the twisting of kinoplasm (cf. Kamiya and Seifriz, 1954). During the rotation the chloroplasts clearly show grana in face view while side views only display a bright, smooth, refractile aspect. The lenticular aspect of chloroplasts on side presumably acts as a refracting system that obscures the grana.

Even while imbedded in the cortical gel, the chloroplasts may show a small degree of movement. However, the motion may oppose the flow of kinoplasm. Indeed, chloroplast protuberances may extend opposite in direction to the kinoplasm flow and the chloroplast also may move in the opposite direction. It should be noted, however, that the chloroplasts and the cytoplasmic network and kinoplasm, in general, are located on slightly different planes not readily distinguished. The small chloroplasts in hair cells are more prone to movement than the larger chloroplasts in spinach mesophyll cells.

It is now generally accepted that light can orient chloroplasts within higher plant cells (cf. Granick, 1955). Intense light is supposed to cause the movement of chloroplasts in a way to expose the least surface area of the chloroplasts. Voerkel (1933) reported that blue light (408–510 mμ) filtered through $CuSO_4$ solution caused movement of chloroplasts. We usually do not see such movements when examining cells with filtered, intense light from a xenon arc burner suitable for cine photomicrography. However, if interference heat filters are removed from the light path, the spinach chloroplasts move to the side walls of mesophyll cells. The interference heat filters decrease the transmitted light to 10% of the incident intensity at 350 and 770 mμ. Beyond these extreme wavelengths the intensity is even less. From 400 to 670 mμ the transmitted intensity is 50–76% and the transmission is further attenuated as 350 and 770 mμ are approached. Thus, it appears that infrared radiation may engender chloroplast movement.

The outer, irregularly shaped jacket that surrounds the granular portion of a chloroplast (Figs. 1 to 3) presumably corresponds to the mobile peristromium of Senn (1908) as described by Guilliermond (1941). The ameboid movements of the jackets do not result in locomotion of the chloroplasts. Only the grana fluoresce red under ultraviolet light. The chloroplast jackets and protuberances, therefore, do not contain chlorophyll or protochlorophyll.

The joining of mitochondria with chloroplast protuberances does not abolish the form changes which we find are characteristic of mitochondria. For example, the following sequence was filmed in 84 sec. A short mitochondrion joined with a protuberance of a tomato hair chloroplast by end-to-end fusion. Movement of the original mitochondrial portion in the kinoplasm pulled the chloroplast about. The combined protuberance became beaded with thick portions separated by fine thread connections. The middle, thick section of the protuberance formed an arm. The original terminal portion of the protuberance was then retracted into the rest of the protuberance. Segmentation of the protuberance then occurred and the mitochondrial-like particle moved away in the kinoplasm.

Granick (1955, 1961) presents the view that in higher plants the number of chloroplasts per cell is less if the chloroplasts are large. Our preliminary results indicate that for mature spinach mesophyll cells there are more and larger chloroplasts in the larger cells. In a single cell, however, there is a distribution of chloroplast sizes. The question arises, from where do the additional chloroplasts originate in the larger mature cells? We do not see chloroplast division even though two granular structures may be in close proximity to each other in a single jacket. By appropriate treatments, such as application of pressure or 2,4-dinitrophenol, separate jacketed chloroplasts can be made to join together to appear as two or more structures bearing grana in a single jacket. Therefore, the mere presence of two structures with grana in a single jacket is not sufficient evidence for chloroplast division.

On rare occasions, we have seen bodies which may be proplastids in mature mesophyll cells of light-grown spinach. Their movements in the kinoplasm are too rapid to permit the identification of protochlorophyll in the particles by fluorescence microscopy. There are, however, bodies resembling proplastids in electron micrographs of spinach chloroplast preparations isolated from mature spinach by Kahn and Wettstein (1961).

In spinach mesophyll cells, jacketed chloroplasts (Fig. 1) are easily distinguished from non-chloroplast bodies. However, in tobacco and tomato hair cells and in spinach leaf epidermal cells, the chloroplasts are small pleomorphic bodies (Figs. 2 and 3) which often do not appear green. The pleomorphism of the chloroplast jackets of these small organelles may allow for them to be mistaken for mitochondria or non-green plastids. Many forms corresponding to "colorless" plastids or leucoplasts are, in fact, chloroplasts with mobile jackets, as can be verified by fluorescence microscopy. When these organelles are classed as chloroplasts by their fluorescence, we see no other particles except the drops of kinoplasm which might be classified as proplastids or leucoplasts.

Summation

The mature, higher plant cell is a complex, dynamic multiphase system. It is stratified from outside to inside as follows: cell wall; non-mobile or extremely slowly moving cortical gel with chloroplasts and, occasionally, mitochondria and sphaerosomes imbedded in it; at least two major layers of kinoplasm, i.e., streaming cytoplasm; the slower moving layer adjacent to the cortical gel contains the major portion of the mitochondria and the faster moving layer nearer the vacuole contains the bulk of the sphaerosomes; and finally the main vacuole. Kinoplasm strands anastomose and sheets may enclose small vesicles or vacuoles, thereby forming a syncytial pattern, the cytoplasmic network. The endoplasmic reticulum may be manifested by portions of the cytoplasmic network. Other strands and sheets of kinoplasm may traverse the central vacuole from wall to wall. Mitochondria, sphaerosomes, Golgi bodies, the nucleus, and chloroplasts, when dislodged from the cortical gel, are moved about the cell in kinoplasm. They follow paths marked by the cytoplasmic network. Although the Golgi bodies and nucleus occur only in kinoplasm in the form of the cytoplasmic network, both the mitochondria and sphaerosomes are not restricted to such paths. Mitochondria and sphaerosomes may be seen passing from layer to layer of the slow and fast moving layers of kinoplasm. Usually the mitochondria and sphaerosomes stream around the chloroplasts which bulge out of the cortical gel. However, sometimes they stream across the chloroplast face on either the cell wall side or near the vacuole. The chloroplasts often are irregular in shape with grana restricted to a regular, oval portion within a jacket not containing chlorophyll. Sometimes protuberances are extended from the jackets and segment into mitochondria-like bodies which behave and appear as mitochondria *in situ* by light microscopy. Mitochondria may coalesce with the chloroplast jackets and completely lose their individual identity or only momentarily fuse or stick with the jackets.

The dynamic behavior of living cells and their organelles may necessitate new conceptions of the basis of structure as it is related to function. Anticipating the need for a more flexible view, Mercer (1960) points out that the dynamic pleomorphism of cell constituents represents the structural counterpart of biochemical turnover and metabolic pools. He clearly warns: "If structure has a statistical existence then many of our classical concepts of cell physiology and cell morphology may be hopelessly inadequate."

Acknowledgments

This work has been supported in part by contract AT-(11-1)-34 from the U.S. Atomic Energy Commission, research grant E536-(C10) from the U.S. Public Health Service, and special research fellowship GSP-17,795 and Research Career Development Award GM-K3-17,795 (U.S. Public Health Service, Division of General Medicine) to S. I. Honda. We have greatly benefited from our discussions with Drs. G. G. Laties and J. B. Biale.

References

Avers, C. J. (1961). *Am. J. Botany* **48**, 137.
Avers, C. J. (1962). *Am. J. Botany* **49**, 996.
Avers, C. J., and King, E. E. (1960). *Am. J. Botany* **47**, 220.
Benda, C. (1902). *Ergeb. Anat. Entwicklungsgeschichte* **12**, 743.

Bonner, W. D., Jr. (1961). *In* "Haematin Enzymes" (J. E. Falk, R. Lemberg, and R. K. Morton, eds.), Part 2, pp. 479-500. Pergamon Press, London.

Chance, B., and Hackett, D. P. (1959). *Plant Physiol.* **34**, 33.

Claude, A. (1943). *Science* **97**, 451.

Dalton, A. J. (1961). *In* "The Cell" (J. Brachet and A. E. Mirsky, eds.), Vol. II, pp. 603-619. Academic Press, New York.

Dalton, A. J., and Felix, M. D. (1954). *Am. J. Anat.* **94**, 171.

Dangeard, P. (1958). *Protoplasmatologia* **3**, A1.

Ephrussi, B., and Slonimski, P. (1955). *Nature* **176**, 1207.

Fawcett, D. W., and Ito, S. (1958). *J. Biophys. Biochem. Cytol.* **4**, 135.

Fernandez-Moran, H. (1962). *Res. Publ. Assoc. Res. Nervous Mental Disease* **40**, 235.

Frédéric, J. (1958). *Arch. Biol. (Liège)* **69**, 167.

Gey, G. O., Shapras, P., Bang, F. B., and Gey, M. K. (1956). *In* "Symposium on the Fine Structure of Cells," pp. 38-54. Interscience, New York.

Glauert, A. M., and Hopwood, D. A. (1960). *J. Biophys. Biochem. Cytol.* **7**, 479.

Granick, S. (1955). *In* "Handbuch der Pflanzenphysiologie" (W. Ruhland, ed.), Vol. I, pp. 507-564. Springer, Berlin.

Granick, S. (1961). *In* "The Cell" (J. Brachet and A. E. Mirsky, eds.), Vol. II, pp. 489-602. Academic Press, New York.

Guilliermond, A. (1941). "The Cytoplasm of the Plant Cell," 247 pp. Chronica Botanica, Waltham, Massachusetts.

Hackett, D. P. (1959). *Ann. Rev. Plant Physiol.* **10**, 113.

Honda, S. I., Hongladarom, T., and Wildman, S. G. (1961). *Federation Proc.* **20**, No. 1, 146.

Honda, S. I., Hongladarom, T., and Wildman, S. G. (1962). "Organelles in Living Plant Cells" (16-mm sound film). Educational Film Sales and Rentals, University Extension, University of California, Berkeley, California.

Hongladarom, T., Honda, S. I., and Wildman, S. G. (1961). *1st Ann. Meeting Am. Soc. Cell Biol.,* Abstracts, p. 92.

Jones, L. E., Hildebrandt, A. C., Riker, A. J., and Wu, J. H. (1960). *Am. J. Botany* **47**, 468.

Kahn, A., and von Wettstein, D. (1961). *J. Ultrastruct. Res.* **5**, 557.

Kamiya, N. (1960). *Ann. Rev. Plant Physiol.* **11**, 323.

Kamiya, N., and Seifriz, W. (1954). *Exptl. Cell Res.* **6**, 1.

Mercer, F. V. (1960). *Ann. Rev. Plant Physiol.* **11**, 1.

Novikoff, A. B. (1961). *In* "The Cell" (J. Brachet and A. E. Mirsky, eds.), Vol. II, pp. 299-421. Academic Press, New York.

Palade, G. E., and Siekevitz, P. (1955). *Federation Proc.* **14**, 262.

Perner, E. S. (1953). *Protoplasma* **42**, 457.

Perner, E. S. (1958). *Protoplasmatologia,* **3**, A2.

Pollister, A. W., and Pollister, P. F. (1957). *Intern. Rev. Cytol.* **6**, 85.

Porter, K. R. (1961). *In* "The Cell" (J. Brachet and A. E. Mirsky, eds.), Vol. II, pp. 621-675. Academic Press, New York.

Robertson, J. D. (1960). *J. Physiol. (London)* **153**, 58 P.

Rose, G. G., and Pomerat, C. M. (1960). *J. Biophys. Biochem. Cytol.* **8**, 423.

Rudzinska, M. A., and Trager, W. (1959). *J. Biophys. Biochem. Cytol.* **6**, 103.

Senn, G. (1908). "Die Gestalts- und Lageveranderung der Pflanzen- Chromatophoren." Engelman, Leipzig, Germany.

Siekevitz, P., and Watson, M. C. (1956). *J. Biophys. Biochem. Cytol.* **2**, 653.

Siekevitz, P., and Watson, M. C. (1957). *Biochim. Biophys. Acta* **25**, 274.

Sjöstrand, F. S. (1956). *Intern. Rev. Cytol.* **5**, 455.
Sorokin, H. (1941). *Am. J. Botany* **28**, 476.
Sorokin, H. (1955). *Am. J. Botany* **42**, 225.
Spencer, D., and Wildman, S. G. (1962). *Australian J. Biol. Sci.* **15**, 599.
Steffen, K. (1955). *In* "Handbuch der Pflanzenphysiologie" (W. Ruhland, ed.), Vol. I, pp. 574-613. Springer, Berlin.
Tobioka, M., and Biesele, J. J. (1956). *J. Biophys. Biochem. Cytol., Suppl.* **2**, 319.
Voerkel, S. H. (1933). *Planta* **21**, 156.
Watson, M. L., and Siekevitz, P. (1956). *J. Biophys. Biochem. Cytol.* **2**, 639.
Wildman, S. G., and Cohen, M. (1955). *In* "Handbuch der Pflanzenphysiologie" (W. Ruhland, ed.), Vol. I, pp. 243-268. Springer, Berlin.
Wildman, S. G., Hongladarom, T., and Honda, S. I. (1962). *Science* **138**, 434.
Yotsuyanagi, Y. (1953). *Cytologia* **18**, 202.
Yotsuyanagi, Y. (1955). *Nature* **176**, 1208.
Yotsuyanagi, Y. (1959). *Compt. Rend. Acad. Sci.* **248**, 274.
Ziegler, D. M., Linnane, A. W., Green, D. E., Dass, C. M. S., and Ris, H. (1958). *Biochim. Biophys. Acta* **28**, 524.

Discussion

Dr. Robineaux: One can observe fragmentation of mitochondria when the mitochondria are going into a stream. What do you think about the active movement of mitochondria?

Doubtless you are familiar with the work of Frédéric on the mitochondria of fibroblasts. What do you think about possible active movements of mitochondria in plants?

Dr. Honda: I feel that there is no active movement in the sense of locomotion. There may be local exchanges of substances in the way Dr. Mahlberg suggested, or local contractions in the mitochondrion as a result of phosphorylation. I doubt whether this should be called active movement.

Dr. Robineaux: Did you ever observe the disappearance of the mitochondria?

Dr. Honda: Yes; this happens occasionally.

Dr. Robineaux: In mitosis?

Dr. Honda: These cells are mature vacuolated cells; they do not divide.

Dr. Wohlfarth-Bottermann: Your pictures are the best I have seen of the endoplasmic reticulum in living cells; but, first, may I ask whether you have studied the endoplasmic reticulum in the electron microscope to make sure this "network" is really endoplasmic reticulum?

Second, I saw in your pictures a suggestion that sphaerosomes and mitochondria are flowing along the "cytoplasmic network." Do you believe that these sphaerosomes or mitochondria are flowing *in* this endoplasmic reticulum or on the outer side?

Dr. Honda: We have made some electron micrographs of cells in which the characteristic membrane structures do correspond to endoplasmic reticulum, but we cannot say that the network is entirely endoplasmic reticulum in all cells. The streaming of the different particles with the network we believe is generally on the surface. They may be trapped within, but this is very abnormal. Such things happen when the cell is treated with pressure or dinitrophenol.

Dr. Allen: I want to ask exactly what you mean by the term "kinoplasm." I think you are aware of the history of the meaning of this term.

Dr. Honda: Yes. We just use the term in the sense of streaming cytoplasm.

Saltatory Particle Movements in Cells

LIONEL I. REBHUN

Department of Biology, Princeton University, Princeton, New Jersey, and the Marine Biological Laboratory, Woods Hole, Massachusetts[1]

Introduction

In this paper we shall deal with a kind of intracellular particle movement that does not follow the statistics associated with Brownian motion. In the latter type of movement, the total path length traversed by a particle in a Newtonian fluid is proportional to the square root of the time considered for all sufficiently long time intervals (Mysels, 1959). In the movement considered here, particles may show little or no motion for intervals of minutes and then suddenly undergo translatory motions involving distances of 20–30 μ in a period of a few seconds, only to become quiescent again. Not all types of particles participate in this movement in a given cell type and, indeed, particles may at one time undergo Brownian movement and suddenly undergo a process converting this to sudden, discontinuous motion, i.e., saltatory motion. There is a wide range of particle types in different cells which may at times in the cell cycle participate in saltatory movement.

In addition to the startling characteristics of saltatory movement itself, cells undergoing mitosis often show a remarkable ability to organize the saltatory movements in such a way that the particles become differentially located around the centrosomes. That is, particles will move into the centers by a kind of in and out movement with the path traversed toward the center, of greater length than that away, the net result being an aggregation of particles surrounding the astral centers. Such aggregations may persist on the nuclear surface through interphase to the following mitosis in some instances (*Spisula*) and in other cases the aggregations break up (*Pectinaria*), re-forming again at the succeeding mitosis.

The evidence presented later leads us to the conclusion that the movements are not autonomous but are induced by the interaction of the particles with some other system in the cytoplasm. We have thus

---

[1] This work has been supported at various times by the NSF (G 13422), the NIH (GM-07426), and the American Cancer Society (P 51 and P 267).

been led to search for some widely distributed system in the cell, present in interphase in essentially unoriented form (since the particle movements are unoriented in interphase) but which undergoes an orientation radial to the asters at mitosis, and which may act as the movement-inducing system. We have in a previous paper tentatively concluded that the endoplasmic reticulum (ER) or some elements associated with it may represent such a system (Rebhun, 1963). Evidence presented later also suggests that mechanisms of saltatory particle movement have many characteristics in common with mechanisms of spindle action (see also, Rebhun, 1963; Ostergren, *et al.*, 1960). If this is so, then we are forced to consider the possibility that fibrils such as have been seen both in spindles *and* asters (Kane, 1962; Harris and Mazia, 1962) may be involved in these movements although no evidence for the extensive, unoriented deployment of these elements in the cytoplasm during interphase has been described. These matters will be discussed at length later. Much of this material has been discussed in a recent paper (Rebhun, 1963) and so a complete description of experimental procedures and of electron micrographs will not be given, but the reader will be referred to that paper for details.

Observations and Literature Review

The personal observations reported here have been made on particle movements in cells of four different organisms with primary emphasis on metachromatic granules in eggs of the surf clam, *Spisula solidissima*. These eggs were treated with methylene blue or with toluidine blue in dilute solution in sea water and then observed. The cytoplasm can be seen to contain many particles, which, in the case of toluidine blue, are stained metachromatically. In the unfertilized egg the particles are uniformly distributed throughout the cytoplasm. On observing them in detail, either directly through the microscope or by means of time-lapse movies taken at rates of 1 frame/sec or 1 frame/2 sec, it can be seen that the particles undergo saltatory movements which have the following detailed characteristics:

1. The motion of observed particles shows a discontinuous velocity distribution in the sense that a given particle may show little or no observable movement for minutes and then may suddenly move for distances up to 30 μ at velocities up to 5 μ/sec, this motion ceasing as suddenly as it started.

2. The movement of a given particle is not influenced by its proximity to other particles. Thus, of two neighboring particles separated by distances as little as 1 μ, one particle may undergo saltatory movement and the second remain stationary. However, occasionally, groups of 2

to 5 particles move in one extended continuous movement with no observable relative motion between them, i.e., they all move together. This is especially true of the aggregates of particles often occurring at mitosis (Rebhun, 1959).

3. Only the metachromatic granules show this behavior in the *unfertilized* egg, as far as can be observed. The various refractile bodies, presumably yolk, lipid, and mitochondria, do not show this movement and are not influenced in their movement by metachromatic granules which are moving very close to them.

4. With methylene blue, the particles may grow considerably larger if eggs are left in stain for long rather than for short periods, and some of them may attain diameters of 2–3 μ. Observation of these particles compared to particles in eggs stained for short periods (such particles being about 0.5 μ in diameter) reveal no difference in saltatory behavior, including distances moved and maximum velocities attained. Further, the larger particles are actually denser than the smaller ones, moving to a more centrifugal position in the egg on stratification in a centrifuge.

5. For most of the movement, especially if the distance traversed is large, the velocity appears to be uniform so that the force applied to the particle must be applied continuously during the motion in overcoming viscous drag. That is, the motion does not appear to be one in which an impulse is applied during a short interval with the particle then gradually coming to a halt.

6. Where verifiable, i.e., with particles large enough for clear observation, no change in form or size of particles can be observed during motion.

The previously mentioned characteristics of the motion are identical to those of yolk granule movements in the unfertilized egg of the annelid, *Pectinaria* [the older name, *Cistenides,* was used previously (Rebhun, 1963)], except that many fewer granules undergo these movements in this egg at any one time. Again, in melanin granules in the migrating melanocytes of the 3–4 day embryo of the minnow, *Fundulus,* the same set of characteristics describe the melanin particle movements. In addition, countermoving streams of particles may often be observed in these cells, with an occasional particle reversing its direction in the middle of the stream and moving opposite to that of the rest of the particles. In the final case with which we have had direct experience, that of echinochrome granules in the unfertilized eggs of the sea urchin, *Arbacia,* the same set of characteristics of the movement may be seen, as has been partially reported by Parpart (1953) and more fully in this Symposium. Considerably more detail concerning these processes may be obtained in previous papers (Rebhun, 1959, 1960, 1963).

Observations of saltatory particle movements, though not usually thoroughly documented, are numerous in the literature and exist for a wide variety of cell types. A cell for which the observations are detailed is the newt heart fibroblast, as studied by Taylor (1957). In this case the fibroblasts were observed in tissue culture. Taylor was interested in studying the viscosity of the cytoplasm at different mitotic stages by using statistical analysis of the movements of particles showing Brown-

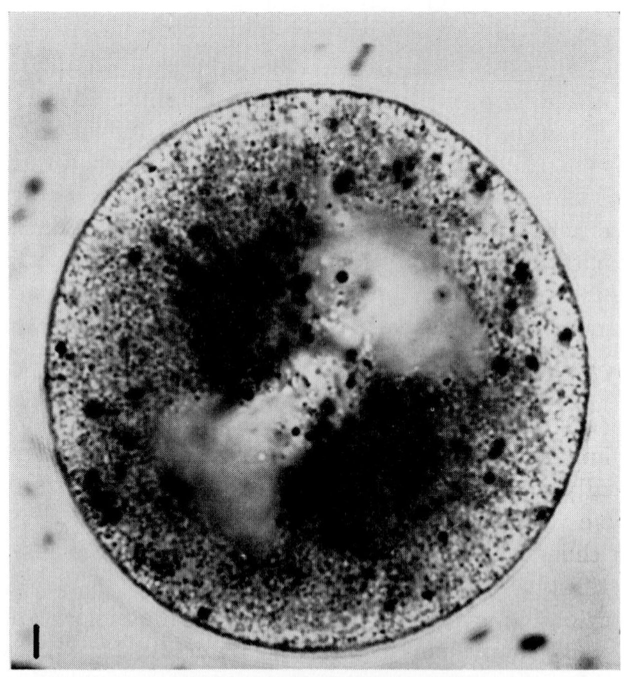

FIG. 1. Fertilized egg of *Spisula* at about 2–3 min subsequent to germinal vesicle breakdown. A tight aggregation of methylene blue particles can be seen in the asters of the forming first polar body spindle.

ian motion. The particles studied were the so-called lipid granules. These granules, however, not only showed Brownian movement, but on occasion would engage in long saltatory movement. Indeed, as may be verified in the film by Bloom and Zirkle (undated), the particles show the same behavior characteristics (1 to 6) as described for the cells previously. In this case, however, when not engaged in saltatory movement, the particles can be seen to be engaged in Brownian movement, whereas in the case of egg particles, this is not always easy to see, although it probably occurs.

In protozoa, movements similar to those described previously have been reported by Andrews (1955; see also Seifritz, 1952), and have been seen by others (Allen, personal communication). Most important is the observation of Andrews that ingested carmine or ink particles undergo such movements. An important case of saltatory particle movements in plant material is that which occurs in extruded cytoplasm of *Chara* and *Nitella*, reported by Jarosch (1956) and Kamiya (1959). A final probable case of such movement occurs in HeLa cells (Rose, 1957). In

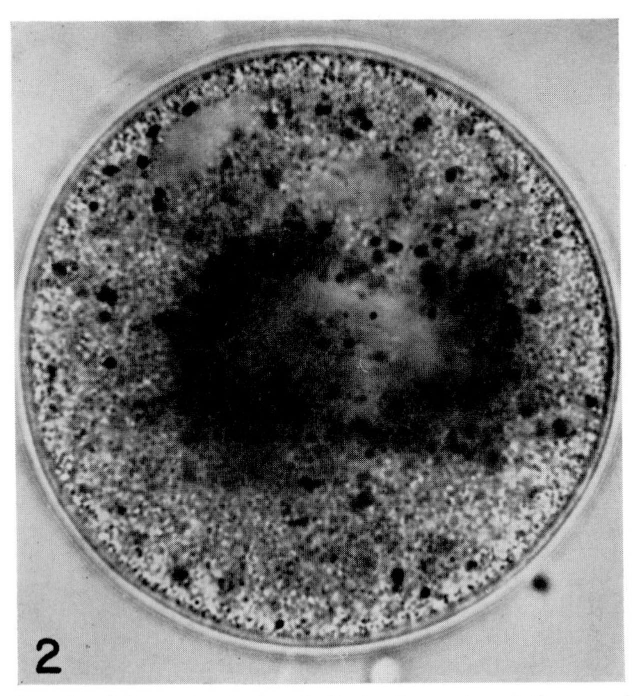

FIG. 2. Completed first polar body spindle in the center of the egg of *Spisula* at about 20 min after fertilization. The degree of astral aggregation of methylene blue particles seen in Figs. 1 and 2 is higher than in most eggs at this stage but is not rare.

this case Rose was describing microkinetospheres and their ability to move against the general stream of material entering the cell by pinocytosis. His description of their "peculiar" movements is suggestive of the movements we have been describing. Indeed, our own informal observation of many films of cells in tissue culture, lead us to conclude that almost all such cells possess a class of particles whose movements appear to be saltatory in the sense of characteristics 1 to 6 described previously.

If the vitally stained eggs of *Spisula* are fertilized, further important behavior characteristics appear. The particles begin to show an orientation in their movement which was previously absent, Figs. 1 and 2. At first, this orientation is variable from cell to cell, but generally, by the polar body stages (*Spisula* eggs only undergo maturation after fertilization) most of the particles have moved into the asters; specifically, at this stage they surround the central aster, few, if any, actually moving out with the polar bodies. By the time the first cleavage spindle has formed in the center of the egg most of the particles are in the astral regions surrounding both centrosomes but not penetrating into them,

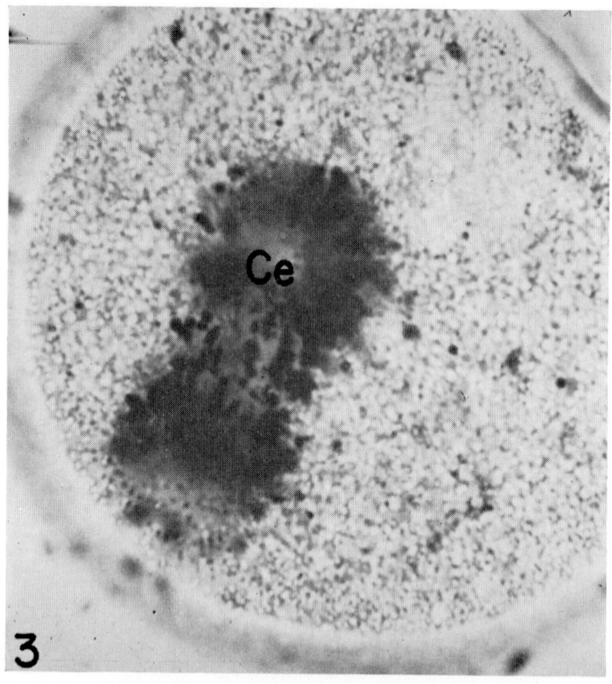

FIG. 3. First cleavage spindle in *Spisula,* rotated about 80° from its normal orientation by pressure on the cover slip. Note the strings of particles on the spindle. Note also the clear centrosomal region (Ce). The degree of aggregation of methylene blue particles seen is universal in normal eggs at this stage.

Fig. 3. At the conclusion of telophase, and after formation of the blastomere nuclei, the particles remain aggregated on the peripheral pole of the nucleus (that is, the pole between the nucleus and the egg surface). During prophase of the next division the mass of particles separates into two masses, each one surrounding the aster of the newly formed spindle, Fig. 4. This cycle is repeated until at least the fifth cleavage.

This description concerns the average location of the particles at various stages but does not describe the motions of the particles during their transport into the asters. During this period the number of particles moving at any one time increases considerably, especially in *Pectinaria,* and the movements gradually become oriented relative to the centrosomes, although always maintaining a saltatory character. That is, individual particles still show the characteristics 1 to 6 (described previously) but now many more are involved in the movement, and the movements are directed. However, though the movement of particles is oriented radial to the astral centers, some particles may move away

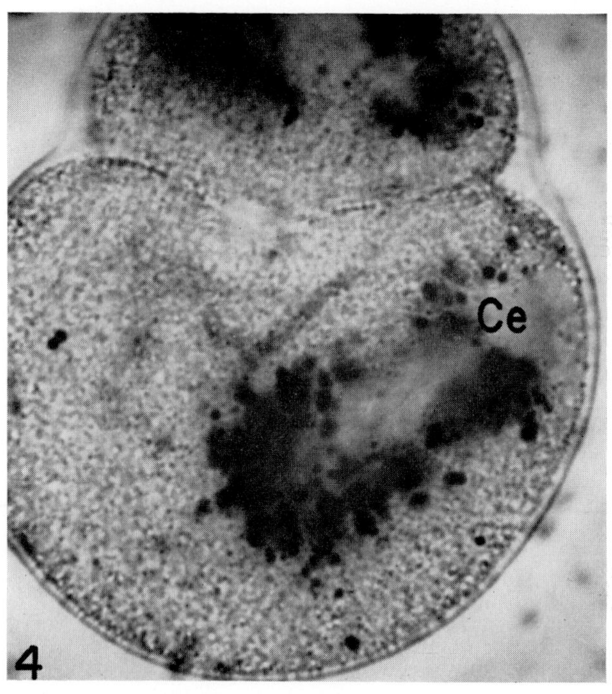

FIG. 4. Beginning of second cleavage in *Spisula* egg. Note clear peripheral centrosome (Ce).

from the center while other particles are moving toward it; that is, contiguous particles may move in opposite, but radial, directions in the aster. However, the average distance traversed into the centers is greater than that out of them and the net result is the accumulation of the particles around the centrosomes. This is, however, a dynamic accumulation in the sense that particles still show bidirectional movement of small amplitude even after the accumulation has resulted in a tight ingathering of particles. The result is the same for *Pectinaria* as for *Spisula,* except that in *Pectinaria* the particles disperse after the reformation of the blastomere nuclei in telophase, whereas in *Spisula,* as indicated previously, the particles remain aggregated on the peripheral pole of the blastomere nucleus.

These observations may be summarized in the following way:

7. At times in the cell cycle when a mitotic apparatus is present, the number of particles moving with saltatory movements at any one time increases and the movements possess the characteristics 1 to 6, previously listed.

8. The particles statistically orient their movements radial to the centrosomes and aggregate tightly about them by a process in which the independent movements of particles into the asters have greater path lengths than those away from the centers. More detail is contained in previous reports (Rebhun, 1959, 1960, 1963).

This ingathering of particles into the aster has been seen by several other investigators, the first report, to our knowledge, having been that of Fischel (1899), who dealt, in part, with the astral distribution of neutral red particles in sea urchin eggs. Iida (1942) also reported the astral location of neutral red particles in sea urchin eggs and referred the movement of the particles to the growth characteristics of the aster. Kojima (1959a,b) in two excellent papers studied the neutral red particles in three species of sea urchins and an echiuroid, from the point of view of astral orientation and the possible involvement of the particles in furrowing activity of the egg, and in ciliary development in later stages of embryogenesis. A considerable amount of work on particles in marine invertebrate eggs stainable with basic metachromatic dyes and showing the movements 1 to 8 described previously has been done by the Belgian group at Brussels (Pasteels, 1955, 1958; Pasteels and Mulnard, 1957; Mulnard, 1958; Dalcq, 1957; Dalcq *et al.*, 1956). This group has stressed the relationships of several types of particles in eggs that show metachromatic reactions to toluidine blue and which show relations to and movements toward the mitotic apparatus similar to those already described. A natural pigment granule present in the echiuroid, *Urechis,* has been described by Taylor (1931) as aggregating in the cleavage asters, by a process of "sliding down the astral rays." The description, though not complete, strongly suggests the process we are discussing here. Such particles were identified as containing heme by Horowitz and Baumberger (1941). Finally and most important, Chambers (1917) reported that 2–4 μ oil droplets in the egg could also be transported toward the centrosome after being pushed into the aster with a microneedle, from which he inferred the presence of centripetal currents in the aster.

It is thus clear that the phenomenon of differential distribution of particles into the asters in cleavage of marine invertebrate eggs is quite widespread and that several different particle types may be involed in these movements in different eggs. In a given egg in mitosis, however, it is possible for more than one type of particle to participate in saltatory movements into the asters. For example, in *Spisula*, at later stages some (not all) cortical granules may move into the astral centers by independent saltatory movements although refractile granules (presumably lipid, yolk, and mitochondria) definitely do not. Again, in *Arbacia* (sea urchin), some of the echinochrome granules which do not move into the cortex during the immediate postfertilization stage, generally become located in the aster at the time of cleavage along with the vitally stained granules (the same astral location of neutral red particles exists in *Arbacia* as Kojima finds in other sea urchins). However, other particles in the cell do not show this differential distribution at these times, although yolk granules in unfertilized *Arbacia* eggs may undergo saltatory movement of smaller amplitude than that of echinochrome granules (Parpart, personal communication).

The best-studied case of the influence of the mitotic apparatus on the saltatory movement of particles in cells other than egg cells is that of the lipid granules in newt heart fibroblasts. In this case, as in those already discussed, both properties, 7 and 8 (listed previously), can be verified for many of the cells in mitosis, although in some of the cells many particles remain in the cytoplasm still undergoing unoriented saltatory movement during mitosis. This behavior may be compared to particles in other fibroblasts which undergo the extensive ingathering of particles shown by eggs during cleavage. However, some differential localization of particles, in many cases extensive, does take place in all these cells, and by processes whose description is identical to that of the egg granules described previously (see Taylor, 1957; Bloom *et al.*, 1955; Bloom and Zirkle, undated film).

We have seen no other reports of the differential movement of particles into the asters in living cells but wish to point out the existence of some phenomena which may be related to our discussion, namely, those of dictyokinesis and of chondrokinesis (see extensive review in Wilson, 1925). In the former case, particles identified as Golgi bodies (dictyosomes) are shown to assume an astral location in mitosis similar to that assumed by the particles discussed previously. In chondrokinesis, the mitochondria are described as undergoing similar differential distributions in some cells. The identification of the particles involved as mitochondria is not always certain, and it is not clear that the differential distribution attained by the particles is due to saltatory movement. If so, however, this latter process might be an example of mitochondria in some cells undergoing saltatory movements. Other then these possible cases, no other examples of mitochondrial aggregation by saltatory movements are known to us. [The aggregation of the long mitochondria of insect spermatocytes about and parallel to the spindle (e.g., Michel, 1943; Barer and Joseph, 1957) during mitosis it not accomplished by saltatory movement.]

Another possible case is that of the aggregation of acid phosphatase particles about the spindle ends (and to some degree also on the spindle equator) of cells in mitosis in regenerating rat liver. This phenomenon has been studied in detail by Dougherty in our laboratory (it appears to have been first reported by Holt, 1957), and Dougherty has shown with the electron microscope that two kinds of particles show this differential localization—dense bodies and microbodies (Dougherty, 1963). Which of these bodies correspond to the acid phosphatase particles seen in the light microscope is not clear. These particles are located at the bile canaliculi during interphase, however, and there is therefore a problem as to the transport mechanism from canaliculi to spindle pole. Again, this looks like a case in which saltatory movement may be involved, but one about which nothing *definite* can be said.

Discussion

We have reviewed actual and possible cases in which saltatory particle movement occurs and in which this movement may be oriented by processes in the cell so that a nonuniform location of the particles occurs. In the cases described here, the process, mitosis, results in the transport of the particles into the asters surrounding the centrosome. In the case described by Parpart in this Symposium, the processes set off by fertilization in *Arbacia* result in the movement of the echinochrome particles into the cortex. In all the cases described, where actual particle movements have been studied, almost the same verbal description characterizing the movement can be given and the major characteristics have been summarized in propositions 1 to 8. The problem which now confronts us is that of attempting to see if some reasonable hypothesis may be found which might help orient us in our understanding of this complex set of events. Several hypotheses concerning the genesis of the movement immediately present themselves: (*a*) that the movements are autonomous in the sense that some process within the particles themselves initiates the movement, (*b*) that the movements are induced in the particles by the transfer of momentum to the particles by some external force generating system, and (*c*) that a combination of external and internal forces are involved. None of these ideas can be completely eliminated on the basis of the data which are on hand. The most potent argument in support of movement arising from internal processes comes from the observation that the particles do, indeed, seem to move independently, that is, that the first part of proposition 2, listed previously, holds. It can be seen that the major stumbling block to any theory of externally induced movement is precisely this fact that adjacent, and in fact, contiguous particles are influenced in entirely different ways. Nevertheless, it is difficult to imagine what kind of processes could be involved in particles as different as lipid droplets, melanin granules, echinochrome granules, heme granules, and microkinetospheres, which would result in the same detailed type of movement, since most of the granules mentioned have considerable claim to being considered as metabolically inert. Most cogently, oil droplets in eggs and carmine and ink particles in protozoa undergo similar movements. Also, hypotheses such as those involving propulsion of a particle by some form of jet process originating, for example, by local influx and efflux of ions, water or other molecules, due to some permeability change in a surface membrane, should certainly predict form changes on the part of the particles, especially in processes involving movements for distances as great as 30 μ and such changes are not observed. Again, one would expect considerable differences in velocities and distances attained between large and small particles, which is not observed. Finally, it is not rare to find groups of 2 to 5 particles moving with the same velocity and for the same distance but not changing the relative distances between them, which may be of the order of a micron. It is very difficult to see what kind of process would correlate the movements of such a group of particles with such precision if the movement arose from some internal process. We feel that these considerations effectively, if not absolutely, rule out internal processes as being totally responsible for the movement.

A modification of this might suggest that the movement is due to the combination of some process in the cell as a whole with some process in the particle. For example, it might be suggested that some force field, such as an electric field set up by metabolic processes centered on some element in the cell; for example, the centriole might interact with charges that develop on the particles, which would result in movement in one or another direction depending on charge sign. Similarly, reversal of direction might be correlated with change of sign on the particle. In addition to the completely *ad hoc* nature of such a hypothesis the same arguments as used before against theories suggesting processes originating within the particle as causing movement may be used here. In addition, the unoriented movements during interphase would be most difficult to explain on this basis. We again conclude the unlikely nature of this specific possibility and others like it.

We are thus left with the middle alternative, namely, that the particle moves because of some process which originates external to the particle itself. Again we may distinguish hypotheses that postulate the interaction of the particle with some general process occurring in the cell and those that involve a direct interaction of the particle with some physically existing structure in the cell which transmits momentum to it. An example of the first kind might be the hydrodynamic forces postulated by Bjerknes as causing chromosome movement (see Schrader, 1953). He showed that oscillating or pulsating spheres in a liquid medium can set up standing waves, and depending on the phases of the spheres (that is, whether they pulsate or oscillate in or out of phase), particles of a density greater than the medium move in one direction relative to the spheres whereas particles less dense than the medium move in the opposite direction. Aside from the fact that the particles, in actuality, do not all move at once, which would seem to be required by this hypothesis, it is difficult to imagine particles changing density as rapidly as would be required by the suddenness of the onset and termination of the movement. Also, other than Pfeiffer's (1956) report of pulsating centrosomes, we know of no other reports of the existence of oscillating spheres in the cell.

We are finally brought to what we consider to be the most likely hypothesis for mechanisms of saltatory movement, namely, the hypothesis that some system in the cell can develop forces which it can transmit to the particles, inducing them to move. It is clear that we have by no means completely eliminated (or enumerated) all other hypotheses, but we feel that arguments similar to those used previously, plus the positive attributes of our suggestions, discussed later, make the suggestion of the external-force transmitting system the hypothesis most likely to succeed. Before inquiring into the requirements which such a system must satisfy, we shall describe two further sets of observations which add to the criteria we shall list.

If fertilized eggs of *Spisula* are centrifuged, the contents of the egg stratify to definite layers; from centrifugal to centripetal poles the layers are cortical granules, yolk, mitochondria, lower hyaline zone (containing primarily ribosomal material), upper hyaline zone (containing mainly elements of the endoplasmic reticulum, Golgi bodies, and multivesicular bodies) and, most centripetal, the lipid layer. Spindles and nuclei stratify into the boundary between the upper hyaline and the lower hyaline layers (see Figs. 5 to 9). If eggs which have been lightly stained with vital dyes are centrifuged so that they are stratified, the stained particles also gather primarily at the boundary between the upper and lower hyaline regions, Fig. 5. However, if eggs are heavily stained so that the stained particles become large (see point 4, p. 505), they stratify into the yolk region. Observation of these particles reveals differences in their saltatory movements. Thus, the particles stratifying to the hyaline zones can still be observed to undergo saltatory movement, generally parallel to the boundary between upper and lower hyaline zone. These particles will tend to gather tightly about the re-forming asters when redistribution gets underway, if the cell has been centrifuged in a stage containing a spindle. On the other hand, the particles going to the yolk layers do not undergo saltatory movements. However, eggs generally recover from the effects of centrifugation, and after variable periods from ½ to 1 hr, may cleave again. In this latter case the heavy particles will again participate in saltatory movements and will gather about the aster centers. Thus, particles stratified to the region bordering the upper hyaline zone continue to exhibit saltatory movement and may participate in redistribution into the asters as these re-form in the hyaline zone. Particles stratified into the yolk layer lose their ability to undergo saltatory movement, but subsequently, and much later, regain that ability as the contents of the egg redistribute and cleavage begins to occur (see Rebhun 1959, 1960, 1963 for further details on some of these processes).

The relationship of these observations to those on echinochrome granules as reported by Parpart (Parpart, 1953 and this Symposium) for unfertilized eggs is clear. In fact, their significance was not realized by us until we had considered them in relation to the work of Parpart.

The second set of phenomena, which we feel are related to saltatory movements, occur in certain phenomena to be seen on the mitotic spindle. In 1960, Östergren *et al.* put forward a rather unique hypothesis for the mechanism of spindle action, which postulated as its main element, the existence of a "pumping" activity on the part of the spindle, which was controlled by and which interacted with the chromosomal fibers resulting in the relative motion of spindle and chromosomal fiber (with attached chromosome) during anaphase separation. Stripped of the language used by these authors, the hypothesis suggests that spindles work because they can develop forces that can move particles relative to themselves without the spindle itself necessarily changing in form or shape during that movement; i.e., spindles are able to develop sliding or shearing forces at their surfaces. Östergren *et al.* (1960) showed that many particles other than chromosomes with chromosomal fibers may be moved to the poles by the activity of the spindle; for example, acentric chromosomal fragments, pieces of nucleoli and other cytoplasmic particles accidentally caught on the spindle are so moved. They also suggested that the particle aggregations seen in newt heart fibroblasts could have their origin in the same type of force generated in the asters as is generated in the spindle. The major suggestion which we would like to glean from this work is that the spindle may work by generating a shearing or sliding force at its surface and that mechanisms for saltatory particle movement may be related to mechanisms for spindle activity.

Our own observational data had also suggested the possible relation of spindle activity to saltatory activity prior to the appearance of the work by Östergren *et al.* (1960) since we had often seen metachromatic particles moving on the surface of spindles during cleavage in *Spisula*.

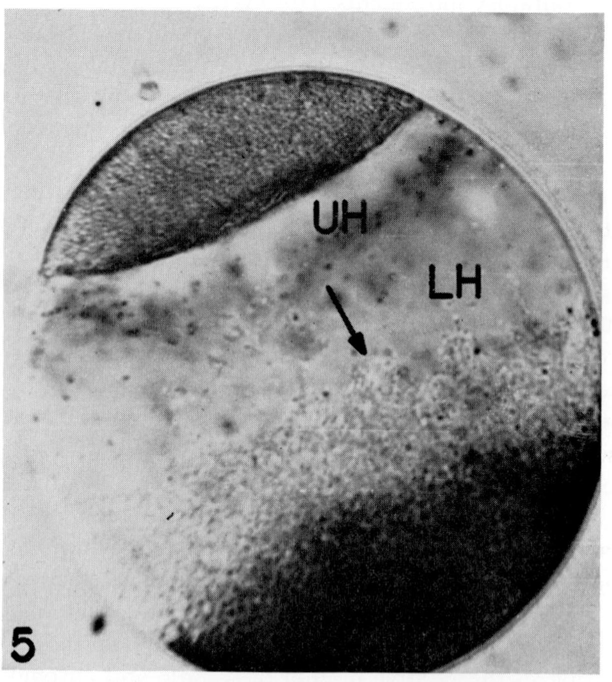

FIG. 5. A *Spisula* egg stratified just prior to first cleavage. Note the toluidine blue particles at the junction of the upper hyaline zone (UH) and lower hyaline zone (LH). The arrow points in the centrifugal direction.

Specifically, particles may often be seen on the spindle surface engaging in what appear to be directed pole-to-pole movements (see Fig. 3). These have many of the characteristics of saltatory movement in that the movements are discontinuous in time, involve continuous motions of the order of 30 μ (the spindle length in first cleavage spindles in *Spisula* eggs), do not differ for large or small particles, and may involve particles moving simultaneously in opposite directions. These characteristics are so similar to the ones described previously for astral particles as to have suggested to us that spindle mechanisms and saltatory particle mechanisms might, indeed, be identical (Rebhun, 1963).

A system in the cell involved in saltatory movement should, therefore, satisfy the following criteria:

(*1*) It must satisfy the orientation requirements set by the behavior of the particles both in interphase and during mitosis. That is, it should be a system unoriented in interphase, but with parts long enough to be physically extant over distances up to 30 μ and it should be oriented radially to the astral centers during mitosis.

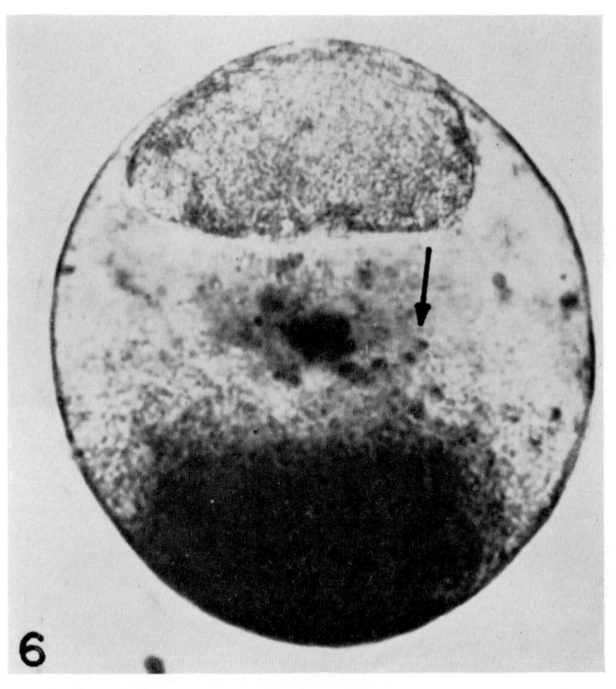

FIG. 6. A *Spisula* egg 10–15 min after stratification (prior to first cleavage). Note the aggregation of toluidine blue particles about a center in the lower part of the lower hyaline zone.

(*2*) It should be the only system in the cell exhibiting these properties.

(*3*) There should be some evidence that this system might be capable of performing the tasks required of it.

(*4*) With the orientation properties of the system in centrifuged eggs one should be capable of explaining the phenomena seen there, especially, the reversible cessation of particle movements in the case of echinochrome granules in *Arbacia* and in the case of heavy (but not light) vitally stained granules in *Spisula*.

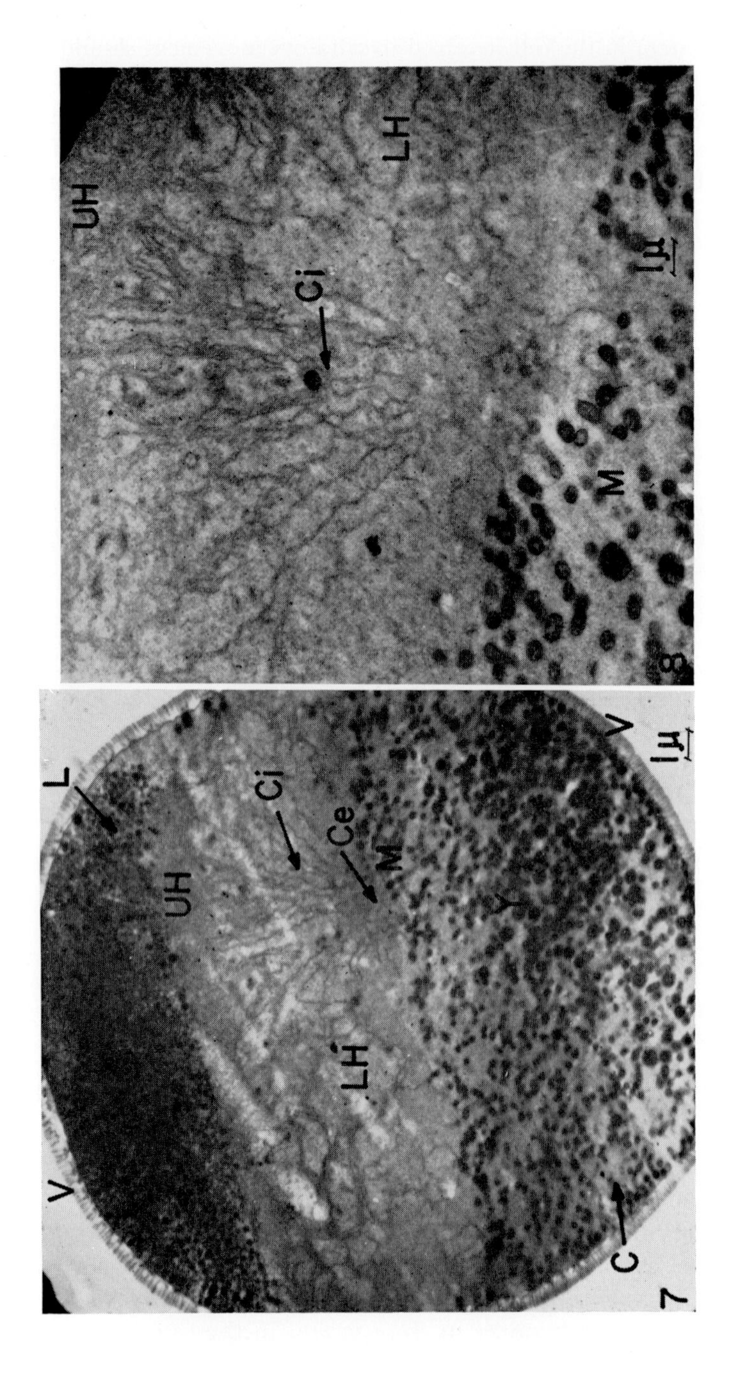
UH
LH
Ci
M
1μ
8
L
V
UH
Ci
Ce
M
Y
LH
V
C
1μ
7

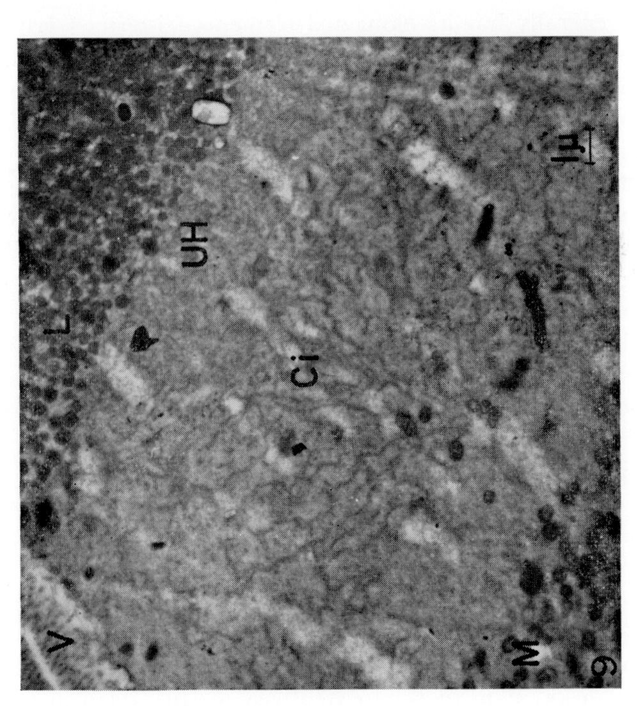

Figs. 7 to 9. Stratified *Spisula* eggs 10–15 min after stratification. Note the radial chains of cisternae (Ci) coming from the upper hyaline zone. In Fig. 7 can be seen the lipid (L), upper and lower hyaline zones, mitochondrial layer (M), and yolk layer (Y). The cortical granules (C) have not formed a distinct layer but are concentrated to some extent in the lower (centrifugal) hemisphere. In Figs. 7 and 9, V represents the vitelline membrane. (For more details, see Rebhun 1960, 1963.)

(*5*) The system should also be present in the spindle, or on its surface. This criterion is more speculative than the previous four but we feel that the observations that generate it are sound.

With these criteria in mind, we shall now examine results from ultrastructural studies of cells to see if any system present in the cell does satisfy the previously mentioned requirements, and, especially, whether criteria (*3*) and (*5*) can be verified.

Ultrastructural Studies of Cells and General Discussion

More complete description of our work on the ultrastructure of egg cells with respect to this problem has already been published (Rebhun, 1960, 1962, 1963). In these papers it is shown in some detail that the endoplasmic reticulum does show the orientation properties required in criterion (*1*) listed previously. Thus, elongated elements of this system are present in the unfertilized eggs of both *Spisula* and *Pectinaria,* and, in both, this system shows an orientation radial to the centers when the spindle begins to form. In addition, in centrifugally stratified eggs, the endoplasmic reticulum of *Spisula* eggs occupies the upper hyaline zone and generally has its elongated elements (cisternae) lying parallel to the boundary of the upper and lower hyaline zones. As eggs begin to redistribute after centrifugation the elements of the endoplasmic reticulum orient radially to the centers of the asters (see Figs. 7–9). It is clear then that the ER forms one component of the visible aster and does show distribution characteristics required of it if it is to be the system involved in saltatory movements. Further, the yolk zone is entirely free of elements of the ER after centrifugation. Finally, in the case of *Arbacia* eggs, our electron micrographs show that the ER stratifies to the hyaline zone and is completely absent from the zone occupied by the echinochrome granules. This agrees with the distribution of the ER as seen in other sea urchins by Pasteels *et al.* (1958). Unfortunately, the studies of ER have not yet been extended to redistributing eggs after centrifugation. However, from the observations we have obtained the ER appears to be a system satisfying criteria (*1*) and (*4*).

Other work indicates a similar distribution of ER relative to asters in egg material. For example, Harris (1961, 1962) and Harris and Mazia (1962) report a radial orientation of ER in asters of eggs of the urchin, *Strongylocentrotus,* similar to that seen here. In division figures in onion root tips, Porter and Machado (1960) also report orientations of ER radial to the asters. A similar, though complex orientation of a smooth ER appears to occur in spermatocyte divisions (Ito, 1960) in *Drosophilia.*

As mentioned, the criteria (*1*) and (*4*) described in the preceding section are satisfied by the ER. However, we have obtained no evidence for the existence of ER in or on the spindle as a constant component in the egg material we have studied, although on occasion small pieces of this system are present (this is in contrast to the presence of "spindle lamellae" in root tip cells (Porter and Machado, 1960) and in carcinoma cells (Buck, 1961). Thus, if we are to retain criterion (5), we find difficulties in the absence of ER from egg cell spindles. Further, criterion (2) may not hold; that is, there has now been a second system described in cells which has many of the orientation characteristics that we require. This is the system of spindle filaments which have been described so well in spindles of amebae by Roth and Daniels (1962) and Roth (this Symposium) and in spindles *and* asters of sea urchin eggs by Kane (1962) (in isolated mitotic apparatuses), by Harris (1961, 1962), and by Harris and Mazia (1962). These filaments, or "tubules," are about 150 A in diameter and occur both in chromosomal fibers and in pole-to-pole fibers in the spindle. More important for our consideration, however, is the fact that they extend into the asters of the mitotic apparatus, and, according to Harris and Mazia (1962) may extend out as far as the egg cortex. They have the proper orientation characteristics needed to account for saltatory movement and are present in force in the spindle. However, there is no evidence that they exist as an extensively deployed component in interphase cells, although they appear prior to nuclear membrane breakdown (Harris and Mazia, 1962). It may be that a search for them in interphase cells would reveal them, although their presumed lack of orientation, if they are present, would create great technical difficulties in identifying them.

We are thus left with the disturbing situation in which the only ultrastructural elements, visible with present electron-microscopical techniques, which might be responsible for saltatory movements do not simultaneously satisfy all the criteria we have set up for the system. It is, of course, always possible that a further system as yet not visualized in the cell may be the responsible element. However, a possible solution may be contained in observations such as those of Jarosch and Kamiya described next.

In a series of papers (Jarosch, 1956, 1957, 1958, 1960) a fibrillar cytoplasmic system was described which consisted of moving filaments in the cytoplasm. These filaments appear to be capable of exerting a force on the medium surrounding them, which results in their forward propulsion. Such systems have been also described by Kuroda and Jarosch in this Symposium. Jarosch and later Kamiya (1959) postulated deep involvement of these filaments in certain movements in the cytoplasm of the exuded cytoplasmic droplets they were concerned with. Specifically, they attempted to account for saltatory movement as a movement of particles along the filaments but in a direction opposite to that in which the filaments are moving. Further, they observed that the filaments appeared to attach to and detach from the nuclear surfaces and that the filaments with their associated countermoving streams appeared to be correlated with the rotations of the nuclei. Similar phenomena were observed for moving chloroplasts. Thus, they considered that the nuclear rotations and chloroplast movements were due to the concerted efforts of filaments attached to the nuclear or chloroplast surface exerting their influence in a fixed direction. The observations on which these ideas are based appear to be well founded and suggest, therefore, that the nuclear surface may form reversible associations with a filamentous component of the cytoplasm which may be able to exert shearing forces on the medium in which it is immersed. Since, however, most recent work indicates that the nuclear surface is derived from, is continuous with, and can contribute to the endoplasmic reticulum, and thus is part of that system (see Porter, 1961), it is not too far-fetched to suggest the possibility that an ER-filament association may be possible, and that force transmission to the particles from the ER may, indeed, be due to this associated filament system. Thus, if saltatory particle movements in these cytoplasmic droplets are due to interaction with a filament capable of exerting a shearing force on the medium, then this may also be the mechanism involved in other cells, if it is justifiable to assume all saltatory motion to have a common mechanism.

We have thus arrived at a hypothesis for saltatory particle movement which has been obtained by consideration of the most diverse phenomena, namely, that these movements are due to sliding or shearing forces exerted on particles in cells by the activity of filaments which may occur in the spindle, in the cytoplasm, or are associated with the endoplasmic reticulum. Many auxiliary hypotheses are used in this work. The highly hypothetical nature of this discussion is made clearer, if we reiterate all the hypotheses used with an estimate of their worth. A similar and extended discussion has been previously published (Rebhun, 1963).

a. Because the details of saltatory movement are so similar in different cells, we have assumed that the mechanism of saltatory movement is likely to be the same. We feel that this hypothesis is quite strong.

b. An analysis of the details of the movement characteristics of the particles leads us to consider that the most reasonable hypothesis for the mechanism of saltatory movement is that of the transfer of momentum to the particle by some outside agency, with the particle itself being quite passive in the process. This hypothesis we feel to be very likely, but not as strong a hypothesis as (*a*).

c. A consideration of some of the phenomena associated with spindle action, especially as it relates to saltatory movement, leads us to consider the possibility that saltatory movement and spindle action are mechanistically similar. This is clearly a hypothesis on a lower level of support than (*a*) and (*b*). This leads to the next hypothesis.

d. Some structure present in the spindle is also present in the cytoplasmic system responsible for saltatory movement. Since the only visible structure consistently present in the spindle is the 150 A filament (or tubule) and since this is present in asters, we have suggested that this filament or one related to it may be responsible for saltatory movement. An even more tenuous suggestion is the final one necessary to tie in the observation on the ER in this process.

e. The ER is involved in saltatory movement because of its association with the true force-producing structure, the filaments. Evidence for such an association comes from observations of filaments associated with rotating nuclei in isolated cytoplasmic droplets of *Chara* and *Nitella*.

The essence of the last three hypotheses is the suggestion that there is available to cells a structure, filamentous in nature, which can exert a force on its surroundings in such a way as to propel itself if it is free to move, or, if not, acting in concert with other similar structures, to propel the surrounding medium or objects in it. Such ideas have originated with Jarosch and Kamiya in the papers cited previously, and though the universality of these ideas is by no means clear, we feel that they deserve consideration in the present context. Finally, it is not suggested that the filaments in different cells are made of the same materials or are even necessarily organized in the same way on a molecular level. We merely require that they act alike, namely, exert forces capable of moving themselves or external objects parallel to their axes, possibly by processes such as those described by Jarosch in this Symposium.

It remains to be seen whether the concepts introduced will prove adequate to interpret these phenomena, or whether concepts drawn from a less structural and morphological sphere will ultimately have to be invoked in the explanation of the mechanisms of saltatory particle movements.

Some discussion on saltatory motion can be found in the Free Discussion of Part IV.

References

Andrews, E. A. (1955). *Biol. Bull.* **114**, 113-117.

Barer, R., and Joseph, S. (1957). *In* "Mitochondria and other Cytoplasmic Inclusions," Society for Experimental Biology Symposium, Vol. X. Cambridge Univ. Press, Cambridge, England.

Bloom, W., and Zirkle, R. (undated). "Mitosis of Newt Cells in Tissue Culture," film-print No. 64. Univ. of Chicago Bookstore, Chicago, Illinois.

Bloom, W., Zirkle, R., and Uretz, R. (1955). *Ann. N.Y. Acad. Sci.* **59**, 503-513.

Buck, R. C. (1961). *J. Biophys. Biochem. Cytol.* **11**, 227-236.

Chambers, R. (1917). *J. Exptl. Zool.* **23**, 483-505.

Dalcq, A. (1957). *Bull. Soc. Zool. France* **82**, 296-316.

Dalcq, A., Pasteels, J. J., and Mulnard, J. (1956). *Bull. Acad. Roy. Belg. Classe Sci., Ser. 5,* **42**, 771-777.

Dougherty, W. J. (1963). Ph.D. Dissertation, Dept. Biol., Princeton Univ., Princeton, New Jersey.

Fischel, A. (1899). *Anat. Heften* **37**, 463-504.

Harris, P. (1961). *J. Biophys. Biochem. Cytol.* **11**, 419-432.

Harris, P. (1962). *J. Cell Biol.* **14**, 475-488.

Harris, P., and Mazia, D. (1962). *In* "The Interpretation of Ultrastructure" (R. J. C. Harris, ed.), Vol. I. Academic Press, New York.

Holt, S. J. (1957). *9th Intern. Congr. Cell Biol., St. Andrews, Scotland.*

Horowitz, N. H., and Baumberger, J. P. (1941). *J. Biol. Chem.* **141**, 407-415.

Iida, T. T. (1942). *Dobutsugaku Zasshi* **54**, 364-366.

Ito, S. (1960). *J. Biophys. Biochem. Cytol.* **7**, 433-442.

Jarosch, R. (1956). *Phyton (Buenos Aires)* **6**, 87-107.

Jarosch, R. (1957). *Biochim. Biophys. Acta* **25**, 204-205.

Jarosch, R. (1958). *Protoplasma* **50**, 93-108.

Jarosch, R. (1960). *Phyton (Buenos Aires)* **15**, 43-66.

Kamiya, N. (1959). *Protoplasmatalogia* **8**, 13a.

Kane, R. L. (1962). *J. Cell Biol.* **15**, 279-287.

Kojima, M. K. (1959a). *Embryologia (Nagoya)* **4**, 191-209.

Kojima, M. K. (1959b). *Embryologia (Nagoya)* **4**, 211-218.

Michel, K. (1943). Film No. C443, Institut für den Wissenschaftlichen Film, Göttingen.

Mulnard, J. (1958). *Arch. Biol. (Liège)* **69**, 645-685.

Mysels, K. J. (1959). "Introduction to Colloid Chemistry." Interscience, New York.

Östergren, G., Molé-Bajer, J., and Bajer, J. (1960). *Ann. N. Y. Acad. Sci.* **90**, 391-408.

Parpart, A. K. (1953). *Biol. Bull.* **105**, 368.

Pasteels, J. J. (1955). *Bull. Acad. Roy. Belg. Classe Sci., Ser. 5,* **41**, 761-768.

Pasteels, J. J. (1958). *Arch. Biol. (Liège)* **69**, 591-619.

Pasteels, J. J., and Mulnard, J. (1957). *Arch. Biol. (Liège)* **68**, 115-163.

Pasteels, J. J., Castiaux, P., and Vandermeersche, G. (1958). *Arch. Biol. (Liège)* **69**, 627-643.

Pfeiffer, H. H. (1956). *Protoplasma* **46**, 585-596.

Porter, K. R. (1961). *In* "The Cell" (J. Brachet and A. E. Mirsky, eds.), Vol. II. Academic Press, New York.

Porter, K. R., and Machado, R. D. (1960). *J. Biophys. Biochem. Cytol.* **7**, 167-180.

Rebhun, L. I. (1959). *Biol. Bull.* **117**, 518-545.

Rebhun, L. I. (1960). *Ann. N. Y. Acad. Sci.* **90**, 357-380.

Rebhun, L. I. (1961). *J. Ultrastruct. Res.* **5**, 208-225.

Rebhun, L. I. (1963). *In* "The Cell in Mitosis" (L. Levine, ed.). Academic Press, New York.

Rose, G. G. (1957). *J. Biophys. Biochem. Cytol.* **3**, 697-704.

Roth, L. E., and Daniels, E. W. (1962). *J. Biophys. Biochem. Cytol.* **12**, 57-78.

Schrader, F. (1953). "Mitosis," 2nd ed. Columbia Univ. Press, New York.

Seifritz, W. (1952). *In* "Deformation and Flow in Biological Systems" (A. Frey-Wyssling, ed.). North Holland, Amsterdam.

Taylor, C. (1931). *Physiol. Zool.* **4**, 423-460.

Taylor, E. W. (1957). Ph.D. Dissertation, Committee on Biophysics, Univ. of Chicago, Chicago, Illinois.

Wilson, E. B. (1925). "The Cell in Development and Heredity," 3rd ed. Macmillan, New York.

Motile Systems with Continuous Filaments[1]

L. E. ROTH

Committee on Cell Biology and Department of Biochemistry and Biophysics, Iowa State University, Ames, Iowa

Introduction

Filaments are integral components of several movement systems. The intent of this report is to present an overview of those filament systems in which individual filaments are continuous throughout their functional length. For example, the individual filaments in cilia extend essentially continuously from base to tip. By contrast, the filament system of striated muscle is composed of filaments not continuous over the functional unit, the sarcomere.

As concepts are derived by consideration of the comparative morphology, biochemistry, and physiology of such filamentous systems, understanding of other types of movements may be enhanced. Similarities and differences between the functions of these systems and the function of muscles may become clearer.

Structural Features of Filament Systems

The movements of chromosomes in mitosis are produced by a well-organized mitotic apparatus (MA). The following descriptive survey includes the four types of nuclear divisions—anastral, astral, intranuclear, and amitotic—and other filament systems that are similar. My purpose is not to describe in ultimate detail; the reader will be directed to numerous publications for this information. The intent is to compare and contrast filament systems and to utilize information available about one to gain further understanding about the others.

The most detailed structural information available to date on mitosis comes from electron-microscopic studies of the anastral division process in the giant, multinucleate amebae (Roth and Daniels, 1962; Daniels and Roth, 1963). The anastral MA has no centrioles nor does it converge to a pole and, therefore, has no asters (Fig. 1). In cross section, fibrils are observed that are irregular in size and shape (Fig. 2, F) and are composed of filaments that measure 15 mμ in diameter and

[1] Supported by a research grant C-5581 from the National Cancer Institute, United States Public Health Service.

have a circular cross section with a dense cortex and light center (Fig. 3, F). Surrounding each filament is a randomly arranged grouping of fine material (Fig. 3, M). This appearance is found in all filaments ob-

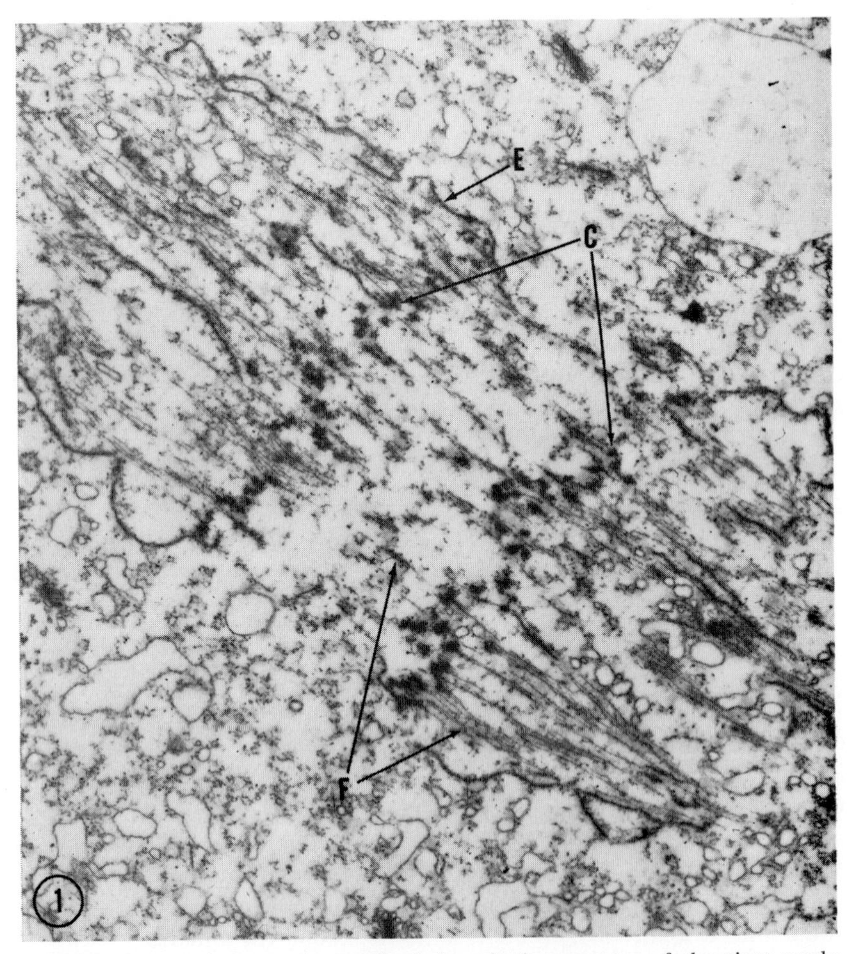

FIG. 1. Survey electron micrograph of the mitotic apparatus of the giant ameba, *Pelomyxa carolinensis*, about 5 min after the beginning of the anaphase. The chromosome plates (C) are about 4 μ apart, filaments (F) are present both in the interzone (continuous filaments) and poleward (continuous and chromosomal filaments), and fragments of the nuclear envelope (E) are within and exterior of the mitotic apparatus. Fixation in osmium tetroxide solution at pH 8.0 with 0.002 *M* cobalt added. Magnification: × 8700. (Micrograph by R. A. Jenkins.)

served, regardless of whether they might be continuous or chromosomal filaments, located on the poleward or interzonal sides of the chromosome plate, or in the metaphase or anaphase.

Numerous ribosomelike particles are scattered between the filaments in a manner not readily suggesting a structural association of particle and filament (Fig. 3, R). Since it is first observed in the prometaphase when the envelope has only a few interruptions, this distribution is undoubtedly the cause of the intense basophilia found in the mitotic apparatus of many different cells (e.g., Boss, 1955; Rustad, 1959).

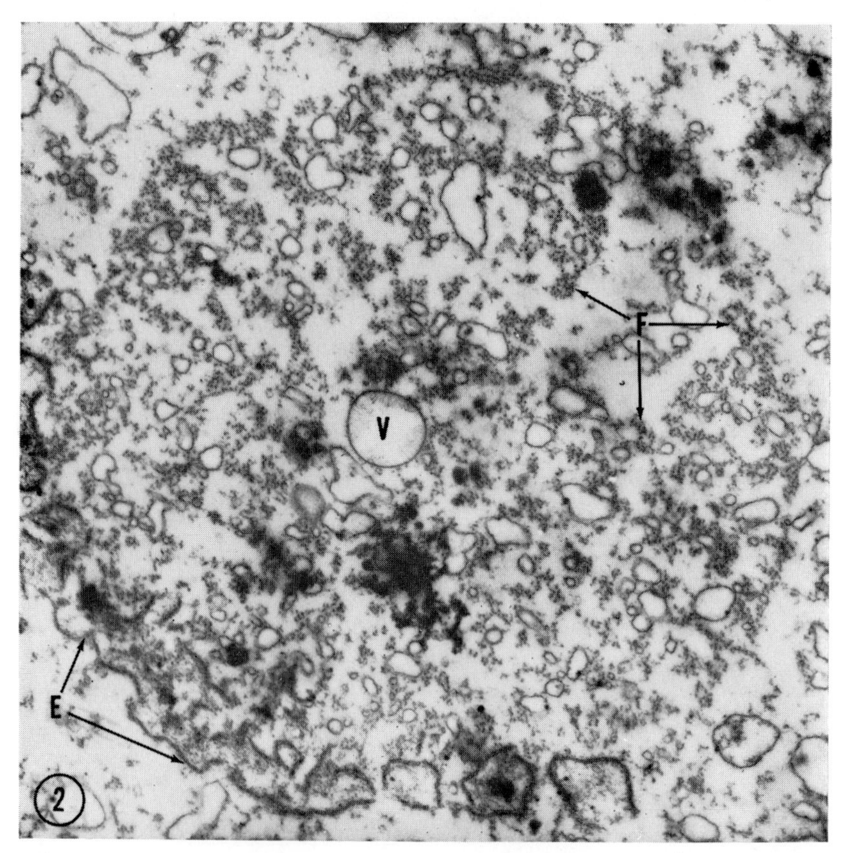

FIG. 2. Cross-sectional survey micrograph on the poleward side of an anaphase mitotic apparatus when the chromosome separation is about 9 μ. Filaments are arranged into irregular fibrils (F), fragments of the envelope (E) are leading the chromosomes (not in section) at the periphery of the MA, and small vesicles including a recently formed pinocytic vesicle (V) are contained within the MA. Fixation in an osmium tetroxide solution at pH 8.0 with 0.002 *M* cobalt. Magnification: × 12,000. (Micrograph by R. A. Jenkins.)

By the telophase, the filaments have become detached from the chromosomes and can be observed in the cytoplasm some distance from the chromosomes that are now surrounded by a continuous nuclear

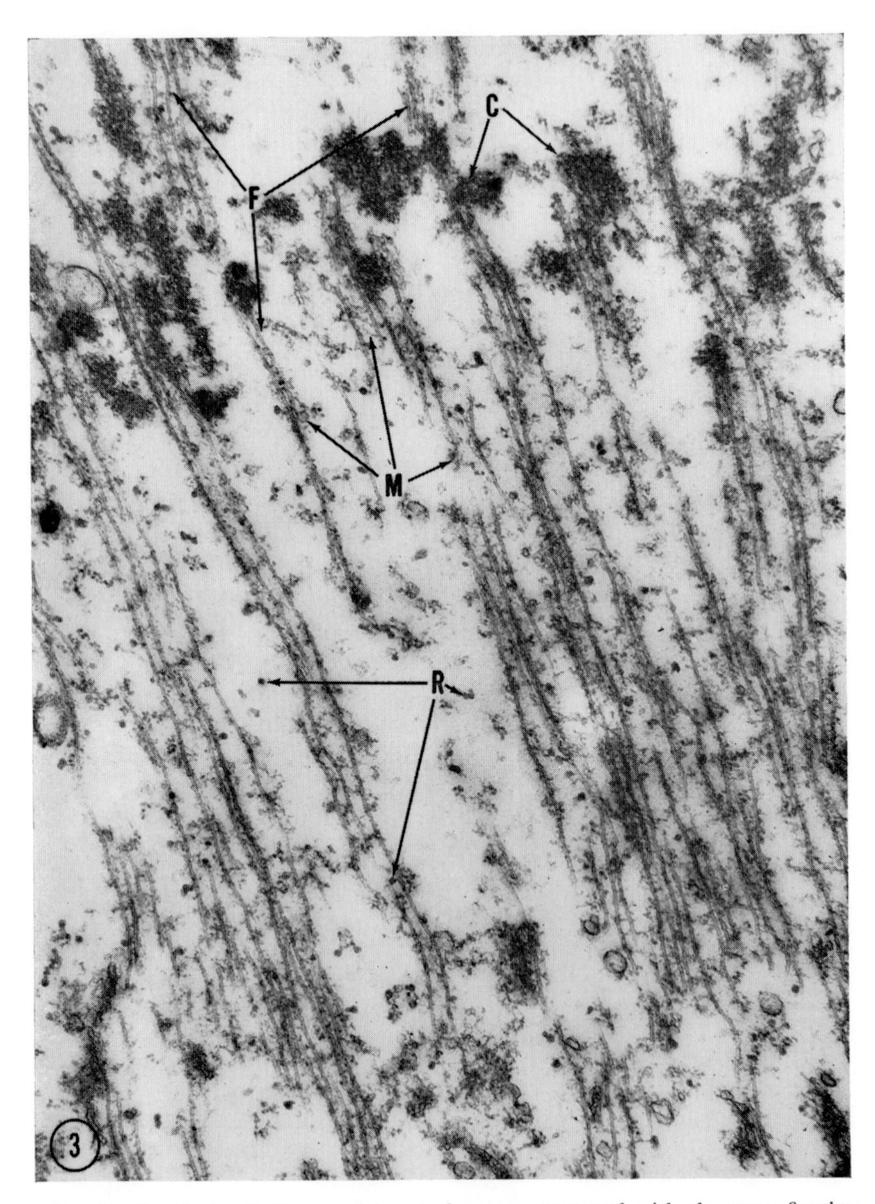

FIG. 3. Detail of mitotic apparatus at the same stage and with the same fixation as Fig. 1. Filaments (F) both on the interzonal and poleward sides of the chromosome plates (C) have the same diameter and appearance and have fine material (M) on their surfaces. Ribosome-like particles (R) are scattered throughout the MA usually without indications of attachment to filaments. Magnification: $\times$ 37,000. (Micrograph by R. A. Jenkins.)

envelope. The length of these bundles suggests that they are the continuous filaments that become perhaps 65 μ long in these cells owing to a marked spindle elongation. At this time, the filaments are arranged in very close bundles and regular configurations, since the material surrounding filaments is no longer present. With the exception of this difference, telophase filaments show no observable difference in diameter or appearance from those in the anaphase and metaphase.

Mitosis of the astral type has been observed in several different cells. The most careful and detailed observations are those made by Harris (1961, 1962) on sea urchin blastomeres; they show that many of the previously mentioned features are also present. The filament diameter and appearance are essentially identical, and ribosomelike particles are similarly distributed. The adhering material seen in anastral mitoses has not been so adequately observed, but is probably present though masked by the many other particles present. No observations of such cells in the telophase have shown filament bundles. Kane (1962) has also studied these cells by isolation with his method of controlled pH in hexandiol and by electron microscopy and has seen similar structures with the exception that the filament diameter is reported to be 21 mμ rather than the 15-mμ diameter reported by Harris and also found in the giant ameba MA. The possibility of his experimental procedure causing alteration of the filaments exists, and the hypothesis follows that such filaments may be amenable to experimental alteration.

Intranuclear mitosis takes place in the micronuclei of protozoan ciliates. Here the nuclear envelope remains continuous throughout division, and the entire process, which is otherwise essentially an anastral mitosis, takes place within the confines of the envelope. The MA in this case is composed of filaments of identical size and appearance to those found in astral and anastral mitosis. The major difference is the complete lack of ribosomal particles (Roth and Shigenaka, 1963). Movement of chromosomes is largely by spindle elongation.

Amitotic division takes place in the separation of ciliate macronuclei. Filaments are contained within the macronuclear volume and are concentrated either at the nuclear constriction (Roth, 1959; Roth and Minick, 1961) or extend in random directions (Roth and Shigenaka, 1963). However, another, more highly organized macronuclear apparatus has been found in the oligotrichous, rumen ciliates (Roth and Shigenaka, 1963); this filament system is external to and encloses both the macronucleus and micronucleus. Although present during the interphase, this system is more highly developed during division. Whether this filament system actually produces a major part of the force that causes division of the macro- and micronuclei is still not established although no adequate mechanism of macronuclear division has yet been suggested. This system has the appearance of an extranuclear division apparatus.

The demonstration of MA filaments by electron microscopy is possible only by the choice of somewhat unusual conditions of osmium tetroxide fixation (Roth and Daniels, 1962; Roth and Jenkins, 1962). Filaments of the MA will be preserved if divalent cations—calcium, magnesium, strontium, or cobalt at 0.002 *M*—are added to an osmium tetroxide fixation fluid in the pH range of 7.2 to 8.0 or without divalent cations if fixation is carried out at a slightly acid pH such as 6.0 (see also Harris, 1962). Such stabilizing conditions are, by contrast, not necessary for preservation of the filaments of cilia, infraciliature, or centrioles. Functional lability is, in fact, a property of the MA as Östergren (1949) demonstrated so well in his observations of the mitotic process in *Luzula* where the chromosomes have diffuse kinetochores. Thus the evidence shows that the MA filaments are more labile than other filaments of similar morphology to which we now turn our attention.

The cytoplasmic infraciliature of many protozoa is composed of filaments. These filaments usually have a diameter of 15 to 21 mμ and are found in hexagonally packed bundles extending from one group of cilia to another (Fig. 4, F); sheetlike groups are also found beneath the pellicular membranes particularly surrounding the orifice of the contractile vacuole (Fig. 5, F). In a few cases, large concentrations of these elements within a small volume of the cytoplasm are present so there is an intimate association of bundles from many different parts of the cell (Fig. 6); this region is probably the "motorium" described first by Sharp (1914) in this genus. The infraciliature is typically seen connected to filaments lying close to the nuclear envelope.

Flagellate protozoa have similar filaments in their cytoplasm. However, the groupings are usually smaller with only a few filaments in sheetlike arrangements coursing beneath the membranes of the pellicle and associated in small numbers with the base of flagella (Roth, 1958).

Amebae typically lack such filaments in their cytoplasm. Rather, the only filaments demonstrated are those already described in the mitotic apparatus.

Cilia are now widely known to be composed of filaments. The shaft structure is rigidly standardized with nine peripheral filaments of essentially doublet cross sections and two central filaments of essentially circular cross sections. For our purposes it is important to pay particular attention to the structure of the ciliary tip. In the tips of protozoan cilia

(Fig. 7), the peripheral filaments have single, circular cross sections (Roth, 1957; Gibbons and Grimstone, 1960; Roth and Shigenaka, 1963). The peripheral filaments do not all terminate at the same level so that cross sections at the tip frequently show diminished numbers. However, the important consideration is the structure of the tips of the peripheral filaments where circular cross sections appear similar to those of the infraciliature and the MA.

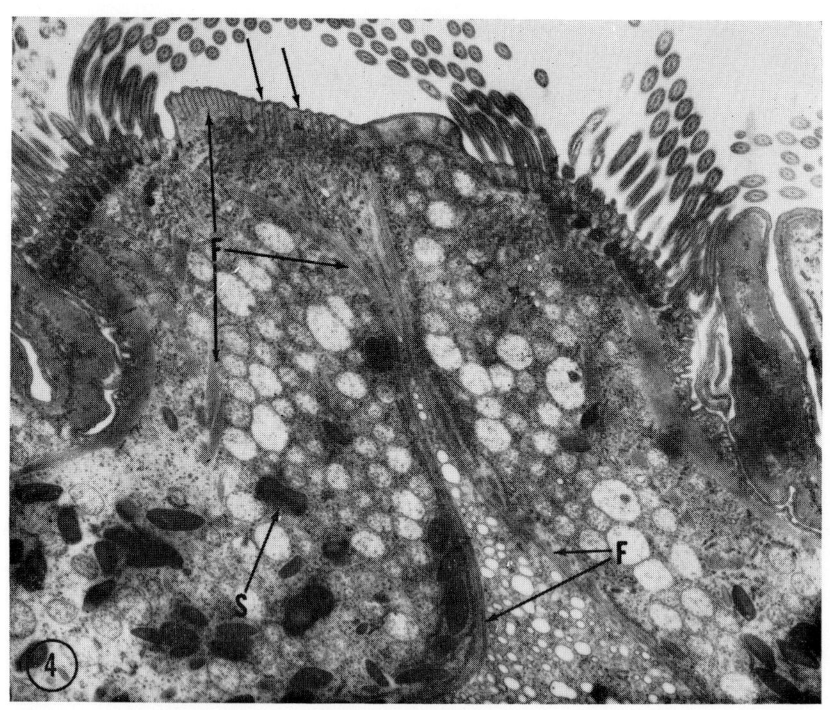

FIG. 4. Survey electron micrograph of a section through the anterior end of the oligotrichous protozoan *Diplodinium ecaudatum*. Filaments (F) form bundles extending to groups of cilia, sheets that are located at the region of ingestion of food particles and fluids, and a barrier wall that separates the digestion region from the remainder of the cytoplasm. Ingesta would move in the direction of the unlettered arrows through the esophageal tube into an enlarged volume of cytoplasm. The dense granules (S) are food-storage bodies that are localized in the anterior, peripheral portions of the cell. Magnification: $\times$ 8300.

Another system of movement closely related to cilia is found in the stalk of several peritrich protozoa. When disturbed, these organisms retract quickly by coiling their stalks tightly; recovery to the extended condition usually takes several seconds so that contraction is not rapidly repeated. The fibrils contained within these stalks are extensions from

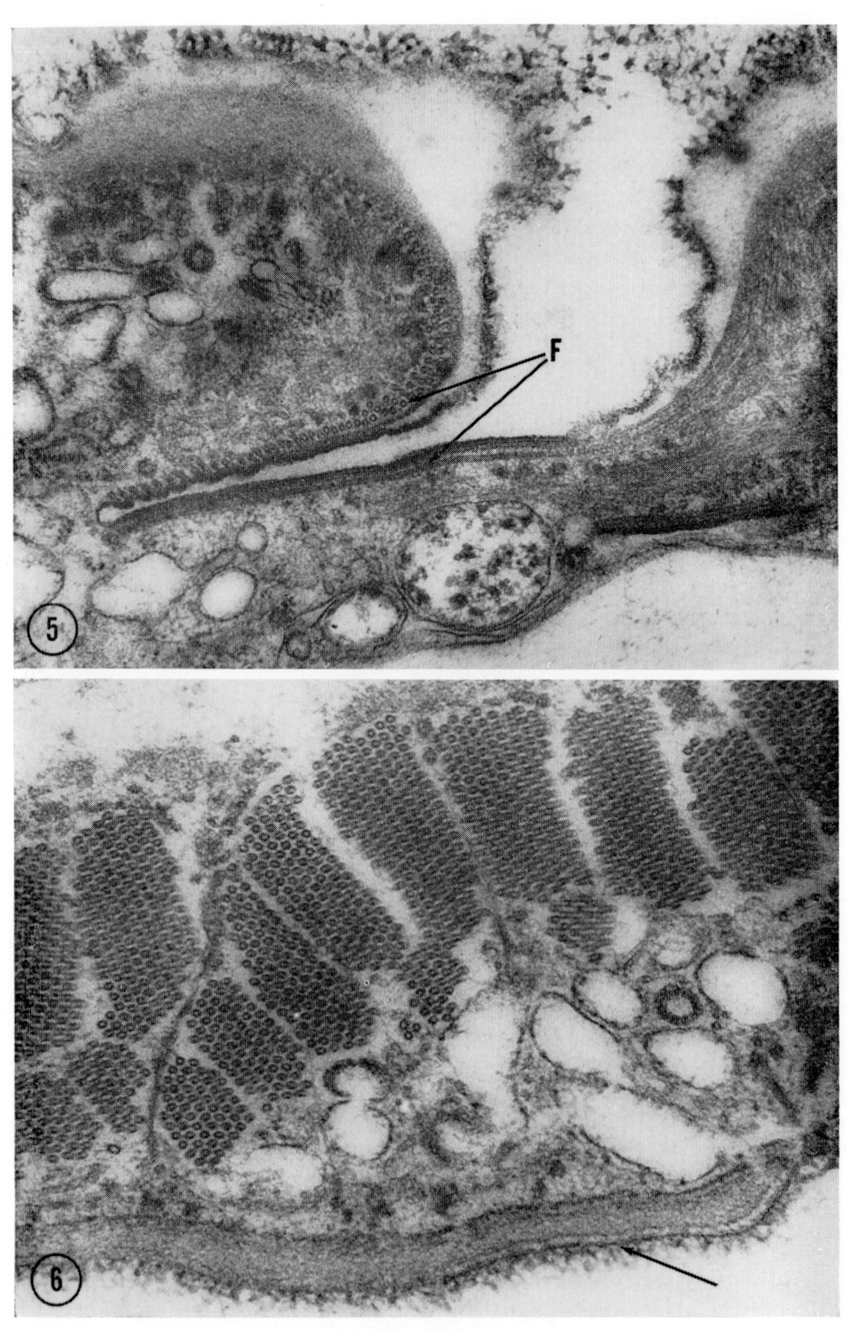

Fig. 5. The orifice of the contractile vacuole in *Diplodinium ecaudatum* showing the filament (F) system around it. Magnification: × 48,000.

Fig. 6. Cross sections of filaments of the infraciliature in *Diplodinium ecaudatum*. Individual filamentous bundles are nearly ubiquitous in the cytoplasm, but in this

typical basal bodies and, though they have marked striations, are similar to ciliary shafts (Randall, 1956; Rouiller *et al.*, 1956).

The structural resemblance of metazoan centrioles to the triplet-filament complex of ciliary basal bodies adds another major link to this series of similarities. New impetus has been given to this long-established concept by the extremely close morphology of basal body (e.g., Gibbons and Grimstone, 1960) and centriole (Gall, 1961; Slautterback, 1963).

With mention of the centriole, we have again returned to considera-

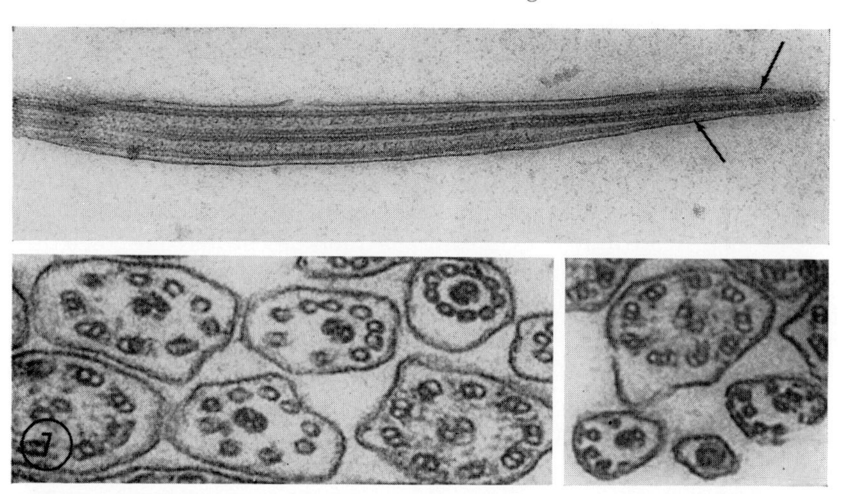

FIG. 7. The structure of the tips of cilia. The upper micrograph from *Isotricha* sp. is a longitudinal section in a plane including both central filaments which fuse and then end in a terminal enlargement. The peripheral filaments are still doublets at the left end of the upper micrograph but soon have circular cross sections and end before the terminal enlargement; peripheral filaments do not all end at the same level as indicated by the unlettered arrows. Magnification: × 57,000. The lower micrographs are from *Diplodinium ecaudatum* and form a series, with a few exceptions, of more distally located cross sections. Lower left, magnification: × 137,000; lower right, magnification: × 118,000. (Lower micrographs by Y. Shigenaka.)

tion of the mitotic apparatus. In spite of similarities, a major difference between basal bodies and centrioles is evident since filaments of the astral MA do not usually contact the centriole; rather, a distance of separation exists with the filaments appearing to end at the periphery of the centrosome which is a clear homogeneous region.

The foregoing survey demonstrates that filaments from many cells and from several different cell structures are very similar in appearance,

---

region many bundles are closely packed immediately below the pellicle (unlettered arrow). This organism is the one in which Sharp (1914) originally described a "motorium." Magnification: × 57,000. (Micrograph by Y. Shigenaka.)

size, and arrangement and also suggests that chemical and functional similarities exist.

Evidence for chemical similarity is available from two areas. First, divalent cations have been implicated as integrally involved. Bishop (1962a) in his valuable detailed review paper lists numerous species in which motility of isolated sperm flagella is activated by the divalent cations of calcium and magnesium and inhibited by proper concentrations of compounds that bind cations by chelation. In another presentation, Bishop (1962b) summarizes the evidence relating to glycerol-KCl-extracted sperm and suggests not only that calcium and magnesium are involved in inducing flagellar movement but that they also perhaps regulate this function according to their position and binding. Divalent cations in this regard probably regulate the ATPase activity of the filament protein. Jahn (1962) has discussed similar effects on cilia particularly in regard to reversal of beat. Although the effect of divalent ions in this case is attributed to membrane binding, he has proposed that limitations caused by bond angles can restrict the ions capable of occupying binding sites and thus explain the biological specificity of certain ions. In regard to the MA, Mazia (1959) has described an effect of calcium on the isolated MA of sea urchin blastomeres where the fibrous appearance is increased by application of cations. As described earlier, stabilization of the MA for electron microscopy by such ions when fixation at pH 8.0 is used is another example of divalent cation involvement with filament systems. Still another filament system in which calcium ions are vitally involved is that in the stalk of peritrich protozoa; the modified ciliary shaft filaments are described by Hoffmann-Berling (1958) to contract when traces of calcium are added and to relax either spontaneously or upon addition of the chelating agent, ethylenediaminetetraacetate (EDTA).

Protein chemists state that one of the common features of the action of inorganic electrolytes on proteins is their binding to the charged groups of proteins (Putnam, 1953). More specifically, the alkali earth cations—calcium in particular—are thought to combine with the free dicarboxylic acid residues of proteins (Höber, 1945), thus suggesting the importance of aspartic and glutamic acids which are present in the protein of the MA in larger quantity than most other amino acids (Mazia, 1956). Binding at these sites is necessary for stabilization at pH 8.0 but not at pH 6.0; the latter pH is probably nearer and sufficiently close to the isoelectric point of MA protein where a lesser hydration of the protein exists (Höber, 1945), and stabilization in addition to osmium tetroxide fixation is not required. In muscle function, the importance of calcium and magnesium ions is clearly established (Weber, 1958). Therefore, divalent cations are apparently necessary components of all motile systems.

Further chemical similarities are suggested by evidence pointing toward the presence of actomyosin-like proteins. Weber (1958) summarizes this subject well and concludes that marked similarities exist although major differences are found in regard to the involvement of adenosine triphosphate (ATP) and calcium ions. Later evidence suggests close similarities in the amino acid composition of protein from the MA (Mazia, 1956), from the flagella of *Chlamydomonas* (Jones and Lewin, 1960), and the cilia of *Tetrahymena* (Watson and Hopkins, 1962); high amounts of aspartic and glutamic acids are present in all three cases. The ability of these proteins to act as nucleoside triphosphatases has been widely established (Weber, 1958), although varying degrees of specificity for this capability are found. For example, Mazia *et al.* (1961) have shown that MA protein can split the terminal phosphate from ATP but not from adenosine diphosphate (ADP), guanosine triphosphate (GTP), cytosine triphosphate (CTP), or uridine triphosphate (UTP). Thus the proteins involved in the movement systems are quite similar though not identical.

Summary

Filaments with the general appearance of a dense cortex and light center and having a diameter of 15 to 21 mμ are present in systems of several different types and in many different cells. Evidence pointing to their chemical similarity is sufficient to suggest that they are quite closely related in their composition and, therefore, that their functional capabilities are similar.

Formation of Filaments

How are such structural filaments formed? If the small amount of available data on the formation of these systems is compiled, a hypothesis of synthetic stages can be formulated. In regard to the MA, significant studies by Went (1960) have shown that synthesis of filament protein has largely been completed before the prophase begins. However, it is obvious by observation of prophase cells that no filament formation has yet taken place. Rather, filaments are formed at the time when the nuclear envelope is first interrupted in the typical astral and anastral divisions. At this time electron microscopy of giant amebae nuclei has shown filaments with a cross-sectional appearance not greatly different from that in later stages: circular cross sections are present with fine material along their surfaces (Roth and Daniels, 1962). Forming filaments in the prometaphase differ very little from functioning

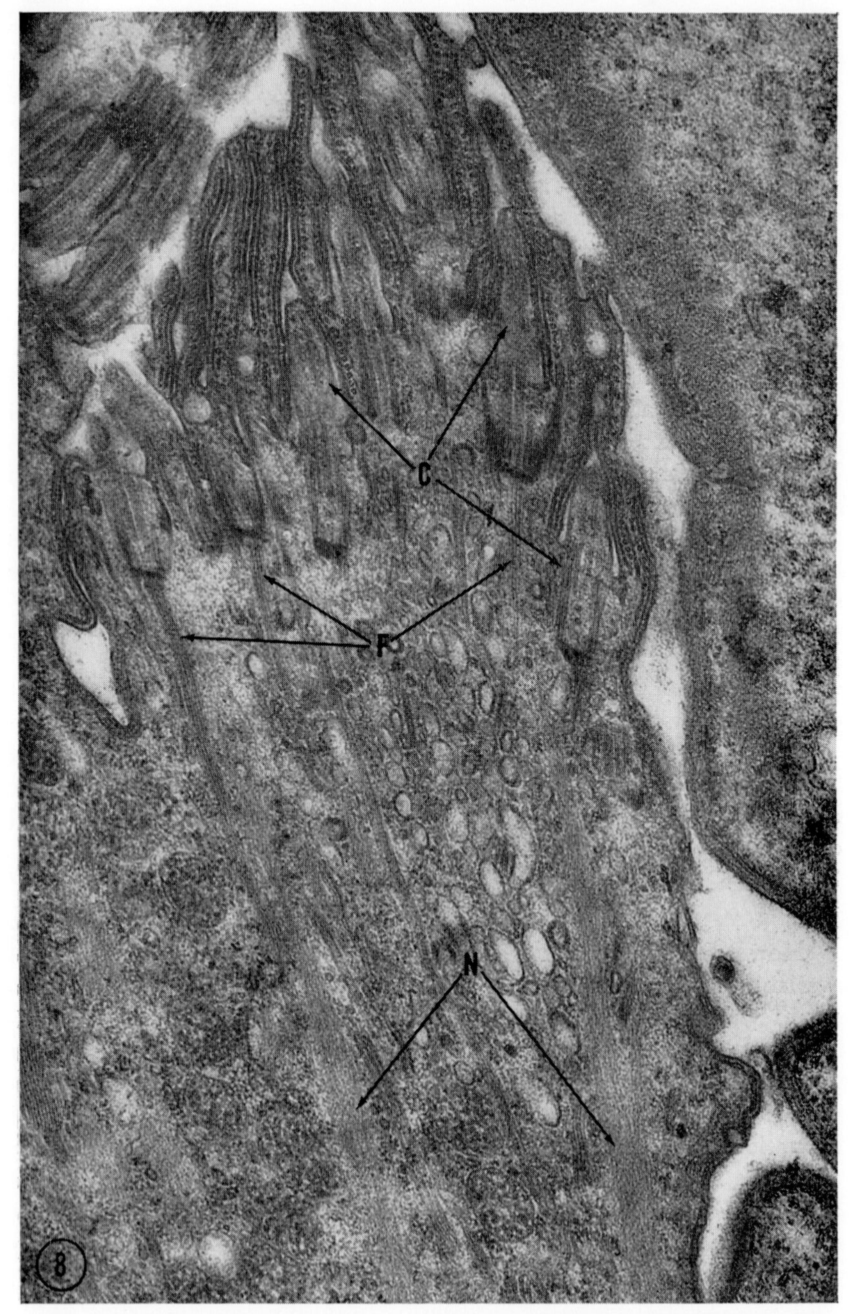

FIG. 8. Section through the tuft of forming cilia present in *Diplodinium ecaudatum* during cytokinesis. The short shafts (less than 1 μ long) of the cilia (C) already

filaments in the anaphase to the extent of our present observations. Thus, synthesis of MA protein takes place in the interphase whereas filament formation takes place later. Therefore, the two processes are separable events both in time and probably also in location.

Centriole formation has been studied in two recent significant studies. Mazia *et al.* (1960) have shown evidence that replication of the cell centers involves the formation of a pro-body of smaller size than the parent structure. Gall (1961), in a detailed electron-microscopic study of centriole formation in spermatogenesis, presented good structural evidence that a procentriole is initially formed close to but not in contact with the centriole. These presentations together suggest that the procentriole is shorter but equal in diameter and filamentous structure to the parent centriole. Thus centriole formation is currently thought to be composed of at least two processes: formation of a short procentriole followed by an elongation of the filamentous structures to form the centriole.

Observations on formation of the protozoan infraciliature are meager. Recent studies of the rumen protozoa (Roth and Shigenaka, 1963), however, have given some evidence that the formation of these filaments takes place by an aggregation and orientation in the cytoplasm from masses of otherwise unoriented material. In dividing *Diplodinium,* an oligotrichous ciliate with a highly developed infraciliary system, a tuft of cilia is formed when cytokinesis takes place. Organisms with a constriction furrow are easily selected and allow detailed observation of the process of formation of the shafts of cilia and the infraciliary filaments. The infraciliature in some of the cases studied in composed of filaments extending for a few microns from the basal bodies of the forming cilia but then fade into a region of oriented but nonfilamentous material (Fig. 8, N). An initial molecular synthesis apparently takes place followed by an aggregation into filaments which finally elongate to the length of the interphase infraciliature. The basal body of the cilium may influence formation of such filaments but direct elongation is unlikely since the number of infraciliary filaments per basal body may be fifty or more.

The formation of cilia is another case of filament formation that should be considered. The basal bodies, because of their extremely close structural similarity to the centriole, are probably formed like centrioles.

---

have filaments attached to their basal bodies. Near the basal bodies, filaments (F) are well defined whereas deeper in the cytoplasm aligned material (N) is present but not filaments. This phenomenon occurs in division stages in several portions of the cytoplasm and probably represents the formation of infraciliary filaments. Magnification: × 36,000.

A pro-body would be formed and elongate to a full basal body, move to and become oriented perpendicular to the plasma membrane, and then greatly elongate to form a shaft of somewhat different, filamentous composition than itself. Unfortunately little evidence is available in support of this hypothesis. Many questions arise: Can the basal body of a mature cilium give rise to such a probasal body; how do the triplet filaments in the base function in forming the doublets in the shaft; and how do the central filaments arise?

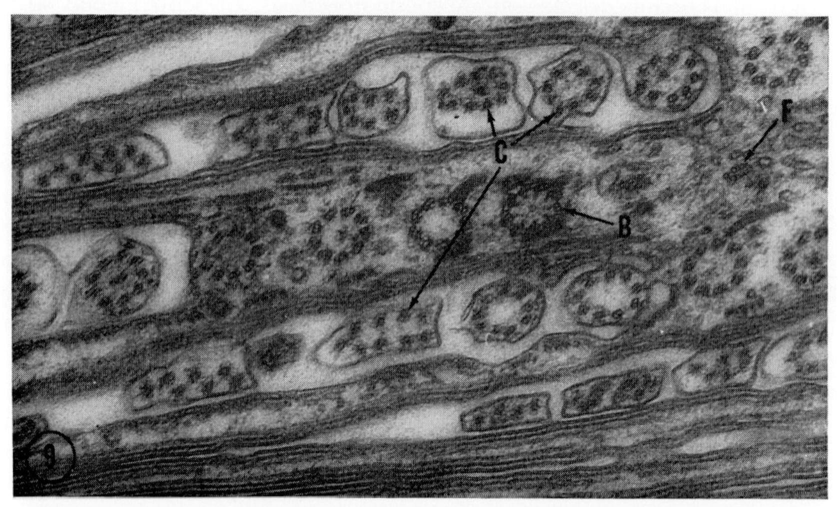

FIG. 9. A section perpendicular to that in Fig. 8 showing cross sections of the forming cilia. Notice that the short shafts (C) are composed of filaments with circular cross sections and are not yet arranged in a circle distally. The basal bodies (B) are not different from those of mature cilia and the infraciliary filaments (F) are well formed. Magnification: × 57,000. (Micrograph by Y. Shigenaka.)

The first stages of ciliary shaft formation have been observed in studies of rumen protozoa (Bretschneider, 1962; Roth and Shigenaka, 1963). The first elements in the shaft of the very short forming cilium are about nine filaments of circular cross sections with material randomly adhering to their surfaces (Fig. 9, F). At this stage the filaments have not yet been oriented into a circular configuration but are randomly arranged within the short tapering part of the cilium thus far formed. The membrane of the cell has been extended so there is always an enclosure of these forming filaments in the cytoplasm of the cell. The appearance of filaments with doublet cross sections, their arrangement into a circular configuration, and the addition of central filaments takes place at a later time by processes not known. Central filaments form without apparent connection with other filaments whereas peripheral filaments seem to form by direct contact with the basal triplets. Thus, the formation of the shafts of cilia can probably be arranged into several parts: first, synthesis of the molecular components; second, simple molecular aggregation to form circular cross sections; third, aggregation of a modified type to form doublet filaments; and fourth, the arrangement of filaments into the nine-and-two complex.

Summary

The formation of filament systems probably involves four events which may be partially separable in time and location. (1) Synthesis of protein molecules must take place. (2) Polymerization of these molecules and (3) cross-bonding of polymer chains result in the formation of filaments. Degrees of intimacy of filament formation with parent filaments have been observed; formation involving direct contact takes place but considerably greater independence must also be possible. (4) Grouping of filaments takes place, e.g., the rather crude grouping to form irregular fibrils in the MA. In cilia, however, modification of the filament-forming process from the first formation of circular filaments to doublet filaments takes place as an additional step followed by a very precise filament arrangement into the nine-and-two pattern. Therefore, four steps of the formation process are recognized: synthesis, polymerization, polymer cross-bonding, and filament grouping.

Production of Movement by Filaments

Progress is being made in understanding the mechanism by which filaments produce movement in striated muscles. Neither the molecular conformation theory nor the molecular interaction (sliding) theory can be either confirmed or excluded at present. In regard to other cell movements, Weber (1958) lucidly summarizes biochemical and comparative physiological evidence to establish important apparent differences in the contraction of muscle and the mechanisms of other cell movements. Structural differences also suggest such differences since the systems previously described, in contrast to muscle, are characterized by filaments continuous throughout their functional length; in addition, neither two different filament types nor regular arrays are necessarily present.

Where rapid and repeated movements are to be produced as in cilia, configurational changes of proteins could account for rapid lengthening and shortening of segments of filaments. Semiconduction along the filaments may take place as postulated in recent papers (Eley, 1962; Green and Fleischer, 1962) from the volume honoring Albert Szent-Györgi who first proposed that such conduction is important in biological systems.

Semiconduction can be suggested in biological systems whenever a close and regular linear arrangement of molecules occurs such as in filaments. However, while configurational changes and semiconduction can explain rapidly repeated movements, they seem inadequate to explain the anaphase movements of chromosomes.

The filament-related anaphase movement has two components: shortening of the chromosome-to-pole distance and elongation of the pole-to-pole distance. Both are active processes. Elongation is so marked in the giant amebae that the final pole-to-pole distance is five times as long as it was in the metaphase; no change is observed in the length of chromosomal filaments in these cells (Short, 1946; Berkeley, 1948). Similarly, chromosome-to-pole shortening can be so marked that the kinetochore-to-pole distance at the end of the anaphase is one-fifth the metaphase distance (Hughes, 1952). Configurational changes of individual protein molecules cannot explain such great changes in length. Rather, removal of molecules from chromosomal filaments and insertion of new molecules into continuous filaments is likely.

A dynamic process seems to be involved. A dynamic polymerization and cross-bonding may take place whereby a filament has added molecules inserted to greatly increase its length or, conversely, molecules removed from it to shorten it. Such a mechanism would be easily explained if the addition and subtraction of molecules took place at the polar ends of individual filaments, but the evidence is against this being the location involved. Rather, polarizing microscope studies as reported and summarized by Inoué (1953, 1959, 1963) are interpreted to show anaphase changes near centers of organization—centrioles, kinetochores, or phragmoplasts—but not at the ends of filaments. Thus contraction of chromosomal filaments may be considered as the removal of molecules at points along the chromosomal filament. Spindle elongation or the lengthening of continuous filaments can be the insertion of molecules into largely the interzonal portion of the continuous filaments which would, therefore, be increased in mass; some evidence for an increase of mass in the interzone has been given from two interference microscopic studies (Rustad, 1959; Longwell and Svihla, 1960). Therefore, the dynamic mechanism involved can best be explained as the insertion and removal of molecules from filaments not at their ends but within their lengths.

In many cells where some amount of both anaphase movements take place, the interzone must be suggested to favor polymerization whereas the region poleward from the chromosomes would favor depolymerization; the possibility exists, therefore, that molecules removed from chromosomal filaments may become incorporated later into continuous filaments. Interestingly, movement by spindle elongation frequently *follows* chromosome-to-pole movement in those cells where both movements are important. In many cells the two movements take place more simultaneously, but I am not aware of any cases where spindle elongation precedes chromosome-to-pole movement.

SUMMARY

Filaments are dynamic structures. Just as membranes are regarded as being dynamic and capable of great surface enlargement in short periods of time, so, also, filaments may need to be similarly regarded as dynamic and capable of considerable elongation and shortening. It seems appropriate to suggest that, as a membrane may increase its surface area, similarly a filament may increase its length. The dynamic mechanism involved may account for filament formation and, furthermore, may also produce the mitotic movements caused by filaments.

Movements without Filaments

In the holotrichous protozoan *Isotricha prostoma,* the extranuclear division apparatus does not contain filaments of the type discussed. Instead, the thick fibers or sheets present (Fig. 10, F) are composed of a fine material in parallel array (Fig. 11). Similar material has been reported to occur in *Physarum* (Wohlfarth-Bottermann, 1962) and also has been shown at this Symposium to be present in amebae at determined sites and times relating to movement (Wohlfarth-Bottermann, 1963). In ciliate protozoa, such as *Stentor* (Randall, 1957) and *Diplodinium* (Fig. 5), both filamentous bundles and nonfilamentous fibers are present in the pellicle in juxtaposition. Such nonfilamentous bundles could be involved in a similar mechanism of movement, differing only in the manner of molecular linkage.

Summary

Filament systems having structural and chemical similarities have been described and discussed in order to derive hypotheses of their formation and function. Formation is suggested to include four steps: synthesis, polymerization, polymer cross-bonding, and filament grouping.

Some cell movements involving filaments can be explained by local changes in polymerized molecules. For example, in cilia configurational changes of the protein molecules or changes in cross-bonding of helically arranged polymers could explain shortening by a small percentage of its length.

Some other cell movements, though perhaps involving very similar molecules and arrangements, cannot be explained by such mechanisms.

The functioning of the filaments of the MA is such an example since the changes in filament length are too great to be explained in this way.

Filaments of the MA and other filaments at the time of their forma-

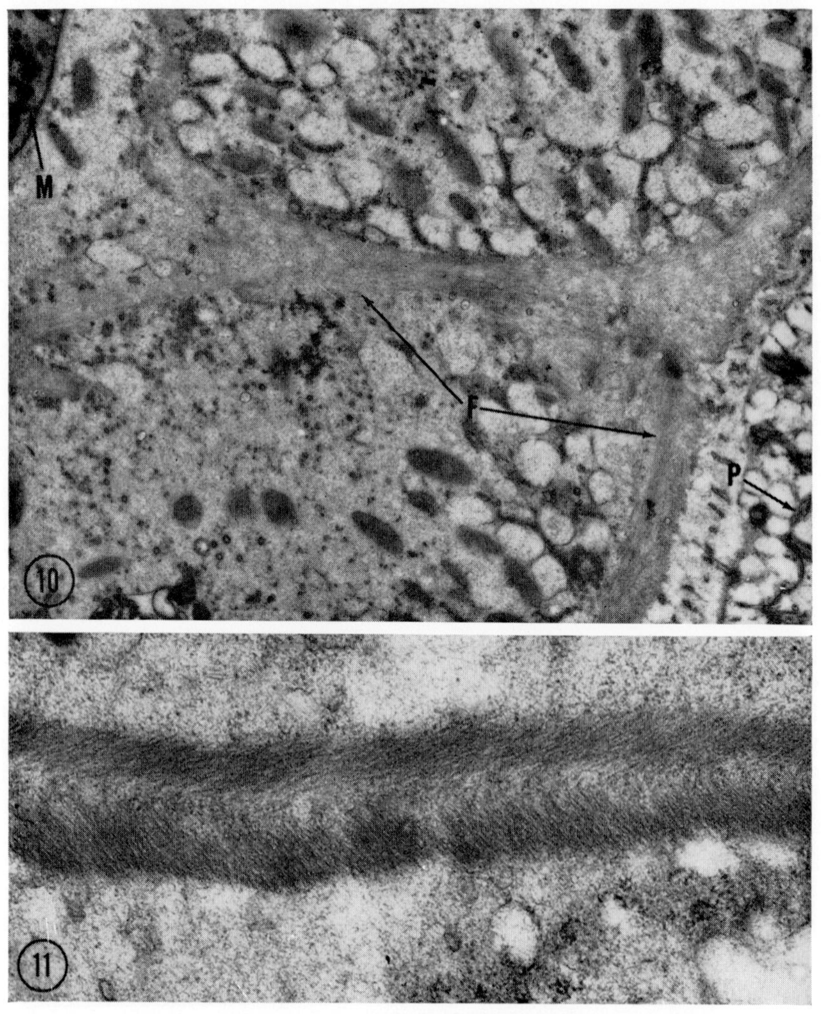

FIG. 10. A survey electron micrograph of *Isotricha prostoma* including portions of the pellicle (P) and macronucleus (M). A system of fibrils (F) is present forming a layer slightly below the plasma membranes and another surrounding the nuclei; the layers are interconnected. Magnification: × 11,000.

FIG. 11. Higher magnification of the fibril in *Isotricha* (Fig. 10). Each section of the pellicular layer shows two sublayers in which parallel, fine elements are found; the orientation of elements in the sublayers is perpendicular. Magnification: × 41,000. (Micrograph by J. Chakraborty.)

tion are similar. The terms "function" and "formation" in regard to the MA filaments may be synonymous. Slightly varying conditions may be found in cells to alter the equilibrium that controls the addition (to lengthen) or removal (to shorten) of molecules from filaments in metastable states. Filament function in the mitotic apparatus is thus considered to be a dynamic process.

Cell movements must be explained at the molecular level. The function of the mitotic apparatus can only begin to be explained as a dynamic molecular phenomenon. To understand ameboid movement, terms such as "contraction" must give way to terms that are chemically more precise; to understand all cell movements, we must apply the principles of the physical chemistry of proteins (Höber, 1945; Putnam, 1953; Weber, 1958 are exemplary).

Varying bonding of similar molecules can produce the several levels of organization known in motile systems: fibrils composed of filaments, fibrils composed of finer elements in birefringent array, isometric arrays, and free molecules constituting the precursor pool. Movement is controlled molecular bonding.

References

Berkeley, E. (1948). *Biol. Bull.* **94**, 169.

Bishop, D. (1962a). *Physiol. Rev.* **42**, 1.

Bishop, D. (1962b). *In* "Spermatozoan Motility" (D. W. Bishop, ed.), pp. 251-268. Am. Assoc. Advan. Sci., Washington, D.C.

Boss, J. (1955). *Exptl. Cell Res.* **8**, 81.

Bretschneider, L. H. (1962). *Koninkl. Ned. Akad. Wetenschap. Proc., Ser. C* **65**, 423.

Daniels, E. W., and Roth, L. E. (1963). *J. Cell Biol.* In press.

Eley, D. D. (1962). *In* "Horizons in Biochemistry" (M. Kasha and B. Pullman, eds.), pp. 341-380. Academic Press, New York.

Gall, J. G. (1961). *J. Biophys. Biochem. Cytol.* **10**, 163.

Gibbons, I. R., and Grimstone, A. V. (1960). *J. Biophys. Biochem. Cytol.* **7**, 697.

Green, D. E., and Fleischer, S. (1962). *In* "Horizons in Biochemistry" (M. Kasha and B. Pullman, eds.), pp. 381-420. Academic Press, New York.

Harris, P. (1961). *J. Biophys. Biochem. Cytol.* **11**, 419.

Harris, P. (1962). *J. Cell Biol.* **14**, 475.

Höber, R. (1945). "Physical Chemistry of Cells and Tissues," pp. 293ff. Blakiston, Philadelphia, Pennsylvania.

Hoffmann-Berling, H. (1958). *Biochim. Biophys. Acta* **27**, 247.

Hughes, A. F. W. (1952). "The Mitotic Cycle," p. 128. Academic Press, New York.

Inoué, S. (1953). *Chromosoma* **5**, 487.

Inoué, S. (1959). *In* "Biophysical Science—A Study Program" (J. L. Oncley, ed.), pp. 402-408. Wiley, New York.

Inoué, S. (1963). *In* "Primitive Motile Systems in Cell Biology" (R. D. Allen and N. Kamiya, eds.), pp. 549–598. Academic Press, New York.

Jahn, T. (1962). *J. Cellular Comp. Physiol.* **60**, 217.

Jones, R. F., and Lewin, R. A. (1960). *Exptl. Cell Res.* **19**, 408.
Kane, R. E. (1962). *J. Cell Biol.* **15**, 279.
Longwell, A. C., and Svihla, G. (1960). *Exptl. Cell Res.* **20**, 294.
Mazia, D. (1956). *Advan. Biol. Med. Phys.* **4**, 95.
Mazia, D. (1959). *In* "Sulfur in Proteins" (R. Benesch *et al.*, eds.), p. 367. Academic Press, New York.
Mazia, D., Harris, P. J., and Bibring, T. (1960). *J. Biophys. Biochem. Cytol.* **7**, 1.
Mazia, D., Chaffee, R. R., and Iverson, R. M. (1961). *Proc. Natl. Acad. Sci. (U.S.)* **47**, 788.
Östergren, G. (1949). *Hereditas* **35**, 445.
Putnam, F. W. (1953). *In* "The Proteins" (H. Neurath and K. Bailey, eds.), Vol. IB, pp. 807-892. Academic Press, New York.
Randall, J. T. (1956). *Nature* **178**, 9.
Randall, J. T. (1957). *Symp. Soc. Exptl. Biol.* **9**, 185.
Roth, L. E. (1957). *J. Biophys. Biochem. Cytol.* **3**, 985.
Roth, L. E. (1958). *J. Ultrastruct. Res.* **1**, 223.
Roth, L. E. (1959). *Proc. Intern. Conf. Electron Microscopy, 4th, West Berlin, 1958*, **2**, 241-244.
Roth, L. E., and Daniels, E. W. (1962). *J. Cell Biol.* **12**, 57.
Roth, L. E., and Jenkins, R. A. (1962). *Proc. Intern. Conf. Electron Microscopy, 5th, Philadelphia, 1962*, **2**, NN-2.
Roth, L. E., and Minick, O. T. (1961). *J. Protozool.* **8**, 12.
Roth, L. E., and Shigenaka, Y. (1963). *J. Cell. Biol.* In press.
Rouiller, C., Faure-Fremiet, E., and Gauchery, M. (1956). *Exptl. Cell Res.* **11**, 527.
Rustad, R. C. (1959). *Exptl. Cell Res.* **16**, 575.
Sharp, R. (1914). *Univ. Calif. (Berkeley) Publ. Zool.* **13**, 43.
Short, R. B. (1946). *Biol. Bull.* **90**, 8.
Slautterback, D. B. (1963). *J. Cell. Biol.* **18**, 367.
Watson, M. R., and Hopkins, J. M. (1962). *Exptl. Cell Res.* **28**, 280.
Weber, H. H. (1958). "The Motility of Muscle and Cells." Harvard Univ. Press, Cambridge, Massachusetts.
Went, H. A. (1960). *Ann. N. Y. Acad. Sci.* **90**, 422.
Wohlfarth-Bottermann, K. E. (1962). *Protoplasma* **54**, 514.
Wohlfarth-Bottermann, K. E. (1963). *In* "Primitive Motile Systems in Cell Biology" (R. D. Allen and N. Kamiya, eds.), pp. 79–110. Academic Press, New York.

Discussion

Dr. Inoué: I am very pleased with Dr. Roth's conclusion that this film of mitotic apparatus represents a dynamic state of organization. I wonder what electron-microscopic evidence one can find to support this notion.

Dr. Roth: I think the information has to come mostly from other studies, but the electron-microscopic evidence which I think is most important is that in the giant ameba, material adheres to the exterior of filaments in prometaphase, metaphase, and anaphase. Those filaments that are in the interzone and, therefore, are presumed to be continuous filaments are the ones that we see in telophase and that pack more closely because material is now absent from their surfaces. This evidence suggests that material has now been incorporated into the filaments and that they have concluded their formation at the same time they have concluded their function.

Dr. Allen: Are the continuous filaments thicker than the chromosomal filaments?

DR. ROTH: We can see no difference in the diameter of continuous and chromosomal filaments.

DR. TERU HAYASHI: To a naive nonelectron microscopist, the films you showed of the mitotic apparatus, especially in those regions where they were not too closely packed, looked very much like the sort of thing one sees with sarcoplasmic reticulum.

DR. ROTH: You are suggesting perhaps they are not filaments, but rather endoplasmic reticulum?

DR. TERU HAYASHI: Yes.

DR. ROTH: I think they are not. You may, if you wish, follow the argument which many people do, that any membranous structure is part of the endoplasmic reticulum; if you do, you may wish to call them endoplasmic reticulum. Cross-sectional views show uniform packing of uniform elements, more than is characteristic of the endoplasmic reticulum.

DR. TERU HAYASHI: What are the dimensions of the separate units? You mentioned diameter.

DR. ROTH: The filament diameter in the mitotic apparatus is 15 mμ, except in Dr. Kane's isolated mitotic apparatus, where it is 20. The infraciliary filaments also seem to have 15–20 mμ diameters. Perhaps there is either 15 or 20 mμ, but we are not ready to say that yet.

DR. REBHUN: I would not stress the differences in the 15 vs. 20 mμ fibers, especially if there are differences of embedding media and different types of fixation. With respect to the endoplasmic reticulum, in Dr. Harris' work you do find the 15-mμ fibers in the asters as well as chains of endoplasmic reticulum, and it looks certainly as if they are different systems.

The thing I would stress is that in your work no changes in diameter can be seen in the chromosomal filaments at any stage in mitosis.

DR. ROTH: That is exactly right.

DR. REBHUN: This does not necessarily mean that there is no shortening of the filaments. They might be, for example, shortening by material "dissolving" from the ends. If so, you cannot tell whether there is active shortening which has nothing to do with the generation of force, or not. It would be very difficult to decide whether the shortening were actually coupled to the performance of work.

DR. ROTH: First of all, a very active process is involved in the spindle elongation of the giant ameba; that the filaments are performing some work seems to be somewhat well established there, if one believes the work that Berkeley (1948) and Short (1946) report in very nice detail in their papers on giant amebae. Elongation, if impeded on one end, results in a "barreling-out" of the spindle in the interzone, and they stress very strongly that the elongation must be an active process.

In relation to your first remark, I suggest only that 15 and 21 mμ filaments may not be the same, or they may be. Such a difference in size is very close to the limit of error in magnification of microscopes, but, nevertheless, in the dividing *Diplodinium*, instead of having only one diameter, we seem to have a range of diameters. If a lot of measurements are made, they turn out to have a slightly bimodal distribution clustering around 15 and 21 mμ when sections are studied from the same or identically handled specimens.

DR. ZIMMERMAN: A comment I would like to make is that you must remember Dr. Kane's work on mitotic apparatus was done on isolated preparations. It is amazing to me that he got any kind of evidence indicative of fibers, because all previous work on the isolated mitotic apparatus had given no such indications. The fact that Dr. Harris' work gave measurements very comparable to yours indicates in all probability that the measurements are valid.

DR. ROTH: Yes. I have very great respect for Dr. Kane's work. It is interesting because it may mean that these filaments can be altered. Perhaps they transform or can be transformed from one diameter to another. Perhaps by his experimental procedure he transformed them in some way. With this in mind, we can look back at the suggestion of bimodal diameter distribution in *Diplodinium* and speculate that the filaments are formed at one diameter but function at another.

DR. ZIMMERMAN: It could be a difference in fixation.

DR. ROTH: The fixation in his case could be the altering factor. However, I am suggesting the organism may be able to alter the filaments; there may be two forms of filaments also, the smaller one being that one which is first formed.

DR. INOUÉ: Dr. Kane has been able to alter artificially the filaments into three different forms; one, a so-called tubular filament of the type you see; another, a solid-appearing filament; and, third, he can put these in solution and cause small droplets to form. I think it could be a simple ionic environment that introduces differences which do not result from the fixation process itself.

DR. ROTH: I think Dr. Inoué will emphasize in his paper the metastability or lability of the filaments of the mitotic apparatus. I would agree that this metastability follows from the many studies that have gone before, particularly the one by Östergren on *Luzula* where the chromosomes may be pulled through filaments in a typical mitotic process. It is possible that the tubular filament does not correspond exactly to the *in vivo* functional unit of the mitotic apparatus. This may possess a metastability that we can alter by fixation procedures and perhaps which the cell alters during function.

Organization and Function of the Mitotic Spindle[1]

SHINYA INOUÉ

Department of Cytology, Dartmouth Medical School, Hanover, New Hampshire

Introduction

The occurrence of birefringence in contractile elements of living cells may have a deeper significance than the mere localization of oriented molecules or micelles. This assertion is supported by two generalizations, the first made by Engelmann in 1875, stating that contractility in living systems depended upon the presence of birefringence (also Engelmann, 1906; but see Schmidt, 1937, p. 251). Whether strictly true or not, the statement takes on added significance in the light of the second generalization made by Dr. Aaron Katchalaski at the February, 1963 Meeting of the Biophysical Society. On thermodynamic grounds, he argued that in order to convert chemical energy (a scalar quantity) *directly* into contractile mechanical force (a vectorial quantity), structural anistotropy is required. In other words, the molecules giving rise to contractile force cannot be randomly arranged, but must form an organized structure whose properties vary in different directions. In general, structural anisotropy is associated with optical anisotropy, i.e., a different response of the material to light vibrating or traveling in different directions. When the velocity of light propagation in a material varies in different directions, the material is said to be birefringent or to exhibit double refraction.

The birefringence of muscle fibers, for example, is strong enough to be seen with conventional polarizing microscopes. The mitotic spindle, on the other hand, exhibits a considerably lower retardation than do muscle fibers and can be detected only by the use of refined polarization microscopic techniques. The contrast due to weakly birefringent specimens must be maximized by reducing all possible sources of stray light and by judicious use of a compensator (Schmidt, 1934; Swann and Mitchison, 1950; Inoué and Dan, 1951; Mitchison, 1953; Inoué, 1961). Conventional polarizing microscopes are not satisfactory for the study of

[1] Supported in part by grants from the National Science Foundation (G 19487) and National Cancer Institute, U.S. Public Health Service (CA 04552).

detailed structures of the mitotic spindle, not only because of excessive stray light, but also because of the presence of a troublesome diffraction anomaly which limits effective lateral resolution (Inoué and Kubota, 1958; Kubota and Inoué, 1958). Some years ago we developed the polarization rectifier which corrects both of these defects (Inoué and Hyde,

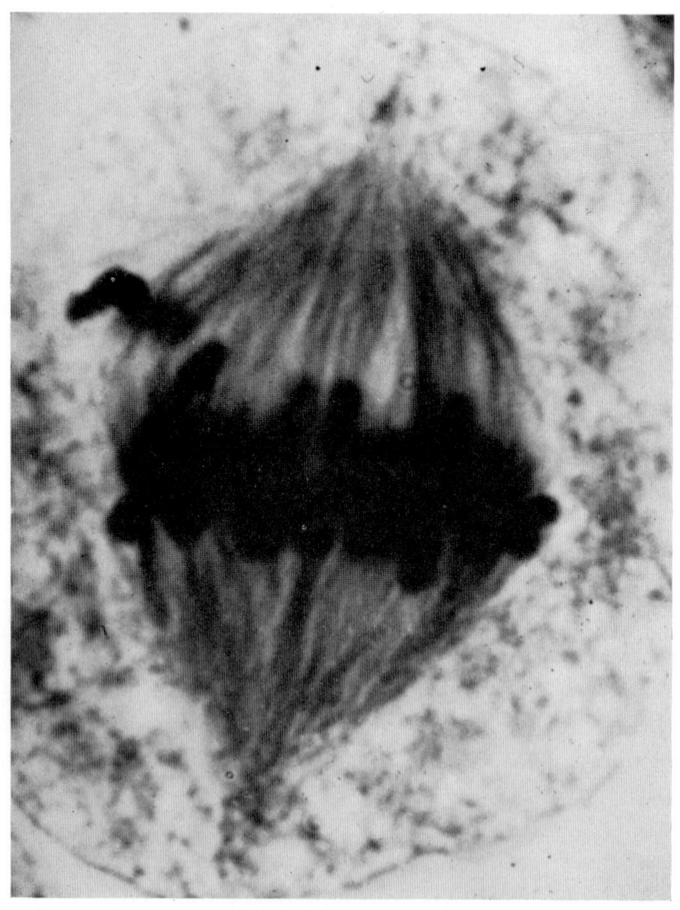

FIG. 1. The spindle of meiosis I as seen in fixed and stained preparation of a hybrid wheat pollen mother cell. (Courtesy, Dr. G. Östergren, University of Lund, Sweden.)

1957). The polarizing microscope used in these studies (Figs. 57, 58) has been described elsewhere (Inoué, 1961).

In this paper the term "mitotic apparatus" will be used in a broad sense, as Mazia and Dan (1952) defined it, to include astral rays, centrioles, chromosomal and continuous spindle fibers, and so on, i.e., all that which makes up the assembly that comes out as one piece when the

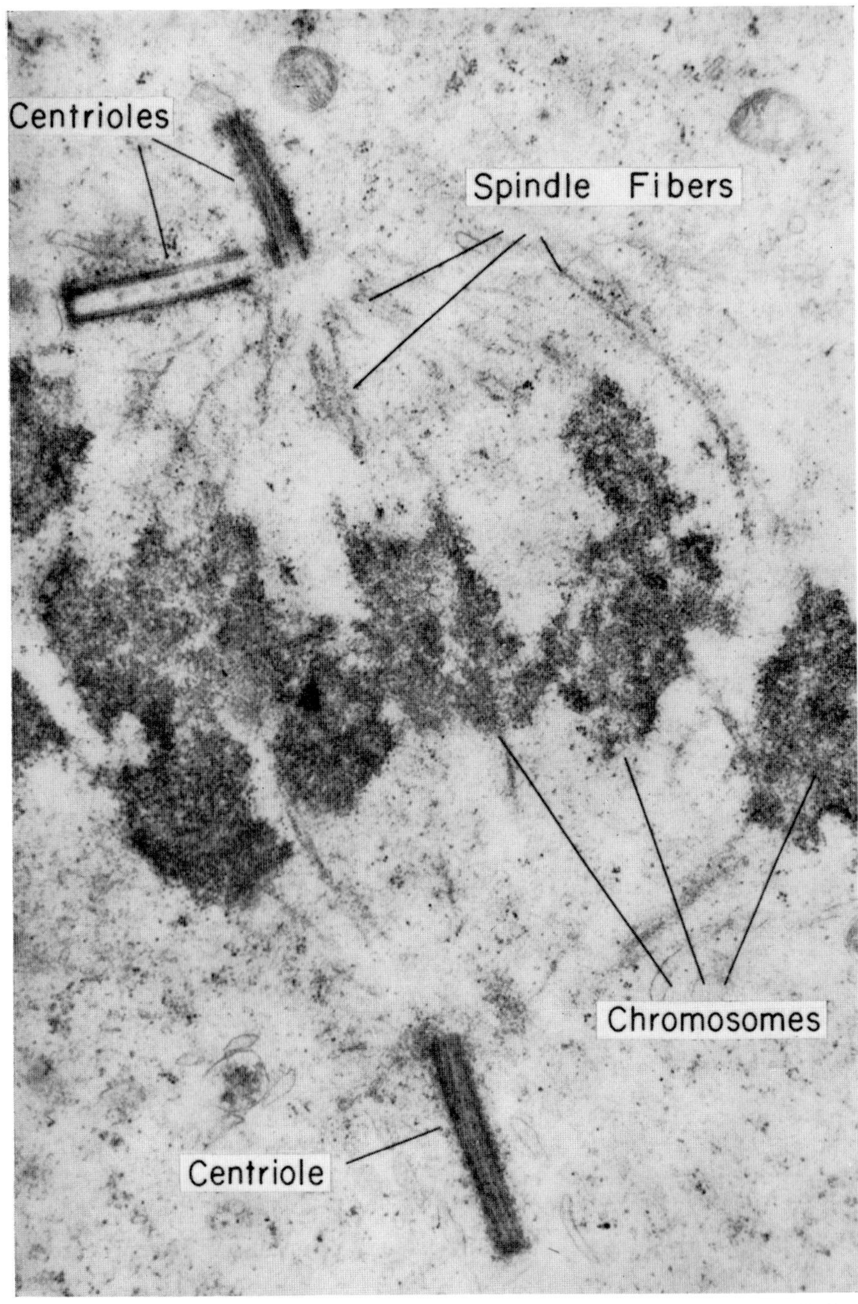

FIG. 2. Electron micrograph of a mitotic spindle from the spermatocyte of the domestic fowl. (From Bloom and Fawcett, 1962, p. 36.)

"spindle" is isolated from the cell in division (Mazia and Dan, 1952; Kane, 1962).

The fibrous structures of the mitotic apparatus (Fig. 1) as generally depicted in textbooks of cytology have been observed in fixed and stained slides since the last century, as also recently by electron microscopy (Fig. 2; also see article by Roth in this Symposium). The functions of these fibrous structures have been variously discussed (see summaries by Schrader, 1953; Mazia, 1961), but there have also been numerous debates as to whether the fibers really existed in living cells. The grounds for the debate have been chiefly that (1) with careful fixation the fibers sometimes could not be demonstrated (see Porter and Machado, 1960, as an example at the electron-microscopic level), whereas with poor fixation they appeared coarser and (2) chromosomes exhibited strange mitotic movements which could not be explained by the simple pulling or pushing by spindle fibers (Östergren, 1949; Schrader, 1953).

As to the reality of these fibers in living cells, I have proved over the years by the use of sensitive polarizing microscopes (Inoué, 1961) that fibers as depicted by the best classic work (e.g., Bělař, 1929a,b) in fact do exist in a wide variety of healthy dividing cells (Inoué, 1953; Inoué movie, 1960b; Inoué and Bajer, 1961). Additional evidence along this line is also presented in this paper.

The main features which are stressed in this paper are the following: (1) The spindle fibers are not static structures but exist in a dynamic state of flux; (2) the fibers are oriented and organized by "centers" (Boveri, 1888; Wilson, 1928) such as centrioles and kinetochores and other organelles; these will be referred to as "orienting centers"; (3) depending on the activity of such centers and the physiological state of the cell, the spindle fibers can be readily built up, broken down, or reorganized. The same material can, therefore, be made into one kind of fiber or another or transformed from one type of fiber to another, depending on which center happens to be active at that time.

Description of Spindle Birefringence

Very broadly speaking, there are two types of spindles, one with discrete centrioles and the other without (Fig. 3). Most commonly in animal cells one encounters the former type and in plant cells the latter. In the animal cell spindle with centrioles the birefringence of the spindle fibers is stronger, and the fibers converge, toward the centrioles and toward the kinetochores. The birefringent fibers between kinetochores and the centrioles we shall call the "chromosomal fibers." The birefringent fibers running from pole to pole we shall call "continuous fibers."

In spindles of plant cells, lacking centrioles, asters are generally missing, and there may or may not be convergence toward the poles. Whether or not there is convergence, the birefringence of the chromosomal fibers is invariably weak at the pole. The birefringence is strongest adjacent to the kinetochores and gradually becomes weaker toward the poles. Continuous fibers are generally present.

If a living cell, for example, a grasshopper spermatocyte (Fig. 4), is observed under a phase-contrast microscope, one may see the centrioles in an appropriate focus. The spindle region is outlined by mitochondria

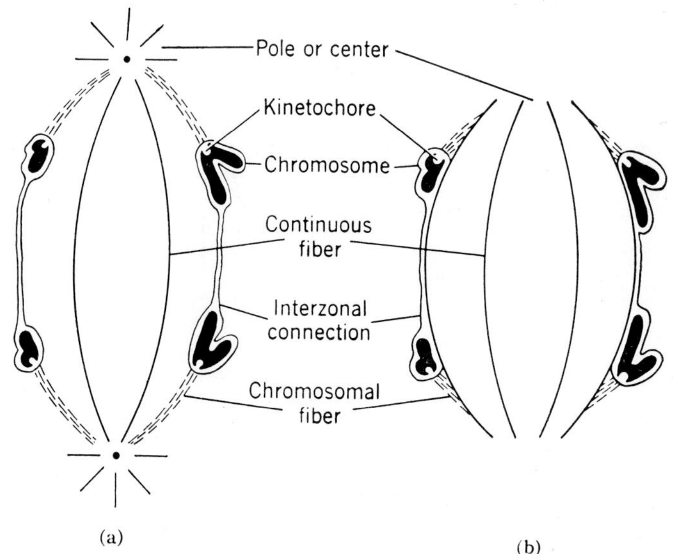

FIG. 3. Schematic diagram of mitotic spindle (a) with centrioles and (b) without. (Modified from Schrader, 1953.)

which are often elongated and surround the spindle as a sheath. The region which should show spindle fibers, however, generally appears structureless under the phase-contrast microscope. "Structureless" here means that there is not an adequate optical path difference (owing in this case to difference of refractive index between the fibers and their surroundings) to produce detectable contrast. It is significant that in living cells in general the spindle fibers do not exhibit such refractive index difference unless the cell is fixed, exposed to acid (Lewis, 1923), or otherwise maltreated. Exceptions to this generalization are found in the flagellate spindle (Cleveland, 1953, 1963), in mite eggs (Cooper, 1941), and occasionally in various other healthy cells just at the onset of anaphase.

Observing a living cell with a sensitive polarizing microscope, one can clearly see these fibers in regions which appear empty under the phase-contrast microscope. In Fig. 5 we see a plant cell, a pollen mother cell of the Easter lily, *Lilium longiflorum,* in anaphase. The spindle converges toward the poles somewhat, but the birefringence of the chromosomal fibers is strongest adjacent to the kinetochores and weaker

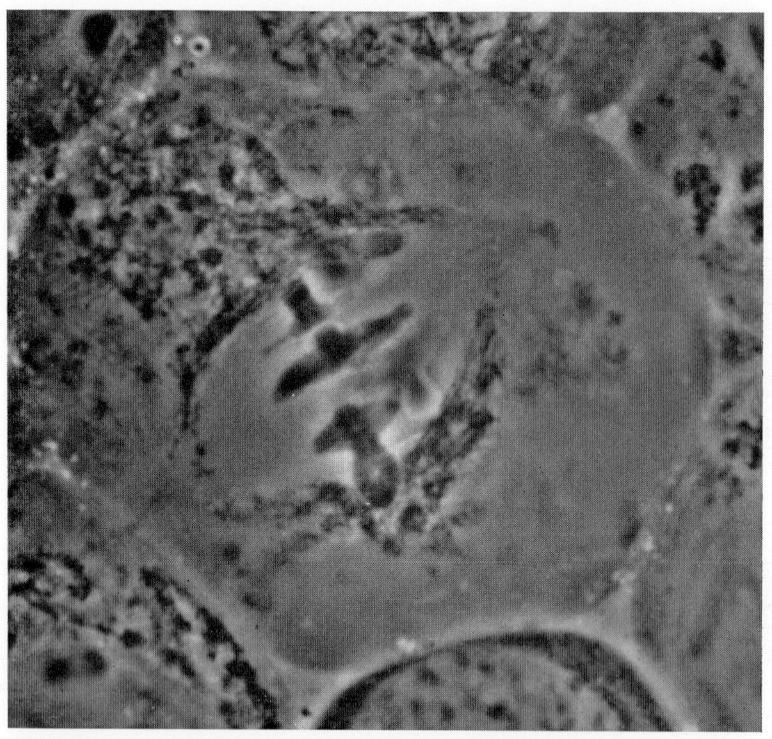

FIG. 4. Living grasshopper spermatocyte (*Chloealtis genicularibus*) as seen with a phase-contrast microscope. Notice early anaphase chromosomes still at equator and mitochondria outlining the spindle. Half-spindles appear empty, and continuous and chromosomal fibers are not visible. (Courtesy, Professor K. Shimakura, Faculty of Agriculture, Hokkaido University, Sapporo, Japan.)

toward the poles. The continuous fibers of this stage are less birefringent and not readily detectable. During anaphase movement, the birefringence of the chromosomal fibers adjacent to the kinetochore remains strong and the fibers become shorter but not detectably thicker. In late anaphase, continuous fibers can be seen again by their increased birefringence.

As an example of an animal cell, Fig. 6 shows the first maturation division spindle in a oocyte of a marine worm, *Chaetopterus pergamen-*

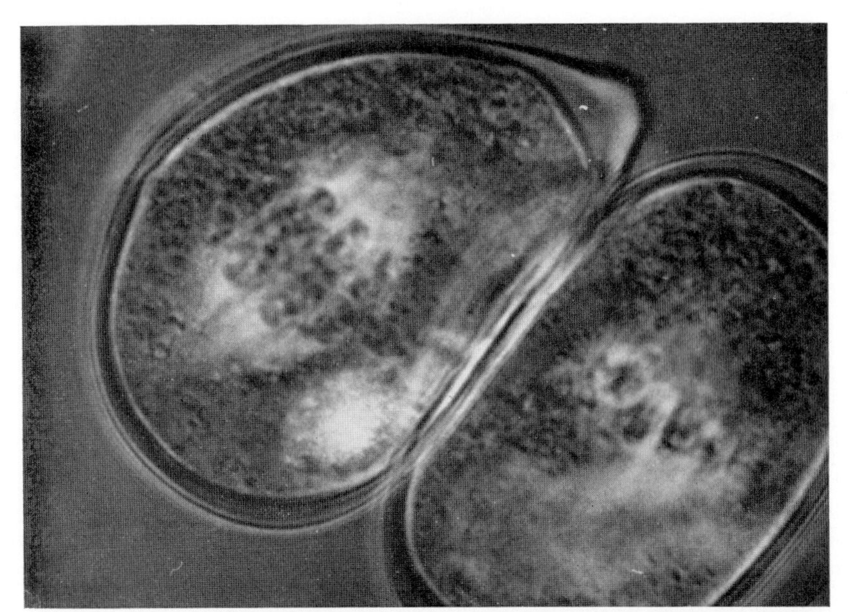

FIG. 5. Living pollen mother cell of the Easter lily, *Lilium longiflorum,* as seen with a sensitive polarizing microscope. The chromosomal fibers show strong birefringence adjacent to the helical chromosomes. (Original photo.)

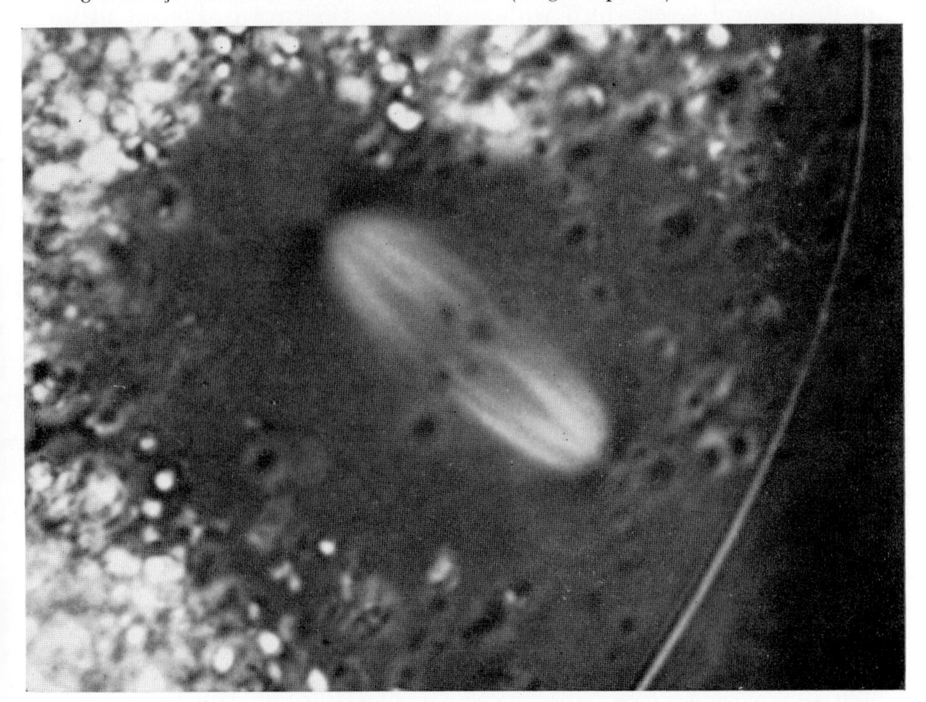

FIG. 6. Birefringent spindle fibers in a living oocyte of a marine worm, *Chaetopterus pergamentaceus.* Egg centrifuged to move away the birefringent yolk granules. (From Inoué, 1953.)

taceus. This cell has been centrifuged to remove the strongly birefringent yolk granules. Not only do the chromosomal fibers converge to the kinetochores of the chromosomes, but they also converge to the poles, where the birefringence of the spindle fibers and astral rays is strong. Owing to the action of the compensator, the positive birefringence of the spindle fibers and astral rays is depicted brighter than the background in one quadrant and darker in the opposite quadrant.

Changes in Birefringence during Mitosis

The appearance, disappearance, and general behavior of the spindle fibers and the change in their birefringence during mitosis are evidence for their dynamic nature. Figures 7 through 14 show a series of low-magnification photographs of the developing eggs of a West Coast jellyfish, *Halistaura cellularia.* These eggs are so clear in ordinary light that one can distinctly see the individual sperm on the other side of the egg.

The contrast within the cytoplasm seen with polarized light is due to walls of vacuoles which fill most of the cytoplasm and the material in-between the vacuoles. The peripheral birefringence is apparently due to the cortical layer which without the birefringence is difficult to differentiate optically.

The nucleus is very difficult to see in interphase, but, as the cell prepares for prophase, it suddenly becomes clear for it swells and a birefringent (spindle) material appears around it. The breakdown of the nuclear membrane is accompanied by the development of a small spindle and asters which appear black or white, depending on the orientation of their fibers relative to the compensator axes. After the metaphase plate is formed, the chromosomes are gradually pulled apart. The spindle length increases, and the over-all birefringence diminishes, especially between the separating chromosomes. This diminution of birefringence was observed as early as 1937 by W. J. Schmidt in Giessen and later studied more in detail by Michael Swann in Edinburgh. Schmidt concluded in 1939, after he decided that the birefringence was due not to chromosomes but to spindle fibers, that the drop of birefringence in the "half-spindle" represented a folding of protein chains. Swann (1951) measured the distribution of birefringence in dividing sea urchin eggs and concluded that a disorienting substance was released from the chromosomes. We should notice, however, that under this magnification one is not able to resolve spindle fibers nor to determine the exact location of the chromosomes. We can, in fact, show with improved resolution that Swann's postulate was based on erroneous assumptions concerning the microscopic structure of the spindle (Inoué and Bajer, 1961).

Spermatocytes of the grasshopper, *Dissosteira carolina*, are shown in Figs. 15 to 26. By prometaphase one clearly sees the strongly birefringent chromosomal fibers attached to the kinetochores, the convergence of the spindle fibers and the birefringent asters at the poles, and weakly birefringent astral rays attached at the poles.

As seen in the time-lapse motion picture to be described below, it is important to note that these fibers are not simply static, but in metaphase and especially prometaphase, their birefringence fluctuates as though one were observing northern lights. This fluctuation is a vivid expression of the dynamic equilibrium and presumably reflects the variation in the amount of oriented material going in and out of the fibers all the time.

At anaphase (Figs. 18 to 21), we see the birefringent chromosomal fibers, the chromosomes, and some sign of the continuous fibers inbetween. The mitochondria which have lined up outside of the spindle also become strongly positively birefringent in late anaphase (Figs. 19 to 24, 26).

Figures 27 to 30 show a pollen mother cell of *Lilium longiflorum*, previously centrifuged to move away the birefringent light-scattering granules. In prometaphase, the birefringent continuous fibers can be seen to run between the poles past the chromosomes (Inoué, 1953). As metaphase approaches, the chromosomal fibers take on a stronger birefringence, especially adjacent to the kinetochores; this obtains throughout division (Figs. 27 and 28). The continuous fibers, whose birefringence is very weak during early anaphase, become more strongly birefringent again in late anaphase and eventually become transformed into phragmoplast fibers in telophase (Figs. 29, 30). Within the phragmoplast thus formed, little vacuoles accumulate, oscillate parallel to the fibers, and finally merge to form the new cell plate (Fig. 30).

We thus see the transformation of the birefringence from continuous fibers to chromosomal fibers, back to continuous fibers again, and finally into phragmoplast fibers (Inoué, 1953).

In the African blood lily, *Haemanthus katherinae*,[2] we see birefringence around the nucleus in the so-called "clear zone" before the nuclear membrane breaks down (Fig. 31). As soon as the nuclear membrane breaks down, this birefringence cannot be distinguished from that now present within the nuclear area (Fig. 32). In other words, this whole area becomes birefringent, occupying the same area and with the same magnitude and positive sign as the spindle, showing longitudinal orientation

[2] This part of the work was done in collaboration with Dr. Andrew Bajer of Kraków, Poland; see Inoué and Bajer, 1961.

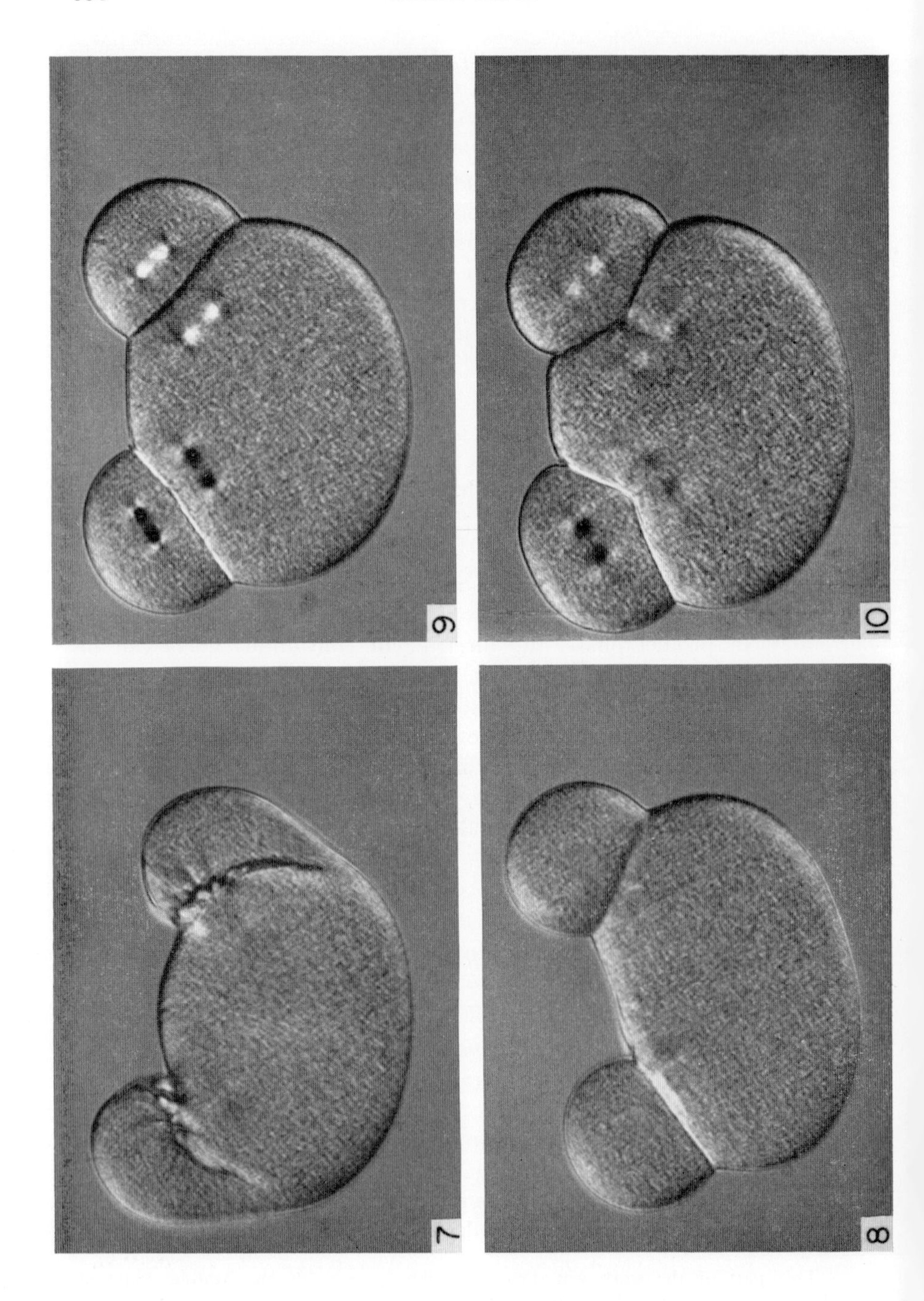

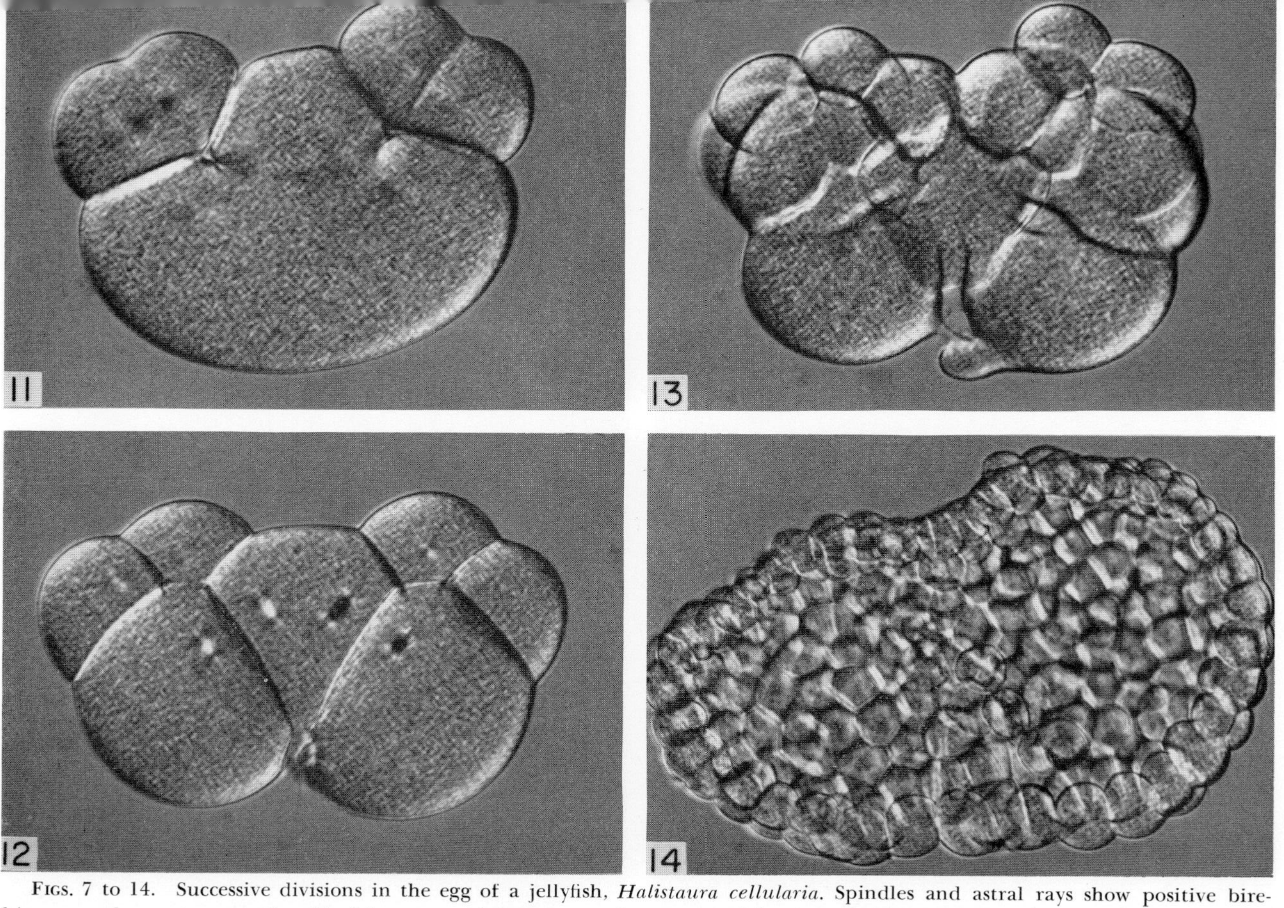

FIGS. 7 to 14. Successive divisions in the egg of a jellyfish, *Halistaura cellularia*. Spindles and astral rays show positive birefringence, the cortex negative birefringence with respect to its tangent. The first cleavage of this egg was suppressed during telophase by cold. (Original photos.)

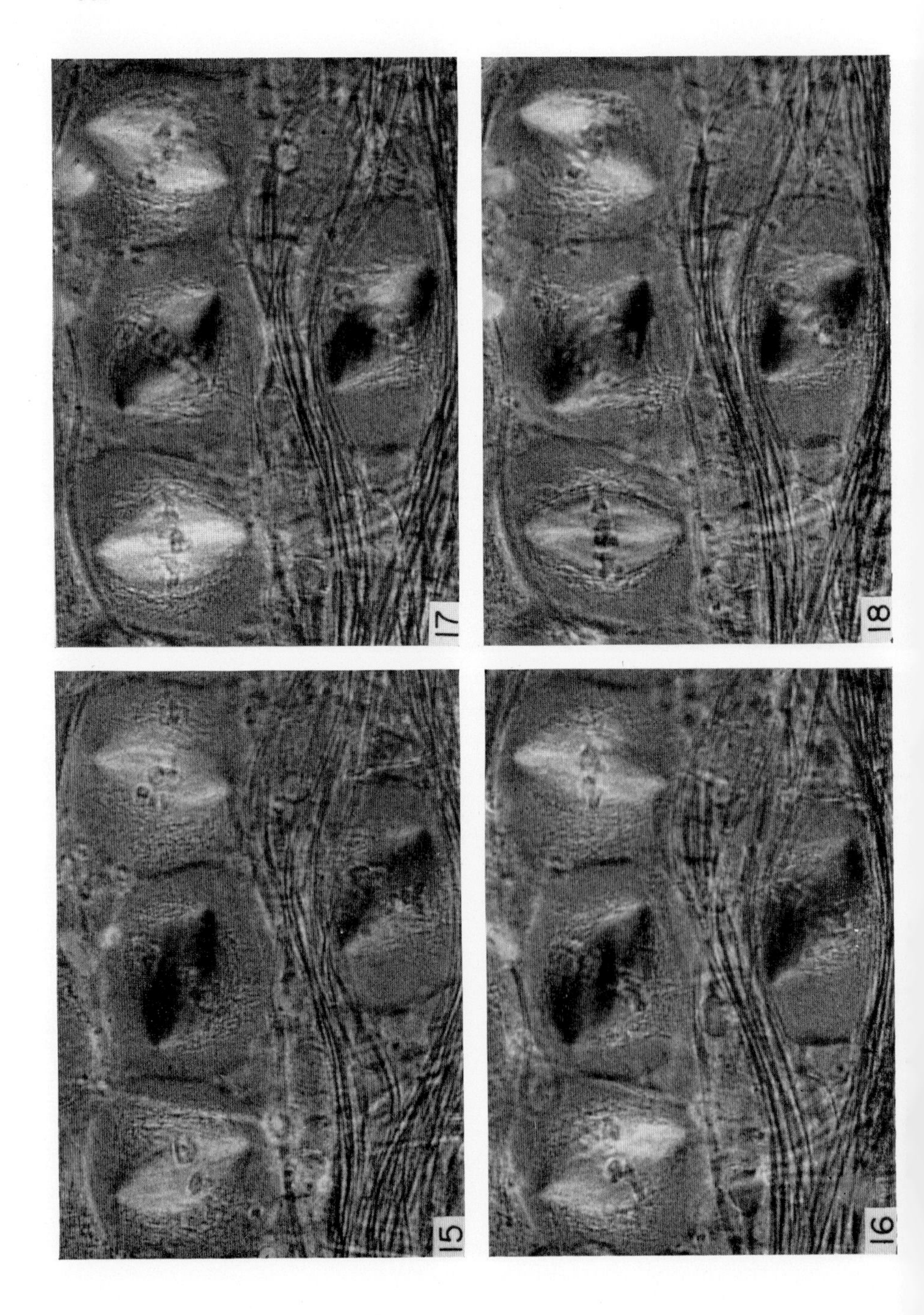
17
18
15
16

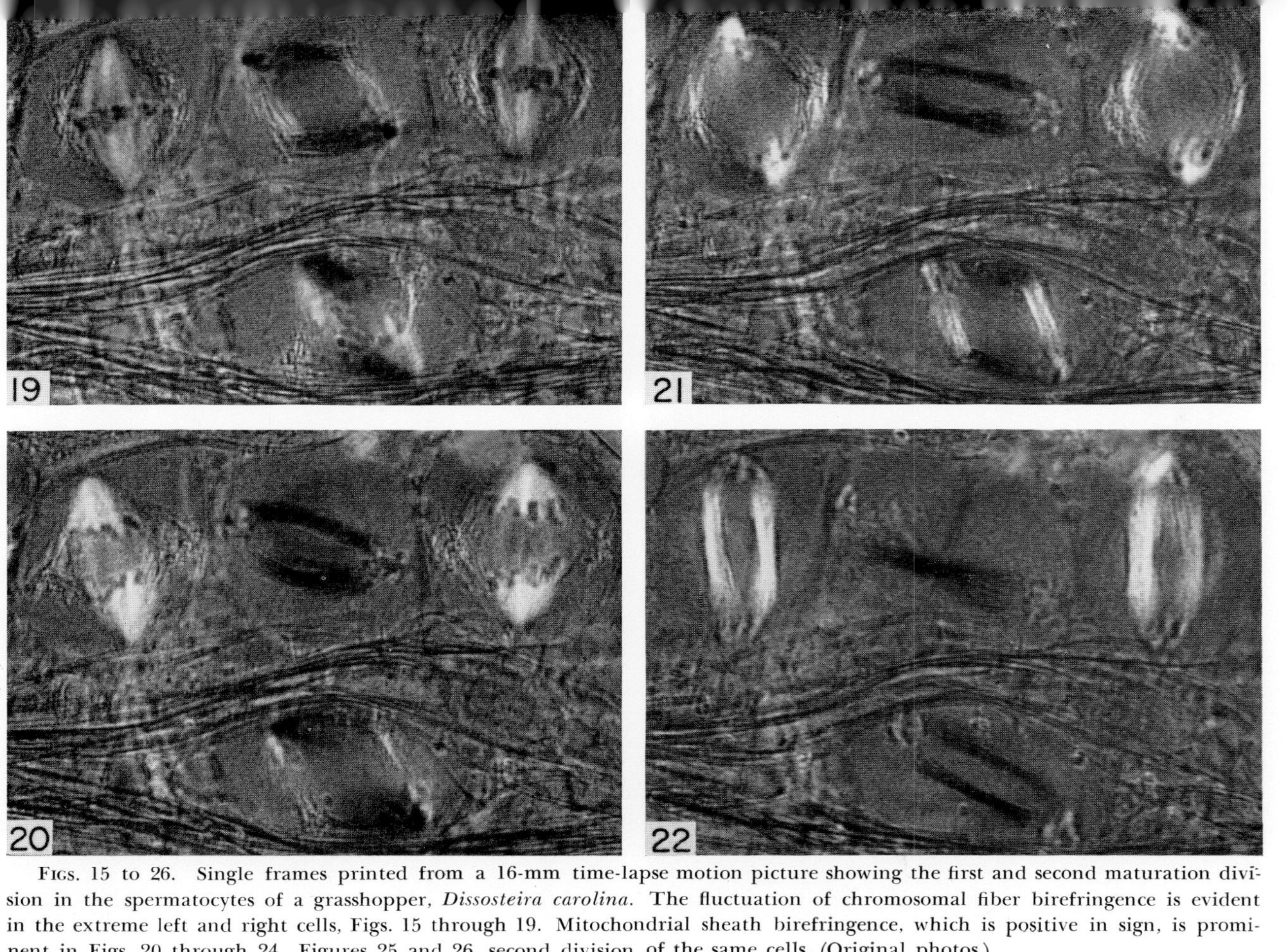

FIGS. 15 to 26. Single frames printed from a 16-mm time-lapse motion picture showing the first and second maturation division in the spermatocytes of a grasshopper, *Dissosteira carolina.* The fluctuation of chromosomal fiber birefringence is evident in the extreme left and right cells, Figs. 15 through 19. Mitochondrial sheath birefringence, which is positive in sign, is prominent in Figs. 20 through 24. Figures 25 and 26, second division of the same cells. (Original photos.)

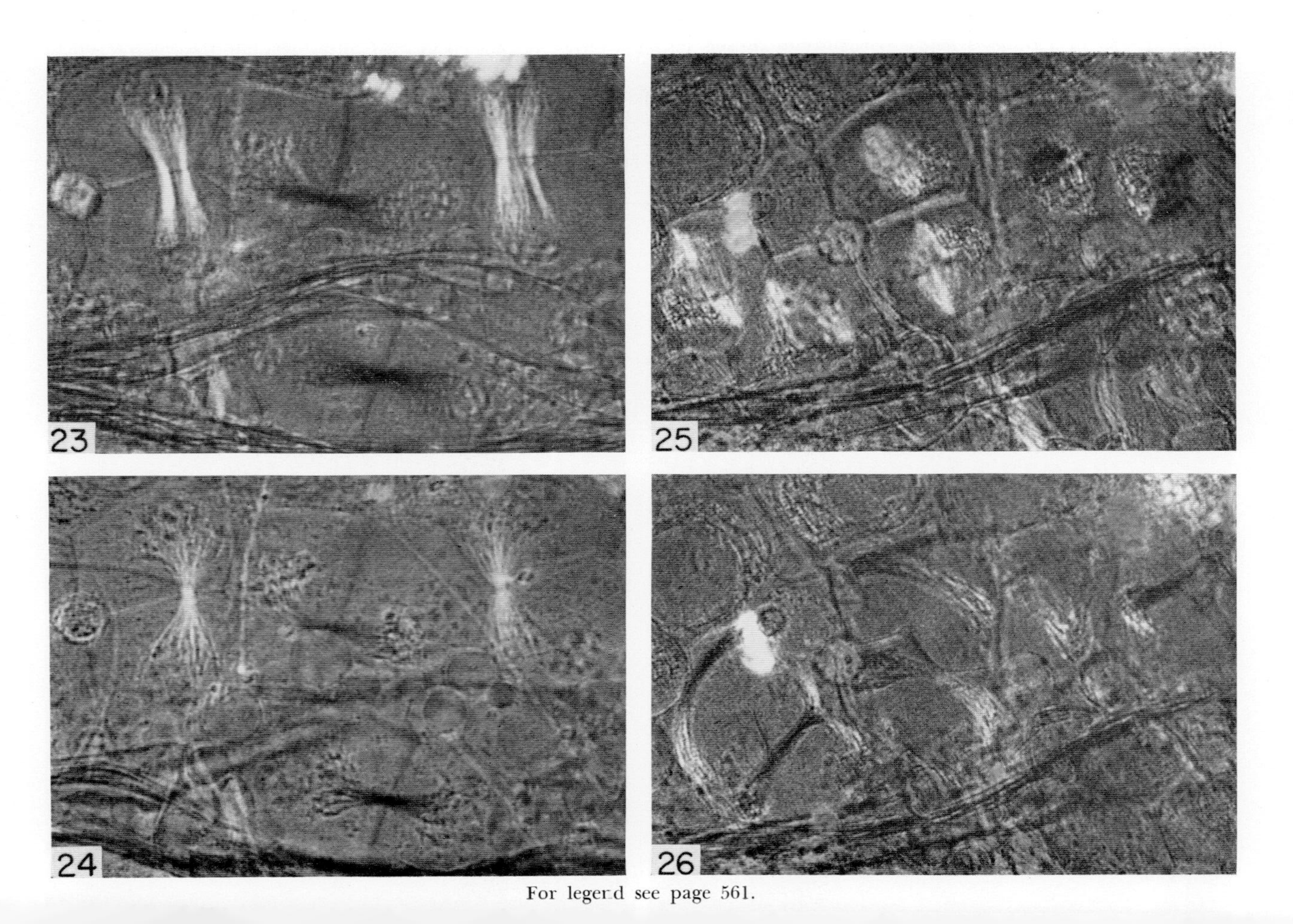

For legend see page 561.

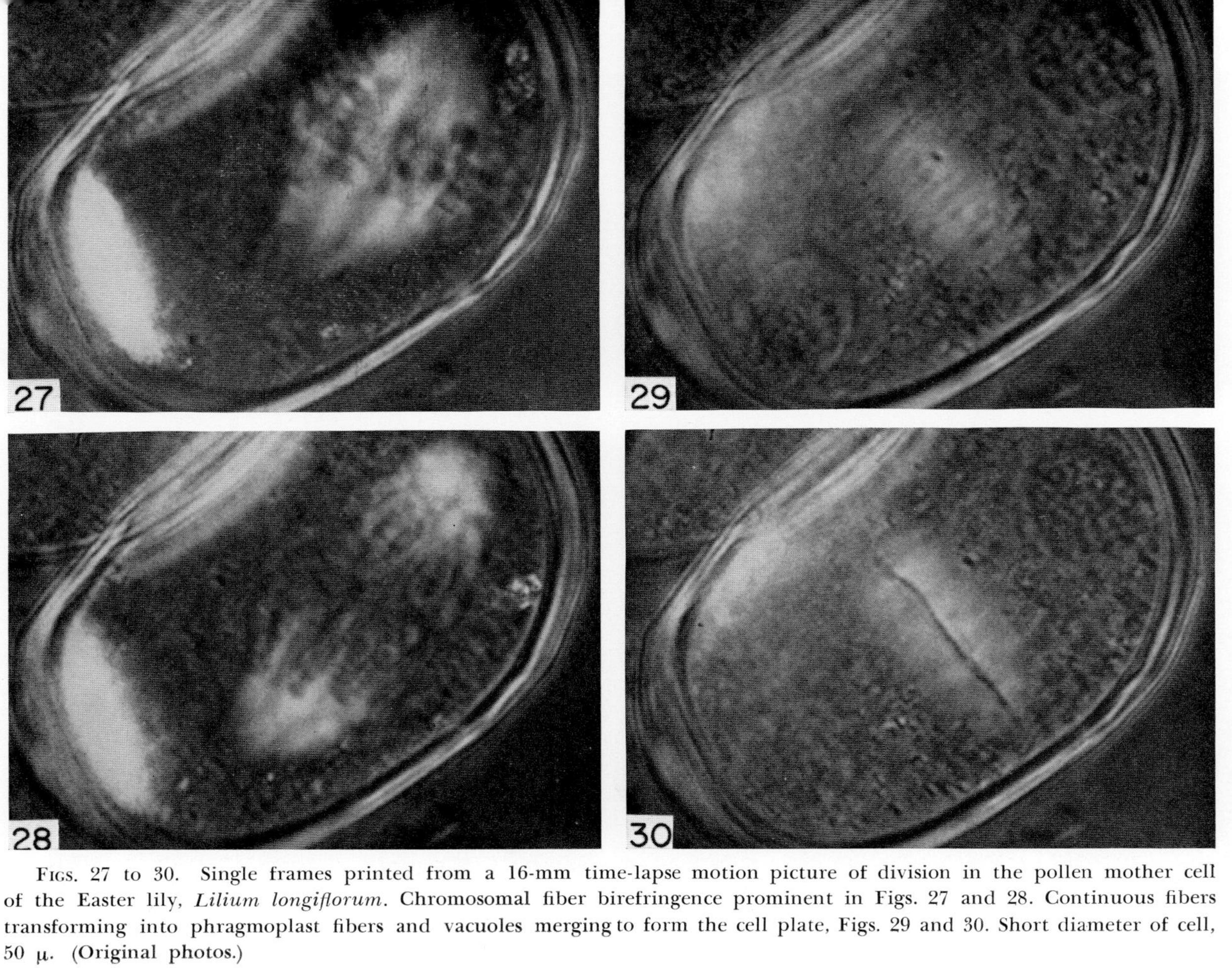

FIGS. 27 to 30. Single frames printed from a 16-mm time-lapse motion picture of division in the pollen mother cell of the Easter lily, *Lilium longiflorum*. Chromosomal fiber birefringence prominent in Figs. 27 and 28. Continuous fibers transforming into phragmoplast fibers and vacuoles merging to form the cell plate, Figs. 29 and 30. Short diameter of cell, 50 μ. (Original photos.)

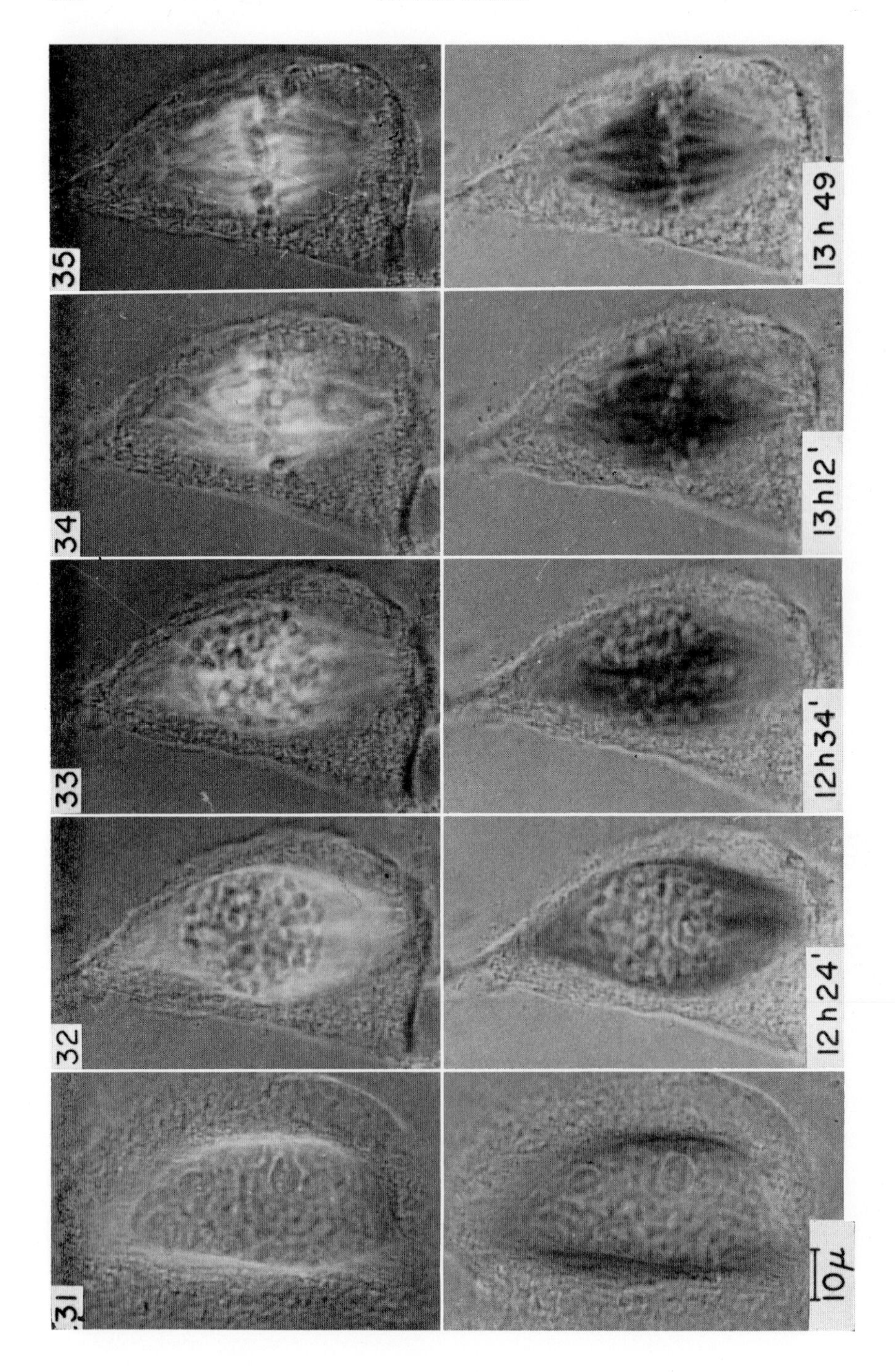
35
34
33
32
31
13 h 49'
13h12'
12h34'
12h24'
10μ

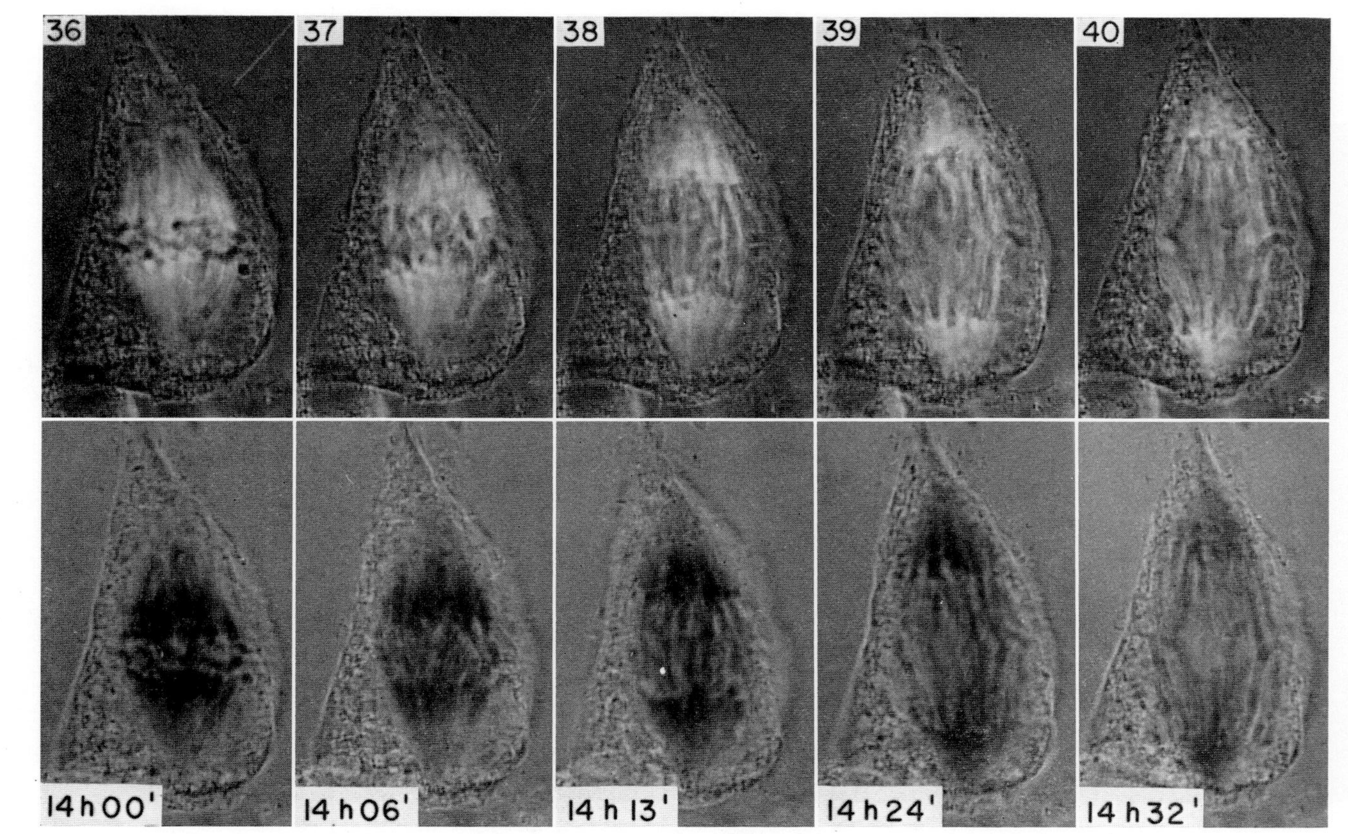

Figs. 31 to 40. Development of the birefringent spindle fibers and anaphase movement in the endosperm cell of the African blood lily, *Haemanthus katherinae*. Pairs of photos are taken with opposite compensation. Same magnification as Figs. 41 to 48. See Fig. 31 for 10 μ scale. (From Inoué and Bajer, 1961.)

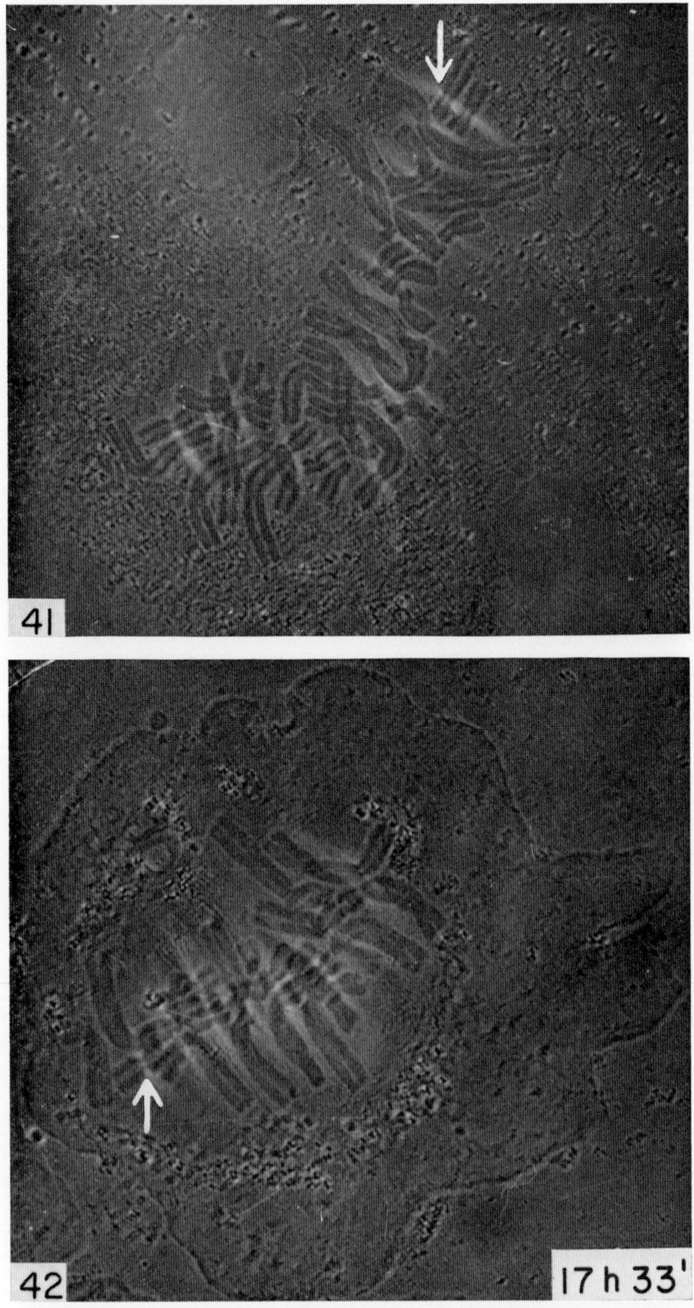

FIGS. 41 to 48. Chromosomes and birefringent spindle fibers in flattened cells of *Haemanthus katherinae*. The strong birefringence of the chromosomal fibers adjacent to the kinetochore is indicated by arrows in Figs. 41 and 42. Continuous fibers visible

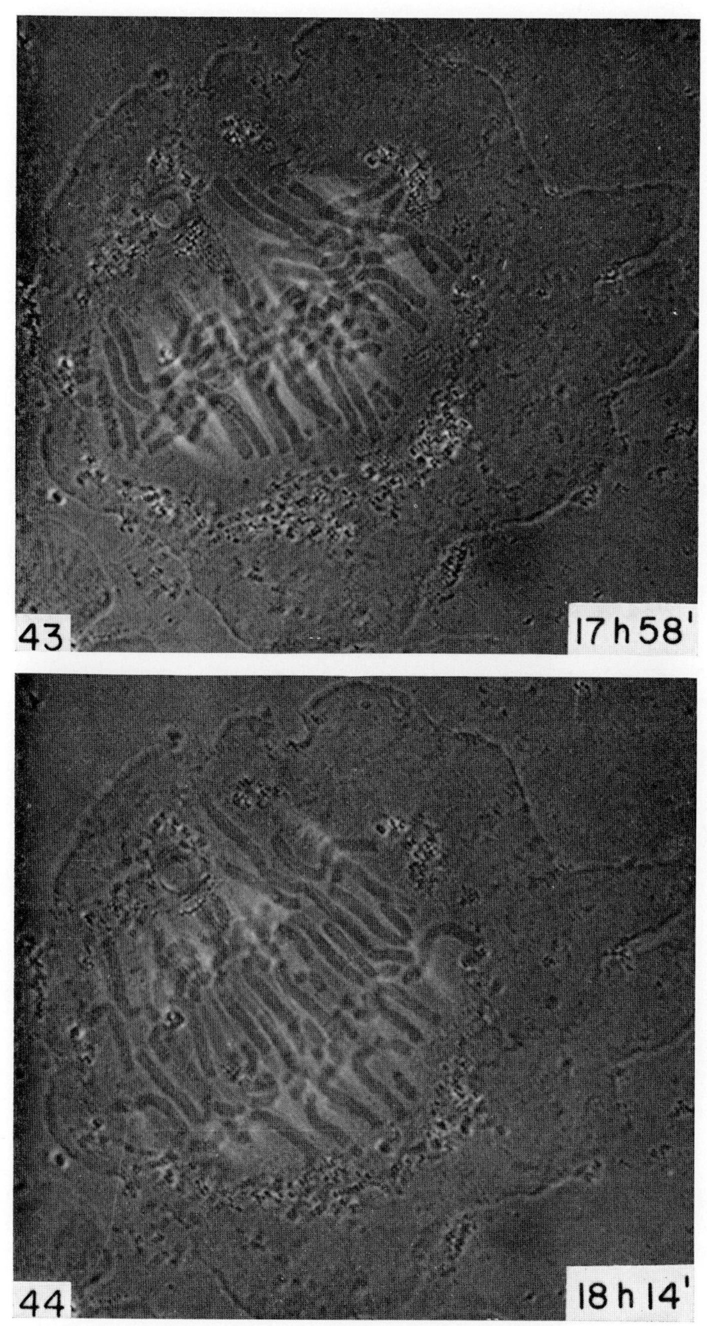

in Figs. 45 and 46. Phragmoplast fibers visible in Figs. 47 and 48. Same magnification as Figs. 31 to 40. A 10 μ scale is shown on Fig. 31. (From Inoué and Bajer, 1961.)

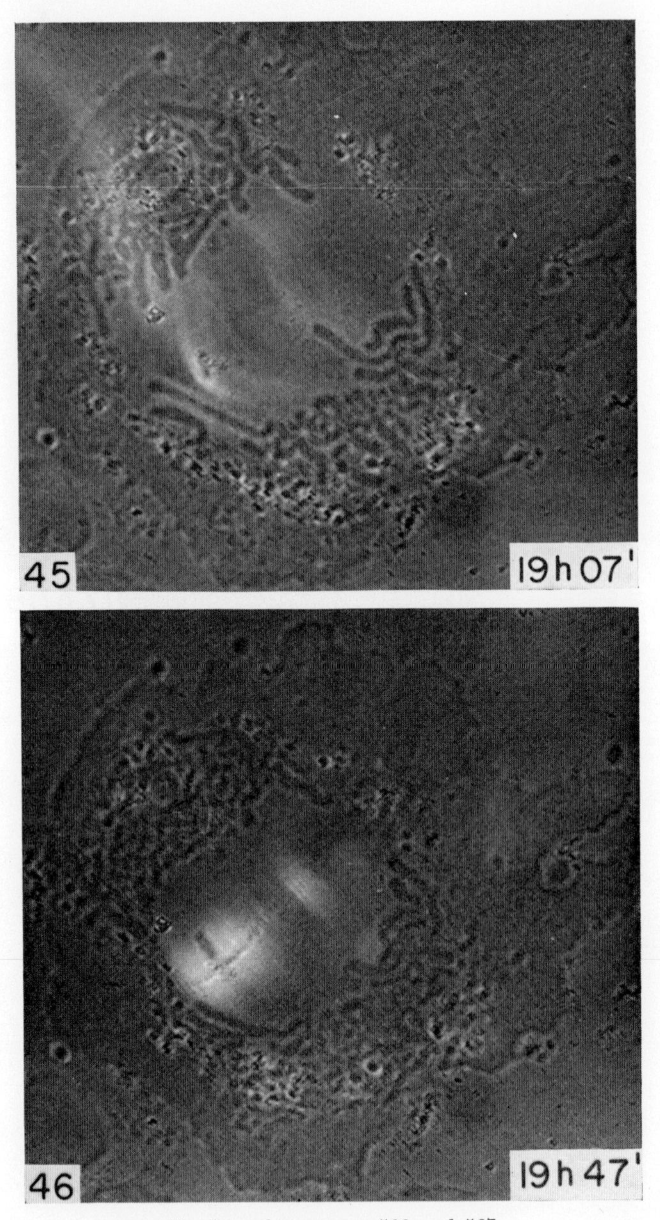

For legend see pages 566 and 567.

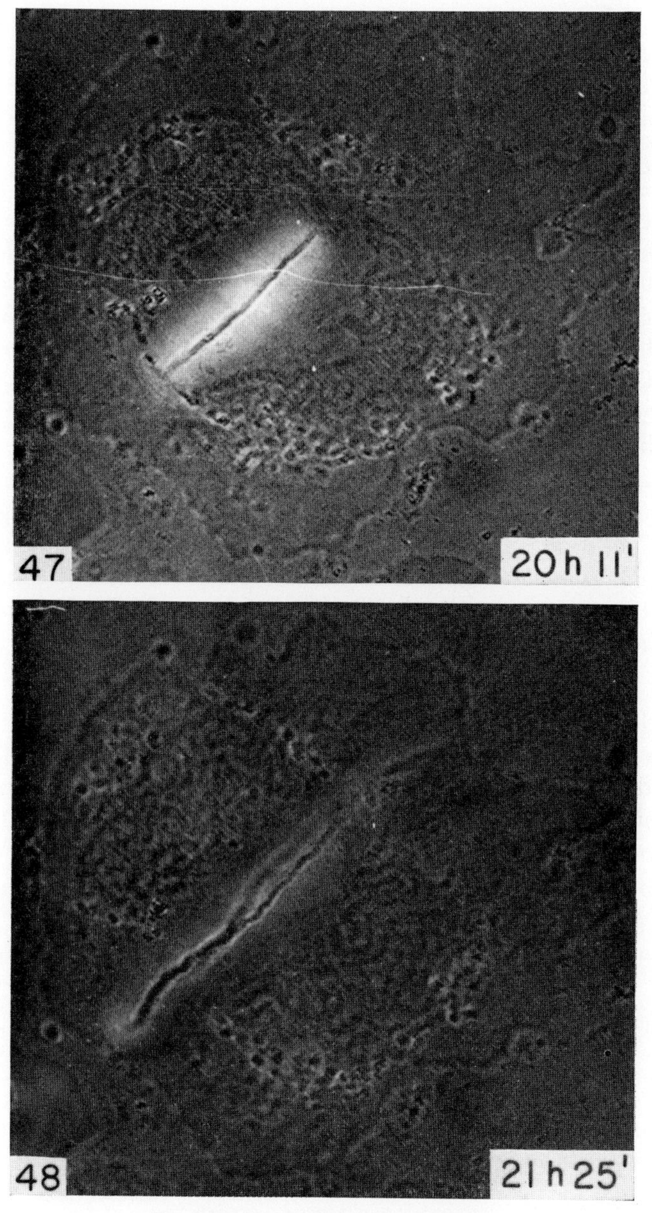

For legend see pages 566 and 567.

of positive birefringent material in agreement with the postulate of Wada (1950). Continuous fibers thus appear, then chromosomal fibers, then gradually the chromosomes are moved up and down until the metaphase plate is formed (Figs. 33 to 35). Here is another example of transformation of the clear-zone fiber to continuous spindle fibers and to chromosomal fibers (Inoué and Bajer, 1961).

Continuing the series (Figs. 36 to 40) throughout anaphase, the birefringence of the chromosomal fibers remains strongest adjacent to the kinetochore region. In these unflattened cells, the individual spindle fibers, although visible, are difficult to distinguish, but if we flatten the endosperm cells (Figs. 41 to 48) we clearly see the individual continuous fibers, the pairs of chromosomes, and their chromosomal fibers. The flattened endosperm cells continue to divide, and in the movie, the unwinding of the chromosome pairs and their pulling apart by the chromosomal fibers can be clearly seen.

The continuous fibers in late anaphase (Fig. 45) become more strongly birefringent in the mid-region and turn into phragmoplast fibers (Figs. 46 to 48); small vacuoles accumulate at the equator amidst the phragmoplast fibers and form the cell plate. As the cell continues to divide, the phragmoplast birefringence extends toward the cell periphery where the new cell plate is laid down.

Motion Picture of Living Cells in Division

A time-lapse motion picture film[3] showing spindle fiber birefringence during division of various animal and plant cells was shown at this Symposium. The pictures taken with the sensitive polarizing microscope developed by the author demonstrate the following points.

1. In the eggs of the jellyfish, *Halistaura cellularia,* flattening the cell places the spindle in a single plane of focus and reveals the synchrony of cell division. The nucleus normally lies at the periphery of the cell; flattening the cell may displace the nucleus and, therefore, the final position of the spindle. Regardless of which direction or where the spindle comes to lie, the cell divides at right angles to the axis of the

[3] The film required exposure times of from 5 to 20 sec/frame because of the low brightness of the specimen. The interval between frames was identical with the exposure. This film is now available at cost (Inoué, 1960b).

FIGS. 49 and 50. First and second cleavage in the egg of a jellyfish, *Aglantha digitale*. The positive birefringence of the spindle fibers and the negative birefringence of the vacuolar walls provide contrast in these illustrations. The vacuolar walls are deformed and point to the astral centers at these stages. (Original photos.)

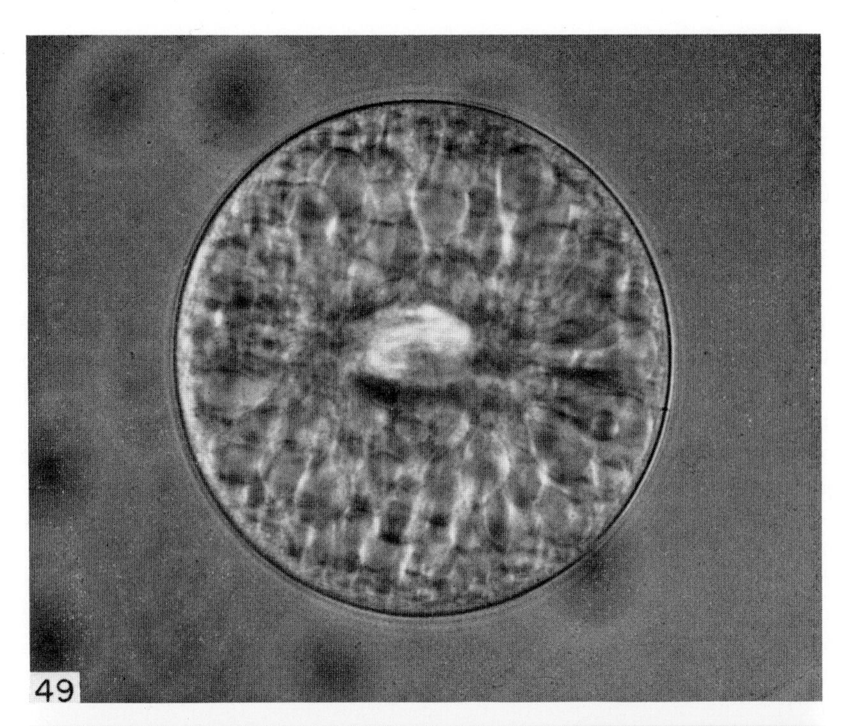
49

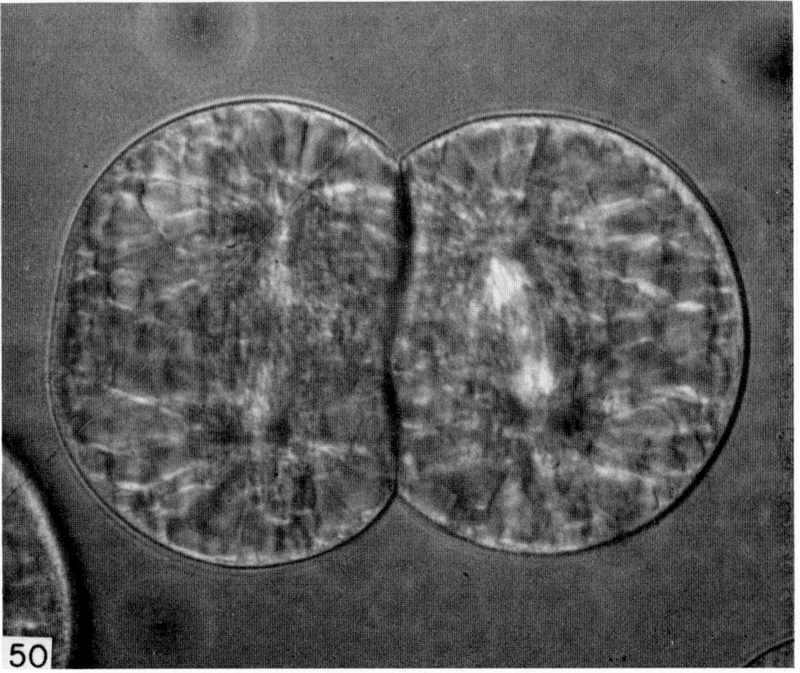
50

spindle, and the furrow begins to form at the surface nearest the spindle. Comparison with the sequence of cleavage in a noncompressed egg (Figs. 7 to 14) demonstrates that the birefringence of the spindle and its relation to the cleavage furrow is not altered by compression.

2. In the eggs of another jellyfish, *Aglantha digitale* (Figs. 49 and 50), which are even more transparent than those of the previous species, as the spindle and asters develop, the cytoplasmic vacuoles become tear-shaped, their apices pointing toward the astral centers. Prior to the separation of the chromosomes, a striking oscillation (or "rocking") of the whole spindle of the kind reported in nematode eggs (Ziegler, 1895) and in chick fibroblasts (Hughes and Swann, 1948) is observed. In the *Aglantha* eggs, the vacuoles change back and forth between tear-drop and round shape as the spindle rocks, as if the astral rays were attached to them and exerting a pull. After mitosis, the vacuoles regain their spherical shape.

3. In spermatocytes of the grasshopper, *Dissosteira carolina* (Figs. 15 to 26), the "northern lights" flickering of individual chromosomal fibers is clearly demonstrated in prometaphase (Figs. 15 to 19). Just prior to anaphase, the birefringence of the chromosomal fibers becomes stronger. While the chromosomes are pulled apart and the chromosomal fibers shorten, their birefringence remains strong until late anaphase (Figs. 18 to 21). The birefringence of the astral rays (Figs. 15 to 22) and of the continuous fibers in the interzonal region (Figs. 19 to 21) is much weaker than that of the chromosomal fibers. The mitochondrial sheath acquires a very strong positive birefringence in telophase (Figs. 19 to 23 and 26) and often becomes twisted in late telophase (Fig. 24).

4. In pollen mother cells of *Lilium longiflorum* (Figs. 27 to 30), chromosomal fiber birefringence becomes strongest just at the onset of anaphase, the birefringence being strongest adjacent to the kinetochores (Fig. 27). The chromosomal fibers retain their strong birefringence as they lead the chromosomes to the poles (Fig. 28). Continuous fiber birefringence is initially weak but becomes stronger during the later stages of anaphase, especially near the equatorial region (Fig. 29) where it finally transforms into the phragmoplast (Fig. 30). At this time, vacuoles come together in the equator and merge to form the cell plate—a structure which is also birefringent but with a different character.

5. Endosperm cells of *Haemanthus katherinae*, unlike most other types of plant cells, are not surrounded by conventional cell-wall material and may thus be flattened like an animal cell in tissue culture (see footnote 2 on p. 557). In the flattened cells, the relatively large chromosomes are sufficiently dispersed (Figs. 41 to 48) so that the individual chromosomes, their kinetochores, the chromosomal and continuous and the

phragmoplast fibers are all clearly visible in polarized light. The chromosomes are untwisted and pulled at their kinetochores by the chromosomal fibers.

As stated previously, the fibers in all these cells are not visible with phase-contrast microscopy in untreated dividing cells.

The motion picture of living cells in division shows not only the reality of the fibers but how the spindle fibers appear to transform from one type of fiber to another, depending on the sequence of events. The various spindle movements and the fluctuation in the fiber birefringence which also reflect the dynamic state of these fibers in the dividing cells are demonstrated.

Effect of Low Temperature[4]

Low-temperature experiments further demonstrate the dynamic state of the spindle fibers.

A pollen mother cell of *L. longiflorum* in early anaphase was chilled to 3°C (Fig. 51) resulting in complete disappearance of spindle birefringence. On raising the temperature to 27°C, the birefringence reappeared, at first as continuous fibers; after about 8 min, both the continuous and chromosomal fibers were fully reorganized. This process of spindle re-formation after chilling is faster than, but otherwise similar to, spindle formation at prometaphase in normal mitosis.

After re-establishment of the spindle birefringence and structural organization, the chromosomes began to move again and completed mitosis.

If we cool a cell at a later stage in anaphase (Fig. 52), by which time the chromosomes have become stretched but not yet separated, the chromosomes are found to "recoil" as the spindle birefringence disappears. In this case, when the temperature is raised, birefringent chromosomal fibers reappear first. Then the chromosomes become stretched, the kinetochore distance increases, and the cell completes mitosis, but only after the size and birefringence of the chromosomal fibers have fully recovered.

There thus exists a direct correlation between spindle birefringence and deformation and movement of the chromosomes (also see Inoué, 1952a on the effect of colchicine).

This temperature treatment can be performed on many types of cells over and over again during a single division. Thus one can interrupt the mitotic process experimentally without impeding the ability of the spindle fibers to become reorganized again.

---

[4] Original report.

Figure 53A shows the birefringent spindle in a *Chaetopterus pergamentaceus* egg. When this cell is chilled, the spindle becomes thin and then quickly loses its birefringence. Upon returning the cell to room temperature, a miniature spindle reappears in the original site, quickly migrates to the cell surface, and grows back again (Fig. 53B), first form-

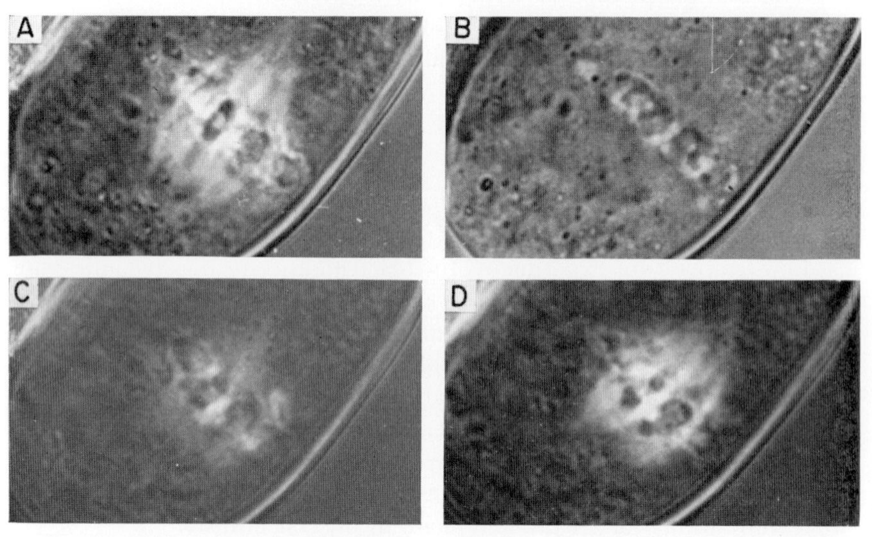

FIG. 51. Cold treatment of a pollen mother cell of *Lilium longiflorum* in early anaphase. The birefringence of the spindle fibers disappears at 3°C and returns rapidly at 27°C. A, Before treatment; B, after 6.5 min at 3°C; C, after 3.5 min at 27°C; D, after 8 min at 27°C. (Original photos.)

ing continuous fibers with chromosomes randomly arranged (Fig. 53C and D), and then eventually forming the metaphase plate and more prominent chromosomal fibers (Fig. 53E and F). This cycle takes about 15 min at 22°C. In the *Chaetopterus* egg which is in a metaphase arrest, such a cycling can be done on the same cell for as many as ten times without losing the spindle material (Inoué, 1952b).

Thermodynamic Studies

By measuring the equilibrium birefringence at various temperatures, one can gain insight into the mechanism of orientation of the molecules making up the fibers.

When the temperature of the cell is changed quickly, the birefringence of the spindle reaches an equilibrium value after a few minutes. The birefringence of the spindle fibers is then measured and can be plotted as a function of temperature, as illustrated in Fig. 54.

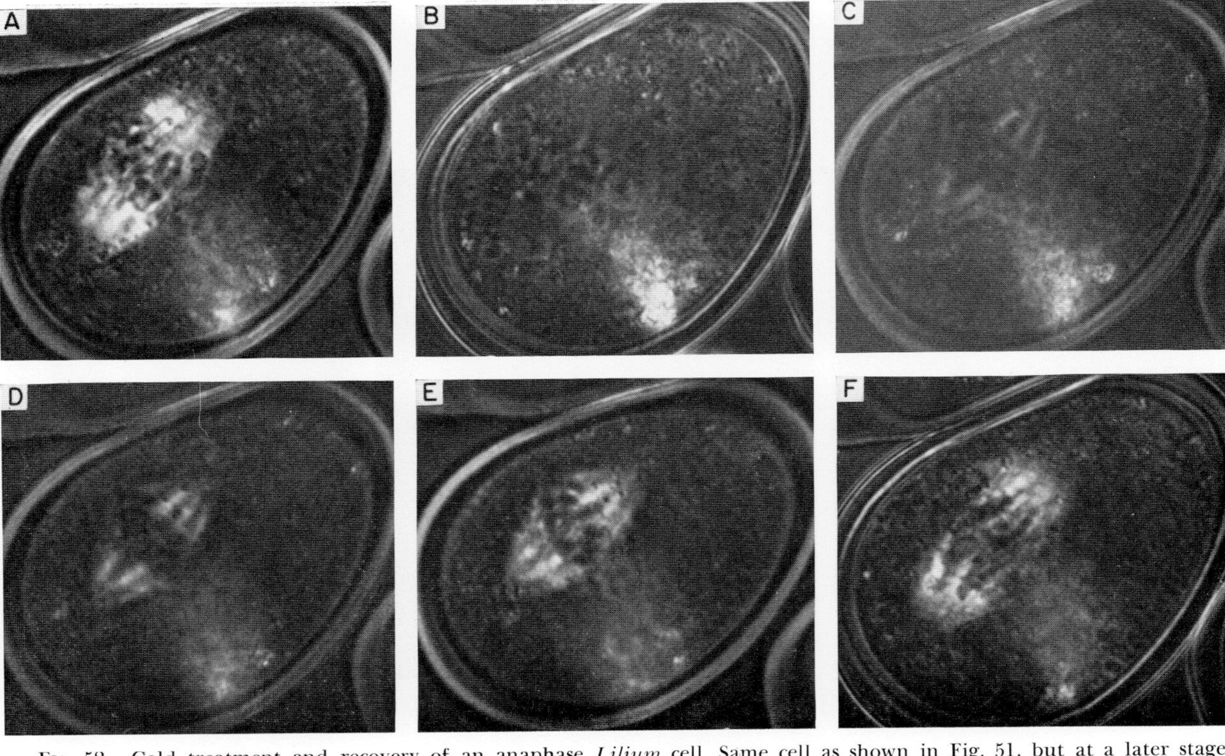

FIG. 52. Cold treatment and recovery of an anaphase *Lilium* cell. Same cell as shown in Fig. 51, but at a later stage. The chromosomes are extended again after spindle birefringence and organization are restored (F). A, Before treatment; B, after 4 min at 4°C; C, recovery after 1 min; D, recovery after 2 min; E, recovery after 4 min; F, recovery after 13 min. (Original photos.)

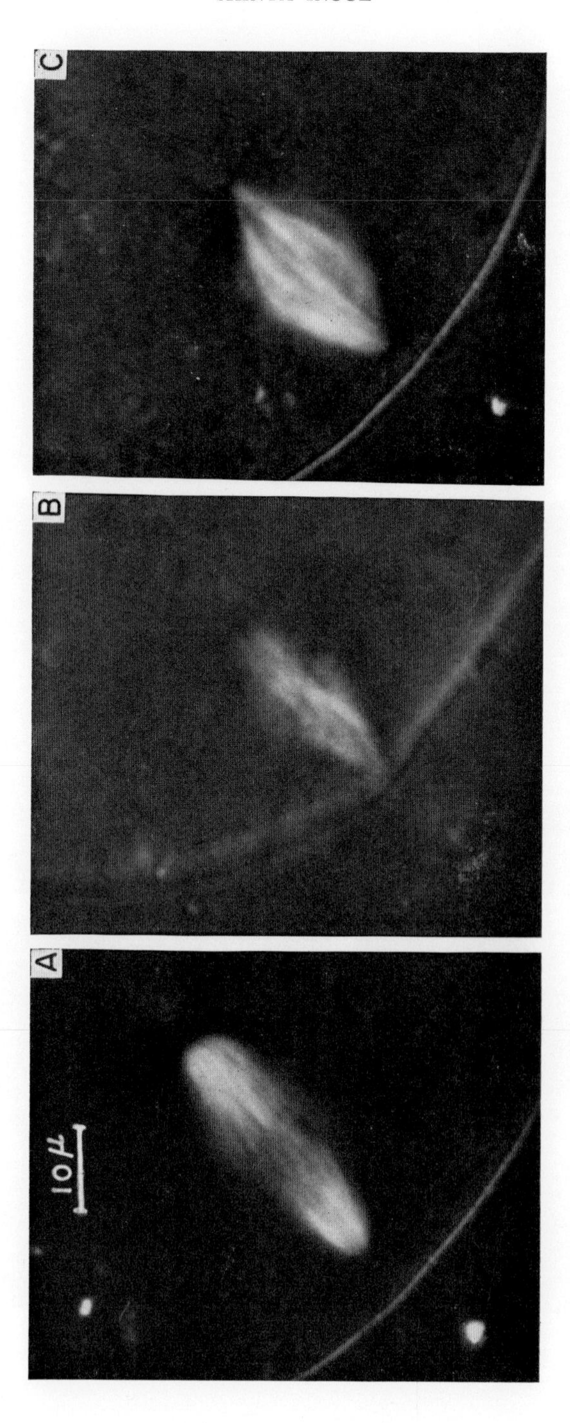
C
B
A
10μ

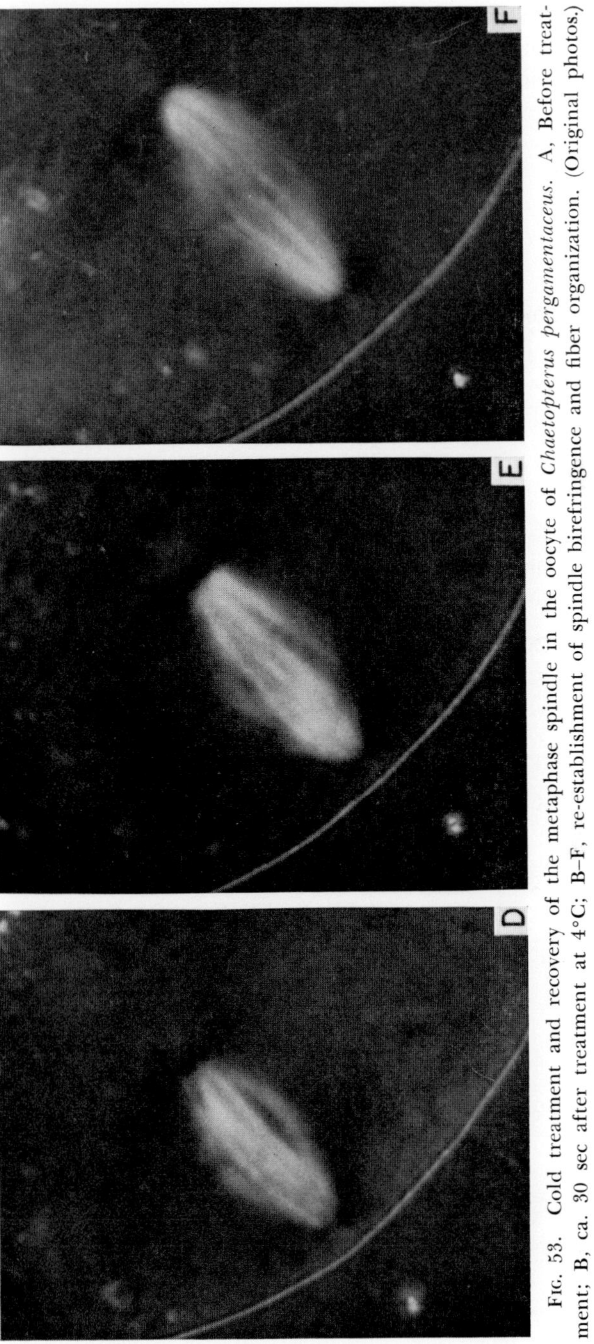

FIG. 53. Cold treatment and recovery of the metaphase spindle in the oocyte of *Chaetopterus pergamentaceus*. A, Before treatment; B, ca. 30 sec after treatment at 4°C; B–F, re-establishment of spindle birefringence and fiber organization. (Original photos.)

As shown in the figure, the birefringence increases as the temperature is raised, at least up to an optimum. This is, at first glance, contrary to intuitive thermodynamics, since one would expect randomization of the spindle elements with the increased temperature. But it does agree

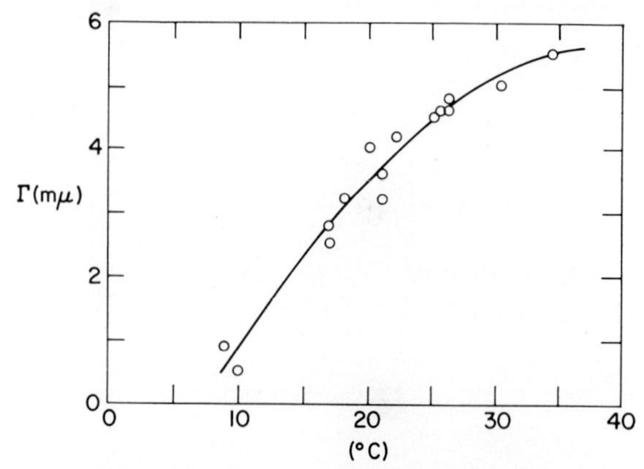

FIG. 54. Equilibrium retardation (Γ) of *Chaetopterus* spindle at various temperatures. (From Inoué, 1959.)

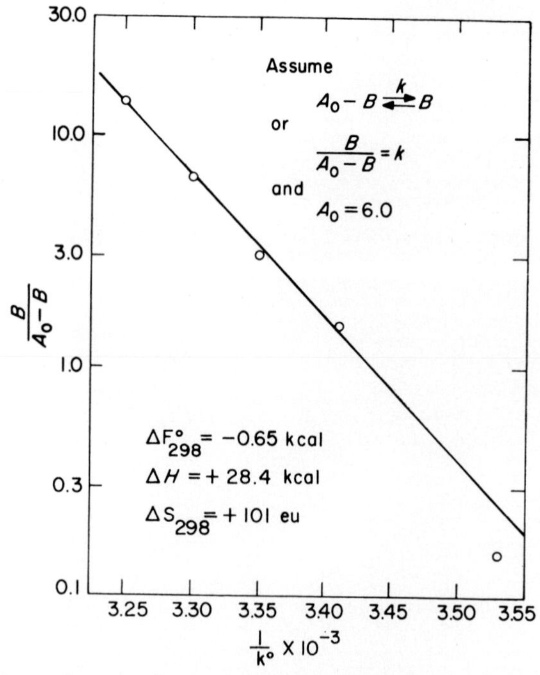

FIG. 55. Log plot of spindle orientation equilibrium versus inverse absolute temperature. (From Inoué, 1959, with values corrected in 1960a.)

with "viscosity" measurements on dividing cells where the gel structure is weakened either with high hydrostatic pressure or with low-temperature treatment (Marsland *et al.*, 1960).

Figure 55 shows a thermodynamic plot of the data shown in Fig. 54. Assuming that the total concentration (A_0) of orientable material is constant and that the birefringence (B) represents the con-

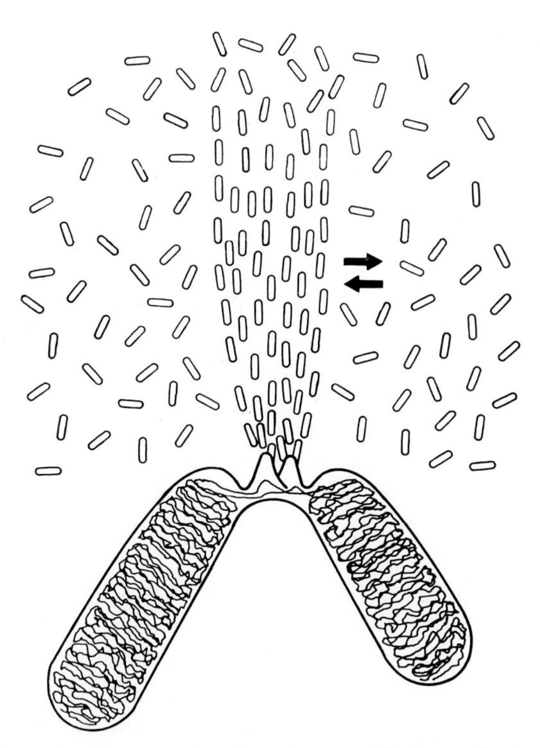

FIG. 56. Schematic diagram of orientation equilibrium of chromosomal fiber. Fiber orientation is organized by kinetochores of chromosomes. The oriented material is in a temperature-sensitive equilibrium with the nonoriented material surrounding it. (Original.)

centration of material oriented at that temperature, one can draw an equation: $A_0 - B \rightleftharpoons B$. Plotting the log of the equilibrium constant $k = (B/A_0) - B$ against the reciprocal of the absolute temperature, one would expect a linear relationship *if* one were dealing with an isolated equilibrium system in which the equilibrium between the oriented and non-oriented material is a function of temperature.

A straight-line relationship has in fact been observed and the change in free energy, the heat evolved per mole, and the entropy change have been calculated by Dr. Manuel Morales (Inoué, 1959, see correction in 1960a).

The result of this analysis showing a great increase of entropy at higher temperatures is similar to that for globular to filamentous actin transformation (Asakura *et al.,* 1960) and to that for tobacco mosaic virus A protein association (Lauffer, 1958; Ansevin and Lauffer, 1963) and may suggest a common mechanism (see Inoué, 1959; also Kauzmann, 1959, Singer, 1962, for pertinent reviews).

The orientation equilibrium of the spindle elements is schematically shown in Fig. 56. On the chromosomes, the kinetochores act as orienting centers which determine the localization of the oriented material. There is a pool of unoriented or perhaps slightly oriented material around the chromosomal fibers, depending on the stage, which can be transformed into oriented chromosomal fibers. Conversely, the chromosomal fiber material may become disoriented. This equilibrium is temperature-sensitive.

Orienting Centers—Ultraviolet Microbeam Experiments[5]

Experiments were done to test further the notion of orienting centers and orientation equilibrium. If we somehow stop the action of the orienting center, then the oriented material should go into the disoriented state. If we could locally disorient the material distal to the orienting center, then the material proximal to it should not be affected because the orienting center remains, but distally the orientation and the birefringence should disappear.

If we remove the disorienting influence, we should again recover the orientation by the incorporation of the nonoriented material into the oriented state. These predictions have been fulfilled by an experiment in which the spindle fibers were irradiated with a small spot of ultraviolet light.

Figure 57 shows a part of the instrumentation, the high extinction polarizing microscope built onto a stable optical bench. As shown in the schematic diagram (Fig. 58), the visible light source is at the top of the instrument. The light coming through the polarizer, a compensator, and condenser, illuminates the object. Through the objective, analyzer, and eyepiece, we see the birefringence of the specimen. A mercury arc lamp (Osram, HBO-200) shines the ultraviolet light onto a small first surface mirror placed in front of the visible source diaphragm. The

[5] This section reports original work carried out in collaboration with Dr. Hidemi Sato of the Department of Cytology, Dartmouth Medical School.

FIG. 57. Rectified polarizing microscope designed by author with help of the staff of the Research Center, American Optical Co. and the Institute of Optics, University of Rochester. Compare with schematic diagram, Fig. 58. (Original photo.) See Inoué, 1961, for further details.

image of the mirror is projected into the specimen plane and delineates the area of the specimen receiving the ultraviolet exposure.

Figure 59 shows a prophase *Haemanthus* endosperm cell in which a small spindlelike body spontaneously developed. This body was irradiated in its mid-region (Figs. 59 and 60), where the birefringence instantly disappeared. After irradiation, the two halves that remained birefringent quickly came together (Fig. 61) and then merged to form another single spindle-shaped body (Fig. 62).

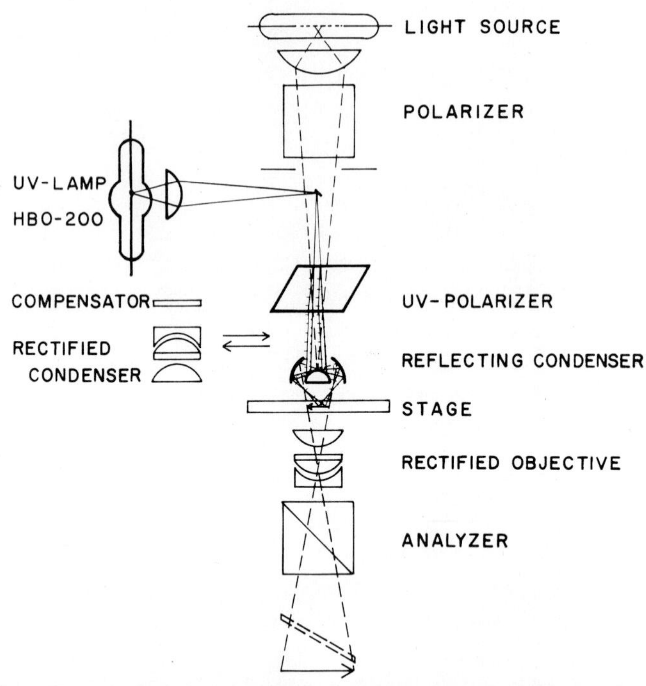

FIG. 58. Schematic diagram of (polarized) ultraviolet-microbeam irradiating system which was incorporated into the rectified polarizing microscope shown in Fig. 57. (Original.)

When a cell in early anaphase is irradiated so that the ultraviolet beam covers the basal regions of some chromosomal fibers and their kinetochores, the whole length of the chromosomal fibers including the distal unirradiated part is lost and does not reappear for a long time (Figs. 63 and 64).

If instead we irradiate only the chromosomal fibers and avoid the kinetochores (Figs. 65 and 66), we find the birefringence has disappeared from the irradiated region as well as from the distal part, but is still intact between the kinetochores and the irradiated region (Fig. 67). In the course of a few minutes the birefringence of the irradiated and distal regions returns (Fig. 68).

These experiments not only support the notion of the dynamic equilibrium and the activity of the organizing centers, but provide us with a means for establishing the locations of organizing centers in general.

For example, in the case of the phragmoplast, the birefringence is stronger next to the cell plate, but the presence of an organizing center in that region of the phragmoplast had never been suspected.

However, when we irradiated a narrow diagonal area of the phragmoplast, the birefringence was lost from the full length of the phragmoplast fibers where the cell plate itself was irradiated (Figs. 69 to 71). If only the distal parts of the fibers, and not the cell plate, were irradiated, the portions of the fibers nearer the cell plate remained birefringent. The birefringence of the distal parts eventually returned.

Figure 72 shows a late phragmoplast in which the birefringence is now confined to the peripheral regions; in the middle, the cell plate has already been laid down. When the center of one phragmoplast is irradiated, as shown in Fig. 73, we find that all birefringence is gone from the phragmoplast fibers on that side (Fig. 74). Here then exists an unsuspected orienting center, the activity of which moves progressively with the phragmoplast as it moves toward the cell periphery.

As shown by these experiments, the phragmoplast fibers behave very much like chromosomal and continuous spindle fibers, not only in their birefringence but in terms of their response to ultraviolet irradiation. This provides further support for the notion that the composition and molecular organization of the phragmoplast fibers is similar or identical to those of the chromosomal fibers, continuous fibers, and the clear-zone fibers.

Conclusion

In summary, then, we have seen in a wide variety of living, dividing cells "fibers" which could explain, at least topologically, the movement of chromosomes. This does not explain the molecular mechanism of chromosome movement yet, but at least we know there is an anisotropic distribution of material which could one way or another account for the development of mechanical forces required for pulling (or pushing) the chromosomes, most likely dependent on the shift of orientation equilibrium.

It may well be that as material is removed from the chromosomal fibers, while the organizing centers are still active, the fibers shorten in length simply to reach a new orientation equilibrium. The continuous

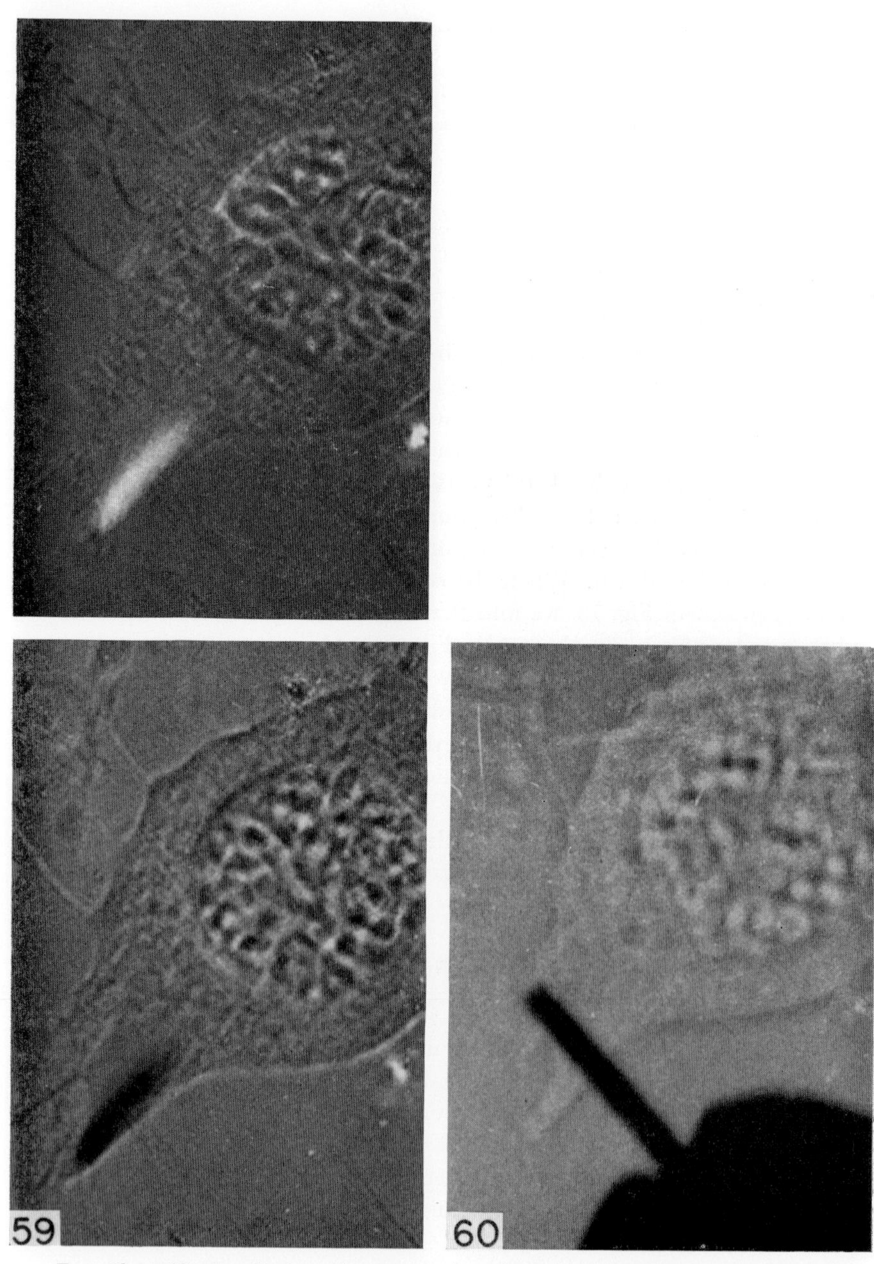

FIGS. 59 to 62. Irradiation of cytoplasmic spindle spontaneously formed in prophase of a *Haemanthus* endosperm cell. Figure 59, before irradiation; Fig. 60, the dark bar indicates the area irradiated with ultraviolet; Fig. 61, 1 min after irradiation

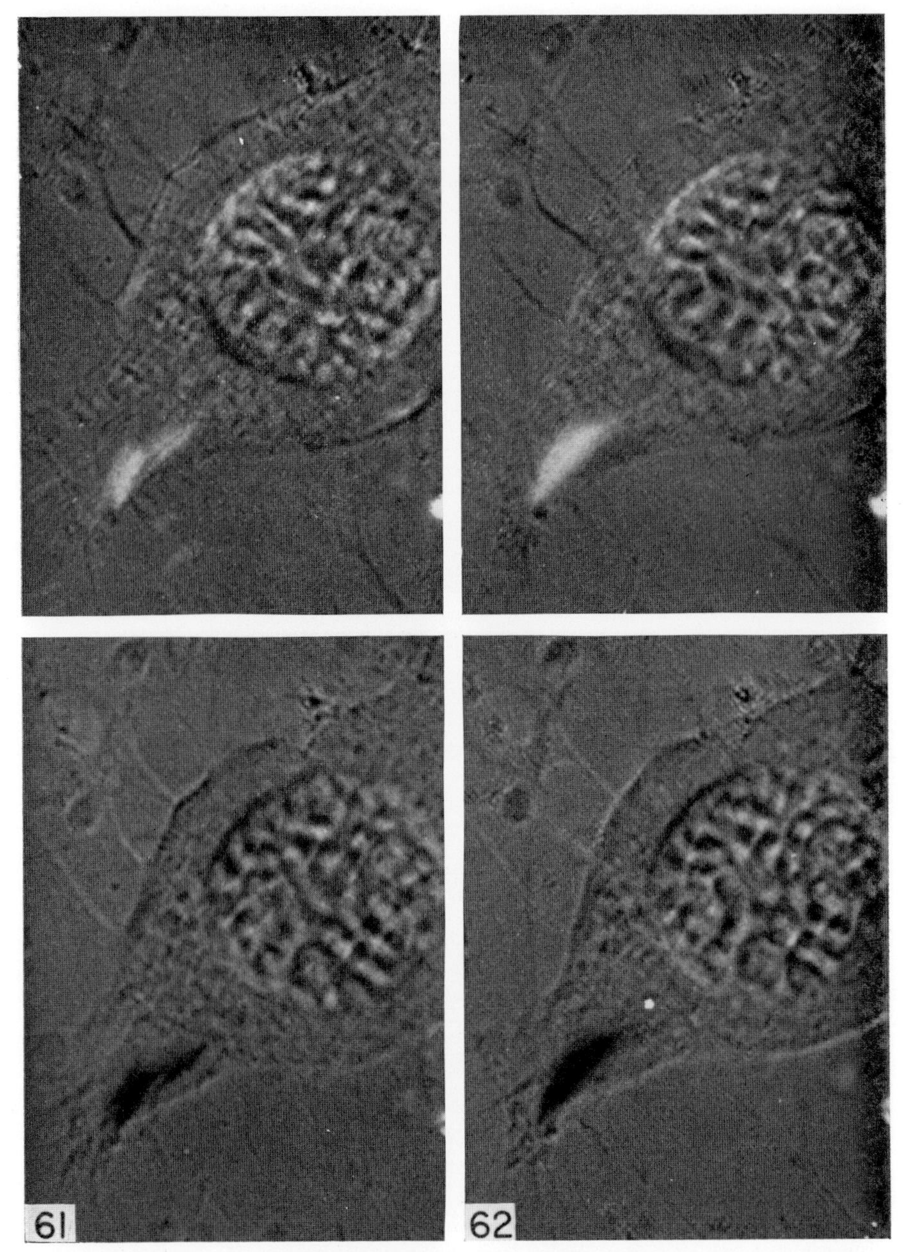

—the irradiated area lost birefringence, the nonirradiated parts have moved together to close the gap; Fig. 62, ca. 5 min after irradiation, the spindle has become reorganized into a single body again. (Original photos.)

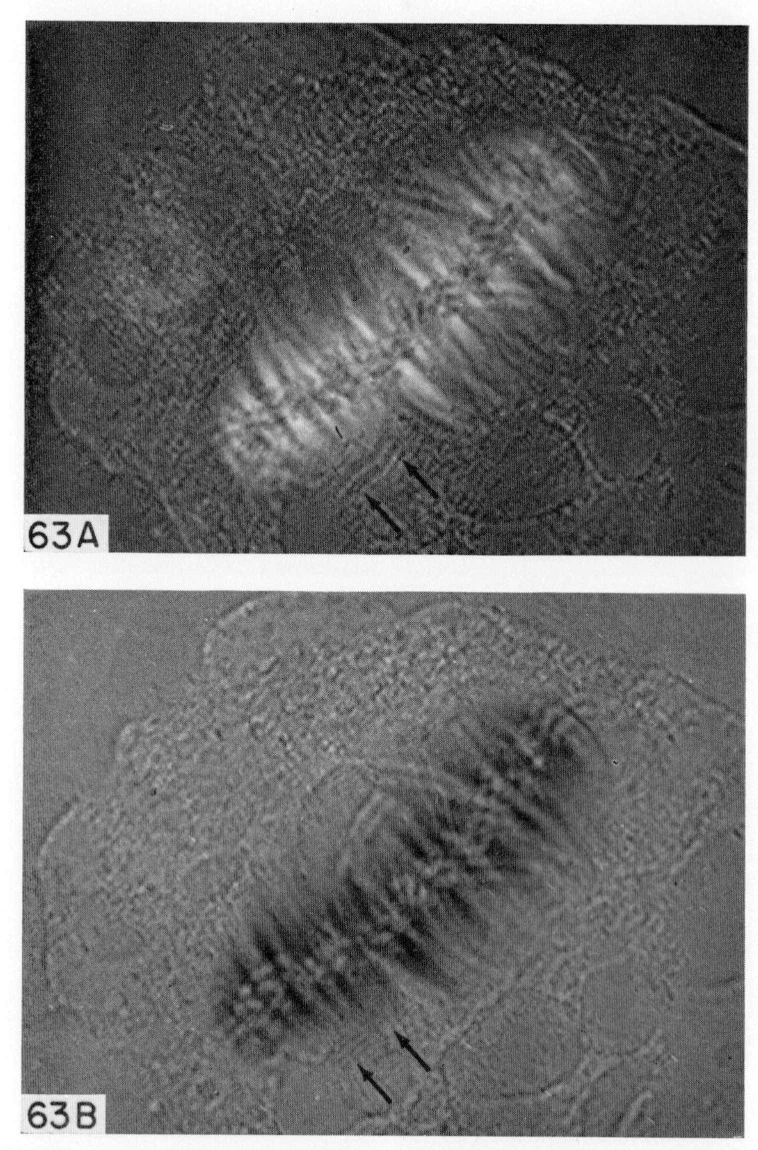

FIGS. 63 and 64. Ultraviolet-microbeam irradiation of chromosomal fibers and kinetochores in one-half spindle of *Haemanthus* endosperm cells. The irradiating ultraviolet beam covered a square area between the tips of the arrows and the

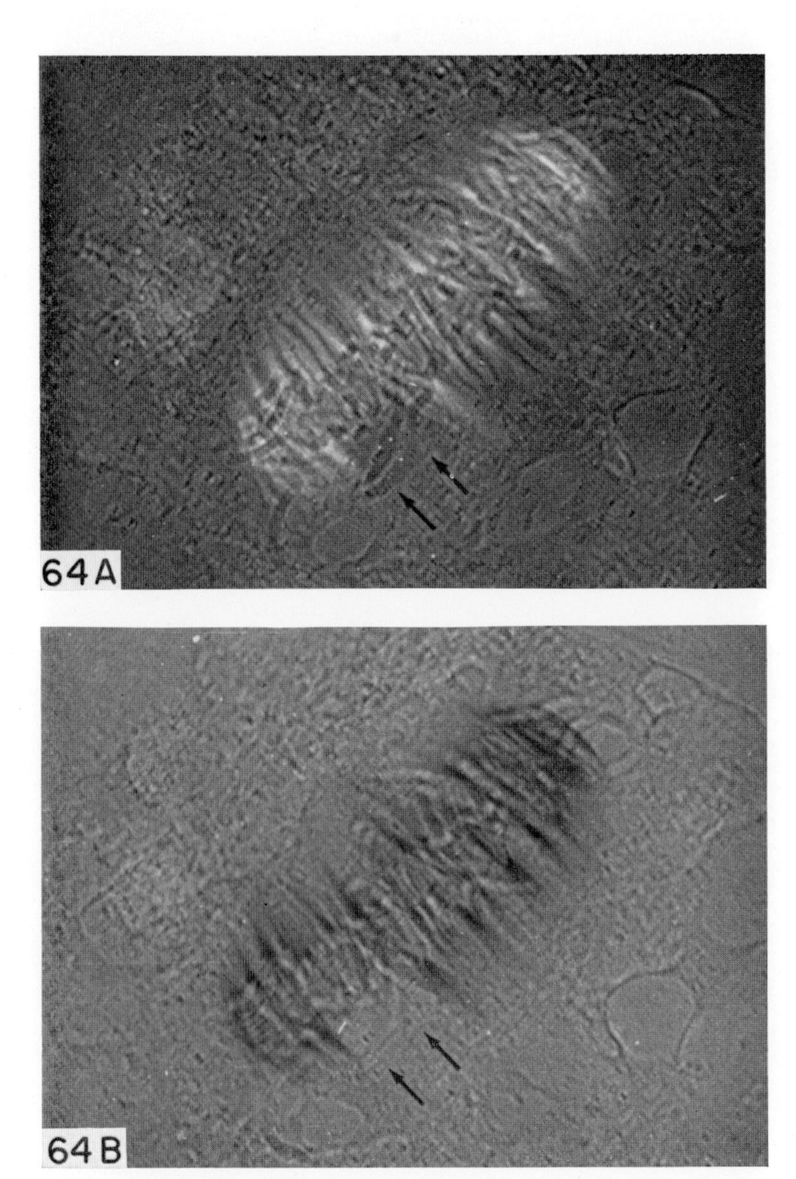

nearby kinetochores. Figure 63, before, Fig. 64, after irradiation. No recovery of the chromosomal fibers can be seen. (Original photos.)

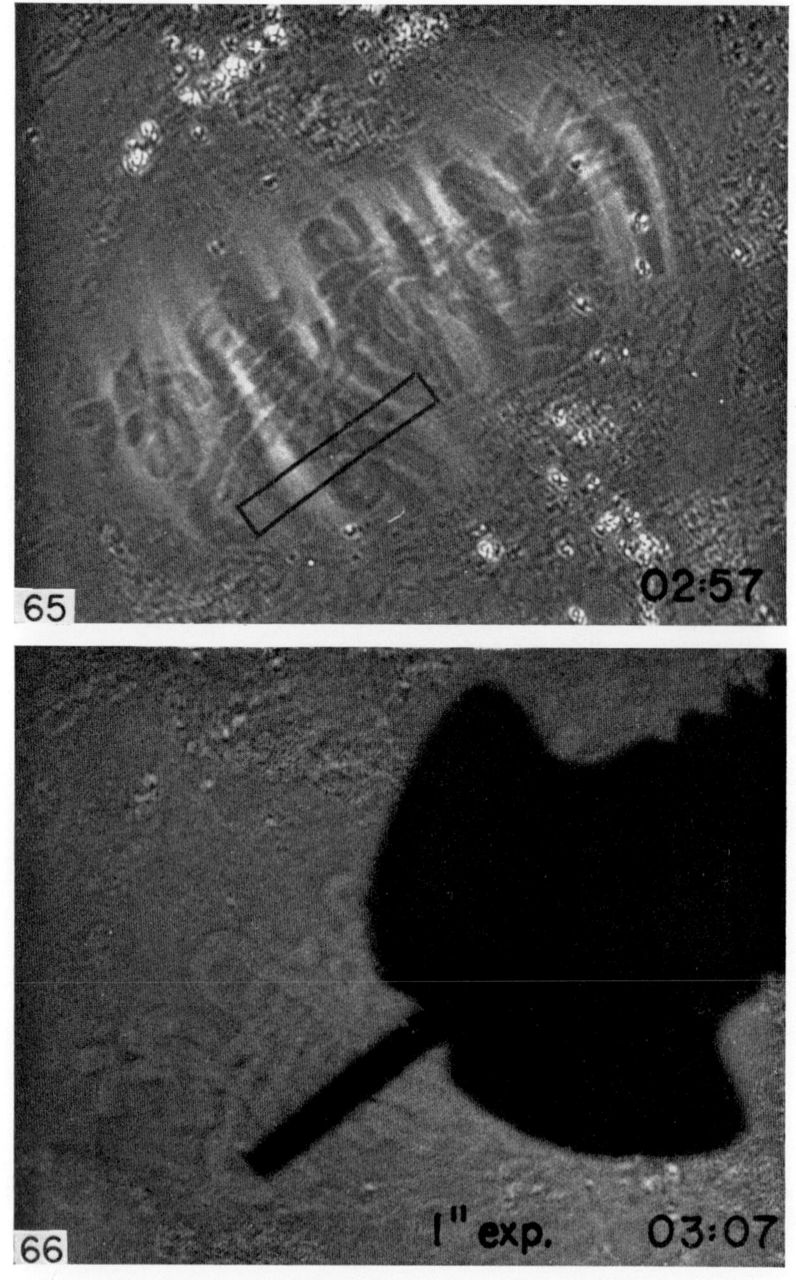

FIGS. 65 to 68. Ultraviolet-microbeam irradiation of chromosomal fiber only. When kinetochores are not irradiated (Figs. 65 and 66), birefringence disappears from the area irradiated and also the distal portions of the chromosomal fibers (Fig. 67).

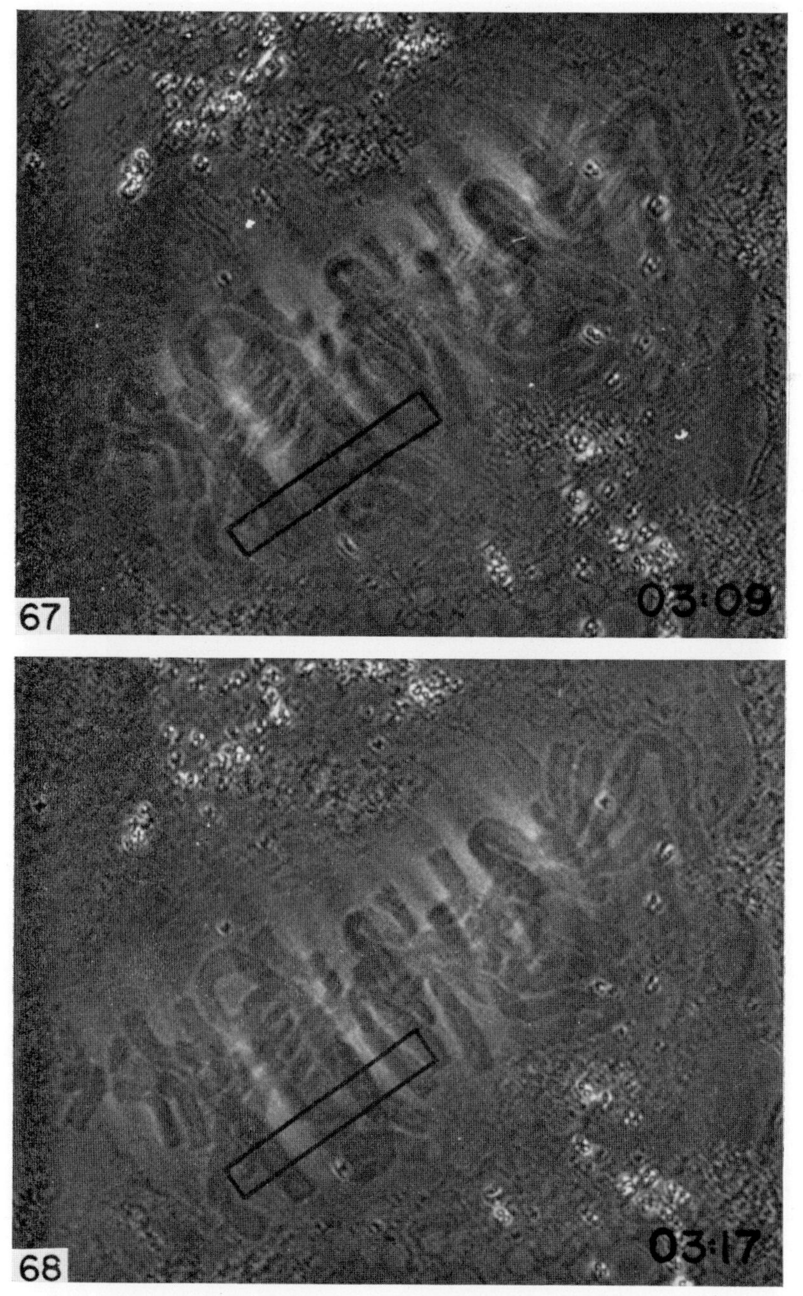

They both rapidly grow back (Fig. 68) if the ultraviolet exposure is not too great. (Original photos.)

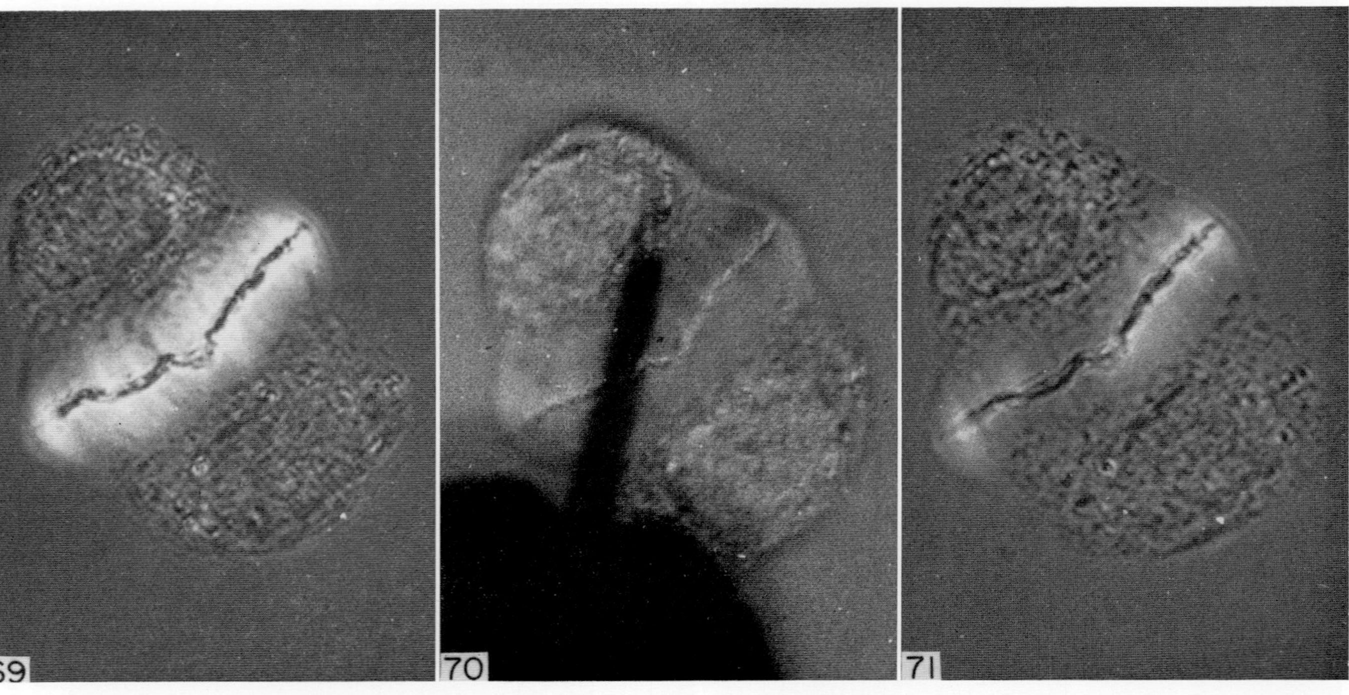

FIGS. 69 to 71. Ultraviolet-microbeam irradiation of phragmoplast in the endosperm cell of *Haemanthus katherinae*. The area irradiated ran diagonally over the phragmoplast as shown by the shadow of the ultraviolet mirror in Fig. 70. Figure 69, before irradiation. Figure 71 shows no birefringence of the phragmoplast fibers where the cell plate was irradiated. (Original photos.)

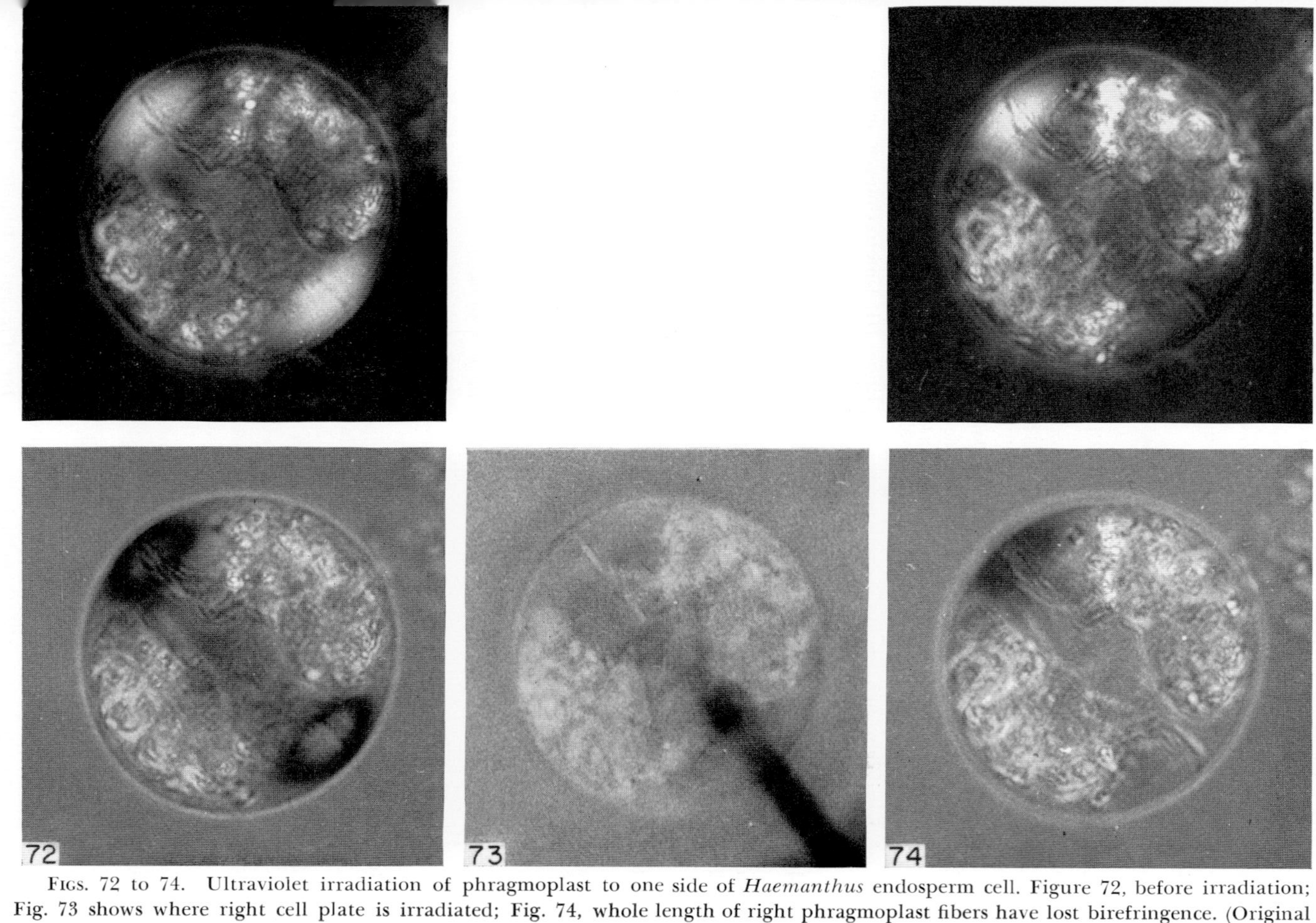

FIGS. 72 to 74. Ultraviolet irradiation of phragmoplast to one side of *Haemanthus* endosperm cell. Figure 72, before irradiation; Fig. 73 shows where right cell plate is irradiated; Fig. 74, whole length of right phragmoplast fibers have lost birefringence. (Original photos.)

fibers could then acquire more material and support the compressive force imposed on the spindle by the chromosomal fibers. It is even possible that by adding more material into the oriented state the fibers can be made to elongate as well. In any event, there is no question that inside the living cells these birefringent fibers are present; they are organized by centers and exist in a highly dynamic state.

References

Ansevin, A. T., and Lauffer, M. A. (1963). Polymerization-depolymerization of tobacco mosaic virus protein. I. *Biophys. J.* **3** (3), 239-251.

Asakura, S., Kasai, M., and Oosawa, F. (1960). The effect of temperature on the equilibrium state of actin solutions. *J. Polymer Sci.* **44**, 35-49.

Bělař, K. (1929a). Beiträge zur Kausalanalyse der Mitose. II. Untersuchungen an den Spermatocyten von *Chorthippus (Stenobothrus lineatus)* Panz. *Wilhelm Roux' Arch. Entwicklungsmech. Organ.* **118**, 359-484.

Bělař, K. (1929b). Beiträge zur Kausalanalyse der Mitose. III. Untersuchungen an den Staubfadenhaarzellen und Blattmeristemzellen von *Tradescantia virginica. Z. Zellforsch. Mikroskop. Anat.* **10**, 73-134.

Bloom, W., and Fawcett, D. W. (1962). "A Textbook of Histology," 8th ed. Saunders, Philadelphia, Pennsylvania.

Boveri, T. (1888). "Zellen Studien." Fischer, Jena, Germany.

Cleveland, L. R. (1953). Studies on chromosomes and nuclear division. IV. Photomicrographs of living cells during meiotic divisions. *Trans. Am. Philos. Soc.* [N.S.] **43**, 805-869.

Cleveland, L. R. (1963). Functions of Flagellate and Other Centrioles in Cell Reproduction. *In* "The Cell in Mitosis" (L. Levine, ed.), pp. 3-30. Academic Press, New York.

Cooper, K. W. (1941). Visibility of the primary spindle fibers and the course of mitosis in the living blastomeres of the mite *Pediculopsis graminum* Reut. *Proc. Natl. Acad. Sci. (U.S.)* **27**, 480-483.

Engelmann, Th. W. (1875). Contractilität und Doppelbrechung. *Arch. Ges. Physiol.* **11**, 432-464.

Engelmann, Th. W. (1906). Zur Theorie der Contractilität. *Sitzber. Kgl. Preuss. Akad. Wiss.*, pp. 694-724.

Hughes, A. F., and Swann, M. M. (1948). Anaphase movements in living cell. A study with phase contrast and polarized light on chick tissue culture. *J. Exptl. Biol.* **25**, 45-70.

Inoué, S. (1952a). The effect of colchicine on the microscopic and submicroscopic structure of the mitotic spindle. *Exptl. Cell Res. Suppl.* **2**, 305-318.

Inoué, S. (1952b). Effects of temperature on the birefringence of the mitotic spindle. *Biol. Bull.* **103**, 316.

Inoué, S. (1953). Polarization optical studies of the mitotic spindle. I. The demonstration of spindle fibers in living cells. *Chromosoma* **5**, 487-500.

Inoué, S. (1959). Motility of cilia and the mechanism of mitosis. *Rev. Mod. Phys.* **31**, 402-408.

Inoué, S. (1960a). On the physical properties of the mitotic spindle. *Ann. N.Y. Acad. Sci.* **90**, Article 2, 529-530.

Inoué, S. (1960b). "Birefringence in Dividing Cells." Time-lapse motion picture. Available at cost from Geo. W. Colburn Laboratory, Inc., Chicago 6, Illinois.

Inoué, S. (1961). Polarizing microscope. Design for maximum sensitivity. *In* "Encyclopedia of Microscopy" (George Clark, ed.), pp. 480-485. Reinhold, New York.

Inoué, S., and Bajer, A. (1961). Birefringence in endosperm mitosis. *Chromosoma* **12**, 48-63.

Inoué, S., and Dan, K. (1951). Birefringence of the dividing cell. *J. Morphol.* **89**, 423-455.

Inoué, S., and Hyde, W. L. (1957). Studies on depolarization of light at microscope lens surfaces. II. The simultaneous realization of high resolution and high sensitivity with the polarizing microscope. *J. Biophys. Biochem. Cytol.* **3**, 831-838.

Inoué, S., and Kubota, H. (1958). Diffraction anomaly in polarizing microscopes. *Nature* **182**, 1725-1726.

Kane, R. E. (1962). The mitotic apparatus: isolation by controlled pH. *J. Cell Biol.* **12**, 47-56.

Katchalsky, A. (1963). Mechanochemistry of cell movement. *In* "Symposium on Non-Muscular Contractions in Biological Systems," 7th Annual Meeting of the Biophysical Society.

Kauzmann, W. (1959). Some factors in the interpretation of protein denaturation. *Advan. Prot. Chem.* **14**, 1-64.

Kubota, H., and Inoué, S. (1958). Diffraction image in the polarizing microscope. *J. Opt. Soc. Am.* **49**, 191-198.

Lauffer, M. A. (1958). Polymerization-depolymerization of tobacco mosaic virus protein. *Nature* **181**, 1338-1339.

Lewis, M. R. (1923). Reversible gelation in living cells. *Bull. Johns Hopkins Hosp.* **34**, 373-379.

Marsland, D., Zimmerman, A. H., and Auclair, W. (1960). Cell division: experimental induction of cleavage furrows in the eggs of *Arbacia punctulata*. *Exptl. Cell Res.* **21**, 179-196.

Mazia, D. (1961). Mitosis and the physiology of Cell Division. *In* "The Cell" (J. Brachet and A. E. Mirsky, eds.), Vol. 3, pp. 77-412. Academic Press, New York.

Mazia, D., and Dan, K. (1952). The isolation and biochemical characterization of the mitotic apparatus of dividing cells. *Proc. Natl. Acad. Sci.* (*U. S.*) **38**, 826-838.

Mitchison, J. M. (1953). A polarized light analysis of the human red cell ghost. *J. Exptl. Biol.* **30**, 379-432.

Östergren, G. (1949). *Luzula* and the mechanism of chromosomal movements. *Hereditas* **35**, 445-468.

Porter, K. R., and Machado, R. D. (1960). Studies on the endoplasmic reticulum: IV. Its form and distribution during mitosis in cells of onion root tip. *J. Biophys. Biochem. Cytol.* **7**, 167-180.

Schmidt, W. J. (1934). Polarisationsoptische Analyse des submikroskopischen Baues von Zellen und Geweben. *In* "Handbuch der biologischen Arbeitsmethoden" (E. Abderhalden, ed.), Sec. 5, Part 10, p. 435. Urban und Schwarzenberg, Berlin and Vienna.

Schmidt, W. J. (1937). "Die Doppelbrechung von Karyoplasma, Zytoplasma und Metaplasma," Protoplasma-Monographien, Vol. 2. Gebrüder Borntraeger, Berlin.

Schmidt, W. J. (1939). Doppelbrechung der Kernspindel und Zugfasertheorie der Chromosomenbewegung. *Chromosoma* **1**, 253-264.

Schrader, F. (1953). "Mitosis. The Movements of Chromosomes in Cell Division," 2nd ed. Columbia Univ. Press, New York.

Singer, S. J. (1962). The properties of proteins in nonaqueous solvents. *Advan. Prot. Chem.* **17**, 1-68.

Swann, M. M. (1951). Protoplasmic structure and mitosis. II. The nature and cause of birefringence changes in the sea-urchin egg at anaphase. *J. Exptl. Biol.* **28**, 434-444.

Swann, M. M., and Mitchison, J. M. (1950). Refinements in polarized light microscopy. *J. Exptl. Biol.* **27**, 226-237.

Wada, B. (1950). The mechanism of mitosis based on studies of the submicroscopic structure and of living state of the *Tradescantia* cell. *Cytologia* **16**, 1-26.

Wilson, E. B. (1928). "The Cell in Development and Heredity," 3rd ed. Macmillan, New York.

Ziegler, H. E. (1895). Untersuchungen über die ersten Entwicklungsvorgänge der Nematoden. *Z. Wiss. Zool.* **60**, 351-410.

DISCUSSION

DR. TERU HAYASHI: Do you get the same equilibrium birefringence values when you lower the temperature as when you raise it, or do you come down a different path?

DR. INOUÉ: Within the limit of experimental error, the *equilibrium* birefringence seems to fall on the same curve. There is, however, a hysteresis in *approaching* the equilibrium value.

DR. ALLEN: I would like to pursue the point you made at the beginning and ask whether it really is necessary to have optical anisotropy to have contraction. It is a common laboratory occurrence that supposedly isotropic gels of gelatin contract; of course, when they contract, they do so isodiametrically.

DR. INOUÉ: When gelatin contracts isodiametrically, we do know there are filaments and that the filaments actually shorten. The structural anisotropy is already built in. It depends whether the filaments were randomly arranged or preferentially arranged whether an anisodiametric or isodiametric contraction occurs. Kunitz showed in the 1920's that you can take gelatin or agar and line up the molecules anisotropically; in this case, the responses to swelling and contraction will also be anisodiametric. This was published in *J. Gen. Physiol.* **13**, 565-606.

DR. ALLEN: I brought this up after thinking about Dr. Wolpert's results. He isolates a system of proteins (supposedly contractile) from the ameba at a low temperature, lets it warm up in the presence of adenosine triphosphate, and finds streaming apparently based on contractile processes. If birefringence is a prerequisite for contraction, then one must assume that the proteins somehow line themselves up as an anisotropic array. If so, this must constitute a self-organizing system.

DR. INOUÉ: Well, if one has a large number of filaments, as I think you yourself well know, they tend to line up as they precipitate out. This is well known in tobacco mosaic virus. Also, if you take a solution of G-actin and give it conditions that will form F-actin with very long filaments, it will line up spontaneously. So I think simply by increasing the concentration to the point where the form or precipitate comes out, one can expect some degree of alignment, unless there is a constraint that makes them not line up.

DR. ALLEN: By his method of preparation, I would guess the concentration of this material could not be greater than 3–4% by weight. What role do you think the centriole plays in lining up spindle fiber elements?

DR. INOUÉ: I think the plants which generally show no centriole have answered this question for us. In the clear-zone material, simply by having anisodiametric condensation, one gets a spontaneous alignment, and very often it is true that in plant cells the orientation of the spindle filaments takes on the orientation or spatial distribution imposed by the nucleus or the cytoplasm.

DR. REBHUN: It appears from your pictures that neither the shape nor the birefringence of the chromosomal fibers change, at least during the early part of anaphase. Is this correct?

DR. INOUÉ: Yes, that is correct.

DR. REBHUN: About how long does this last? In other words, how long can the chromosomal fibers move without changing either of these parameters?

DR. INOUÉ: The length is very difficult to judge. In *Haemanthus,* we have not been able to measure the length because the birefringence tapers off into an imperceptible value. In the case of *Lilium,* one can, by frame-by-frame analysis, measure the position of the spindle poles. In this case, the apparent change of position of the tip is illusory, and I can hardly detect any length change of the over-all spindle in *Lilium* until very late anaphase or telophase.

DR. REBHUN: How about the chromosomal fibers?

DR. INOUÉ: The chromosomal fibers obviously must be shorter, and they do not go beyond the poles.

DR. REBHUN: And they do not change birefringence?

DR. INOUÉ: Near the kinetochore, as far as I can tell, they do not change to any measurable extent. We do not have the exact numbers.

DR. MARSLAND: There appears to be a remarkable similarity between the behavior of this material and other gel structures. As you probably know, we can cause a dissolution of the spindle with pressure, and it reconstitutes itself after decompression. The type of gel with which we are dealing becomes more disorganized with lowered temperatures and more highly organized with increased temperatures. I would like to point out, also, that perhaps a gel strand is a structure which can contract without changing its diameter. As a gel contracts, it necessarily loses material by syneresis. This type of contraction is probably one example where you can get contraction of a structure without any change in the diameter of the fiber.

DR. INOUÉ: Well, perhaps some of the chemists would also like to comment on your remark. If syneresis is taking place, simply expelling water without loss of protein matrix, then the refractive index should go up if the mass is not changing.

DR. MARSLAND: There might be quite a residuum of protein material which has not entered into gel structure which might be expressed in syneresis.

DR. INOUÉ: I am sorry; I did not understand.

DR. MARSLAND: When gelation occurs, it does not necessarily involve all of the available material, and part of the fluid which presumably is trapped in the framework of the gel may still have quite a residuum of protein material. So that the change in birefringence might be very small.

DR. INOUÉ: I did not mean in birefringence; I meant in total refractility. You are saying that there may be contraction.

DR. MARSLAND: I also meant total refractility. As you are losing one thing, aren't you condensing another? In other words, the syneretic exudate may have a high protein content and huge refractility.

DR. INOUÉ: Does it sound likely?

DR. MARSLAND: Maybe I am wrong, but I do not think so.

CHAIRMAN BISHOP: It sounds like an argument for the Free Discussion. Any other questions?

DR. KAUZMANN: Is this form or intrinsic birefringence, or are they the same thing?

DR. INOUÉ: I can only answer this question operationally. If one fixes a cell with one of the classic fixatives, in which the fiber birefringence appears to remain, let's say in Bouin's solution, and then changes the refractive index of the medium with organic solvents, the birefringence then disappears at the matching refractive index and returns again at a higher index. Therefore, according to this operational definition there is form birefringence, but since fixation involves very drastic changes, that is the extent of the answer I can give.

CHAIRMAN BISHOP: I can understand your reticence in naming molecules, but could you indicate which part of the spindle structure might contain a myosin-like structure?

DR. INOUÉ: May I ask you whether anybody has shown a myosin-like material?

CHAIRMAN BISHOP: I am not at all sure that what Dr. Mazia has found is myosin, but can you identify what he claims is myosin-like, or others have discussed as being myosin-like, with respect to the position of the fibrils here?

DR. INOUÉ: Dr. Mazia has isolated spindles by various means and put the material into solution, in general using rather strong reagents. Then he finds a material, which is predominantly protein mixed with some ribonucleic acid. Presumably this is the material which is making up the fibers; but I have a feeling that what we are seeing in this kind of preparation is material which has already become rather insoluble, or denatured and vulcanized. I doubt that this final step would be reversible.

But if we take the intermediate state, where the reaction is still reversible, for example, the spindle isolated by Dr. Kane, in distilled water at pH of about 6.5, if the pH is raised by 0.2 of a unit the spindle will instantly dissolve and go into solution. If one observes the spindle with a metaphase plate, the chromosomes will be seen to fall as a flake onto the slide. In such a preparation I think one can start worrying about what the state of aggregation is and whether this is myosin-like or not in terms of the over-all amino acid analysis. I believe Dr. Andrew G. Szent-Györgyi pointed out there is no resemblance to myosin. There was some similarity to actin, but I gather the data are not critical enough to say yes or no. Is that right?

DR. ANDREW G. SZENT-GYÖRGYI: Yes, you are right.

DR. GOLDACRE: If you can disorient the spindle by lowering the temperature, how do you account for formation of tubes, as mentioned by Dr. Evans Roth?

DR. INOUÉ: I do not quite follow your suggestion or question, Dr. Goldacre.

DR. GOLDACRE: I gather the electron microscope shows tubular structure in the spindles.

DR. INOUÉ: In cross section there is an outer dense region, and in longitudinal section, two lines. If we call it a tube, this is one of the ways the spindle filaments can appear under the electron microscope. Conversely, Dr. Kane has changed the conditions (the salt concentrations, etc.) of the isolated spindle before fixation and then he sees solid structures. I am at a loss to say just what temperature and disorientation may have to do with the tubular aspect.

DR. GOLDACRE: Apparently you may be building up tubes by allowing temperature to rise after the spindle material has been disoriented and disaggregated by lowering the temperature.

DR. INOUÉ: The over-all morphology and the volume of the spindle determine birefringence changes to such an extent that I don't think we can explain the temperature effect simply by a transformation of the over-all birefringent material into tubes. I gather you are suggesting this would be less birefringent.

Dr. Goldacre: I just cannot understand how a tubular structure could be built up by raising the temperature of disoriented material.

Dr. Inoué: Shouldn't we be more sure we are dealing with a tube before we worry about it?

Dr. Goldacre: I suppose so.

Dr. Wolpert: Do you have any idea whether a reversible process of random disorientation leading to orientation could give rise to small movements in cytoplasm? In other words, could it exert a force for such a process as saltatory motion of particles?

Dr. Inoué: Well, I have only limited experience on this matter. If you form liquid crystals of detergent, let's say, and watch the front of the crystals forming, this does not appear to displace granules which seem to be floating around freely in the medium.

Dr. Hoffmann-Berling: May I comment on the energy characteristics? Couldn't they be those of another process which regulates spindle production?

Dr. Inoué: Dr. Hoffmann-Berling would like to caution us, I believe, that what we observe as entropy change and free-energy change may not be the primary change in the spindle material itself. I agree with you whole-heartedly, and, in fact, I had this data lying around for 6 or 7 years before we actually put this interpretation on because of this danger. However, it appears that there are many encouraging results in isolated systems such as I mentioned, the G-F actin transformation and the tobacco mosaic virus A protein polymerization.

Dr. Hoffmann-Berling: A low entropy and high ΔH.

Dr. Inoué: High ΔH and huge ΔS. So I am somewhat encouraged that perhaps we may be dealing with something rather close to the spindle itself.

Dr. Edwin Taylor: While you are on the subject, I wish you would clarify something that has always bothered me. That is in making this calculation, Dr. Morales either knew the molecular weight of the spindle protein or I do not know what these numbers could refer to.

Dr. Inoué: Dr. Kauzmann, would you answer this?

Dr. Kauzmann: If the model used by Dr. Inoué is correct, then he does not need to know the molecular weight of the spindle protein. If this protein can exist in only two states—one of which is birefringent and the other of which is not—and if the relative amounts of the two states is proportional to the amount of birefringence, then one can deduce an equilibrium constant for the two states. From this equilibrium constant one can find the difference in free energy (ΔF) of the two states (equilibrium constant $= K = \exp(-\Delta F/RT)$) without knowing the molecular weights. This is one of the wonderful consequences of thermodynamics. Remember, however, that the reality of this ΔF depends upon the adequacy of the assumed model. If there were several kinds of protein in the spindle, each having a different equilibrium constant, or if there were a series of intermediate states, then the ΔF value that Dr. Inoué has calculated would not have much meaning.

Dr. Taylor: What reaction does this ΔF refer to?

Dr. Kauzmann: It could refer to the direct aggregate reaction or it could refer to some controlling equilibrium that indirectly determines the birefringence.

Concerning the point that has been raised about the pH, could one not look for pH changes by introducing an inert, indicator dye and look for color changes that would result from changes in the pH?

Dr. Inoué: Yes. Intracellular pH has been measured by colors of indicators, and I think such an experiment would be worth trying. There is also the possibility, for instance, of calcium ion release which I gather affects active polymerization a great deal; there are all kinds of other things that could go on.

DR. ALLEN: I just wanted to ask you how well you can separate the birefringence of the spindle from that of the mitochondrial sheath that is above and below it in some cases. Doesn't the sheath act as a compensator sometimes?

DR. INOUÉ: Yes, sometimes. The mitochondrial sheath in telophase has a much stronger birefringence than the continuous fibers in grasshopper spermatocytes. However, this is not always true with other types of cells. I can generally distinguish the birefringence of spindle structures both morphologically and by optical sectioning from the birefringence of the surrounding material.

Screw-Mechanical Basis of Protoplasmic Movement

ROBERT JAROSCH

Biological Research Division, Austrian Nitrate Works, Linz/Donau, Austria

During the last few decades, most biologists seem to have agreed that the basic mechanism of biological movements is contraction. The term "contraction" is generally understood to be an active shortening of contractile elements, whatever these may be. This concept alone, however, is insufficient to explain the great variety of biological movements. I should like to outline in this paper a different view of the process of contraction, but first I would like briefly to review the theory of contraction from a historical viewpoint.

Historical Sketch of Contraction Hypotheses

As early as 1854, Schultze (1854, 1863) spoke of protoplasm as a "contractile substance." Afterward there followed a period during which a wide variety of speculations on the nature of the contractile process were put forward, which, however, did not lead to any progress in understanding. Progress did not, in fact, become possible until 1902 when Fischer (1906a, b) elucidated the primary molecular structure of proteins, and until the secondary structure was studied by X-ray diffraction (Herzog and Jancke, 1920). The α–β transformation of the polypeptide chain (Astbury and Street, 1931) appeared to biologists to hold the first clue to the explanation of contractility, and several hypotheses were advanced. Finally, a little over a decade ago, Pauling *et al.* (1951) pointed out an important aspect of protein structure: many proteins contain polypeptide chain regions that are in the relatively rigid α-helix form. More modern explanations of contraction have thus tended to look for explanations of the contraction process in structural aspects of proteins more complex than the polypeptide chain sequence itself.

The Basic Structure of the α-Proteins

The α-helix (Fig. 1) behaves as a rigid, elastic screw, since the hydrogen bonds between adjacent gyres do not allow the monomers to rotate freely around the axis of the polypeptide chain. The helix has

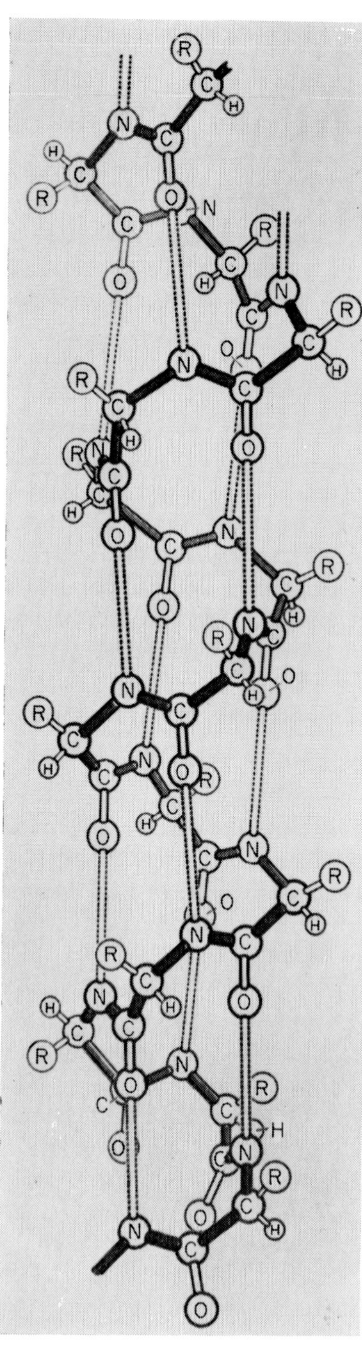

FIG. 1. The α-helix (from Kendrew, 1961). Hydrogen bonds between adjacent gyres stabilize the molecule so that a free rotation of the monomers around the axis of polypeptide chain is inhibited.

constant dimensions: pitch about 5.4 A, diameter about 11.0 A, pitch angle about 26°. However, Corey and Pauling (1953) and Pauling and Corey (1954) pointed out the possibility of slight variations both in the length of the hydrogen bonds (2.68–2.92 A) and in the bond angle at the α-carbon atom (108.9–110.8°). This means a possible maximal variation of about 7% in the pitch. "The changes in the nature of the side-chain

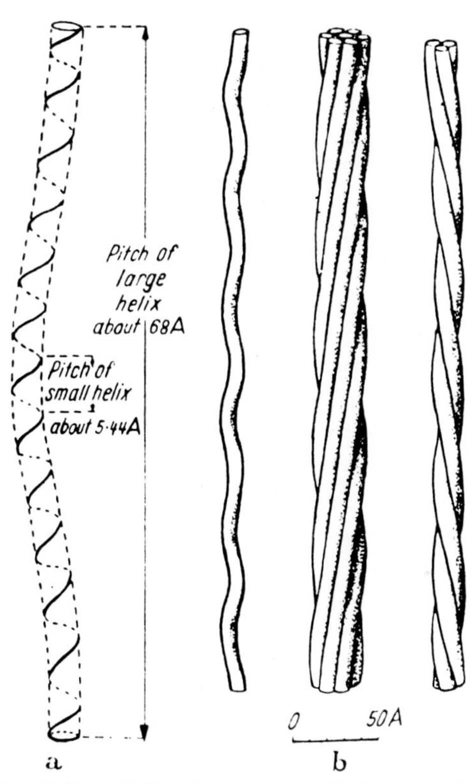

FIG. 2. The axis of the α-helix also takes a spiral course (helix of the second order) in keratin so that three or seven spirals form a strand. (After Pauling and Corey, 1953.)

groups might well cause the hydrogen bond distance to vary, either directly through the interaction of side chains with the carbonyl and the imino groups, or indirectly by steric hindrance or van der Waals' attraction" (Corey and Pauling, 1955).

Crick (1952), and independently Pauling and Corey (1953) came to the further conclusion that in keratin, the axes of the α-helices are wound in a screwlike manner (Fig. 2), due, in part, to special sequences of amino acids. "Let us consider an α-helix composed of a polypeptide in which a unit of four amino acid residues of different types is continually repeated. Two of the hydrogen bonds might be longer than the other two, by about 0.2 A. This difference in length would cause a curvature of the axis of the α-helix" (Pauling and Corey, 1953). If we speak of the α-helix as a screw of the first order, then we can call wound helices, "second-order screws." Thus, in considering the structures involved in movement, we may have to consider screws of the second, third, fourth, and even higher orders.

The existence of higher-order screws in cells appears to be demonstrated by the major and minor coils in chromosomes [where possibly as many as seven orders may be observed (Amano *et al.*, 1956; Bopp-Hassenkamp, 1959; Howanitz, 1953)] and in the bacterial flagellum (cf. Figs. 7 and 8) and, occasionally, in spirochetes, where two higher orders are apparent.

Since one aspect of the basic structure of α-proteins seems to be that they are elastic screws, we would expect among their dynamic functions not only simple shortening by contraction, but also other processes that are connected with the special mechanics of coiled structures. Various aspects of the mechanics and dynamics of coils can be studied in model experiments using helices made of steel wire (Jarosch, 1963a,b).

Since much of what we shall say concerning coiled structures is suggested directly by certain phenomena occurring in bacterial flagellae and in the behavior of filaments and bundles of filaments in extruded droplets of plant cytoplasm, we shall discuss both model experiments and observations simultaneously. In this presentation, we shall first discuss processes occurring in single helical structures and then those occurring in multiple helices. It should be pointed out that phenomena seen in the model experiments occur in steel-wire screws and, thus, these phenomena are independent of the ultimate nature of the elastic material of which the helices are composed. However, since proteins involved in protoplasmic movement undoubtedly possess α-helical regions, much of what we say can probably be referred back to this configuration, as the one from which higher-order helices are built.

The Motive Principle in Protein Helices

If the pitch of an elastic screw is changed, a torsional force is built up in the screw. This force, in turn, causes a rotation. In a revolving screw, the single gyres progress as "apparent waves" and may displace the screw relative to its environment, as, for example a corkscrew moving relative to a cork. This principle, which was already known to Archimedes, is the basis for the wide technological use of the screw. In an elastic screw, decreasing the pitch always results in a rotation, in which the gyres travel as apparent waves toward the fixed end of the screw (Fig. 3b,d). An increase in the pitch, by contrast, always causes the apparent waves to travel away from the fixed end (Fig. 3a,c,e). These rules apply regardless of the coiling-sense of the screw.

In a coil of steel wire, which has dimensions proportional to those of an α-helix 1 μ in length and with a pitch angle of 26°, experiments show that a change in pitch of 7% (caused by stretching) results in about 17.5 revolutions of the free end of the screw. In a longer screw, more

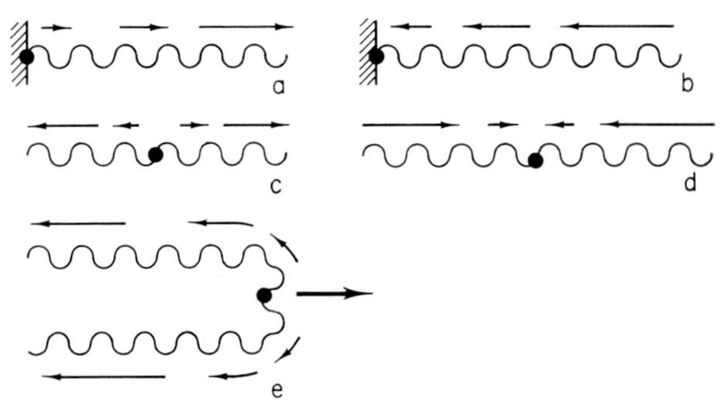

FIG. 3. Direction of the progress of apparent waves (small arrows) in the screw resulting from rotation caused by changes in pitch. The fixed position is shown with a black spot. With an increase in the pitch (a,c,e), the apparent waves progress away from the fixed end. With a decrease in the pitch (b,d) they proceed toward the fixed end. (e) Diagram of a chromosome during an anaphase movement.

rotations occur. From Pauling and Corey's data on α-helices, the number of residues per turn can vary from 3.60 to 3.67. The difference of 0.07 of a residue between these two extremes for a given α-helix is 1.9% of one turn. Thus, an α-helix 1 μ long will change length by approximately 190 A for this pitch change, and since the pitch of the α-helix is about 5.4 A, this represents about 190/5.4 or 35.2 rotations. However, in a real fluid medium, the torsion resulting in rotation will cause movement of the gyres against the medium, resulting in frictional forces. Thus, one would not expect to find the full 35.2 rotations actually occurring. In our work we, therefore, use the value 17.5 derived from model experiments, although this value is undoubtedly too small, since some of the torsion introduced into the model is dissipated in plastic deformation of the steel wire—a phenomenon we would not expect in elastic protein helices.

Together with the α-helices, the helices of higher order will also rotate, whereas the rigidity of the screws decreases with increasingly higher orders. This means that screws of the higher orders may tend to bend rather than participate in rotation around the long axis, the latter behavior being the main response of helices of lower order. If we ignore this tendency to bend, then we may derive an expression for the number of rotations R of a superhelix composed of wound α-helices corresponding to the maximum number of rotations of the lower-order helix of which it is composed. For a helix of third order this expression is

$$R = \frac{17.5l}{\sin \alpha_2 \sin \alpha_3} \quad (1)$$

where l is the actual length of the axis of the third-order helix in microns, 17.5 is the number of rotations of an α-helix 1 μ in length if it undergoes a maximally allowable change in pitch (as discussed previously), and α_2 and α_3 are the inclination angles of the second- and third-order helices, respectively.

The displacement, S, of a cytoplasmic structure is mainly dependent on the pitch, P, of the highest-order screw that rotates about its axis. Under the most favorable conditions, that is, if the environment behaves as a firm body,

$$S = PR \quad (2)$$

Frequently, however, especially in higher-order screws, the displacement will be less than this value [cf. Hancock (1953) and Taylor (1951, 1952) for discussion of wave propagation in screwlike structures] because the properties of the environment become more similar to a fluid at the microscopic and macroscopic levels.

The idea I wish to develop is that the cause of biological movements may be torsion pressure within protein helices. The basic process that produces this tension—i.e., the change in pitch of the helix—occurs before the screw rotates. This may explain the fact that after interruption of some biological movements by injury, the movement takes some time to decay (e.g., cf. Metzner, 1920). Under certain circumstances a change of pitch in protein helices may be possible even in dead cells. This, at least, is one possible interpretation of the results of Hoffmann-Berling (1953, 1955).

Now I should like to discuss two possible applications of the mechanics of screws in nature.

Muscular Contraction

According to the recent work of Huxley's group (Hanson and Huxley, 1955; Huxley, 1956, 1957), contraction and relaxation of striated muscle are accompanied by a sliding motion of protofibrils relative to one another. The cause of this motion remains unexplained. If, however, one assumes that the actin protofibrils are screws which are attached on one side of the Z-membrane, then contraction can be explained as caused by a decrease in pitch (Figs. 3b and 4a), whereas relaxation might result from an increase in pitch (Figs. 3a and 4b). This concept is supported by older electron-microscopic observations which indicate that the axial

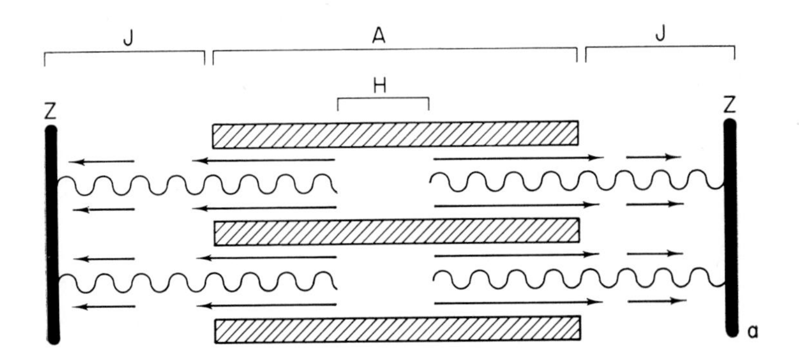

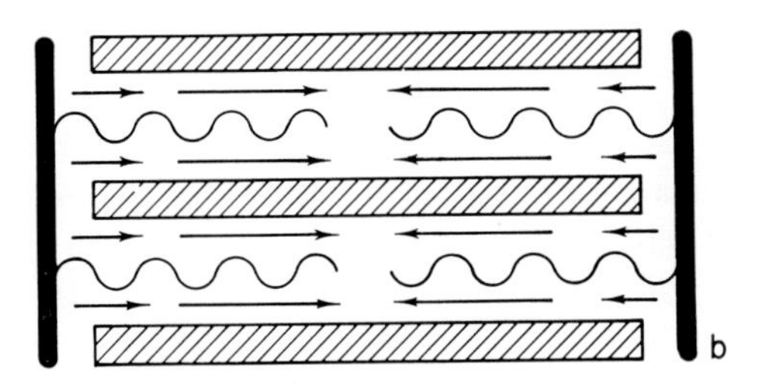

FIG. 4. A diagram showing the movement in muscle. (a) Condition at the beginning of contraction after a decrease in pitch of actin screw (cf. Fig. 3b). (b) Condition at the beginning of relaxation phase after an increase in pitch (cf. Fig. 3a).

period (which apparently corresponds to the pitch of the actin helices) is less in the contracted than in the relaxed state [250 versus 450 A, according to Draper and Hodge (1949)].

Movement of Chromosomes

The movements of chromosomes may also be related to changes in their coiling pitch. It is known that chromosomes exhibit a cycle of spiralization which reaches a peak in the condensed metaphase state, after which the chromosomes begin to elongate again. If the coils of both arms, which are fixed at the centromere, rotate when their pitch increases on uncoiling, then the centromere would be pushed ahead, in a manner similar to the screw action in propelling a ship. It is possible that such considerations are important in anaphase movements (Fig. 3c and e).

Temperature Dependence of Cytoplasmic Streaming

If the cause of movement is torsional force in protein helices, then during the relaxation of this torsion, the resulting apparent waves will encounter resistance in the surrounding aqueous medium. The viscosity of water, and, therefore, this resistive force are both temperature dependent. It is probably for this reason that cytoplasmic movements are proportional to temperature. On the other hand, the motive force [which we interpret as the expression of the torsional force and which in slime molds has a torsional component (Kamiya and Seifriz, 1954; Kamiya, 1959)] is *not* temperature dependent. This has been experimentally demonstrated by Kamiya (1953, 1959), in the slime mold and in Characeae by Hayashi (1960), who used the balance-centrifugal acceleration method of Kamiya and Kuroda (1958) to measure the motive force for cytoplasmic rotation. Hayashi's data strongly suggest that the effect of temperature on the velocity of cytoplasmic streaming is basically an effect on cytoplasmic viscosity.

If the change in pitch always occurs when the torsion tension has decreased to a certain minimal value, then the frequency of the change in pitch must be temperature dependent, because an increasing of temperature will cause a quicker release by quicker rotations. This explains the shortening in the motion period, e.g., of shuttle streaming (Kamiya, 1953, 1959) and gliding movements, as, for example, of *Bacillaria paradoxa* (Jarosch, 1958b) when the temperature is increased.

The ability to take up torsional forces will arise proportionally with the length of a screw. The previously mentioned motion period will, therefore, also arise proportionally with the screw length. The periods, e.g., of the "shunting movements" in gliding diatoms and Oscillatoriacae (cf. Jarosch, 1962) seem, therefore, to be dependent on the organism's length.

Helices of Different Pitch Wound Together into a Superhelix

So far we have dealt mostly with the behavior of a single helix. If two coils with slightly different pitch are wound together (let the pitches be p_1 and p_2, respectively, $p_2 < p_1$), a superhelix is obtained with a considerably larger pitch, P (Fig. 5). The same is true for a superhelix made of three (Fig. 6) or more single helices. The superhelix can be thought of as the resultant of the superposition of single helices. An exact mathematical treatment of these combinations has not yet been possible.

Many model experiments with wound steel-wire helices have, how-

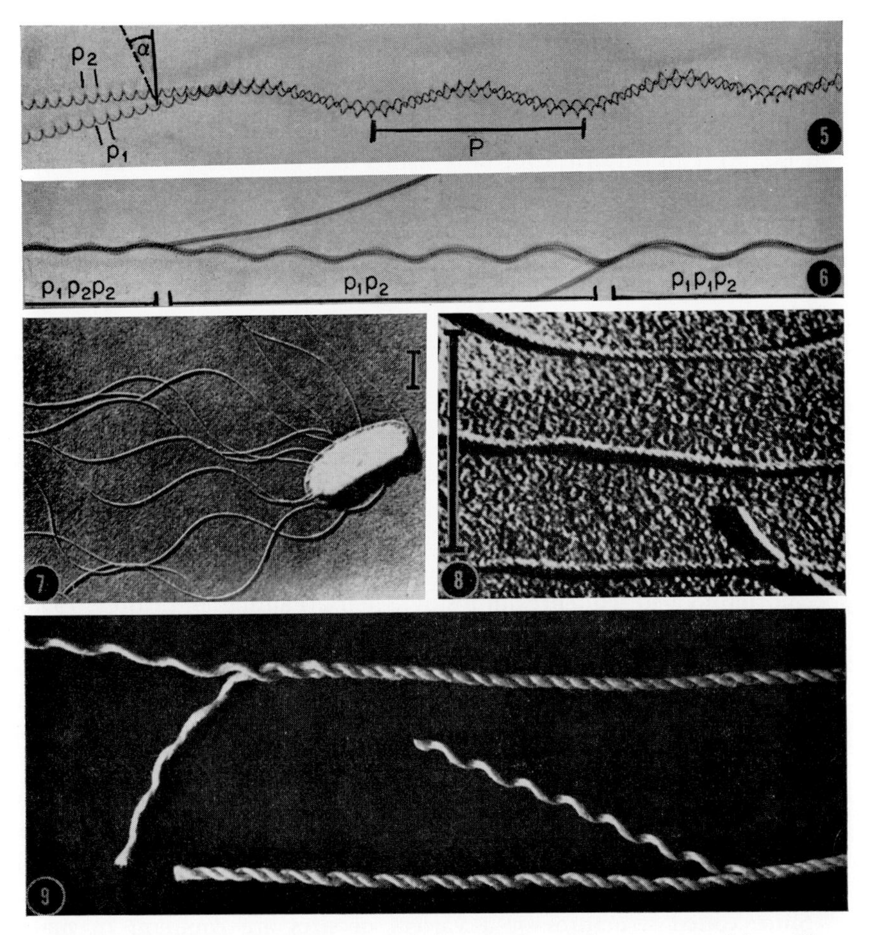

FIGS. 5 to 9. Bacterial flagella and their imitation with a screw model. Figure 5—a superscrew with a pitch P is formed when two screws of pitches p_1 and slightly smaller, p_2, are wound together; α, a pitch angle. Figure 6—a model of a superscrew, which consists of three screws in its left and right portions and of two screws in the middle portion. There are only two different pitches (p_1 and smaller, p_2). Figure 7—a bacterium *(Brucella bronchiseptica)* with flagella (superscrews). Scale: 0.5 μ. (From Labaw and Mosley, 1955.) Figure 8—the same flagella as shown in Fig. 7 enlarged. Each flagellum consists of three fibrils wound around one another. Scale: 0.5 μ. (From Labaw and Mosley, 1955.) Figure 9—string model demonstrating the course of the fibrils. Top, two fibrils; bottom, three fibrils.

ever, yielded the following empirical approximation for a superhelix formed from two separate helices:

$$P = 2\bar{p}\left[\frac{1.5K}{(p_1/p_2) - 1} + 3.1\right] \quad (3)$$

where $\bar{p} = \dfrac{(p_1 + p_2)}{2}$, the average pitch of the two helices.

Graphs of P against $\bar{p}$ were made for helices with different pitches, and Eq. (3) is an empirical formula which best fits these data. It is valid only within the limits $p_1/p_2 < 2$ and $P/p_1 < 60$ (Jarosch, 1963b).

The value of K is dependent on the average of the pitch angles, $\bar{\alpha}$, and is given in Table I (based on empirical measurements).

TABLE I
VALUES OF K

$\bar{\alpha}$	K	$\bar{\alpha}$	K
10°	2.14	35°	0.30
15°	1.30	40°	0.23
20°	0.84	45°	0.20
25°	0.50	50°	0.18
30°	0.33	60°	0.15

Whereas protein helices of second order may originate from certain specific amino acid sequences in a single helix (Pauling and Corey, 1953), higher-order helices may more reasonably be assumed to be the product of two or more helices wound together. The possible variations in pitch of such helices increases with the order of the helix. Thus, the α-helix may vary only by about 7% in pitch whereas Pauling and Corey (1953) show that a pitch decrease of 5.5% in an α-helix may cause the pitch of a second-order helix to double. Thus, helices of higher order may have a nearly unlimited variability in pitch (e.g., compare the chromonena helices of chromosomes, which may be completely extended or completely coiled).

Since coiled structures of a variety of sizes are common in many plants, it might be expected that the protoplasm of plants would show more striking differences in the pitch of its protein helices than would the proteins of animals. Detailed analyses have been presented so far only for bacteria.

The flagella of bacteria (Fig. 7) behave in a manner analogous to steel wires in the model experiments (Fig. 5). The pitch of the bacterial flagellum (P) is characteristic for each species (Reichert, 1909; Pijper and Abraham, 1954; Pijper, 1957). The two or three intertwined helical

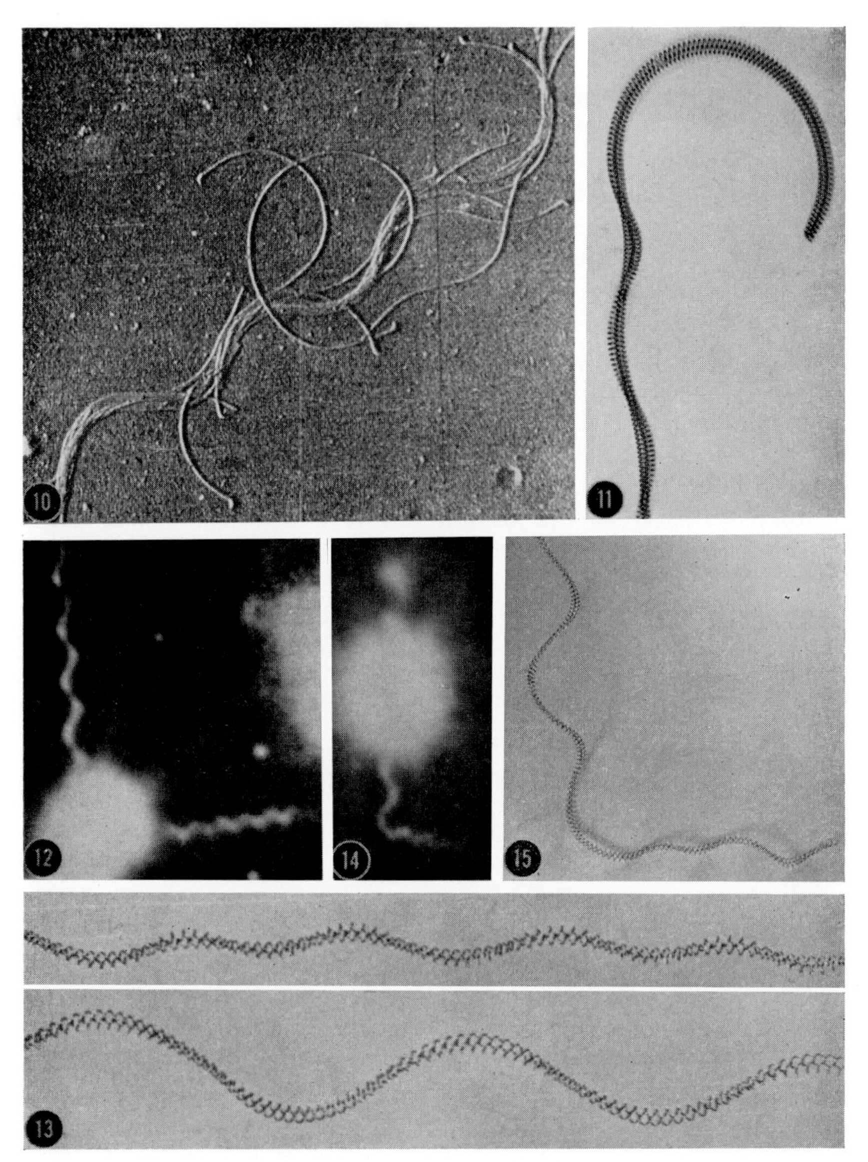

FIGS. 10 to 15. Imitation of phenomena displayed by bacterial flagella with a screw model. Figure 10—detached bacterial flagella with big arcs. (From Houwink and van Iterson, 1950.) Figure 11—the screw model in Fig. 5 can easily be laid in a plane making it form an arc such as is shown in Fig. 10. Figure 12—"biplicity" in two bundles of flagella in a bacterium. (From Pijper, 1957.) Figure 13—imitation of the biplicity with one and the same screw model. Biplicity with opposite winding direction (bottom) is brought about by torsion in winding direction at the primary condition. Slight pull brings back the primary condition (top). Figure 14—biplicity of one and the same flagellar bundle (From Pijper, 1957.) Figure 15—imitation of the biplicity in Fig. 14 with the screw model.

fibrillae have been demonstrated most clearly by Starr and Williams (1952) and by Labaw and Mosley (1954, 1955) (cf. Fig. 8). These fibrillae are apparently helices of the third order. The helix of the flagellum itself would consequently be of the fourth order. A model demonstrates the course of the two (Fig. 9, upper) or three (Fig. 9, lower) fibrillae more clearly. The difference between the model of Fig. 5 and the bacterial flagellum would seem to be primarily the proportionately larger diameter of the fibrils relative to the diameter of their constituent helices.

If values for P and α are taken from the paper by Labaw and Mosley (1954) and introduced into Eq. (3), a difference in pitch of 3.5% is obtained for the two fibrillar helices of the flagellum. It can probably be taken as evidence for the actual existence of pitch differences in component helices of the flagellum that certain characteristic properties of the bacterial flagellum can be imitated in a wire coil model. These properties are: (1) the arc which is visible frequently at the end of a flagellum (Figs. 10 and 11), and (2) the phenomenon of "biplicity," that is, the sudden appearance of one-half the usual pitch (Fig. 12). The latter can be seen in the same flagellum (Fig. 14) and can easily be imitated in the coil model by application of torsion in the direction of the coil winding (Figs. 13, 15).

Dynamics of a Stable Superhelix

In working with steel-wire models, such as that in Fig. 5, it is difficult to produce torsional forces by a pitch change, as they probably are produced in reality, and we have therefore done our experiments by twisting the model to produce internal torsional forces. There are two extreme cases in superhelices: those in which twisting the first-order helix either with or against its coiling sense will *not* reverse the sense of the superhelix; and those in which it will. Let us consider the latter case first. If we produce a torsional force by twisting the first-order helices *against* the direction in which they are wound, apparent waves of the superhelix will travel *away* from the fixed end of the screw. If we produce such a force by twisting *in* the direction of winding of the first-order coil, it may suddenly change the direction of coiling of the secondary helix (as can easily be seen in model experiments), and again the waves travel *away* from the fixed point. Thus, in this case, the direction of apparent wave motion is independent of the direction of torsional force. The *primary superhelix* (defined here as a superhelix wound in the same sense as the individual helices of which it is constituted) changes into a *secondary superhelix* (one in which the sense of winding is opposite to the sense of the separate helices). In the "biplicity" phenomenon of bacterial flagella, the flagellum with the doubled pitch appears to be a stable secondary superhelix. This behavior means that a periodic change in pitch can cause a movement which goes always in the same direction, i.e., away from the fixed end of the helix. This, apparently, is the situation in flagella: continuous movement in the

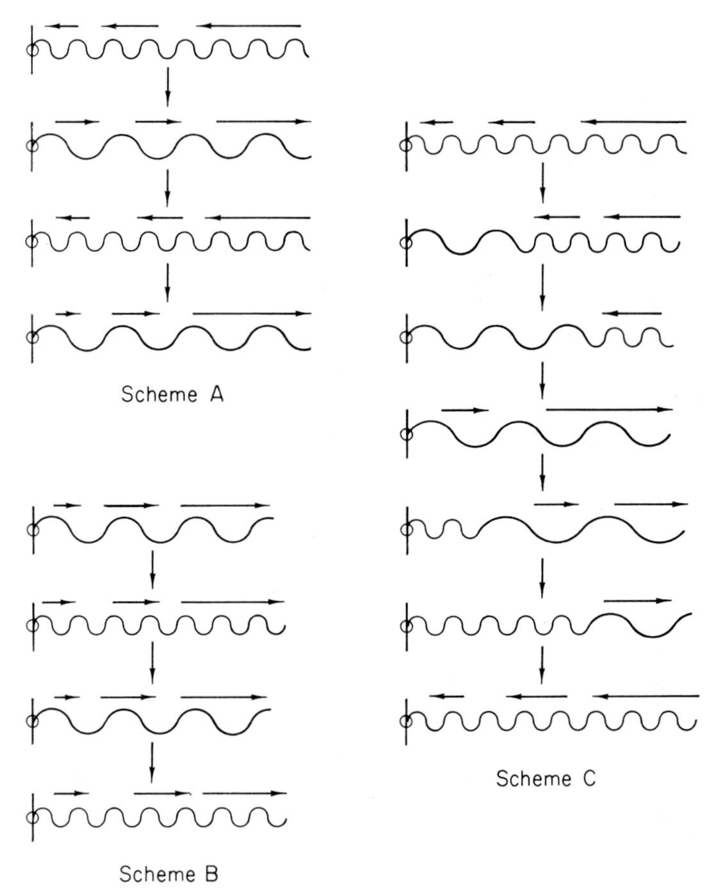

FIG. 16. Possible schemes for shifting of protein helices with periodic increases and decreases in pitch. For further details see the text.

direction toward the fixed end of the helix is mechanically impossible. The first case where the superhelix does not reverse was considered earlier (cf. Fig. 3a,b).

If there is a common mechanism to protoplasmic movements, we believe that it is based on changes in the pitch of protein helices. Figure 16 shows three possible propulsion schemes based on this simple idea. In scheme A, there is no change in the sense of the superhelix, so that the direction of propulsion reverses when the pitch increases and decreases. This applies to muscular contraction. In scheme B there is a change in the direction of coiling, so that propulsion is always in the same direction (away from the point of fixation). While scheme B appears to fit the facts of flagellar movement, both schemes A and B probably apply to the motions observed in the cytoplasm of plant cells; in addition, there are probably transitions between these schemes. For example, if the change in pitch occurs very slowly (scheme C), then the conditions in the scheme B may not be fulfilled and the primary superhelices will respond during the transition by rotations. Because of these rotations, the torsion tension required for reversal of the sense cannot be achieved. This might well be the case in the gliding movements of diatoms, blue-green algae (Jarosch, 1962), and similar organisms. Exactly which scheme is operative depends not only on the structure of the helix but also on the energetics of the change in pitch.

If these considerations are correct, then there does not seem to be any fundamental difference between flagellar and gliding movement. The superhelix of the bacterial flagellum, when wound around the cell mass, becomes a gliding organ if it cannot rotate around the axis of the highest helical order, but instead bends in a flexible manner around the bacterium. In the eubacteria and spirochetes, this rotation proceeds in the direction opposite to body rotation and causes swimming. In the filose bacteria and Cyanophyceae the rotation is no longer observed; we assume in these cases that gliding is caused by rotation of helices of the next lower order.

If superhelices, which are under high torsional forces, have free play, certain characteristic phenomena are observed in model experiments with wire. (1) If the superhelix is fixed at both ends, a braidlike configuration of the individual helices may form and grow in extent; formations of this sort may occur in accidental intertwining of blue-green algae and in filamentous bacteria. (2) If the superhelix is fixed only at one end and if there occurs a sudden increase in torsional force, the helix may snap back toward the fixation point and twist around it. During the change of direction of certain gliding movements and during bacterial movements, these phenomena, especially in flagella, occur widely. A more detailed discussion of these phenomena will not be given here.

Behavior of Unstable Bundles of Helices

In contrast to flagella, which are stable bundles of helices, the cytoplasm of many cells apparently contains unstable bundles of helices. The behavior of these is very complicated and variable. The combination of single helices into bundles seems to be a physical phenomenon common to all active (i.e., rotating) single helices which, because of the rotation, will twist around one another. This leads to the formation of helices of higher order. Parallel rotating superhelices (e.g., the single flagella of a bacterium) must become entrained with respect to phase (cf. Gray 1951; Taylor 1951). This in turn leads to a real wave motion of the bundle (helical waves), whereas the rotating single helices exhibit only apparent wave motion.

If the protoplasmic fibrils of Characeae discussed earlier in this Symposium by Dr. Kuroda and previously by us (Jarosch, 1956a, 1957, 1958a, 1960; cf. Yotsuyanagi, 1953 and Kamiya, 1959), can be considered as bundles of helices, their properties can be explained rather simply in terms of the mechanics of helices. In drops of cytoplasm freshly squeezed from living cells, the bundles of fibrils first appear bent and flexible (Fig. 18). In this state the fibrils themselves are hardly visible, but they can be recognized by small attached particles. These fibrillar bundles may form rings, which, however, soon exhibit straight stiffened regions, with the result that polygonal shapes are formed. The sides and corners of these fibrillar bundles move in a wavelike fashion, whereas the particles attached to the ring do not participate in the movement (Fig. 20). In older preparations, the polygons come to rest, showing diffuse regions of brightness in dark-field illumination and birefringence in polarized light (Fig. 19). Particle movements along the fibers and bundles of fibers are complex but may also be interpreted in terms of rotating helices. A particle attached to a single rotating helix will take part in this rotation (Fig. 17a). However, one which is attached to a bundle of rotating helices may either be pushed parallel to the axis of the bundle, or else exhibit some rotation without translation (Fig. 17b and c). The motions parallel to the axis of the bundle occur as a kind of reaction movement similar to those in, e.g., bundles of bacterial flagella, on the surfaces of gliding organisms, and in protoplasmic filaments. However, it is at first difficult to understand how the fibrils may exhibit strong active propulsion in a lengthwise direction, yet small attached particles may move only slightly. This can be explained if we assume that the propulsive activity of the fibrils is located in a higher-order coil than that to which the particle is attached (see Fig. 22). A particle which becomes detached correspondingly obtains an impulse counter to the direction of fibrillar movement, because, by being detached it may enter the counter-moving streamlet.

Coil models can imitate certain aspects of behavior of bundles of filaments described previously. For example, if a normally wound wire coil is bent, it appears to be curved smoothly (Fig. 21, outer coil). However, a maximally contracted wire coil (i.e., α equals nearly zero) shows straight stiffened regions and, at the site of the bend, a rather sharp arc (Fig. 21, inner coil). This sharp arc can often be observed clearly in completely stiffened fibrillar bundles in cytoplasmic droplets. In Fig. 21 the two coils are illuminated by a kind of dark-field illumination; it can be seen that the normally (outer) wound coil does not shine so brightly as the tightly wound one. It seems reasonable to assume, then, that in cytoplasmic droplets the straight stiffening of the fibrils is due to some kind of "supercontraction" of the superhelices of higher orders. It can be

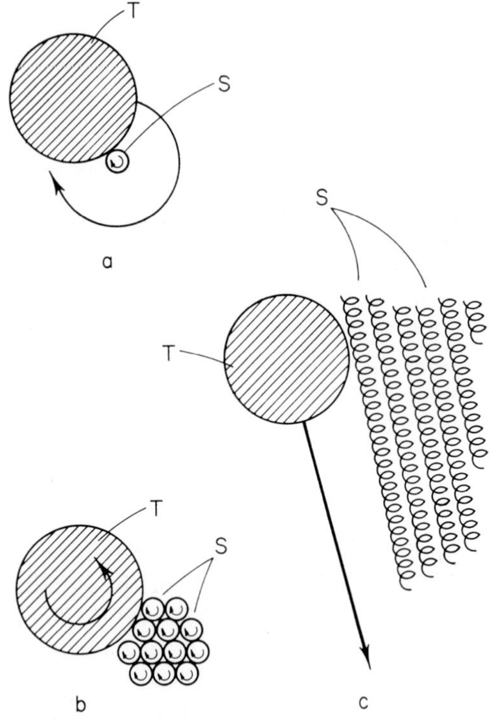

FIG. 17. The behavior of a particle (T) attaching to rotating screws (S). (a) Attached to a single screw and (b, c) to a bundle of screws.

thought of as the mechanical consequence of a persisting tendency to contract, even when the single turns are closely apposed. The stiffening force will probably increase with the internal tension.

Thus, in cytoplasm squeezed from Characeae the single helices probably mostly decrease in pitch resulting in increased stiffness. Over an extended period (some hours) it can be observed that the wave motion becomes progressively slower. At about the time that the light scattering of the polygons increases very slight wave motion of the corners of the polygons sometimes still persists. Here again the existing torsional force appears to run out slowly.

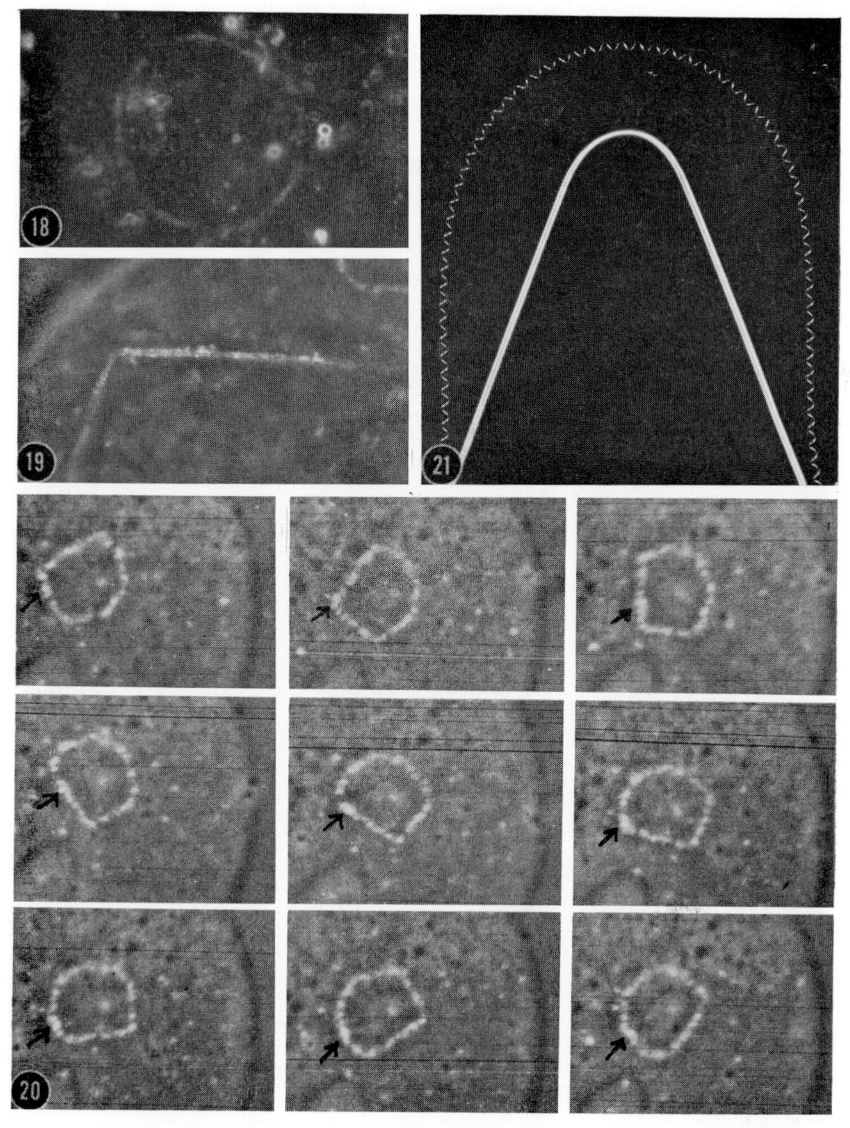

FIGS. 18 to 21. Imitation of the phenomena shown by protoplasmic fibrils of Characeae with a screw model. Figure 18—a ring of fibril bundle undulating in the form of arc in the freshly squeezed out protoplasmic drop. Figure 19—Linearly stiffened fibril bundles in an old plasma drop. Figure 20—wavelike propagation of a corner in a ring of fibril bundle. A particle imbedded in the ring, to which the arrow points, scarcely changes its position. The nine cine pictures were taken at intervals of 0.25 sec. Figure 21—imitation of the rectilinear stiffening of fibrils with a "supercontracted" coil.

The wave motion of the rings of fibrillar bundles in characeous cytoplasm can be imitated in a striking manner if a superhelix model similar to that shown in Fig. 5 is made into a ring by joining the ends and is made to revolve around the helical axis. The similarity is so striking that

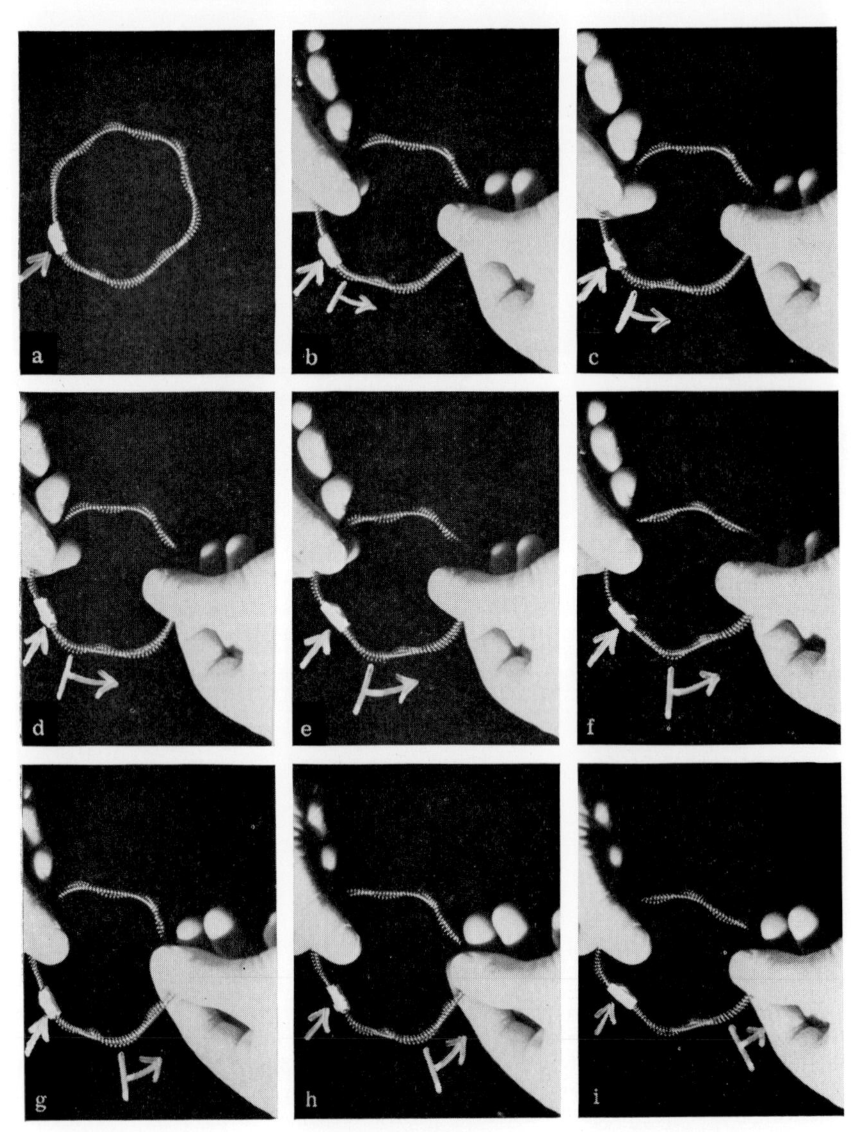

FIG. 22. Imitation of the undulatory motion (see Fig. 20) by revolving a ringlike superhelix model around the helical axis. The simple arrow marks a gummed tape as a "particle." The arrow with cross beam marks a moving "wave."

it is not difficult to become convinced that there are helices in protoplasm which rotate and give rise to protoplasmic streaming phenomena. Figures 22a–i are successive photos of such a coiled model and show the wave motion of the sides and corners and fixed position of the "particle."

It is unfortunate that the phenomena which I have tried to explain in terms of screw-mechanical principles lie mostly at or just below the limit of resolution of the light microscope. Although observations on the organization of protein structures at this level must be regarded with caution, it should be mentioned that Strugger (1956, 1957; Strugger and Lindner, 1959) has claimed the existence of submicroscopic helices as protoplasmic structural elements on the basis of electron-microscopic observations on plant cells. Even if his analysis of the ribosomes as cross sections through these helices seems to be incorrect, the value of his observations themselves should be acknowledged (cf. also Weissenfels, 1957, 1958).

We have been assuming that much of plant cytoplasm streaming may be related to rotating superhelices and that the organization of such streaming is due to superhelix formation. For example, the transition from agitation to order by streaming in cells (Jarosch, 1956b; Kamiya, 1959) may very well be due to the organization of previously randomly arranged, rotating helices into bundles of superhelices, in a manner analogous to that discussed for fibril and polygon formation. If so, it may be of some value to calculate the number of resolutions per second to be expected of superhelices. Starting with the equation for the velocity of wave motion;

$$\text{Velocity} = \text{frequency} \times \text{wavelength}$$

we can write for a rotating submicroscopic helix $V = rP$. In this expression the displacement velocity, V, corresponds to the streaming velocity of the protoplasm; r designates the number of rotations per second. The pitch, P, can be measured in electron micrographs. Its value, if we accept Strugger's results for the purposes of this argument, averages about 300 A ($= 0.03\ \mu$). Protoplasmic streaming at a rate of 2 μ/sec, therefore, yields for r a value of 2/0.03 or 67 rotations/sec. Though this rotation velocity is high, it might well be the basis for the "dynamical organization" of the protoplasm, as described, for instance, for the amoeba cell by Abé (1961, 1962). What we are suggesting can be demonstrated by rotating the model of a superhelix, of the kind shown in Fig. 5, at high speed around its long axis, for instance by connecting it to an electric motor. If this is done, *standing waves* will result, the distances of their nodes correspond to the repeat distances of the rotating superscrew (Fig. 23a, b, c). Using the proper repeat distances, varying patterns of standing waves may be produced. Frequently these standing waves are not quite stationary. Then their movement resembles the wave-like motions observed at protoplasmic surfaces. They are quite unstable, as each mechanical alteration changes the state of equilibrium (Fig. 23c).

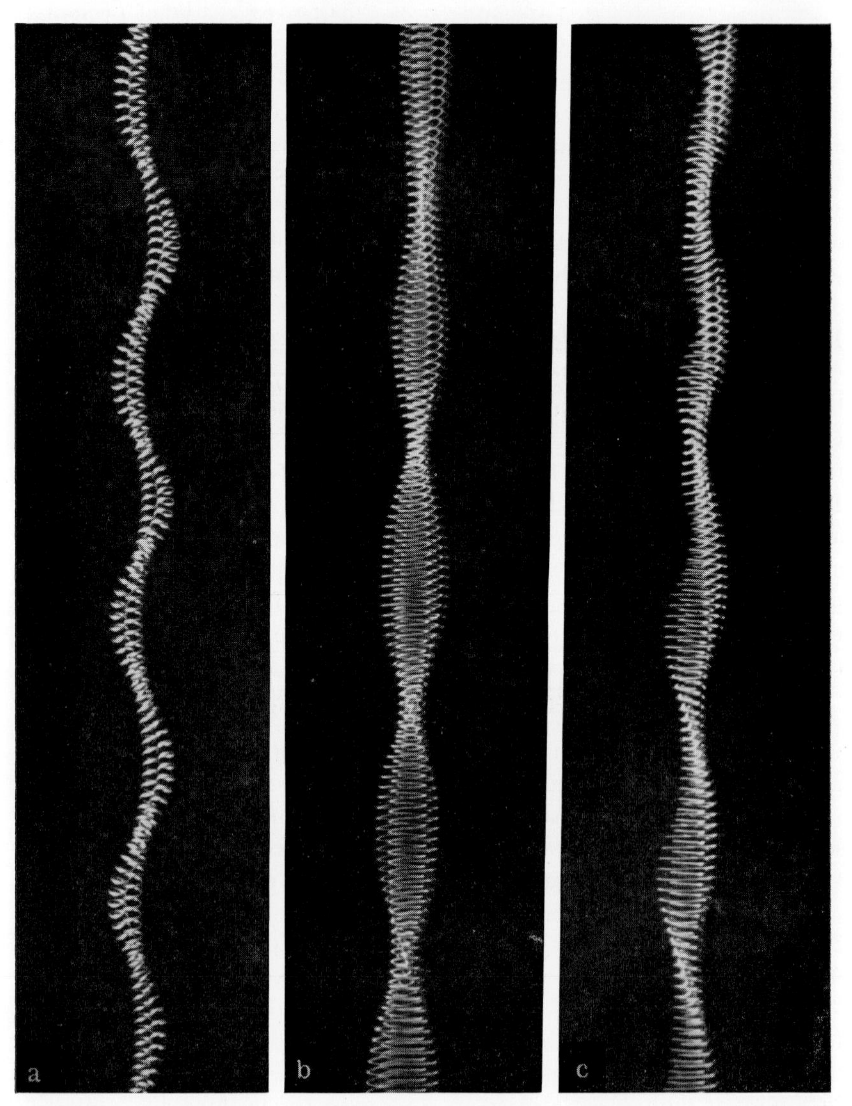

FIG. 23. (a) Resting superhelix similar to that shown in Fig. 5. (b) "Standing waves" on a quickly rotating superhelix (about 17 rotations/sec). The distances of neighboring nodes correspond to the pitch. (c) "Disturbed standing waves" caused by touching the rotating superhelix with the fingers.

If such rotating helices do exist in cells, then these standing waves may set up fixed patterns. These patterns may even play a decisive role in cell morphogenesis determining, for instance, in growing plant cells the site of addition of cell wall material (compare Kiermayer and Jarosch, 1962). It may also be that a pattern of this kind determines the localization of the pseudopods in amebae. Perhaps high-speed movies of cells would reveal some of these phenomena.

In summary, a general theory has been developed to suggest that torsional forces within protein helices giving rise to alterations in pitch may explain a variety of motions in living organisms from flagellated and gliding bacteria to the striated muscle of higher organisms. Although the idea is new and therefore untested, it has the advantage that it is generally applicable to a wide variety of so far unexplained motions which we associate with life.

Acknowledgment

I am indebted to Dr. R. D. Allen and Dr. L. Rebhun for bringing this article to its present form. I am also very grateful to Mr. K. Bachmann for the translation of the manuscript.

References

Abé, T. H. (1961). *Cytologia* **26**, 378.

Abé, T. H. (1962). *Cytologia* **27**, 111.

Amano, S., Dohi, S., Tanaka, H., Uchino, F., and Hanaoka, M. (1956). *Cytologia* **21**, 241.

Astbury, W. T., and Street, A. (1931). *Phil. Trans. Roy. Soc.* **A230**, 75.

Bopp-Hassenkamp, G. (1959). *Protoplasma* **50**, 243.

Corey, R. B., and Pauling, L. (1953). *Proc. Roy. Soc.* **B141**, 10.

Corey, R. B., and Pauling, L. (1955). *Rend. Ist. Lombardo Sci. Lettere* **B89**, 10.

Crick, F. H. C. (1952). *Nature* **170**, 882.

Draper, M. H., and Hodge, A. J. (1949). *Australian J. Exptl. Biol. Med. Sci.* **27**, 465.

Fischer, E. (1906a). *Ber. Deut. Chem. Ges.* **39**, 453, 2893.

Fischer, E. (1906b). "Untersuchungen über Aminosäuren, Polpeptide und Proteine." Springer, Berlin.

Gray, J. (1951). *Nature* **168**, 929.

Hancock, G. J. (1953). *Proc. Roy. Soc.* **A217**, 96.

Hanson, J., and Huxley, H. E. (1955). *Symp. Soc. Exptl. Biol.* **9**, 228.

Hayashi, T. (1960). *Sci. Papers Coll. Gen. Educ. Univ. Tokyo* **10**, 245.

Herzog, R. O., and Jancke, W. (1920). *Ber. Deut. Chem. Ges.* **53**, 2162.

Hoffmann-Berling, H. (1953). *Biochim. Biophys. Acta* **10**, 629.

Hoffmann-Berling, H. (1955). *Biochim. Biophys. Acta* **16**, 146.

Houwink, A. L., and van Iterson, W. (1950). *Biochim. Biophys. Acta* **5**, 10.

Howanitz, W. (1953). *Wasman J. Biol.* **11**, 1.

Huxley, H. E. (1956). *Endeavour* **15**, 177.

Huxley, H. E. (1957). *Biophys. Biochem. Cytology* **3**, 631.

Jarosch, R. (1956a). *Phyton (Buenos Aires)* **6**, 87.

Jarosch, R. (1956b). *Protoplasma* **47**, 478.

Jarosch, R. (1957). *Biochim. Biophys. Acta* **25**, 204.
Jarosch, R. (1958a). *Protoplasma* **50**, 93.
Jarosch, R. (1958b). *Protoplasma* **50**, 277.
Jarosch, R. (1960) *Phyton (Buenos Aires)* **15**, 43.
Jarosch, R. (1962). *In* "Physiology and Biochemistry of Algae" (R. A. Lewin, ed.), Academic Press, New York.
Jarosch, R. (1963a). *Protoplasma: Höfler-Festschrift.* **57**, 448.
Jarosch, R. (1963b). *J. Theoret. Biol.* In press.
Kamiya, N. (1953). *Ann. Rept. Sci. Works Fac. Sci. Osaka Univ.* **1**, 53.
Kamiya, N. (1959). *Protoplasmatologia* **8** (3a).
Kamiya, N., and Kuroda, S. (1958). *Protoplasma* **50**, 144.
Kamiya, N., and Seifriz, W. (1954). *Exptl. Cell Res.* **6**, 1.
Kendrew, J. C. (1961). *Sci. Am.* **205**, 96.
Kiermayer, O., and Jarosch, R. (1962). *Protoplasma* **54**, 382.
Labaw, L. W., and Mosley, V. M. (1954). *Biochim. Biophys. Acta* **15**, 325.
Labaw, L. W., and Mosley, V. M. (1955). *Biochim. Biophys. Acta* **17**, 322.
Metzner, P. (1920). *Jahrb. Wiss. Botan.* **59**, 325.
Pauling, L., Corey, R. B., and Branson, H. H. (1951). *Proc. Natl. Acad. Sci. U.S.* **37**, 205.
Pauling, L., and Corey, R. B. (1953). *Nature* **171**, 59.
Pauling, L., and Corey, R. B. (1954). *Fortschr. Chem. Org. Naturstoffe* **8**, 310.
Pijper, A. (1957). *Ergeb. Mikrobiol. Immunitätsforsch. Exptl. Therap.* **30**, 37.
Pijper, A., and Abraham, G. (1954). *J. Gen. Microbiol.* **10**, 425.
Reichert, K. (1909). *Zentr. Bakteriol. Parasitenk. Abt. 1, Orig.* **51**, 14.
Schultze, M. (1854). "Über den Organismus der Polythalamien." Leipzig, Germany.
Schultze, M. (1863). "Das Protoplasma der Rhizopoden und der Pflanzenzellen." Leipzig, Germany.
Starr, M. P., and Williams, R. C. (1952). *J. Bacteriol.* **63**, 701.
Strugger, S. (1956). *Naturwissenschaften* **43**, 451.
Strugger, S. (1957). *Ber. Deut. Botan. Ges.* **70**, 91.
Strugger, S., and Lindner, H. (1959). *Protoplasma* **50**, 607.
Taylor, G. (1951). *Proc. Roy. Soc.* **A209**, 447.
Taylor, G. (1952). *Proc. Roy. Soc.* **A211**, 225.
Weissenfels, N. (1957). *Naturwissenschaften* **44**, 241.
Weissenfels, N. (1958). *Z. Naturforsch.* **13b**, 182.
Yotsuyanagi, Y. (1953). *Cytologia* **18**, 146, 202.

Discussion

Dr. Rebhun: I would like to point out that there are really two elements in what you have discussed. One is that you developed a certain number of models and phenomena for elastic helices. The second is that in your paper you tried to link these with processes that occur in the α-helix.

I think that these two things are quite separate, because, while I think your mechanical models are extremely interesting, I feel there could be considerable disagreement as to whether your application of these to the biological systems in terms of the α-helix is proper. For example, if the thin filaments in muscle are actually actin filaments, there is considerable evidence that they do not consist of α-helices elongated throughout the length of the fibril, but of F-actin composed, at least from recent microscopic work, of globular elements. I think, therefore, that your analysis might refer to the higher order of helix without necessarily having to relate it to an α-helix.

DR. JAROSCH: This work describes mechanics which might apply to helices of higher order, even if the smallest order is not an α-helix but any kind of screw. But certainly the values described for the changes in the proportion of α-helix show a remarkable agreement with the values demanded by the screw-mechanical concept.

DR. ANDREW G. SZENT-GYÖRGYI: I must comment on some of the experimental data on muscle which Dr. Jarosch mentioned. First, actin is perhaps the only fibrous muscle protein that does not show the 5.1 A periodicity. It is very hard to consider it a fully coiled α-helix. Second, the changes in the periodicity were observed at an early stage of electron microscopy at a time when effects of sectioning artifacts and shrinkage were not clearly realized and taken into account. It really should be done over before one could accept it without reservations. If the observation is true, it would be one of the most important pieces of information concerning changes in the structure of contracting muscle.

DR. JAROSCH: The computations which were made here using these data taken from the literature are not supposed to be more than an example, using some values and obtaining a result which is of the right order of magnitude.

DR. HOFFMANN-BERLING: I would like to apply your models to the movements of bacterial flagella. It is an unsolved problem whether bacterial flagella contract throughout their whole length as the flagella of higher organisms do. However, if they do, energy is required to produce or to reverse the mechanical alterations at the flagellar tip. That energy must be transported from the bulk of the cytoplasm over the length of the flagellum to the tip. The flagella of higher organisms are surrounded by a membrane, and the transport can be imagined to occur inside the flagellum. However, bacterial flagella are extremely narrow and do not seem to have membranes.

What I ask is: Do your models offer some explanation for how energy could be transported *mechanically* from one end of the flagellum to the other? Do you see what I mean? In your models, you twist a screw, fixed at both ends. Can you imagine a different model, where one end of the screw is unfixed and where, by twisting, you can make waves travel to the unfixed end?

DR. JAROSCH: Concerning the energy that transports along the flagellum, there is a problem, not only in the explanation with screw models but in any other explanation. Dr. Gray has shown that the energy content is the same along the whole flagellum.

DR. INOUÉ: I would like to answer part of Dr. Hoffmann-Berling's comments about introducing mechanical distortions in the middle of a flagellum. We have done similar experiments with protozoan axostyles by altering their mechanical properties simply by irradiating with a microbeam of blue light. We can change the wavelengths and the wave form traveling down the axis, which is ordinarily a sawtooth wave. Then one can watch what happens at the irradiated spot and one, indeed, finds the waves altered; as soon as the waves come to a normal part again, they get transmitted in the same way as before, getting exactly the type of deformation that Dr. Jarosch is proposing, namely, a differential contraction wave or stiffening wave which is traveling down the length. There is no time to present the data in detail, but the observations do agree with Dr. Jarosch's model and Miss Kuroda's observation.

DR. ALLEN: I just wanted to say that I am very glad that Dr. Jarosch decided to give his presentation on the screw-mechanical basis of protoplasmic movement rather than on the topic he was originally invited to discuss, namely, the movements in isolated droplets of *Nitella* cytoplasm. There are many motions in living material which have spiral components to them. To remind ourselves of just a few, I might mention the helical waves in flagella, the twisting of slime mold strands that Dr.

Kamiya showed in his film, the spiral path of the ameba (Schaeffer), the spiral contraction of the *Vorticella* stalk, the twisting during contraction of acephaline gregarines (which Christopher Watters has shown us in his film) and many other phenomena.

So far as I know, no general theory of contraction that has ever been proposed has ever taken these twisting motions into account. I think that Dr. Jarosch's model, whether it is exactly correct or only in the right direction, is a very important thing for us to think about when considering general mechanisms of contractility.

FREE DISCUSSION

DR. EDWIN TAYLOR: Since I was invited here not just to listen but to discuss, I would like to make a few comments about saltatory motion and streaming. In discussing saltatory movement the first question is whether the particles are self-propelled or moved by an external agent. The motion of small particles in cytoplasm is controlled entirely by viscous drag rather than inertia. The energy input to obtain the velocities observed in saltatory movements is relatively large and if one considers the requirements in terms of various possible mechanisms one is soon reduced to a mechanochemical cycle. Although a rough calculation suggests that the motion might be barely possible by jet propulsion this implies a particular structure for the particles. The observations of Dr. Rebhun and ourselves indicated that the movements could be exhibited by a variety of particles with no obvious structural features in common. In addition, a sphere, 1 μ in diameter, will undergo rotary Brownian motion with a relaxation time of a few seconds. Thus there would be some difficulty in maintaining the axis of the jet for a time sufficient to propel the particle over large distances in a straight line.

I was going to make some remarks about other ways of obtaining flow but they have been partially pre-empted by Dr. Jarosch. Many of the motions which he proposed can be obtained, I believe, with transverse waves rather than torsional oscillations.

The sperm provides a good example of how a small object can be moved by the propagation of a sine wave. This is, I think, a gel contraction system but it doesn't work by the obvious mechanism of a gross contraction producing a pressure. The hydrodynamic problem has been solved by Sir G. I. Taylor, and we can use his solution to see whether this is a feasible mechanism. The propagation of waves along a flat plate leads to motion in the direction opposite to the direction of propagation. If the plate is held fixed one would expect to get motion of the fluid in the opposite direction with a comparable velocity. Using data for sea urchin sperm, we have a flow velocity of 100 μ/sec for an oscillation of 1 μ.

Therefore, this mechanism is feasible and it perhaps would be of interest to see if there is any evidence for vibrations of the wall in *Nitella,* for example. I do not wish to propose this as yet as another theory of streaming but mainly to point up the moral that there are other ways of producing flow besides gel contraction.

I would like to make one comment on the controversy regarding location of the propulsive force in the ameba. It is very difficult to decide where the force is exerted if one only studies the flow of cytoplasm in the steady state. The problem is somewhat similar to flow along an elastic pipe. If a constant pressure is applied at one end to start the flow a wave must travel along the walls of the pipe at a velocity determined by the Young's modulus and diameter of the pipe. There is a well-known system in which a contractile gel changes its shape in a rhythmic fashion and produces waves along an elastic pipe. These waves are easily detected and they are known as the pulse. There are some obvious similarities between circulation of the blood and shuttle streaming in slime molds, and some of the work on circulation, particularly the hydrodynamic analysis of the flow, may be of use to us. The velocity of the expected pulse wave could be estimated from a rough guess at the Young's modulus of the ectoplasm in the ameba case and the wave might be looked for by stroboscopic photography.

DR. INOUÉ: I would like to make a very brief comment on sine wave motion. As we were taught in high school, the plane sine wave does not introduce displacement of material as a whole. If you take a particle on the surface of water which is undergoing a sine wave translation, the particle translates in a circle with its center standing still. So you have to add something else in addition to the plane sine wave.

DR. EDWIN TAYLOR: The system you are referring to is one in which the motion is determined almost entirely by inertia whereas for the motion of small objects in a viscous fluid the system is determined by viscous forces, the inertial terms are negligible. Under these circumstances a solution of the hydrodynamic equations, as given by G. I. Taylor, shows that the propagation of a sine wave along an object, such as a sperm tail, causes it to move in the direction opposite to the propagation of the wave.

CHAIRMAN BISHOP: I should think the difference here would be the fact that this particle is not part of a wave or attached to a wave, but it is a discrete distance away from the surface if it is being propelled.

DR. HAYES: I would like to take advantage of being a discussant here to make a few general comments on the taxonomy of waves. It is clear that taxonomy is part of what you are interested in. I think it is important to know what kind of a wave you are talking about when you are talking about a wave. So let me suggest a taxonomy for waves.

The common wave that we are used to in mechanical systems—and there are analogous waves also in electrical systems—can be classed as dynamic. These have the properties that there is something which is called inertia (this is not mass inertia in an electrical system) which gives rise to inertial forces; these are balanced against elastic forces of some kind. In the dynamic system there is always an invariant or an approximate invariant of some kind. In the vibrating string on a violin or in, say, a shallow-water wave coming up on a beach, the invariant is the velocity of the wave.

We have the basic relation which, of course, holds for all waves, that the velocity V, the frequency N, and the wavelength L are related, with $V = NL$. With the most classic of these dynamic waves, the velocity is an invariant; but before we go on to other types of waves, let me point out that this is not always true. Deep water waves have $V\sqrt{L}$ invariant; capillary waves have $V\sqrt{L}$ invariant. So it is not always the velocity of propagation of the waves which is the invariant.

Another type of wave, which I don't think is important in biology, but which may be, is the kinematic wave. A kinematic wave is essentially a modulation of some formation on a stream that is flowing with approximately constant velocity. If you have a river and you periodically change the amount of salt concentration that you are adding to the river and you measure the salt concentration at various points in the river, you will find that it will have a wave motion. But this wave motion depends only on what you put in and the speed of the wave V is simply the speed of the carrier. This speed is invariant. These are important; if you modulate the speed, you can get amplification. This is the basis for a type of electronic tube.

Let me introduce another type of wave which we can call helical (or Jaroschian!). This is the sort of wave that you get when you rotate a helix. It is clear that the invariant here is the wavelength L, which is precisely the wavelength of the helix. The velocity V is strictly proportional to N.

There is another type of wave which you would observe if you had, say, a collection of oscillators with each oscillator oscillating essentially by itself almost independently of its neighbors, but with some sort of a weak interaction between neighbors. The main effect of the weak interaction would not be to transfer energy, but only to interrelate phase, that is, to hook in one oscillator with respect to its neighbor into exactly the same frequency but with a given time or phase delay. This may be the mechanism which appears in a peristaltic motion, or in the slime mold motion. Let's call this kind of wave "local oscillatory." The frequency is the invariant, and the speed of the wave is governed by the nature of the interaction between them and that which governs the relative phase; this interaction could be very weak.

Waves of any of these types can transport debris. The helical wave can transport a particle along it, as we have seen. The dynamic wave can transport debris. People who ride waves on surf boards or who body surf are debris that are being carried with the phase velocity of a wave. The observable in a kinematic wave is essentially simply debris. Debris can also be carried by waves of the local-oscillatory type.

I think the main point is that what type of a wave is involved in a wave motion may be extremely important. In order to type the wave you must have an understanding of the mechanism of the wave.

Dr. Ling: What I would like to comment on is the molecular mechanism of contractile phenomena.

First, I think many of you are aware of the important work by Kauzmann, Doty, and many others in studying the transformation of an α-helix into a random coil during protein denaturation. In this aspect especially with the aid of synthetic polypeptides, the experimental advance has been most rapid. The theoretical treatment has been handled very capably by many people, including Dr. Shellman and Dr. Peller at Princeton University, a few years ago, and many others. This model considers that if a hydrogen bond is formed at one point, then the entropy for the next hydrogen bond is such that it will be favoring the formation of adjacent hydrogen bonds, and an α-helix may thus be formed by this cooperative mechanism with a purely entropic, near neighbor interaction.

This model is without doubt very much a part of the underlying mechanism; but there is also a good deal of evidence to show this cannot be the entire story. I will only mention two now-established examples.

One is, as Harrington and Shellman have shown, oxidized ribonuclease which at the isoelectric point does not form an α-helix at all. Yet oxidized ribonuclease has all the side chains which are potentially capable of forming what Kauzmann has referred to as hydrophobic bonds and other bonds. On the other extreme we have poly-*L*-alanine. It can be shown, as Doty has pointed out, the side chain is only 2 A in length and the distance between the centers of the nearest alanine side chains is 5.7 A. Consequently, the helices cannot form hydrogen bonds or any other tertiary bond. Yet it forms an α-helix so strong that aqueous denaturants cannot break it up. So this points out there must be other forces involved, other than the formation of tertiary bonds which make the α-helix either stronger or weaker.

Again, as a biologist, I want to make a general conclusion from the many important discussions in this Symposium: the fact that the control mechanism is extremely important in biology. From Dr. Hoffmann-Berling's data, you can make a simple calculation and show that in 1 kg of muscle, ATPase is capable of hydrolyzing approximately 1 mmole of adenosine triphosphate (ATP) per second.

You can also make another estimation and show that in the resting muscle the rate of ATP regeneration is not more than 1 mmole/100 sec. Thus, if all the ATPase acts at full throttle, the resting cell cannot contain any significant amount of ATP; yet it does contain about 5 mmoles ATP. How can this be achieved? This can be achieved, of course, through control.

Dr. Hoffmann-Berling referred to calcium and to relaxing factors as agents that are perhaps involved in such control mechanisms. The question is how are these to influence the protein conformation changes such as illustrated in the α-helix—random coil transformation.

Recently we presented another model in which the near-neighbor interaction is considered to involve not merely an entropy term but also an enthalpy term due to the inductive effect. We are inclined to think that it is this inductive effect which is the basic unit of the control mechanism.

CHAIRMAN STEWART: I seem to have been suddenly appointed Chairman. Would any of the other discussants like to say anything?

DR. ANDREW G. SZENT-GYÖRGYI: At this Symposium there have been a number of references made and experiments presented which indicate that some of the filamentous structures and proteins may be similar to actomyosin, and that the motility of many cells may be based on a mechanism similar to contraction of muscle. I would like to point out that even if this analogy proves to be correct, this will not solve your problems. In fact, we do not know very well how muscle contracts and have no direct experiment telling us what, if any, structural changes in actomyosin are responsible for the contraction of muscle. Ironically, cellular motility would be easier to explain in molecular terms if the proteins responsible for motility behaved like synthetic polymers. The way these polymers contract, develop tension, and perform work is quite well understood and has been experimentally and theoretically treated in a beautiful manner by a number of workers, especially by Katchalsky, by Kuhn, and by Flory.

These polymers can shorten and convert chemical energy into mechanical work, and what is more, can do it much more simply and efficiently than muscle. The conversion of mechanical energy and chemical energy is freely reversible. It is possible to show, for instance, that mechanical work exerted on the polymer results in a change of the free energy associated with the activity of charged groups of the polymer and the surrounding ions.

One would, of course, like to know in what sense the muscle proteins and the contraction of muscle is different from the contraction of synthetic polymers. In first approximation perhaps the most outstanding difference is that muscle is a two-component system. The different sites on actin and on myosin must interact with each other to produce shortening under the influence of ATP. The two-component system has the property that allows a simple control in separating the resting state and activity in an extremely clearcut fashion. Resting state corresponds to the state where interaction between the two proteins is prevented, in case of muscle by the relaxing factor system. Activation brings about the interaction between actin and myosin which starts the contraction process. Thus the system does not depend on large concentration changes of certain ions. One does not have to postulate changes in pH or other such mechanisms which may be rather difficult to conceive in biological systems. In addition, there is no need for the compartmentalization of small ions. The gain in the ability to control rest and activity is offset by the structural and chemical complexity of muscle as compared with synthetic polymers.

We do not know and have no direct experiment to show how the actomyosin system changes during contraction. As a matter of fact, the most widely accepted theory for muscle contraction (it is not accepted by me personally, but by most people in the field) is the sliding theory of contraction, proposed by Huxley and Hanson. It is based on their studies of the fine structure of certain striated muscles. Their model suggests that there is no over-all irreversible change in the structure of the filaments of muscle which are formed from the aggregates of myosin and of actin. Thus there is still discussion going on as to whether contraction of muscle is the result of some type of intramolecular folding or is it due to some type of sliding.

I would very carefully examine the various pieces of evidence to see whether or not similarities with the synthetic polymers exist, especially since these systems seem to be simpler and theoretically much better understood than the contraction of muscle. There is one case, described by Hoffmann-Berling, the contraction of extracted *Vorticella* stalk, which behaves rather like the synthetic polyacids.

DR. ALLEN: I would like to ask Dr. Szent-Györgyi whether it might not be reasonable to suppose, in view of the fact that electron microscopically no one has seen a change in length of any fibrous component in the contracting structure, that the possibility exists of there being at least three states of such fibers—contracted, relaxed, and fixed?

DR. ANDREW G. SZENT-GYÖRGYI: I agree.

DR. ALLEN: And, whichever living state the muscle is in, it goes into the fixed state on being killed?

DR. ANDREW G. SZENT-GYÖRGYI: That we do not know for certain. We have rather scanty observation on the fine structure of contracted muscle. We do not know when we see filaments which are there in a certain geometric pattern whether we see everything which was associated with the filaments in the living state. We do not know how far extraction, shrinkage, and other changes may have caused alterations. Do we see all of the myosin, all of the actin which were present, or just see what remained? These are problems usually associated with electron-microscopic techniques. The more the interpretations depend on electron microscopy, the more carefully one has to consider these eventualities. At present, evaluation of the results may involve such arguments as these. I do not think this would be the right place or time to go into the subject in more detail.

DR. WOLPERT: Leaving contractility aside for a moment, one characteristic of protoplasm is that it undergoes what have been called sol-gel transformations. I wonder if you could tell us something of sol-gel transformation in actomyosin.

DR. ANDREW G. SZENT-GYÖRGYI: I think there is a very good correlation between sol-gel transformations and the behavior of actomyosin. In resting muscle I think it is fairly well agreed that there is very little bonding between actin and myosin. That is the reason why resting muscle is highly extensible, has a lower elastic modulus, and lower viscosity. The effect of excitation is that an interaction is established between actin and myosin. That interaction increases the viscosity and elastic modulus and reduces the extensibility of the muscle. So if you have a two-component system, the components of which become locked or hooked to each other, then you will have a "gelation" or what I think was understood here when sol-gel transformation was discussed.

DR. WOLPERT: Can you get such interactions also in isolated systems?

DR. ANDREW G. SZENT-GYÖRGYI: Yes, the viscosity of actomyosin in salt solutions of high ionic strength is very high. The addition of ATP, which reduces the interaction between actin and myosin, causes viscosity drop.

DR. ALLEN: Dr. Goldacre presented me with a list of some fourteen observations which he considers are impossible to explain in terms of the frontal-contraction model. If I may be permitted, I would like to follow in the footsteps of Wilson, and pursue these fourteen points.

First, Dr. Goldacre would predict from the frontal-contraction theory that polypodial amebae should pull themselves to pieces. This would be true if there were no control system, either mechanical or chemical, to regulate the movement of pseudopodia. So far, no measurements have been made of how the force varies with the extent of outgrowth of a pseudopodium, but I think one has to assume either that the motive force gets weaker or the resistance to that force becomes greater as the pseudopod elongates. I would point out, however, that in one situation, amebae do pull themselves apart in just the manner that Dr. Goldacre predicts: in cytokinesis.

Second, Dr. Goldacre claims to have demonstrated a positive hydrostatic pressure within the tail, and not a negative one, as *he* predicts from the frontal-contraction model, and he would like me to explain this. Before accepting such a demonstration as a fact, I would like to know more about the conditions of the experiment and see in what manner the outflow occurred. However, I would regard the demonstration of turgidity within the cell as a side issue, for this could (1) exist without playing any role whatever in movement, or (2) it could play a subordinate, or (3) a dominant role, as Dr. Goldacre believes. The frontal-contraction model would function perfectly in a turgid cell, either by itself, or in addition to such effects as pressure might have on streaming processes. I do not deny that pressure might play some role in cytoplasmic streaming in the ameba, but I believe that the complexity of movement is such that its mechanism cannot be explained solely on this basis.

Dr. Kanno's experiment (see the Free Discussion of Part III, comment by Dr. Abé), in which he withdrew about two-thirds of an ameba's cytoplasm by means of a pipette inserted into the tail, without reversing the normal forward cytoplasmic streaming, is perhaps the strongest direct evidence we have that streaming is the result of directional forces rather than pressure.

Dr. Goldacre's third point was his claim to having measured the contractile tension developed by the shortening tail. The idea of using a fork-shaped microspring balance to measure forces in cells is so elegant and original that it is a pity not to have these methods, observations, and data published in full so that they can be made proper use of in many systems besides the ameba. However, even if Dr. Goldacre succeeds in recording his observations for publication, I would not regard the measurement of tension in the shortening tail as evidence of active contraction necessarily taking place there. The exact same result would be expected if the tail collapsed passively as the result of tension applied to its inner surface through elastic elements in the endoplasm. We know from polarized light studies to be published soon that such elastic structure does exist and furthermore exhibits fluctuating photoelasticity, showing that it is under fluctuating tension.

Dr. Goldacre's fourth point was that his experiments with *p*-chloronitrobenzene vapor indicate to him that locomotion is possible when the fountain zone is completely free of plasmagel. My comment is that although these cells show some very interesting phenomena (such as the peeling off of some kind of a new surface from the anterior surface of the cell membrane) which I think deserve much further study, I do not follow Dr. Goldacre's conclusion that the fountain zone is completely free of plasmagel. In his interesting experiments, the normal cell structure has been so severely altered that I, for one, could not distinguish with assurance what was sol and gel, or even determine what the pattern of cytoplasmic streaming was. I think all that one can conclude from his pictures is that sporadic streaming of some kind can take place even when a substantial part of the pseudopodial tip contains hyaline material. However, it is not clear whether this hyaline material corresponds to the syneretic exudate or to granule-free whole ground substance in the "gel" or "sol" state.

Dr. Goldacre's fifth point concerns the spherical pseudopods formed as the result of heparin injection. I would really prefer to reserve judgment until I saw with my own eyes or on film just how such a spherical pseudopod was formed. Given an initially turgid cell (which I mentioned earlier is a possibility), it is quite possible that such pseudopods might result from the local dissolution of gel structure. Such an interpretation would certainly follow from the old ideas of Heilbrunn and his students regarding the action of heparin-like substances on cytoplasm.

I would point out, however, that spherical pseudopods are not normal structures. To explain the species-specific form of pseudopodia (sometimes as in *Chaos*, these are almost perfectly cylindrical when first formed), it is absolutely necessary to assume an efficient feedback mechanism coordinating the rate of streaming with the rate of "gelation." On the other hand, if one assumes this "gelation" is instead a contraction, then no such mechanism is required. This point has been overlooked.

Dr. Goldacre's sixth point concerns the detection of phosphate ion in the ameba's tail by cytochemical tests. This is an interesting observation, but I would prefer to reserve further comment until I find out more about the specificity of the stain from cytochemists and learn more from Dr. Goldacre about the reproducibility of the results. Superficially, such data would seem to be strong support for Dr. Goldacre's contention that the tail actively contracts.

The seventh point Dr. Goldacre would like me to explain is the contraction of plasmagel induced by membrane contact. Dr. Goldacre's theory is based in part on the fact that when the cell membrane is brought into mechanical contact with the ectoplasmic layer, the latter responds to stimulation and apparently contracts. I have seen this myself. Dr. Goldacre's interpretation stresses the importance of the membrane in this contact. An alternative possibility is that the ectoplasmic layer (in fact, perhaps any gel portion of the cell) is "sensitized" to stimulation. I have a strip of motion picture film which shows a rather violent rounding-up response of an *Amoeba dubia* to the impact of a gold sphere several microns in diameter falling on the *inside* surface of the ectoplasmic tube. I therefore prefer the simpler classical view that "protoplasm is irritable."

Dr. Goldacre's eighth point is that the nucleus in *Amoeba proteus* is free to rotate as it "flows" forward; therefore it is not caught in a gel plug under tension. Dr. Goldacre is correct on the observations, but the same observations lead us to somewhat different conclusions! I doubt whether the nucleus (in *A. proteus*) or nuclei (in *Chaos*) ever become rigidly caught up in the axial endoplasm; instead, they tend to ride in the more fluid shear zone. I do not see that we can conclude very much from either these observations or these different interpretations concerning the mechanism of movement.

The ninth point is Dr. Goldacre's claim that locomotion is possible with a gel/sol ratio of less than 1, first at 40°C, second after heparin injection. The first example conflicts with the very careful observations of Mast and Prosser, who pointed out that 32°C is close to the limit at which amebae can move. As the temperature increases to this point, the gel/sol ratio (*computed on an area basis*) (which we call the A_t/A_s ratio) approaches 1.0. We have data which are in general agreement from *Chaos chaos*. Before abandoning these data, I would like to have more information. At the present time, I see no discrepancy between the data and the predictions of my model. As far as the effects of heparin are concerned, I would accept Dr. Goldacre's theory as an explanation, since the presence of heparin in a living cell is hardly a normal situation. My model deals only with normal pseudopod formation and its role in ameboid locomotion. The effects of diverse chemical agents are very difficult to interpret at this stage in our knowledge.

The tenth point concerns the existence of the so-called "plasmagel sheet," the name given by Mast to the temporary border between the granular cytoplasm and hyaline cap fluid. No one doubts that such a border appears in some but not all cells, but there is considerable doubt concerning the various functions that have been attributed to this structure. Dr. Marsland, for example, would have us believe that it acts as a filter, despite the fact that it does not separate the parts of the cell which would, according to this view, represent unfiltered and filtered cytoplasm: the endoplasm and ectoplasm, respectively. Until we know more about the ultrastructure of that part of the cell, it is premature to assign to it any function—particularly when it is not a constant feature of ameboid cells. Until more is known, I prefer to take the simpler view that this border represents an interface between a granular gel of relatively high refractive index and a hyaline material (presumably a syneretic exudate) of relatively low refractive index.

The "films" peeling off the inner surface of the membrane in *p*-nitrochlorbenzene may bear some resemblance to the plasmagel sheet, but should be distinguished from it until more is known about the ultrastructure or function of each. Operationally, the two structures should be defined separately.

Point number eleven is not quite clear to me, but I gather that Dr. Goldacre wants to hear my reaction to his observations of broken sections of membrane contaminating his preparations of dissociated cytoplasm. I have two comments on that. First, Dr. Goldacre and I disagree here on the observations themselves.

The film I showed had a long sequence of an ameba trapped in a capillary unable to locomote, but showing sporadic "fountain streaming" such as Mast described years ago. In these sequences, the plasmalemma or all membrane was clearly visible in all parts of the cell. The second sequence showed dissociated cytoplasm from a broken ameba photographed under identical optical conditions and showing no membranes whatever in the streaming cytoplasm. I do not deny that it is possible to prepare dissociated cytoplasm contaminated by considerable amounts of membrane; the fact is, however, that our preparations were not so contaminated.

The second comment is on the interpretations: The argument about the presence or absence of membranes is a side issue. The motions of the cytoplasmic "units of streaming," or "hairpin loops of cytoplasm" have so far been explainable only in terms of a contraction occurring at the bend of each loop. The loops themselves represent, as nearly as we can tell, a radial breakdown in pseudopodial structure, such that the bends of the loop correspond in structure and function to what we have called the "fountain zone" of the intact cell.

The twelfth point concerns the production of pseudopodia from an initially spherical cell by frontal contraction. This is a process that may well be more easily explained at the present time in terms of a weak spot appearing in a contracting skin. However, it is also quite possible that frontal contraction could explain it, but it would require the rapid formation of oriented contractile material in a circular pattern at the base of a prospective pseudopod to provide an anchor for the postulated hairpin loops to push against. In principle, such a mechanism would be possible; however, we have no evidence as to the presence or absence of the required ultrastructure.

The thirteenth point concerns the "plasmagel network" in the anterior half of the cell. When Dr. Goldacre first raised this point in Leiden in 1961, I misunderstood him; I thought he was referring to dorsoventral strands passing through the endoplasm such as Dr. Abé has now described in *Amoeba striata*. Now that I have seen Dr. Goldacre's diagrams, I realize that he is referring to the branched endoplasmic streams frequently found in *Amoeba proteus* and other species as the result of the fusion of neighboring pseudopodia. As the cell advances after such a fusion, shared ectoplasmic walls pass backward (relative to the cell) and take on the aspect of gel islands. We agree on the observation, but I doubt that it has any significance as far as mechanism of movement is concerned. It only means that each endoplasmic branch leads to an advancing portion of the anterior end—a fountain zone where either a gelation or a contraction occurs—depending on which way you wish to look at it.

The last point is a request for an alternative interpretation of Dr. Goldacre's observation that when ATP is injected into an ameba, the region injected always becomes a tail, never a front. My comment is that I doubt whether we can conclude anything at present from the injection of ATP into a living cell. It is already known that ATP can have several other chemical effects besides giving the energy of its terminal pyrophosphate bond to a mechanochemical system. One such function might be to solate the tail endoplasm, decreasing the resistance of that part of the cell to tension from another part. Perhaps the ATP is not used until it gets to the front. Perhaps it is not used at all, but the cell merely responds to stimulation. I would regard a clear experimental demonstration that the ATP injected had been used in the tail as much more convincing than a behavioral response.

It might be worth pointing out in passing that Podolsky has shown that physiological concentrations of ATP injected into living muscle fibers cause no contraction, whereas extremely small concentrations of calcium ion are effective. In view of the fact that divalent cations are effective in causing the contraction of ameba cytoplasm, one might wonder whether the contractions observed might have been caused by impurities.

I would like to conclude by recalling a statement made by Dr. Goldacre to the effect that a single fact can disprove a theory. In principle this should be true, but the more complicated the phenomenon to be explained, the easier it is to save the theory by an assumption. It may be possible to settle this question by forever going back and checking assumptions. However, since what is at issue is the localization of the motive force, would it not seem a more direct approach to ask physical questions that yield answers dealing with forces and the results of the application of forces? We have been using this approach in our work, and I am sorry now that organizing this meeting prevented me from making a contribution in this area rather than in pure description.

DR. GOLDACRE: Might I make a comment on some of those?

CHAIRMAN STEWART: Yes; I think it would be only fair.

DR. GOLDACRE: It seems that Dr. Allen does not believe some of the experimental results I have put forward in diagrams only. I can assure him I can produce photographs, if he would like, of any of these things, and I would be happy to demonstrate these phenomena in his own laboratory, including the presence, for example, of cell membrane made visible with toluidine blue in capillary tubes, broken according to his method.

With regard to why amebae do not pull into pieces when they are polypodial, Dr. Allen said this does in fact happen in cell division. I would like to know why it doesn't happen all the time in a normal polypodial ameba.

The second point, the fact that the negative pressure which is predicted by the fountain-zone theory in the ameba would require medium to be sucked in whenever it is broken, whereas, in fact, the granules run out. That is the observation. That can easily be demonstrated, and I will be quite happy to do so.

I showed a colored photograph of the blue phosphomolybdate reaction in the ameba's tail. Dr. Allen says that he needs some evidence of this. Well, what can one give beyond the photograph in a lecture? It would be possible to give a demonstration, if you like, in your own laboratory.

Next the contraction due to membrane-plasmagel contact: It has been suggested this has nothing to do with the membrane, but that the plasmagel alone is sensitive. Well, the observation is that if one prods an ameba with, for example, a blunt needle, nothing happens unless the cell membrane is pushed the whole of the way across the hyaline layer. If one pushes it halfway across, nothing happens. This seems to me to imply that membrane contact is necessary for response.

DR. KITCHING: It is interesting that Dr. Mast showed that light shone on the plasmagel would stop a pseudopod, whereas light shone on the hyaline cap failed to do so.

DR. GOLDACRE: Probably reaction to a light stimulus is a special case. But in every case where the membrane is brought into contact with the plasmagel by any method, by mechanical means or electrical means or hydraulic means, you get this contraction and the formation of a tail at the point of contact.

The gel/sol ratio seems to me to be an important point. Dr. Allen said that Dr. Mast showed that no streaming took place when the gel-sol ratio became less than 1. Well, Dr. Mast's gel/sol ratios were not volume ratios but thickness ratios, and in a cylinder, a thickness ratio of 1 according to Dr. Mast would be a volume ratio of 3. So that his figures don't really apply. On your theory, it is the relative *volumes* which are important.

DR. ALLEN: It is meaningful only to speak in terms of cross-sectional *areas*. This is why we tried to avoid confusion by using the term A_t/A_s ratio.

DR. GOLDACRE: Yes; volume ratio is the same thing in a tube. Dr. Mast never found any gel/sol ratios below a value of 3. Only if Dr. Mast had found a limiting *volume* ratio of 1, instead of 3, could there have been this interesting correlation with the fountain-zone theory. In the case of heparin, it seems quite clear there is no gel there at all, yet locomotion proceeds.

Dr. Allen, also, does not admit the existence of the plasmagel sheet which I showed on the cine film. I find it difficult to know what else to do to convince him of that. If one can photograph it and show it on cine film and see it sieving granules out of the flowing stream, there is reason to think it is there. Its existence excludes the possibility of having a forward-flowing rod of gelled endoplasm coming up the middle of the ameba. I do not think there is any doubt of the existence of this sheet.

Might I ask you what kind of evidence would convince you of its existence?

DR. ALLEN: The plasmagel sheet? Well, there is an interface between the hyaline cap fluid and the granular cytoplasm. There is no doubt about this. But one cannot decide on the basis of an interface what the physical properties on the two sides of the interface are. Mast assumed incorrectly they were both sol. Therefore he postulated a structured border.

DR. GOLDACRE: The pressing together of the granules as if their suspension medium was passing through a sieve shows, I think, that the endoplasm is liquid here.

DR. GRIFFIN: May I comment on that? I think there is a little misunderstanding. It is shown very clearly in your film and in Dr. Allen's film of intact *Chaos* in a capillary that something peels off the membrane at the front. From my recollection of Dr. Mast's papers, he did not see this phenomenon. It is quite rare in normal *Amoeba proteus.* He postulated a partially permeable alveolar sheet to account for the anterior separation of hyaline and granular materials and later wrote that he had actually seen such an anterior alveolar barrier in *Pelomyxa palustris.* There is no evidence in either of these amebae that a plasmagel sheet such as he postulated actually is present at the front during locomotion, although it looks as though *P. palustris* moves by a mechanism like that proposed by Dr. Mast.

DR. GOLDACRE: I think we must consider only *Amoeba proteus* in this connection, because that is what we are both concerned with. I do not think Dr. Mast *postulated* but rather *observed* the plasmagel sheet.

DR. GRIFFIN: You showed something peeling off the membrane. I think we all agree on that; but I do not think that should be called a plasmagel sheet. It may be something else.

DR. GOLDACRE: It is something that has the properties of a sieve.

DR. GRIFFIN: I think that is true; and dissolves very rapidly in protoplasm.

CHAIRMAN STEWART: A very common difficulty in biology involves trying to oversimplify that which is very complicated. It seems to me this is why Dr. Abé's beautiful description of *Amoeba striata* is pertinent. There is a lot more going on than either of the proponents of these two opposite points of view is saying. Neither of these points of view can account so far for all the facts. We do not yet know how an ameba streams.

DR. KITCHING: I rather feel that we are beating a dead horse. Does it matter frightfully whether the major part of the contraction in the plasmagel is in the front or back? A question which I think is far more important, which nobody is willing to talk about, is why the advancing plasmasol fails to gelate right at the front end, but, as Dr. Goldacre says, peels off. For some reason or other, it fails to form a thick plasmagel there and does it only at the side. Something important is going on at the front end, something to do with stimulus provided by the environment. This is a part of ameboid movement which has been neglected so far.

DR. GOLDACRE: I think the films that peel off at the front are not gelated plasmasol but extremely thin films, too thin to resolve in the light microscope. They are comparable with the cell membrane itself. It seems to me they are synthesized by action on the cell itself, and they are repeatedly peeling off and could perhaps represent protein synthesis in sheets.

DR. ALLEN: I would like to disagree with the suggestion of Dr. Kitching that the localization of the motive force is not an important problem. The aim of all modern cell biology is to get molecular explanations of cell dynamics. If one does not try to find a site at which the motive force is applied, one cannot possibly hope to utilize techniques that are capable of tracking molecular events in living cells. I think that one has to just consider simple hypotheses. These hypotheses then should be tested experimentally, in this case by physical experiments since the hypotheses so far deal only with force and deformation. If one does not take such a viewpoint, then one essentially gives up all hope of finding molecular explanations.

To my way of thinking, it is almost equivalent to becoming a vitalist to concede without demonstrating that the motive force is distributed randomly throughout a system.

DR. GOLDACRE: That is one point on which I can agree with you.

CHAIRMAN STEWART: That sounds like an excellent place to end the conference!

Subject Index

D